WILEY SERIES IN PROBABILITY
AND MATHEMATICAL STATISTICS

ESTABLISHED BY WALTER A. SHEWHART AND SAMUEL S. WILKS

Editors

Vic Barnett, Ralph A. Bradley, J. Stuart Hunter, David G. Kendall, Rupert G. Miller, Jr., Adrian F. M. Smith, Stephen M. Stigler, Geoffrey S. Watson

Probability and Mathematical Statistics

ADLER • The Geometry of Random Fields
ANDERSON • The Statistical Analysis of Time Series
ANDERSON • An Introduction to Multivariate Statistical Analysis, *Second Edition*
ARNOLD • The Theory of Linear Models and Multivariate Analysis
BARNETT • Comparative Statistical Inference, *Second Edition*
BHATTACHARYYA and JOHNSON • Statistical Concepts and Methods
BILLINGSLEY • Probability and Measure, *Second Edition*
BOROVKOV • Asymptotic Methods in Queuing Theory
BOSE and MANVEL • Introduction to Combinatorial Theory
CAINES • Linear Stochastic Systems
CASSEL, SARNDAL, and WRETMAN • Foundations of Inference in Survey Sampling
CHEN • Recursive Estimation and Control for Stochastic Systems
COCHRAN • Contributions to Statistics
COCHRAN • Planning and Analysis of Observational Studies
CONSTANTINE • Combinatorial Theory and Statistical Design
DOOB • Stochastic Processes
DUDEWICZ and MISHRA • Modern Mathematical Statistics
EATON • Multivariate Statistics: A Vector Space Approach
ETHIER and KURTZ • Markov Processes: Characterization and Convergence
FABIAN and HANNAN • Introduction to Probability and Mathematical Statistics
FELLER • An Introduction to Probability Theory and Its Applications, Volume I, *Third Edition*, Revised; Volume II, *Second Edition*
FULLER • Introduction to Statistical Time Series
FULLER • Measurement Error Models
GRENANDER • Abstract Inference
GUTTMAN • Linear Models: An Introduction
HALL • Introduction to The Theory of Coverage Problems
HAMPEL, RONCHETTI, ROUSSEEUW, and STAHEL • Robust Statistics: The Approach Based on Influence Functions
HANNAN • Multiple Time Series
HANNAN • The Statistical Theory of Linear Systems
HARRISON • Brownian Motion and Stochastic Flow Systems
HETTMANSPERGER • Statistical Inference Based on Ranks
HOEL • Introduction to Mathematical Statistics, *Fifth Edition*
HUBER • Robust Statistics
IMAN and CONOVER • A Modern Approach to Statistics
IOSIFESCU • Finite Markov Processes and Applications
JOHNSON and BHATTACHARYYA • Statistics: Principles and Methods, *Revised Printing*
LAHA and ROHATGI • Probability Theory
LARSON • Introduction to Probability Theory and Statistical Inference, *Third Edition*
LEHMANN • Testing Statistical Hypotheses, *Second Edition*
LEHMANN • Theory of Point Estimation
MATTHES, KERSTAN, and MECKE • Infinitely Divisible Point Processes

Probability and Mathematical Statistics (Continued)

MUIRHEAD • Aspects of Multivariate Statistical Theory
PURI and SEN • Nonparametric Methods in General Linear Models
PURI and SEN • Nonparametric Methods in Multivariate Analysis
PURI, VILAPLANA, and WERTZ • New Perspectives in Theoretical and Applied Statistics
RANDLES and WOLFE • Introduction to the Theory of Nonparametric Statistics
RAO • Linear Statistical Inference and Its Applications, *Second Edition*
RAO • Real and Stochastic Analysis
RAO and SEDRANSK • W.G. Cochran's Impact on Statistics
RAO • Asymptotic Theory of Statistical Inference
ROBERTSON, WRIGHT and DYKSTRA • Order Restricted Statistical Inference
ROGERS and WILLIAMS • Diffusions, Markov Processes, and Martingales, Volume II: Îto Calculus
ROHATGI • An Introduction to Probability Theory and Mathematical Statistics
ROHATGI • Statistical Inference
ROSS • Stochastic Processes
RUBINSTEIN • Simulation and The Monte Carlo Method
SCHEFFE • The Analysis of Variance
SEBER • Linear Regression Analysis
SEBER • Multivariate Observations
SEN • Sequential Nonparametrics: Invariance Principles and Statistical Inference
SERFLING • Approximation Theorems of Mathematical Statistics
SHORACK and WELLNER • Empirical Processes with Applications to Statistics
STOYANOV • Counterexamples in Probability

Applied Probability and Statistics

ABRAHAM and LEDOLTER • Statistical Methods for Forecasting
AGRESTI • Analysis of Ordinal Categorical Data
AICKIN • Linear Statistical Analysis of Discrete Data
ANDERSON and LOYNES • The Teaching of Practical Statistics
ANDERSON, AUQUIER, HAUCK, OAKES, VANDAELE, and WEISBERG • Statistical Methods for Comparative Studies
ARTHANARI and DODGE • Mathematical Programming in Statistics
ASMUSSEN • Applied Probability and Queues
BAILEY • The Elements of Stochastic Processes with Applications to the Natural Sciences
BAILEY • Mathematics, Statistics and Systems for Health
BARNETT • Interpreting Multivariate Data
BARNETT and LEWIS • Outliers in Statistical Data, *Second Edition*
BARTHOLOMEW • Stochastic Models for Social Processes, *Third Edition*
BARTHOLOMEW and FORBES • Statistical Techniques for Manpower Planning
BECK and ARNOLD • Parameter Estimation in Engineering and Science
BELSLEY, KUH, and WELSCH • Regression Diagnostics: Identifying Influential Data and Sources of Collinearity
BHAT • Elements of Applied Stochastic Processes, *Second Edition*
BLOOMFIELD • Fourier Analysis of Time Series: An Introduction
BOX • R. A. Fisher, The Life of a Scientist
BOX and DRAPER • Empirical Model-Building and Response Surfaces
BOX and DRAPER • Evolutionary Operation: A Statistical Method for Process Improvement
BOX, HUNTER, and HUNTER • Statistics for Experimenters: An Introduction to Design, Data Analysis, and Model Building
BROWN and HOLLANDER • Statistics: A Biomedical Introduction
BUNKE and BUNKE • Statistical Inference in Linear Models, Volume I

(*continued on back*)

Order Restricted
Statistical Inference

To my friend Chul-Gyu.

Tim Robertson

Applied Probability and Statistics (Continued)

CHAMBERS • Computational Methods for Data Analysis
CHATTERJEE and HADI • Sensitivity Analysis in Linear Regression
CHATTERJEE and PRICE • Regression Analysis by Example
CHOW • Econometric Analysis by Control Methods
CLARKE and DISNEY • Probability and Random Processes: A First Course with Applications, *Second Edition*
COCHRAN • Sampling Techniques, *Third Edition*
COCHRAN and COX • Experimental Designs, *Second Edition*
CONOVER • Practical Nonparametric Statistics, *Second Edition*
CONOVER and IMAN • Introduction to Modern Business Statistics
CORNELL • Experiments with Mixtures: Designs, Models and The Analysis of Mixture Data
COX • Planning of Experiments
COX • A Handbook of Introductory Statistical Methods
DANIEL • Biostatistics: A Foundation for Analysis in the Health Sciences, *Fourth Edition*
DANIEL • Applications of Statistics to Industrial Experimentation
DANIEL and WOOD • Fitting Equations to Data: Computer Analysis of Multifactor Data, *Second Edition*
DAVID • Order Statistics, *Second Edition*
DAVISON • Multidimensional Scaling
DEGROOT, FIENBERG and KADANE • Statistics and the Law
DEMING • Sample Design in Business Research
DILLON and GOLDSTEIN • Multivariate Analysis: Methods and Applications
DODGE • Analysis of Experiments with Missing Data
DODGE and ROMIG • Sampling Inspection Tables, *Second Edition*
DOWDY and WEARDEN • Statistics for Research
DRAPER and SMITH • Applied Regression Analysis, *Second Edition*
DUNN • Basic Statistics: A Primer for the Biomedical Sciences, *Second Edition*
DUNN and CLARK • Applied Statistics: Analysis of Variance and Regression, *Second Edition*
ELANDT-JOHNSON and JOHNSON • Survival Models and Data Analysis
FLEISS • Statistical Methods for Rates and Proportions, *Second Edition*
FLEISS • The Design and Analysis of Clinical Experiments
FOX • Linear Statistical Models and Related Methods
FRANKEN, KÖNIG, ARNDT, and SCHMIDT • Queues and Point Processes
GALLANT • Nonlinear Statistical Models
GIBBONS, OLKIN, and SOBEL • Selecting and Ordering Populations: A New Statistical Methodology
GNANADESIKAN • Methods for Statistical Data Analysis of Multivariate Observations
GREENBERG and WEBSTER • Advanced Econometrics: A Bridge to the Literature
GROSS and HARRIS • Fundamentals of Queueing Theory, *Second Edition*
GUPTA and PANCHAPAKESAN • Multiple Decision Procedures: Theory and Methodology of Selecting and Ranking Populations
GUTTMAN, WILKS, and HUNTER • Introductory Engineering Statistics, *Third Edition*
HAHN and SHAPIRO • Statistical Models in Engineering
HALD • Statistical Tables and Formulas
HALD • Statistical Theory with Engineering Applications
HAND • Discrimination and Classification
HOAGLIN, MOSTELLER and TUKEY • Exploring Data Tables, Trends and Shapes
HOAGLIN, MOSTELLER, and TUKEY • Understanding Robust and Exploratory Data Analysis
HOCHBERG and TAMHANE • Multiple Comparison Procedures

Order Restricted Statistical Inference

TIM ROBERTSON

University of Iowa

F. T. WRIGHT

University of Missouri-Rolla

R. L. DYKSTRA

University of Iowa

JOHN WILEY & SONS

Chichester · New York · Brisbane · Toronto · Singapore

To Joanie, Judy, and Pat and to our friend and mentor, Dan Brunk.

Library of Congress Cataloging-in-Publication Data:

Robertson, T. (Tim), 1937–
Order restricted statistical inference

(Wiley series in probability and mathematical statistics. Probability and mathematical statistics section)
Bibliography: p.
Includes index.
1. Order statistics. 2. Regression analysis.
I. Wright, F. T. II. Dykstra, R. (Richard)
III. Title. IV. Series: Wiley series in probability and mathematical statistics. Probability and mathematical statistics.
QA278.7.R63 1988 519.5 87-27896

ISBN 0 471 91787 7

British Library Cataloguing in Publication Data:

Robertson, Tim
Order restricted statistical inference.
—(Wiley series in probability and mathematical statistics).
1. Mathematical statistics 2. Probabilities
I. Title II. Wright, F. T. III. Dykstra, R. L.
519.5′4 QA276

ISBN 0 471 91787 7

Phototypesetting by Thomson Press (India) Ltd., New Delhi
Printed in Great Britain by Anchor Brendon Ltd., Tiptree, Essex

Contents

Preface

Many types of problems are concerned with identifying meaningful structure in real world situations. Structure involving orderings and inequalities is often useful since it is easy to interpret, understand, and explain. For example, the probability of a particular response may increase with the treatment level; a regression function may be nondecreasing, or convex, or both; the failure rate of a component may increase as it ages; or the treatment response may stochastically dominate the control. The fact that the utilization of such ordering information increases the efficiency of statistical inference procedures is well documented. The one-tailed, two-sample t-test provides a familiar example in which the procedure which utilizes the prior information dominates procedures which ignore this information. Tests for identifying this structure often require good estimates under inequality constraints and serve as motivation for much of the material in this book.

The origins of order restricted statistical inference are usually dated back to the early 1950s. At about the same time, a number of researchers started working independently on similar problems. These researchers included Dan Brunk and his coworkers and V. J. Chacko in the United States, David Bartholomew in the United Kingdom, and Constance van Eeden in Amsterdam. Herman Chernoff was also working on closely related ideas.

After these early efforts, the field developed rapidly during the 1960s and early 1970s. The theory of restricted estimation and testing was developed under a variety of assumptions, and many outstanding researchers made important contributions to this literature. In 1972, Richard Barlow, David Bartholomew, J. M. Bremner, and Dan Brunk published the book, *Statistical Inference Under Order Restrictions*, which assembled much of the early work and served as a basis for subsequent research in this area.

There has been a considerable amount of development in the field since the time of that publication. The book by Barlow *et al.* contains approximately three hundred references related to this area. This book, written about fifteen years later, contains over eight hundred such references and points to the ubiquitous nature of this topic. Order restricted inference has been, and in the authors' opinion will continue to be, a fertile area for research.

Obviously, not all of the developments up to this time can be discussed in this book. We have chosen to emphasize those procedures that are based on the likelihood principle. Beyond that, the choices of topics that have been included were based on our interests and reflect our biases.

While this book was written primarily as a research reference, it could be used as a text for a special topics course. We recommend covering Chapters 1, 2, 5 and 6 and then choosing topics according to interest.

The aforementioned book by Barlow, Bartholomew, Bremner, and Brunk has had a significant impact on the development of this area of investigation. We are keenly aware of our debt to those authors for their seminal contributions. Even beyond that, we are indebted to them because their book has served as a foundation for this work. We followed the outline they used as well as their practice of including, at the end of each chapter, a 'complements' section with historical comments and a brief mention of some topics not included in the chapter. Of course, we realize that we have missed some of the original sources and subsequent developments, for which we offer our apologies. We also wish to express our thanks to the authors of that book and its publisher, John Wiley and Sons, for permission to use, with only minor changes, the material in Chapters 7 and 8 of this book. We have viewed our task as that of updating a book which is otherwise difficult to improve.

We are grateful to the authors and publishers who kindly gave permission to reproduce the following tables and a figure. Table 7.6.1 is taken from Barlow and Proschan (1969); Table 7.6.2 is taken from Proschan (1963); and Table 7.6.3 and Figure 7.4.1 are taken from Barlow *et al.* (1972).

The writing of this book, as well as much of the authors' research on this subject, has been supported by the Office of Naval Research under contracts N00014-78-C-0655, N00014-80-C-0321, N00014-80-C-0322, and N00014-83-K-0249. In addition, the encouragement for this project given by Edward J. Wegman and Lyle D. Broemeling of the Office of Naval Research is greatly appreciated. Partial support, during the latter stages of the book, was also given by the Army Research Office through the Mathematical Sciences Institute at Cornell University while the last author was visiting there. The authors' work in this area has benefited greatly from our interactions with David Bartholomew, Dan Brunk, David Hanson, students, and other colleagues too numerous to mention here. Some of the ideas for presenting this material arose from seminars, courses, or parts of courses, taught at the University of Iowa, the University of California at Davis, and the University of Missouri-Rolla. We are grateful for the opportunities to have taught those courses and for the ideas and comments given to us by faculty and students who participated in those courses and seminars.

C. I. C. Lee and B. Singh carefully read and commented on parts of an earlier manuscript, and H. Mukerjee and P. C. Wollan did thorough jobs of reviewing the manuscript for the publisher. Through their suggestions these colleagues have made invaluable contributions to this work. However, all responsibility for deficiencies in this book belongs to the authors. We also wish to thank the staff at John Wiley and Sons for the careful way they handled the manuscript; Paul Wright for a professional job of typing the manuscript and preparing the figures; and the University of Missouri-Rolla for the use of their word processing facilities.

We express our appreciation to our families and particularly to our wives, Joan, Judy, and Pat, for their love, encouragement, support, and patience.

IOWA CITY, IOWA
ROLLA, MISSOURI
1987

Introduction and Overview

Estimators which utilize prior ordering information are discussed in Chapter 1 with emphasis given to maximum likelihood estimators. In the situations considered, the parameters are viewed as a regression function defined on a finite set that is endowed with an ordering. The order restrictions on the parameters are specified by requiring the regression function to be isotonic with respect to the ordering on the index set. That is, the functional values are assumed to be ordered in the same way as their arguments. The isotonic regression problem arises from the maximum likelihood estimation of normal means under an isotonic restriction. Section 2 contains numerical and graphical computation algorithms for the isotonic regression in the simply (totally) ordered case as well as an example in multidimensional scaling. Isotonic regression over a quasi-ordered index set and computation algorithms are discussed in Sections 3 and 4. A recently developed algorithm for computing the least squares estimate of a bivariate regression function which is nondecreasing in both independent variables is included. A generalized isotonic regression problem, whose solution yields maximum likelihood estimates for ordered parameters from an exponential family, is formulated in Section 5. Error reduction properties of the isotonic regression are discussed in Section 6. A particularly nice result in that section states that the isotonic regression has smaller pointwise mean square error than the sample means in the simply ordered case. Fenchel duality is mentioned briefly and is applied to obtain the maximum likelihood estimates of the probability vectors of stochastically ordered multinomial populations, cf. Section 7.

In Chapter 2, the likelihood ratio tests of order restricted hypotheses involving normal means are presented. Bartholomew's $\bar{\chi}^2$ and $\bar{E}^2$ tests of homogeneity with an order restricted alternative are discussed as well as their analogues for testing an order restriction as a null hypothesis. The null distributions of these statistics are mixtures of chi-square or beta distributions depending on whether the variances are known or not. The mixing coefficients, called level probabilities, are determined by the distribution of the number of distinct values in the isotonic regression estimator under the assumption that the means are equal. A recursive relation for computing the level probabilities is given in Section 4 along with expressions for arbitrary sample sizes and the well-known Bartholomew–Miles recursive formula for the case of equal sample sizes and a simple order. Formulas for the power of these tests in special cases are derived in Section 5 and used to study the increase in efficiency obtained by using the order restricted tests rather than omnibus tests. Properties of the tests are presented in Section 6, including some recent results on the monotonicity of the power functions. Questions concerning the bias of these tests are also considered. Section 7 briefly mentions testing problems with restrictions imposed by requiring the parameter vector to lie in polyhedral, circular or arbitrary convex cones.

The distributions of the LRT statistics discussed in Chapter 2 are rather complex, particularly if the weights, which are proportional to the sample sizes, are not equal. Thus approximations are of interest. In Section 1 of Chapter 3, the equal-weights null distributions are considered as approximations to the null distributions for unequal weights. For many practical purposes these approximations are adequate. If the number of populations is large, then even the equal-weights null distributions are tedious to use. Two-moment chi-square and beta approximations to the equal-weights null distributions are given in Section 2. Considering the simple order and the simple tree ordering, approximations to these null distributions are presented for the case in which one of the weights is larger than the others, and for arbitrary weights sets. In the latter case, one type of approximation is based on the pattern of large and small weights. Section 6 contains stochastic bounds for the null distributions of these statistics. Two-moment approximations for the powers of these tests at slippage alternatives are given in Section 7. Because this material is rather technical in nature and the equal-weights approximations are adequate in many situations, this chapter could be omitted on a first reading.

Chapter 4 contains some extensions and generalizations of the tests discussed in Chapter 2. Likelihood ratio tests for ordered hypotheses involving the parameters of k independent samples from an exponential family are discussed in Section 1, and several particular exponential families are considered. In Section 2 the use of contrast statistics for testing ordered hypotheses is studied. Optimal contrast tests are derived and the likelihood ratio tests for ordered normal means with known variances are shown to be adaptive contrast tests. The power of a contrast test may be relatively small at the edges of the cone of parameter vectors that satisfy the ordering. This drawback is alleviated to a large degree by the multiple contrast tests presented in Section 3. The power of the likelihood ratio tests, multiple contrast tests and single contrast tests are compared in the fourth section of the chapter. The theory of distribution-free tests of ordered hypotheses is treated in Section 5, and some asymptotic relative efficiencies are presented there. One-sided tests for a multivariate normal mean vector are considered in the last section of Chapter 4.

Inferences for multinomial parameters are considered in Chapter 5. In Sections 1 and 2, maximum likelihood estimates and likelihood ratio tests subject to order restrictions are considered. A weaker inequality restriction called 'lower starshaped' or 'decreasing on the average' is treated in Section 3. Because the cone corresponding to this ordering is an orthant, the distribution theory is simpler than in the simply ordered case. The 'duality' between isotonic and stochastic ordering restrictions is further exploited in Section 4 where likelihood ratio tests for and against a stochastic ordering between multinomial populations are developed. The concepts involved in testing situations in which both the null and alternative hypotheses involve non-trivial order restrictions are illustrated in Section 5 on the problems of testing symmetry and unimodality versus symmetry, and symmetry and unimodality versus unimodality. While the idea of making order restricted inferences in contingency tables seems quite natural, there is a limited number of such results available. Some results for contingency tables which are related to isotonic regression are mentioned in the sixth section of this chapter.

A detailed discussion of duality is given in Chapter 6. The theory of Fenchel duality is considered in Section 2 and that theory is applied to the generalized isotonic regression problem, I-projections with linear inequality constraints, and log-linear models. Lagrangian duality or Kuhn–Tucker theory, which is discussed in Section 4, is related to Fenchel duality in Section 5. Lagrangian duality is used to obtain the maximum likelihood estimates of stochastically ordered distributions with possibly right censored data. In Section 8, the use of Wald's test, Rao's test and the Lagrange multiplier test for hypotheses with linear inequality constraints is briefly discussed.

Chapter 7 treats inferences concerning distributions subject to shape restrictions. In the situations considered, the restricted maximum likelihood estimates are the isotonic regression of the usual unconstrained estimates. Estimation of a monotone density is studied in Section 2. The asymptotic properties of this isotonic estimator are obtained via the error reduction properties of the isotonic regression and the large sample properties of the empirical CDF. The results for monotone densities are extended to unimodal densities in the third section. Estimation of the failure rate and the generalized failure rate function when they are constrained to be monotone is investigated in Sections 4 and 5, respectively. Testing exponentiality with the alternative restricted to have a monotone failure rate is considered in Section 6. Properties of this test, such as unbiasedness, monotonicity of its power function and an asymptotic minimax property, are mentioned.

Projections in the Hilbert space of real valued, square integrable functions are considered in Chapter 8. Projections that are onto the set of functions that are isotonic with respect to a quasi-order on the domain space, or more generally, onto the set of functions which are measurable with respect to a σ-lattice of subsets of the domain are considered. Properties of these projections, which may be thought of as conditional expectations given a σ-lattice, are studied in Section 2 and characterization theorems for conditional expectations given a σ-lattice are treated in Section 3.

The last chapter contains a brief discussion of topics in the area of inequality constrained inference which at this time are not fully developed or could not be fully treated because of space limitations. In Section 1, optimization problems with isotonic constraints but norms other than the L_2 norm are considered. Special emphasis is given to the L_p norms with $p \neq 2$. The nature of the limiting results for the isotonic regression is indicated in Section 2. For a finite index set, these results follow from error reduction properties and limiting results for the unconstrained estimator. For an infinite index set, such as $[0, 1]$ or $[0, 1]^k$, the results are more interesting mathematically. The observation points must become dense in the index set in a particular way and the true regression function must satisfy certain regularity conditions. Martingale types of convergence results are discussed for projections in a Hilbert space setting. A few results on Bayesian order restricted inference are mentioned in Section 3. Order constrained Bayes estimates of multinomial probabilities and the parameters in an exponential family are discussed. Bayesian solutions to a bioassay problem and a reliability problem are also presented. Because the isotonic regression averages the unrestricted estimator over subsets of the index set, it can be thought of as a smoothing operator. A discussion of 'isotonization' as a smoothing process is found in Section 4. Restricted estimation and testing problems for the intensity of a nonhomogeneous Poisson process are outlined in Section 6, and some work on confidence intervals and simultaneous confidence bounds are presented in Section 7.

Tables to be used in implementing the $\bar{\chi}^2$ and $\bar{E}^2$ tests discussed in Chapter 2 are given in the Appendix and the Bibliography contains those references which are cited in the text as well as related references. The pages on which a reference is cited are given in brackets at the end of the reference.

Symbols and Acronyms

Av, average or weighted average, 18, 22, 23, 24, 70
α, main effect in a two-way layout, 66
$\hat{\alpha}$, maximum likelihood estimate of α, 66
$\bar{\alpha}$, maximum likelihood estimate of α subject to an isotonic restriction, 67
$\mathscr{A}$, collection of vectors or functions antitonic with respect to a quasi-order, 49
b, breadth of a partially ordered set, 144
β, main effect in a two-way layout, 66
$\hat{\beta}$, maximum likelihood estimate of β, 66
$\bar{\beta}$, maximum likelihood estimate of β subject to an isotonic restriction, 67
CSD, cumulative sum diagram, 7
γ, interaction term in a two-way layout, 66
$\hat{\gamma}$, maximum likelihood estimate of γ, 66
Δ, distance of a vector from the linear subspace of constant vectors, 90
dom f, effective domain of a function f, 266
e, number of exterior elements in a partially ordered set, 148
epi f, epigraph of a convex function f, 266
eph h, epigraph of a concave function h, 266
$\bar{E}^2_{01}$, likelihood ratio test statistic for testing homogeneity with an ordered alternative and variances unknown, 63
$\bar{E}^2_{01}(\nu)$, generalization of $\bar{E}^2_{01}$ for an independent estimator of σ^2 with ν degrees of freedom, 64
$\bar{E}^2_{12}$, likelihood ratio test statistic for testing an ordered null hypothesis with variances unknown, 65
$\bar{E}^2_{12}(\nu)$, generalization of $\bar{E}^2_{12}$ for an independent estimator of σ^2 with ν degrees of freedom, 65
f^*, convex conjugate of a convex function f, 267
g^*, isotonic regression of g, 4, 14
GCM, greatest concave majorant, 7
h^*, concave conjugate of a concave function h, 267
IPFP, iterative proportional fitting procedure, 288
$I(p|q)$, I divergence of p with respect to q, 281

$\mathscr{I}$, collection of vectors or functions isotonic with respect to a quasi-order, 5
J, Jonckheere–Terpstra test statistic, 201
$\mathscr{K}^*$, dual or polar of the cone $\mathscr{K}$, 5
$\mathscr{K}^{*\mathbf{w}}$, dual or polar of the cone $\mathscr{K}$ with respect to the weight vector $\mathbf{w}$, 45
$\mathscr{K}(A)$, polyhedral cone, 109
$\mathscr{K}_J$, face of a polyhedral cone, 109
$\mathscr{K}_J^I$, relative interior of a face of a polyhedral cone, 109
LRT, likelihood ratio test, 59
L_2, collection of square integrable functions, 369
$L_2(\mathscr{U})$, collection of square integrable functions which are measurable with respect to $\mathscr{U}$, 371
$\text{ls}(\mathscr{K})$, linearity space of $\mathscr{K}$, 109
Λ, likelihood ratio, 61
$\mathscr{L}$, collection of lower layers, 21
$\mathscr{L}_{mk}$, collection of decomposition of X into unordered level sets, 76
MLE, maximum likelihood estimator, 6
MLSA, minimum lower sets algorithm, 24
$\bar{\mu}$, projection of μ onto the linear subspace of constant vectors, 90
$\hat{\mu}$, estimator of common mean under homogeneity, 61
μ^*, estimator of a mean vector subject to an isotonic constraint, 61
PAVA, pool adjacent violators algorithm, 8
P_k, orthant probability, 75
$P(l, k; w)$, level probability ($P(l, k)$ if $w_1 = w_2 = \cdots = w_k$), 69
$P_S(l, k; w)$ or $P_S(l, k)$, level probability for a simple order, 77
$P_T(l, k; w)$ or $P_T(l, k)$, level probability for a simple tree ordering, 82
$P_L(l, k; w)$ or $P_L(l, k)$, level probability for a simple loop ordering, 84
$P_h(l, k; w)$ or $P_h(l, k)$, level probability for a unimodal ordering, 410
$P_{R\times C}(l, RC; w)$ or $P_{R\times C}(l, RC)$, level probability for a matrix ordering, 410
$P_w(\cdot|\mathscr{K})$, projection onto $\mathscr{K}$ with respect to a weighted Euclidean distance, 98
$P_{\Sigma^{-1}}(\cdot|\mathscr{K})$, projection onto $\mathscr{K}$ with distance $d(\mathbf{x}, \mathbf{y}) = (\mathbf{x}-\mathbf{y})'\mathbf{\Sigma}^{-1}(\mathbf{x}-\mathbf{y})$, 218
$\pi_{01}(\mu)$, power of $\bar{\chi}_{01}^2$ at μ, 89
$\pi_{12}(\mu)$, power of $\bar{\chi}_{12}^2$ at μ, 89
$\pi_{01}^*(\mu)$, power of $\bar{E}_{01}^2$ at μ, 90
$\pi_{12}^*(\mu)$, power of $\bar{E}_{12}^2$ at μ, 90
$\pi_c(\mu)$, power of a contrast test at μ, 177
ϕ, PDF of the standard normal distribution, 87
Φ, CDF of the standard normal distribution, 89
$\psi(\mathbf{x})$, with ψ defined on the reals; $\psi(\mathbf{x})$ is the vector valued function whose ith component is $\psi(x_i)$, 272
$Q(l, k)$, limiting level probability for a simple order, 125
$R(l, k+1)$, limiting level probability for a simple tree ordering, 130
$R(\mathscr{U})$, collection of real valued functions measurable with respect to $\mathscr{U}$, 371
$S_{01}(S_{01}(v))$, a nondecreasing function of $\bar{E}_{01}^2(\bar{E}_{01}^2(v))$, 64

$S_{12}(S_{12}(v))$, a nondecreasing function of $\bar{E}_{12}^2(\bar{E}_{12}^2(v))$, 65
sgn(x), sign of x, 202
$\sum$, summation
$\sum$, used for a disjoint union of sets, 71
T_c, contrast statistic, 177
$T_c(R)$, contrast of ranks, 204
$\mathscr{U}$, collection of upper layers, 21
X, index set, 6
$\bar{\chi}_{01}^2$, LRT statistic for testing homogeneity of normal means with an ordered alternative and known variances, 61
$\bar{\chi}_{12}^2$, LRT statistic for testing an ordered restricted null hypothesis with known variances, 62
$\bar{\chi}_{01}^2(R)$ and $\bar{\chi}_{12}^2(R)$, rank analogues of $\bar{\chi}_{01}^2$ and $\bar{\chi}_{12}^2$, 204
$\bar{\chi}_{01}^2(S)$ and $\bar{\chi}_{12}^2(S)$, chi-bar-square tests based on scores, 205
$\nabla f(x_0)$, gradient of f evaluated at x_0, 267
$\lesssim$, quasi-order on X, 12
$\wedge$, infimum, $a \wedge b$ is the smaller of a and b, 6
$\vee$, supremum, $a \vee b$ is the larger of a and b, 6
$\cap$, intersection
$\cup$, union
$\langle x, y \rangle$, inner product between x and y, 99, 266
$\|x\|$ norm of x, 99

CHAPTER 1

Isotonic Regression

1.1 INTRODUCTION

Regression is concerned with the fitting of curves or functions to a set of points or more generally to the joint distribution of a random vector. Of course, statistical methods for assessing the quality of the fit are of importance, too. The functions chosen to fit the points or the joint distribution are, for historical reasons, called *regression functions.* A regression function may be used in a number of ways, such as for prediction, for studying the degree of association between the variables, or for providing new insights into the phenomenon which generated the data.

Consider a finite set of points, $(x_1, y(x_1)), (x_2, y(x_2)), \ldots, (x_k, y(x_k))$, with $y(x_i)$ a real number and x_i either a real number or a vector of real numbers. The notation $y(x_i)$ has been chosen to emphasize that this value of the dependent variable, y, corresponds to the particular value of the independent variable, $x = x_i$. In some situations, the shape of the regression function to be used depends on the assumptions made regarding the way the points were obtained. For instance, if the points are a realization of a random sample from a bivariate normal distribution, then, because the conditional mean function for such a distribution is linear, a linear regression function is typically employed. In other situations, one may restrict attention to a suitable class of regression functions and then select the one that best fits the data using some measure of the quality of the fit, such as least squares or least absolute deviations. One approach to selecting the restricted class of functions is to consider those with a specified functional form, such as polynomial, logarithmic, or exponential. However, the approach taken in this monograph is to assume that the regression functions considered satisfy various order or shape restrictions; that is inequality restrictions that constrain the values of these functions. For the case in which the independent variable is real, the collection of nondecreasing functions and the collection of convex functions are important examples of order restricted classes of regression functions.

In the simplest examples, the regression function is assumed to be constant as a function of x and the quality of fit is measured by least squares. In this case it is well known that the best fit is obtained by taking the constant value to be the arithmetic mean of the y values. On the other hand, if one assumes that the regression function is constant and measures the quality of fit by absolute deviations, then the best fit is obtained by taking the constant to be the median of the y values.

On the other end of the spectrum, if no assumption is made regarding the shape of the regression function then its resulting value at a data point is the mean or median of the y values at that particular data point, depending upon whether quality of fit is measured by least squares or least absolute deviations. This unrestricted regression function is likely to be very erratic while the constant regression function may, in a sense, be too smooth. Most approaches to regression lie somewhere in between these two extremes. Undoubtedly, the most common approach is to assume that the regression function is affine or, more generally, a polynomial. However, such an assumption is typically made for mathematical convenience rather than because the investigator has information that the 'true' regression function has this particular form.

On the other hand, in many situations, a researcher may strongly believe that the underlying regression function has a particular shape or form which can be characterized by certain order restrictions, and, consequently, it is quite natural to select a regression function from the appropriate order restricted class of functions.

Example 1.1.1 Some of the dental study data discussed by Potthoff and Roy (1964) is presented in Table 1.1.1. In this study, which was conducted at the University of North Carolina Dental School, the size, in millimeters, of the pituitary fissure was measured for a number of subjects. The data in Table 1.1.1 are for three girls aged 8 years, three girls aged 10, three girls aged 12, and two girls aged 14.

It is reasonable to assume that the pituitary fissure increases with age and if the regression function (as a function of age) is chosen from the class of all nondecreasing functions, then this provides an example of *isotonic regression.*

Table 1.1.1 Size of pituitary fissure (Potthoff and Roy, 1964)

AGE	8	10	12	14
	21	24	21.5	23.5
	23.5	21	22	25
	23	25	19	
Mean	22.5	23.333	20.833	24.25
Median	23	24	21.5	24.25

The word *isotonic* means having the same tone, and in this instance refers to the fact that the functional values increase with age, or that these two variables, age and size of pituitary fissure, have the same 'tone'. If the function is assumed to be nonincreasing it would be proper to term it *antitonic*. The word *monotonic* means either isotonic or antitonic. Other more sophisticated shapes may be included in this theory by placing a partial order restriction on the independent variable and requiring that the regression function be isotonic with respect to this partial order (see Section 1.3). Not all order restrictions on a regression function can be described by assuming the function is isotonic with respect to a partial order on the independent variable. However, it is interesting to note that the methods developed for isotonic inferences also provide solutions for some of these more general inequality restricted inferences.

Neither the means nor the medians are increasing for the data in Table 1.1.1 so that some smoothing of the unrestricted regression function is necessary. The least squares line for the data is increasing and has a slope of 0.0652 and an intercept of 21.891. This function is rather 'flat' and, in fact, is quite close to the overall mean of 22.591. Using least squares the best nondecreasing fit to the data has the value 22.222 at ages 8, 10, 12 and the value 24.25 at age 14. We will discuss several algorithms for computing these isotonic estimates later.

Perhaps the most familiar problem in order restricted inference is that of testing the equality of two normal means under the assumption that the first mean is no larger than the second. One could ignore the ordering assumptions and use the standard two-sample t-test for the equality of the two means. However, it is intuitively obvious that this is not the thing to do, and it is well known that the one-sided test is significantly more powerful than the two-sided test. In fact, under the appropriate assumptions the one-sided test is uniformly most powerful in this situation (see Lehmann, 1959). Taking shapes or order restrictions into account can improve the efficiency of a statistical analysis by reducing the error or expected error of estimates or by increasing the power of test procedures, provided, of course, that the hypothesized order restriction actually holds. Evidence supporting this claim is given in Theorems 1.6.1 and 1.6.2 and in the analysis of the power of the restricted tests discussed in Chapters 2 and 3.

One of the facts that unifies the theory of order restricted inference is that the solutions to a wide class of restricted optimization problems can be found by restricted least squares. To be specific, suppose $X = \{x_1, x_2, \ldots, x_k\}$, w is a positive weight function defined on X and $\mathscr{F}$ is a restricted family of functions on X. For an arbitrary function g, defined on X, let g^* be the least squares projection of g onto the collection $\mathscr{F}$. Specifically, g^* is the solution to the restricted least squares problem

$$\text{minimize} \sum_{x \in X} [g(x) - f(x)]^2 w(x) \text{ subject to } f \in \mathscr{F}.$$

If the family $\mathscr{F}$ satisfies certain conditions, then g^* provides the solution for a variety of other optimization problems for which the objective functions are quite different from that of least squares including the estimation of ordered normal variances, ordered binomial parameters (bioassay), ordered Poisson means, ordered multinomial parameters as well as a variety of problems from other areas, such as reliability theory and density estimation (cf. Section 1.5).

1.2 ISOTONIC REGRESSION: THE SIMPLY ORDERED CASE

Let X be the finite set $\{x_1, x_2, \dots, x_k\}$ with the simple order $x_1 \lesssim x_2 \lesssim \cdots \lesssim x_k$. A function, f, on X is isotonic (with respect to this ordering) if $f(x_1) \leqslant f(x_2) \leqslant \cdots \leqslant f(x_k)$. Throughout this chapter w is assumed to be a positive weight function defined on X.

Definition 1.2.1 Suppose g is a given function on X. A function g^* on X is an *isotonic regression* of g with weights w if and only if g^* is isotonic and g^* minimizes

$$\sum_{x \in X} [g(x) - f(x)]^2 w(x) \tag{1.2.1}$$

in the class of all isotonic functions f on X.

Consider the special case in which $k = 3$ and $w(x) \equiv 1$. Functions on X are then just triples of real numbers $(f(x_1), f(x_2), f(x_3))$ and can be thought of as points in Euclidean three space, R^3, and the objective function given by (1.2.1) is just the

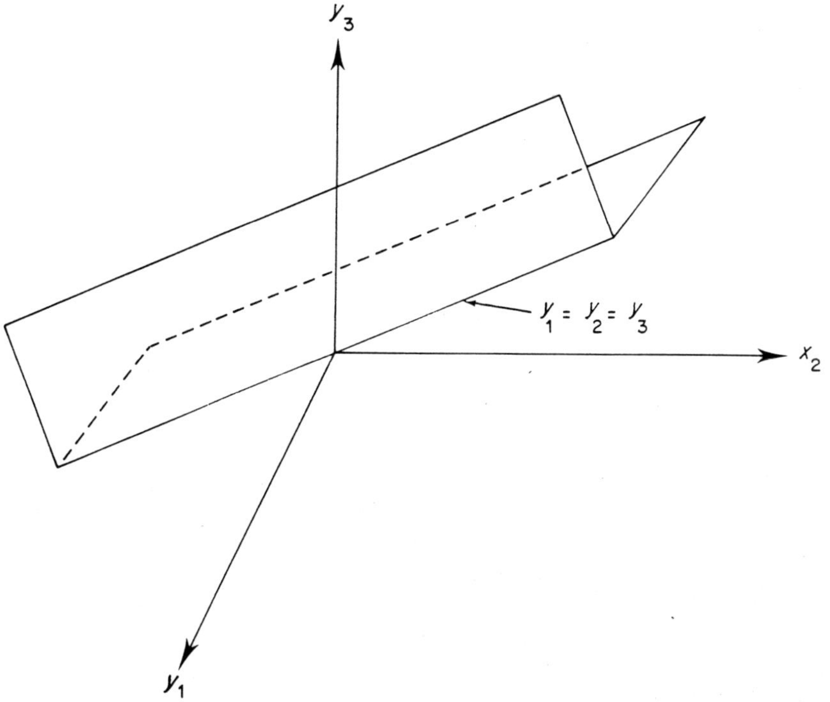

Figure 1.2.1 The isotonic functions with the usual simple order and $k = 3$

square of the Euclidean distance between g and f. The set, $\mathscr{I}$, of isotonic points (functions) is a 60° wedge which has two *faces* and an *edge*. The edge is given by $f(x_1) = f(x_2) = f(x_3)$ and the two faces are characterized by $f(x_1) = f(x_2) < f(x_3)$ and $f(x_1) < f(x_2) = f(x_3)$. A picture of the set, $\mathscr{I}$, of isotonic functions is given in Figure 1.2.1 and a two-dimensional view of it looking down the 'edge' is given in Figure 1.2.2.

Within this context the isotonic regression problem becomes: given the point $g = (g(x_1), g(x_2), g(x_3))$ find the point g^* which is the closest point of $\mathscr{I}$ to g. Clearly, R^3 can be divided into the four sets $\mathscr{I}$, A, B, and C depicted in Figure 1.2.2. The calculation of g^* depends upon which one of these sets contains the point g. It is clear that if $g \in \mathscr{I}$ then $g^* = g$. It will be shown later that A is the set of points $\mathbf{y} = (y_1, y_2, y_3)$ such that $y_2 > y_3$ and $y_1 < (y_2 + y_3)/2$ and that if $g \in A$ then $g^* = (g(x_1), [g(x_2) + g(x_3)]/2, [g(x_2) + g(x_3)]/2)$. Similarly, if $g \in B$ then $g^* = ([g(x_1) + g(x_2)]/2, [g(x_1) + g(x_2)]/2, g(x_3))$ and if $g \in C$ then $g^*(x_i) = [g(x_1) + g(x_2) + g(x_3)]/3$ for $i = 1, 2, 3$. Note that if $g \in A$ then g^* is the projection of g onto the subspace $\{\mathbf{y} = (y_1, y_2, y_3); y_2 = y_3\}$ and similar statements can be made for $g \in B$ and $g \in C$. It is also interesting to note that the set

$$\mathscr{I}^* = C \cap \{\mathbf{y} : \sum y_i = 0\}$$

is the set of all points, $\mathbf{y}$, which make an obtuse angle with every point in $\mathscr{I}$ (i.e. the

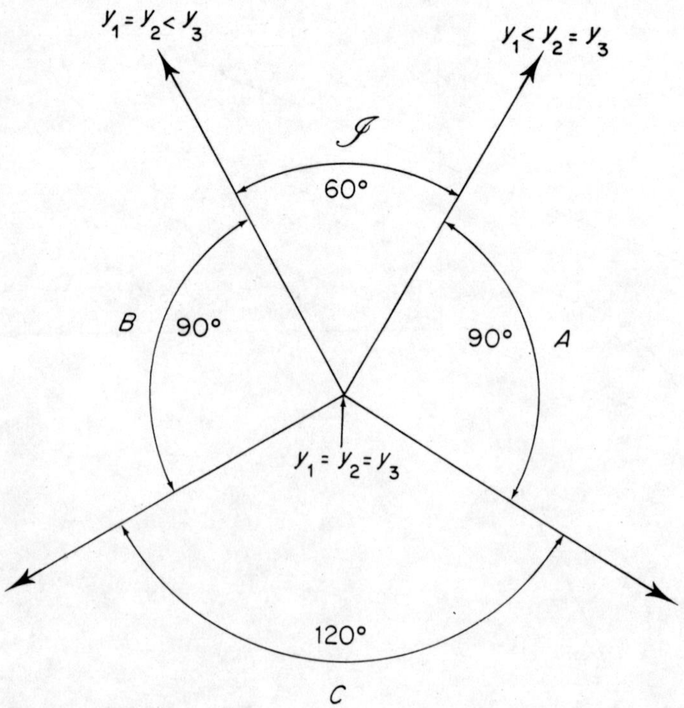

Figure 1.2.2 A view of the isotonic functions looking down the edge

inner product, $\langle \mathbf{x}, \mathbf{y} \rangle$ of $\mathbf{x}$ with $\mathbf{y}$ is nonpositive for all $\mathbf{x} \in \mathscr{I}$). This set is called the *Fenchel dual* or *polar* of $\mathscr{I}$ and will play an important role in the theory developed in this monograph.

The following example was treated in the work by Brunk (1955).

Example 1.2.1 Suppose one has independent random samples from k normal populations with means $\mu(x_1)$, $\mu(x_2), \ldots, \mu(x_k)$ and a common variance σ^2. For the ith sample, let n_i denote the sample size, let Y_{i1}, $Y_{i2}, \ldots, Y_{in_i}$ denote the observations and let $\bar{Y}_i = \sum_{j=1}^{n_i} Y_{ij}/n_i$ denote the sample mean for $i = 1, 2, \ldots, k$. It is well known that the vector $(\bar{Y}_1, \bar{Y}_2, \ldots, \bar{Y}_k)$ is the unrestricted maximum likelihood estimator (MLE) of $\boldsymbol{\mu} = (\mu(x_1), \mu(x_2), \ldots, \mu(x_k))$, the population mean vector. Suppose, however, that it is known that $\mu(x_1) \leqslant \mu(x_2) \leqslant \cdots \leqslant \mu(x_k)$. Then, one may want the estimate to also satisfy this constraint, but because of sampling variability the sample means $\bar{Y}_1, \bar{Y}_2, \ldots, \bar{Y}_k$ may not be ordered in this way. How can we obtain estimates which satisfy this restriction?

Taking the negative of the logarithm of the likelihood function, one sees that the restricted MLE minimizes $\sum_{i=1}^{k} \sum_{j=1}^{n_i} [Y_{ij} - \mu(x_i)]^2$ subject to $\mu(x_1) \leqslant \mu(x_2) \leqslant \cdots \leqslant \mu(x_k)$. By adding and subtracting $\bar{Y}_i$ and squaring, this sum can be rewritten as

$$\sum_{i=1}^{k} \sum_{j=1}^{n_i} (Y_{ij} - \bar{Y}_i)^2 + 2 \sum_{i=1}^{k} \sum_{j=1}^{n_i} (Y_{ij} - \bar{Y}_i)[\bar{Y}_i - \mu(x_i)] + \sum_{i=1}^{k} [\bar{Y}_i - \mu(x_i)]^2 n_i.$$

However, the first sum does not involve the $\mu(x_i)$ and the second sum is zero. Hence, the restricted MLE minimizes $\sum_{i=1}^{k} [\bar{Y}_i - \mu(x_i)]^2 n_i$ subject to $\mu(x_1) \leqslant \cdots \leqslant \mu(x_k)$. If we let $X = \{x_1, x_2, \ldots, x_k\}$ with $x_1 \lesssim x_2 \lesssim \cdots \lesssim x_k$, g and w be defined on X by $g(x_i) = \bar{Y}_i$ and $w(x_i) = n_i$, then it is clear that the restricted MLE of $\boldsymbol{\mu} = (\mu(x_1), \mu(x_2), \ldots, \mu(x_k))$ is provided by the isotonic regression g^* of g with weights w.

Returning to the discussion of isotonic regression on a simply ordered set, for arbitrary k, g, and w, let $l = \min_{x \in X} g(x)$ and $u = \max_{x \in X} g(x)$. If f is any isotonic function, then the function f' defined on X by $f'(x) = [l \vee f(x)] \wedge u$ (with $a \wedge b$ the smaller of a and b and $a \vee b$ the larger of a and b) is also isotonic and, moreover, $|g(x) - f'(x)| \leqslant |g(x) - f(x)|$ for all $x \in X$. Hence,

$$\sum_{x \in X} [g(x) - f'(x)]^2 w(x) \leqslant \sum_{x \in X} [g(x) - f(x)]^2 w(x).$$

Thus, the solution to (1.2.1) satisfies $l \leqslant f(x) \leqslant u$ for all $x \in X$. Again, thinking of these functions as points in R^k and (1.2.1) as a function on R^k, the isotonic regression minimizes a continuous function over a closed and bounded subset of R^k, and hence exists. This argument holds in the more general context studied in Section 1.3. The uniqueness of the isotonic regression is given in Theorem 1.2.1, for a simple order restriction. The remainder of this section is devoted to computation algorithms for g^* for a simple order restriction. The following very elegant algorithm for g^* is due to W.T. Reid (cf. Brunk, 1956).

Graphical representation—greatest convex minorant

Plot the points $P_j = (W_j, G_j)$; $j = 0, 1, 2, \ldots, k$ with $W_j = \sum_{i=1}^{j} w(x_i)$ and $G_j = \sum_{i=1}^{j} g(x_i)w(x_i)$ for $j = 1, 2, \ldots, k$ and $P_0 = (0, 0)$. The plot of these points is called the *cumulative sum diagram* (CSD) for the given function g with weights w. Note that the slope of the segment joining P_{j-1} to P_j is $g(x_j)$; $j = 1, 2, \ldots, k$. Let G^* be the *greatest convex minorant* (GCM) of the CSD on the interval $[0, W_k]$. The value, $G^*(t)$, is the supremum of the values, at t, of all convex functions which lie entirely below the CSD. It is straightforward to see that the function defined in this way is convex on $[0, W_k]$. A graph of the CSD and GCM for a given set of points is given in Figure 1.2.3.

Because of the convexity of G^* on $[0, W_k]$, G^* must be left differentiable at each of the points $W_1, W_2, \ldots, W_k$. Let $g^*(x_i)$ be the left derivative of G^* at W_i for $i = 1, 2, \ldots, k$, and note that if for some i the GCM at W_i lies strictly below the CSD, then the slopes of the GCM to the left and right of W_i are the same. In other words, for $i = 1, 2, \ldots, k-1$,

$$G^*(W_i) < G_i \Rightarrow g^*(x_{i+1}) = g^*(x_i). \tag{1.2.2}$$

For the plot in Figure 1.2.3, $g^*(x_1) = g^*(x_2) < g^*(x_3) = g^*(x_4) = g^*(x_5)$.

Theorem 1.2.1 *If X is simply ordered, the left derivative or left-hand slope, g^*, of the*

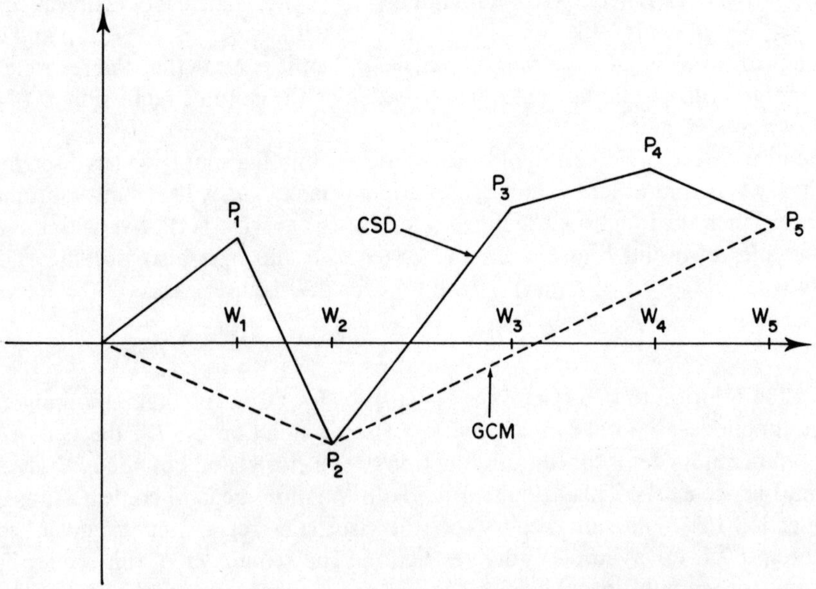

Figure 1.2.3 The cumulative sum diagram (CSD) and the greatest convex minorant (GCM)

GCM furnishes the isotonic regression of g. Indeed, if f is isotonic on X, then

$$\sum_{x\in X}[g(x)-f(x)]^2w(x)\geqslant\sum_{x\in X}[g(x)-g^*(x)]^2w(x)+\sum_{x\in X}[g^*(x)-f(x)]^2w(x). \tag{1.2.3}$$

The isotonic regression is unique.

Proof: Subtracting and adding $g^*(x)$ in the left side of (1.2.3) and squaring the terms, the left-hand side of (1.2.3) is equal to the right-hand side plus twice the sum

$$\sum_{x\in X}[g(x)-g^*(x)][g^*(x)-f(x)]w(x), \tag{1.2.4}$$

so that it suffices to show that the sum in (1.2.4) is nonnegative. Abel's partial summation formula (see Hille, 1959, p. 106) applied to the sum in (1.2.4) yields

$$\begin{aligned}
&\sum_{i=1}^{k}[g(x_i)-g^*(x_i)][g^*(x_i)-f(x_i)]w(x_i)\\
&\quad=\sum_{i=1}^{k}\{[f(x_i)-f(x_{i-1})]-[g^*(x_i)-g^*(x_{i-1})]\}[G_{i-1}-G^*(\mathrm{W}_{i-1})]\\
&\quad\quad+[g^*(x_k)-f(x_k)][G_k-G^*(\mathrm{W}_k)],
\end{aligned}$$

with $x_0=f(x_0)=g^*(x_0)=G_0=G^*(x_0)=0$. The last term is zero since $G^*(\mathrm{W}_k)=G_k$. Now by (1.2.2), $\sum_{i=1}^{k}[g^*(x_i)-g^*(x_{i-1})][G_{i-1}-G^*(\mathrm{W}_{i-1})]=0$ and it follows that the sum is nonnegative since f is isotonic and $G_{i-1}-G^*(\mathrm{W}_{i-1})\geqslant 0$ for all i. To verify uniqueness, suppose that g_1 is another isotonic function on X such that

$$\sum_{x\in X}[g(x)-g_1(x)]^2w(x)\leqslant\sum_{x\in X}[g(x)-f(x)]^2w(x)$$

for all isotonic f. Then since both g_1 and g^* are isotonic

$$\sum_{x\in X}[g(x)-g_1(x)]^2w(x)=\sum_{x\in X}[g(x)-g^*(x)]^2w(x).$$

Letting f be equal to g_1 in (1.2.3) and using the last equality, one sees that $\sum_{x\in X}[g^*(x)-g_1(x)]^2w(x)\leqslant 0$, which implies that $g_1(x)=g^*(x)$ for all $x\in X$. □

The most widely used algorithm for computing the isotonic regression for a simple order is the *pool-adjacent-violators algorithm* (PAVA) first published by Ayer and coworkers in 1955.

The pool-adjacent-violators algorithm (PAVA)

It is clear that if, for some i, $g(x_{i-1})>g(x_i)$ then the graph of the GCM on the interval $[\mathrm{W}_{i-2},\mathrm{W}_i]$ is a straight line segment. Thus if we connect the two points P_{i-2} and P_i by a straight line segment the GCM of this new plot is the same as the

GCM of the original plot. Thus, the GCM can be constructed by a sequence of operations in which one replaces adjacent line segments in the previous step by one line segment if there is a 'violator'. For example, consider the plot given in Figure 1.2.4. The slope of the segment joining P_0 and P_1 is greater than the slope of the segment joining P_1 and P_2 so this constitutes a violation. Hence, these two segments are replaced by the one joining P_0 and P_2. Now the slope of the segment joining P_0 and P_2 is greater than the slope of the segment joining P_2 and P_3, so these two segments are replaced by one segment joining P_0 to P_3. Thus, for this data set, the GCM is determined and consequently so is the isotonic regression.

The isotonic regression, g^*, determines a partition of X into sets of consecutive elements on which it is constant. These sets are called the *level sets* or *solution blocks*. On each of these level sets, the value of g^* is the weighted average of the values of g over the level set using weights w. In other words, the value of g^* for points in the level set is the slope of the side of the GCM corresponding to this block.

The PAVA starts with g. If g is isotonic (i.e. the CSD is convex) then $g^* = g$. Otherwise, there must exist a subscript i such that $g(x_{i-1}) > g(x_i)$. These two values are then replaced by their weighted average, namely $\text{Av}(\{i-1, i\}) = [g(x_{i-1})w(x_{i-1}) + g(x_i)w(x_i)]/[w(x_{i-1}) + w(x_i)]$. The two weights $w(x_{i-1})$ and $w(x_i)$ are replaced by $w(x_{i-1}) + w(x_i)$. If this new set of $k-1$ values is isotonic (i.e. $g(x_1) \leqslant \cdots \leqslant g(x_{i-2}) \leqslant \text{Av}(\{i-1, i\}) \leqslant \cdots \leqslant g(x_k)$), then $g^*(x_{i-1}) = g^*(x_i) = \text{Av}(\{i-1, i\})$ and $g^*(x_j) = g(x_j)$ otherwise. If this new set of values is not isotonic, then this process is repeated using the new values and weights until an isotonic set of values is obtained. The value of $g^*(x_i)$ is the weighted average over the

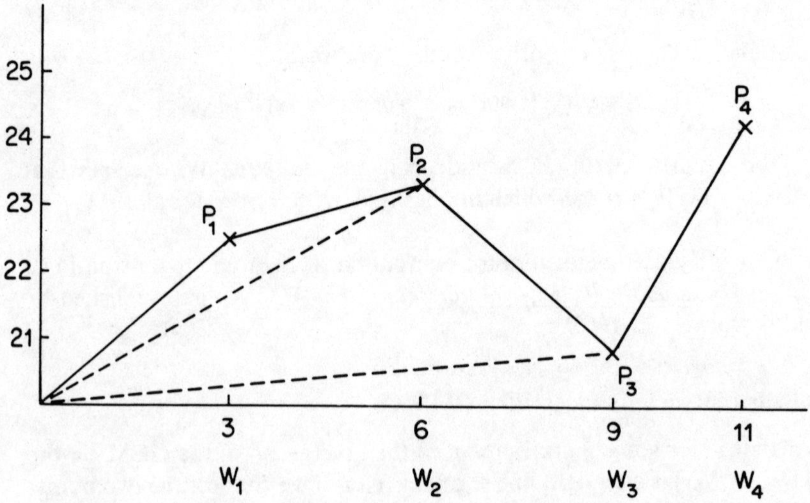

Figure 1.2.4 Graphical interpretation of pooling adjacent violators

block in which x_i is contained. Perhaps the easiest way to visualize this algorithm is to construct a tree as in the following example.

Example 1.2.2 Let $k = 4$, let g be the set of four means in Table 1.1.1, and let w be the corresponding sample sizes (that is $w(x_1) = w(x_2) = w(x_3) = 3$ and $w(x_4) = 2$). We wish to find the isotonic regression g^* of g with weights w. The values at the uppermost branches of the tree in Figure 1.2.5 are the four means from Table 1.1.1. The only violator is between the second and third values and they are both replaced by the average, $(23.333 \times 3 + 20.833 \times 3)/6 = 22.083$. A new violator has been created, i.e. the first value is now larger than the second. Hence, these two are combined. It should be noted that once values are combined in the process, they remain together throughout the process. Therefore these two violators are replaced by the average $(22.5 \times 3 + 22.083 \times 6)/9 = 22.222$. The remaining two values are isotonic so that $g^*(x_i) = g^*(x_2) = g^*(x_3) = 22.222$ and $g^*(x_4) = 24.25$. Note that in the second step the weight on the second value, 22.083, is $w(x_2) + w(x_3) = 6$. The weights accumulate in this fashion through the entire process. Note that during the first step the two means 23.333 and 20.833 are replaced by 22.083 which is equal to the mean of the six pituitary fissure distances corresponding to subjects aged 10 or 12. In other words, 22.083 is the mean of the 'pooled' data. This is the origin of the phrase *pool-adjacent-violators algorithm*. In Example 1.2.1, it was noted that the isotonic regression gives the restricted MLEs for simply ordered normal means. Therefore, if these data represent independent random samples from four normal populations with nondecreasing means and a common variance, then the values obtained by the PAVA are the restricted MLEs.

Two other algorithms, namely the *minimum-lower-sets algorithm* and the *min–max formula*, have been used extensively in deriving properties of the isotonic

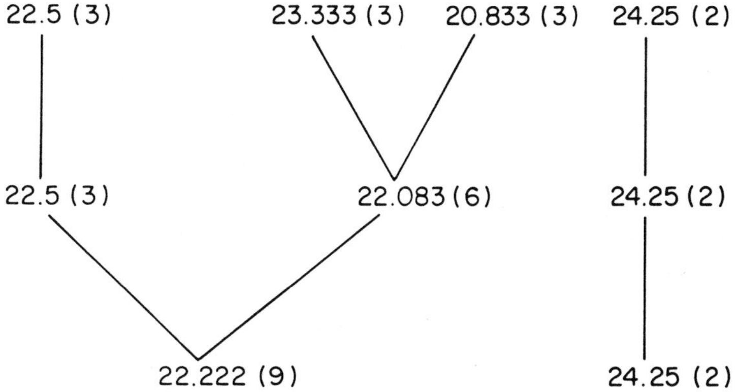

Figure 1.2.5 An example of the PAVA with weights in parentheses

regression. As with the PAVA, they are both straightforward consequences of the graphical representation of the isotonic regression for the simply ordered case. However, these algorithms hold in the more general context discussed in the next section and we will postpone their description until then.

Example 1.2.3 Multidimensional scaling: an application of isotonic regression over a simply ordered set In an interesting experiment carried out by Abelson and Sermat (1962) (see also Shepard, 1962b, p. 245) the stimuli were thirteen photographs of various facial expressions obtained from one actress. The facial expressions were meant to portray thirteen different emotions such as terror, anger, anxiety, maternal love, light sleep, etc. Subjects were presented with two photographs at a time and asked to judge the dissimilarity of the portrayed facial expressions on a nine-step scale. Thus for each of the $\binom{13}{2} = 78$ pairs, (i, j), of facial expression they obtained a dissimilarity index δ_{ij}. The object of multidimensional scaling is to find thirteen points in a low-dimensional Euclidean space which represent those thirteen facial expressions in such a way that if $\delta_{i_1,j_1} \leqslant \delta_{i_2,j_2}$ then the distance between the points representing expressions i_1 and j_1 is no larger than the distance between the points representing expressions i_2 and j_2. It is always possible to find such a set of points if the dimension of the space is sufficiently large. However, the points are a picture of the facial expressions so it is also desirable to find these points in a space which has a low dimension like 2. If one fixes the dimension to be 2, for example, then the technique proposed by Kruskal (1964a), and described in the next paragraph, determines a set of points which may not preserve the ordering, but which hopefully comes close.

Given a candidate set of points in R^2 let d_{ij} be the distance between the point representing the ith expression and the point representing the jth expression. Compute the isotonic regression d^*_{ij} of d_{ij} given the simple ordering which is obtained from the dissimilarity scores δ_{ij}. The 'raw stress' of those thirteen points is defined to be

$$\sum_{i=2}^{13} \sum_{j=1}^{i-1} (d_{ij} - d^*_{ij})^2.$$

Of course, if the points have distances which are ordered in exactly the same way as the dissimilarity scores then $d^*_{ij} = d_{ij}$ and the raw stress is zero. A normalized stress is defined to be

$$\left\{ \frac{\sum_{i=2}^{13} \sum_{j=1}^{i-1} (d_{ij} - d^*_{ij})^2}{\sum_{i=2}^{13} \sum_{j=1}^{i-1} d^2_{ij}} \right\}^{1/2}.$$

Programming techniques are then used to determine n points so as to minimize the normalized stress. Isotonic regression enters the problem only in determining the stress of a given set of points.

1.3 ISOTONIC REGRESSION OVER A QUASI-ORDERED SET

Example 1.3.1 The entries in Table 1.3.1 are the first-year grade point averages (GPA) of 2397 students who entered the University of Iowa in the fall of 1978. The independent variables, in the margins, are the student's composite scores on the ACT Assessment (ACTC) and the student's high school percentile rank (HSR). The entries in the table are the average first-year GPA of the students having the HSR and ACTC scores listed in the margins. The numbers in parentheses are the number of students in that particular category. If one wishes to use the entries in the table to predict the first-year GPA for an incoming student, then there are two problems. First, there are a number of categories with no entries, particularly in the lower right-hand portion of the table. This may not seem like a serious problem because students with a high ACT score and a low HSR are rare. However, they do exist and prediction of their success in college is much more difficult than predicting the success of students whose scores fall near the diagonal. The second problem is the reversals or violators in the table. To elaborate, if A and B are candidates for admission to the university and if A has a lower score than B in both predictor variables, then it would not seem reasonable to predict a higher first-year GPA for A than for B. However, such reversals among the averages do occur in the table in several places. One solution is to find the isotonic regression of the values in the table requiring that the regression be isotonic in both predictor variables. The required order restriction on the predictor variables is not a simple order restriction because it does not require isotonicity between values corresponding to predictor variables which are higher in HSR but lower in ACTC scores or vice versa. The order restriction imposed on the predictor variables is a partial order restriction. The appropriate isotonic regression exists in this setting. Conditions which characterize the isotonic regression for such a partial order will be given in this section, and in fact arbitrary quasi-orders are considered. Other properties of the isotonic regression, which will be useful later, are obtained also. Additional properties of the isotonic regression are given in Section 1.4 and, in a more general context, in Chapter 8.

Definition 1.3.1 A binary relation $\lesssim$ on X is a *simple order* on X if

1. it is *reflexive*: $x \lesssim x$ for $x \in X$;
2. it is *transitive*: $x, y, z \in X$, $x \lesssim y$ and $y \lesssim z$ imply $x \lesssim z$;
3. it is *antisymmetric*: $x, y \in X$, $x \lesssim y$ and $y \lesssim x$ imply $x = y$; and
4. *every two elements of X are comparable*: $x, y \in X$ implies that either $x \lesssim y$ or $y \lesssim x$.

A binary relation $\lesssim$ on X is a *partial order* if it is reflexive, transitive, and antisymmetic, but there may be noncomparable elements. A *quasi-order* is reflexive and transitive. It need not be antisymmetric and it may admit

Table 1.3.1 First-year GPA of freshmen entering the University of Iowa as freshmen in the fall of 1978

$91 \leqslant HSR \leqslant 99$	1.57 (4)	2.11 (5)	2.73 (18)	2.96 (39)	2.97 (126)	3.13 (219)	3.41 (232)	3.45 (47)	3.51 (4)
$81 \leqslant HSR \leqslant 90$	1.80 (6)	1.94 (15)	2.52 (30)	2.68 (65)	2.69 (117)	2.82 (143)	2.75 (70)	2.74 (8)	(0)
$71 \leqslant HSR \leqslant 80$	1.88 (10)	2.32 (13)	2.32 (51)	2.53 (83)	2.58 (115)	2.55 (107)	2.72 (24)	2.76 (4)	(0)
$61 \leqslant HSR \leqslant 70$	2.11 (6)	2.23 (32)	2.29 (59)	2.29 (84)	2.50 (75)	2.42 (44)	2.41 (19)	(0)	(0)
$51 \leqslant HSR \leqslant 60$	1.60 (11)	2.06 (16)	2.12 (49)	2.11 (63)	2.31 (57)	2.10 (40)	1.58 (4)	2.13 (1)	(0)
$41 \leqslant HSR \leqslant 50$	1.75 (6)	1.98 (12)	2.05 (31)	2.16 (42)	2.35 (34)	2.48 (21)	1.36 (4)	(0)	(0)
$31 \leqslant HSR \leqslant 40$	1.92 (7)	1.84 (6)	2.15 (5)	1.95 (27)	2.02 (13)	2.10 (13)	1.49 (2)	(0)	(0)
$21 \leqslant HSR \leqslant 30$	1.62 (1)	2.26 (2)	1.91 (5)	1.86 (14)	1.88 (11)	3.78 (1)	1.40 (2)	(0)	(0)
$HSR \leqslant 20$	1.38 (1)	1.57 (2)	2.49 (5)	2.01 (7)	2.07 (7)	(0)	0.75 (1)	(0)	(0)
High school percentile rank	ACTC 1–12	13–15	16–18	19–21	22–24	25–27	28–30	31–33	34–36

noncomparable elements. Every simple order is a partial order and every partial order is a quasi-order.

Definition 1.3.2 A real valued function, f, on X is *isotonic* with respect to the quasi-ordering $\lesssim$ on X if $x, y \in X$ and $x \lesssim y$ imply $f(x) \leqslant f(y)$.

Definition 1.3.3 Let g be a given function on X and w a given positive function on X. An isotonic function g^* on X is an *isotonic regression* of g with weights w if and only if

$$\sum_{x \in X} [g(x) - g^*(x)]^2 w(x) \leqslant \sum_{x \in X} [g(x) - f(x)]^2 w(x)$$

for all functions f on X which are isotonic.

Example 1.3.1(continued) Let i index the ACTC score and let j index the HSR so that $X = \{(i, j);\ i = 1, 2, \ldots, 9,\ j = 1, 2, \ldots, 9\}$. Thus, for example, the pair $(1, 1)$ indexes the category of individuals who scored between 1 and 12 on the ACT and achieved a HSR no higher than 20. Define the binary relation $\lesssim$ on X by $(i, j) \lesssim (k, l)$ if and only if $i \leqslant k$ and $j \leqslant l$. It is easy to verify that $\lesssim$ is a partial order on X, but it is not a simple order because there are noncomparable pairs such as $(1, 2)$ and $(2, 1)$. Define the functions $g(i, j)$ and $w(i, j)$ on X to be the average first-year GPA and number of students for the category corresponding to the pair (i, j). An isotonic regression g^* of g with weights w would provide a function on X which is nondecreasing in both predictor variables. The values in the lower right corner of Table 1.3.1 with zero weights do cause some difficulties. There have been a number of solutions posed for this problem, including computing the isotonic regression on the reduced table obtained by deleting zero weight values from X, and then filling in the reduced table in an isotonic way. This question will be discussed after the discussion of computation algorithms for g^*. An isotonic

Table 1.3.2 Least squares doubly nondecreasing regression function with the same categories as in Table 1.3.1

1.87	2.17	2.73	2.96	2.97	3.13	3.41	3.45	3.51
1.87	2.17	2.52	2.68	2.69	2.79	2.79	2.79	2.79[a]
1.87	2.17	2.32	2.53	2.57	2.57	2.72	2.76	2.76[a]
1.87	2.17	2.29	2.29	2.46	2.46	2.46	2.46[a]	2.46[a]
1.73	2.06	2.12	2.13	2.25	2.25	2.25	2.25	2.25[a]
1.73	1.98	2.05	2.13	2.25	2.25	2.25	2.25[a]	2.25[a]
1.73	1.94	1.98	1.98	2.02	2.05	2.05	2.05[a]	2.05[a]
1.62	1.94	1.96	1.96	1.96	2.05	2.05	2.05[a]	2.05[a]
1.38	1.57	1.96	1.96	1.96	1.96	1.96	1.96[a]	1.96[a]

[a] Not uniquely determined.

regression of g with weight function, w, is given in Table 1.3.2. It is interesting to note that the constant values in the table (they are blocked off) give indications of 'nearly equivalent' pairs of values of HSR and ACT scores.

Returning to the general problem, the existence of an isotonic regression is proved in the corollary to Theorem 1.3.4 and in a more general context in Chapter 8 (see the corollary to Theorem 8.2.1). Theorem 1.3.2 states that the isotonic regression is unique, and gives a necessary and sufficient condition for an isotonic function to be the isotonic regression of a given function g with weights w.

It is useful to think of the collection, $\mathscr{I}$, of isotonic functions as a subset of a Euclidean space which has as its dimension the number of points in X. In this setting the isotonic regression, g^*, is the closest point of $\mathscr{I}$ to g with distances measured by the weighted Euclidean distance $d_w(\cdot,\cdot)$ given by

$$d_w(g,f)^2 = \sum_{x\in X} [g(x) - f(x)]^2 w(x). \tag{1.3.1}$$

The set $\mathscr{I}$ of isotonic functions has a number of interesting properties only some of which are needed to prove the theorems in this chapter. $\mathscr{I}$ is a *closed convex cone* which is also a *lattice*. Specifically,

1. $\mathscr{I}$ is closed in the topology induced by the metric given by (1.3.1);
2. $\mathscr{I}$ is *convex*: if $f, g\in\mathscr{I}$ and $0 \leqslant \alpha \leqslant 1$, then $\alpha f + (1-\alpha)g \in \mathscr{I}$;
3. $\mathscr{I}$ is a *cone*: if $f\in\mathscr{I}$ and $a \geqslant 0$ then $af\in\mathscr{I}$; and
4. $\mathscr{I}$ is a *lattice*: if $f, g\in\mathscr{I}$, then the functions $f\vee g$ and $f\wedge g$ defined on X by $(f\vee g)(x) = f(x)\vee g(x)$ and $(f\wedge g)(x) = f(x)\wedge g(x)$ are also in $\mathscr{I}$.

Theorem 1.3.1 *Suppose $\mathscr{C}$ is any convex set of functions on X and g and w are given functions on X with $w(x) > 0$ for all $x\in X$. If $g^*\in\mathscr{C}$ and g^* solves*

$$\textit{minimize} \sum_{x\in X} [g(x) - f(x)]^2 w(x) \textit{ subject to } f\in\mathscr{C}, \tag{1.3.2}$$

then for every $f\in\mathscr{C}$,

$$\sum_{x\in X} [g(x) - g^*(x)][g^*(x) - f(x)]w(x) \geqslant 0 \tag{1.3.3}$$

and

$$\sum_{x\in X} [g(x) - f(x)]^2 w(x) \geqslant \sum_{x\in X} [g(x) - g^*(x)]^2 w(x) + \sum_{x\in X} [g^*(x) - f(x)]^2 w(x). \tag{1.3.4}$$

Conversely, if $u\in\mathscr{C}$ and

$$\sum_{x\in X} [g(x) - u(x)][u(x) - f(x)]w(x) \geqslant 0 \tag{1.3.5}$$

for all $f\in\mathscr{C}$ then u solves (1.3.2). There is at most one such function.

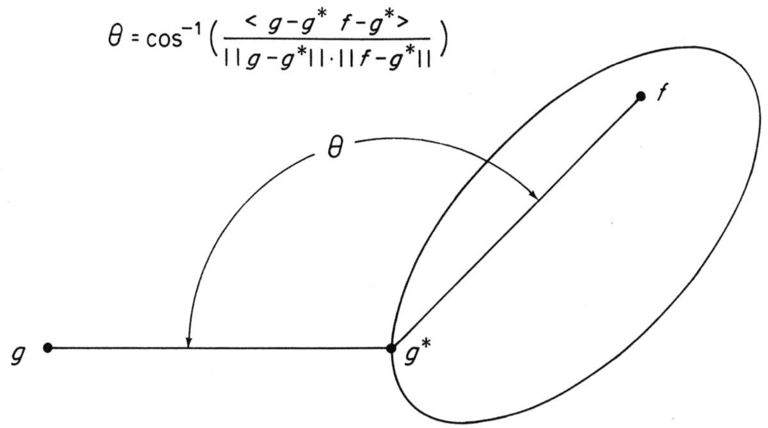

Figure 1.3.1 Projection onto a closed, convex set

Thinking of functions on X as points in a Euclidean space with metric given by (1.3.1), the expression

$$\sum_{x \in X} u(x)\, v(x)\, w(x)$$

is positive or negative according as the 'angle' between u and v is acute or obtuse. Theorem 1.3.1 says that $g^* \in \mathscr{C}$ solves (1.3.2) (i.e. g^* is the closest point of $\mathscr{C}$ to g) if and only if the angle between $g - g^*$ and $f - g^*$ is obtuse for each $f \in \mathscr{C}$ (see Figure 1.3.1).

Proof of Theorem 1.3.1: First suppose g^* solves (1.3.2) and that f is an arbitrary member of $\mathscr{C}$. Then since $\mathscr{C}$ is convex $(1-\alpha)g^* + \alpha f \in \mathscr{C}$ for $0 \leqslant \alpha \leqslant 1$ and since g^* solves (1.3.2),

$$\sum_{x \in X} \{g(x) - [(1-\alpha)g^*(x) + \alpha f(x)]\}^2 w(x) \tag{1.3.6}$$

must assume its minimum as a function of α at $\alpha = 0$. However, (1.3.6) is a quadratic in α so that its derivative at $\alpha = 0$ must be nonnegative. Its derivative at $\alpha = 0$ is just twice the left-hand side of (1.3.3). This proves (1.3.3), and (1.3.4) follows by adding and subtracting $g^*(x)$ in the left-hand side of (1.3.4) and squaring the terms.

Conversely, suppose $u \in \mathscr{C}$ and u satisfied (1.3.5). Then for any $f \in \mathscr{C}$

$$\begin{aligned}
&\sum_{x \in X} [g(x) - f(x)]^2 w(x) \\
&\quad \geqslant \sum_{x \in X} [g(x) - u(x)]^2 w(x) + \sum_{x \in X} [u(x) - f(x)]^2 w(x) \\
&\quad \geqslant \sum_{x \in X} [g(x) - u(x)]^2 w(x)
\end{aligned}$$

so that u solves (1.3.2).

In order to complete the proof suppose u_1 and u_2 are two functions in $\mathscr{C}$ satisfying (1.3.5). Then

$$\sum_{x \in X} [g(x) - u_1(x)][u_1(x) - u_2(x)]w(x) \geqslant 0$$

and

$$\sum_{x \in X} [g(x) - u_2(x)][u_2(x) - u_1(x)]w(x) \geqslant 0.$$

Adding the left sides of these inequalities one obtains

$$-\sum_{x \in X} [u_1(x) - u_2(x)]^2 w(x) \geqslant 0$$

which implies that $u_1 = u_2$. □

If the closed convex set is also a cone, as in the case of the collection of all functions isotonic with respect to a quasi-order, then the conditions given in the next result characterize the projection.

Theorem 1.3.2 *If $\mathscr{K}$ is a convex cone of functions on X and g and w are given functions on X with $w > 0$ then a function g^* on X solves (1.3.2) if and only if $g^* \in \mathscr{K}$ and*

$$\sum_{x \in X} [g(x) - g^*(x)]g^*(x)w(x) = 0 \tag{1.3.7}$$

and

$$\sum_{x \in X} [g(x) - g^*(x)]f(x)w(x) \leqslant 0 \qquad \textit{for all } f \in \mathscr{K}. \tag{1.3.8}$$

Proof: If $g^* \in \mathscr{K}$ and g^* satisfies (1.3.7) and (1.3.8), then by subtracting the left sides of those expressions g^* satisfies (1.3.5), and by Theorem 1.3.1, g^* solves (1.3.2). Conversely, suppose g^* solves (1.3.2). Then by setting $f(x) = cg^*(x)$ in (1.3.3) first with $c > 1$ and then with $0 < c < 1$, one sees that g^* satisfies (1.3.7). Using this property together with (1.3.3), g^* satisfies (1.3.8). □

Note that any constant function on X is trivially isotonic with respect to any quasi-order on X.

Theorem 1.3.3 *If $\mathscr{K}$ is any convex cone of functions on X and if $\mathscr{K}$ contains all the constant functions on X and if g^* solves (1.3.2) then*

$$\sum_{x \in X} g(x)w(x) = \sum_{x \in X} g^*(x)w(x). \tag{1.3.9}$$

Proof: This follows easily by first letting $f(x) \equiv 1$ and then letting $f(x) \equiv -1$ in (1.3.8). □

Theorem 1.3.4 *If g_1 and g_2 are isotonic functions on X such that $g_1(x) \leqslant g(x) \leqslant g_2(x)$ for $x \in X$, and if g^* is an isotonic regression of g, then also*

$g_1(x) \leqslant g^*(x) \leqslant g_2(x)$ for $x \in X$. In particular, if a and b are constants such that $a \leqslant g(x) \leqslant b$ for $x \in X$, then also $a \leqslant g^*(x) \leqslant b$ for $x \in X$.

Proof: Suppose g^* is the isotonic regression of g and that there exist $x_0 \in X$ such that $g^*(x_0) < g_1(x_0)$. Define the function g_1^* on X by $g_1^*(x) = g^*(x) \vee g_1(x)$ for $x \in X$. Then g_1^* is isotonic and $|g_1^*(x) - g(x)| \leqslant |g^*(x) - g(x)|$ for all $x \in X$ with $|g_1^*(x_0) - g(x_0)| < |g^*(x_0) - g(x_0)|$. Thus,

$$\sum_{x \in X} [g_1^*(x) - g(x)]^2 w(x) \leqslant \sum_{x \in X} [g^*(x) - g(x)]^2 w(x),$$

a contradiction, so that $g^*(x) \geqslant g_1(x)$ for $x \in X$. The argument that $g^*(x) \leqslant g_2(x)$ for $x \in X$ is similar using g_2^*, where $g_2^* = g^* \wedge g_2(x)$, rather than g_1^*. □

Corollary *An isotonic regression of g exists.*

Proof: From Theorem 1.3.4 the isotonic regression must belong to the set of all isotonic functions bounded below by $\min_{x \in X} g(x)$ and above by $\max_{x \in X} g(x)$. Thinking of functions on X as points in a Euclidean space and thinking of $\sum_{x \in X} [g(x) - f(x)]^2 w(x)$ as a continuous function on that closed, bounded subset of this Euclidean space, it must achieve its minimum value. □

Suppose g and w are functions defined on X, set

$$\mathrm{Av}(A) = \frac{\sum_{x \in A} w(x) g(x)}{\sum_{x \in A} w(x)}$$

for those A which are nonempty subsets of X, and let $[g^* = c]$ denote $\{x \in X : g^*(x) = c\}$. While $\mathrm{Av}(A)$ depends on g, this will not be made explicit in the notation.

Theorem 1.3.5 *If c is any real number and if the set $[g^* = c]$ is nonempty then $c = \mathrm{Av}([g^* = c])$.*

Proof: Clearly,

$$\sum_{x \in X} [g(x) - g^*(x)]^2 w(x) = \sum_{[g^* \neq c]} [g(x) - g^*(x)]^2 w(x) + \sum_{[g^* = c]} [g(x) - c]^2 w(x).$$

The expression $\sum_{[g^* = c]} [g(x) - t]^2 w(x)$ is a quadratic in t and has an absolute minimum at $t = \mathrm{Av}([g^* = c])$. If $c \neq \mathrm{Av}([g^* = c])$ then one could form a new function which is equal to g^* on $[g^* \neq c]$ and is a little closer to $\mathrm{Av}([g^* = c])$ on $[g^* = c]$. Clearly since X is finite if the new function does not differ too much from c on $[g^* = c]$, then it will also be isotonic on X. This new function will give a smaller sum of squares than g^* does, which is a contradiction. This contradiction establishes Theorem 1.3.5. □

Theorem 1.3.5 reduces the problem of computing g^* to finding the sets on which g^* is constant (i.e. its level sets). For a simple order the sides of the GCM determine the level sets. They are also determined for a simple order by looking at the roots of the tree constructed in the PAVA. For a partial or quasi-order the problem of determining the level sets is much more complex and algorithms for the computation of the isotonic regression will be given later in the chapter.

Example 1.3.2 (The simple tree) An important partial order is the simple tree. It arises in sampling situations where one wishes to compare several treatments with a control or standard making use of prior information that all of the treatment means are at least as large as the control mean. (The case in which all of the treatment means are no larger than the control mean is included in the theory by changing the signs of all the means.) For example, in a drug study, several drugs may be compared to a zero-dose control.

Let $X = \{x_0, x_1, x_2, \ldots, x_k\}$ and define the partial order $\lesssim$ on X by $x_0 \lesssim x_i$; $i = 1, 2, \ldots, k$ with no relationship between x_i and x_j for $i, j \geqslant 1$. This partial order restriction is called the *simple tree*. A function f on X is isotonic (with respect to this partial order) if and only if $f(x_i) \geqslant f(x_0)$; $i = 1, 2, \ldots, k$. We now describe an algorithm for computing the isotonic regression g^* of a given function g with weights $w(w > 0)$. If $g(x_0) \leqslant g(x_i)$; $i = 1, 2, \ldots, k$, then $g^* = g$. Otherwise, arrange the values $g(x_1), g(x_2), \ldots, g(x_n)$ in increasing order ($g(x_0)$ is not to be included). Denote these values by $g(x_{(1)}) \leqslant g(x_{(2)}) \leqslant \cdots \leqslant g(x_{(k)})$ and let $w(x_{(i)})$ denote the weight corresponding to $g(x_{(i)})$ for $i = 1, 2, \ldots, k$. Next, find the smallest positive integer j for which

$$A_j = \frac{w(x_0)g(x_0) + \sum_{i=1}^{j} w(x_{(i)})g(x_{(i)})}{w(x_0) + \sum_{i=1}^{j} w(x_{(i)})} < g(x_{(j+1)}). \tag{1.3.10}$$

Such an integer will exist unless $A_{k-1} \geqslant g(x_{(k)})$ and in this case set $j = k$. Now $g^*(x_0) = A_j$ and the value of $g^*(x_i)$ is either A_j or $g(x_i)$ depending upon whether $g(x_i)$ is included in $\sum_{i=1}^{j} w(x_{(i)})g(x_{(i)})$ or not.

The function g^* constructed in the preceding paragraph is shown to be the isotonic regression of g by verifying the three properties given in Theorem 1.3.2. However, another property of the collection, $\mathscr{I}$, of isotonic functions on X needs to be established first. Define the $k+2$ functions, $e_1, e_2, \ldots, e_{k+2}$, on X by

$$e_i(x_i) = 1 \qquad \text{and} \qquad e_i(x_j) = 0 \qquad \text{for } j \in \{0, 1, \ldots, k\} - \{i\}$$

for $i = 1, 2, \ldots, k$, and $e_{k+1} = -e_{k+2}$ are defined by $e_{k+1}(x_j) = 1$ for $j = 0, 1, 2, \ldots, k$. Note that if $f \in \mathscr{I}$, then f can be written as a nonnegative linear combination of $e_1, e_2, \ldots, e_{k+2}$. Specifically, if $f(x_0) < 0$, then for any $x \in X$,

$f(x)=(-f(x_0))e_{k+2}(x)+\sum_{i=1}^{k}[f(x_i)-f(x_0)]e_i(x)$ and if $f(x_0)\geq 0$, then $f(x)=f(x_0)e_{k+1}(x)+\sum_{i=1}^{k}[f(x_i)-f(x_0)]e_i(x)$. A cone with this property (i.e. there exists a finite collection of elements such that every element in the cone is a nonnegative linear combination of these elements) is said to be *finitely generated* and the elements of the finite collection are called *generators*. It is well known that a convex cone is finitely generated if and only if it can be written as a finite intersection of closed half-spaces (i.e. it is *polyhedral*; cf. Rockafellar, 1970, pp. 171). In summary, every function f in $\mathscr{I}$ can be written $f=\sum_{i=1}^{k+2}\alpha_i e_i$ with $\alpha_i\geq 0$ for $i=1,2,\ldots,k+2$.

We verify that the function g^* described above is the isotonic regression of g with respect to the simple tree by verifying the properties of Theorem 1.3.2. The generators are used in verifying (1.3.8). First note that $g^*\in\mathscr{I}$. Now, let B be the subset of X consisting of all x such that $g^*(x)=A_j$ where j is the smallest positive integer such that (1.3.10) holds. In order to verify (1.3.7), note that $g^*(x)=A_j$ for all $x\in B$ and $g^*(x)=g(x)$ for $x\notin B$. The next observation is used to establish (1.3.7) and to show that (1.3.8) holds for f, a generator of $\mathscr{I}$. For any f which is constant on B, left f_B denote its value on B and note that

$$\sum_{x\in X}[g(x)-g^*(x)]f(x)w(x)=\sum_{x\in B}[g(x)-A_j]f(x)w(x)$$

$$=f_B\left[\sum_{x\in B}g(x)w(x)-A_j\sum_{x\in B}w(x)\right]=0$$

by (1.3.10). Because g^* is constant on B, (1.3.7) follows from this observation. Inequality (1.3.8) will first be established for f equal to the generators of $\mathscr{I}$. For e_{k+1} and e_{k+2} the desired conclusion follows from the last observation because they are constant on B. Next consider e_i with $i\leq k$. From the definition of e_i,

$$\sum_{x\in X}[g(x)-g^*(x)]e_i(x)w(x)=[g(x_i)-g^*(x_i)]w(x_i). \tag{1.3.11}$$

This is zero if $x_i\notin B$. If $x_i\in B$, then note that from the way A_j is defined $A_{j-1}\geq g(x_{(j)})$, or otherwise j would not have been the smallest positive integer having the given property. However, A_j is the weighted average of A_{j-1} and $g(x_{(j)})$ so that $A_{j-1}\geq A_j\geq g(x_{(j)})\geq\cdots\geq g(x_{(1)})$. Thus $g(x_i)\leq A_j=g^*(x_i)$ if $x_i\in B$ and $x_i\neq x_0$. Therefore (1.3.11) is nonpositive for $i=1,2,\ldots,k+2$. Now suppose f is an arbitrary member of $\mathscr{I}$. Then since the e_i generate $\mathscr{I}$, $f=\sum_{j=1}^{k+2}\alpha_j e_j$ and

$$\sum_{x\in X}[g(x)-g^*(x)]f(x)w(x)=\sum_{i=0}^{k}[g(x_i)-g^*(x_i)]f(x_i)w(x_i)$$

$$=\sum_{i=0}^{k}\sum_{j=1}^{k+2}\alpha_j[g(x_i)-g^*(x_i)]e_j(x_i)w(x_i)$$

$$=\sum_{j=1}^{k+2}\alpha_j\sum_{i=0}^{k}[g(x_i)-g^*(x_i)]e_j(x_i)w(x_i)$$

which is nonpositive since α_j and (1.3.11) are nonpositive. Thus by Theorem 1.3.2 the function constructed above must be the isotonic regression of g with weights w.

Theorem 1.3.6 *For an arbitrary real valued function, Ψ, defined on the reals,*

$$\sum_{x \in X} [g(x) - g^*(x)]\Psi[g^*(x)]w(x) = 0.$$

Proof: By Theorem 1.3.5, for any value c in the range of g^*,

$$\sum_{[g^* = c]} [g(x) - c]w(x) = 0.$$

The desired conclusion follows from the equality

$$\sum_{x \in X} [g(x) - g^*(x)]\Psi[g^*(x)]w(x) = \sum_{c} \Psi(c) \sum_{[g^* = c]} [g(x) - c]w(x). \quad \square$$

1.4 COMPUTATION ALGORITHMS

The calculation of g^*, given g, the weights w, and the quasi-order on X, can be accomplished via quadratic programming. An extensive literature on methods of computing quadratic programming solutions for such problems exists. The problem of computing the isotonic regression is special and a number of algorithms have been proposed for this specific problem. The graphical representation and the PAVA discussed in Section 1.2 are very elegant algorithms, but they apply only to simple order restrictions. The algorithms presented here all involve averaging g over suitably selected subsets of X; the term *amalgamation of means* has frequently been used in connection with such computations.

Suppose the function g, the weights w, and the quasi-order, $\lesssim$, are given on X.

Definition 1.4.1 A subset L of X is a *lower set* with respect to the quasi-order $\lesssim$ if $y \in L$, $x \in X$, and $x \lesssim y$ imply $x \in L$. A subset U of X is an *upper set* if $x \in U$, $y \in X$, and $x \lesssim y$ imply $y \in U$. We denote the class of all lower sets by $\mathscr{L}$ and the class of all upper sets by $\mathscr{U}$.

Note that each of the classes $\mathscr{L}$ and $\mathscr{U}$ is closed under arbitrary unions and intersections and both contain the empty set ϕ and the set X. A set U is an upper set if and only if its complement, U^c, is a lower set. The classes $\mathscr{L}$ and $\mathscr{U}$ and the collection, $\mathscr{I}$, of all isotonic functions on X are related by the following result whose proof is a straightforward exercise. Extending the notation being used, $[f < a]$ represents $\{x \in X: f(x) < a\}$ for f defined on X and real numbers a.

Theorem 1.4.1 *A function f on X is isotonic if and only if any one of the following conditions is satisfied:*

1. $[f < a] \in \mathscr{L}$ *for every real* a;
2. $[f \leqslant a] \in \mathscr{L}$ *for every real* a;
3. $[f > a] \in \mathscr{U}$ *for every real* a; *or*
4. $[f \geqslant a] \in \mathscr{U}$ *for every real* a.

It is also straightforward to verify that a subset U of X is an upper set if and only if its indicator function, I_U, is isotonic. A subset L of X is a lower set if and only if $-I_L$ is isotonic.

Definition 1.4.2 A subset B of X is a *level set* if and only if there exists a lower set L and an upper set U such that $B = L \cap U$.

The following result is easily verified.

Theorem 1.4.2 *A subset B of X is a level set if and only if there exists an isotonic function f on X and a real number a such that $B = [f = a]$.*

A key result for the justification of the first two algorithms given in this section is the following property of the isotonic regression. Suppose that g and w are fixed and recall that $\mathrm{Av}(B)$ denotes the weighted average of g over the nonempty subset B of X.

Theorem 1.4.3 *If g^* is the isotonic regression of g then for each real number a,*

$$\mathrm{Av}(L \cap [g^* \geqslant a]) \geqslant a,$$
$$\mathrm{Av}(L \cap [g^* > a]) > a,$$
$$\mathrm{Av}(U \cap [g^* \leqslant a]) \leqslant a,$$

and

$$\mathrm{Av}(U \cap [g^* < a]) < a,$$

for any $L \in \mathscr{L}$ or $U \in \mathscr{U}$ such that the set over which the weighted average is taken is not empty.

Proof: Note first that by Theorem 1.3.6 with Ψ being the indicator function of the open interval (a, b), one has, for $-\infty \leqslant a < b \leqslant \infty$,

$$\sum_{x \in [a < g^* < b]} [g(x) - g^*(x)]w(x) = 0. \tag{1.4.1}$$

Further, if $U \in \mathscr{U}$ then

$$\sum_{x \in U} [g(x) - g^*(x)]w(x) = \sum_{x \in X} [g(x) - g^*(x)]I_U(x)w(x) \leqslant 0 \tag{1.4.2}$$

by (1.3.8) since I_U is isotonic. Now suppose a is real, $L \in \mathscr{L}$, and $L \cap [g^* > a] \neq \phi$. Then

$$\sum_{x \in L \cap [g^* > a]} [g(x) - a]w(x) > \sum_{x \in L \cap [g^* > a]} [g(x) - g^*(x)]w(x)$$
$$= \sum_{x \in [g^* > a]} [g(x) - g^*(x)]w(x)$$
$$- \sum_{x \in L^c \cap [g^* > a]} [g(x) - g^*(x)]w(x) \geqslant 0$$

since the first sum is zero by (1.4.1) and the second sum is nonpositive by (1.4.2). Thus

$$\sum_{x \in L \cap [g^* > a]} g(x)w(x) > a \sum_{x \in L \cap [g^* > a]} w(x)$$

which is equivalent to the second conclusion. The other parts of the theorem are established similarly. □

To simplify the notation, throughout the remainder of the section L and U will only be used to denote lower sets and upper sets respectively.

The max–min formulas for isotonic regression

Theorem 1.4.4 *The isotonic regression of g is given by*

$$g^*(x) = \max_{U:x \in U} \min_{L:x \in L} \text{Av}(L \cap U)$$
$$= \max_{U:x \in U} \min_{L:L \cap U \neq \phi} \text{Av}(L \cap U)$$
$$= \min_{L:x \in L} \max_{U:x \in U} \text{Av}(L \cap U)$$
$$= \min_{L:x \in L} \max_{U:L \cap U \neq \phi} \text{Av}(L \cap U).$$

Proof: Suppose $a = g^*(x)$. Then by Theorem 1.4.3, for L with $L \cap [g^* \geqslant a] \neq \phi$, $\text{Av}(L \cap [g^* \geqslant a]) \geqslant a$. Thus, letting $U_a = [g^* \geqslant a]$, $\min_{L:L \cap U_a \neq \phi} \text{Av}(L \cap U_a) \geqslant a$. However, $[g^* \leqslant a] \in \mathscr{L}$ and $[g^* \leqslant a] \cap [g^* \geqslant a] \neq \phi$ so that $\text{Av}([g^* \leqslant a] \cap [g^* \geqslant a]) = \text{Av}([g^* = a]) = a$ by Theorem 1.3.5. Thus, $\min_{L:L \cap U_a \neq \phi} \text{Av}(L \cap U_a) = a$, which implies that

$$\max_{U:x \in U} \min_{L:L \cap U \neq \phi} \text{Av}(L \cap U) \geqslant a.$$

On the other hand, if $x \in U$, then $[g^* \leqslant a] \cap U \neq \phi$ and $\text{Av}([g^* \leqslant a] \cap U) \leqslant a$ by Theorem 1.4.3. Hence, $\min_{L:L \cap U \neq \phi} \text{Av}(L \cap U) \leqslant a$ and since this is true for all U such that $x \in U$, $\max_{U:x \in U} \min_{L:L \cap U \neq \phi} \text{Av}(L \cap U) \leqslant a$. The second conclusion has been proved. The other parts of the theorem are argued similarly. □

In order to illustrate this theorem let $X = \{x_1, x_2, \ldots, x_k\}$ with $\lesssim$ defined on X by $x_1 \lesssim x_2 \lesssim \cdots \lesssim x_k$ (i.e. a simple order). Then, there are exactly $k+1$ lower sets and exactly $k+1$ upper sets. Of course, the empty set, ϕ, and the set X are both lower and upper sets. The other lower sets are sets of the form $\{x_1, x_2, \ldots, x_i\}$; $i = 1, 2, \ldots, k-1$, and the other upper sets are sets of the form $\{x_i, x_{i+1}, \ldots, x_k\}$; $i = 2, \ldots, k$. For this simple order Theorem 1.4.4 yields

$$\begin{aligned} g^*(x_i) &= \max_{j:j\leqslant i}\min_{h:h\geqslant i} \quad \mathrm{Av}(\{x_j, x_{j+1}, \ldots, x_h\}) \\ &= \max_{j:j\leqslant i}\min_{h:h\geqslant j} \quad \mathrm{Av}(\{x_j, x_{j+1}, \ldots, x_h\}) \\ &= \min_{h:h\geqslant i}\max_{j:j\leqslant i} \quad \mathrm{Av}(\{x_j, x_{j+1}, \ldots, x_h\}) \\ &= \min_{h:h\geqslant i}\max_{j:j\leqslant h} \quad \mathrm{Av}(\{x_j, x_{j+1}, \ldots, x_h\}). \end{aligned} \tag{1.4.3}$$

If $\lesssim$ is an arbitrary quasi-order on X and w_n is a sequence of weight functions and g_n is a sequence of functions defined on X which satisfy $w_n(x_i) \to w(x_i) > 0$ and $g_n(x_i) \to g(x_i)$ for $i = 1, 2, \ldots, k$, then using the expressions in Theorem 1.4.4, g_n^* converges pointwise to g^*. This establishes the following result.

Corollary *The isotonic regression viewed as a function on the 2k-tuples of real numbers $(w_1, w_2, \ldots, w_k, g_1, g_2, \ldots, g_k)$ with $w_i > 0$ for $i = 1, 2, \ldots, k$ is continuous. Furthermore, with $w > 0$ fixed, the isotonic regression viewed as a mapping from R^k into R^k is continuous.*

Minimum lower sets algorithm

Definition 1.4.3 A function M defined on the nonempty subsets of X which has the property that $M(A+B)$ is between $M(A)$ and $M(B)$ whenever A and B are disjoint nonempty subsets of X is called a *Cauchy mean value function* (note that the notation $A+B$ is used for $A \cup B$ whenever A and B are disjoint). A *strict Cauchy mean value function* has the additional property that $M(A+B)$ is strictly between $M(A)$ and $M(B)$ whenever $M(A) \neq M(B)$.

Note that the weighted average, Av, which is used in the algorithms in this chapter is a strict Cauchy mean value function. A strict Cauchy mean value function, M, has the property that if $M(A+B) \leqslant M(A)$ then $M(B) \leqslant M(A+B)$ and if $M(A+B) \geqslant M(A)$ then $M(B) \geqslant M(A+B)$. The *minimum lower sets algorithm* for isotonic regression is described next.

Select a lower set L_1 such that $\mathrm{Av}(L_1) \leqslant \mathrm{Av}(L)$ for all L. Suppose L_1' is another lower set having this property. Then $L_1 = (L_1 \cap L_1') + (L_1 - L_1')$ and $L_1 \cap L_1'$ is a lower set. Thus, since Av is a strict Cauchy mean value function and since $\mathrm{Av}(L_1) \leqslant \mathrm{Av}(L_1 \cap L_1')$ it follows that $\mathrm{Av}(L_1 - L_1') \leqslant \mathrm{Av}(L_1)$. Now $\mathrm{Av}(L_1') \leqslant \mathrm{Av}(L_1)$ and $L_1 \cup L_1' = (L_1 - L_1') + L_1'$ so that $\mathrm{Av}(L_1 \cup L_1') \leqslant \mathrm{Av}(L_1)$

and $L_1 \cup L_1'$ is another lower set of minimum average. Therefore, the union of all lower sets of minimum average is the largest lower set of minimum average. Let L_1, and also B_1, denote this lower set. This level set is the set on which g^* assumes its smallest value:

$$g^*(x) = \mathrm{Av}(B_1) = \min\{\mathrm{Av}(L): L \in \mathscr{L}\} \quad \text{for } x \in B_1.$$

Now consider the averages of level sets of the form $L \cap L_1^c$, level sets consisting of lower sets with L_1 subtracted. Select again the largest of these level sets of minimum average, say $B_2 = L_2 \cap L_1^c$. The level set B_2 is the set on which g^* assumes its next smallest value:

$$g^*(x) = \mathrm{Av}(B_2); \quad x \in B_2.$$

This process is continued until X is exhausted.

There is an analogous *maximum upper sets algorithm*, in which the largest upper set of maximum average is selected first.

Theorem 1.4.5 *Let a and b be adjacent values of g^* with $a < b$, that is $[g^* = a] \neq \phi$, $[g^* = b] \neq \phi$, and $[g^* = c] = \phi$ for all $a < c < b$. Then $\mathrm{Av}([g^* = b]) \leqslant \mathrm{Av}(L \cap [g^* > a])$ for all L such that $L \cap [g^* > a] \neq \phi$ and $\mathrm{Av}([g^* = b]) < \mathrm{Av}(L \cap [g^* > a])$ for all L properly containing $[g^* \leqslant b]$. Thus $[g^* = b]$ is the largest level set of the form $L \cap [g^* > a]$ which has the minimum average among such level sets, as described in the minimum lower sets algorithm.*

Proof: By Theorem 1.3.5, $\mathrm{Av}([g^* \leqslant b] \cap [g^* > a]) = \mathrm{Av}([g^* = b]) = b$. Also if $L \cap [g^* > a] \neq \phi$, $\mathrm{Av}(L \cap [g^* > a]) = \mathrm{Av}(L \cap [g^* \geqslant b]) \geqslant b$ by Theorem 1.4.3. Thus $[g^* = b]$ has the minimum average among all nonempty level sets of the form $L \cap [g^* > a]$. Moreover, if L contains $[g^* \leqslant b]$ properly, then $L \cap [g^* > b] \neq \phi$ and $\mathrm{Av}(L \cap [g^* > b]) > b$ again by Theorem 1.4.3. Since $L \cap [g^* \leqslant b] \in \mathscr{L}$, it follows from Theorem 1.4.3 that $\mathrm{Av}(L \cap [g^* \leqslant b] \cap [g^* > a]) \geqslant b$. Furthermore, $\mathrm{Av}(L \cap [g^* > a]) = \mathrm{Av}(\{L \cap [g^* \leqslant b] \cap [g^* > a]\} + \{L \cap [g^* > b]\}) > b = \mathrm{Av}([g^* = b])$, which yields the last conclusion of the theorem. □

It might be instructive for the reader to think about the interpretation of the minimum lower sets algorithm for a simple order in the context of the graphical representation given in Section 1.2.

Example 1.4.1 (The simple tree continued) Recall that in the 'simple tree' example, $X = \{x_0, x_1, \ldots, x_k\}$ with $\lesssim$ defined on X by $x_0 \lesssim x_i$; $i = 1, 2, \ldots, k$. The lower sets consist of ϕ, $\{x_0\}$, and $\{x_0\} \cup A$ with A any nonempty subset of $\{x_1, x_2, \ldots, x_k\}$. Thus there are 2^{k+1} lower sets. On the other hand, if an upper set contains x_0 it must contain $\{x_1, x_2, \ldots, x_k\}$ as a subset. Thus, there are 2^{k+1} upper sets consisting of ϕ, X, and all of the nonempty subsets of $\{x_1, x_2, \ldots, x_k\}$. Note that this is also a consequence of the fact that every upper set is the complement of a lower set and vice versa. It is a straightforward exercise to verify that the minimum lower sets algorithm applied to the simple tree yields the algorithm described previously in Example 1.3.2.

For a simple order on a set X with k elements there are k lower sets to be considered in finding the first level set. If there are i elements of X in the first level set then there are $k - i$ level sets to be considered in the next iteration. On the other hand, in order to apply the minimum lower sets algorithm in a problem involving a simple tree ordering on a set X with k elements requires consideration of 2^{k-1} lower sets to find the first level set, but after that first set is found the isotonic regression is determined since $g^*(x) = g(x)$ for x not in the first level set.

It is clear that for partial orders the number of lower (upper) sets to be considered in applying the minimum lower sets algorithm (or the min–max formula) can be much larger than for a simple order. This large number of lower sets can cause real problems in the computation of the isotonic regression for partial orders.

For example, consider the type of partial order discussed in Example 1.3.1. We refer to such a partial order as a *matrix order*. Specifically, let $X = \{(i, j); i = 1, 2, \ldots, a; j = 1, 2, \ldots, b\}$ and define the partial order $\lesssim$ on X by $(i, j) \lesssim (h, l)$ if and only if $i \leqslant h$ and $j \leqslant l$. Visualizing X as an a by b array with $(1, 1)$ in the lower left-hand corner and (a, b) in the upper right-hand corner, a subset of X is a lower set if and only if whenever it contains a point it contains all the elements of X to the lower left of that point. The boundary of a lower set corresponds to a path from the upper left-hand corner of the grid to the lower right-hand corner of the grid. By counting such paths it follows that the number of lower sets is $\binom{a+b}{a}$ (including the empty set). Now if $a = b = 20$ and if consideration of each lower set were to require one microsecond of computer time, then finding the first level set in applying the minimum lower sets algorithm would require 2312 minutes or 38.5 hours of CPU time. (One microsecond seems conservative in light of the fact that computation of the average value over that set would take at least two multiplications, two additions, and a division and the comparison would require a subtraction.) Moreover, if the first level set is small (as it would be with good data) the second cycle is nearly as difficult as the first.

A considerable amout of research concerning the computation of the isotonic regression with respect to a partial order has been conducted. Many of these algorithms are for very specific partial orders such as the simple order, the simple tree order, or the more general tree order. Some of these algorithms try to mimic the PAVA by pooling pairs which violate the order restriction. However, early 'poolings' may have to be broken later in the process and these algorithms can require a significant amount of checking and readjustment. Thus computer programs which implement these algorithms require intricate branching logic and are complicated to program. Moreover, these algorithms can be very difficult to describe. We close this section with a description of an iterative algorithm for the matrix partial order discussed above. The concept involved in this algorithm can be applied in a number of problems involving complicated partial orders.

An iterative algorithm for the matrix partial order

Real functions with domain, X, may be thought of as $a \times b$ matrices, i.e.

$$G = (g_{ij}) = (g((i, j))).$$

Note that a function G (a matrix) is isotonic with respect to this partial order if and only if the elements of G along each row and column are nondecreasing. The algorithm requires only the ability to solve the isotonic regression problem with the usual nondecreasing order (in one dimension) along rows and columns. These row and column smoothings can be easily accomplished with the PAVA.

Step 1. Let $\hat{G}^{(1)} = (\hat{g}_{ij}^{(1)})$ denote the isotonic regression of G over rows, i.e. $\hat{G}^{(1)}$ minimizes $\sum_{i=1}^{a} \sum_{j=1}^{b} (g_{ij} - f_{ij})^2 w_{ij}$ subject to $f_{1j} \leqslant f_{2j} \leqslant \cdots \leqslant f_{aj}$ for $j = 1, 2, \ldots, b$. Let $R^{(1)} = (r_{ij}^{(1)}) = (\hat{g}_{ij}^{(1)} - g_{ij})$ be the first set of 'row increments'.
Step 2. Let $\tilde{G}^{(1)} = (\tilde{g}_{ij}^{(1)})$ denote the isotonic regression over columns of $G + R^{(1)}$, i.e. $\tilde{G}^{(1)}$ minimizes $\sum_{i=1}^{a} \sum_{j=1}^{b} (g_{ij} + r_{ij}^{(1)} - f_{ij})^2 w_{ij}$ subject to $f_{i1} \leqslant f_{i2} \leqslant \cdots \leqslant f_{ib}$ for $i = 1, 2, \ldots, a$. Call $C^{(1)} = \tilde{G}^{(1)} - (G + R^{(1)})$ the first set of 'column increments'. Note that $\tilde{G}^{(1)} = G + R^{(1)} + C^{(1)}$.
Step 3. At the beginning of the nth cycle, $\hat{G}^{(n)}$ is obtained by isotonizing $G + C^{(n-1)}$ over rows. The nth set of row increments is defined by $R^{(n)} = \hat{G}^{(n)} - (G + C^{(n-1)})$ so that $\hat{G}^{(n)} = G + C^{(n-1)} + R^{(n)}$. Next, obtain $\tilde{G}^{(n)}$ by isotonizing $G + R^{(n)}$ over columns. The nth set of column increments is given by $C^{(n)} = \tilde{G}^{(n)} - (G + R^{(n)})$ or, equivalently, $\tilde{G}^{(n)} = G + R^{(n)} + C^{(n)}$.

The utility of the algorithm lies in the following theorem which will be proved in Section 1.7, after the necessary mathematical tools have been developed.

Theorem 1.4.6 *Both $\hat{G}^{(n)}$ and $\tilde{G}^{(n)}$ converge to the isotonic regression, G^*, with respect to the matrix partial order as $n \rightarrow \infty$.*

As one would expect, this procedure works equally well when the order restrictions are modified to require nonincreasing rows and nonincreasing columns. One has only to change the one-dimensional smoothings to operate in the appropriate direction.

Of course Theorem 1.4.6 would not necessarily hold if some of the values of the weight function were allowed to be zero, since then the individual row and column isotonic regressions would not be uniquely defined.

This problem may be avoided by the following procedure: (1) remove those cells which have zero weights from the grid and disregard all orderings which involve these cells, (2) carry out the row–column isotonic regression procedure on the reduced grid until adequate convergence is attained, and (3) insert any values into the zero-weight cells which are consistent (satisfy the order restrictions) with values given by the row–column procedure. This solution will not be viable in certain pathological cases in which removal of the zero-weight cells yields a problem with insufficient constraints between the positive weight cells.

An alternative approach is to assign very small weights (say 10^{-5}) to those cells with zero weights. These small weights will have a negligible effect on the solution, yet unique convergence of the row–column procedure is guaranteed. Moreover, if the PAVA is being used for the row and column smoothings, this will avoid the problem of division by zero. However, it is possible to greatly slow down the rate of convergence when very small weights are assigned.

In order to illustrate the algorithm, consider the data in Table 1.3.1. The isotonic regression in Table 1.3.2, correct to four significant digits, was obtained after 500 iterations (250 row smoothings and 250 column smoothings) at a cost of 9 seconds of CPU time. Since the cost of this algorithm is essentially linear in the number of points in the grid, even large arrays can be isotonized at a reasonable cost. A Fortran program to implement the algorithm is given in Bril *et al.* (1984).

This algorithm extends naturally to regression problems involving more than two idependent variables. Thus, suppose one has three independent variables and the isotonic regression must be nondecreasing in rows, columns, and layers. The algorithm would proceed as follows:

Step 1. First smooth $G=(g_{ijk})$ over all rows. Let $\hat{G}^{(1)}$ denote the row-smoothed values and $R^{(1)}=(r_{ijk}^{(1)})=\hat{g}_{ijk}-g_{ijk})$ the first set of 'row increments'.

Step 2. Smooth $G+R^{(1)}$ over all columns. Let $\tilde{G}^{(1)}$ denote the column-smoothed values and $C^{(1)}=(c_{ijk}^{(1)})=\{\tilde{g}_{ijk}-(g_{ijk}+r_{ijk}^{(1)})\}$ the first set of 'column increments'.

Step 3. Smooth $G+R^{(1)}+C^{(1)}$ over all layers. Let $\bar{G}^{(1)}$ denote the layer-smoothed values and $L^{(1)}=(l_{ijk}^{(1)})=\{\bar{g}_{ijk}^{(1)}-(g_{ijk}+r_{ijk}^{(1)}+c_{ijk}^{(1)})\}$ the first set of 'layer increments'.

Step 4. Now smooth $G+C^{(1)}+L^{(1)}$ over rows again to obtain the second set of row-smoothed values $\hat{G}^{(2)}$. The row increments are updated and become $R^{(2)}=(r_{ijk}^{(2)})=\{\hat{g}_{ijk}^{(2)}-(g_{ijk}+c_{ijk}^{(1)}+l_{ijk}^{(1)})\}$.

Step 5. Continue. Sequentially smooth over columns, layers, and rows, each time updating the increments.

Additional independent variables are handled in a similar fashion.

Let $\mathscr{K}_r$ be the collection of all functions F on X (i.e. matrices) which are nondecreasing over rows (that is $f_{1j}\leqslant f_{2j}\leqslant\cdots\leqslant f_{aj}$ for $j=1,2,\ldots,b$ and let $\mathscr{K}_c$ be the collection of all functions on X which are nondecreasing over columns. Then $\mathscr{K}_r$ and $\mathscr{K}_c$ are the collections of all functions on X which are isotonic with respect to obvious partial orders on X. Let $\mathscr{K}$ be the collection of all functions on X which are isotonic with respect to the matrix partial order on X. The clearly $\mathscr{K}=\mathscr{K}_r\cap\mathscr{K}_c$ and this algorithm gives a technique for finding the least square projection onto $\mathscr{K}$ in terms of least squares projections onto $\mathscr{K}_r$ and $\mathscr{K}_c$. A schematic diagram of the algorithm is given in Figure 1.4.1.

The algorithm described here for finding the isotonic regression for the matrix partial order can be generalized to a problem of finding the least squares projection onto any finite intersection of closed, convex cones. Thus if $\mathscr{K}$ is expressible as $\mathscr{K} = \mathscr{K}_1 \cap \mathscr{K}_2 \cap \cdots \cap \mathscr{K}_l$, where the $\mathscr{K}_i$ are closed convex cones in a metric space, then one may perform sequential projections onto the individual $\mathscr{K}_i$ (making the proper adjustments on the projected vectors) and obtain convergence to the projection onto $\mathscr{K}$ (see Dykstra, 1983).

Using duality, which will be discussed later in this chapter, the aforementioned algorithm can also be used to find solutions to least squares problems where the constraint region is expressible as a direct sum of closed convex cones rather than an intersection. Thus, the algorithm can also be to used solve least squares problems over constraint regions of the form $\mathscr{K} = \mathscr{K}_1 \oplus \mathscr{K}_2 \oplus \cdots \oplus \mathscr{K}_l = \{x_1 + x_2 + \cdots + x_l : x_i \in \mathscr{K}_i;\, i = 1, 2, \ldots, l\}$.

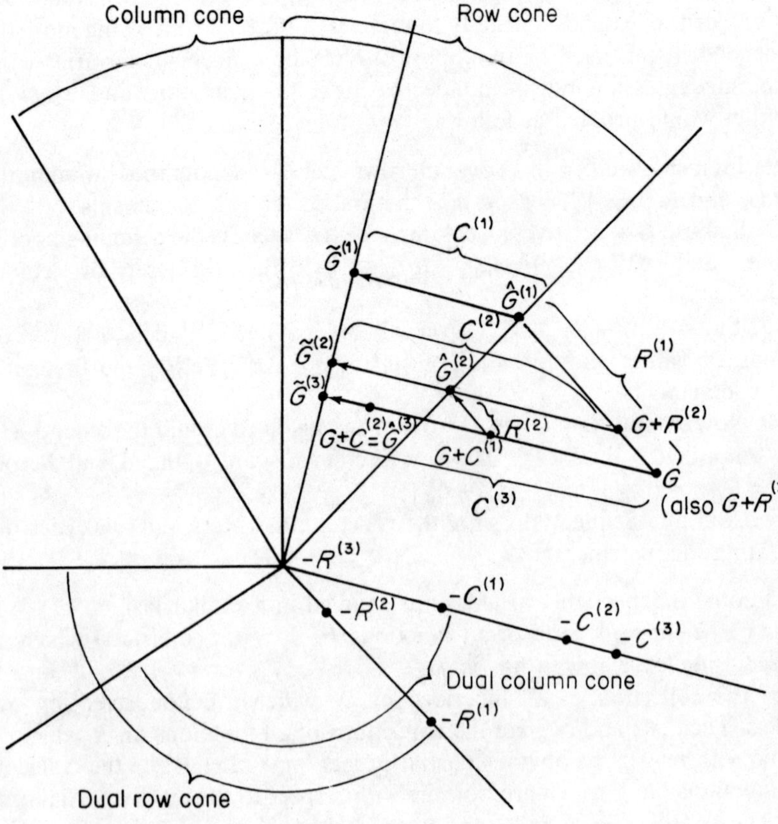

Figure 1.4.1 Schematic diagram of the iterative algorithm

1.5 GENERALIZED ISOTONIC REGRESSION

Suppose Φ is a convex function which is finite on an interval I containing the range of the function g on X and has the value $+\infty$ elsewhere. Then it is well known that Φ is continuous on the interior of I, I^0, and, in fact, Φ is both left and right differentiable at each point in I^0. Let ϕ be a nondecreasing function on I which coincides with any determination of the derivative of Φ on I^0 (i.e. for each $x \in I^0$, $\phi(x)$ is a number between the left derivative of Φ at x and the right derivative of Φ at x). The value of ϕ at the left endpoint of I may be $-\infty$ and its value at the right endpoint of I may be $+\infty$.

For each $u, v \in I$ define the function $\Delta_\Phi(u, v)$ by

$$\Delta_\Phi(u, v) = \Phi(u) - \Phi(v) - (u - v)\phi(v),$$

with $\Delta_\Phi(u, u) = 0$ and $\Delta_\Phi(u, v) = \infty$ if either u or v is not a member of I. If $v < u$ then $\Phi(u) - \Phi(v)$ is the change in the graph of Φ over the interval $[v, u]$ and $(u - v)\phi(v)$ is the change in the graph of a tangent line to Φ at v over that same interval (see Figure 1.5.1).

Now $\Delta_\Phi(u, v)$ could be $+\infty$ at the endpoints of I, but it is impossible for $\Delta_\Phi(u, v)$ to be $-\infty$. Moreover, by a straightforward consideration of all possible cases it can be seen that

$$\Delta_\Phi(u, v) \geqslant 0 \qquad \text{for all } u, v \in I,$$

with strict inequality provided $u \neq v$ and Φ is strictly convex (this is known as the subgradient inequality; see Rockafellar, 1970, p. 214). Note that the value of $\Delta_\Phi(u, v)$ is the excess of the rise of the graph of Φ between v and u over the rise of a tangent line to the graph of Φ at v. Clearly the addition of a linear function to Φ does not affect Δ_Φ. Also it follows directly from the definition that

$$\Delta_\Phi(r, t) = \Delta_\Phi(r, s) + \Delta_\Phi(s, t) + (r - s)[\phi(s) - \phi(t)] \tag{1.5.1}$$

if $r, s, t \in I$.

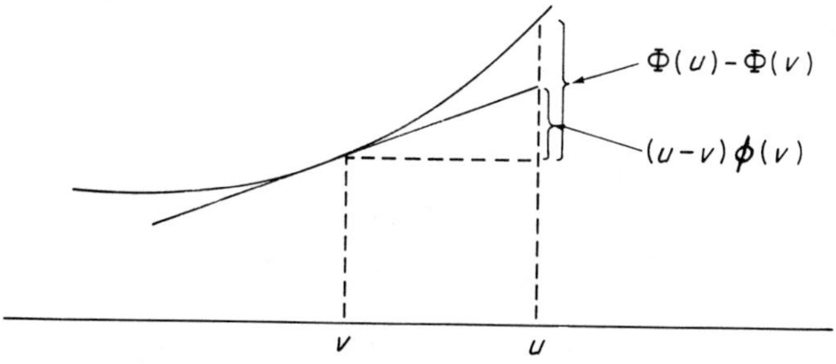

Figure 1.5.1 A comparison of the increase of a convex function with that of its tangent line

Theorem 1.5.1 *If f is isotonic on X and if the range of f is in I then*

$$\sum_{x\in X} \Delta_\Phi[g(x), f(x)]w(x) \geqslant \sum_{x\in X} \Delta_\Phi[g(x), g^*(x)]w(x) + \sum_{x\in X} \Delta_\Phi[g^*(x), f(x)]w(x). \tag{1.5.2}$$

Consequently g^* minimizes

$$\sum_{x\in X} \Delta_\Phi[g(x), f(x)]w(x) \tag{1.5.3}$$

in the class of all isotonic f with range in I and maximizes

$$\sum_{x\in X} \{\Phi[f(x)] + [g(x) - f(x)]\phi[f(x)]\}w(x). \tag{1.5.4}$$

The minimizing (maximizing) function is unique if Φ is strictly convex.

Proof: Using (1.5.1) with $r = g(x)$, $s = g^*(x)$, and $t = f(x)$, the difference between the left-hand side and the right-hand side of (1.5.2) is

$$\sum_{x\in X} [g(x) - g^*(x)]\{\phi[g^*(x)] - \phi[f(x)]\}w(x).$$

However, $\sum_{x\in X}[g(x) - g^*(x)]\phi[g^*(x)]w(x) = 0$ by Theorem 1.3.6. Moreover, ϕ is nondecreasing so that $\phi[f(x)]$ is isotonic and $\sum_{x\in X}[g(x) - g^*(x)]\phi[f(x)]w(x) \leqslant 0$ by Theorem 1.3.2. This yields the first conclusion of the theorem. The fact that g^* minimizes (1.5.3) follows from the fact that $\Delta_\Phi \geqslant 0$ and the next conclusion follows from the fact that the first term of $\Delta_\Phi[g(x), f(x)]$, which is $\Phi[g(x)]$, does not depend on f. The uniqueness conclusion follows from the fact that the second term on the right of (1.5.2) is strictly positive unless $g^* = f$. □

Corollary *Let $\psi_1, \psi_2, \ldots, \psi_p$ be arbitrary real valued functions and let $h_1, h_2, \ldots, h_m$ be isotonic functions on X. Then g^* minimizes (1.5.3) in the class of all isotonic functions, f, on X with range in I satisfying all of the side conditions*

$$\sum_{x\in X} [g(x) - f(x)]\psi_j[f(x)]w(x) = 0; \qquad j = 1, 2, \ldots, p$$

$$\sum_{x\in X} [g(x) - f(x)]h_j(x)w(x) \leqslant 0; \qquad j = 1, 2, \ldots, m.$$

Proof: By Theorems 1.3.6 and 1.3.2, g^* satisfies all of these side conditions. □

Theorem 1.5.1 and its corollary can be used to show that the isotonic regression (i.e. the least squares projection of a function onto the collection of all isotonic functions) provides a solution for a wide variety of restricted estimation problems in which the objective functions do not look at all like least squares. The technique for solving these problems involves finding the appropriate choice of the function Φ in Theorem 1.5.1.

Example 1.5.1 Maximum likelihood estimation of ordered binomial parameters (bioassay) A problem frequently encountered in bioassay is the problem of estimating an ordered set of binomial parameters. These estimates may in turn be used to estimate the minimum stimulus which will produce a specified response. Suppose that for each of several levels $x_1 \lesssim x_2 \lesssim \cdots \lesssim x_k$ of a stimulus (e.g. doses of a drug) the probabilities of a positive response are $p(x_1), p(x_2), \ldots, p(x_k)$ and suppose p is known to be nondecreasing on $X = \{x_1, x_2, \ldots, x_k\}$. Suppose that for $x \in X$ there are $n(x)$ independent trials at stimulus level x producing $a(x)$ positive responses. The 'usual' estimate of $p(x)$ is $\bar{y}(x) = a(x)/n(x)$ and if $n(x)$ is large for each x, $\bar{y}(x_1), \bar{y}(x_2), \ldots, \bar{y}(x_k)$ can be expected to be increasing. However, in practice this is frequently not the case and another estimate is required. Theorem 1.5.1 will be used to show that the isotonic regression of $\bar{y}$ with weights n provides the nondecreasing maximum likelihood estimate of p.

If f denotes an arbitrary function on X whose values are bounded between 0 and 1, then the likelihood function evaluated at f is given by

$$\prod_{x \in X} [f(x)]^{a(x)}[1 - f(x)]^{n(x) - a(x)}$$

and the negative log-likelihood can be written as

$$-\sum_{x \in X} \{\bar{y}(x) \ln f(x) + [1 - \bar{y}(x)] \ln [1 - f(x)]\} n(x). \tag{1.5.5}$$

Let $I = [0, 1]$ and define Φ on $I^0 = (0, 1)$ by $\Phi(y) = y \ln y + (1 - y) \ln (1 - y)$. Define $\Phi(0) = \Phi(1) = 0$ by continuity. The first derivative of Φ is given by $\phi(y) = \ln y - \ln (1 - y)$ for $y \in (0, 1)$ with $\phi(0) = -\infty$ and $\phi(1) = \infty$. The second derivative of Φ on $(0, 1)$ is given by $[y(1 - y)]^{-1}$ which is positive so that Φ is convex on $(0, 1)$. Writing out $\sum_{x \in X} \Delta_\Phi[\bar{y}(x), f(x)] n(x)$ and omitting all terms which do not depend upon f one obtains (1.5.5). Thus, by Theorem 1.5.1 the isotonic regression, $\bar{y}^*$, of $\bar{y}$ with weights $w = n$ provide the maximum likelihood estimates of $p(x_1), p(x_2), \ldots, p(x_k)$ subject to the constraint that the estimates must be nondecreasing in x. Note that this analysis will work for any problem in which the estimate is required to be isotonic with respect to a quasi-order on X.

Example 1.5.2 Another interesting data set involving the 'matrix' partial order discussed in Example 1.3.1, but this time with binomial parameters, involves the prediction of success in college. In this problem it is desired to estimate the probability of obtaining a 'B or better' GPA for entering college students. The estimate is to be based upon the student's high school GPA and composite ACT score.

The data for 1490 college students is given in Table 1.5.1. The two entries are the total cell frequencies and the observed relative frequencies. One would like the estimate of the probability of 'B or better' to be nondecreasing in both variables, the high school GPA and the ACT score. Note that there are a number of 'violators' in the table, even for these relatively large sample sizes. From

Table 1.5.1 The probability of making a B 'or better' GPA (top number = total cell frequency; bottom number = relative frequency)

ACT	HIGH SCHOOL GRADE POINT AVERAGE				
COMPOSITE	0–1.55	1.56–2.25	2.26–2.95	2.96–3.65	3.66–4.00
28^+	0	7	10	47	44
	0.0000	0.2857	0.2000	0.5745	0.8864
23–27	7	56	88	180	84
	0.0000	0.1250	0.1818	0.2833	0.5238
18–22	23	166	152	149	33
	0.0435	0.0301	0.0724	0.1946	0.1212
13–17	27	149	96	61	4
	0.0000	0.0470	0.0313	0.0492	0.5000
0–12	10	57	33	7	0
	0.0000	0.0000	0.0606	0.0000	0.0000

Table 1.5.2 The probability of making 'B or better' GPA estimated by least squares isotonic regression (weights = cell frequencies). Same categories as in Table 1.5.1

0.0333[a]	0.2353	0.2353	0.5745	0.8864
0.0333	0.1250	0.1818	0.2833	0.5238
0.0333	0.0377	0.0724	0.1881	0.1881
0.0000	0.0377	0.0377	0.0492	0.1881
0.0000	0.0000	0.0377	0.0377	0.0377[a]

[a]Any value which satisfies the order restrictions will suffice here since this cell has zero weight.

Example 1.5.1 the maximum likelihood estimates, subject to the contraints that they are nondecreasing in each of the independent variables, are provided by the isotonic regression of the observed relative frequencies using the cell frequencies as weights. These estimates are given in Table 1.5.2. They were computed using the iterative algorithm of Theorem 1.4.6.

Exponential families

Suppose ν is a σ-finite measure on the Borel subsets of the real line and consider a regular exponential family of distributions defined by probability densities of the form

$$f(y; \theta, \tau) = \exp\left[p_1(\theta)p_2(\tau)K(y; \tau) + S(y; \tau) + q(\theta, \tau)\right] \tag{1.5.6}$$

for $y \in A$, $\theta \in (\underline{\theta}, \bar{\theta})$, $\tau \in T$. The densities are with respect to ν and $-\infty \leqslant \underline{\theta} < \bar{\theta} \leqslant \infty$. The parameter τ is thought of as a nuisance parameter, and the derivatives in the following regularity conditions are taken with respect to θ. Suppose that

$$p_1 \text{ and } q(\cdot, \tau) \text{ both have continuous second derivatives on } (\underline{\theta}, \bar{\theta}) \text{ for all } \tau \in T, \tag{1.5.7}$$

$$p_1'(\theta) > 0 \text{ for all } \theta \in (\underline{\theta}, \bar{\theta}),\ p_2(\tau) > 0 \text{ for all } \tau \in T, \tag{1.5.8}$$

and

$$q'(\theta, \tau) = -\theta p_1'(\theta) p_2(\tau) \text{ for all } \theta \in (\underline{\theta}, \bar{\theta}) \text{ and } \tau \in T. \tag{1.5.9}$$

If Y is any random variable having density function $f(y; \theta, \tau)$ then using Theorem 9 on page 52 of Lehmann (1959), the integral $\int_A f(y; \theta, \tau)\nu(dy) = 1$ can be twice differentiated with respect to θ, under the integral sign, obtaining $E[K(Y; \tau)] = \theta$ and $V[K(Y; \tau)] = [p_1'(\theta) p_2(\tau)]^{-1}$.

Suppose one has independent random samples from k populations belonging to the above exponential family with the sample sizes denoted by n_i and the observations by Y_{ij}, i.e. Y_{ij} has density $f(\cdot; \theta(x_i), \tau_i)$ for $j = 1, 2, \ldots, n_i$ and $i = 1, 2, \ldots, k$. If one makes no assumptions regarding the shape of θ, then differentiating the log-likelihood function and applying the assumptions on p_1, p_2 and q, one sees that the maximum likelihood estimator of $\theta(x_i)$ is

$$\hat{\theta}(x_i) = n_i^{-1} \sum_{j=1}^{n_i} K(Y_{ij}; \tau_i). \tag{1.5.10}$$

Suppose $\lesssim$ is a quasi-order on $X = \{x_1, x_2, \ldots, x_k\}$; suppose one has prior information that θ is isotonic with respect to this quasi-order; and assume an estimate of θ which also has this property is desired.

Theorem 1.5.2 *Assume independent random samples from each of k populations characterized by exponential densities of the form (1.5.6) and satisfying (1.5.7), (1.5.8), and (1.5.9). The maximum likelihood estimate of θ subject to the constraint that the estimate be isotonic with respect to the quasi-order $\lesssim$ on X is given by the isotonic regression of $\hat{\theta}$ with weights, $w(x_i) = n_i p_2(\tau_i)$, and $\hat{\theta}(x_i)$ given by (1.5.10).*

Proof: Suppose f is a function on X with values in $(\underline{\theta}, \bar{\theta})$. Using (1.5.6) the likelihood function at f is given by

$$L[\mathbf{y}; f] = \exp\left\{ \sum_{i=1}^{k} p_1[f(x_i)] p_2(\tau_i) n_i \hat{\theta}(x_i) + \sum_{i=1}^{k} \sum_{j=1}^{n_i} S(y_{ij}; \tau_i) + \sum_{i=1}^{k} n_i q[f(x_i), \tau_i] \right\}.$$

Taking the logarithm and discarding all terms which do not involve f, the optimization problem is to find the isotonic function, f, that maximizes

$$\sum_{i=1}^{k} \{p_1[f(x_i)] p_2(\tau_i) n_i \hat{\theta}(x_i) + n_i q[f(x_i), \tau_i]\} \tag{1.5.11}$$

Fix $\theta_0 \in (\underline{\theta}, \overline{\theta})$, apply (1.5.9), and integrate by parts to obtain

$$q(\theta, \tau) = \int_{\theta_0}^{\theta} q'(t, \tau)\,dt + C = \int_{\theta_0}^{\theta} [-tp_1'(t)p_2(\tau)]\,dt + C$$

$$= -p_2(\tau)\left[\theta p_1(\theta) - \theta_0 p_1(\theta_0) - \int_{\theta_0}^{\theta} p_1(t)\,dt\right] + C \qquad (1.5.12)$$

where the constant C may depend upon θ_0 and τ but does not depend upon θ. Substituting this expression for $q(\theta, \tau)$ into (1.5.11) and again dropping all terms which do not involve f, (1.5.11) becomes

$$\sum_{i=1}^{k} \left\{ p_1[f(x_i)]\hat{\theta}(x_i) - f(x_i)p_1[f(x_i)] + \int_{\theta_0}^{f(x_i)} p_1(t)\,dt \right\} n_i p_2(\tau_i)$$

$$= \sum_{i=1}^{k} \left\{ \int_{\theta_0}^{f(x_i)} p_1(t)\,dt + [\hat{\theta}(x_i) - f(x_i)]p_1[f(x_i)] \right\} n_i p_2(\tau_i).$$

Identifying $\hat{\theta}$ with g, $n_i p_2(\tau_i)$ with $w(x_i)$, and letting

$$\Phi(x) = \int_{\theta_0}^{x} p_1(t)\,dt$$

for $x \in (\underline{\theta}, \overline{\theta})$, the above expression is just (1.5.4). The function Φ is convex by (1.5.8). □

Many of the following problems were originally solved using Theorem 1.5.1 with the appropriate choice of Φ, which was frequently difficult to find. In each of our examples the measure ν will be either a counting measure or Lebesgue measure.

Example 1.5.1 (continued) The maximum likelihood estimation of ordered binomial parameters discussed in Example 1.5.1 involves sampling from an exponential family of distributions. Let Y_{ij} be 1 or 0 depending upon whether the jth response to the stimulus at level x_i is positive or negative. Then Y_{i1}, $Y_{i2}, \ldots, Y_{in(x_i)}$ is a random sample from a distribution with probability density

$$f([y; \theta(x_i)] = \theta(x_i)^y[1 - \theta(x_i)]^{1-y}$$

$$= \exp\left\{ y \ln\left[\frac{\theta(x_i)}{1 - \theta(x_i)}\right] + \ln[1 - \theta(x_i)] \right\}$$

for $y = 0, 1$. Let ν be the counting measure on $\{0, 1\}$. This is an exponential family of the form (1.5.6) with $(\underline{\theta}, \bar{\theta}) = (0, 1)$, $\tau = 1$, $S(y; \tau) = 0$, $q(\theta; \tau) = \ln[1 - \theta]$, $K(y; \tau) = y$, $p_2(\tau) = 1$, and $p_1(\theta) = \ln[\theta/(1 - \theta)]$. It is straightforward to verify that these functions satisfy (1.5.7), (1.5.8), and (1.5.9). Thus, from Theorem 1.5.2 the restricted maximum likelihood estimate of θ is the isotonic regression of $\hat{\theta}$ with $\hat{\theta}(x_i) = n_i^{-1} \sum_{j=1}^{n_i} Y_{ij}$ and weights $w(x_i) = n_i$. This is the same as the solution found previously using Theorem 1.5.1.

Example 1.5.2 The 'geometric' extremum problem Suppose

$$f(y; \theta) = \theta^y (1 - \theta); \qquad y = 0, 1, 2, \ldots,$$

where $\theta \in (0, 1)$. This family of functions is not of the form (1.5.6) where the functions satisfy (1.5.9). However, one can use Theorem 1.5.2 to solve the order restricted problem associated with this family by reparameterizing in terms of η where $\theta = \eta/(1 + \eta)$ and $\eta > 0$. This transformation is one-to-one and in fact $\eta = \theta/(1 - \theta)$. Thus,

$$f(y; \eta) = \left(\frac{\eta}{1 + \eta}\right)^y \left(\frac{1}{1 + \eta}\right) = \exp\{y[\ln \eta - \ln(1 + \eta)] - \ln(1 + \eta)\}. \quad (1.5.13)$$

Let $(\underline{\eta}, \bar{\eta}) = (0, \infty)$, $\tau = 1$, $S(y; \tau) = 0$, $q(\eta; \tau) = -\ln(1 + \eta)$, $K(y; \tau) = y$, $p_2(\tau) = 1$, and $p_1(\eta) = \ln \eta - \ln(1 + \eta)$. Then (1.5.13) is of the form (1.5.6) where straightforward differentiation of the functions shows that they satisfy (1.5.7), (1.5.8), and (1.5.9). Thus, if one sets $\hat{\eta}(x_i) = n_i^{-1} \sum_{j=1}^{n_i} Y_{ij}$, then the order restricted maximum likelihood estimate of η is, by Theorem 1.5.2, the isotonic regression of $\hat{\eta}$ with weights $w(x_i) = n_i$. Since the transformation, $\theta = \eta/(1 + \eta)$, is increasing in η, the order restricted maximum likelihood estimate of θ is given by $\hat{\eta}/(1 + \hat{\eta})$.

Example 1.5.3 The 'gamma' extremum problem Suppose

$$f(y; \theta, \tau) = \frac{y^{\tau - 1} e^{-y/\theta}}{\theta^\tau \Gamma(\tau)}; \qquad y > 0,$$

where $\theta > 0$, $\tau > 0$. Then $f(y; \theta, \tau)$ can be rewritten as

$$f(y; \theta, \tau) = \exp\left[\left(-\frac{1}{\theta}\right)\left(\frac{y}{\tau}\right)\tau + (\tau - 1)\ln y - \tau \ln \theta - \ln \Gamma(\tau)\right].$$

With $(\underline{\theta}, \bar{\theta}) = (0, \infty)$, $S(y; \tau) = (\tau - 1)\ln y$, $q(\theta, \tau) = -\tau \ln \theta - \ln \Gamma(\tau)$, $K(y; \tau) = y/\tau$, $p_2(\tau) = \tau$, and $p_1(\theta) = -1/\theta$, this is another example of an exponential family, and the restricted maximum likelihood estimate is given by the isotonic regression of $\hat{\theta}$ where $\hat{\theta}(x_i) = n_i^{-1} \sum_{j=1}^{n_i} Y_{ij}/\tau_i$ with weights $w(x_i) = n_i \tau_i$.

Example 1.5.4 The 'normal means' extremum problem Suppose

$$f(y; \theta, \tau) = \frac{1}{\sqrt{2\pi\tau}} e^{-(y - \theta)^2/(2\tau)}; \qquad -\infty < y < \infty,$$

with $\tau > 0$. Then $f(y; \theta, \tau)$ can be rewritten as

$$f(y; \theta, \tau) = \exp\left[\theta\left(\frac{1}{\tau}\right)y - \frac{y^2}{2\tau} - \frac{\theta^2}{2\tau} - \frac{\ln(2\pi\tau)}{2}\right].$$

Thus, letting $(\underline{\theta}, \overline{\theta}) = (-\infty, \infty)$, $S(y; \tau) = -y^2/(2\tau)$, $q(\theta, \tau) = -\theta^2/(2\tau) - \ln(2\pi\tau)/2$, $K(y; \tau) = y$, $p_2(\tau) = 1/\tau$, and $p_1(\theta) = \theta$, one sees that this is an exponential family. The restricted maximum likelihood estimates are given by the isotonic regression with $w(x_i) = n_i/\tau_i$. Note that $w(x_i)$ is the 'precision' of $\hat{\theta}(x_i)$ as an estimate of $\theta(x_i)$.

Example 1.5.5 The 'normal variances' extremum problem Suppose

$$f(y; \theta, \tau) = \frac{1}{\sqrt{2\pi\theta}} e^{-(y-\tau)^2/(2\theta)}; \qquad -\infty < y < \infty,$$

with $\theta > 0$. Then $f(y; \theta, \tau)$ can be rewritten as

$$f(y; \theta, \tau) = \exp\left[-(2\theta)^{-1}(y-\tau)^2 - \frac{\ln(2\pi\theta)}{2}\right]$$

so that if one lets $(\underline{\theta}, \overline{\theta}) = (0, \infty)$, $S(y, \tau) = 0$, $q(\theta; \tau) = -\ln(2\pi\theta)/2$, $K(y; \tau) = (y-\tau)^2$, $p_2(\tau) = 1$, and $p_1(\theta) = -(2\theta)^{-1}$, then this is an exponential family. The restricted maximum likelihood estimates are given by the isotonic regression of $\hat{\theta}$ with weights $w(x_i) = n_i$, where

$$\hat{\theta}(x_i) = n_i^{-1} \sum_{j=1}^{n_i} (Y_{ij} - \tau_i)^2.$$

Example 1.5.6 The 'Poisson' extremum problem It is quite straightforward to see that the Poission family with densities of the form

$$f(x; \lambda) = e^{-\lambda}\frac{\lambda^x}{x!}; \qquad x = 0, 1, 2, \ldots,$$

with $\lambda > 0$, is an exponential family. Thus, Theorem 1.5.2 can also be applied in this setting to give restricted maximum likelihood estimates. However, in practice the Poisson distribution frequently arises in a different kind of sampling situation.

Suppose that a system (i.e. a computer, airplane, etc.) is put into operation and that y_1 failures or errors occur during the initial time period $[0, T_1]$. The device is then modified to improve performance and y_2 failures occur during the second time interval of length T_2. The process continues for time intervals of length $T_1, T_2, \ldots, T_k$. A common model for analyzing this experimental situation is to assume that the number Y_i of failures during the ith time interval has a Poisson distribution with mean $\lambda_i T_i$, $i = 1, 2, \ldots, k$, and that these random variables are independent. The density of Y_i, with $\theta = \lambda_i$ and $\tau = T_i$, is

$$\exp\left[(\log\theta)\tau\left(\frac{y}{\tau}\right) + y\log\tau - \log y! - \theta\tau\right],$$

and this is an exponential family with $(\underline{\theta}, \bar{\theta}) = (0, \infty)$, $p_1(\theta) = \log \theta$, $p_2(\tau) = \tau$, $K(y; \tau) = y/\tau$, $S(y; \tau) = y \log \tau - \log y!$, and $q(\theta, \tau) = -\theta\tau$. The regularity conditions, (1.5.7), (1.5.8), and (1.5.9), hold and so the maximum likelihood estimates of the λ_i subject to $\lambda_1 \geqslant \lambda_2 \geqslant \cdots \geqslant \lambda_k$ are provided by the isotonic regression of g with weights $w_i = T_i$, where $g(x_i) = y_i/T_i$ for $i = 1, 2, \ldots, k$.

Example 1.5.7 Maximizing a product (maximum likelihood estimation of multinomial parameters) Consider an experiment, the outcome of which must be one of k mutually exclusive events having probabilities $p(x_1), p(x_2), \ldots, p(x_k)$. Suppose n independent trials of this experiment are performed and that $\hat{p}(x_i)$ is the relative frequency of the event having probability $p(x_i)$ (i.e. the function $\hat{p}$ on $X = \{x_1, x_2, \ldots, x_k\}$ is the unrestricted maximum likelihood estimate of p). Suppose $\lesssim$ is a quasi-order on X and consider the problem of finding the maximum likelihood estimate of p which is isotonic with respect to this quasi-order. The problem is different from the ones considered before in that, in addition to being isotonic and bounded between zero and one, the estimate must satisfy $\sum_{x \in X} f(x) = 1$.

The following generalization of this problem is treated. Suppose that in addition to the quasi-order $\lesssim$ on X, one is given a pair, c and r, of positive functions on X and a number s. The problem is to find an isotonic function f on X which maximizes

$$\prod_{x \in X} [f(x)]^{r(x)} \tag{1.5.14}$$

subject to the additional constraint

$$\sum_{x \in X} c(x) f(x) = s. \tag{1.5.15}$$

Set

$$R = \sum_{x \in X} r(x), \qquad g(x) = \frac{sr(x)}{Rc(x)}$$

and

$$w(x) = \frac{r(x)}{g(x)} = \frac{Rc(x)}{s}.$$

The problem is to maximize

$$\sum_{x \in X} g(x) \log f(x) w(x) \tag{1.5.16}$$

in the class of all isotonic functions, f, satisfying

$$\sum_{x \in X} [g(x) - f(x)] w(x) = 0. \tag{1.5.17}$$

Let $\Phi(u) = u \log u$ for $u \in (0, \infty)$ and note that

$$\Delta_\Phi(g, f) = g \log g - g \log f - (g - f).$$

From Theorem 1.5.1, g^*, the isotonic regression of g, maximizes

$$\sum_{x \in X} [g(x) \log f(x) + g(x) - f(x)] w(x)$$

in the class of isotonic functions. However, by Theorem 1.3.6, $\sum_{x \in X}[g(x) - g^*(x)]w(x) = 0$, and consequently g^* maximizes

$$\sum_{x \in X} g(x) \log f(x) w(x)$$

in the class of all isotonic functions which satisfy (1.5.17). Hence, the isotonic regression of g with weights w maximizes (1.5.14) among the collection of isotonic functions which satisfy (1.5.15).

Note that computing g^* involves weighted averages of the form

$$\frac{\sum_{x \in A} g(x) w(x)}{\sum_{x \in A} w(x)}$$

for subsets A of X. This weighted average is equal to

$$\frac{s \sum_{x \in A} r(x)}{R \sum_{x \in A} c(x)}.$$

It follows from any one of the appropriate algorithms that the solution, g^*, is equal to s/R multiplied by the isotonic regression of r/c with weights of c.

Returning to the multinomial problem, $s = 1$, $r(x) = n\hat{p}(x)$, and $c(x) = 1$. Thus $R = \sum_{t \in X} r(t) = n$ and it follows that the restricted maximum likelihood estimate is the isotonic regression of $\hat{p}$ with weights $w(x) = 1$.

1.6 REDUCTION OF ERROR (THE ISOTONIC REGRESSION AS AN IMPROVED 'DATA' SET)

Example 1.6.1 (The normal means problem revisited) Suppose we have independent random samples from each of k normal populations with unknown means $\mu(x_1), \mu(x_2), \ldots, \mu(x_k)$ and known variances $\sigma^2(x_1), \sigma^2(x_2), \ldots, \sigma^2(x_k)$, and let the sample size and sample mean from the population associated with x_i be denoted by $n(x_i)$ and $\bar{Y}(x_i)$, respectively. If $\lesssim$ is a quasi-order on X and if it is known that μ is isotonic with respect to this quasi-order then it follows from Example 1.5.4 that the isotonic regression, $\bar{Y}^*$, of $\bar{Y}$ with weights $w(x_i) = n(x_i)/\sigma^2(x_i)$ provides the maximum likelihood estimate of μ subject to the isotonic restriction.

Now from Theorem 1.3.1 it follows that

$$\sum_{x \in X} [\bar{Y}(x) - f(x)]^2 w(x) \geqslant \sum_{x \in X} [\bar{Y}(x) - \bar{Y}^*(x)]^2 w(x) + \sum_{x \in X} [\bar{Y}^*(x) - f(x)]^2 w(x)$$

for all isotonic f. Since the first sum on the right-hand side is nonnegative it follows that

$$\sum_{x \in X} [\bar{Y}(x) - f(x)]^2 w(x) \geq \sum_{x \in X} [\bar{Y}^*(x) - f(x)]^2 w(x)$$

for all isotonic f. In particular, since the true mean function μ is assumed to be isotonic,

$$\sum_{x \in X} [\bar{Y}(x) - \mu(x)]^2 w(x) \geq \sum_{x \in X} [\bar{Y}^*(x) - \mu(x)]^2 w(x). \tag{1.6.1}$$

Now $w(x_i) = n(x_i)/\sigma^2(x_i)$ is the Fisher information in the sample from the population associated with x_i with regard to the parameter $\mu(x_i)$. The number $w(x_i)$ is also referred to as the 'precision' of the estimator $\bar{Y}(x_i)$ of $\mu(x_i)$. Thus, for any estimate, θ, of μ the quantity

$$\sum_{x \in X} [\theta(x) - \mu(x)]^2 w(x)$$

is a natural way to measure the total squared error of the function θ as an estimator of the function μ. The inequality (1.6.1) states that, as estimators of μ, the square error of $\bar{Y}^*$ is no more than that of $\bar{Y}$ and this is true for all isotonic μ. Moreover, the inequality in (1.6.1) is strict unless $\bar{Y}^*$ is identically equal to $\bar{Y}$, which is the case if and only if $\bar{Y}$ is isotonic. This rather obvious result is also quite remarkable as it speaks to error and not expected error.

The random function $\bar{Y}$ is sufficient for the parameter function μ. In other words, $\bar{Y}$ contains all the information in the data that is important for estimating μ. Thus, intuitively, $\bar{Y}$ could be thought of as the 'data' in this problem and the above result could be interpreted as saying that 'isotonizing' improves the quality of the data in the sense of mean square error as long as the true parameter function, μ, is isotonic. This point of view could be particularly appropriate when the primary concern is not with estimating μ itself but with estimating some functional of μ, such as in the bioassay situation where one might be interested in estimating the minimum effective dosage level. The above discussion supports the use of the appropriate functional of $\bar{Y}^*$ rather than the functional of $\bar{Y}$. It will be shown in this section that 'isotonizing' estimates reduce error in several other senses, giving further evidence attesting to the claim that it should improve the quality of an estimation procedure.

An analogous conclusion can be reached in virtually all of the examples discussed in the preceding section. In fact, within the context of the exponential family discussed in Theorem 1.5.2, the unrestricted maximum likelihood estimate, $\hat{\theta}$, of θ is sufficient for θ by the factorization theorem. Moreover, the Fisher information in the sample from the population associated with x_i is $n_i p_1'[\theta(x_i)] p_2(\tau_i)$, which is just $n_i p_2(\tau_i) = w(x_i)$ if the exponential family is parameterized in its natural form (i.e. $p_1(\theta) = \theta$). Theorem 1.3.1 applied in this context yields

$$\sum_{x \in X} [\hat{\theta}(x) - \theta(x)]^2 w(x) \geq \sum_{x \in X} [\hat{\theta}^*(x) - \theta(x)]^2 w(x) \tag{1.6.2}$$

with strict inequality unless $\hat{\theta}^* = \hat{\theta}$, with $\hat{\theta}^*$ the isotonic regression of $\hat{\theta}$ with weight function w.

Theorem 1.6.1 *Suppose $\lesssim$ is a quasi-order on X. If $\hat{\theta}$ is any function on X and if $\hat{\theta}^*$ is the isotonic regression of $\hat{\theta}$ with weights w then*

$$\sum_{x \in X} \Phi[\hat{\theta}^*(x) - \theta(x)]w(x) \leqslant \sum_{x \in X} \Phi[\hat{\theta}(x) - \theta(x)]w(x) \tag{1.6.3}$$

for any convex function Φ on $(-\infty, \infty)$ and any isotonic function θ on X.

Proof: Since Φ is convex,

$$\Phi(v) \geqslant \Phi(u) + (v-u)\phi(u)$$

for ϕ any determination of the derivative of Φ. Letting $v = \hat{\theta}(x) - \theta(x)$ and $u = \hat{\theta}^*(x) - \theta(x)$, applying this inequality for each x and summing over x, one obtains

$$\sum_{x \in X} \Phi[\hat{\theta}(x) - \theta(x)]w(x) \geqslant \sum_{x \in X} \Phi[\hat{\theta}^*(x) - \theta(x)]w(x) + \sum_{x \in X} [\hat{\theta}(x) - \hat{\theta}^*(x)]\phi[\hat{\theta}^*(x) - \theta(x)]w(x) \tag{1.6.4}$$

Now consider the second sum on the right-hand side of (1.6.4). Let $B_1, B_2, \ldots, B_m$ be the level sets obtained from the minimum lower sets algorithm for computing $\hat{\theta}^*$. The function $\hat{\theta}^*$ is constant on each of these sets and has the value $\text{Av}(B_i)$ on B_i. Fix i and consider the function $\hat{\theta}$ restricted to B_i. If one restricts the quasi-order to B_i and constructs the isotonic regression of the restriction of $\hat{\theta}$ to B_i with respect to this restricted quasi-order, then it is clear from the way B_i was selected that the restricted isotonic regression is constant on B_i and has the value $\text{Av}(B_i)$. Thus,

$$\sum_{x \in X} [\hat{\theta}(x) - \hat{\theta}^*(x)]\phi[\hat{\theta}^*(x) - \theta(x)]w(x) = \sum_{i=1}^{m} \sum_{x \in B_i} [\hat{\theta}(x) - \text{Av}(B_i)]\phi[\text{Av}(B_i) - \theta(x)]w(x).$$

Now $-\theta$ is antitonic on B_i with respect to the partial order restricted to B_i. Since ϕ and θ are nondecreasing, $\phi[\text{Av}(B_i) - \theta(x)]$ is also antitonic with respect to this restricted partial order. It follows from (1.3.8) that

$$\sum_{x \in B_i} [\hat{\theta}(x) - \text{Av}(B_i)]\phi[\text{Av}(B_i) - \theta(x)]w(x) \geqslant 0.$$

This, together with (1.6.4), yields the desired result. □

Letting $\Phi(x) = |x|^p$ for $p \geqslant 1$, one obtains the following corollary.

Corollary A *Under the assumptions of Theorem 1.6.1,*

$$\left[\sum_{x\in X}|\hat{\theta}^*(x)-\theta(x)|^p w(x)\right]^{1/p} \leqslant \left[\sum_{x\in X}|\hat{\theta}(x)-\theta(x)|^p w(x)\right]^{1/p} \tag{1.6.5}$$

for all $p\geqslant 1$ and for all isotonic θ.

Letting $p\to\infty$, one obtains the following result.

Corollary B *Under the assumptions of Theorem 1.6.1,*

$$\sup_{x\in X}|\hat{\theta}^*(x)-\theta(x)| \leqslant \sup_{x\in X}|\hat{\theta}(x)-\theta(x)| \tag{1.6.6}$$

for all isotonic functions θ on X.

Applying these results to a situation in which $\hat{\theta}$ is an estimate of θ, one sees that isotonizing the estimate reduces error in a number of ways including total absolute error ($p=1$) and maximum absolute error ($p=\infty$). This is, of course, provided θ is isotonic.

The estimates $\hat{\theta}^*(x_1), \hat{\theta}^*(x_2), \ldots, \hat{\theta}^*(x_k)$ are generally biased. In fact, for the simple order $x_1 \lesssim x_2 \lesssim \cdots \lesssim x_k$,

$$\hat{\theta}^*(x_1) = \min_{1\leqslant j\leqslant k}\left[\frac{\sum_{i=1}^{j}\hat{\theta}(x_i)w(x_i)}{\sum_{i=1}^{j}w(x_i)}\right].$$

Thus, if the basic estimates are unbiased, as they frequently are, then $\hat{\theta}^*(x_1)$ is the minimum of several random quantities, one of which, $\hat{\theta}(x_1)$, is unbiased for $\theta(x_1)$. Thus, $\hat{\theta}^*(x_1)\leqslant\hat{\theta}(x_1)$ and typically $\hat{\theta}^*(x_1)<\hat{\theta}(x_1)$ with positive probability so that in such cases it is a biased estimate. In light of this observation it is rather surprising that the isotonic regression reduces the expected squared error at each x in the normal means problem for a simple order.

Example 1.6.2 (The normal means problem revisited) Suppose $\bar{Y}(x_i)$ is the mean of a sample size $n(x_i)$ from a normal population with unknown mean $\mu(x_i)$ and known variances $\sigma^2(x_i)$, $i=1,2,\ldots,k$. Assume that the samples from the various populations are independent and that it is known that μ is isotonic with respect to the simple order, $x_1 \lesssim x_2 \lesssim \cdots \lesssim x_k$ on X. Let $\bar{Y}^*$ be the isotonic regression of $\bar{Y}$ with weights $w(x)=n(x)/\sigma^2(x)$ and recall that, from Example 1.5.4, $\bar{Y}^*$ is the maximum likelihood estimate of μ subject to the isotonic constraint.

Theorem 1.6.2 *Under the above conditions, for each $x\in X$,*

$$E[(\bar{Y}(x)-\mu(x))^2] \geqslant E[(\bar{Y}^*(x)-\mu(x))^2]. \tag{1.6.7}$$

Proof: The proof is an induction on k. The result is clearly true for $k = 1$. Assume the result holds true for $k-1$ and for this simple order. Let $\mathrm{Av}(s,t)$ denote $\mathrm{Av}[\{x_s, x_{s+1}, \ldots, x_t\}]$ with $1 \leqslant s \leqslant t \leqslant k$. Two cases are considered.

Case 1. Let $\mu(x_i) \leqslant E[\mathrm{Av}(1,k)]$ and $i < k$.

Let $\bar{Y}'$ be the isotonic regression of $\bar{Y}(x_1), \bar{Y}(x_2), \ldots, \bar{Y}(x_{k-1})$ with weights $w(x_1), w(x_2), \ldots, w(x_{k-1})$. By the min–max formulas given in (1.4.3),

$$\bar{Y}^*(x_i) = \min_{t:t \geqslant i} \max_{s:s \leqslant i} \mathrm{Av}(s,t) \leqslant \min_{t:k>t \geqslant i} \max_{s:s \leqslant i} \mathrm{Av}(s,t)$$
$$= \bar{Y}'(x_i). \tag{1.6.8}$$

It follows that if $\bar{Y}'(x_i) > \bar{Y}(x_i)$ then

$$\bar{Y}^*(x_i) = \max_{s:s \leqslant i} \mathrm{Av}(s,k) \geqslant \mathrm{Av}(1,k). \tag{1.6.9}$$

Using the independence and normality of the variables $\bar{Y}(x_1)$, $\bar{Y}(x_2), \ldots, \bar{Y}(x_k)$, the collection of random variables $\{\bar{Y}_i - \mathrm{Av}(1,k): i = 1, 2, \ldots, k\}$ can be shown to be independent of $\mathrm{Av}(1,k)$ by computing covariances. For $i < k$, $\bar{Y}'(x_i) = [\bar{Y}(x_i) - \mathrm{Av}(1,k)]' + \mathrm{Av}(1,k)$ with $[\bar{Y}(x_i) - \mathrm{Av}(1,k)]'$ the isotonic regression of $\bar{Y}(x_1) - \mathrm{Av}(1,k)$, $\bar{Y}(x_2) - \mathrm{Av}(1,k), \ldots, \bar{Y}(x_{k-1}) - \mathrm{Av}(1,k)$ with weights $w(x_1), w(x_2), \ldots, w(x_{k-1})$. Also $\bar{Y}^*(x_i) = [\bar{Y}(x_i) - \mathrm{Av}(1,k)]^* + \mathrm{Av}(1,k)$. Hence, $\bar{Y}'(x_i) - \bar{Y}^*(x_i)$ is a function of $\{\bar{Y}_j - \mathrm{Av}(1,k): j = 1, 2, \ldots, k\}$ and consequently

$$E\{[\bar{Y}'(x_i) - \bar{Y}^*(x_i)]\mathrm{Av}(1,k)\} = E[\bar{Y}'(x_i) - \bar{Y}^*(x_i)]E[\mathrm{Av}(1,k)]. \tag{1.6.10}$$

Applying the induction hypothesis,

$$\begin{aligned} E\{[\bar{Y}(x_i) - \mu(x_i)]^2\} &\geqslant E\{[\bar{Y}'(x_i) - \mu(x_i)]^2\} \\ &= E\{[\bar{Y}'(x_i) - \bar{Y}^*(x_i)]^2\} + E\{[\bar{Y}^*(x_i) - \mu(x_i)]^2\} \\ &\quad + 2E\{[\bar{Y}'(x_i) - \bar{Y}^*(x_i)][\bar{Y}^*(x_i) - \mu(x_i)]\}. \end{aligned} \tag{1.6.11}$$

The inner product in (1.6.11) can be divided into two terms, namely

$$\begin{aligned} &E\{[\bar{Y}'(x_i) - \bar{Y}^*(x_i)][\bar{Y}^*(x_i) - \mathrm{Av}(1,k)]\} \\ &\quad = E\{[\bar{Y}'(x_i) - \bar{Y}^*(x_i)][\bar{Y}^*(x_i) - \mathrm{Av}(1,k)]I_{[\bar{Y}(x_i) > \bar{Y}^*(x_i)]}\} \end{aligned} \tag{1.6.12}$$

and

$$\begin{aligned} E\{[\bar{Y}'(x_i) - \bar{Y}^*(x_i)][\mathrm{Av}(1,k) - \mu(x_i)]\} = E[\bar{Y}'(x_i) - \bar{Y}^*(x_i)] \\ E[\mathrm{Av}(1,k) - \mu(x_i)]. \end{aligned} \tag{1.6.13}$$

The quantity in (1.6.12) is nonnegative by (1.6.9) and the quantity in (1.6.13) is nonnegative by (1.6.8) and the assumption that $\mu(x_i) \leqslant \mathrm{Av}(1,k)$. This completes the argument for this case. Note that if this case does not hold then either $\mu(x_i) > E[\mathrm{Av}(1,k)]$ or $i = k$. Each of these possibilities are treated in case 2.

Case 2. Let $\mu(x_i) \geqslant E[\mathrm{Av}(1,k)]$ and $i > 1$.

This part of the proof consists of multiplying by -1, reversing the roles of k and 1, $k-1$ and 2, etc., and using the first case. □

Remark *Note that in the proof of Theorem 1.6.2 if $k \geqslant 2$ then $\bar{Y}'(x_i)$ and $\bar{Y}^*(x_i)$ are unequal on a set of positive probability so that $E\{[\bar{Y}'(x_i) - \bar{Y}^*(x_i)]^2\} > 0$. Thus the inequality in Theorem 1.6.2 is strict for $k \geqslant 2$.*

The proof given for Theorem 1.6.2 depends very much on the properties of normal variables and upon the fact that the partial order is simple. It would be of interest to know if this theorem is valid under other conditions. In particular, does Theorem 1.6.2 hold for partial orders or for samples from other types of populations? A partial answer to this question is provided by the following discussion which is based upon some observations of Chu-In Charles Lee.

Let $X = \{x_0, x_1, \ldots, x_k\}$; $\lesssim$ be the simple tree partial order on X which was discussed in Example 1.3.2; let $Y(x_0)$, $Y(x_1), \ldots, Y(x_k)$ be independent standard normal variables; and let $Y^*(x_0)$, $Y^*(x_1), \ldots, Y^*(x_k)$ be the isotonic regression of the $Y(x_i)$ with equal weights. Consider the behavior of $Y^*(x_0)$ as $k \to \infty$. Using the minimum lower sets algorithm for computing $Y^*(x_0)$, one can write

$$Y^*(x_0) = \min_{S \subset \{1,2,\ldots,k\}} \frac{Y(x_0) + \sum_{j \in S} Y(x_j)}{1 + \mathrm{card}\,(S)}$$

where card (S) denotes the number of points in the set S. Thus,

$$Y^*(x_0) \leqslant \frac{Y(x_0) + \min_{1 \leqslant i \leqslant k} Y(x_i)}{2}$$

Now $Y(x_1)$, $Y(x_2), \ldots, Y(x_k)$ is a random sample of size k from a standard normal distribution so that, as $k \to \infty$,

$$\min_{1 \leqslant i \leqslant k} Y(x_i) \to -\infty$$

with probability one. It follows that $Y^*(x_0) \to -\infty$ as $k \to \infty$ with probability one. Thus, the mean square error of $Y^*(x_0)$ as an estimate of the mean of the population indexed by x_0, namely zero, goes to infinity as $k \to \infty$ and, for sufficiently large k,

$$E\{[Y^*(x_0) - 0]^2\} > E\{[Y(x_0) - 0]^2\} = 1.$$

For $k = 1(1)\ 10(2)\ 20,\ 25$, Table 1.6.1 gives Monte Carlo estimates of $E\{[Y^*(x_0) - 0]^2\}$ based on 100 000 iterations. Note that for $k \geqslant 8$, $E\{[Y^*(x_0)]^2\} > E\{[Y(x_0)]^2\} = 1$. Table 1.6.1 also contains estimates of $E\{[Y^*(x_i)]^2\}$ for $i \geqslant 1$. Why should mean square error be reduced by the isotonic

Table 1.6.1 Monte Carlo estimates of $E\{[Y^*(x_0)]^2\}$ and $E\{[Y^*(x_1)]^2\}$ for the simple tree partial order

k	$E\{[Y^*(x_0)]^2\}$	$E\{[Y^*(x_1)]^2\}$	k	$E\{[Y^*(x_0)]^2\}$	$E\{[Y^*(x_1)]^2\}$
1	0.7500	0.7500	9	1.1086	0.7647
2	0.7308	0.7179	10	1.1695	0.7772
3	0.7662	0.7164	12	1.2632	0.7876
4	0.8191	0.7214	14	1.3610	0.7998
5	0.8817	0.7290	16	1.4509	0.8117
6	0.9387	0.7408	18	1.5300	0.8185
7	0.9982	0.7492	20	1.6072	0.8288
8	1.0553	0.7578	25	1.7769	0.8479

regression at each point for a simple order and not for the tree partial order? One difference between these two order restrictions is in the number of lower sets. For the simple order the number of lower sets is linear in the number of populations, while for the tree partial order it is exponential in the number of populations. If this is the explanation then many other partial orders should be like the simple tree in this respect. However, at this time, this question has not been thoroughly investigated.

1.7 EXTREMUM PROBLEMS ASSOCIATED WITH DUAL CONVEX CONES

This section gives a very brief introduction to the important topic of dual convex cones. The subject and its ramifications in order restricted inference will be explored in depth in Chapter 6. For $\mathscr{C}$, a convex set of functions on X, it is convenient to denote the projection of g onto $\mathscr{C}$, with weights w, by $P_w(g|\mathscr{C})$, or if w is understood, by $P(g|\mathscr{C})$. That is $P_w(g|\mathscr{C})$ is the solution to (1.3.2).

The Fenchel *dual*, $\mathscr{K}^{*w}$ (also called the *polar*), of a nonempty, closed, convex, cone, $\mathscr{K}$, of functions on X is defined to be the set of all functions on X which make an obtuse angle with all functions in $\mathscr{K}$. Specifically,

$$\mathscr{K}^{*w} = \left\{ f : \sum_{x \in X} f(x) g(x) w(x) \leqslant 0 \text{ for all } g \in \mathscr{K} \right\}$$

(cf. Figures 1.4.1 and 1.7.1). If w is understood, then the dual will be denoted by $\mathscr{K}^*$. Figure 1.7.1 is very instructive in that it is suggestive of the relationship between $\mathscr{K}$ and $\mathscr{K}^*$, as well as that of $P(g|\mathscr{K})$ and $P(g|\mathscr{K}^*)$, in the general case. In particular, the figure with corners at g, $P(g|\mathscr{K})$, $P(g|\mathscr{K}^*)$, and the common vertex of $\mathscr{K}$ and $\mathscr{K}^*$ is, in a sense, a 'rectangle' and $P(g|\mathscr{K}^*) = g - P(g|\mathscr{K})$, as will be shown later. The proof of the following theorem is a straightforward exercise.

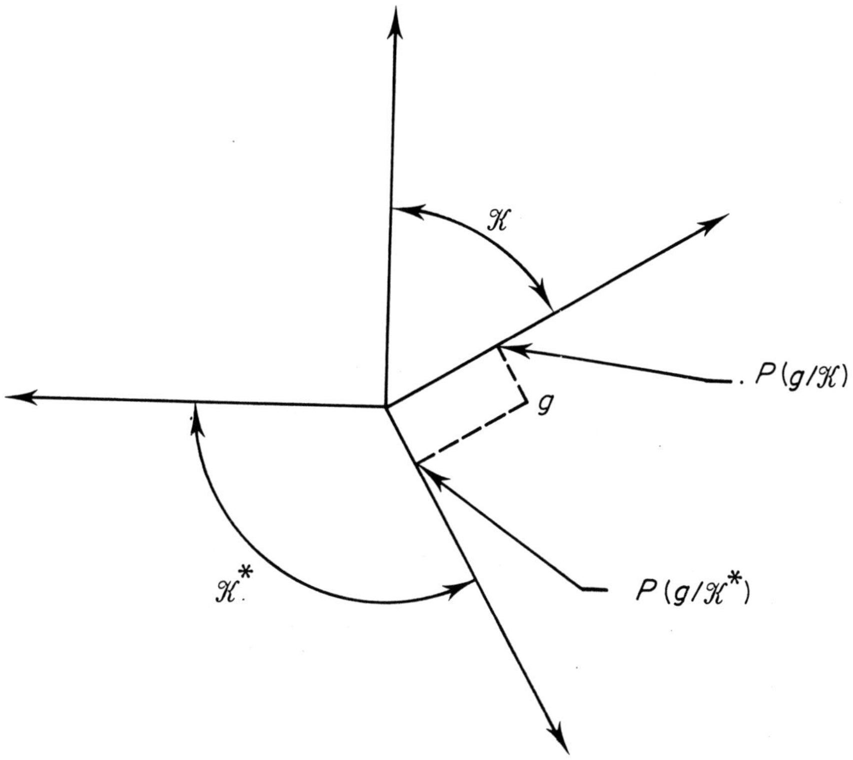

Figure 1.7.1 A cone, its dual and related projections

Theorem 1.7.1 *The dual $\mathcal{K}^*$ of a nonempty, convex, cone, $\mathcal{K}$, of functions on X is a nonempty, closed, convex, cone of functions on X.*

Theorem 1.7.2 *If $\mathcal{K}$ is a nonempty, closed, convex, cone of functions on X and $\mathcal{K}^{*w}$ is its dual then*

$$P_w(f|\mathcal{K}^{*w}) = f - P_w(f|\mathcal{K}). \tag{1.7.1}$$

Proof: It will be shown that the function $g = f - P_w(f|\mathcal{K})$ has the properties given in Theorem 1.3.2 which characterizes $P_w(f|\mathcal{K}^{*w})$. In the first place $g \in \mathcal{K}^{*w}$, because if $h \in \mathcal{K}$, then

$$\sum_{x \in X} g(x)h(x)w(x) = \sum_{x \in X} [f(x) - P_w(f|\mathcal{K})(x)]h(x)w(x) \leqslant 0$$

by (1.3.8) applied to $P_w(f|\mathcal{K})$. Next,

$$\sum_{x \in X} [f(x) - g(x)]g(x)w(x) = \sum_{x \in X} P_w(f|\mathcal{K})(x)[f(x) - P_w(f|\mathcal{K})(x)]w(x) = 0$$

by (1.3.7). Finally, suppose $h \in \mathscr{K}^{*w}$ and note that

$$\sum_{x \in X} [f(x) - g(x)]h(x)w(x) = \sum_{x \in X} P_w(f|\mathscr{K})(x)h(x)w(x) \leqslant 0,$$

since $P_w(f|\mathscr{K}) \in \mathscr{K}$ and $h \in \mathscr{K}^{*w}$. The proof is completed. □

Theorem 1.7.3 *For any nonempty, closed, convex, cone, $\mathscr{K}$, of functions on X, $(\mathscr{K}^{*w})^{*w} = \mathscr{K}$.*

Proof: It follows easily from the definition of $(\mathscr{K}^{*w})^{*w}$ that $\mathscr{K} \subset (\mathscr{K}^{*w})^{*w}$. The other containment depends upon the assumption that $\mathscr{K}$ is closed. Suppose $f \in (\mathscr{K}^{*w})^{*w}$ and $f \notin \mathscr{K}$. Since $\mathscr{K}$ is closed,

$$\|f - P(f|\mathscr{K})\|_w^2 = \sum_{x \in X} [f(x) - P(f|\mathscr{K})(x)]^2 w(x) > 0.$$

But $f - P(f|\mathscr{K}) \in \mathscr{K}^{*w}$ by (1.3.8) and $f \in (\mathscr{K}^{*w})^{*w}$ so that

$$\begin{aligned} 0 &\geqslant \sum_{x \in X} f(x)[f(x) - P(f|\mathscr{K})(x)]w(x) \\ &= \sum_{x \in X} [f(x) - P(f|\mathscr{K})(x)]^2 w(x) + \sum_{x \in X} P(f|\mathscr{K})(x)[f(x) - P(f|\mathscr{K})(x)]w(x) \\ &= \|f - P(f|\mathscr{K})\|^2 > 0 \end{aligned}$$

by (1.3.7). This contradiction establishes that $(\mathscr{K}^{*w})^{*w} \subset \mathscr{K}$. □

Duality has been used in order restricted inference in a number of ways. For example, speaking heuristically, the dual of a simple order restriction is a stochastic ordering. This observation has led to the solution of several problems in which the restriction is a stochastic ordering restriction by relating the problem through its dual to the simple order restriction and using Theorem 1.7.2 (cf. Example 1.7.1). There are other problems in which the dual has a much simpler structure than that of the original cone and Theorem 1.7.2 leads to the solution. The study of marginal homogeneity in contingency tables (cf. Section 4.6) provides an example. The solution to several of these problems is obtained via the corollary to the next theorem.

Theorem 1.7.4 *Let $\mathscr{H}$ be an arbitrary class of functions on X and let Φ be convex and ϕ a determination of its derivative. Let g_1 be a given function on X and w be a positive weight function. Then, a function h solves*

$$\textit{minimize} \sum_{x \in X} \{\Phi[f(x)] - f(x)g_1(x)\}w(x) \textit{ subject to } f \in \mathscr{H}$$

provided there exists a function g_2 on X such that

$$\sum_{x \in X} [h(x) - g_2(x)]\{\phi[h(x)] - g_1(x)\}w(x) = 0 \tag{1.7.2}$$

and

$$\sum_{x \in X} [f(x) - g_2(x)]\{\phi[h(x)] - g_1(x)\} w(x) \geqslant 0 \qquad (1.7.3)$$

for all $f \in \mathscr{H}$. The minimizing function is unique if Φ is strictly convex.

Proof: If $f \in \mathscr{H}$ then

$$\Phi(f) - fg_1 - [\Phi(h) - hg_1] = \Delta_\Phi(f, h) + (f - h)\phi(h) - (f - h)g_1$$

where Δ_Φ is defined as in Section 1.5. Since $\Delta_\Phi \geqslant 0$, one may conclude that

$$\begin{aligned}\Phi(f) - fg_1 - [\Phi(h) - hg_1] &\geqslant (f - h)[\phi(h) - g_1] \\ &= [(f - g_2) + (g_2 - h)][\phi(h) - g_1]\end{aligned}$$

and the inequality is strict if Φ is strictly convex and $f(x) \neq h(x)$. It now follows from (1.7.2) and (1.7.3) that

$$\sum_{x \in X} \{\Phi[f(x)] - f(x)g_1(x)\} w(x) \geqslant \sum_{x \in X} \{\Phi[h(x)] - h(x)g_1(x)\} w(x)$$

with strict inequality if Φ is strictly convex and if there is an $x \in X$ such that $f(x) \neq h(x)$. □

Corollary A *Let g be a given function on X and w a positive weight function. Suppose Φ is a convex function and let ϕ be a determination of its derivative. Let $\mathscr{I}$ be the class of all functions on X which are isotonic with respect to a given quasi-order on X, and let $\mathscr{I}^{*w}$ be the dual of $\mathscr{I}$. Then the isotonic regression $g^* = P_w(g \mid \mathscr{I})$ solves*

$$\textit{minimize} \sum_{x \in X} \Phi[f(x)] w(x) \textit{ subject to } g - f \in \mathscr{I}^{*w} \qquad (1.7.4)$$

The minimizing function is unique if Φ is strictly convex.

Proof: Apply Theorem 1.7.4 with $g_1 = 0$, $g_2 = g$, and $\mathscr{H} = \{f : g - f \in \mathscr{I}^{*w}\}$. It suffices to verify that $h = g^*$ satisfies (1.7.2) and (1.7.3). The former becomes

$$\sum_{x \in X} [g^*(x) - g(x)]\phi[g^*(x)] w(x) = 0$$

by Theorem 1.3.6. The latter is, for $f \in \mathscr{I}$,

$$\sum_{x \in X} [f(x) - g(x)]\phi[g^*(x)] w(x) \geqslant 0$$

since $\phi(g^*) \in \mathscr{I}$ (ϕ is nondecreasing) and $-(f - g) \in \mathscr{I}^{*w}$. □

Corollary B *Suppose that Φ is a differentiable convex function with derivative ϕ. Let $\phi^{-1}(t)$ be defined as* $\inf\{u : \phi(u) \geqslant t\}$, *and assume that g_1 is a function defined on X such that the range of g_1 is in the domain of ϕ^{-1}. Finally, suppose that $\mathscr{I}$ is the* *collection of all functions isotonic with respect to a particular quasi-order and that w is a positive weight function. Then a function which minimizes the expression*

$$\sum_{x\in X} \{\Phi[f(x)] - f(x)g_1(x)\} w(x) \tag{1.7.5}$$

for $f \in \mathscr{I}$ *is given by* $\phi^{-1}(g_1^*) = \{\phi^{-1}[g_1^*(x_1)], \ldots, \phi^{-1}[g_1^*(x_k)]\}$, *where* $g_1^* = P_w(g_1 | \mathscr{I})$ *is the isotonic regression of* g_1 *onto* $\mathscr{I}$. *The solution is unique if* Φ *is strictly convex.*

Proof: Note that $\phi^{-1}(g_1^*) \in \mathscr{I}$ since ϕ^{-1} is nondecreasing and hence preserves the ordering used in defining $\mathscr{I}$. Moreover, it must be the case that $\phi\{\phi^{-1}[g_1^*(x)]\} = g_1^*(x)$. Then if g_2 is taken to be identically zero, conditions (1.7.2) and (1.7.3) of Theorem 1.7.4 reduce, respectively, to

$$\sum_{x\in X} \phi^{-1}[g_1^*(x)][g_1^*(x) - g_1(x)]w(x) = 0$$

and

$$\sum_{x\in X} f(x)[g_1^*(x) - g_1(x)]w(x) \geqslant 0 \text{ for } f \in \mathscr{I}.$$

These expressions follow from Theorem 1.3.6 and (1.3.8) of Theorem 1.3.2 and prove the corollary. □

Let $\mathscr{A}$ be the set of functions on $X = \{x_1, x_2, \ldots, x_k\}$ which are isotonic with respect to the partial order $x_1 \gtrsim x_2 \gtrsim \cdots \gtrsim x_k$ (i.e. $f \in \mathscr{A}$ if and only if f is antitonic with respect to the usual simple order, $x_1 \lesssim x_2 \lesssim \cdots \lesssim x_k$). As in Example 1.3.2 (i.e. the simple tree) this cone is finitely generated. In fact any set of functions which are isotonic with respect to a quasi-order on X is finitely generated. One set of generators consists of the constant functions 1 and -1 together with the collection of all indicator functions of nonempty upper sets associated with the quasi-order. For the simple order on X, described above, those generators are e_1, $e_2, \ldots, e_{k+1}$, where $e_1(x) = 1$, $e_2(x) = -1$ for all $x \in X$ and for $i = 3, 4, \ldots, k+1$,

$$e_i(x_j) = 1 \quad \text{for } 1 \leqslant j \leqslant i-2 \qquad \text{and} \qquad e_i(x_j) = 0 \quad \text{for } i-2 < j \leqslant k.$$

If $f \in \mathscr{A}$ and $f(x_k) < 0$, then

$$f = -f(x_k)e_1 + \sum_{j=3}^{k+1} [f(x_{j-2}) - f(x_{j-1})]e_j.$$

If $f \in \mathscr{A}$ and $f(x_k) \geqslant 0$, then

$$f = f(x_k)e_1 + \sum_{j=3}^{k+1} [f(x_{j-2}) - f(x_{j-1})]e_j.$$

Since the coefficients are all nonnegative the functions $e_1, e_2, \ldots, e_{k+1}$ generate $\mathscr{A}$.

Using the generators of $\mathscr{A}$ it is straightforward to characterize the dual, $\mathscr{A}^{*w}$, of $\mathscr{A}$. It is simply the set of all functions making an obtuse angle with each of the generators of $\mathscr{A}$. In this case,

$$\mathscr{A}^{*w} = \left\{ h: \sum_{j=1}^{i} h(x_j)w(x_j) \leqslant 0;\ i = 1, 2, \ldots, k-1 \text{ and } \sum_{j=1}^{k} h(x_j)w(x_j) = 0 \right\} \quad (1.7.6)$$

In order to see this, note that if $h \in \mathscr{A}^{*w}$ then h is in the right-hand side of (1.7.6) because it must make an obtuse angle with each of the generators of $\mathscr{A}$. The fact that $\sum_{j=1}^{k} h(x_j)w(x_j) = 0$ follows from taking the inner product of h with e_1 and e_2. Conversely, if h is in the right-hand side of (1.7.6) and if $f \in \mathscr{A}$ then $f = \sum_{i=1}^{k+1} \alpha_i e_i$ with $\alpha_i \geqslant 0$ and

$$\sum_{i=1}^{k} f(x_i)h(x_i)w(x_i) \leqslant 0.$$

Since f was an arbitrary member of $\mathscr{A}$, $h \in \mathscr{A}^{*w}$.

The relationship between the properties which characterize the dual of $\mathscr{A}$ and the concept of a stochastic ordering are noted in the next example. This observation coupled with Theorem 1.7.2 and the corollaries to Theorem 1.7.4 lead to the solution of several restricted estimation problems in which the restriction involves a stochastic ordering.

Example 1.7.1 One sample maximum likelihood estimates of multinomial parameters subject to a stochastic ordering constraint As in Example 1.5.7 suppose one has an experiment, the outcome of which must be one of k mutually exclusive events which have probabilities $p(x_1), p(x_2), \ldots, p(x_k)$ and suppose these probabilities are to be estimated. If n independent trials of this experiment are performed, then $p(x_i)$ could be estimated by the relative frequency, $\hat{p}(x_i)$ of the event which has probability $p(x_i)$. However, in some situations, one may believe that the $p(x_i)$ satisfy certain order restrictions and may desire estimates which also satisfy these restrictions. For instance, suppose that the x_i represent ordered categories, that is $x_1 \lesssim x_2 \lesssim \cdots \lesssim x_k$, and suppose it is believed that the distribution associated with the $p(x_i)$ is stochastically smaller than that associated with a known set of probabilities, $q(x_i)$, with $q(x_i) \geqslant 0$ and $\sum_{i=1}^{k} q(x_i) = 1$. Then one may require the $p(x_i)$ to satisfy

$$\sum_{j=1}^{i} p(x_j) \geqslant \sum_{j=1}^{i} q(x_j); \quad i = 1, 2, \ldots, k-1, \qquad (1.7.7)$$

with of course $\sum_{i=1}^{k} p(x_i) = \sum_{i=1}^{k} q(x_i) = 1$. The next result gives the maximum likelihood estimate of p subject to the constraint (1.7.7) which will be denoted by $p \gg q$.

Theorem 1.7.5 *If $\hat{p}(x_i) > 0$; $i = 1, 2, \ldots, k$, then the MLE of p subject to the restriction that $p \gg q$ is given by*

$$\bar{p} = \hat{p} P_{\hat{p}}(q/\hat{p} \mid \mathscr{A}). \qquad (1.7.8)$$

Proof: The MLE, $\bar{p}$, of p solves

$$\text{minimize} - \sum_{i=1}^{k} \hat{p}(x_i) \ln p(x_i) \text{ subject to } p \gg q. \tag{1.7.9}$$

Set $s(x_i) = p(x_i)/\hat{p}(x_i)$, $w(x_i) = \hat{p}(x_i)$, $g(x_i) = q(x_i)/\hat{p}(x_i)$ and $\Phi(y) = -\ln y$. Then by writing (1.7.9) in terms of s and by dropping terms that do not involve s, the function $\bar{s}$ defined by $\bar{s}(x_i) = \bar{p}(x_i)/\hat{p}(x_i)$ solves

$$\text{minimize} \sum_{i=2}^{k} w(x_i)\Phi[s(x_i)]$$

subject to

$$\sum_{j=1}^{i} w(x_j)[g(x_j) - s(x_j)] \leqslant 0; \quad i = 1, 2, \ldots, k-1,$$

and

$$\sum_{j=1}^{k} w(x_j)[g(x_j) - s(x_j)] = 0.$$

However, by (1.7.6) this constraint can be expressed as $g - s \in \mathscr{A}^{*w}$. Applying Corollary A of Theorem 1.7.4, one sees that the solution is given by $\bar{s} = P_w(g|\mathscr{A})$ or $\bar{p} = \hat{p}P_{\hat{p}}(q/\hat{p}|\mathscr{A})$. This is the desired result. □

Using duality, the concept of isotonic regression can be seen to provide a solution to a class of new problems, including the two-sample analogue to Example 1.7.1, a taut string problem, a production planning problem, and an inventory problem. Examples such as these will be taken up in Chapters 5 and 6.

1.8 PROOF OF THEOREM 1.4.6 AND SOME MISCELLANEOUS RESULTS

This section contains a proof of Theorem 1.4.6 which states that the iterative algorithm for the matrix partial order converges. A few miscellaneous results regarding the isotonic regression are also included. These results are not needed in Chapter 2 and so the reader may wish to go on to the second chapter and begin the study of tests of order restricted hypotheses.

Proof of Theorem 1.4.6: Denote the inner product norm by

$$\|F\| = \langle F, F \rangle^{1/2} = \left(\sum_{i=1}^{a} \sum_{j=1}^{b} f_{ij}^2 w_{ij} \right)^{1/2}$$

and observe that

$$\|\hat{G}^{(n)}\|^2 \geqslant \|\tilde{G}^{(n)}\|^2 \geqslant \|\hat{G}^{(n+1)}\|^2 \text{ for all } n. \tag{1.8.1}$$

Recall that $\mathscr{K}_r$ and $\mathscr{K}_c$ denote the row and column cones, respectively, and the dual cones are

$$\mathscr{K}_r^* = \left\{H: \sum_{i=1}^{a} h_{ij} f_{ij} w_{ij} < 0 \text{ for } j = 1, 2, \ldots, b \text{ and each } F \in \mathscr{K}_r\right\}$$

and

$$\mathscr{K}_c^* = \left\{H: \sum_{j=1}^{b} h_{ij} f_{ij} w_{ij} \leqslant 0 \text{ for } i = 1, 2, \ldots, a \text{ and each } F \in \mathscr{K}_c\right\}.$$

Now $-R^{(n+1)} = (G + C^{(n)}) - \hat{G}^{(n)} = G + C^{(n)} - P_w(G + C^{(n)} | \mathscr{K}_r) = P_w(G + C^{(n)} | \mathscr{K}_r^*)$ by Theorem 1.7.2. Thus $-R^{(n+1)}$ minimizes $\| G + C^{(n)} - F \|^2$ for $F \in \mathscr{K}_r^*$. Similarly, $-C^{(n)}$ minimizes $\| G + R^{(n)} - F \|^2$ for $F \in \mathscr{K}_r^*$. Thus, since $-R^{(n)} \in \mathscr{K}_r^*$ and $-C^{(n-1)} \in \mathscr{K}_c^*$,

$$\| G + R^{(n)} - (-C^{(n-1)}) \|^2 \geqslant \| G + R^{(n)} - (-C^{(n)}) \|^2 \geqslant \| G + C^{(n)} - (-R^{(n+1)}) \|^2,$$

which is equivalent to (1.8.1).

Next it is shown that $\{C^{(n)}\}$ and $\{R^{(n)}\}$ are bounded. If not, let (i_0, j_0) be a minimal point in X (with regard to the matrix order, $\lesssim$) such that either $\{r_{i_0 j_0}^{(n)}\}$ or $\{c_{i_0 j_0}^{(n)}\}$ is unbounded. Say there exists a subsequence $\{n_i\}$ such that $r_{i_0 j_0}^{(n_i)} \to -\infty$. The fact that $-R^{(n)} \in \mathscr{K}_r^*$ implies that $\sum_{i=1}^{i_0} r_{ij_0}^{(n)} w_{ij_0} \leqslant 0$ for all n; $r_{i_0 j_0}^{(n_i)} \to \infty$ would contradict the assumption that (i_0, j_0) is minimal. The fact that $r_{i_0 j_0}^{(n_i)} \to -\infty$ together with $\tilde{G}^{(n)} = G + R^{(n)} + C^{(n)}$ and the fact that $\tilde{G}^{(n)}$ is bounded in norm (cf. (1.8.1)) implies that $c_{i_0 j_0}^{(n_i)} \to \infty$. This, in turn, contradicts the assumption that (i_0, j_0) is minimal since $\sum_{j=1}^{j_0} c_{i_0 j}^{(n)} w_{i_0 j} \leqslant 0$ for all n.

Using the error norm property of projections onto convex sets (see Theorem 1.3.1 and its implication (1.6.1)), one concludes that

$$\begin{aligned} \| C^{(i)} - C^{(i-1)} \|^2 &= \| G + C^{(i)} - (G + C^{(i-1)}) \|^2 \geqslant \| R^{(i+1)} - R^{(i)} \|^2 \\ &= \| G + R^{(i+1)} - (G + R^{(i)}) \|^2 \geqslant \| C^{(i+1)} - C^{(i)} \|^2 \quad \text{for all } i. \end{aligned} \tag{1.8.2}$$

The next step is to show that

$$\| R^{(i+1)} - R^{(i)} \|^2 \to 0, \text{ and hence } \| C^{(i+1)} - C^{(i)} \|^2 \to 0 \quad \text{as } i \to \infty. \tag{1.8.3}$$

If (1.8.3) were not the case then there would exist $(i_0, j_0) \in X$ and $\varepsilon > 0$ such that

$$|r_{i_0 j_0}^{(i+1)} - r_{i_0 j_0}^{(i)}| > \varepsilon \text{ for infinitely many } i. \tag{1.8.4}$$

However, since $\{R^{(n)}\}$ is bounded there exists a finite M such that

$$|r_{i_0 j_0}^{(i)} - r_{i_0 j_0}^{(j)}| < M \quad \text{for all } i, j. \tag{1.8.5}$$

Write

$$\begin{aligned} &\| R^{(i+1)} - R^{(i)} \|^2 - \| C^{(i+1)} - C^{(i)} \|^2 \\ &= \| G + R^{(i)} + C^{(i)} - (G + R^{(i+1)} + C^{(i+1)}) \|^2 \\ &\quad + 2\langle G + R^{(i)} + C^{(i)} - (G + R^{(i)} + C^{(i)}), C^{(i+1)} - C^{(i)} \rangle \end{aligned} \tag{1.8.6}$$

and note that the left side of (1.8.6) converges to zero since, by (1.8.2), both terms converge to the same quantity. The last term on the right-hand side is nonnegative by Theorem 1.3.2. Thus

$$(R^{(i+1)} - R^{(i)}) + (C^{(i+1)} - C^{(i)}) \to 0 \quad \text{as } i \to \infty. \tag{1.8.7}$$

In a similar fashion, beginning with

$$\| C^{(i+1)} - C^{(i)} \|^2 - \| R^{(i+1)} - R^{(i+1)} \|^2,$$

it can be shown that

$$(R^{(i+2)} - R^{(i+1)}) + (C^{(i+1)} - C^{(i)}) \to 0 \quad \text{as } i \to \infty. \tag{1.8.8}$$

Subtracting (1.8.7) from (1.8.8) yields

$$(R^{(i+2)} - R^{(i+1)}) - (R^{(i+1)} - R^{(i)}) \to 0 \quad \text{as } i \to \infty.$$

Thus for sufficiently large N_0 and fixed n_0, the terms

$$(r_{i_0 j_0}^{(N_0+i+1)} - r_{i_0 j_0}^{(N_0+i)}), \quad i = 1, 2, \ldots, n_0,$$

can be made arbitrarily close to

$$(r_{i_0 j_0}^{(N_0+1)} - r_{i_0 j_0}^{(N_0)}).$$

This, however, contradicts the fact that both (1.8.4) and (1.8.5) are true.

Since $\{R^{(n)}\}$ and $\{C^{(n)}\}$ are bounded, there must exist convergent subsequences. Suppose $R^{(n_i)} \to R$ and $C^{(n_i)} \to C$. Then, using (1.8.3),

$$\tilde{G}^{(n_i)} = G + R^{(n_i)} + C^{(n_i)}$$

and

$$\hat{G}^{(n_i+1)} = G + R^{(n_i+1)} + C^{(n_i)}$$

both converge to $G + R + C = G'$. Since $\hat{G}^{(n)} \in \mathscr{K}_r$ and $\tilde{G}^{(n)} \in \mathscr{K}_c$ and these cones are closed, $G' \in \mathscr{K}_r \cap \mathscr{K}_c$. Furthermore,

$$\begin{aligned} \langle G - G', G' \rangle &= \langle G + R - G', G' \rangle - \langle R, G' \rangle \\ &= \lim_{i \to \infty} \langle G + R^{(n_i)} - \tilde{G}^{(n_i)}, \tilde{G}^{(n_i)} \rangle + \lim_{i \to \infty} \langle G + C^{(n_i)} - \hat{G}^{(n_i+1)}, \hat{G}^{(n_i+1)} \rangle \\ &= 0 + 0 \end{aligned}$$

by (1.3.7). Similarly, if $V \in \mathscr{K}_r \cap \mathscr{K}_c$,

$$\begin{aligned} \langle G - G', V \rangle &= \langle G + R - G', V \rangle - \langle R, V \rangle \\ &= \lim_{i \to \infty} \langle G + R^{(n_i)} - \tilde{G}^{(n_i)}, V \rangle + \lim_{i \to \infty} \langle G + C^{(n_i)} - \hat{G}^{(n_i+1)}, V \rangle \\ &\leqslant 0 + 0 \end{aligned}$$

by (1.3.8). Hence, using Theorem 1.3.2, G' is the isotonic regression of G with respect to the matrix order. Moreover, since $-C$ minimizes $\|G+R-F\|^2$ for $F\in\mathscr{K}_c^*$ and $-R$ minimizes $\|G+C-F\|^2$ for $F\in\mathscr{K}_r^*$, one may conclude from the norm reducing property of projections that

$$\|C^{(n)}-C\|^2 = \|G+C^{(n)}-(G+C)\|^2 \geqslant \|R^{(n+1)}-R\|^2$$
$$= \|G+R^{(n+1)}-(G+R)\|^2 \geqslant \|C^{(n+1)}-C\|^2 \text{ for all } n.$$

Thus $R^{(n)}\to R$ and $C^{(n)}\to C$ as $n\to\infty$, which implies that $\hat{G}^{(n)}=G+R^{(n)}+C^{(n-1)}$ and $\tilde{G}^{(n)}=G+R^{(n)}+C^{(n)}$ both converge to $G'=G+R+C=P_w(G|\mathscr{K}_r\cap\mathscr{K}_c)$ as $n\to\infty$. □

It is important to note that the solution $G'=G+R+C$ does not uniquely determine R and C. In fact, if one begins with column smoothings rather than row smoothing, different limits for R and C are obtained even though the same limiting solution, G', is obtained. The algorithm also extends in a natural fashion to situations in which one has more than two independent variables and the proof of convergence follows similar lines to that given above.

Miscellaneous results regarding the isotonic regression

Lemma *Let f and h be isotonic functions on X and let ρ be a nonnegative function of a real argument. Then the function which assigns to an arbitrary $x\in X$ the value $f(x)\rho[h(x)]$ is equal to a sum of a function of h and an isotonic function on X.*

Proof: Let I_A denote the indicator of the set A and let C denote the range of h, that is $C=\{h(x):x\in X\}$. Set

$$f_m=\min_{x\in X} f(x) \qquad \text{and} \qquad f^m=\max_{x\in X} f(x)$$

and consider the function Ψ of a real argument defined by

$$\Psi(t)=(f^m-f_m)\sum_{c\in C}\rho(c)I_{(-\infty,c]}(t).$$

It will be shown that the function f_1 defined by

$$f_1(x)=f(x)\rho[h(x)]-\{f_m\rho[h(x)]+\Psi[h(x)]\}$$

is isotonic on X. Since the function in braces is a function of h this will prove the lemma. It is thus necessary to show that $x\lesssim y$ implies that $f_1(x)\leqslant f_1(y)$. However,

$$f_1(y)=[f(y)-f_m]\rho[h(y)]-(f^m-f_m)\sum_{c\in C:c\geqslant h(y)}\rho(c),$$

and a similar expression is obtained for $f_1(x)$. Since h is isotonic,

$$\begin{aligned} f_1(y) - f_1(x) &= [f(y) - f_m]\rho[h(y)] - [f(x) - f_m]\rho[h(x)] \\ &\quad + (f^m - f_m) \sum_{c \in C: h(x) \leqslant c < h(y)} \rho(c) \\ &= [f(y) - f_m]\rho[h(y)] + [f_m - f(x)]\rho[h(x)] \\ &\quad + (f^m - f_m) \sum_{c \in C: h(x) < c < h(y)} \rho(c) \geqslant 0. \quad \square \end{aligned}$$

Theorem 1.8.1 *A necessary and sufficient condition that an isotonic function, u, on X be the isotonic regression of g is that for every isotonic function f on X and every nonnegative function ρ of a real argument*

$$\sum_{x \in X} [g(x) - u(x)]\rho[u(x)]f(x)w(x) \leqslant 0.$$

Proof: Suppose the inequality holds for all such ρ and f. Setting $\rho = 1$, it is found that (1.3.8) is satisfied. Setting $\rho = 1$ and first $f = 1$ and then $f = -1$ yields (1.3.9). Now let m and M be the lower and upper bounds on the values of u, respectively. Setting $f = 1$ and $\rho(t) = (t - m) \vee 0$ and using (1.3.9) it is found that

$$\sum_{x \in X} [g(x) - u(x)]u(x)w(x) \leqslant 0.$$

Setting $f = 1$ and $\rho(t) = (M - t) \vee 0$ and using (1.3.9) it is found that

$$\sum_{x \in X} [g(x) - u(x)]u(x)w(x) \geqslant 0,$$

so that (1.3.7) is satisfied. An application of Theorem 1.3.2 completes the proof of sufficiency.

The necessity is established next. By the preceding lemma,

$$f(x)\rho[g^*(x)] = \Psi[g^*(x)] + f_1(x)$$

with f_1 an isotonic function. Applying Theorem 1.3.6,

$$\sum_{x \in X} [g(x) - g^*(x)]\Psi[g^*(x)]w(x) = 0,$$

and applying Theorem 1.3.2,

$$\sum_{x \in X} [g(x) - g^*(x)]f_1(x)w(x) \leqslant 0.$$

Hence,

$$\sum_{x \in X} [g(x) - g^*(x)]\rho[g^*(x)]f(x)w(x) \leqslant 0. \quad \square$$

Corollary *If ρ is a nonnegative function of a real argument, if Ψ is an arbitrary function of a real argument, and if g^* is the isotonic regression of g, then the isotonic regression of $\rho(g^*)g + \Psi(g^*)$ is $\rho(g^*)g^* + \Psi(g^*)$, provided the latter is isotonic.*

Theorem 1.8.1 yields a theorem of Robertson (1965) which is useful in finding estimates under stochastic ordering restrictions (cf. Section 4.4).

Theorem 1.8.2 *Let a and b be positive functions on X and let g^* be the isotonic regression of a/b with weights b with respect to a quasi-order on X. Then $1/g^*$ is the antitonic regression of b/a with weights a, i.e. the isotonic regression of b/a with weights a and with respect to the reverse quasi-order on X.*

Proof: Since $a/b > 0$, by Theorem 1.3.4, $g^* > 0$. Let $u = 1/g^*$. By Theorem 1.8.1 it suffices to show that for every antitonic function f on X and every nonnegative function ρ on the reals,

$$\sum_{x \in X} \left[\frac{b(x)}{a(x)} - \frac{1}{g^*(x)} \right] \rho \left[\frac{1}{g^*(x)} \right] f(x) a(x) \leqslant 0.$$

However, the left member can be rewritten as

$$\sum_{x \in X} \left[\frac{a(x)}{b(x)} - g^*(x) \right] \left[\frac{1}{g^*(x)} \right] \rho \left[\frac{1}{g^*(x)} \right] [-f(x)] b(x),$$

and since $-f(x)$ is isotonic this sum is nonpositive by Theorem 1.8.1. □

1.9 COMPLEMENTS

The phrase 'monotone regression, appears in Lombard and Brunk (1963) and also in Kruskal (1965; p. 253). Isotonic, i.e. order preserving, has the advantage that it specifies direction whereas monotonic may be thought of as order preserving or order reversing (antitonic).

The use of the GCM for representing isotonic regression is due to W. T. Reid (cf. Brunk, 1956). It was used in the problem of estimating the failure rate of a distribution by Grenander (1956). The PAVA and the max–min formulas were given by Ayer and coworkers (1955) in a study of maximum likelihood estimation of completely ordered binomial parameters (bioassay). The least squares property of the solution, in effect identifying it with isotonic regression, is also noted in that paper. The bioassay problem was studied independently by Eeden (1956), who admitted partial orders and gave methods of solution. The PAVA is implicit in Eeden's research. It is also Miles' method $\mathscr{A}_1$ (1959) (cf. also Bartholomew, 1959a). The minimum lower sets algorithm is given in Brunk, Ewing and Utz (1957) and in Brunk (1955). The iterative algorithm for the matrix partial order is developed in Dykstra and Robertson (1982a). This type of iterative algorithm has been extended to a large number of restricted optimization problems (cf. Dykstra, 1983).

The problem of developing algorithms for the isotonic regression has received a great deal of attention from researchers in the area and Barlow *et al.* (1972) describe several that are not discussed here. The *up-and-down blocks algorithm*, developed by J. B. Kruskal (1964b) is essentially a version of PAVA and applies only in the case of a simple order. Many of the other algorithms which have been developed for quasi or partial orders attempt to mimic the PAVA by finding simple orders which are in some sense consistent with the partial order thus allowing a PAVA-type algorithm to be applied and thereby allowing poolings in pairs. After a pair is pooled it may not stay pooled in the final estimate. Thus, such algorithms usually require complicated checking and branching logic and can be difficult to understand and to program.

The *minimum violator algorithm*, due to W. A. Thompson Jr (1962), the *minimax order algorithm*, due to Alexander (1970), and an algorithm due to Eeden (1956, 1957a, 1957b, 1958) are described in Section 2.3 of Barlow *et al.* (1972). The algorithm given in Example 1.3.2 for the simple tree is a special case of Thompson's minimum violator algorithm. This minimum violator algorithm applies to partial orders in which each element of X, with one exception, has exactly one immediate predecessor. At least two researchers have observed recently that *Alexander's minimax order algorithm* does not converge (cf. Lee, 1983, and Murray, 1983). Another of Eeden's algorithms is based upon the following important observation. If there are two or more disjoint subsets of X such that the order restriction does not relate any members of X contained in distinct members of this collection then the isotonic regression may be computed on these sets separately.

Hoadley (1971) gives an algorithm for simple order and Gebhardt (1970) gives algorithms for the simple order and for the matrix partial order discussed in Section 1.4. Other algorithms are discussed in Dykstra (1981, 1983) and in Lee (1983).

The idea discussed in Example 1.7.3 of using isotonic regression to measure the quality of fit in multidimensional scaling is due to Kruskal (1964a).

Among the more interesting properties of the isotonic regression g^* is that of being the *isotonic function most highly correlated with* g. A precise statement of this property in a more general setting is given in Corollary E to Theorem 8.2.7.

The fact that the isotonic regression solves a wide class of restricted optimization problems, of which those for finite X are discussed in Section 1.5, was discovered independently by W. T. Reid and by Brunk, Ewing, and Utz (1957). Application of the results of this paper to maximum likelihood estimation of ordered parameters was made in Brunk (1958). (In spite of the apparent discrepancy in dates, the work of Brunk, Ewing, and Utz, 1957, preceded that of Brunk, 1955). Further statistical applications are noted in Brunk (1965). Restricted optimization problems for *norms other than the* L_2 *norm* are discussed in Section 9.1.

The application of Theorem 1.5.1 to the exponential family discussed in Theorem 1.5.2 was discovered by a student, Zehua Chen, in a class on order restricted inference at the University of Iowa in 1985.

Chacko (1966) considered the problem of statistical inferences regarding a set of multinomial parameters. The generalization of the problem of finding the MLE of ordered multinomial parameters derived in Example 1.5.7 is discussed in Robertson (1965) using Theorem 1.8.2.

The reduction of total square error property for the isotonic regression contained in (1.6.1) was apparently first observed by an anonymous referee (cf. Ayer *et al.*, 1955, p. 644) for the bioassay problem. Eeden obtained this result in a more general setting. The reduction of maximum absolute error contained in Corollary B to Theorem 1.6.1 was obtained by Robertson and Wright (1974b) in a general context for estimators representable with max–min formulas involving Cauchy mean value functions. It was obtained in a related problem by Barlow and Ubhaya (1971). The reduction of mean square error at each $x \in X$ for the normal means problem with a simple order was deduced by Lee (1981). The observation that this property does not hold, in general, for partial order restrictions is also due to Lee.

Fenchel duality and its role in order restricted inference was explored by Barlow and Brunk (1972). This paper is an important milestone in the theory of order restricted inference. They also observed the dual relationship between the simple and stochastic ordering problems. Maximum likelihood estimates under a stochastic ordering restriction were first researched by Brunk *et al.* (1966).

Stochastic approximation has to do with finding roots of regression equations. Mukerjee (1981) explores the idea of using monotonicity properties of the regression function to improve the quality of the estimate of the roots.

Order restricted estimates are not, in general, *admissible*. The question of admissibility of these estimates is discussed in Blumenthal and Cohen (1968), Sackrowitz (1970, 1982), Cohen and Sackrowitz (1970), and Sackrowitz and Strawderman (1974).

Some Bayesian approaches to estimating ordered parameters are discussed in Section 9.3 and the *robust estimation* of ordered location parameters is treated in Magel and Wright (1984b).

Order restricted estimates can play a role in resolving the problem of *negative estimates of variance* in ANOVA. A discussion of this theory and some examples are given in Thompson (1962).

In this chapter, estimation of an isotonic function defined on a finite index set has been treated. The problem of estimating an isotonic function defined on an interval in R^k has been studied in the literature. Section 9.2 discusses the large sample properties of such estimates and some smoothing techniques for such estimates are mentioned in Section 9.3.

CHAPTER 2

Tests of Ordered Hypotheses: The Normal Means Case

2.1 INTRODUCTION

Many of the methods of statistical inference are derived from the problem of comparing several normal populations. As a preliminary step in such investigations, it is often desirable to test the null hypothesis that the means are equal. The classical tests have been developed for the extreme situations in which the alternative is either unrestricted or else has very stringent restrictions placed upon it, such as assuming that these means are a linear function of a known variable. In applications, a researcher may be reluctant to specify a functional form for the means, but may believe *a priori* that their ordering is known, or more generally that the means are isotonic with respect to a known quasi-order on the index set. In this chapter, the likelihood ratio tests (LRTs) for homogeneity of normal means with order restricted alternatives are developed and studied. For alternatives which satisfy, or nearly satisfy, the order restriction, these tests have greater power than the omnibus procedures, and at the same time they provide protection against the possibility of specifying an incorrect functional form.

If the quasi-ordering imposed on the alternative is in question, one may wish to test this order restriction as the null hypothesis with an unrestricted alternative. This procedure shares the disadvantage of goodness-of-fit tests in that the conclusion the researcher may desire to establish is the null hypothesis. Consequently, one must exercise caution in specifying the level of the test. In other situations, this procedure may be used as a test for the presence of a trend in normal means for which the probability of rejecting the trend, when it is present, is controlled. LRTs in this setting are also discussed in this chapter. Tests based on a single contrast as well as the maximum of several contrast have been proposed for the two testing situations mentioned above. In Chapter 4, these contrast tests are described and their power characteristics compared with those of the LRTs. As is often the case, the procedures for normal means suggest large sample normal approximations for nonnormal distributions as well as distribution-free procedures based on ranks. Such procedures are discussed in Chapter 4 also.

The restricted MLEs obtained in Chapter 1 are needed for the statistics used in these LRTs and so these testing problems are formulated in terms of the notation of that chapter. Let $X = \{x_1, x_2, \ldots, x_k\}$ be an index set with $\lesssim$ a quasi-order on X and let $\mu(x_i)$ be the mean of a normal population with variance $\sigma^2(x_i)$ for $i = 1, 2, \ldots, k$. For example, if the x_i denote increasing dosage levels of a drug and it is believed that the mean response should increase with dosage levels in this range, then one would set $X = \{1, 2, \ldots, k\}$ with $\lesssim$ the usual simple order on X. If several treatments are being compared with a control and it is believed that the treatment means are at least as large as the control mean, then one would set $X = \{0, 1, \ldots, k\}$ and let $\lesssim$ be the simple tree ordering discussed in Chapter 1. Consider a regression setting with two independent variables and observations at each point on the $R \times C$ grid obtained by fixing the first independent variable at one of R increasing levels and the second independent variable at one of C increasing levels. If the regression function is nondecreasing in each variable with the other fixed and $\mu(i, j)$ denotes the mean response with the first variable fixed at level i and the second at level j, then one would set $X = \{(i, j): i = 1, 2, \ldots, R, j = 1, 2, \ldots, C\}$ with $\lesssim$ the matrix order which was discussed in Chapter 1.

To simplify the notation, one can, by relabelling, take $X = \{1, 2, \ldots, k\}$ and assume that $\lesssim$ is defined on X. One can also denote the means by $\mu_1, \mu_2, \ldots, \mu_k$, the standard deviations by $\sigma_1, \sigma_2, \ldots, \sigma_k$, the mean vector by $\boldsymbol{\mu}$ and the vector of standard deviations by $\boldsymbol{\sigma}$. With this convention the hypothesis that the means satisfy the order restriction can be written as

$$H_1: \boldsymbol{\mu} \text{ is isotonic with respect to } \lesssim.$$

The symbol H, with or without subscripts, will be used to denote a hypothesis as well as the region of the parameter space consisting of points that satisfy the hypothesis. The hypothesis of homogeneity is

$$H_0: \mu_1 = \mu_2 = \cdots = \mu_k$$

and H_2 places no restrictions on the means. In this chapter tests of H_0 versus $H_1 - H_0$, that is the mean vector is isotonic but not constant, and of H_1 versus $H_2 - H_1$ are developed. As in the classical one-way analysis of variance, the variances are assumed to be known or known up to a multiplicative constant.

2.2 THE $\bar{\chi}^2$ AND $\bar{E}^2$ TESTS

The $\bar{\chi}^2$ tests: variances known

For $i = 1, 2, \ldots, k$, suppose that $\bar{Y}_i$ is the mean of a random sample of size n_i from a normal population with unknown mean μ_i and known variance

σ_i^2, and suppose that the samples are independent. Results obtained under these assumptions provide large sample approximations for the case of unknown variances, one-parameter exponential families, and rank tests. Because $\bar{\mathbf{Y}} = (\bar{Y}_1, \bar{Y}_2, \ldots, \bar{Y}_k)$ is sufficient for $\boldsymbol{\mu}$, the LRTs of H_0 versus $H_1 - H_0$ and of H_1 versus $H_2 - H_1$ can be determined from $L(\bar{y}_1, \bar{y}_2, \ldots, \bar{y}_k; \boldsymbol{\mu}, \boldsymbol{\sigma})$, the likelihood function.

Testing H_0 versus $H_1 - H_0$

Under H_0, the MLE of $\mu_1 = \mu_2 = \cdots \mu_k$ is given by

$$\hat{\mu} = \frac{\sum_{i=1}^k w_i \bar{Y}_i}{\sum_{i=1}^k w_i} \qquad \text{with } w_i = n_i/\sigma_i^2. \tag{2.2.1}$$

Under H_1, the MLE of $\boldsymbol{\mu}$ is $\boldsymbol{\mu}^*$, the isotonic regression of $\bar{\mathbf{Y}}$ with weight vector $\mathbf{w} = (w_1, w_2, \ldots, w_k)$ and the quasi-order $\lesssim$ which determines H_1 (cf. Definition 1.3.3). The unrestricted MLE of $\boldsymbol{\mu}$ is $\bar{\mathbf{Y}}$.

The likelihood ratio for testing H_0 versus $H_1 - H_0$ is

$$\Lambda = \frac{\max_{\mu \in H_0} L(\bar{\mathbf{y}}; \boldsymbol{\mu}, \boldsymbol{\sigma})}{\max_{\mu \in H_1} L(\bar{\mathbf{y}}; \boldsymbol{\mu}, \boldsymbol{\sigma})} \tag{2.2.2}$$

and

$$\begin{aligned} -2\log\Lambda &= \sum_{i=1}^k w_i(\bar{y}_i - \hat{\mu})^2 - \sum_{i=1}^k w_i(\bar{y}_i - \mu_i^*)^2 \\ &= \sum_{i=1}^k w_i(\mu_i^* - \hat{\mu})^2 + 2\sum_{i=1}^k w_i(\bar{y}_i - \mu_i^*)(\mu_i^* - \hat{\mu}). \end{aligned} \tag{2.2.3}$$

Applying Theorem 1.3.6, the last term in (2.2.3) is seen to be zero. Hence, the LRT rejects H_0 for large values of

$$\bar{\chi}_{01}^2 = \sum_{i=1}^k w_i(\mu_i^* - \hat{\mu})^2. \tag{2.2.4}$$

If the level sets of $\boldsymbol{\mu}^*$ have been determined to be $B_1, B_2, \ldots, B_m$, the weight corresponding to B_l is $W_l = \sum_{i \in B_l} w_i$, and the value of $\boldsymbol{\mu}^*$ on B_l is $U_l = \mathrm{Av}(B_l)$, then

$$\hat{\mu} = \frac{\sum_{l=1}^m \sum_{i \in B_l} w_i \bar{Y}_i}{\sum_{l=1}^m W_l} = \frac{\sum_{l=1}^m W_l U_l}{\sum_{l=1}^m W_l},$$

which will also be denoted by $\bar{U}$, and

$$\bar{\chi}_{01}^2 = \sum_{l=1}^m \sum_{i \in B_l} w_i(\mu_i^* - \hat{\mu})^2 = \sum_{l=1}^m W_l(U_l - \bar{U})^2.$$

Hence, $\bar{\chi}_{01}^2$ can be computed using U_l and W_l, $l = 1, 2, \ldots, m$. To illustrate the use of this computational formula, consider the case with $k = 5$, $\lesssim$ the usual simple order on $X = \{1, 2, \ldots, 5\}$ and the data given in the example below.

Example 2.2.1

Population, i	1	2	3	4	5
Sample size, n_i	10	12	9	15	11
Sample means, $\bar{y}_i$	63.9	58.2	62.3	75.4	68.5

Suppose that the variances are known to have the common value $\sigma^2 = 100$. Using the PAVA, the level sets are found to be $B_1 = \{1, 2\}$, $B_2 = \{3\}$, and $B_3 = \{4, 5\}$; the corresponding weights are $W_1 = 0.22$, $W_2 = 0.09$, and $W_3 = 0.26$; and $U_1 = 60.791$, $U_2 = 62.3$ and $U_3 = 72.481$. Hence,

$$\bar{U} = \frac{22 \times 60.791 + 9 \times 62.3 + 26 \times 72.481}{57} = 66.361$$

and

$$\begin{aligned}\bar{\chi}^2_{01} &= 0.22(60.791 - 66.361)^2 + 0.09(62.3 - 66.361)^2 + 0.26(72.481 - 66.361)^2 \\ &= 18.048.\end{aligned}$$

Testing H_1 versus H_2–H_1

The likelihood ratio for testing H_1 versus H_2–H_1 is

$$\Lambda = \frac{\max_{\mu \in H_1} L(\bar{\mathbf{y}}; \boldsymbol{\mu}, \boldsymbol{\sigma})}{\max_{\mu \in H_2} L(\bar{\mathbf{y}}; \boldsymbol{\mu}, \boldsymbol{\sigma})}$$

and

$$-2 \log \Lambda = \sum_{i=1}^{k} w_i (\bar{y}_i - \mu_i^*)^2.$$

Thus, in this case the LRT rejects H_1 for large values of

$$\bar{\chi}^2_{12} = \sum_{i=1}^{k} w_i (\bar{Y}_i - \mu_i^*)^2. \tag{2.2.5}$$

Using the data in Example 2.2.1, $\bar{\chi}^2_{12}$ is computed to be

$$\begin{aligned}&0.1(63.9 - 60.791)^2 + 0.12(58.2 - 60.791)^2 + 0.15(75.4 - 72.481)^2 \\ &\quad + 0.11(68.5 - 72.481)^2 = 4.794.\end{aligned} \tag{2.2.6}$$

It is instructive to note that, applying Theorem 1.3.6, the omnibus test statistic for H_0 versus H_2–H_0 is

$$\chi^2_{02} = \sum_{i=1}^{k} w_i (\bar{Y}_i - \hat{\mu})^2 = \bar{\chi}^2_{01} + \bar{\chi}^2_{12}. \tag{2.2.7}$$

The $\bar{E}^2$ tests: variances unknown

For $i = 1, 2, \ldots, k$ and $j = 1, 2, \ldots, n_i$, suppose that Y_{ij} is a normally distributed random variable with unknown mean μ_i and variance of the form

$$\sigma_i^2 = a_i \sigma^2 \tag{2.2.8}$$

with $a_1, a_2, \ldots, a_k$ known and σ^2 unknown, and assume that the Y_{ij} are independent. Of course, for the case of equal but unknown variances, one would set $a_1 = a_2 = \cdots = a_k = 1$. Under assumption (2.2.8), the weights are defined by $w_i = n_i/a_i$ for $i = 1, 2, \ldots, k$. The likelihood function is

$$(2\pi\sigma^2)^{-N/2} \prod_{i=1}^{k} a_i^{-n_i/2} \exp\left[-\frac{1}{2\sigma^2} \sum_{i=1}^{k} a_i^{-1} \sum_{j=1}^{n_i} (y_{ij} - \mu_i)^2 \right]$$

with $N = \sum_{i=1}^{k} n_i$. To maximize $L(\mathbf{y}; \boldsymbol{\mu}, \sigma)$ with $\boldsymbol{\mu} \in H \subset R^k$ and $\sigma > 0$, one first observes that for fixed $\sigma > 0$ the $\hat{\boldsymbol{\mu}}_H = (\hat{\mu}_{H1}, \hat{\mu}_{H2}, \ldots, \hat{\mu}_{Hk}) \in H$ that maximizes $L(\mathbf{y}; \boldsymbol{\mu}, \sigma)$ is the same for each σ, and for a fixed vector $\hat{\boldsymbol{\mu}}_H$ the $\hat{\sigma}^2$ that maximizes $L(\mathbf{y}; \boldsymbol{\mu}, \sigma)$ is given by

$$\hat{\sigma}^2 = \sum_{i=1}^{k} a_i^{-1} \sum_{j=1}^{n_i} (y_{ij} - \hat{\mu}_{Hi})^2 / N. \tag{2.2.9}$$

Maximization of the likelihood function with $\sigma = 1$ is the unknown variances problem solved previously with σ_i^2 replaced by a_i. It follows that the solutions given there yield estimates of $\boldsymbol{\mu}$ under H_0, H_1, and H_2.

Testing H_0 versus $H_1 - H_0$

Denoting the estimate of σ^2 given in (2.2.9) with $H = H_i$ by $\hat{\sigma}_{H_i}^2$, it follows that the likelihood ratio test of H_0 versus $H_1 - H_0$ rejects for small values of $\Lambda = (\hat{\sigma}_{H_1}^2 / \hat{\sigma}_{H_0}^2)^{N/2}$.

Equivalently, the LRT rejects H_0 for large values of

$$1 - \Lambda^{2/N} = \frac{\sum_{i=1}^{k} a_i^{-1} \sum_{j=1}^{n_i} (y_{ij} - \hat{\mu})^2 - \sum_{i=1}^{k} a_i^{-1} \sum_{j=1}^{n_i} (y_{ij} - \mu_i^*)^2}{\sum_{i=1}^{k} a_i^{-1} \sum_{j=1}^{n_i} (y_{ij} - \hat{\mu})^2}.$$

The numerator is equal to

$$\sum_{i=1}^{k} w_i (\mu_i^* - \hat{\mu})^2 + 2 \sum_{i=1}^{k} w_i (\mu_i^* - \hat{\mu})(\bar{y}_i - \mu_i^*)$$

and, as before, the last term is zero by Theorem 1.3.6. Hence, the LRT rejects for large values of

$$\bar{E}_{01}^2 = \frac{\sum_{i=1}^{k} w_i (\mu_i^* - \hat{\mu})^2}{\sum_{i=1}^{k} a_i^{-1} \sum_{j=1}^{n_i} (Y_{ij} - \hat{\mu})^2}. \tag{2.2.10}$$

It should be noted that

$$\sum_{i=1}^{k} a_i^{-1} \sum_{j=1}^{n_i} (Y_{ij} - \hat{\mu})^2 = \sum_{i=1}^{k} w_i(\bar{Y}_i - \hat{\mu})^2 + \sum_{i=1}^{k} a_i^{-1} \sum_{j=1}^{n_i} (Y_{ij} - \bar{Y}_i)^2$$
$$= \sum_{i=1}^{k} w_i(\bar{Y}_i - \mu_i^*)^2 + \sum_{i=1}^{k} a_i^{-1} \sum_{j=1}^{n_i} (Y_{ij} - \bar{Y}_i)^2$$
$$+ \sum_{i=1}^{k} w_i(\mu_i^* - \hat{\mu})^2,$$

so that

$$\bar{E}_{01}^2 = \frac{\sum_{i=1}^{k} w_i(\mu_i^* - \hat{\mu})^2}{\sum_{i=1}^{k} w_i(\mu_i^* - \hat{\mu})^2 + \sum_{i=1}^{k} w_i(\mu_i^* - \bar{Y}_i)^2 + \sum_{i=1}^{k} a_i^{-1} \sum_{j=1}^{n_i} (Y_{ij} - \bar{Y}_i)^2}.$$

For later use, with v a nonnegative integer, define

$$\bar{E}_{01}^2(v) = \frac{\sum_{i=1}^{k} w_i(\mu_i^* - \hat{\mu})^2}{\sum_{i=1}^{k} w_i(\mu_i^* - \hat{\mu})^2 + \sum_{i=1}^{k} w_i(\mu_i^* - \bar{Y}_i)^2 + \sigma^2 Q(v)} \tag{2.2.11}$$

with $\{Q(v)\}$ a sequence of random variables which are independent of the $\bar{Y}_i$ and $Q(v) \sim \chi_v^2$, where χ_v^2 denotes a central chi-square variable with v degrees of freedom. These random variables are constructed in such a way that as $v \to \infty$, one may assume that the other quantities in $\bar{E}_{01}^2(v)$ stay fixed. While $\sigma^2 Q(N-k)$ and $\sum_{i=1}^{k} a_i^{-1} \sum_{j=1}^{n_i} (Y_{ij} - \bar{Y}_i)^2$ need not be the same, these two random variables have the same distribution and both are independent of the other quantities in $\bar{E}_{01}^2$. Thus, while $\bar{E}_{01}^2$ and $\bar{E}_{01}^2(N-k)$ need not be the same, these two quantities have the same distribution. The statistic $S_{01}(v) = v\bar{E}_{01}^2(v)/[1 - \bar{E}_{01}^2(v)]$ is a strictly increasing function of $\bar{E}_{01}^2(v)$ and, as will be shown, as $v \to \infty$ its distribution approaches that of $\bar{\chi}_{01}^2$, which will be denoted symbolically by $S_{01}(v) \xrightarrow{D} \bar{\chi}_{01}^2$ as $v \to \infty$. Write

$$S_{01}(v) = \frac{v\sigma^2 \bar{\chi}_{01}^2}{\sigma^2 \bar{\chi}_{12}^2 + \sigma^2 Q(v)} = \frac{\bar{\chi}_{01}^2}{\bar{\chi}_{12}^2/v + Q(v)/v}. \tag{2.2.12}$$

Fix $\mathbf{w} = (w_1, w_2, \ldots, w_k)$. As $v \to \infty$, $Q(v)/v \xrightarrow{P} 1$. (The mean and variance of $Q(v)/v$ are 1 and $2/v$, respectively.) Moreover, since $\bar{\chi}_{12}^2$ is fixed, $\bar{\chi}_{12}^2/v \to 0$ at every point in the underlying probability space. Consequently, $S_{01}(v) \xrightarrow{D} \bar{\chi}_{01}^2$ as $v \to \infty$. Hence, for large $N-k$, the distribution of $S_{01} = (N-k)\bar{E}_{01}^2/(1 - \bar{E}_{01}^2)$ is approximated by that of $\bar{\chi}_{01}^2$. It is more convenient to table the critical values for S_{01} because they lend themselves to interpolation for large degrees of freedom. The following example illustrates the calculation of $\bar{E}_{01}^2$ and S_{01}.

Example 2.2.2 Assume that the means in Example 2.2.1 were calculated from samples from normal populations with unknown but equal variances and that, in addition,

$$\sum_{j=1}^{n_i} Y_{ij}^2 = 41382.1,\ 42117.9,\ 35531.4,\ 86471.5,\ 52483.7;\ i = 1, 2, \ldots, 5.$$

In that example, the numerator of $\bar{E}_{01}^2$ was calculated to be 1804.8 and the denominator is given by

$$\sum_{i=1}^{5} \sum_{j=1}^{n_i} y_{ij}^2 - N\hat{\mu}^2 = 257\,986.6 - 57(66.361)^2 = 6971.008.$$

(Rounding in $\hat{\mu}$ makes some difference in this value. The same comment holds throughout this example.) Thus, $\bar{E}_{01}^2 = 0.2589$ and $S_{01} = 18.166$.

Testing H_1 versus H_2–H_1

The LRT of H_1 versus H_2–H_1 rejects H_1 for small values of

$$\Lambda = \left(\frac{\hat{\sigma}_{H_2}^2}{\hat{\sigma}_{H_1}^2} \right)^{N/2},$$

or equivalently for large values of

$$1 - \Lambda^{2/N} = \frac{\sum_{i=1}^{k} w_i (\bar{Y}_i - \mu_i^*)^2}{\sum_{i=1}^{k} a_i^{-1} \sum_{j=1}^{n_i} (Y_{ij} - \mu_i^*)^2}, \tag{2.2.13}$$

which will be denoted by $\bar{E}_{12}^2$. Because

$$\bar{E}_{12}^2 = \frac{\sigma^2 \bar{\chi}_{12}^2}{\sigma^2 \bar{\chi}_{12}^2 + \sum_{i=1}^{k} a_i^{-1} \sum_{j=1}^{n_i} (Y_{ij} - \bar{Y}_i)^2},$$

with $Q(\nu)$ defined as before, define

$$\bar{E}_{12}^2(\nu) = \frac{\bar{\chi}_{12}^2}{\bar{\chi}_{12}^2 + Q(\nu)} \quad \text{and} \quad S_{12}(\nu) = \frac{\nu \bar{E}_{12}^2(\nu)}{1 - \bar{E}_{12}^2(\nu)}. \tag{2.2.14}$$

Because $S_{12}(\nu)$ is an increasing function of $\bar{E}_{12}^2(\nu)$, rejecting for large values of $S_{12}(\nu)$ is equivalent to rejecting for large values of $\bar{E}_{12}^2(\nu)$. Furthermore, $S_{12}(\nu) = \bar{\chi}_{12}^2/[Q(\nu)/\nu]$, which converges in distribution to $\bar{\chi}_{12}^2$ as $\nu \to \infty$. Of course, $\bar{E}_{12}^2 \sim \bar{E}_{12}^2(N-k)$ which, for large $N-k$, can be approximated by $\bar{\chi}_{12}^2$. Because the critical values of $S_{12} = (N-k)\bar{E}_{12}^2/(1-\bar{E}_{12}^2)$ lend themselves better to interpolation for large degrees of freedom, they will be tabled rather than those for $\bar{E}_{12}^2$.

The statistic $\bar{E}_{12}^2$ is easily computed using the formula

$$\bar{E}_{12}^2 = \frac{\sigma^2 \bar{\chi}_{12}^2}{N\hat{\sigma}_{H_2}^2 + \sigma^2 \bar{\chi}_{12}^2} \tag{2.2.15}$$

where $\sigma^2 \bar{\chi}_{12}^2$ is $\bar{\chi}_{12}^2$ computed with weights $w_i = n_i/a_i$ for $i = 1, 2, \ldots, k$. To illustrate the use of (2.2.15), consider the data in Example 2.2.2. From (2.2.6), $\sigma^2\bar{\chi}_{12}^2$ is found to be 479.4 and

$$N\hat{\sigma}_{H_2}^2 = \sum_{i=1}^{5}\left(\sum_{j=1}^{n_i} y_{ij}^2 - n_i\bar{y}_i^2\right) = 4683.86,$$

and thus $\bar{E}_{12}^2 = 0.0928$ and $S_{12} = 5.322$.

Tests for main effects in a two-way layout

Tests of ordered hypotheses in the general linear model are discussed in Chapter 4. However, because the balanced two-way layout is a natural generalization of the topics discussed thus far in this chapter and because of the simplicity of orthogonal designs it will be treated here. The usual model, with interaction, is

$$Y_{ijl} = \mu + \alpha_i + \beta_j + \gamma_{ij} + \varepsilon_{ijl}, \qquad i = 1, 2, \ldots, R,$$
$$j = 1, 2, \ldots, C, \qquad l = 1, 2, \ldots, n, \tag{2.2.16}$$

with the ε_{ijl} independent and identically distributed normal variables with mean zero and variance σ^2, and for $r \in \{1, 2, \ldots, R\}$ and $c \in \{1, 2, \ldots, C\}$,

$$\sum_{i=1}^{R} \alpha_i = \sum_{j=1}^{C} \beta_j = \sum_{i=1}^{R} \gamma_{ic} = \sum_{j=1}^{C} \gamma_{rj} = 0.$$

It is convenient to let μ_{ij} denote $\mu + \alpha_i + \beta_i + \gamma_{ij}$. Tests of the hypothesis $H_0{:}\alpha_1 = \alpha_2 = \cdots \alpha_R = 0$ versus $H_1 - H_0$ with $H_1{:}\boldsymbol{\alpha}$ is isotonic with respect to $\lesssim$ where $\lesssim$ is a quasi-order on $\{1, 2, \ldots, R\}$, and tests of H_1 versus $H_2 - H_1$ with the hypothesis H_2 placing no restrictions on $\boldsymbol{\alpha} = (\alpha_1, \alpha_2, \ldots, \alpha_R)$ are considered. The usual notation for averaging over an index will be used; for instance,

$$\bar{Y}_{ij\cdot} = \frac{\sum_{l=1}^{n} Y_{ijl}}{n},$$

$$\bar{Y}_{\cdot j\cdot} = \frac{\sum_{i=1}^{R}\sum_{l=1}^{n} Y_{ijl}}{nR},$$

and

$$\bar{Y}_{\ldots} = \frac{\sum_{ijl} Y_{ijl}}{nRC}.$$

Define

$$\hat{\mu} = \bar{Y}_{\ldots}, \qquad \hat{\alpha}_i = \bar{Y}_{i\cdot\cdot} - \bar{Y}_{\ldots}, \quad \hat{\beta}_j = \bar{Y}_{\cdot j\cdot} - \bar{Y}_{\ldots}, \qquad \hat{\gamma}_{ij} = \bar{Y}_{ij\cdot} - \bar{Y}_{i\cdot\cdot} - \bar{Y}_{\cdot j\cdot} + \bar{Y}_{\ldots}, \tag{2.2.17}$$

and consequently $\hat{\mu}_{ij} = \hat{\mu} + \hat{\alpha}_i + \hat{\beta}_j + \hat{\gamma}_{ij} = \bar{Y}_{ij\cdot}$.

By straightforward calculations, it is easily shown that

$$T_S = T_\mu + T_\alpha + T_\beta + T_\gamma + T_W \tag{2.2.18}$$

with

$$T_S = \sum_{ijl} (Y_{ijl} - \mu_{ij})^2$$

$$T_\mu = nRC(\hat{\mu} - \mu)^2$$

$$T_\alpha = nC \sum_{i=1}^{R} (\hat{\alpha}_i - \alpha_i)^2$$

$$T_\beta = nR \sum_{j=1}^{C} (\hat{\beta}_j - \beta_j)^2$$

$$T_\gamma = n \sum_{i=1}^{R} \sum_{j=1}^{C} (\hat{\gamma}_{ij} - \gamma_{ij})^2$$

and

$$T_W = \sum_{ijl} (Y_{ijl} - \hat{\mu}_{ij})^2.$$

For the case of known or unknown σ^2, the MLEs of μ, $\boldsymbol{\beta}$ and γ, which are the same under H_0, H_1, and H_2 are given by $\hat{\mu}, \hat{\boldsymbol{\beta}}$ and $\hat{\gamma}$; and under H_1, the MLE of $\boldsymbol{\alpha}$ is $\hat{\boldsymbol{\alpha}}_{H_1} = \hat{\boldsymbol{\alpha}}^*$, the isotonic regression of $\hat{\boldsymbol{\alpha}}$ with equal weights and the quasi-order determining H_1; under H_2, the MLE of $\boldsymbol{\alpha}$ is $\hat{\boldsymbol{\alpha}}_{H_2} = \hat{\boldsymbol{\alpha}}$; and under H_0, α_i is constrained to be zero for $i = 1, 2, \ldots, R$, that is $\hat{\boldsymbol{\alpha}}_{H_0} = (0, 0, \ldots, 0)$. Because $\bar{Y}_{...}$ is constant with respect to i,

$$\hat{\boldsymbol{\alpha}}_{H_1} = (\bar{Y}^*_{1..} - \bar{Y}_{...}, \bar{Y}^*_{2..} - \bar{Y}_{...}, \ldots, \bar{Y}^*_{R..} - \bar{Y}_{...}).$$

Employing the techniques used to obtain (2.2.9), the MLE of σ^2 under $H_\delta, \delta = 0, 1, 2$, is found to be

$$\hat{\sigma}^2_{H_\delta} = \frac{nC \sum_{i=1}^{R} (\hat{\alpha}_i - \hat{\alpha}_{H_\delta i})^2 + \sum_{ijl} (Y_{ijl} - \bar{Y}_{ij.})^2}{nRC}.$$

If σ^2 is known, the LRT of H_0 versus H_1–H_0 rejects for large values of

$$-2 \log \Lambda = \frac{nC[\sum_{i=1}^{R} (\bar{Y}_{i..} - \bar{Y}_{...})^2 - \sum_{i=1}^{R} (\bar{Y}^*_{i..} - \bar{Y}_{i..})^2]}{\sigma^2}$$

$$= \frac{nC \sum_{i=1}^{R} (\bar{Y}^*_{i..} - \bar{Y}_{...})^2}{\sigma^2}, \tag{2.2.19}$$

and the LRT of H_1 versus H_2–H_1 rejects for large values of

$$-2 \log \Lambda = \frac{nC \sum_{i=1}^{R} (\bar{Y}_{i..} - \bar{Y}^*_{i..})^2}{\sigma^2}. \tag{2.2.20}$$

Because $\bar{Y}_{i\cdot\cdot} \sim \mathcal{N}(\mu + \alpha_i, \sigma^2/(nC))$, (2.2.19) has the same distribution as $\bar{\chi}^2_{01}$ and (2.2.20) has the same distribution as $\bar{\chi}^2_{12}$ with $k = R$, $\mu_i = \mu + \alpha_i$, and $w_i = nC/\sigma^2$ for $i = 1, 2, \ldots, R$.

If σ^2 is unknown, the LRT of H_δ versus $H_{\delta+1}-H_\delta$ rejects for small values of $\Lambda^{2/N} = \hat{\sigma}^2_{H_{\delta+1}}/\hat{\sigma}^2_{H_\delta}$. Thus, for H_0 versus H_1-H_0, the LRT rejects for large values of

$$1 - \Lambda^{2/N} = \frac{nC[\sum_{i=1}^R (\bar{Y}_{i\cdot\cdot} - \bar{Y}_{\cdots})^2 - \sum_{i=1}^R (\bar{Y}_{i\cdot\cdot} - \bar{Y}^*_{i\cdot\cdot})^2]}{nC\sum_{i=1}^R (\bar{Y}_{i\cdot\cdot} - \bar{Y}_{\cdots})^2 + T_W},$$

and repeating the argument used to establish (2.2.19), this becomes

$$\frac{nC\sum_{i=1}^R (\bar{Y}^*_{i\cdot\cdot} - \bar{Y}_{\cdots})^2}{nC\sum_{i=1}^R (\bar{Y}_{i\cdot\cdot} - \bar{Y}_{\cdots})^2 + T_W}. \tag{2.2.21}$$

The LRT of H_1 versus H_2-H_1 rejects for large values of

$$\frac{nC\sum_{i=1}^R (\bar{Y}_{i\cdot\cdot} - \bar{Y}^*_{i\cdot\cdot})^2}{nC\sum_{i=1}^R (\bar{Y}_{i\cdot\cdot} - \bar{Y}^*_{i\cdot\cdot})^2 + T_W}. \tag{2.2.22}$$

Because $\bar{Y}_{i\cdot\cdot} \sim \mathcal{N}(\mu + \alpha_i, \sigma^2/(nC))$, $T_W \sim \sigma^2\chi^2_{(n-1)RC}$, and T_W is independent of $(\bar{Y}_{1\cdot\cdot}, \bar{Y}_{2\cdot\cdot}, \ldots, \bar{Y}_{R\cdot\cdot})$, (2.2.21) has the same distribution as $\bar{E}^2_{01}((n-1)RC)$ and (2.2.22) has the same distribution as $\bar{E}^2_{12}((n-1)RC)$ with $k = R, \mu_i = \mu + \alpha_i$ and $w_i = nC$ for $i = 1, 2, \ldots, R$. Similar conclusions can be drawn regarding restricted tests of hypotheses regarding the parameter vector $\boldsymbol{\beta}$.

2.3 THE NULL HYPOTHESIS DISTRIBUTIONS OF $\bar{\chi}^2$ AND $\bar{E}^2$

For a given quasi-order $\lesssim$ on $X = \{1, 2, \ldots, k\}$, H_0 will be shown to be least favorable within H_1 for $\bar{\chi}^2_{12}$ and $\bar{E}^2_{12}$, i.e. the maximum, over $\boldsymbol{\mu} \in H_1$, of the probability of rejecting occurs in H_0. Then the distributions of $(\bar{\chi}^2_{01}, \bar{\chi}^2_{12})$ and $(\bar{E}^2_{01}, \bar{E}^2_{12})$ are obtained for $\boldsymbol{\mu} \in H_0$ by conditioning on $B_1, B_2, \ldots, B_m$, the level sets of $\boldsymbol{\mu}^*$. For the two-way layout as well as other applications to be made later, it is convenient to have the distribution of $\bar{E}^2_{01}(v)$ and $\bar{E}^2_{12}(v)$ for arbitrary degrees of freedom, v. Chernoff (1954) considered LRTs for some special situations in which the parameters are restricted by linear constraints. He showed that $-2\log\Lambda$, with Λ the likelihood ratio, has a $\bar{\chi}^2$ distribution. This type of distribution is defined in this section. The proofs of this section are patterned after the works by Bartholomew (1959a, b) and Robertson and Wegman (1978), and the book by Barlow *et al.* (1972).

Least favorable configurations for $\bar{\chi}^2_{12}$ and $\bar{E}^2_{12}(v)$

Let $\boldsymbol{\eta} \in H_1$ and let $V_1, V_2, \ldots, V_k$ be independent with V_i normally distributed with zero mean and variance σ_i^2/n_i. For $\boldsymbol{\mu} = \boldsymbol{\eta}$,

$$\bar{\chi}^2_{12} \stackrel{D}{=} \sum_{i=1}^k w_i[V_i + \eta_i - (\mathbf{V} + \boldsymbol{\eta})^*_i]^2$$

and for $\boldsymbol{\mu} = \mathbf{0}$,

$$\bar{\chi}_{12}^2 \overset{D}{=} \sum_{i=1}^{k} w_i(V_i - V_i^*)^2$$

with $(\mathbf{V} + \boldsymbol{\eta})_i^*$ and V_i^* the ith component of the isotonic regression of $\mathbf{V} + \boldsymbol{\eta}$ and $\mathbf{V}$, respectively, with weights w_i. Because $\mathbf{V}^* + \boldsymbol{\eta} \in H_1$,

$$\sum_{i=1}^{k} w_i[V_i + \eta_i - (\mathbf{V} + \boldsymbol{\eta})_i^*]^2 \leqslant \sum_{i=1}^{k} w_i[V_i + \eta_i - (V_i^* + \eta_i)]^2 = \sum_{i=1}^{k} w_i(V_i - V_i^*)^2$$

by the definition of $(\mathbf{V} + \boldsymbol{\eta})^*$. Hence, H_0 is least favorable within H_1 for $\bar{\chi}_{12}^2$. The test based on $E_{12}^2(v)$ is equivalent to that based on $S_{12}(v) = \bar{\chi}_{12}^2/[Q(v)/v]$ with $Q(v)$ defined as before. Because $Q(v)$ is independent of $\bar{\chi}_{12}^2$, it follows that H_0 is also least favorable within H_1 for $\bar{E}_{12}^2(v)$.

Distribution of $(\bar{\chi}_{01}^2, \bar{\chi}_{12}^2)$ and $(\bar{E}_{01}^2(v), \bar{E}_{12}^2(v))$ under H_0

The null distributions of the $\bar{\chi}^2$ and $\bar{E}^2$ tests statistics depend on the quasi-order, $\lesssim$, $X = \{1, 2, \ldots, k\}$ and the weight vector, $\mathbf{w}$, through the level probabilities defined below.

Definition 2.3.1 Let $\boldsymbol{\mu} \in H_0$. The level probabilities are defined by

$$P(l, k; \mathbf{w}) = P[M = l], \qquad l = 1, 2, \ldots, k, \tag{2.3.1}$$

with $Y_1, Y_2, \ldots, Y_k$ independent normal variables with mean μ_i and variance w_i^{-1} and M the number of level sets in $\mathbf{Y}^*$, the isotonic regression of $\mathbf{Y} = (Y_1, Y_2, \ldots, Y_k)$ with weight vector $\mathbf{w}$. Clearly, the $P(l, k; \mathbf{w})$ do not depend on the common value of μ_i, and they are unchanged if $\mathbf{w}$ is multiplied by a positive constant. (Since $\boldsymbol{\mu} \in H_0$ implies that $\mu_1 = \mu_2 = \cdots = \mu_k$ and $(\mathbf{y} + \mathbf{a})^* = \mathbf{y}^* + \mathbf{a}$ for any constant vector $\mathbf{a}$, one may without loss of generality assume that $\mu_i = 0$ in the definition of the level probabilites.)

Theorem 2.3.1 *Let $\boldsymbol{\mu} \in H_0$. For c_1 and c_2 real numbers and v a nonnegative integer,*

$$P[\bar{\chi}_{01}^2 \geqslant c_1, \bar{\chi}_{12}^2 \geqslant c_2] = \sum_{l=1}^{k} P(l, k; \mathbf{w}) P[\chi_{l-1}^2 \geqslant c_1] P[\chi_{k-l}^2 \geqslant c_2] \tag{2.3.2}$$

where χ_i^2 is a standard chi-square variable with i degrees of freedom ($\chi_0^2 \equiv 0$) and

$$P[\bar{E}_{01}^2(v) \geqslant c_1, \bar{E}_{12}^2(v) \geqslant c_2]$$

$$= \sum_{l=1}^{k} P(l, k; \mathbf{w}) P[B_{(l-1)/2,(v+k-l)/2} \geqslant c_1] P[B_{(k-l)/2, v/2} \geqslant c_2] \tag{2.3.3}$$

where $B_{a,b}$ is a beta variable with parameters a and b ($B_{0,b} \equiv 0$).

Before giving the proof of Theorem 2.3.1 the following corollary, which is an immediate consequence, is stated.

Corollary *For $\boldsymbol{\mu}\in H_0$ and v a nonnegative integer,*

$$P[\bar{\chi}^2_{01} \geqslant c] = \sum_{l=1}^{k} P(l, k; \mathbf{w}) P[\chi^2_{l-1} \geqslant c]$$

$$P[\bar{\chi}^2_{12} \geqslant c] = \sum_{l=1}^{k} P(l, k; \mathbf{w}) P[\chi^2_{k-l} \geqslant c] \qquad (2.3.4)$$

$$P[\bar{E}^2_{01}(v) \geqslant c] = \sum_{l=1}^{k} P(l, k: \mathbf{w}) P[B_{(l-1)/2,(v+k-l)/2} \geqslant c]$$

and

$$P[\bar{E}^2_{12}(v) \geqslant c] = \sum_{l=1}^{k} P(l, k; \mathbf{w}) P[B_{(k-l)/2,v/2} \geqslant c].$$

Setting $v = N - k$ in the third and fourth equations in (2.3.4) *gives the distributions of $\bar{E}^2_{01}$ and $\bar{E}^2_{12}$.*

There are several lemmas needed for the proof of Theorem 2.3.1. The first gives a characterization of the level sets. The partition, $B_1, B_2, \ldots, B_m$, of X denotes the ordered level sets of an isotonic function $\mathbf{y}$ provided $\mathbf{y}$ is constant on each B_i and the value of $\mathbf{y}$ increases with i. In Section 1.4, it was shown that for a vector $\mathbf{y}$, the MLSA determines lower layers, $L_0, L_1, \ldots, L_m$, with $\phi = L_0 \subset L_1 \subset \cdots \subset L_m = X$ and that $\mathbf{y}^*$, the isotonic regression of $\mathbf{y}$, is constant on the $B_i = L_i - L_{i-1}$ (which are nonempty). Furthermore, the values of $\mathbf{y}^*$ on the B_i increase with i. Hence, $B_1, B_2, \ldots, B_m$ are the ordered level sets of $\mathbf{y}^*$. The proof of the next lemma uses the fact that Av is a strict Cauchy mean value function (see Definition 1.4.3).

Lemma A *For $\mathbf{y} = (y_1, y_2, \ldots, y_k)$, the ordered level sets of $\mathbf{y}^*$ are $B_1, B_2, \ldots, B_m$ ($B_i = L_i - L_{i-1}$) if and only if*

$$\text{for } L_{i-1} \subset L \subset L_i \text{ with } L - L_{i-1} \neq \phi,\quad L_i - L \neq \phi \text{ and } i \in \{1, 2, \ldots, m\},$$
$$\mathrm{Av}(B_i) \leqslant \mathrm{Av}(L - L_{i-1}) \qquad (2.3.5a)$$

and

$$\mathrm{Av}(B_1) < \mathrm{Av}(B_2) < \cdots < \mathrm{Av}(B_m). \qquad (2.3.5b)$$

Proof of Lemma A: Recall, that at the ith step, the MLSA chooses L_i to be the largest lower layer L which properly contains the one chosen in the last step, $L_{i-1}(L_0 = \phi)$, and minimizes $\mathrm{Av}(L - L_{i-1})$ over such lower layers. Hence, if $B_1, B_2, \ldots, B_m$ are the ordered level sets of $\mathbf{y}^*$, then with $L_i = B_1 \cup B_2 \cup \cdots \cup B_i$,

$$\text{for } L \supset L_{i-1} \text{ with } L - L_{i-1} \neq \phi \text{ and } i \in \{1, 2, \ldots, m\},$$
$$\mathrm{Av}(B_i) \leqslant \mathrm{Av}(L - L_{i-1}) \qquad (2.3.6a)$$

and

$$\text{for } L \supset L_i \text{ with } L - L_i \neq \phi \text{ and } i \in \{1, 2, \ldots, m-1\},$$

$$\mathrm{Av}(B_i) < \mathrm{Av}(L - L_{i-1}). \tag{2.3.6b}$$

Conversely, if (2.3.6a) holds, then each L_i minimizes $\mathrm{Av}(L - L_{i-1})$ over L properly containing L_{i-1}. As was noted in Section 1.4, the collection of lower sets properly containing L_{i-1} and minimizing $\mathrm{Av}(L - L_{i-1})$ is closed under unions. Hence, if L' is another such lower set, then so is $L_i \cup L'$. If $L' - L_i \neq \phi$ then (2.3.6b) does not hold. Therefore (2.3.6b) implies $L' \subset L_i$ and consequently (2.3.6a, b) imply that $B_1, B_2, \ldots, B_m$ are the ordered level sets of $\mathbf{y}^*$.

It will be shown that (2.3.5) and (2.3.6) are equivalent. Clearly, (2.3.6a) implies (2.3.5a). With $1 \leqslant i \leqslant m-1$ and $L = L_{i+1}$, (2.3.6b) shows that $\mathrm{Av}(B_i) < \mathrm{Av}(B_i + B_{i+1})$. Because Av is a strict Cauchy mean value function, $\mathrm{Av}(B_i) < \mathrm{Av}(B_{i+1})$. Hence, (2.3.6b) implies (2.3.5b).

The proof is completed by showing that (2.3.5) implies (2.3.6). Let $1 \leqslant i \leqslant m$ and $L \supset L_{i-1}$ with $L - L_{i-1} \neq \phi$. Using $\sum$ to denote a union of disjoint sets, one can write

$$L - L_{i-1} = \sum_{j=i}^{m} (L \cap B_j)$$

Note that the sets in this union are of the following three types: empty, $L \cap B_j = B_j$, and $\phi \neq L \cap B_j$ is properly contained in B_j. Not all of the $L \cap B_j$ are empty. If $L \cap B_j = B_j$, then $\mathrm{Av}(L \cap B_j) \geqslant \mathrm{Av}(B_i)$ because of (2.3.5b). If $B_j - (L \cap B_j) \neq \phi$ and $L \cap B_j \neq \phi$, then $L' = (L \cup L_{j-1}) \cap L_j$ is a lower set with $L_{j-1} \subset L' \subset L_j$, $L' - L_{j-1} \neq \phi$, $L_j - L' \neq \phi$, and $L' - L_{j-1} = L \cap B_j$. Hence, by (2.3.5a, b), $\mathrm{Av}(L \cap B_j) = \mathrm{Av}(L' - L_{j-1}) \geqslant \mathrm{Av}(B_i)$. Because Av is a Cauchy mean function $\mathrm{Av}(L - L_{i-1}) \geqslant \mathrm{Av}(B_i)$, that is (2.3.6a) holds. Next, it is shown that (2.3.5a, b) imply (2.3.6b). Suppose $1 \leqslant i \leqslant m-1$ and $L \supset L_i$ with $L - L_i \neq \phi$. Write $L - L_{i-1} = (L - L_i) + B_i$ and note that applying (2.3.6a) and (2.3.5b), $\mathrm{Av}(L - L_i) \geqslant \mathrm{Av}(B_{i+1}) > \mathrm{Av}(B_i)$. By the strict Cauchy mean property, $\mathrm{Av}(L - L_{i-1}) > \mathrm{Av}(B_i)$. □

Lemma B *Let* $\mathbf{Z} = (Z_1, Z_2, \ldots, Z_s)'$ *with the* Z_i *independent standard normal variables and let* $\mathbf{A} = ((a_{ij}))$ *be a* $t \times s$ *matrix of reals. The conditional distribution of* $\sum_{i=1}^{s} Z_i^2$, *given that each component of* $\mathbf{AZ}$ *is nonnegative, is* χ_s^2, *provided the conditioning event has positive probability.*

Proof of Lemma B: Making the polar transformation

$$Z_1 = R \sin\theta_1, \quad Z_2 = R\cos\theta_1 \sin\theta_2, \ldots,$$
$$Z_i = R\cos\theta_1 \cos\theta_2 \cdots \cos\theta_{i-1} \sin\theta_i, \ldots, \qquad Z_s = R\cos\theta_1\cos\theta_2\cdots\cos\theta_{s-1},$$

$\sum_{i=1}^{s} Z_i^2 = R^2 \sim \chi_s^2$. Furthermore, the event conditioned on is determined by

$\theta_1, \theta_2, \ldots, \theta_{s-1}$ which are independent of R (cf. Kendall and Stuart, 1958, p. 247). □

Lemma C *Let $V_1, V_2, \ldots, V_s$ be independent normal variables with common mean and variances $b_1^{-1}, b_2^{-1}, \ldots, b_s^{-1}$, respectively; let $\mathbf{V} = (V_1, V_2, \ldots, V_s)'$; let $\bar{V} = \sum_{i=1}^{s} b_i V_i / \sum_{i=1}^{s} b_i$; and let $\mathbf{A} = ((a_{ij}))$ be a $t \times s$ matrix of reals with zero row sums. The conditional distribution of $\sum_{i=1}^{s} b_i(V_i - \bar{V})^2$, given that each component of $\mathbf{AV}$ is nonnegative, is χ^2_{s-1}, provided the event conditioned on has positive probability.*

Proof of Lemma C: Because the row sums of $\mathbf{A}$ are zero, the common value of the means may be taken to be zero. Let $U_i = b_i^{1/2} V_i$ for $i = 1, 2, \ldots, s$ and set $\mathbf{U} = (U_1, U_2, \ldots, U_s)'$. Make an orthogonal transformation from $\mathbf{U}$ to $\mathbf{Z} = (Z_1, Z_2, \ldots, Z_s)'$ with $Z_s = (\sum_{i=1}^{s} b_i)^{1/2} \bar{V}$, that is $\mathbf{Z} = \mathbf{PU}$ with $\mathbf{P} = ((p_{ij}))$ an $s \times s$ orthogonal matrix with $p_{si} = b_i^{1/2} / (\sum_{j=1}^{s} b_j)^{1/2}$ for $i = 1, 2, \ldots, s$. It is easily shown that

$$\sum_{i=1}^{s} b_i (V_i - \bar{V})^2 = \sum_{i=1}^{s-1} Z_i^2$$

and the Z_i are independent standard normal variables. The proof is completed by showing that $\mathbf{AV} = \mathbf{D}(Z_1, Z_2, \ldots, Z_{s-1})'$ with $\mathbf{D}$ a $t \times (s-1)$ matrix of reals and applying Lemma B. Now $\mathbf{AV} = \mathbf{ABP'Z}$ with $\mathbf{B}$ the $s \times s$ diagonal matrix which has the ith diagonal element given by $b_i^{-1/2}$. Hence, it suffices to show that $(\mathbf{ABP'})_{is} = 0$ for $i = 1, 2, \ldots, t$. However,

$$(\mathbf{ABP'})_{is} = \sum_{j=1}^{s} a_{ij} p_{sj} b_j^{-1/2} = \left(\sum_{j=1}^{s} b_j \right)^{-1/2} \sum_{j=1}^{s} a_{ij} = 0. \quad \square$$

Lemma D *Suppose $T_1, T_2, \ldots, T_m$ are random variables and $E_1, E_2, \ldots, E_m$ are nonnull events with $(T_1, I_{E_1}), (T_2, I_{E_2}), \ldots, (T_m, I_{E_m})$ independent where I_A denotes the indicator of the event A. If for $i = 1, 2, \ldots, m$, the conditional distribution of T_i given E_i is $\chi^2_{r_i}$, then the conditional distribution of $\sum_{i=1}^{m} T_i$ given $\bigcap_{i=1}^{m} E_i$ is χ^2_r with $r = \sum_{i=1}^{m} r_i$.*

Proof of Lemma D: Because of the independence assumption, $\bigcap_{i=1}^{m} E_i$ is nonnull and given $I_{E_1} = 1, I_{E_2} = 1, \ldots, I_{E_m} = 1, T_1, T_2, \ldots, T_m$ are independent chi-square variables with the appropriate degrees of freedom. The desired conclusion follows from the reproductive property of the chi-squared distributions. □

Proof of Theorem 2.3.1: Let $\boldsymbol{\mu} \in H_0$, c_1 and c_2 be real numbers and let $G(B_1, B_2, \ldots, B_m)$ denote the event on which $B_1, B_2, \ldots, B_m$ are the ordered level sets of $\bar{\mathbf{Y}}^*$. Write

$$P[\bar{\chi}^2_{01} \geq c_1, \bar{\chi}^2_{12} \geq c_2] = \sum_{m=1}^{k} \sum_{B_1, B_2, \ldots, B_m} P[\bar{\chi}^2_{01} \geq c_1, \bar{\chi}^2_{12} \geq c_2, G(B_1, B_2, \ldots, B_m)] \tag{2.3.7}$$

with the second summation taken over all possible decompositions of X into m ordered nonempty level sets, $B_1, B_2, \ldots, B_m$. Fix m and possible level sets $B_1, B_2, \ldots, B_m$ or equivalently fix m and lower sets $\phi \subset L_1 \subset L_2 \subset \cdots \subset L_m = X$ with $L_i - L_{i-1} \neq \phi$. Define $G_1 = G_1(B_1, B_2, \ldots, B_m)$ and $G_2 = G_2(B_1, B_2, \ldots, B_m)$ to be the events on which (2.3.5a) and (2.3.5b), respectively, are satisfied. Now $G_1 \cap G_2 = G(B_1, B_2, \ldots, B_m)$ and on this event, $\bar{\chi}^2_{01} = \chi^2_{BB}$ and $\bar{\chi}^2_{12} = \chi^2_{WB}$ with

$$W_l = \sum_{i \in B_l} w_i$$

$$U_l = \mathrm{Av}(B_l)$$

$$\bar{U} = \frac{\sum_{l=1}^m W_l U_l}{\sum_{l=1}^m W_l}$$

$$\chi^2_{BB} = \sum_{l=1}^m W_l(U_l - \bar{U})^2$$

and

$$\chi^2_{WB} = \sum_{l=1}^m \sum_{i \in B_l} w_i(\bar{Y}_i - U_l)^2.$$

It is well known that $\{\bar{Y}_i - U_l : i \in B_l\}$ is independent of U_l (this is easily seen by computing the covariances) and hence $\{\bar{Y}_i - U_l : i \in B_l, l = 1, 2, \ldots, m\}$ is independent of $\{U_l : l = 1, 2, \ldots, m\}$. (One easy way to see this is by using the fact that pairwise independence implies independence for jointly normal variables.) Because I_{G_1} and χ^2_{WB} are functions of the former collection of variables and I_{G_2} and χ^2_{BB} are functions of the latter collection,

$$P[\bar{\chi}^2_{01} \geqslant c_1, \bar{\chi}^2_{12} \geqslant c_2, G_1, G_2] = P[\bar{\chi}^2_{BB} \geqslant c_1, G_2]P[\bar{\chi}^2_{WB} \geqslant c_2, G_1].$$

Applying Lemma C with $s = m, b_l = W_l$ and $V_l = U_l$,

$$P[\bar{\chi}^2_{BB} \geqslant c_1, G_2] = P[\bar{\chi}^2_{BB} \geqslant c_1 | G_2]P(G_2) = P[\chi^2_{m-1} \geqslant c_1]P(G_2).$$

In considering the other factor, define $T_l = \sum_{i \in B_l} w_i(\bar{Y}_i - U_l)^2$ and with $\mathscr{L}$ the collection of lower sets in X,

$$E_l = \left[\min_{L \in \mathscr{L}: L_{l-1} \subset L \subset L_l, L - L_{l-1} \neq \phi, L_l - L \neq \phi} \sum_{i \in L - L_{l-1}} w_i(\bar{Y}_i - U_l) \geqslant 0 \right].$$

Note that $\sum_{l=1}^m T_l = \chi^2_{WB}$ and $\bigcap_{l=1}^m E_l = G_1$. For each l, apply Lemma C with $s = \text{card}\,(B_l)$, the V_j taken to be the $\bar{Y}_i$ with $i \in B_l$ and the b_j the corresponding w_i. Then applying Lemma D,

$$P[\bar{\chi}^2_{WB} \geqslant c_2, G_1] = P[\chi^2_{k-m} \geqslant c_2]P(G_1).$$

Therefore, because $P(G_1)P(G_2) = P(G_1 \cap G_2) = P[G(B_1, B_2, \ldots, B_m)]$,

$$P[\bar{\chi}^2_{01} \geqslant c_1, \bar{\chi}^2_{12} \geqslant c_2] = \sum_{m=1}^k P[\chi^2_{m-1} \geqslant c_1]P[\chi^2_{k-m} \geqslant c_2] \sum_{B_1, B_2, \ldots, B_m} P[G(B_1, B_2, \ldots, B_m)], \quad (2.3.8)$$

and the first conclusion, (2.3.2), follows from the fact that the inside summation on the right-hand side is $P(m, k; \mathbf{w})$. The following has been shown: if $G_1 \cap G_2$ is nonnull, then, conditioned on $G_1 \cap G_2$, $\bar{\chi}_{01}^2$ and $\bar{\chi}_{12}^2$ are independent chi-squared variables with $m-1$ and $k-m$ degrees of freedom, respectively.

The second conclusion is established next. From (2.2.11) and (2.2.14),

$$\bar{E}_{01}^2(v) = \frac{\bar{\chi}_{01}^2}{\bar{\chi}_{01}^2 + \bar{\chi}_{12}^2 + Q(v)} \qquad \text{and} \qquad \bar{E}_{12}^2(v) = \frac{\bar{\chi}_{12}^2}{\bar{\chi}_{12}^2 + Q(v)}$$

with $Q(v) \sim \chi_v^2$ independently of $(\bar{\chi}_{01}^2, \bar{\chi}_{12}^2)$. As in (2.3.7), partition the probability space over the possible values of M, that is $\{1, 2, \ldots, k\}$, and the possible ordered level sets, $B_1, B_2, \ldots, B_m$. Then, condition on $G_1 \cap G_2$ for those level sets for which $G_1 \cap G_2$ is nonnull. Conditioning on $G_1 \cap G_2, \bar{\chi}_{01}^2, \bar{\chi}_{12}^2$, and $Q(v)$ are independent chi-squared variables with $m-1$, $k-m$, and v degrees of freedom, respectively. It is well known that the sum of two independent chi-squared variables is independent of their ratio (cf. Hogg and Craig, 1978, p. 138), and hence, because $\bar{E}_{12}^2(v)$ is a function of $\bar{\chi}_{12}^2/Q(v)$ and $\bar{E}_{01}^2(v)$ is a function of $\bar{\chi}_{12}^2 + Q(v)$ and $\bar{\chi}_{01}^2$, they are conditionally independent. It is also clear that, conditioning on $G_1 \cap G_2, \bar{E}_{01}^2(v) \sim B_{(m-1)/2,(v+k-m)/2}$ and $\bar{E}_{12}^2(v) \sim B_{(k-m)/2,v/2}$. The second conclusion follows from these facts as the first follows from (2.3.8). □

It should be noted that in the proof of Theorem 2.3.1, it was shown that, conditional on l level sets, $\bar{\chi}_{01}^2$ and $\bar{\chi}_{12}^2$ are stochastically independent, $\bar{\chi}_{01}^2 \sim \chi_{l-1}^2$ and $\bar{\chi}_{12}^2 \sim \chi_{k-l}^2$.

2.4 THE LEVEL PROBABILITIES, $P(l, k; \mathbf{w})$

Theorem 2.3.1 gives the general form of the null distributions of the $\bar{\chi}^2$ and the $\bar{E}^2$ test statistics, but to make use of these results the values of the $P(l, k; \mathbf{w})$ are needed. It is not possible in all cases to determine simple closed form expressions for these probabilities. In this section, a general method for calculating the $P(l, k; \mathbf{w})$ is discussed and it is applied to a few important orderings. Approximations and bounds for the $P(l, k; \mathbf{w})$ and the related distributions are considered in the next chapter.

A general recursive algorithm for calculating the $P(l, k; \mathbf{w})$

The recursive approach to calculating the level probabilities consists of two parts. First the $P(k, k; \mathbf{w})$ are computed. Then for $1 < l < k$ a recursive formula for $P(l, k; \mathbf{w})$, in terms of $P(j, j; \mathbf{w})$ and $P(1, j; \mathbf{w})$ for $j < k$, is obtained. Of course, the value of $P(1, k; \mathbf{w})$ can be obtained from the equation

$$\sum_{l=1}^{k} P(l, k; \mathbf{w}) = 1.$$

Recall that M is the number of level sets in $\mathbf{Y}^*$, the isotonic regression of $\mathbf{Y} = (Y_1, Y_2, \ldots, Y_k)$ with weights w_i and quasi-order $\lesssim$, provided $Y_1, Y_2, \ldots, Y_k$ are independent normal variables with a common mean and variances $w_1^{-1}, w_2^{-1}, \ldots, w_k^{-1}$. The common mean may be chosen to be zero. These probabilities depend on $\lesssim$, but this will not be made explicit in the notation. The probabilities $P[Y_1 < Y_2 < \cdots < Y_k]$ play a key role in these calculations. Making the transformation $U_i = Y_{i+1} - Y_i$ for $i = 1, 2, \ldots, k-1$, this probability becomes the orthant probability, P_{k-1}, defined by

$$P_{k-1} = P[U_1 > 0, U_2 > 0, \ldots, U_{k-1} > 0], \qquad (2.4.1)$$

in which $\mathbf{U} = (U_1, U_2, \ldots, U_{k-1})$ has a multivariate normal distribution with zero means and correlation matrix of the form ($\rho_{ii} = 1$)

$$\rho_{i,i+1} = \rho_{i+1,i} = -\left[\frac{w_i w_{i+2}}{(w_i + w_{i+1})(w_{i+1} + w_{i+2})}\right]^{1/2}; \qquad i = 1, 2, \ldots, k-2$$

$$\rho_{i,j} = 0; \qquad |i - j| > 1. \qquad (2.4.2)$$

There is a vast literature on the determination of orthant probabilities (see Gupta, 1963, and the references in the discussion below). Expressions for the first four P_{k-1} are given below for arbitrary weights (for $k = 4$ and 5 the special form of ρ_{ij} is used):

$$P_1 = \frac{1}{2}$$

$$P_2 = \frac{1}{4} + \frac{1}{2\pi} \sin^{-1} \rho_{12}$$

$$P_3 = \frac{1}{8} + \frac{1}{4\pi} (\sin^{-1} \rho_{12} + \sin^{-1} \rho_{23})$$

and

$$P_4 = \frac{1}{16} + \frac{1}{8\pi} (\sin^{-1} \rho_{12} + \sin^{-1} \rho_{23} + \sin^{-1} \rho_{34})$$

$$+ \frac{1}{4\pi^2} \int_0^{-\rho_{34}} \int_0^{-\rho_{12}} [(1 - u^2)(1 - v^2) - \rho_{23}^2]^{-1/2} \, du \, dv \qquad (2.4.3)$$

(see, for example, McFadden, 1960, Abrahamson, 1964, and Childs, 1967). The integral in P_4 can be written as

$$\int_0^{-\rho_{34}} (1 - v^2)^{-1/2} \sin^{-1} \left[\frac{-(1 - v^2)^{1/2} \rho_{12}}{(1 - v^2 - \rho_{23}^2)^{1/2}}\right] dv,$$

and the term involving this integral can be obtained from the tables in Abrahamson. He tables a function, $P(a, b, c)$, for nonnegative values of its arguments. In terms of $P(a, b, c)$,

$$P_4 = P(-\rho_{12}, -\rho_{23}, -\rho_{34}) + \frac{1}{4\pi}(\sin^{-1}\rho_{12} + \sin^{-1}\rho_{23} + \sin^{-1}\rho_{34}).$$

We know of no closed form expression for P_{k-1} for $k > 5$. If the weights are equal, $w_1 = w_2 = \cdots = w_k$, then P_{k-1} is the probability that the elements of a random sample are increasing. Because each permutation of the ranks are equally likely, $P_{k-1} = 1/k!$. This result holds for any set of symmetrically dependent Y_i for which ties occur with probability zero. In particular, it holds if the Y_i are independently and identically distributed continuous variables.

To illustrate the use of P_{k-1} in the calculation of $P(k, k; \mathbf{w})$, the simple loop ordering is considered. It is convenient to adopt the following notation: $[i_1, i_2, \ldots, i_\alpha] \lesssim [j_1, j_2, \ldots, j_\beta]$ if and only if $i_s \lesssim j_t$ for $s = 1, 2, \ldots, \alpha$ and $t = 1, 2, \ldots, \beta$. The simple loop ordering with $k = 4$ is characterized by $1 \lesssim [2, 3] \lesssim 4$, and $\boldsymbol{\mu}$ is isotonic with respect to this ordering if and only if $\mu_1 \leqslant [\mu_2, \mu_3] \leqslant \mu_4$. In this case, $P(4, 4; \mathbf{w}) = P[Y_1 < Y_2 < Y_3 < Y_4] + P[Y_1 < Y_3 < Y_2 < Y_4]$, and the latter two probabilities are P_3 with $\mathbf{w} = (w_1, w_2, w_3, w_4)$ and $\mathbf{w}' = (w_1, w_3, w_2, w_4)$. The $P(l, 4; \mathbf{w})$ for $l < 4$ can be computed using the recurrence formula developed below.

For an arbitrary quasi-order, $P(k, k; \mathbf{w})$ is the sum of $P[Y_{\pi(1)} < Y_{\pi(2)} < \cdots < Y_{\pi(k)}]$ over certain π, permutations of $\{1, 2, \ldots, k\}$. However, if $\boldsymbol{\mu}$ is isotonic with respect to $\lesssim$ implies $\mu_i \leqslant \mu_j$ and $\mu_j \leqslant \mu_i$ for $1 \leqslant i \neq j \leqslant k$, then $[M = k] = \phi$ and $P(k, k; \mathbf{w}) = 0$.

Next, the needed recurrence formula is developed. Let $1 \leqslant m \leqslant k$ and let $B_1, B_2, \ldots, B_m$ be a partition of X into (unordered) level sets. In the proof of Theorem 2.3.1, the concept of the ordered level sets of an isotonic function was used. If $\lesssim$ is the simple tree ordering on $X = \{0, 1, 2, 3\}$, if $\mathbf{y}^*$ denotes the isotonic regression with respect to $\lesssim$, and $y_0^* = y_1^* < y_2^* < y_3^*$, then the ordered level sets are $\{0, 1\}$, $\{2\}$, $\{3\}$. However, if $y_0^* = y_1^* < y_3^* < y_2^*$, then the ordered level sets are $\{0, 1\}$, $\{3\}$, $\{2\}$, but in both cases the decomposition of X is the same. For a fixed m, let $\mathscr{L}_{mk}$ be the collection of all decompositions of X into m (unordered) level sets. For A, a nonempty subset of X, set $W_A = \sum_{i \in A} w_i$, $C_A = \text{card}(A)$, and, for $A = \{i_1, i_2, \ldots, i_j\}$ with $1 \leqslant i_1 < i_2 < \cdots < i_j \leqslant k$ and $j = C_A$, set $\mathbf{w}(A) = (w_{i_1}, w_{i_2}, \ldots, w_{i_j})$. For a given decomposition, define $\lesssim'$ on $\{1, 2, \ldots, m\}$ by $i \lesssim' j$ if there exist $s \in B_i$ and $t \in B_j$ with $s \lesssim t$, and on each B_i let $\lesssim_i$ denote the restriction of $\lesssim$ to B_i. Furthermore, let $P(m, m; W_{B_1}, W_{B_2}, \ldots, W_{B_m})$ be the probability of m level sets with partial order $\lesssim'$ and weight vector $(W_{B_1}, W_{B_2}, \ldots, W_{B_m})$ and let $P(1, C_{B_i}; \mathbf{w}(B_i))$ be the probability of one level set with partial order $\lesssim_i$ and weight vector $\mathbf{w}(B_i)$.

Theorem 2.4.1 *For* $m \in \{2, 3, \ldots, k-1\}$,

$$P(m, k; \mathbf{w}) = \sum_{(B_1, B_2, \ldots, B_m) \in \mathscr{L}_{mk}} P(m, m; W_{B_1}, W_{B_2}, \ldots, W_{B_m}) \prod_{i=1}^{m} P(1, C_{B_i}; \mathbf{w}(B_i)). \quad (2.4.4)$$

Proof: As in the proof of Theorem 2.3.1, let $G(B_1, B_2, \ldots, B_m)$ denote the event on which the ordered level sets are $B_1, B_2, \ldots, B_m$ and let $G_1 = G_1(B_1, B_2, \ldots, B_m)$ and $G_2 = G_2(B_1, B_2, \ldots, B_m)$ be the events on which (2.3.5a) and (2.3.5b), respectively, are satisfied. In that proof, it was shown that $G(B_1, B_2, \ldots, B_m) = G_1 \cap G_2$, $P(G_1 \cap G_2) = P(G_1)P(G_2)$ and

$$P(m, k; \mathbf{w}) = \sum_{B_1, B_2, \ldots, B_m} P(G_1)P(G_2) \quad (2.4.5)$$

with the summation taken over all decompositions of X into ordered level sets. With $B_i = L_i - L_{i-1} \neq \phi$ and

$$E_i = \left[\min_{L \in \mathscr{L}: L_{i-1} \subset L \subset L_i, L - L_{i-1} \neq \phi, L_i - L \neq \phi} \mathrm{Av}(L - L_{i-1}) \geqslant \mathrm{Av}(B_i) \right],$$

it was noted in the proof of Theorem 2.3.1 that $E_1, E_2, \ldots, E_m$ are independent with $G_1 = \bigcap_{i=1}^{m} E_i$. Thus $P(G_1) = \prod_{i=1}^{m} P(E_i)$. Because the collection of lower sets in B_i with quasi-order $\lesssim_i$ is $\{L \cap B_i : L \text{ a lower set in } X\}$, E_i is the set on which the isotonic regression of $\{Y_j : j \in B_i\}$ with weight vector $\mathbf{w}(B_i)$ and quasi-order $\lesssim_i$ has one level set. Hence, $P(G_1) = \prod_{i=1}^{m} P(1, C_{B_i}; \mathbf{w}(B_i))$. For a fixed $(B_1, B_2, \ldots, B_m) \in \mathscr{L}_{mk}$, let $\sum'$ denote the sum over all permutations, π, of $(1, 2, \ldots, m)$ for which $B_{\pi(1)}, B_{\pi(2)}, \ldots, B_{\pi(m)}$ are ordered level sets of some isotonic $\mathbf{y}$. Then (2.4.5) can be written as

$$\sum_{(B_1, B_2, \ldots, B_m) \in \mathscr{L}_{mk}} \prod_{i=1}^{m} P(1, C_{B_i}, \mathbf{w}(B_i)) \sum{}' P[\mathrm{Av}(B_{\pi(1)}) < \mathrm{Av}(B_{\pi(2)}) < \cdots < \mathrm{Av}(B_{\pi(m)})],$$

but the interior sum is $P(m, m; W_{B_1}, W_{B_2}, \ldots, W_{B_m})$. □

While formula (2.4.4) does provide a means of calculating $P(l, k; \mathbf{w})$ it can be tedious to use and requires that one is abie to compute $P(k, k, \mathbf{w})$.

The simply ordered case with arbitrary weights

The most important special case is that of a simple order, i.e. $H_1 : \mu_1 \leqslant \mu_2 \leqslant \cdots \leqslant \mu_k$. Arbitrary weights and $k \leqslant 5$ will be discussed first. Arbitrary k and equal weights are considered in the next subsection. The level probabilities for a simple order will be denoted by $P_S(l, k; \mathbf{w})$. The case $k = 2$ is straightforward, and, in fact, $P_S(1, 2; \mathbf{w}) = P_S(2, 2; \mathbf{w}) = \frac{1}{2}$.

Consider $k=3$, $P_S(3,3;\mathbf{w})=P_2=\frac{1}{4}+1/2\pi\sin^{-1}\rho_{12}$ and, using (2.4.4),

$$\begin{aligned} P_S(2,3;\mathbf{w}) &= P_S(2,2;w_1+w_2,w_3)P_S(1,2;w_1,w_2) \\ &\quad + P_S(2,2;w_1,w_2+w_3)P_S(1,2;w_2,w_3) \\ &= \tfrac{1}{2}\times\tfrac{1}{2}+\tfrac{1}{2}\times\tfrac{1}{2}=\tfrac{1}{2}. \end{aligned}$$

and subtraction, $P_S(1,3;\mathbf{w})=\frac{1}{4}-(1/2\pi)\sin^{-1}\rho_{12}$.

For $k=4$, $P_S(4,4;\mathbf{w})=P_3$ is given by (2.4.3) and

$$\begin{aligned} P_S(3,4;\mathbf{w}) &= P_S(3,3;w_1,w_2,w_3+w_4)P_S(1,2;w_3,w_4) \\ &\quad + P_S(3,3;w_1,w_2+w_3,w_4)P_S(1,2;w_2,w_3) \\ &\quad + P_S(3,3;w_1+w_2,w_3,w_4)P_S(1,2;w_1,w_2). \end{aligned}$$

The second factor in each of these products is $\frac{1}{2}$ and the first factor is obtained by the formula for $P_S(3,3;\mathbf{w})$. For instance,

$$P_S(3,3;w_1,w_2,w_3+w_4)=\frac{1}{4}+\frac{1}{2\pi}\sin^{-1}\rho$$

with

$$\rho=-\left[\frac{w_1(w_3+w_4)}{(w_1+w_2)(w_2+w_3+w_4)}\right]^{1/2}=\frac{\rho_{12}}{[1-\rho_{23}^2]^{1/2}}.$$

Because $\rho_{13}=0$, ρ is the partial correlation coefficient between Y_1 and Y_2 given Y_3, which will be denoted by $\rho_{12\cdot 3}$. Hence,

$$P_S(3,4;\mathbf{w})=\frac{3}{8}+\frac{1}{4\pi}(\sin^{-1}\rho_{12\cdot 3}+\sin^{-1}\rho_{13\cdot 2}+\sin^{-1}\rho_{23\cdot 1}).$$

Again applying (2.4.4), noting that $\mathscr{L}_{24}$ consists of the following three decompositions: $\{1\},\{2,3,4\}$; $\{1,2,3\},\{4\}$; and $\{1,2\},\{3,4\}$, and recalling that $P_S(2,2;\mathbf{w})=P_S(1,2;\mathbf{w})=\frac{1}{2}$,

$$\begin{aligned} P_S(2,4;\mathbf{w}) &= \tfrac{1}{2}P_S(1,3;w_2,w_3,w_4)+\tfrac{1}{2}P_S(1,3;w_1w_2,w_3)+\tfrac{1}{8} \\ &= \frac{3}{8}-\frac{1}{4\pi}(\sin^{-1}\rho_{23}+\sin^{-1}\rho_{12})=\tfrac{1}{2}-P_S(4,4;\mathbf{w}). \end{aligned}$$

Of course, this implies that $P_S(1,4;\mathbf{w})=\frac{1}{2}-P_S(3,4;\mathbf{w})$.

In the case, $k=5$, $P_S(5,5;\mathbf{w})$ is obtained using (2.4.3) with Abrahamson's table or numerical integration. In Bartholomew (1959b) the following expressions for the $P_S(l,5;\mathbf{w})$ are given:

$$\begin{aligned} P_S(4,5;\mathbf{w}) &= \frac{1}{4}+\frac{1}{8\pi}(\sin^{-1}\rho_{12}+\sin^{-1}\rho_{34})+\frac{1}{8\pi}(\sin^{-1}\rho_{12\cdot 3}+\sin^{-1}\rho_{23\cdot 1} \\ &\quad +\sin^{-1}\rho_{34\cdot 2}+\sin^{-1}\rho_{13\cdot 2}+\sin^{-1}\rho_{23\cdot 4}+\sin^{-1}\rho_{24\cdot 3}) \end{aligned}$$

$$P_S(3,5;\mathbf{w}) = \frac{3}{8} + \frac{1}{8\pi}(\sin^{-1}\rho_{12\cdot34} + \sin^{-1}\rho_{13\cdot24} + \sin^{-1}\rho_{14\cdot23} + \sin^{-1}\rho_{23\cdot14}$$
$$+ \sin^{-1}\rho_{24\cdot13} + \sin^{-1}\rho_{34\cdot12}) - \frac{1}{8\pi}(\sin^{-1}\rho_{12} + \sin^{-1}\rho_{23}$$
$$+ \sin^{-1}\rho_{34}) - \frac{1}{4\pi^2}(\sin^{-1}\rho_{12}\sin^{-1}\rho_{34\cdot12} + \sin^{-1}\rho_{23}\sin^{-1}\rho_{14\cdot23}$$
$$+ \sin^{-1}\rho_{34}\sin^{-1}\rho_{12\cdot34})$$

$$P_S(2,5;\mathbf{w}) = \tfrac{1}{2} - P_S(4,5;\mathbf{w})$$

and

$$P_S(1,5;\mathbf{w}) = \tfrac{1}{2} - P_S(5,5;\mathbf{w}) - P_S(3,5;\mathbf{w}).$$

(In Barlow *et al.*, 1972, the lead constant in $P_S(3,5;\mathbf{w})$ is given to be $\frac{1}{2}$, but by examining the constant terms in the lower order expressions one sees that $\frac{3}{8}$ is correct.)

For $2 \leqslant k \leqslant 5$ and a simple order, the probability of an odd number of level sets is the same as the probability of an even number of level sets, i.e.,

$$\sum_{l\,\mathrm{odd}} P_S(l,k;\mathbf{w}) = \sum_{l\,\mathrm{even}} P_S(l,k;\mathbf{w}) = \tfrac{1}{2}.$$

This will also be seen to be the case for larger k and equal weights in the next subsection. Based on this observation and the calculation of the level probabilities for various partial orders, it has been conjectured that this result is true for an arbitrary partial order and $k \geqslant 2$. (See Shapiro, 1987a, for a discussion of the conjecture.) If this result were proved, it could be used in the computation of the level probabilities.

The simply ordered case with equal weights

For the case of a simple order, the level probabilities are much more manageable if the weights are equal. In fact, in this case it is clear that the probability of k levels is $1/k!$. If the weights are equal, then $\mathbf{w}$ will be suppressed in the notation, i.e. the level probabilities will be denoted by $P_S(l,k)$. Because the $P(l,k;\mathbf{w})$ are unchanged if the weights are multiplied by a positive constant, one may assume that the $Y_1, Y_2, \ldots, Y_k$ in the definition of the level probabilities are independent standard normal variables. It will be shown that $P_S(l,k)$ are distribution free over the collection of independent, identically distributed continuous random variables, i.e. the probability that the isotonic regression of $Y_1, Y_2, \ldots, Y_k$ with a simple order and equal weights has l level sets does not depend on the distribution of the Y_i, provided they are independent with a common continuous distribution. An expression for the probability generating function (PGF) of

$\{P_S(l,k)\}$ will be obtained and it will be used to derive a recurrence relationship for the equal-weights level probabilities. The proof given below is an adaptation, due to Barton and Mallows (1961), of that given by Miles (1959).

With $\mathbf{g}=(g_1,g_2,\ldots,g_k)$ and $\mathbf{w}=(w_1,w_2,\ldots,w_k)$ fixed vectors of reals with $w_i>0$, let $r(l,k)$ denote the number of permutations, π, of $(1,2,\ldots,k)$ for which the PAVA applied to $(g_{\pi(1)},g_{\pi(2)},\ldots,g_{\pi(k)})$ with weight vector $(w_{\pi(1)},w_{\pi(2)},\ldots,w_{\pi(k)})$ gives l level sets. The following quite remarkable result, originally due to Miles, shows that under a simple condition on $\mathbf{g}$ and $\mathbf{w}$, $r(l,k)$ does not depend on $\mathbf{g}$ or $\mathbf{w}$ and the generating function of the $\{r(l,k)\}$ has a very simple form.

Lemma *If* $\mathbf{g}$ *and* $\mathbf{w}$ *are such that no two weighted averages of the different sets of* g_i *with weights* w_i *are equal, then the generating function of* $\{r(l,k)\}$ *is*

$$\sum_{l=1}^{k} r(l,k)s^l = s(s+1)\cdots(s+k-1). \tag{2.4.6}$$

Proof: Without loss of generality, assume that the g_i are in increasing order, that is $g_1<g_2<<\cdots<g_k$. (By assumption the g_i are distinct.) The lemma is established by induction and (2.4.6) clearly holds for $k=1$ and 2. Assume the result holds for $1,2,\ldots,k-1$ and consider the $k!$ permutations of the columns of

$$\begin{pmatrix} g_1\ g_2\cdots g_k \\ w_1\ w_2\cdots w_k \end{pmatrix}. \tag{2.4.7}$$

Now $r(l,k)=r_1(l,k)+r_2(l,k)$ with $r_1(l,k)$ the number of permutations, π, which produce l level sets and satisfy $\pi(k)=k$ and $r_2(l,k)$ is the number of permutations which produce l level sets and $\pi(k)\neq k$. There are $(k-1)!$ permutations, π, which satisfy $\pi(k)=k$ and in fact there is a one-to-one correspondence between these permutations and the permutations of $(1,2,\ldots,k-1)$. For each of these $(k-1)!$ permutations g_k is in the last column and the number of level sets given by the PAVA is one more than the number of level sets produced by the first $k-1$ columns. Hence, with $r(\cdot,k-1)$ computed with respect to the first $k-1$ columns of (2.4.7), (note that $r_1(i,k)=0$)

$$\sum_{l=1}^{k} r_1(l,k)s^l = s\sum_{l=1}^{k} r(l-1,k-1)s^{l-1} = s\sum_{l=1}^{k-1} r(l,k-1)s^l.$$

Clearly, $g_1,g_2,\ldots,g_{k-1}$ have no two weighted averages, with weights $w_1,w_2,\ldots,w_{k-1}$, which are equal and so by the inductive hypothesis

$$\sum_{l=1}^{k} r_1(l,k)s^l = s^2(s+1)\cdots(s+k-2). \tag{2.4.8}$$

The remaining permutations can be enumerated as follows: for each i there are $(k-1)!$ permutations of the columns of (2.4.7) which have g_k in the column immediately before g_i. Fix $i\in\{1,2,\ldots,k-1\}$. Using the PAVA to compute the isotonic regression for such a permuted array, one may amalgamate g_k and g_i at the first step. This produces a $k-1$ dimensional array (one column contains $(w_i g_i + w_k g_k)/(w_i + w_k)$ and $w_i + w_k$) which has no two weighted averages equal. Denoting the number of permutations of $(1, 2, \ldots, k-1)$ which produces l level sets in this array by $r_i'(l, k-1)$ and applying the inductive hypothesis,

$$\sum_{l=1}^{k} r_2(l, k)s^l = \sum_{i=1}^{k-1} \sum_{l=1}^{k-1} r_i'(l, k-1)s^l = (k-1)s(s+1)\cdots(s+k-2). \quad (2.4.9)$$

The desired conclusion follows by summing (2.4.8) and (2.4.9). □

Theorem 2.4.2 *Let $Y_1, Y_2, \ldots, Y_k$ be independent, identically distributed and continuous and let $P_S(l, k)$ be the probability that the isotonic regression of $Y = (Y_1, Y_2, \ldots, Y_k)$ with the simple order and equal weights has l level sets. The probability generating function of $\{P_S(l, k)\}$ is given by*

$$P_k(s) = \sum_{l=1}^{k} P_S(l, k)s^l = s(s+1)\cdots(s+k-1)/k!. \quad (2.4.10)$$

Proof: Let $m(y_1, y_2, \ldots, y_k)$ denote the number of level sets in the isotonic regression of $y_1, y_2, \ldots, y_k$ with equal weights and the simple order, $1 \lesssim 2 \lesssim \cdots \lesssim k$; let $P_{\mathbf{Y}}$ be the probability induced by $\mathbf{Y}$; let $P_{\mathbf{Y}'}$ be the probability induced by the order statistics; let Π denote the collection of permutations of $(1, 2, \ldots, k)$; and let A_0 be the set of vectors $(y_1, y_2, \ldots, y_k)$ with $y_1 < y_2 < \cdots < y_k$ which have no two averages (equal weights) equal. Because the Y_i are independent and continuous, $P_{\mathbf{Y}'}(A_0) = 1$. Hence,

$$P_k(s) = \int s^{m(y_1, y_2, \ldots, y_k)} dP_{\mathbf{Y}}(y_1, y_2, \ldots, y_k)$$

$$= \frac{1}{k!} \sum_{\pi \in \Pi} \int_{y_1 < y_2 < \cdots < y_k} s^{m(y_{\pi(1)}, y_{\pi(2)}, \ldots, y_{\pi(k)})} dP_{\mathbf{Y}'}(y_1, y_2, \ldots, y_k)$$

$$= \frac{1}{k!} \int_{A_0} \sum_{\pi \in \Pi} s^{m(y_{\pi(1)}, y_{\pi(2)}, \ldots, y_{\pi(k)})} dP_{\mathbf{Y}'}(y_1, y_2, \ldots, y_k)$$

$$= \frac{1}{k!} \int_{A_0} [s(s+1)\cdots(s+k-1)] dP_{\mathbf{Y}'} = s(s+1)\cdots(s+k-1)/k!. \quad □$$

Corollary A *The probabilities $P_S(l, k)$ satisfy*

$$P_S(1, k) = \frac{1}{k}$$

$$P_S(k,k) = \frac{1}{k!}$$

and

$$P_S(l,k) = \frac{1}{k} P_S(l-1,k-1) + \frac{k-1}{k} P_S(l,k-1) \qquad \textit{for } l = 2,3,\ldots,k-1. \tag{2.4.11}$$

Proof: The formulas for $P_S(1,k)$ and $P_S(k,k)$ follow immediately from (2.4.10) and the recurrence relationship follows from the equality $P_k(s) = sP_{k-1}(s)/k + (k-1)P_{k-1}(s)/k$. □

The numerator of $P_k(s)$ is the generating function of the absolute values of the Stirling numbers of the first kind (cf. Abramowitz and Stegun, 1965, p. 824). Denoting the Stirling numbers of the first kind by S_k^l, the following result is an immediate consequence of this observation.

Corollary B *The probabilities $P_S(l,k)$ are given by*

$$P_S(l,k) = |S_k^l|/k! \qquad \text{for } l = 1,2,\ldots,k.$$

Numerical values of the $P_S(l,k)$ for $k \leqslant 20$ are given in the Appendix. Using these values, the critical values for the $\bar{\chi}^2$ and $\bar{E}^2$ tests with equal weights can be obtained. Tables of these critical values are also included in the Appendix.

The simple tree alternative

Because of its use in making one-sided comparisons of several treatments with a control, the simple tree alternative, $\mu_0 \leqslant [\mu_1, \mu_2, \ldots, \mu_k]$, is of practical importance. The level probabilities for a simple tree ordering are denoted by $P_T(l, k+1; \mathbf{w})$. Using the method discussed earlier, $P_T(k+1, k+1; \mathbf{w})$ can be computed by summing $P(Y_{\pi(0)} < Y_{\pi(1)} < \cdots < Y_{\pi(k)})$ over the $k!$ permutations, π, of $(0, 1, \ldots, k)$ which satisfy $\pi(0) = 0$. The approach given below is less tedious. Let $Y_0, Y_1, \ldots, Y_k$ be independent normal variables with a common mean and variances $w_0^{-1}, w_1^{-1}, \ldots, w_k^{-1}$, respectively. For $i = 1, 2, \ldots, k$, set $U_i = Y_i - Y_0$ and note that

$$P_T(k+1, k+1; \mathbf{w}) = P[U_1 > 0, U_2 > 0, \ldots, U_k > 0].$$

This probability is the orthant probability, P_k, discussed earlier, but in this case the correlation matrix satisfies ($\rho_{ii} = 1$)

$$\rho_{ij} = \left[\frac{w_i w_j}{(w_0 + w_i)(w_0 + w_j)}\right]^{1/2} \qquad \text{for } 1 \leqslant i \neq j \leqslant k. \tag{2.4.12}$$

Thus, in this case,

$$P_1 = \frac{1}{2}$$

$$P_2 = \frac{1}{4} + \frac{1}{2\pi} \sin^{-1} \rho_{12}$$

and

$$P_3 = \frac{1}{8} + \frac{1}{4\pi} (\sin^{-1} \rho_{12} + \sin^{-1} \rho_{13} + \sin^{-1} \rho_{23}). \tag{2.4.13}$$

For $k > 3$, the P_k are simpler than for a simple order, because $\rho_{ij} = \lambda_i \lambda_j$ for $i \neq j$ with $\lambda_i = [w_i/(w_0 + w_i)]^{1/2}$, and the work of Dunnett and Sobel (1955) can be applied to show that

$$P_k \int_{-\infty}^{\infty} \left[\prod_{i=1}^{k} \Phi\left(\frac{\lambda_i x}{(1 - \lambda_i^2)^{1/2}} \right) \right] \phi(x)\, dx \tag{2.4.14}$$

with Φ and ϕ the distribution and density function of a standard normal variable.

Of course, $P_T(2, 2; \mathbf{w}) = P_T(1, 2; \mathbf{w}) = \frac{1}{2}$. For the case $k = 2$, $P_T(3, 3; \mathbf{w}) = P_2$,

$$\begin{aligned} P_T(2, 3; \mathbf{w}) &= P_T(2, 2; w_0 + w_1, w_2) P_T(1, 2; w_0, w_1) \\ &\quad + P_T(2, 2; w_0 + w_2, w_1) P_T(1, 2; w_0, w_2) \\ &= \tfrac{1}{2} \times \tfrac{1}{2} + \tfrac{1}{2} \times \tfrac{1}{2} = \tfrac{1}{2} \end{aligned}$$

and $P_T(1, 3; \mathbf{w}) = \frac{1}{2} - P_T(3, 3; \mathbf{w})$. For $k = 3$, $P_T(4, 4; \mathbf{w}) = P_3$ and Bartholomew (1961a) has shown that

$$P_T(3, 4; \mathbf{w}) = \frac{3}{8} + \frac{1}{4\pi} (\sin^{-1} \rho_{12\cdot 3} + \sin^{-1} \rho_{13\cdot 2} + \sin^{-1} \rho_{23\cdot 1})$$

$$P_T(2, 4; \mathbf{w}) = \tfrac{1}{2} - P_T(4, 4; \mathbf{w})$$

and

$$P_T(1, 4; \mathbf{w}) = \tfrac{1}{2} - P_T(3, 4; \mathbf{w}).$$

If the weights are equal, the level probabilities for the simple tree ordering are less complex. Again, for the equal-weights case, suppress the $\mathbf{w}$, i.e. denote the level probabilities by $P_T(l, k + 1)$. The decompositions in $\mathscr{L}_{lk+1}$ consist of $\{0\} \cup A$, with A a subset of $\{1, 2, \ldots, k\}$ of cardinality $k + 1 - l$ and $l - 1$ singletons. Each term in the recursion formula, (2.4.4), is the same, and assuming the common weight is 1,

$$P_T(l, k + 1) = \binom{k}{l - 1} P_T(1, k - l + 2) P_T(l, l; (k - l + 2), 1, \ldots, 1). \tag{2.4.15}$$

The last factor in (2.4.15) has been tabled by Ruben (1954) for l and k with

$0 \leqslant k \leqslant 49$ and $k-9 \leqslant l \leqslant k+1$. Using either numerical integration or Ruben's table, the $P_T(l, k+1)$ in this case are easily tabulated. Numerical values of $P_T(l, k+1)$ with $k \leqslant 19$ and the associated critical values for the $\bar{\chi}^2$ and $\bar{E}^2$ test statistics are given in the Appendix.

The simple loop alternative

The restriction $\mu_1 \leqslant [\mu_2, \mu_3, \ldots, \mu_{k-1}] \leqslant \mu_k$ is called a simple loop alternative. For $k=3$, it is a simple ordering and for $k=4$, it corresponds to a 2×2 matrix ordering. The level probabilities, which are denoted $P_L(l, k; \mathbf{w})$, will be obtained for $k=4$. Computing orthant probabilities using (2.4.3),

$$P_L(4,4;\mathbf{w}) = P[Y_1 < Y_2 < Y_3 < Y_4] + P[Y_1 < Y_3 < Y_2 < Y_4]$$
$$= \frac{1}{4\pi}\left(\frac{\pi}{2} - \theta_1 - \theta_3 + \frac{\pi}{2} - \theta_2 - \theta_4\right)$$

where

$$\theta_1 = \sin^{-1}\left[\frac{w_1 w_3}{(w_1+w_2)(w_2+w_3)}\right]^{1/2}$$

$$\theta_2 = \sin^{-1}\left[\frac{w_2 w_4}{(w_2+w_3)(w_3+w_4)}\right]^{1/2}$$

and θ_3 is θ_1 and θ_4 is θ_2 with w_2 and w_3 interchanged. Using $\sin(\pi/2 - \theta_1 - \theta_3) = \cos(\theta_1 + \theta_3) = \cos\theta_1 \cos\theta_3 - \sin\theta_1 \sin\theta_3$,

$$P_L(4,4;\mathbf{w}) = \frac{1}{4\pi}(\sin^{-1}\tau_{21,13} + \sin^{-1}\tau_{24,43})$$

with

$$\tau_{21,13} = \left[\frac{w_2 w_3}{(w_1+w_2)(w_1+w_3)}\right]^{1/2} \quad \text{and} \quad \tau_{24,43} = \left[\frac{w_2 w_3}{(w_2+w_4)(w_3+w_4)}\right]^{1/2}.$$

Listing the elements in $\mathscr{L}_{34}$($\mathscr{L}_{lk}$ is defined in the discussion before Theorem 2.4.1) and applying the recursion formula (2.4.4) and (2.4.3), one sees that

$$P_L(3,4;\mathbf{w}) = \frac{1}{2} + \frac{1}{4\pi}(\sin^{-1}\tau_{[12]3,34}$$
$$+ \sin^{-1}\tau_{[13]2,24} + \sin^{-1}\tau_{[24]3,31} + \sin^{-1}\tau_{[34]2,21}),$$

with

$$\tau_{[rs]t,tu} = -\left[\frac{(w_r + w_s)w_u}{(w_r+w_s+w_t)(w_t+w_u)}\right]^{1/2}.$$

The partitions in $\mathscr{L}_{24}$ are

$$\{1,2\},\{3,4\};\{1,3\}\{2,4\};\{1,2,3\},\{4\}, \text{ and } \{1\},\{2,3,4\},$$

and the first two each have probability $\frac{1}{8}$. For the third decomposition, the probability is $\frac{1}{2}P_T(1,3; w_1, w_2, w_3)$, i.e. one-half of the probability of one level set in a simple tree with $k=2$. The probability of the fourth decomposition is one-half of the probability of one level set in an 'inverted tree'. The inverted tree is the negative of a tree and $P_{\mathbf{w}}(\mathbf{x}|-\mathscr{K}) = -P_{\mathbf{w}}(-\mathbf{x}|\mathscr{K})$ (cf. Theorem 1.3.2 or Theorem 8.2.3). Hence, the probability of the fourth decomposition is one-half the probability of one level set in a simple tree with $k=2$ and weight vector (w_4, w_2, w_3), i.e. it is $\frac{1}{2}P_T(1,3; w_4, w_2, w_3)$. It is now straightforward to show that $P_L(2,4;\mathbf{w}) = \frac{1}{2} - P_L(4,4;\mathbf{w})$, which forces $P_L(1,4;\mathbf{w}) = \frac{1}{2} - P_L(3,4;\mathbf{w})$. If the weights are equal, then

$$P_L(4,4) = \tfrac{1}{12}$$

$$P_L(3,4) = \frac{1}{2} - \frac{1}{\pi}\sin^{-1}\frac{1}{\sqrt{3}} = 0.3041$$

$$P_L(2,4) = \tfrac{5}{12}$$

and

$$P_L(1,4) = \frac{1}{\pi}\sin^{-1}\frac{1}{\sqrt{3}} = 0.1959.$$

Other partial orders

The results for the simple loop with $k=4$ are the only exact expressions for the level probabilities with a matrix order. Lemke (1983) obtained Monte Carlo estimates of the equal-weights level probabilities for a matrix order with moderate dimensions. These Monte Carlo estimates are given in the Appendix.

The unimodal partial ordering, i.e. the ordering which imposes the restriction $\mu_1 \leqslant \mu_2 \leqslant \cdots \leqslant \mu_h \geqslant \mu_{h+1} \geqslant \cdots \geqslant \mu_k$ with $1 < h < k$, is of practical importance. The equal-weights level probabilities for a unimodal ordering are given in the Appendix for $k \leqslant 15$. (For $3 \leqslant k \leqslant 7$ they were computed using the recursive approach discussed here, and for $8 \leqslant k \leqslant 15$ they were estimated using Monte Carlo techniques.) If the two modal means are not ordered, i.e. the ordering restriction only requires $\mu_1 \leqslant \mu_2 \leqslant \cdots \leqslant \mu_h$ and $\mu_{h+1} \geqslant \mu_{h+2} \geqslant \cdots \geqslant \mu_k$, then the level probabilities are the convolution of $\{P_S(l, h; w_1, \ldots, w_h)\}$ and $\{P_S(l, k-h; w_{h+1}, \ldots, w_k)\}$. This is a special case of the setting treated in the next paragraph.

If $(X, \lesssim)$ is decomposable, that is $X = X_1 + X_2$ and no element in X_1 is related to an element in X_2, then the isotonic regression can be computed separately on each index set. This implies that, with probability one, the number of level sets, M, is $M_1 + M_2$, where M_i is the number of level sets on $X_i, i = 1, 2,$

and M_1 and M_2 are independent. Hence, $\{P(l,k;\mathbf{w})\}$ is the convolution of the level probabilities for X_1 and X_2, and, of course, the PGF of M is the product of the PGFs of M_1 and M_2.

If $\lesssim$ is a quasi-order, then the relation, $\approx$, defined by $i \approx j$ if and only if $i \lesssim j \lesssim i$, is an equivalence relation and induces a natural partial order, $\lesssim'$, on the equivalence classes $\{E_\alpha : \alpha = 1, 2, \ldots, k'\}$. For convenience, identify α with E_α, think of $\lesssim'$ as a partial order on $\{1, 2, \ldots, k'\}$, let H_1' denote the vectors in $R^{k'}$ isotonic with respect to $\lesssim'$, and for $1 \leqslant \alpha \leqslant k'$, set

$$w'_\alpha = \sum_{i \in E_a} w_i \qquad \text{and} \qquad Y'_\alpha = \sum_{i \in E_\alpha} \frac{w_i Y_i}{w'_\alpha}.$$

Because the lower sets in $(X, \lesssim)$ are of the form $\bigcup_{\alpha \in L'} E_\alpha$ with L' a lower set in $(\{1, 2, \ldots, k'\}, \lesssim')$, the number of level sets in the isotonic regression of $\mathbf{Y}$, is the same as the number of level sets in the isotonic regression of $\mathbf{Y}'$. In particular, if the level probabilities for the partial order, $\lesssim'$, with weights $\mathbf{w}'$, are denoted by $P(l, k'; \mathbf{w}')$, then $P(l, k; \mathbf{w}) = P(l, k'; \mathbf{w}')$. This implies that $P(l, k; \mathbf{w}) = 0$ for $k' < l \leqslant k$.

2.5 THE POWER FUNCTIONS OF THE $\bar{\chi}^2$AND $\bar{E}^2$ TESTS: SOME EXACT COMPUTATIONS AND APPROXIMATIONS

The power functions of the $\bar{\chi}^2$ and $\bar{E}^2$ tests can be computed by methods similar to those used in determining their null distributions. For $l = 1, 2, \ldots, k$, the probability that there are l level sets and the test statistic exceeds the critical value is obtained. These values are summed over l. Because of the complexity of these calculations, exact powers have been computed for only four or fewer populations. These values will be used to provide some indication of the amount of increase in power obtained by using these tests which make use of ordering information. Approximations indicate that there are similar gains for larger k. These results concerning power will be used in Chapter 4 where the LR procedures are compared with other procedures for making use of ordering information. For simplicity, the case of equal weights will be considered; the results in this case give a compelling argument for the use of order restricted tests.

For $k = 3$ there are the following three basic types of partial orderings (other than the trivial one which imposes no restrictions on μ_1, μ_2, and μ_3):

$$H_1^{(1)}: \mu_1 \leqslant \mu_2 \leqslant \mu_3, \qquad H_1^{(2)}: \mu_1 \leqslant [\mu_2, \mu_3], \qquad \text{and} \qquad H_1^{(3)}: \mu_1 \leqslant \mu_2.$$

By relabeling, all of the partial order restrictions on $\{1, 2, 3\}$ can be expressed as one of the above. Because of their practical importance, attention is focused on $H_1^{(1)}$ and $H_1^{(2)}$. For $k = 4$, the simple order, the simple loop, and the simple tree are considered.

Exact computations for $k = 3$ and 4

For the case $k = 3$, let $Y_i \sim \mathcal{N}(\mu_i, w^{-1})$ for $i = 1, 2, 3$ with Y_1, Y_2, Y_3 independent and w a positive real number. Multiplying the Y_i by $\sqrt{w}$ leads to the case of unit weights with mean vector $\sqrt{w}\,\boldsymbol{\mu}$. Thus attention is restricted to the case in which $w_i \equiv 1$. The simple order is considered first.

For $C > 0$, $P[\bar{\chi}^2_{01} \geqslant C, l = 3]$ is computed first. With $\bar{Y} = (Y_1 + Y_2 + Y_3)/3$, this probability is

$$P[Y_2 - Y_1 > 0, Y_3 - Y_2 > 0, \sum_{i=1}^{3} (Y_i - \bar{Y})^2 \geqslant C].$$

By making the orthogonal transformation $V_1 = (Y_2 - Y_1)/\sqrt{2}$ and $V_2 = (2Y_3 - Y_1 - Y_2)/\sqrt{6}$, it can be written as $P[V_1 > 0, \sqrt{3}V_2 - V_1 > 0, V_1^2 + V_2^2 \geqslant C]$. (It should be noted that, with this transformation, $\sum_{i=1}^{3}(Y_i - \bar{Y})^2 = V_1^2 + V_2^2$.) Applying the polar transformation $V_1 = R\cos\theta$ and $V_2 = R\sin\theta$, the conditions $V_1 > 0$ and $\sqrt{3}V_2 - V_1 > 0$ are equivalent to $\cos\theta > 0$, $\sqrt{3}\sin\theta - \cos\theta > 0$ or, equivalently, $\pi/6 < \theta < \pi/2$. Because V_1 and V_2 are independent normal variables with means $\lambda_1 = (\mu_2 - \mu_1)/\sqrt{2}$ and $\lambda_2 = (2\mu_3 - \mu_1 - \mu_2)/\sqrt{6}$ and unit variances, with $\lambda_1 = \Delta\sin\beta$ and $\lambda_2 = \Delta\cos\beta$, this probability is

$$\frac{\exp(-\frac{1}{2}\Delta^2)}{2\pi} \int_{\pi/6+\beta}^{\pi/2+\beta} \int_{C}^{\infty} r\exp(-\tfrac{1}{2}r^2 + r\Delta\sin\theta)\,dr\,d\theta$$

$$= \frac{\exp(-\frac{1}{2}\Delta^2)}{2\pi} \int_{\pi/6+\beta}^{\pi/2+\beta} \psi(\Delta\sin\theta, C)\,d\theta \qquad (2.5.1)$$

where $\psi(x, C) = [x\Phi(x - \sqrt{C}) + \phi(x - \sqrt{C})]/\phi(x)$ with Φ and ϕ being the distribution function and density of a standard normal variable, respectively. For $\bar{E}^2_{01}(v)$, let Q be a chi-squared variable with v degrees of freedom which is independent of (Y_1, Y_2, Y_3). From (2.2.11) one sees that $\bar{E}^2_{01}(v) = \bar{\chi}^2_{01}/(\bar{\chi}^2_{01} + \bar{\chi}^2_{12} + Q)$. Because $l = 3$ implies that $\bar{\chi}^2_{12} = 0$, with $C \in (0, 1)$,

$$P[\bar{E}^2_{01}(v) \geqslant C, l = 3] = P[\bar{\chi}^2_{01} \geqslant CQ/(1 - C), l = 3]$$

$$= \frac{\exp(-\frac{1}{2}\Delta^2)}{2\pi} \int_{\pi/6+\beta}^{\pi/2+\beta} \int_{0}^{\infty} h(u)\psi\left(\Delta\sin\theta, \frac{Cu}{1 - C}\right) du\,d\theta \qquad (2.5.2)$$

where $h(u)$ is the density of a χ^2 variable with v degrees of freedom. Of course, $P[\bar{\chi}^2_{12} \geq C, l=3] = P[\bar{E}^2_{12}(v) \geq C, l=3] = 0$.

Next, $P[\bar{\chi}^2_{01} \geq C, l=2]$ is considered. The event $[l=2]$ is the union of $[Y_1 \geq Y_2, Y_1 + Y_2 < 2Y_3]$ and $[Y_2 \geq Y_3, Y_2 + Y_3 > 2Y_1]$. For the first case with the same transformation as above, $\bar{\chi}^2_{01} = V_2^2$ ($\bar{\chi}^2_{12} = V_1^2$), and

$$\begin{aligned} P[Y_1 \geq Y_2, Y_1 + Y_2 < 2Y_3, \bar{\chi}^2_{01} \geq C] &= P[V_1 \leq 0, V_2 > 0, V_1^2 \geq C] \\ &= \Phi(-\lambda_1)\Phi(\lambda_2 - \sqrt{C}) \\ &= \Phi(-\Delta \sin\beta)\Phi(\Delta\cos\beta - \sqrt{C}). \end{aligned} \tag{2.5.3}$$

Similarly,

$$\begin{aligned} P[Y_1 \geq Y_2, Y_1 + Y_2 < 2Y_3, \bar{\chi}^2_{12} \geq C] &= P[V_1 \leq 0, V_2 > 0, V_1^2 \geq C] \\ &= \Phi(-\sqrt{C} - \lambda_1)\Phi(\lambda_2) \\ &= \Phi(-\sqrt{C} - \Delta\sin\beta)\Phi(\Delta\cos\beta); \end{aligned} \tag{2.5.4}$$

for $C \in (0, 1)$,

$$\begin{aligned} &P[Y_1 \geq Y_2, Y_1 + Y_2 < 2Y_3, \bar{E}^2_{01}(v) \geq C] \\ &= P\left[V_1 \leq 0, V_2 > 0, V_2^2 \geq \frac{C(V_1^2 + Q)}{1 - C}\right] \\ &= \int_0^\infty h(u) \int_{-\infty}^0 \phi(v - \Delta\sin\beta)\Phi\left\{-\left[\frac{C(v^2 + u)}{1 - C}\right]^{1/2} + \Delta\cos\beta\right\} dv\, du; \end{aligned} \tag{2.5.5}$$

and

$$\begin{aligned} &P[Y_1 \geq Y_2, Y_1 + Y_2 < 2Y_3, \bar{E}^2_{12}(v) \geq C] \\ &= P\left[V_1 \leq 0, V_2 > 0, V_1^2 \geq \frac{CQ}{1 - C}\right] \\ &= \Phi(\Delta\cos\beta) \int_0^\infty h(u)\Phi\left[-\Delta\sin\beta - \left(\frac{Cu}{1 - C}\right)^{1/2}\right] du. \end{aligned} \tag{2.5.6}$$

Multiplying by minus one and interchanging the indices 1 and 3, the second case, $[Y_2 \geq Y_3, Y_2 + Y_3 > 2Y_1]$, is transformed into the first case. The λ_1 and λ_2 for the transformed variables, which will be denoted by λ_1' and λ_2', are

$$\lambda_1' = \frac{\mu_3 - \mu_2}{\sqrt{2}} = \Delta\left(\frac{\sqrt{3}}{2}\cos\beta - \frac{1}{2}\sin\beta\right) = \Delta\sin\left(\frac{\pi}{3} - \beta\right)$$

and

$$\lambda_2' = \frac{\mu_3 + \mu_2 - 2\mu_1}{\sqrt{6}} = \Delta\left(\frac{1}{2}\cos\beta + \frac{\sqrt{3}}{2}\sin\beta\right) = \Delta\cos\left(\frac{\pi}{3} - \beta\right).$$

Hence, the contribution to the powers of $\bar{\chi}_{01}^2, \bar{\chi}_{12}^2, \bar{E}_{01}^2(v)$ and $\bar{E}_{12}^2(v)$ for this case are given by (2.5.3), (2.5.4), (2.5.5), and (2.5.6), respectively, with β replaced by $\pi/3 - \beta$.

If $l = 1$, then $\bar{\chi}_{01}^2 = \bar{E}_{01}^2 = 0$. To compute $P[\bar{\chi}_{12}^2 \geqslant C, l = 1]$, one first observes that applying the MLSA, $[l = 1] = [Y_1 \geqslant \bar{Y}, Y_1 + Y_2 \geqslant 2\bar{Y}] = [2Y_1 - Y_2 - Y_3 \geqslant 0, 2Y_3 - Y_1 - Y_2 \leqslant 0]$. Thus, with V_1, V_2, R, and θ defined as before, if $l = 1$, then $\bar{\chi}_{12}^2 = V_1^2 + V_2^2$ and

$$P[\bar{\chi}_{12}^2 \geqslant C, l = 1] = P\left[\sin\left(\theta + \frac{\pi}{3}\right) \leqslant 0, \sin\theta \leqslant 0, R^2 \geqslant C\right]$$

$$= \frac{\exp(-\frac{1}{2}\Delta^2)}{2\pi} \int_{\beta - \pi}^{\beta - \pi/3} \psi(\Delta \sin\theta, C)\mathrm{d}\theta. \qquad (2.5.7)$$

In a similar manner,

$$P[\bar{E}_{12}^2(v) \geqslant C, l = 1] = \frac{\exp(-\frac{1}{2}\Delta^2)}{2\pi} \int_{\beta - \pi}^{\beta - \pi/3} \int_0^\infty h(u)\psi\left(\Delta\sin\theta, \frac{Cu}{1 - C}\right)\mathrm{d}u\,\mathrm{d}\theta. \qquad (2.5.8)$$

In summary, for the simple order with $k = 3$ and $n_i/\sigma_i^2 = 1$ $(i = 1, 2, 3)$, the powers of $\bar{\chi}_{01}^2$ and $\bar{\chi}_{12}^2$ are

$$\pi_{01}(\boldsymbol{\mu}) = \frac{\exp(-\frac{1}{2}\Delta^2)}{2\pi} \int_{\pi/6 + \beta}^{\pi/2 + \beta} \psi(\Delta\sin\theta, C)\,\mathrm{d}\theta$$

$$+ \Phi(-\Delta\sin\beta)\Phi(\Delta\cos\beta - \sqrt{C}) + \Phi\left[-\Delta\sin\left(\frac{\pi}{3} - \beta\right)\right]$$

$$\cdot\Phi\left[\Delta\cos\left(\frac{\pi}{3} - \beta\right) - \sqrt{C}\right] \qquad (2.5.9)$$

and

$$\pi_{12}(\boldsymbol{\mu}) = \frac{\exp(-\frac{1}{2}\Delta^2)}{2\pi} \int_{\beta - \pi}^{\beta - \pi/3} \psi(\Delta\sin\theta, C)\,\mathrm{d}\theta$$

$$+ \Phi(-\Delta\sin\beta - \sqrt{C})\Phi(\Delta\cos\beta) + \Phi\left[-\Delta\sin\left(\frac{\pi}{3} - \beta\right) - \sqrt{C}\right]$$

$$\cdot\Phi\left[\Delta\cos\left(\frac{\pi}{3}\right) - \beta\right], \qquad (2.5.10)$$

respectively. For the simple order with $k = 3$ and $n_i/\sigma_i^2 = 1$ $(i = 1, 2, 3)$, the powers of $\bar{E}_{01}^2(v)$ and $\bar{E}_{12}^2(v)$ are

$$\pi_{01}^*(\boldsymbol{\mu}) = \frac{\exp(-\frac{1}{2}\Delta^2)}{2\pi} \int_{\pi/6+\beta}^{\pi/2+\beta} \int_0^\infty h(u)\psi\left(\Delta \sin\theta, \frac{Cu}{1-C}\right) du\, d\theta$$
$$+ \int_0^\infty \int_{-\infty}^0 h(u)\phi(v - \Delta\sin\beta)\Phi\left\{-\left[\frac{C(v^2+u)}{1-C}\right]^{1/2} + \Delta\cos\beta\right\} dv\, du$$
$$+ \int_0^\infty \int_{-\infty}^0 h(u)\phi\left(v - \Delta\sin\left(\frac{\pi}{3} - \beta\right)\right)\Phi\left\{-\left[\frac{C(v^2+u)}{1-C}\right]^{1/2}\right.$$
$$\left. + \Delta\cos\left(\frac{\pi}{3} - \beta\right)\right\} dv\, du \tag{2.5.11}$$

and

$$\pi_{12}^*(\boldsymbol{\mu}) = \frac{\exp(-\frac{1}{2}\Delta^2)}{2\pi} \int_{\beta-\pi}^{\beta-\pi/3} \int_0^\infty h(u)\psi\left(\Delta\sin\theta, \frac{Cu}{1-C}\right) du\, d\theta$$
$$+ \Phi(\Delta\cos\beta) \int_0^\infty h(u)\Phi\left[-\Delta\sin\beta - \left(\frac{Cu}{1-C}\right)^{1/2}\right] du$$
$$+ \Phi\left[\Delta\cos\left(\frac{\pi}{3} - \beta\right)\right] \int_0^\infty h(u)\Phi\left[-\Delta\sin\left(\frac{\pi}{3} - \beta\right) - \left(\frac{Cu}{1-C}\right)^{1/2}\right] du, \tag{2.5.12}$$

respectively. If $n_i/\sigma_i^2 = w$ for each i, then the power function at $\boldsymbol{\mu}$ can be obtained by evaluating (2.5.9), (2.5.10), (2.5.11), and (2.5.12) at $\sqrt{w}\,\boldsymbol{\mu}$.

The power functions of the classical tests of homogeneity, H_0, with the unrestricted alternative, H_2–H_0, depend on $\boldsymbol{\mu}$ through

$$\Delta = \left[\sum_{i=1}^{k} w_i(\mu_i - \bar{\mu})^2\right]^{1/2} \qquad \text{with } \bar{\mu} = \frac{\sum_{i=1}^k w_i\mu_i}{\sum_{i=1}^k w_i}, \tag{2.5.13}$$

which is the same Δ that appears in the discussion above. However, the power functions of the tests for a restricted alternative depend on $\boldsymbol{\mu}$ through Δ and, in the case of $k = 3$, also through β. These test statistics and their power functions are invariant under a shift by a constant vector. Figure 2.5.1 gives the hyperplane

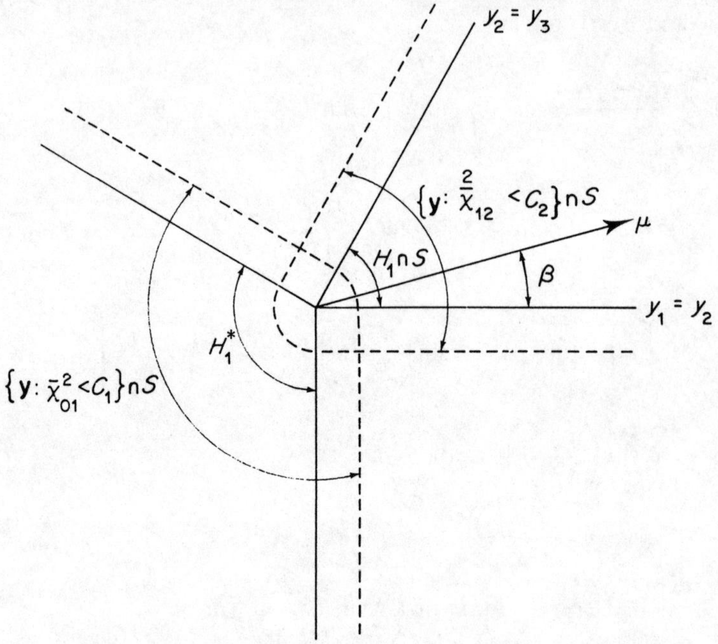

Figure 2.5.1 The hyperplane, $S = \{y \in R^3: y_1 + y_2 + y_3 = 0\}, H_1 \cap S, H_1^*$, and the acceptance regions for $\bar{\chi}^2_{01}$ and $\bar{\chi}^2_{12}$

in R^3 defined by $S = \{(y_1, y_2, y_3): y_1 + y_2 + y_3 = 0\}$, $H_1 \cap S$, H_1^* (the dual of H_1; see Section 1.7) and the acceptance regions for $\bar{\chi}^2_{01}$ and $\bar{\chi}^2_{12}$ intersected with S. The quantity Δ is the distance from $\boldsymbol{\mu}$ to H_0 and β is the angle $\boldsymbol{\mu}$ makes with the ray on which $y_1 = y_2 < y_3$. The restriction $\boldsymbol{\mu} \in H_1$ is equivalent to $0 \leqslant \beta \leqslant \pi/3$. For a fixed value of Δ, Bartholomew (1961a, 1961b), by differentiating with respect to β, found that $\pi_{01}(\boldsymbol{\mu})$, which is periodic in β with period 2π, is increasing for $\beta \in [-5\pi/6, \pi/6]$ and decreasing for $\beta \in [\pi/6, 7\pi/6]$. Thus, $\pi_{01}(\boldsymbol{\mu})$ has its maximum, for fixed $\Delta > 0$, at $\beta = \pi/6$, which is the middle of $H_1(\mu_2 - \mu_1 = \mu_3 - \mu_2 > 0)$ and has its minimum over H_1 at $\beta = 0(\mu_1 = \mu_2 < \mu_3)$ and at $\beta = \pi/3(\mu_1 < \mu_2 = \mu_3)$. It is clear that $\pi_{01}(\boldsymbol{\mu})$ and $\pi^*_{01}(\boldsymbol{\mu})$ are symmetric about $\beta = \pi/6$ and $\beta = 7\pi/6$. (Properties, such as unbiasedness, consistency, and monotonicity will be investigated in the next section.)

The power function of $\bar{\chi}^2_{12}$, $\pi_{12}(\boldsymbol{\mu})$ is also periodic in β with period 2π and symmetric about $\beta = \pi/6$ and $\beta = 7\pi/6$. Chu-In Charles Lee, in a private communication, has shown that for a fixed Δ, $\pi_{12}(\boldsymbol{\mu})$ is increasing for $\beta \in [\pi/6, 7\pi/6]$ and decreasing for $\beta \in [-5\pi/6, \pi/6]$. Hence, $\pi_{12}(\boldsymbol{\mu})$ has its maximum at $\beta = 7\pi/6$, which is the middle of $H_1^*(\mu_2 - \mu_1 = \mu_3 - \mu_2 < 0)$, and has its minimum over H_2–H_1 at $\beta = 0$ and $\beta = \pi/3$. Properties of the power function of $\bar{E}^2_{12}(v)$ follow from those of $\bar{\chi}^2_{12}$ by noting that rejecting for large values of $\bar{E}^2_{12}(v)$ is

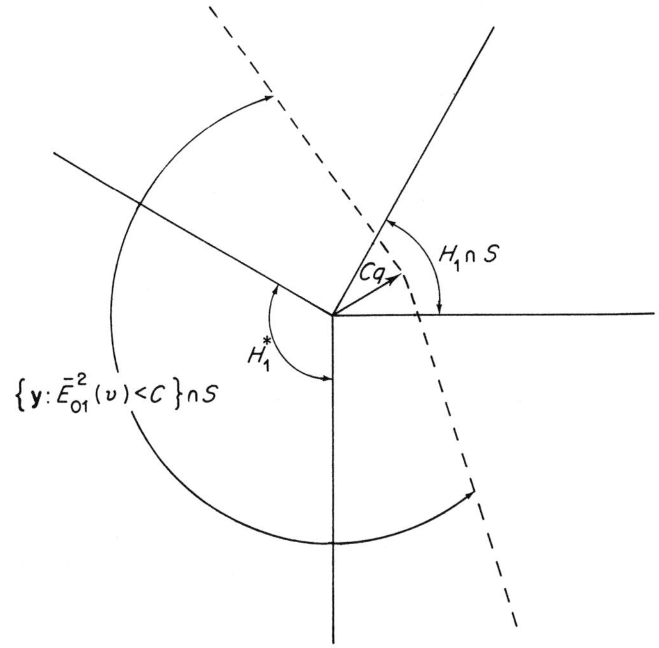

Figure 2.5.2 The hyperplane, $S = \{y \in R^3: y_1 + y_2 + y_3 = 0\}$, and the intersection of S with the acceptance region of $\bar{E}^2_{01}(v)$

equivalent to rejecting for large values of $\bar{\chi}^2_{12}/Q$ and that $\bar{\chi}^2_{12}$ and Q are independent and then conditioning on $Q = q$. In particular, the acceptance region, $\bar{\chi}^2_{12}/Q < C$, for a fixed value of $Q = q$ is the same as that of $\bar{\chi}^2_{12}$ with critical value qC.

The power function for $\bar{E}^2_{01}(v)$ is more complex. Recall that rejecting for large values of $\bar{E}^2_{01}(v)$ is equivalent to rejecting for large values of $\bar{\chi}^2_{01}/(\bar{\chi}^2_{12} + Q)$. For $Q = q$, the acceptance region, $\bar{\chi}^2_{01}/(\bar{\chi}^2_{12} + Q) < C$, is pictured in Figure 2.5.2. While $\bar{\chi}^2_{01}$, $\bar{\chi}^2_{12}$, and $\bar{E}^2_{12}(v)$ have convex acceptance regions, this is not the case for $\bar{E}^2_{01}(v)$. Numerical computations indicate that, for a fixed Δ, the maximum of $\pi^*_{01}(\boldsymbol{\mu})$ occurs at $\beta = \pi/6$ and the minimum over H_1 occurs at $\beta = 0$ and $\beta = \pi/3$.

The determination of the power functions for the simple tree with $k = 3$ and unit weights uses the same technique as in the simply ordered case. Also, as for the simple order, the equal-weights case, $w_1 = w_2 = w_3 = w$, is treated by multiplying $\boldsymbol{\mu}$ by $\sqrt{w}$ and applying the formulas for unit weights. To make use of the notation already developed in this section, it is covenient to denote the control weight and mean by w_1 and μ_1 and to let w_2 and w_3 and μ_2 and μ_3 denote the treatment weights and means, respectively. Bartholomew shows that the power of $\bar{\chi}^2_{01}$ in this case is

$$\pi_{01}(\boldsymbol{\mu}) = \frac{\exp(-\frac{1}{2}\Delta^2)}{2\pi} \int_{-\pi/6+\beta}^{\pi/2+\beta} \psi(\Delta \sin\theta, C)\,d\theta$$

$$+ \Phi(-\Delta\sin\beta)\Phi(\Delta\cos\beta - \sqrt{C}) + \Phi\left[-\Delta\sin\left(\frac{\pi}{3}+\beta\right)\right]$$

$$\cdot\Phi\left[-\Delta\cos\left(\frac{\pi}{3}+\beta\right) - \sqrt{C}\right]. \tag{2.5.14}$$

The power function of $\bar{\chi}^2_{12}$ in this case is found by Mukerjee, Robertson, and Wright (1986b) to be

$$\pi_{12}(\boldsymbol{\mu}) = \frac{\exp(-\frac{1}{2}\Delta^2)}{2\pi} \int_{\pi+\beta}^{4\pi/3+\beta} \psi(\Delta\sin\theta, C)\,d\theta$$

$$+ \Phi(\Delta\cos\beta)\Phi(-\Delta\sin\beta - \sqrt{C}) + \Phi\left[\Delta\cos\left(\frac{2\pi}{3}-\beta\right)\right]$$

$$\cdot\Phi\left[-\Delta\sin\left(\frac{2\pi}{3}-\beta\right) - \sqrt{C}\right]. \tag{2.5.15}$$

Robertson and Wright (1985) found the power function of $\bar{E}^2_{01}(v)$ to be

$$\pi_{01}(\boldsymbol{\mu}) = \frac{\exp(-\frac{1}{2}\Delta^2)}{2\pi} \int_{-\pi/6+\beta}^{\pi/2} \int_0^{\infty} h(u)\psi\left(\Delta\sin\theta, \frac{Cu}{1-C}\right) du\,d\theta$$

$$+ \int_0^{\infty}\int_{-\infty}^{0} h(u)\Phi\left\{\Delta\cos\beta - \left[\frac{C(u+v^2)}{1-C}\right]^{1/2}\right\}\phi(v-\Delta\sin\beta)\,dv\,du$$

$$+ \int_0^{\infty}\int_{-\infty}^{0} h(u)\Phi\left\{-\Delta\cos\left(\frac{\pi}{3}+\beta\right) - \left[\frac{C(u+v^2)}{1-C}\right]^{1/2}\right\}$$

$$\cdot\Phi\left[v - \Delta\sin\left(\frac{\pi}{3}+\beta\right)\right] dv\,du, \tag{2.5.16}$$

and because $\bar{E}^2_{12}(v) \geq C$ is equivalent to $\bar{\chi}^2_{12}/Q \geq C/(1-C)$,

$$\pi^*_{12}(\boldsymbol{\mu}) = \frac{\exp(-\frac{1}{2}\Delta^2)}{2\pi} \int_{\pi+\beta}^{(4\pi/3)+\beta} \int_0^{\infty} h(u)\psi\left(\Delta\sin\theta, \frac{Cu}{1-C}\right) du\,d\theta$$

$$+\Phi(\Delta\cos\beta)\int_0^\infty h(u)\Phi\left\{-\Delta\sin\beta-\left(\frac{Cu}{1-C}\right)^{1/2}\right\}du$$

$$+\Phi\left[\Delta\cos\left(\frac{2\pi}{3}-\beta\right)\right]\int_0^\infty h(u)\Phi\left[-\Delta\sin\left(\frac{2\pi}{3}-\beta\right)-\left(\frac{Cu}{1-C}\right)^{1/2}\right]du. \tag{2.5.17}$$

For the simple tree, the region corresponding to the alternative hypothesis is $0\leqslant\beta\leqslant 2\pi/3$ and the power functions are symmetric about $\beta=\pi/3$ and $\beta=4\pi/3$. Chu-In Charles Lee has also shown that the power function of $\bar{\chi}^2_{01}$ is increasing on $[-2\pi/3,\pi/3]$ and is decreasing on $[\pi/3,4\pi/3]$, but the power functions of $\bar{\chi}^2_{12}$ and $\bar{E}^2_{12}(v)$ are decreasing on $[-2\pi/3,\pi/3]$ and are increasing on $[\pi/3,4\pi/3]$. (Based on numerical computations, we conjecture that the intervals of monotonicity of the power function of $\bar{E}^2_{01}$ and $\bar{\chi}^2_{01}$ agree.)

For a fixed $\Delta>0$, the minimum power over H_1 for the simple order is believed to occur at $\mu_1=\mu_2=\cdots=\mu_{k-1}<\mu_k$ and at $\mu_1<\mu_2=\cdots=\mu_k$. For the simple tree the minimum power, for fixed $\Delta>0$ and $\boldsymbol{\mu}\in H_1$ (while the alternative region depends on $\lesssim$, this is not made explicit in the notation), is believed to occur at $\boldsymbol{\mu}$ with $\mu_1=\mu_\alpha$ for $\alpha\in X-\{j\}$ and $\mu_1<\mu_j, j=2,3,\ldots,k$. In both cases, these points have $\mu_\alpha=\mu_\beta$ for all (α,β) with $\alpha\lesssim\beta$ except for one pair. For a fixed $\Delta>0$ and $\boldsymbol{\mu}$ in the alternative region, the maximum power for the simple order (simple tree ordering) is believed to occur at $\mu_2-\mu_1=\mu_3-\mu_2=\cdots=\mu_k-\mu_{k-1}>0$ $(\mu_2-\mu_1=\mu_3-\mu_1=\cdots=\mu_k-\mu_1>0)$. These points have the differences in the μ_i 'large' for pairs that are ordered by $\lesssim$. In other words, the gain in power due to ordering information is greatest when the means which are ordered differ the most.

Table 2.5.1 gives the powers of the LRTs of homogeneity with an unrestricted alternative and for the simply ordered alternative as well as the simple tree alternative. The powers given there are for $\alpha=0.05$, $k=3$, and degrees of freedom $v=10,20,40,\infty$. For v finite the powers given are those of $\bar{E}^2_{01}(v)$ and for $v=\infty$ they are the powers of $\bar{\chi}^2_{01}$. The powers of the order restricted tests which are labeled as maximum or minimum are the powers at the points in the alternative region which have been conjectured to provide the maximum or minimum power.

The more restrictive the partial order, $\lesssim$, which determines H_1, the greater the prior information that is being provided; notice that the powers for the simple order are greater than for the simple tree and, in fact, the minimum powers for the simple order are greater than the maximum power for the simple tree. Bartholomew (1961a, 1961b) considers the alternative $H_1^{(3)}:\mu_1\leqslant\mu_3$, which is less restrictive than the simple order and the simple tree. The gain in power obtained

Table 2.5.1 Comparison[a] of the powers of unrestricted and restricted LRTs for the simple order and the simple tree order with $k = 3$ and $\alpha = 0.05$

	$\nu = 10$	$\nu = 20$	$\nu = 40$	$\nu = \infty$
		$\Delta = 1$		
Unrestricted	0.111	0.121	0.126	0.133
Simple (max.)	0.224 (101.8%)	0.232 (91.7%)	0.237 (88.1%)	0.244 (83.5%)
Simple (min.)	0.200 (80.2%)	0.209 (72.7%)	0.214 (69.8%)	0.221 (66.2%)
Tree (max.)	0.189 (70.3%)	0.199 (64.5%)	0.204 (61.9%)	0.210 (57.9%)
Tree (min.)	0.150 (35.1%)	0.161 (33.1%)	0.166 (31.7%)	0.173 (30.1%)
		$\Delta = 2$		
Unrestricted	0.319	0.364	0.389	0.415
Simple (max.)	0.550 (72.4%)	0.575 (58.0%)	0.589 (51.4%)	0.605 (45.8%)
Simple (min.)	0.503 (57.7%)	0.534 (46.7%)	0.550 (41.4%)	0.569 (37.1%)
Tree (max.)	0.472 (48.0%)	0.507 (39.3%)	0.523 (34.4%)	0.543 (30.8%)
Tree (min.)	0.411 (28.8%)	0.451 (23.9%)	0.472 (21.3%)	0.494 (19.0%)
		$\Delta = 3$		
Unrestricted	0.629	0.702	0.737	0.771
Simple (max.)	0.842 (33.9%)	0.867 (23.5%)	0.878 (19.1%)	0.892 (15.7%)
Simple (min.)	0.810 (28.8%)	0.843 (20.1%)	0.857 (16.3%)	0.872 (13.1%)
Tree (max.)	0.778 (23.7%)	0.819 (16.7%)	0.836 (13.4%)	0.857 (11.2%)
Tree (min.)	0.734 (17.0%)	0.784 (11.7%)	0.807 (9.5%)	0.831 (7.8%)
		$\Delta = 4$		
Unrestricted	0.873	0.923	0.942	0.957
Simple (max.)	0.970 (11.1%)	0.979 (6.1%)	0.981 (4.1%)	0.987 (3.1%)
Simple (min.)	0.960 (10.0%)	0.974 (5.5%)	0.977 (3.7%)	0.983 (2.7%)
Tree (max.)	0.947 (8.5%)	0.966 (4.7%)	0.971 (3.1%)	0.979 (2.3%)
Tree (min.)	0.931 (6.6%)	0.956 (3.6%)	0.964 (2.5%)	0.974 (1.8%)

[a] The percentage gain in power obtained by using the restricted test rather than the unrestricted test is given in parentheses.

by using order restricted tests for this alternative is less than that for the simple tree.

It should also be noted that the increase in power due to the use of order restricted tests is larger for smaller degrees of freedom. For instance, with $\nu = 10$ and $\Delta = 3$, the classical F test has power 0.629 and for the simple order, the minimum power over the appropriate cone, 0.810, is 28.8 percent larger.

However, with $\nu = \infty$ and $\Delta = 3$, the usual χ^2 test has power 0.771 and for the simple order, the minimum power over the appropriate cone, 0.872, is 13.1 percent larger.

One sees from Table 2.5.1 that the use of order restricted tests can provide considerable gains in power if the mean vector, $\boldsymbol{\mu}$, satisfies the alternative hypothesis. With $\nu = 40$, the minimum of the power of $\bar{E}^2_{01}(\nu)$ over the simple cone is 69.8 percent larger than that of the usual F test if $\Delta = 1$ and the corresponding values for $\Delta = 2, 3$, and 4 are 41.4, 16.3, and 3.7 percent, respectively. Of course, if $\boldsymbol{\mu}$ is outside of the cone the powers of the order restricted tests may be less than the usual tests. For the simple order and the simple tree, Table 2.5.2 gives the powers of $\bar{\chi}^2_{01}$ with $\alpha = 0.05$ and $k = 3$ for $\beta = 30, 45, \ldots, 240°$. (For ease of comparison the values for the usual χ^2 test are given too.) It is clear that for slight departures from the assumption that $\boldsymbol{\mu}$ is isotonic with respect to the usual simple order, that is $\beta = 75, 90°$ (or for the simple tree $\beta = 135°$), the power of the order restricted tests still exceeds that of the χ^2 test. However, this is not necessarily the case for gross departures. In fact, as will be shown in the next section for the simple order and $\boldsymbol{\mu}$ in the interior of H_1^* (see Section 1.7 for a discussion of the dual cone), i.e. for $150° < \beta < 270°$ the power of $\bar{\chi}^2_{01}$ converges to

Table 2.5.2 The power of $\bar{\chi}^2_{01}$ as a function of β for $\alpha = 0.05$, $k = 3$, the simple order, and the simple tree

TEST		SIMPLE ORDER				SIMPLE TREE			
		$\Delta = 1$	$\Delta = 2$	$\Delta = 3$	$\Delta = 4$	$\Delta = 1$	$\Delta = 2$	$\Delta = 3$	$\Delta = 4$
χ^2		0.133	0.415	0.771	0.957	0.133	0.415	0.771	0.957
	$\beta = 30°$	0.244	0.605	0.892	0.987	0.202	0.537	0.854	0.979
	$\beta = 45°$	0.238	0.597	0.889	0.986	0.208	0.542	0.856	0.979
	$\beta = 60°$	0.221	0.569	0.874	0.983	0.210	0.543	0.856	0.979
	$\beta = 75°$	0.193	0.513	0.834	0.973	0.208	0.542	0.856	0.979
	$\beta = 90°$	0.157	0.420	0.742	0.935	0.202	0.537	0.854	0.979
	$\beta = 105°$	0.117	0.297	0.567	0.809	0.191	0.524	0.849	0.978
	$\beta = 120°$	0.079	0.171	0.325	0.518	0.173	0.495	0.831	0.974
$\bar{\chi}^2_{01}$	$\beta = 135°$	0.049	0.076	0.119	0.179	0.148	0.439	0.783	0.959
	$\beta = 150°$	0.028	0.025	0.025	0.025	0.118	0.349	0.679	0.908
	$\beta = 165°$	0.015	0.007	0.003	0.001	0.085	0.237	0.493	0.755
	$\beta = 180°$	0.009	0.002	0.000	0.000	0.056	0.128	0.261	0.445
	$\beta = 195°$	0.006	0.000	0.000	0.000	0.034	0.053	0.086	0.135
	$\beta = 210°$	0.005	0.000	0.000	0.000	0.019	0.016	0.016	0.016
	$\beta = 225°$	0.006	0.000	0.000	0.000	0.012	0.004	0.002	0.001
	$\beta = 240°$	0.009	0.002	0.000	0.000	0.010	0.002	0.000	0.000

Table 2.5.3 Comparison of the powers of the χ^2 and $\bar{\chi}^2_{01}$ test for the simple order, the simple loop, and the simple tree with $k = 4$ and $\alpha = 0.05$

TEST	ALTERNATIVE HYPOTHESIS	$\Delta = 1$	$\Delta = 2$	$\Delta = 3$	$\Delta = 4$
χ^2	H_2–H_0	0.115	0.350	0.710	0.945
	Simple order (max.)	0.238(107.0%)	0.594(69.7%)	0.885(24.6%)	0.985(4.2%)
	Simple order (min.)	0.202(75.7%)	0.531(51.7%)	0.849(19.6%)	0.977(3.4%)
$\bar{\chi}^2_{01}$	Simple loop (max.)	0.229(99.1%)	0.574(64.0%)	0.871(22.7%)	0.982(3.9%)
	Simple loop (min.)	0.198(72.2%)	0.518(48.0%)	0.838(18.0%)	0.974(3.1%)
	Simple tree (max.)	0.187(62.6%)	0.489(39.7%)	0.813(14.5%)	0.967(2.3%)
	Simple tree (min.)	0.140(21.7%)	0.416(18.9%)	0.767(8.0%)	0.955(1.1%)

zero as $\Delta \to \infty$. The same result holds for the simple tree and the interior of the dual of the simple tree corresponds to $210° < \beta < 270°$.

Table 2.5.3 gives a comparison of the power of the usual χ^2 test for homogeneity against all alternatives with the powers of the corresponding $\bar{\chi}^2_{01}$ tests with $k = 4$, $\alpha = 0.05$, the simple order, the simple loop, and the simple tree. The formulas for the power functions of the order restricted tests for these orderings are derived in Bartholomew (1961a). The powers labeled maximum and minimum are those computed for the values of $\boldsymbol{\mu}$, which have been conjectured to provide, for a fixed value of $\Delta > 0$, the maximum and minimum over the appropriate cone. In particular, for the simple loop the maximum power is conjectured to occur for $\boldsymbol{\mu}$ which satisfy $\mu_2 - \mu_1 = \mu_3 - \mu_1 = \mu_4 - \mu_3 = \mu_4 - \mu_2 > 0$ and the minimum is believed to occur for $\boldsymbol{\mu}$ which satisfy $\mu_1 = \mu_2 = \mu_3 < \mu_4$ or $\mu_1 < \mu_2 = \mu_3 = \mu_4$.

Comparing Tables 2.5.1 and 2.5.3, one sees that for the simple order the percentage increases in power obtained by using the order restricted test are greater for $k = 4$ than for $k = 3$. For the simple tree, the percentage increases for the maximum powers are greater for $k = 4$ than for $k = 3$, but the percentage increases for the minimum powers are about the same or slightly less. This suggests that for larger k, the extra effort required to conduct the order restricted test is again worth while provided $\boldsymbol{\mu}$ satisfies, or nearly satisfies, the order restriction, that is $\boldsymbol{\mu}$ is isotonic with respect to $\lesssim$.

As a further study of the case of larger k, the powers of the $\bar{\chi}^2_{01}$ tests with $\alpha = 0.05$, $k = 8, 12$, a simple order, and a simple tree were estimated using Monte Carlo techniques with 10 000 iterations. The estimated powers at the $\boldsymbol{\mu}$ which are conjectured to give the maximum and minimum powers over the appropriate cone are given in Table 2.5.4. The trends observed in the powers as k increased from 3 to 4 seem to continue. The percentage gain in the power of the order restricted tests compared to the omnibus χ^2 tests increases as k increases from 4 to 8 and from 8 to 12 for the simple order. For the simple tree, these percentages for

Table 2.5.4 Comparisons of the powers[a] of the χ^2 and $\bar{\chi}^2_{01}$ tests for the simple order and the simple tree, $\alpha = 0.05$ and $k = 8, 12$

TEST	ALTERNATIVE HYPOTHESIS	$\Delta = 1$	$\Delta = 2$	$\Delta = 3$	$\Delta = 4$
			$k = 8$		
χ^2	H_2–H_0	0.090	0.249	0.535	0.853
$\bar{\chi}^2_{01}$	Simple order (max.)	0.232 (157.8%)	0.574 (130.5%)	0.866 (61.9%)	0.980 (14.9%)
	Simple order (min.)	0.168 (86.7%)	0.452 (81.5%)	0.788 (47.3%)	0.960 (12.5%)
	Simple tree (max.)	0.148 (64.4%)	0.383 (53.8%)	0.701 (31.0%)	0.919 (7.7%)
	Simple tree (min.)	0.096 (6.7%)	0.277 (11.2%)	0.601 (12.3%)	0.876 (2.7%)
			$k = 12$		
χ^2	H_2–H_0	0.080	0.205	0.466	0.776
$\bar{\chi}^2_{01}$	Simple order (max.)	0.234 (192.5%)	0.572 (179.0%)	0.868 (86.3%)	0.978 (26.0%)
	Simple order (min.)	0.162 (102.5%)	0.429 (109.3%)	0.762 (63.5%)	0.949 (22.3%)
	Simple tree (max.)	0.128 (60.0%)	0.318 (55.1%)	0.605 (29.8%)	0.859 (10.7%)
	Simple tree (min.)	0.082 (2.5%)	0.226 (10.2%)	0.494 (6.0%)	0.799 (3.0%)

[a] Powers for $\bar{\chi}^2_{01}$ were estimated by Monte Carlo techniques with 10 000 iterations.

the maximum powers seem to increase and these percentages for the minimum powers seem to remain constant or possibly decrease slightly. (The apparent decreases are due to changes in the powers that are small enough to possibly be due to Monte Carlo error.)

2.6 THE POWER FUNCTIONS OF THE $\bar{\chi}^2$ AND $\bar{E}^2$ TESTS: CONSISTENCY, MONOTONICITY, AND UNBIASEDNESS

As in Section 2.2, let $\bar{Y}_1, \bar{Y}_2, \ldots, \bar{Y}_k$ be the means of independent samples; let $\bar{Y}_i \sim \mathcal{N}(\mu_i, \sigma_i^2/n_i)$ with $\sigma_i^2 = a_i\sigma^2$ for $i = 1, 2, \ldots, k$ where the a_i are known positive constants; and let S^2 be an estimate of σ^2 which is independent of $\bar{\mathbf{Y}} = (\bar{Y}_1, \bar{Y}_2, \ldots, \bar{Y}_k)$ and $\nu S^2/\sigma^2 \sim \chi^2(\nu)$ with ν a positive integer. Consider the hypotheses

$$H_0: \mu_1 = \mu_2 = \cdots = \mu_k, \qquad H_1: \boldsymbol{\mu} = (\mu_1, \mu_2, \ldots, \mu_k) \in \mathscr{K} \qquad \text{and} \qquad H_2: \boldsymbol{\mu} \in R^k$$

with $\mathscr{K}$ a closed, convex cone containing H_0. The collection of vectors isotonic with respect to a quasi-order, $\lesssim$, on $X = \{1, 2, \ldots, k\}$ is an example of such a $\mathscr{K}$.

If σ^2 is known, then, without loss of generality, it may be assumed to be one and in that case, the LRT of H_0 versus $H_1 - H_0$ rejects H_0 for large values of

$$\bar{\chi}^2_{01} = \sum_{i=1}^{k} w_i(P_w(\bar{\mathbf{Y}} \mid \mathscr{K})_i - \hat{\mu})^2 \tag{2.6.1}$$

where $w_i = n_i/a_i$, $P_{\mathbf{w}}(\cdot|\mathscr{K})$ denotes the projection onto $\mathscr{K}$ with distance determined by $\mathbf{w}$ and $\hat{\mu} = \sum_{i=1}^{k} w_i \bar{Y}_i / \sum_{i=1}^{k} w_i$. In the case in which σ^2 is known to be one, the LRT of H_1 versus $H_2 - H_1$ rejects H_1 for large values of

$$\bar{\chi}_{12}^2 = \sum_{i=1}^{k} w_i(\bar{Y}_i - P_{\mathbf{w}}(\bar{\mathbf{Y}}|\mathscr{K})_i)^2. \tag{2.6.2}$$

If σ^2 is unknown the corresponding tests reject for large values of

$$\bar{E}_{01}^2(\nu) = \frac{\bar{\chi}_{01}^2}{\bar{\chi}_{01}^2 + \bar{\chi}_{12}^2 + \nu S^2} \tag{2.6.3}$$

and

$$\bar{E}_{12}^2(\nu) = \frac{\bar{\chi}_{12}^2}{\bar{\chi}_{12}^2 + \nu S^2}, \tag{2.6.4}$$

respectively. In this section, properties of these tests, such as consistency and unbiasedness, as well as monotonicity of their power functions, are investigated. In the variance known case, the definitions of $\bar{\chi}_{01}^2$ and $\bar{\chi}_{12}^2$ given in (2.6.1) and (2.6.2) assume that $\sigma^2 = 1$ and the distribution of $\bar{E}_{01}^2(\nu)$ and $\bar{E}_{12}^2(\nu)$ do not depend on σ^2. Thus, throughout this section, suppose $\sigma^2 = 1$.

Consistency of the tests

The following result, which is like Lemma 5.2 of Perlman (1969), will be used to establish the uniform consistency of these tests. Let $\langle\cdot,\cdot\rangle_{\mathbf{w}}$ and $\|\cdot\|_{\mathbf{w}}$ be the usual weighted inner product and norm.

Lemma *Let c and ε be fixed positive numbers and let $Y_1, Y_2, \ldots, Y_k$ be independent normal variables with zero means and variances $w_1^{-1}, w_2^{-1}, \ldots, w_k^{-1}$, respectively. There exist real numbers $M(\varepsilon, c, k)$ and $M'(\varepsilon, c, k)$ which only depend on ε, c, and k, such that $\|\boldsymbol{\mu} - P_{\mathbf{w}}(\boldsymbol{\mu}|\mathscr{K})\|_{\mathbf{w}} \geq M(\varepsilon, c, k)$ implies that*

$$P[\|\mathbf{Y} + \boldsymbol{\mu} - P_{\mathbf{w}}(\mathbf{Y} + \boldsymbol{\mu}|\mathscr{K})\|_{\mathbf{w}} \geq c] \geq 1 - \varepsilon \tag{2.6.5}$$

and $\|P_{\mathbf{w}}(\boldsymbol{\mu}|\mathscr{K}) - P_{\mathbf{w}}(\boldsymbol{\mu}|H_0)\|_{\mathbf{w}} \geq M'(\varepsilon, c, k)$ implies that

$$P[\|P_{\mathbf{w}}(\mathbf{Y} + \boldsymbol{\mu}|\mathscr{K}) - P_{\mathbf{w}}(\mathbf{Y} + \boldsymbol{\mu}|H_0)\|_{\mathbf{w}} \geq c] \geq 1 - \varepsilon. \tag{2.6.6}$$

Proof: Because $\|\mathbf{Y}\|_{\mathbf{w}}^2 \sim \chi^2(k)$, for $\varepsilon > 0$ there exists an $a = a(\varepsilon, k)$ for which $P[\|\mathbf{Y}\|_{\mathbf{w}} \leq a] \geq 1 - \varepsilon$. Observe that

$$\|\boldsymbol{\mu} - P_{\mathbf{w}}(\boldsymbol{\mu}|\mathscr{K})\|_{\mathbf{w}} \leq \|\boldsymbol{\mu} - P_{\mathbf{w}}(\mathbf{Y} + \boldsymbol{\mu}|\mathscr{K})\|_{\mathbf{w}} \leq \|\mathbf{Y}\|_{\mathbf{w}} + \|\mathbf{Y} + \boldsymbol{\mu} - P_{\mathbf{w}}(\mathbf{Y} + \boldsymbol{\mu}|\mathscr{K})\|_{\mathbf{w}}.$$

Hence, with $M(\varepsilon, c, k) = a + c$, (2.6.5) follows if $\|\boldsymbol{\mu} - P_{\mathbf{w}}(\boldsymbol{\mu}|\mathscr{K})\|_{\mathbf{w}} \geq M(\varepsilon, c, k)$. For the second conclusion, observe that

$$\begin{aligned}\|P_{\mathbf{w}}(\boldsymbol{\mu}|\mathscr{K})-P_{\mathbf{w}}(\boldsymbol{\mu}|H_0)\|_{\mathbf{w}} \leqslant \|P_{\mathbf{w}}(\boldsymbol{\mu}|\mathscr{K})-P_{\mathbf{w}}(\mathbf{Y}+\boldsymbol{\mu}|\mathscr{K})\|_{\mathbf{w}} \\ +\|P_{\mathbf{w}}(\mathbf{Y}+\boldsymbol{\mu}|\mathscr{K})-P_{\mathbf{w}}(\bar{\mathbf{Y}}+\boldsymbol{\mu}|H_0)\|_{\mathbf{w}}. \\ +\|P_{\mathbf{w}}(\mathbf{Y}+\boldsymbol{\mu}|H_0)-P_{\mathbf{w}}(\boldsymbol{\mu}|H_0)\|_{\mathbf{w}}.\end{aligned}$$

However, because projection onto $\mathscr{K}$ is a distance reducing operator (see Theorem 8.2.5), one obtains

$$\|P_{\mathbf{w}}(\boldsymbol{\mu}|\mathscr{K})-P_{\mathbf{w}}(\boldsymbol{\mu}|H_0)\|_{\mathbf{w}} \leqslant 2\|\mathbf{Y}\|_{\mathbf{w}}+\|P_{\mathbf{w}}(\mathbf{Y}+\boldsymbol{\mu}|\mathscr{K})-P_{\mathbf{w}}(\mathbf{Y}+\boldsymbol{\mu}|H_0)\|_{\mathbf{w}}.$$

The desired conclusion follows by choosing $M'(\varepsilon, c, k)=c+2a$. □

The next theorem is an immediate consequence of this lemma.

Theorem 2.6.1 *Let $\bar{\mathbf{Y}}$, $\bar{\chi}^2_{01}$, and $\bar{\chi}^2_{12}$ be defined as in this section. The power function of $\bar{\chi}^2_{01}$, that is $\pi_{01}(\boldsymbol{\mu})$, converges uniformly to one as $\|P_{\mathbf{w}}(\boldsymbol{\mu}|\mathscr{K})-P_{\mathbf{w}}(\boldsymbol{\mu}|H_0)\|_{\mathbf{w}}\to\infty$ and the power of $\bar{\chi}^2_{12}$, that is $\pi_{12}(\boldsymbol{\mu})$, converges uniformly to one as $\|\boldsymbol{\mu}-P_{\mathbf{w}}(\boldsymbol{\mu}|\mathscr{K})\|_{\mathbf{w}}\to\infty$.*

The implications of Theorem 2.6.1 with regard to consistency are considered next. Suppose the sample sizes satisfy

$$n_i=\lambda_i N \text{ with } \lambda_i>0, \quad i=1,2,\ldots,k, \quad \text{and} \quad \sum_{i=1}^{k}\lambda_i=1. \tag{2.6.7}$$

One could assume that $n_i=\lambda_{Ni}N$ with $\lambda_{Ni}\to\lambda_i$, but for simplicity assume (2.6.7). If $\boldsymbol{\mu}$ is fixed and $\mathbf{w}_0=(\lambda_1/a_1,\lambda_2/a_2,\ldots,\lambda_k/a_k)$, then $P_{\mathbf{w}_N}(\boldsymbol{\mu}|\mathscr{K})=P_{\mathbf{w}_0}(\boldsymbol{\mu}|\mathscr{K})$, $P_{\mathbf{w}_N}(\boldsymbol{\mu}|H_0)=P_{\mathbf{w}_0}(\boldsymbol{\mu}|H_0)$ and $\|P_{\mathbf{w}_N}(\boldsymbol{\mu}|\mathscr{K})-P_{\mathbf{w}_N}(\boldsymbol{\mu}|H_0)\|_{\mathbf{w}_N}/N=\|P_{\mathbf{w}_0}(\boldsymbol{\mu}|\mathscr{K})-P_{\mathbf{w}_0}(\boldsymbol{\mu}|H_0)\|_{\mathbf{w}_0}$. Furthermore, the last quantity is positive if and only if $P_{\mathbf{w}_0}(\boldsymbol{\mu}|\mathscr{K})$ is not constant. It is not difficult to show that $P_{\mathbf{w}_0}(\boldsymbol{\mu}|\mathscr{K})$ is constant if and only if $\boldsymbol{\mu}\in\mathscr{K}^{*\mathbf{w}_0}\oplus H_0$ where $K^{*\mathbf{w}_0}$ is the dual of $\mathscr{K}$ (see Section 1.7 for the definition of a dual cone) and $A\oplus B=\{a+b: a\in A, b\in B\}$. Hence, $\bar{\chi}^2_{01}$ is consistent if $\boldsymbol{\mu}\notin\mathscr{K}^{*\mathbf{w}_0}\oplus H_0$ or, equivalently, $P_{\mathbf{w}_0}(\boldsymbol{\mu}|\mathscr{K})$ is not constant, provided the sample sizes satisfy (2.6.7). On the other hand, if (2.6.7) holds and $\boldsymbol{\mu}$ lies in the interior of $\mathscr{K}^{*\mathbf{w}_0}\oplus H_0$, by the strong law of large numbers, $\bar{\mathbf{Y}}\xrightarrow{\text{a.s.}}\boldsymbol{\mu}$. Hence, for almost all points in the underlying probability space and sufficiently large N, $\bar{\mathbf{Y}}\in\mathscr{K}^{\mathbf{w}_0}\oplus H_0$, which implies that $P_{\mathbf{w}_N}(\bar{\mathbf{Y}}|\mathscr{K})=P_{\mathbf{w}_0}(\bar{\mathbf{Y}}|\mathscr{K})=P_{\mathbf{w}_0}(\bar{\mathbf{Y}}|H_0)=P_{\mathbf{w}_N}(\bar{\mathbf{Y}}|H_0)$. Thus, as $N\to\infty$, $\bar{\chi}^2_{01}\xrightarrow{\text{a.s.}}0$ and $\pi_{01}(\boldsymbol{\mu})\to 0$ for such $\boldsymbol{\mu}$.

Suppose $\boldsymbol{\mu}\notin\mathscr{K}$ is fixed and that the sample sizes satisfy (2.6.7). Because $\boldsymbol{\mu}\notin\mathscr{K}$, $\|\boldsymbol{\mu}-P_{\mathbf{w}_0}(\boldsymbol{\mu}|\mathscr{K})\|_{\mathbf{w}_0}>0$ and consequently $\|\boldsymbol{\mu}-P_{\mathbf{w}_N}(\boldsymbol{\mu}|\mathscr{K})\|_{\mathbf{w}_N}\to\infty$. By Theorem 2.6.1, $\pi_{12}(\boldsymbol{\mu})\to 1$ as $N\to\infty$. Therefore $\bar{\chi}^2_{12}$ is consistent. Furthermore, an argument like that given above shows that if $\boldsymbol{\mu}$ is in the interior of $\mathscr{K}$ and the sample sizes satisfy (2.6.7), then $\pi_{12}(\boldsymbol{\mu})\to 0$ as $N\to\infty$.

In considering the $\bar{E}^2$ tests, suppose $\boldsymbol{\mu}$ is fixed and the sample sizes satisfy

(2.6.7). In the usual one-way analysis of variance setting, $v = \sum_{i=1}^{k} n_i - k = N - k$ and so $v \approx N$. In Section 2.2 it was shown that rejecting for $\bar{E}_{01}^2(v) \geqslant C$ is equivalent to rejecting for $S_{01}(v) = v\bar{\chi}_{01}^2/(\bar{\chi}_{12}^2 + vS^2) \geqslant vC/(1-C)$ (recall that σ^2 is assumed to be one). If C is the α-level critical value for $\bar{E}_{01}^2(v)$, then $vC/(1-C)$ converges, as $v \to \infty$, to the α-level critical value for $\bar{\chi}_{01}^2$. Because $\bar{\chi}_{12}^2/v \approx \bar{\chi}_{12}^2/N \xrightarrow{\text{a.s.}} \|\boldsymbol{\mu} - P_{\mathbf{w}_0}(\boldsymbol{\mu}|\mathscr{K})\|_{\mathbf{w}_0} < \infty$ and $S^2 \xrightarrow{P} \sigma^2 = 1$, the consistency of $\bar{E}_{01}^2(N-k)$ follows from that of $\bar{\chi}_{01}^2$ provided $\boldsymbol{\mu} \notin \mathscr{K}^{*\mathbf{w}_0} \oplus H_0$. Using the fact that $E_{12}^2(v) \geqslant C$ is equivalent to $S_{12}(v) = \bar{\chi}_{12}^2/S^2 \geqslant vC/(1-C)$, one can show that $\bar{E}_{12}^2\ (N-k)$ is consistent if the sample sizes satisfy (2.6.7) and $y \notin H$.

Theorem 2.6.1 also has implications concerning the powers of $\bar{E}_{01}^2(v)$ and $\bar{E}_{12}^2(v)$ for the situation in which the sample sizes and the degrees of freedom remain fixed, but $\boldsymbol{\mu}$ moves away from the null hypothesis regions. Suppose $\mathbf{w}$ and v are fixed and $\|\boldsymbol{\mu} - P_{\mathbf{w}}(\boldsymbol{\mu}|\mathscr{K})\|_{\mathbf{w}} \to \infty$. Conditioning on S^2, one sees that $P[S_{12}(v) \geqslant vC/(1-C)] \to 1$. In this setting, the behavior of the power function of $\bar{E}_{01}^2(v)$ is more complex. However, examination of the case $k=3$ is quite informative. In Figure 2.5.2, let the ray in the fourth quadrant which starts at the origin and is parallel to the side of the acceptance region of $\bar{E}_{01}^2(v)$ in the fourth quadrant be denoted, in polar coordinates, by $\theta = -\beta_0$ with $\beta_0 > 0$. Similarly, let the ray in the second quadrant which starts at the origin and is parallel to the side of the acceptance region of $\bar{E}_{01}^2(v)$ in the second quadrant be determined by $\theta = \beta_1$. Any ray $\theta = \beta$ with $-\beta_0 < \beta < \beta_1$ has points with Δ sufficiently large which lie outside of the acceptance region and it is intuitively clear that the power of $\bar{E}_{01}^2(v)$ approaches one along such rays as $\Delta \to \infty$. On the other hand, for any ray $\theta = \beta$ with $\beta_1 < \beta < 2\pi - \beta$, the distance from the point (Δ, θ) to the boundary of the acceptance region increases as $\Delta \to \infty$. Thus, one would expect the power function of $\bar{E}_{01}^2(v)$ to converge to zero as $\Delta \to \infty$ along such rays.

Monotonicity of the power functions

Two basic approaches are used in establishing monotonicity properties of the power functions of the $\bar{\chi}^2$ and $\bar{E}^2$ tests. The simplest approach uses subset containment arguments and the second approach is geometric.

As in the last subsection, let $\bar{\mathbf{Y}}$ be a vector of sample means and S^2 be an independent variance estimator; let $\mathscr{K}$ be a closed convex cone which contains the constant vectors, that is H_0; let $\mathscr{K}^{*\mathbf{w}}$ be the dual of $\mathscr{K}$; and let $\bar{\chi}_{01}^2, \bar{\chi}_{12}^2, \bar{E}_{01}^2(v)$, and $\bar{E}_{12}^2(v)$ be as defined in (2.6.1) through (2.6.4). For a fixed critical value C, let the power functions of these $\bar{\chi}^2$ and $\bar{E}^2$ tests be denoted by $\pi_{01}(\boldsymbol{\mu})$, $\pi_{12}(\boldsymbol{\mu})$, $\pi_{01}^*(\boldsymbol{\mu})$ and $\pi_{12}^*(\boldsymbol{\mu})$, respectively. The next result, which is a restatement of Lemma 8.2 of Perlman (1969) in this setting, is proved using subset containment arguments.

Theorem 2.6.2 *If $\boldsymbol{\mu}' - \boldsymbol{\mu} \in \mathscr{K}$, then $\pi_{12}(\boldsymbol{\mu}) \geqslant \pi_{12}(\boldsymbol{\mu}')$ and $\pi_{12}^*(\boldsymbol{\mu}) \geqslant \pi_{12}^*(\boldsymbol{\mu}')$; if $\boldsymbol{\mu}' - \boldsymbol{\mu} \in -\mathscr{K}^{*\mathbf{w}} \oplus H_0$, then $\pi_{01}(\boldsymbol{\mu}) \leqslant \pi_{01}(\boldsymbol{\mu}')$; and if $\boldsymbol{\mu}' - \boldsymbol{\mu} \in \mathscr{K} \cap (-\mathscr{K}^{*\mathbf{w}_0} \oplus H_0)$, then $\pi_{01}^*(\boldsymbol{\mu}) \leqslant \pi_{01}^*(\boldsymbol{\mu}')$.*

Before the proof of the Theorem 2.6.2 is given the following lemma is established.

Lemma *Let $\mathscr{K}$ be a closed, convex cone,* **w** *a positive weight vector, and $\mathscr{K}^{*\mathbf{w}}$ be the dual of $\mathscr{K}$. If $\boldsymbol{\lambda} \in \mathscr{K}$, then for* $\mathbf{y} \in R^k$,

$$\| P_{\mathbf{w}}(\mathbf{y} + \boldsymbol{\lambda} \,|\, \mathscr{K}^{*\mathbf{w}}) \|_{\mathbf{w}} \leqslant \| P_{\mathbf{w}}(\mathbf{y} \,|\, \mathscr{K}^{*\mathbf{w}}) \|_{\mathbf{w}}.$$

Proof: Because $\mathscr{K}$ is a convex cone, $P_{\mathbf{w}}(\mathbf{y} | \mathscr{K}) + \boldsymbol{\lambda} \in \mathscr{K}$ and hence

$$\begin{aligned} \| P_{\mathbf{w}}(\mathbf{y} + \boldsymbol{\lambda} | \mathscr{K}^{*\mathbf{w}}) \|_{\mathbf{w}} &= \| \mathbf{y} + \boldsymbol{\lambda} - P_{\mathbf{w}}(\mathbf{y} + \boldsymbol{\lambda} | \mathscr{K}) \|_{\mathbf{w}} \leqslant \| \mathbf{y} + \boldsymbol{\lambda} - [P_{\mathbf{w}}(\mathbf{y} | \mathscr{K}) + \boldsymbol{\lambda}] \|_{\mathbf{w}} \\ &= \| \mathbf{y} - P_{\mathbf{w}}(\mathbf{y} | \mathscr{K}) \|_{\mathbf{w}} = \| P_{\mathbf{w}}(\mathbf{y} | \mathscr{K}^{*\mathbf{w}}) \|_{\mathbf{w}}. \quad \square \end{aligned}$$

Proof of Theorem 2.6.2: Suppose $\bar{\mathbf{Y}}$ has mean $\boldsymbol{\mu}$ and $\boldsymbol{\lambda} = \boldsymbol{\mu}' - \boldsymbol{\mu}$. If $\boldsymbol{\lambda} \in \mathscr{K}$ then applying the lemma

$$\begin{aligned} \| \bar{\mathbf{Y}} + \boldsymbol{\lambda} - P_{\mathbf{w}}(\bar{\mathbf{Y}} + \boldsymbol{\lambda} | \mathscr{K}) \|_{\mathbf{w}} &= \| P_{\mathbf{w}}(\bar{\mathbf{Y}} + \boldsymbol{\lambda} | \mathscr{K}^{*\mathbf{w}}) \|_{\mathbf{w}} \leqslant \| P_{\mathbf{w}}(\bar{\mathbf{Y}} | \mathscr{K}^{*\mathbf{w}}) \|_{\mathbf{w}} \\ &= \| \bar{\mathbf{Y}} - P_{\mathbf{w}}(\bar{\mathbf{Y}} | \mathscr{K}) \|_{\mathbf{w}}. \end{aligned} \tag{2.6.8}$$

Noting that $\bar{\mathbf{Y}} + \boldsymbol{\lambda}$ has mean vector $\boldsymbol{\mu}'$, (2.6.8) implies that $\pi_{12}(\boldsymbol{\mu}) \geqslant \pi_{12}(\boldsymbol{\mu}')$. Because $\bar{E}_{12}^2(v) \geqslant C$ is equivalent to $\bar{\chi}_{12}^2/S^2 \geqslant vC/(1 - C)$, by conditioning on S^2, $\pi_{12}(\boldsymbol{\mu}) \geqslant \pi_{12}(\boldsymbol{\mu}')$ follows from (2.6.8).

If $\boldsymbol{\lambda} \in -\mathscr{K}^{*\mathbf{w}_0} \oplus H_0$, then $\boldsymbol{\lambda} = \boldsymbol{\lambda}^{(1)} + \boldsymbol{\lambda}^{(2)}$ with $-\boldsymbol{\lambda}^{(1)} \in \mathscr{K}^{*\mathbf{w}}$ and $\boldsymbol{\lambda}^{(2)} \in H_0$. Now $P_{\mathbf{w}}(\boldsymbol{\lambda}^{(2)} | H_0) = \boldsymbol{\lambda}^{(2)}$ and, because $H_0 \subset \mathscr{K}$, $P_{\mathbf{w}}(\boldsymbol{\lambda}^{(1)} | H_0) = \mathbf{0}$. With $\mathbf{Y}' = \bar{\mathbf{Y}} - P_{\mathbf{w}}(\bar{\mathbf{Y}} | H_0)$, $\|P_{\mathbf{w}}(\bar{\mathbf{Y}} + \boldsymbol{\lambda} | \mathscr{K}) - P_{\mathbf{w}}(\bar{\mathbf{Y}} + \boldsymbol{\lambda} | H_0)\|_{\mathbf{w}} = \|P_{\mathbf{w}}(\mathbf{Y}' + \boldsymbol{\lambda}^{(1)} | \mathscr{K})\|_{\mathbf{w}}$, and recalling that $(\mathscr{K}^{*\mathbf{w}})^{*\mathbf{w}} = \mathscr{K}$ (cf. Theorem 1.7.3) and applying the lemma with $\boldsymbol{\lambda} = -\boldsymbol{\lambda}^{(1)}$, the last expression is bounded below by

$$\| P_{\mathbf{w}}(\mathbf{Y}' | \mathscr{K}) \|_{\mathbf{w}} = \| P_{\mathbf{w}}(\bar{\mathbf{Y}} | \mathscr{K}) - P_{\mathbf{w}}(\bar{\mathbf{Y}} | H_0) \|_{\mathbf{w}}.$$

Thus,

$$\| P_{\mathbf{w}}(\bar{\mathbf{Y}} + \boldsymbol{\lambda} | \mathscr{K}) - P_{\mathbf{w}}(\bar{\mathbf{Y}} + \boldsymbol{\lambda} | H_0) \|_{\mathbf{w}} \geqslant \| P_{\mathbf{w}}(\bar{\mathbf{Y}} | \mathscr{K}) - P_{\mathbf{w}}(\bar{\mathbf{Y}} | H_0) \|_{\mathbf{w}} \tag{2.6.9}$$

and the second conclusion of the theorem follows from (2.6.9).

If $\boldsymbol{\mu}' - \boldsymbol{\mu} \in \mathscr{K} \cap (-\mathscr{K}^{*\mathbf{w}} \oplus H_0)$, then (2.6.8) and (2.6.9) both hold. Recalling that $\bar{E}_{01}^2(v) \geqslant C$ is equivalent to $\bar{\chi}_{01}^2/(\bar{\chi}_{12}^2 + vS^2) \geqslant vC/(1 - C)$ and conditioning on S^2, the third conclusion follows easily. $\square$

Corollary *If $\boldsymbol{\mu} \in \mathscr{K}$, then $\pi_{12}(\delta\boldsymbol{\mu})$ and $\pi_{12}^*(\delta\boldsymbol{\mu})$ are nonincreasing functions of $\delta \in (-\infty, \infty)$; if $\boldsymbol{\mu} \in -\mathscr{K}^{*\mathbf{w}} \oplus H_0$, then $\pi_{01}(\delta\boldsymbol{\mu})$ is a nondecreasing function of $\delta \in (-\infty, \infty)$; and if $\boldsymbol{\mu} \in \mathscr{K} \cap (-\mathscr{K}^{*\mathbf{w}} \oplus H_0)$, then $\pi_{01}^*(\delta\boldsymbol{\mu})$ is a nondecreasing function of $\delta \in (-\infty, \infty)$.*

Proof: If $\boldsymbol{\mu} \in \mathscr{K}$ and $\delta_1 < \delta_2$ then $\delta_2\boldsymbol{\mu} - \delta_1\boldsymbol{\mu} = (\delta_2 - \delta_1)\boldsymbol{\mu} \in \mathscr{K}$. The first conclusion is an immediate consequence of the theorem. The other parts of the corollary are proved similarly. $\square$

One could define an ordering on the elements in R^k by $\boldsymbol{\mu} \ll \boldsymbol{\mu}'$ if $\boldsymbol{\mu}' - \boldsymbol{\mu} \in \mathscr{K}$. Theorem 2.6.2 shows that π_{12} and π_{12}^* are antitonic with respect to this ordering. Marshall, Walkup, and Wets (1967) give a nice discussion of such cone orderings, and Raubertas, Lee, and Nordheim (1986) discuss their use in these testing problems. Replacing $\mathscr{K}$ by $-\mathscr{K}^{*\mathbf{w}} \oplus H_0$ defines another cone ordering. Theorem 2.6.2 shows that π_{01} is isotonic with respect to this second ordering.

It is helpful to see what Theorem 2.6.2 and its corollary say in the simply ordered case. If $\mathscr{K} = \mathscr{I} = \{\mathbf{y} \in R^k : y_1 \leqslant y_2 \leqslant \cdots \leqslant y_k\}$, then for $\mathbf{y} \in \mathscr{I}$,

$$\sum_{j=1}^{i} w_j(y_j - P_{\mathbf{w}}(\mathbf{y}|H_0)_j) \leqslant \sum_{j=1}^{k} w_j(y_j - P_{\mathbf{w}}(\mathbf{y}|H_0)_j) = 0.$$

Using (1.7.6), one sees that $\mathbf{y} - P_{\mathbf{w}}(\mathbf{y}|H_0) \in -\mathscr{I}^{*\mathbf{w}}$ or $\mathbf{y} \in -\mathscr{I}^{*\mathbf{w}} \oplus H_0$. Thus, $\mathscr{I} \subset -\mathscr{I}^{*\mathbf{w}} \oplus H_0$ and, in fact, it is not difficult to exhibit $\mathbf{y} \in (-\mathscr{I}^{*\mathbf{w}} \oplus H_0) - \mathscr{I}$. Figure 2.6.1 shows $\mathscr{I}$ and $-\mathscr{I}^{*\mathbf{w}} \oplus H_0$ intersected with the hyperplane, $S = \{\mathbf{y} \in R^k : y_1 + y_2 + y_3 = 0\}$ for the case $k = 3$ with equal weights. As in the discussion of the $P(l, k; \mathbf{w})$, the parameterization $(y_2 - y_1)/\sqrt{2} = \Delta \sin \beta$ and $(2y_3 - y_2 - y_1)/\sqrt{6} = \Delta \cos \beta$ is used. In the hyperplane S, the power function $\pi_{01}(\boldsymbol{\mu})$ is nondecreasing on the rays starting at the origin with $\theta = \beta$ and $-\pi/6 \leqslant \beta \leqslant \pi/2$ and is nonincreasing on the rays $\theta = \beta$ with $5\pi/6 \leqslant \beta \leqslant 3\pi/2$. (Because these power functions are invariant under a shift by a constant vector, it is sufficient to examine their behavior on S.) One would conjecture that the region of isotonicity of $\pi_{01}(\boldsymbol{\mu})$ is larger than $-\mathscr{I}^* \oplus H_0$. Mukerjee, Robertson, and Wright (1986b) comment that they have shown that in the case $k = 3$ it is

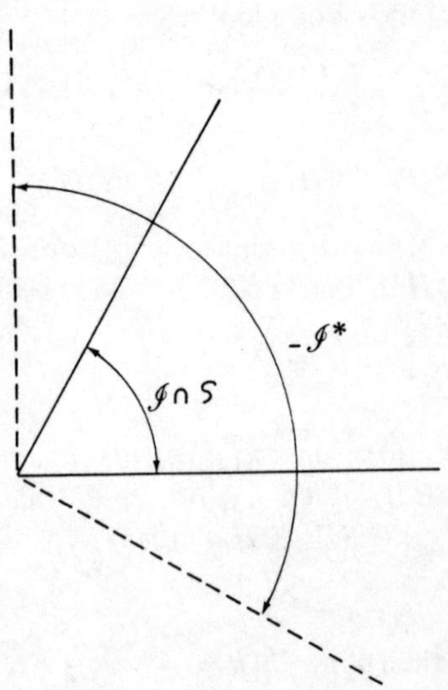

Figure 2.6.1 The hyperplane $S = \{\mathbf{y}: y_1 + y_2 + y_3 = 0\}$, $\mathscr{I} \cap S$, and $-\mathscr{I}^*$

$-\pi/3 \leqslant \beta \leqslant 2\pi/3$ but that the techniques used do not extend to larger k. It is not difficult to show that for $k=3$, the region of monotonicity cannot be extended beyond $[-\pi/3, 2\pi/3]$. The partial derivative of π_{01} with respect to Δ, evaluated at $\Delta = 0$, is

$$\left(\frac{\sqrt{3}}{2} + \frac{\sqrt{C}}{2\pi}\right)\phi(\sqrt{C})\cos\left(\beta - \frac{\pi}{6}\right)$$

which is positive for $(-\pi/3, 2\pi/3)$ and negative for $(2\pi/3, 5\pi/3)$. However, as was noted earlier by applying Theorem 2.6.1, one can show that for $\beta \in (-\pi/2, 5\pi/6)$, $\pi_{01}(\boldsymbol{\mu}) \to 1$ as $\Delta \to \infty$. Hence, π_{01} is not monotonic for $\beta \in (-\pi/2, -\pi/3) \cup (2\pi/3, 5\pi/6)$. For the simple order with arbitrary k the most general result known concerning the monotonicity of $\pi_{01}(\boldsymbol{\mu})$ is that given in Theorem 2.6.2.

In the simply ordered case, $\pi_{12}(\delta\boldsymbol{\mu})$ and $\pi_{12}^*(\delta\boldsymbol{\mu})$ are nondecreasing functions of δ provided $\boldsymbol{\mu} \in -\mathscr{I} = \mathscr{A} = \{\boldsymbol{\mu}: \mu_1 \geqslant \mu_2 \geqslant \cdots \geqslant \mu_k\}$. For $k=3$, $\mathscr{A} \cap S$ consists of those points with $\pi \leqslant \beta \leqslant 4\pi/3$. Because of the results for π_{01}, one might wonder if these monotonicity results could be extended to $\boldsymbol{\mu} \in \mathscr{I}^* \oplus H_0$ (for $k=3$, that is $5\pi/6 \leqslant \beta \leqslant 3\pi/2$). However, the approach used in the proof of Theorem 2.6.2 will not work. The inequality $\|\mathbf{y} + \boldsymbol{\lambda} - P_{\mathbf{w}}(\mathbf{y} + \boldsymbol{\lambda}|\mathscr{K})\|_{\mathbf{w}} \leqslant \|\mathbf{y} - P_{\mathbf{w}}(\mathbf{y}|\mathscr{K})\|_{\mathbf{w}}$ does not hold for arbitrary $\mathbf{y}$ and $\boldsymbol{\lambda}$ in $-\mathscr{K}^{*\mathbf{w}} \oplus H_0$. For an example in which the inequality fails take $\mathbf{y} = \mathbf{0}$, $\mathscr{K} = \mathscr{I}$, and $\boldsymbol{\lambda} \in (-\mathscr{I}^*) - \mathscr{I}$. (This can easily be seen in the case $k=3$ from Figure 2.6.1.) However, such results can be proved using a geometric argument. The key result is a one-sided analog to an inequality due to Anderson (1955).

Let $\boldsymbol{\mu} \in R^k$, let $S_{\mu} = \{b\boldsymbol{\mu}: -\infty < b < \infty\}$ be the subspace generated by $\boldsymbol{\mu}$, and for $A \subset R^k$, let the positive part of A in the $\boldsymbol{\mu}$ direction be defined by

$$A^+ = \{\mathbf{y} \in A: P_{\mathbf{w}}(\mathbf{y}|S_{\mu}) = b\boldsymbol{\mu} \text{ with } b > 0\}. \tag{2.6.10}$$

With $\bar{\mathbf{Y}}$ the vector of means of independent samples and $\bar{Y}_i \sim \mathcal{N}(\mu_i, w_i^{-1})$, let $P_{\mu}(A) = P_{\mu}[\bar{\mathbf{Y}} \in A]$. The following result is proved in Mukerjee, Robertson, and Wright (1986b). (Unimodal densities with elliptical contours are treated there.)

Theorem 2.6.3 *If $A \subset R^k$ is a convex set which satisfies*

$$\mathbf{y} - 2P_{\mathbf{w}}(\mathbf{y}|S_{\mu}) \in A \quad \textit{for each } \mathbf{y} \in A^+, \tag{2.6.11}$$

then $h(\delta) = P_{\delta\mu}(A)$ is nonincreasing for $\delta \geqslant 0$.

Theorem 2.6.3 will be applied with A the acceptance region of $\bar{\chi}_{01}^2$ and of $\bar{\chi}_{12}^2$. For $\mathscr{K}$ a closed, convex cone and $C \geqslant 0$, set

$$A(\mathscr{K}, C) = \{\mathbf{y} \in R^k: \|P_{\mathbf{w}}(\mathbf{y}|\mathscr{K})\|_{\mathbf{w}}^2 \leqslant C\}.$$

The following lemma is used to show $A(\mathscr{K}, C)$ is convex.

Lemma A *For* $\mathbf{x}, \mathbf{y} \in R^k$,

$$\| P_w(\mathbf{x}+\mathbf{y} \mid \mathscr{K}) \|_w \leqslant \| P_w(\mathbf{x} \mid \mathscr{K}) + P_w(\mathbf{y} \mid \mathscr{K}) \|_w \leqslant \| P_w(\mathbf{x} \mid \mathscr{K}) \|_w + \| P_w(\mathbf{y} \mid \mathscr{K}) \|_w.$$

Proof: The second inequality is just the triangular inequality for norms. Applying (1.7.1), $\mathbf{x}+\mathbf{y} = P_w(\mathbf{x} \mid \mathscr{K}) + P_w(\mathbf{x} \mid \mathscr{K}^{*w}) + P_w(\mathbf{y} \mid \mathscr{K}) + P_w(\mathbf{y} \mid \mathscr{K}^{*w}) = P_w(\mathbf{x} \mid \mathscr{K}) + P_w(\mathbf{y} \mid \mathscr{K}) + \mathbf{z}$ with $\mathbf{z} = P_w(\mathbf{x} \mid \mathscr{K}^{*w}) + P_w(\mathbf{y} \mid \mathscr{K}^{*w}) \in \mathscr{K}^{*w}$. Writing $\| P_w(\mathbf{x}+\mathbf{y} \mid \mathscr{K}) \|_w$ as $\| P_w(P_w(\mathbf{x} \mid \mathscr{K}) + P_w(\mathbf{y} \mid \mathscr{K}) + \mathbf{z} \mid \mathscr{K}) \|_w$, recalling that $(\mathscr{K}^{*w})^{*w} = \mathscr{K}$ and applying the lemma given after the statement of Theorem 2.6.2,

$$\begin{aligned} \| P_w(\mathbf{x}+\mathbf{y} \mid \mathscr{K}) \|_w &\leqslant \| P_w(P_w(\mathbf{x} \mid \mathscr{K}) + P_w(\mathbf{y} \mid \mathscr{K}) \mid \mathscr{K}) \|_w \\ &= \| P_w(\mathbf{x} \mid \mathscr{K}) + P_w(\mathbf{y} \mid \mathscr{K}) \|_w. \quad \square \end{aligned}$$

If $\mathbf{x}, \mathbf{y} \in A(\mathscr{K}, C)$ and $0 \leqslant \gamma \leqslant 1$, then

$$\begin{aligned} \| P_w(\gamma \mathbf{x} + (1-\gamma)\mathbf{y} \mid \mathscr{K}) \|_w &\leqslant \| \gamma P_w(\mathbf{x} \mid \mathscr{K}) + (1-\gamma) P_w(\mathbf{y} \mid \mathscr{K}) \|_w \\ &\leqslant \gamma \| P_w(\mathbf{x} \mid \mathscr{K}) \|_w + (1-\gamma) \| P_w(\mathbf{y} \mid \mathscr{K}) \|_w \leqslant \sqrt{C}. \end{aligned}$$

Hence, $A(\mathscr{K}, C)$ is convex. In the next result, the notation $A - \boldsymbol{\mu}$, with $A \subset R^k$ and $\boldsymbol{\mu} \in R^k$, means $\{\mathbf{y} - \boldsymbol{\mu} : \mathbf{y} \in A\}$. Later with $\boldsymbol{\mu}_0$ fixed it will be convenient to write $P_{\boldsymbol{\mu}}(A) = P_{\boldsymbol{\mu}-\boldsymbol{\mu}_0}(A - \boldsymbol{\mu}_0)$.

Lemma B *Let* $\mathscr{K}$ *be a closed, convex cone,* $C \geqslant 0$, $\boldsymbol{\mu}$, $\boldsymbol{\mu}_0 \in R^k$ *and* $A = A(\mathscr{K}, C) - \boldsymbol{\mu}_0$. *The subset A is convex. If* $\boldsymbol{\mu} \in \mathscr{K}$ *and* $\langle \boldsymbol{\mu}, \boldsymbol{\mu}_0 \rangle_w = \sum_{i=1}^k w_i \mu_i \mu_{0i} = 0$, *then for each* $\mathbf{y} \in A^+$, $\mathbf{y} - 2P_w(\mathbf{y} \mid S_{\boldsymbol{\mu}}) \in A$.

Proof: The convexity of A is immediate. Since $P_w(\mathbf{y} - \boldsymbol{\mu}_0 \mid S_{\boldsymbol{\mu}}) = P_w(\mathbf{y} \mid S_{\boldsymbol{\mu}})$ and $\mathbf{y} - \boldsymbol{\mu}_0 - 2P_w(\mathbf{y} \mid S_{\boldsymbol{\mu}}) \in A$ if and only if $\mathbf{y} - 2P_w(\mathbf{y} \mid S_{\boldsymbol{\mu}}) \in A(\mathscr{K}, C)$, one may assume $\boldsymbol{\mu}_0 = \mathbf{0}$. Let $\mathbf{y} \in A^+$. Applying (1.7.1),

$$\begin{aligned} &\| P_w(\mathbf{y} - 2P_w(\mathbf{y} \mid S_{\boldsymbol{\mu}}) \mid \mathscr{K}) \|_w^2 \\ &= \| \mathbf{y} - 2P_w(\mathbf{y} \mid S_{\boldsymbol{\mu}}) \|_w^2 - \| P_w(\mathbf{y} - 2P_w(\mathbf{y} \mid S_{\boldsymbol{\mu}}) \mid \mathscr{K}^{*w}) \|_w^2 \\ &= \langle \mathbf{y}, \mathbf{y} \rangle_w - 4 \langle \mathbf{y} - P_w(\mathbf{y} \mid S_{\boldsymbol{\mu}}), P_w(\mathbf{y} \mid S_{\boldsymbol{\mu}}) \rangle_w - \| P_w(\mathbf{y} - 2b\boldsymbol{\mu} \mid \mathscr{K}^{*w}) \|_w^2 \end{aligned}$$

with $b > 0$. However, $\langle \mathbf{y} - P_w(\mathbf{y} \mid S_{\boldsymbol{\mu}}), P_w(\mathbf{y} \mid S_{\boldsymbol{\mu}}) \rangle_w = 0$ and applying the lemma given after the statement of Theorem 2.6.2, $\| P_w(\mathbf{y} - 2b\boldsymbol{\mu} \mid \mathscr{K}^{*w}) \|_w^2 \geqslant \| P_w(\mathbf{y} \mid \mathscr{K}^{*w}) \|_w^2$. Hence, $\| P_w(\mathbf{y} - 2P_w(\mathbf{y} \mid S_{\boldsymbol{\mu}}) \mid \mathscr{K}) \|_w^2 \leqslant \| \mathbf{y} \|_w^2 - \| P_w(\mathbf{y} \mid \mathscr{K}^{*w}) \|_w^2 = \| P_w(\mathbf{y} \mid \mathscr{K}) \|_w^2 \leqslant C$. Thus, $\mathbf{y} - 2P_w(\mathbf{y} \mid S_{\boldsymbol{\mu}}) \in A(\mathscr{K}, C)$, which is the desired conclusion. $\square$

Theorem 2.6.4 *Let* $\boldsymbol{\mu} \in R^k$. *As functions of* δ, $\pi_{01}(\boldsymbol{\mu} + \delta P_w(\boldsymbol{\mu} \mid \mathscr{K}))$, $\pi_{12}(\boldsymbol{\mu} + \delta P_w(\boldsymbol{\mu} \mid \mathscr{K}^{*w}))$, *and* $\pi_{12}^*(\boldsymbol{\mu} + \delta P_w(\boldsymbol{\mu} \mid \mathscr{K}^{*w}))$ *are nondecreasing for* $\delta \geqslant -1$.

Proof: For π_{01} write $\boldsymbol{\mu} + \delta P_w(\boldsymbol{\mu} \mid \mathscr{K}) = P_w(\boldsymbol{\mu} \mid \mathscr{K}^{*w}) + (\delta + 1) P_w(\boldsymbol{\mu} \mid \mathscr{K})$. Because

$P[\bar{\chi}_{01}^2 = C] = 0$ for $C > 0$ under H_0, the acceptance region for $\bar{\chi}_{01}^2$ can be expressed as $\| P_w(\bar{\mathbf{Y}}|\mathscr{K}) - P_w(\bar{\mathbf{Y}}|H_0)\|_w^2 \leqslant C$. Using the characterization of the projection onto a closed, convex cone, that is (1.3.7) and (1.3.8), one can readily show that $P_w(\mathbf{y}|\mathscr{K}) - P_w(\mathbf{y}|H_0) = P_w(\mathbf{y}|\mathscr{K} \cap S)$ with $S = \{\mathbf{y} \in R^k: \sum_{i=1}^k w_i y_i = 0\}$. Thus, $\bar{\chi}_{01}^2 = \| P_w(\bar{\mathbf{Y}}|\mathscr{K} \cap S)\|_w^2$ and by Theorem 2.6.2,

$$\begin{aligned}\pi_{01}[\boldsymbol{\mu} + \delta P_w(\boldsymbol{\mu}|\mathscr{K})] &= \pi_{01}[\boldsymbol{\mu} + \delta P_w(\boldsymbol{\mu}|\mathscr{K}) - (\delta + 1)P_w(\boldsymbol{\mu}|H_0)] \\ &= \pi_{01}[P_w(\boldsymbol{\mu}|\mathscr{K}^{*w}) + (\delta - 1)P_w(\boldsymbol{\mu}|\mathscr{K} \cap S)].\end{aligned}$$

The desired conclusion for π_{01} follows by applying Theorem 2.6.3 with $A = A(\mathscr{K} \cap S, C) - \boldsymbol{\mu}_0$, $\boldsymbol{\mu}_0 = P_w(\boldsymbol{\mu}|\mathscr{K}^{*w})$, and $\boldsymbol{\mu} = P_w(\boldsymbol{\mu}|\mathscr{K} \cap S)$ and appealing to Lemma B. (Note that

$$\begin{aligned}\langle P_w(\boldsymbol{\mu}|\mathscr{K}^{*w}), P_w(\boldsymbol{\mu}|\mathscr{K} \cap S)\rangle_w &= \langle P_w(\boldsymbol{\mu}|\mathscr{K}^{*w}), P_w(\boldsymbol{\mu}|\mathscr{K}) - P_w(\boldsymbol{\mu}|H_0)\rangle_w \\ &= 0 - P_w(\boldsymbol{\mu}|H_0) \sum_{i=1}^k w_i P_w(\boldsymbol{\mu}|\mathscr{K}^{*w})_i = 0.)\end{aligned}$$

For π_{12} note that $\boldsymbol{\mu} + \delta P_w(\boldsymbol{\mu}|\mathscr{K}^{*w}) = P_w(\boldsymbol{\mu}|\mathscr{K}) + (\delta + 1)P_w(\boldsymbol{\mu}|\mathscr{K}^{*w})$ and that $\bar{\chi}_{12}^2 = \| P_w(\bar{\mathbf{Y}}|\mathscr{K}^{*w})\|_w^2$. The desired result for π_{12} follows by applying Theorem 2.6.3 with $A = A(\mathscr{K}^{*w}, C) - \boldsymbol{\mu}_0$, $\boldsymbol{\mu}_0 = P_w(\boldsymbol{\mu}|\mathscr{K})$, and $\boldsymbol{\mu} = P_w(\boldsymbol{\mu}|\mathscr{K}^{*w})$ and appealing to Lemma B. For π_{12}^*, recall that $\bar{E}_{12}^2(v) \geqslant C$ is equivalent to $\bar{\chi}_{12}^2/S^2 \geqslant vC/(1 - C)$ and condition on S^2. □

Corollary *If $\boldsymbol{\mu} \in \mathscr{K}$, then $\pi_{01}(\delta\boldsymbol{\mu})$ is nondecreasing for $\delta \geqslant 0$. If $\boldsymbol{\mu} \in \mathscr{K}^{*w} \oplus H_0$ then $\pi_{12}(\delta\boldsymbol{\mu})$ and $\pi_{12}^*(\delta\boldsymbol{\mu})$ are nondecreasing for $\delta \geqslant 0$.*

Proof. If $\boldsymbol{\mu} \in \mathscr{K}$, then $\boldsymbol{\mu} + \delta P_w(\boldsymbol{\mu}|\mathscr{K}) = (\delta + 1)\boldsymbol{\mu}$. If $\boldsymbol{\mu} \in \mathscr{K}^{*w} \oplus H_0$, then $\boldsymbol{\mu} = \boldsymbol{\mu}^{(1)} + \boldsymbol{\mu}^{(2)}$ with $\boldsymbol{\mu}^{(1)} \in \mathscr{K}^{*w}$ and $\boldsymbol{\mu}^{(2)} \in H_0$. However, $\pi_{12}(\delta\boldsymbol{\mu}) = \pi_{12}(\delta\boldsymbol{\mu}^{(1)})$ and $\pi_{12}^*(\delta\boldsymbol{\mu}) = \pi_{12}^*(\delta\boldsymbol{\mu}^{(1)})$. Also, $\boldsymbol{\mu}^{(1)} + \delta P_w(\boldsymbol{\mu}^{(1)}|\mathscr{K}^{*w}) = (\delta + 1)\boldsymbol{\mu}^{(1)}$. The corollary is now an immediate consequence of Theorem 2.6.4. □

The techniques used in the proof of Theorem 2.6.4 and its corollary can not be employed on $\bar{E}_{01}^2(v)$ because, as was seen in the last section, its acceptance region is not convex. The results in Theorem 2.6.4 complement those given in Theorem 2.6.2 and its corollary. In the simply ordered case, Theorem 2.6.2 establishes the monotonicity of π_{01} on $\{\delta\boldsymbol{\mu}: -\infty < \delta < \infty\}$ with $\boldsymbol{\mu} \in -\mathscr{I}^{*w} \oplus H_0 \supset \mathscr{I}$ and so Theorem 2.6.4 has nothing new to say in that case. However, Theorem 2.6.2 shows that π_{12} (and π_{12}^*) is monotone on $\{\delta\boldsymbol{\mu}: -\infty < \delta < \infty\}$ if $\boldsymbol{\mu} \in -\mathscr{I} = \mathscr{A}$, but Theorem 2.6.4 shows that π_{12} and π_{12}^* is monotone on $\{\delta\boldsymbol{\mu}: 0 < \delta < \infty\}$ for $\boldsymbol{\mu} \in \mathscr{I}^{*w} \oplus H_0 \supset \mathscr{A}$. For the simple tree ordering $-\mathscr{I}^{*w} \oplus H_0 \subset \mathscr{I}$ and consequently Theorem 2.6.4 provides new information about the monotonicity of the power function π_{01} but not π_{12} nor π_{12}^*. Theorem 2.6.2 and 2.6.4 can also be used to establish the monotonicity of the power functions along directions parallel to the edges or faces of the cones $\mathscr{K}$ and $\mathscr{K}^{*w} \oplus H_0$. Such results will be briefly discussed in the subsection on unbiasedness, but a more detailed discussion is given in Mukerjee, Robertson, and Wright (1986b).

The question of bias

The power function of $\bar{\chi}^2_{01}$ and $\bar{E}^2_{01}(\nu)$ are constant over H_0. If C is the α-level critical value for $\bar{\chi}^2_{01}$ and $\boldsymbol{\mu} \in \mathscr{K}$, then because $\pi_{01}(\delta\boldsymbol{\mu})$ is nondecreasing for $\delta \in (-\infty, \infty)$ (see the corollary to Theorem 2.6.2), $\pi_{01}(\boldsymbol{\mu}) \geqslant \pi_{01}(\mathbf{0}) = \alpha$. Thus, $\bar{\chi}^2_{01}$ is unbiased. The power function of $\bar{E}^2_{01}(\nu)$ was also shown to be nondecreasing on $\{\delta\boldsymbol{\mu}: -\infty < \delta < \infty\}$ provided $\boldsymbol{\mu} \in \mathscr{K} \cap (-\mathscr{K}^{*w} \oplus H_0)$. Hence, the argument given above shows that $\bar{E}^2_{01}(\nu)$ is unbiased if $\mathscr{K} \subset -\mathscr{K}^{*w} \oplus H_0$, which is the case for a simple order. Questions concerning the bias of $\bar{E}^2_{01}(\nu)$ for an arbitrary quasi-order are unresolved. However, for the simple tree ordering $\mathscr{K} \not\subset -\mathscr{K}^{*w} \oplus H_0$, but based on extensive numerical calculation it seems that $\bar{E}^2_{01}(\nu)$ is unbiased for this ordering also. Thus, it seems that $\bar{E}^2_{01}(\nu)$ is unbiased and that new techniques are needed to establish this fact.

If $\bar{\chi}^2_{12}$ is unbiased, then because π_{12} is continuous, $\pi_{12}(\boldsymbol{\mu}) = \alpha$ for $\boldsymbol{\mu} \in \partial\mathscr{K}$. Since π_{12} is analytic and not constant, one would expect that $\bar{\chi}^2_{12}$ is biased. The degree of bias is considered next for the simply ordered case. As noted previously, $\lim_{\delta \to \infty} \pi_{12}(\delta\boldsymbol{\mu}) = 1$ for $\boldsymbol{\mu} \notin H_1$. The complement of H_1 is partitioned into several sets and the behavior of π_{12} is examined on each of these sets. It will follow from the above observation that the supremum of π_{12} over each of these sets is one.

The set, H_1, has $2^{k-1} - 1$ faces and the preimages under $P(\cdot | H_1)$ of these faces form a partition for the complement of H_1. The infimum of π_{12} varies over the members of this partition. The idea is best explained by considering the case $k = 3$ with the usual simple order, that is $H_1 = \{\boldsymbol{\mu} \in R^3: \mu_1 \leqslant \mu_2 \leqslant \mu_3\}$. The complement of H_1 is partitioned into the following three sets:

$$E_1 = \{\mathbf{y} \notin H_1: P_w(\mathbf{y}|\mathscr{I})_1 = P_w(\mathbf{y}|\mathscr{I})_2 = P_w(\mathbf{y}|\mathscr{I})_3\} = (\mathscr{I}^{*w} - \{0\}) \oplus H_0$$
$$E_2 = \{\mathbf{y} \notin H_1: P_w(\mathbf{y}|\mathscr{I})_1 < P_w(\mathbf{y}|\mathscr{I})_2 = P_w(\mathbf{y}|\mathscr{I})_3\}$$
$$E_3 = \{\mathbf{y} \notin H_1: P_w(\mathbf{y}|\mathscr{I})_1 = P_w(y|\mathscr{I})_2 < P_w(\mathbf{y}|\mathscr{I})_3\}.$$

E_1 is paired with H_0, E_2 is paired with the face, $\beta = \pi/3$, and E_3 with the face, $\beta = 0$ (see Figure 2.6.1). If $\boldsymbol{\mu} \in E_1$, then applying the second part of the corollary to Theorem 2.6.4, one finds that $\inf_{\mu \in E_1} \pi_{12}(\boldsymbol{\mu}) = \pi_{12}(\mathbf{0})$, which is the significance level of the test.

Suppose $\boldsymbol{\mu} \in E_2$. Using (1.3.7) and (1.3.8) and a straightforward argument, it is easy to see that for $\delta \geqslant -1$, $P_w(\boldsymbol{\mu} + \delta P_w(\boldsymbol{\mu}|\mathscr{I})|\mathscr{I}) = (1+\delta)P_w(\boldsymbol{\mu}|\mathscr{I})$ and thus the points, $\boldsymbol{\mu} + \delta P_w(\boldsymbol{\mu}|\mathscr{I})$, with $\delta \geqslant -1$, are all in E_2. Incidentally, this also implies that they are all equidistant from $\mathscr{I}$. Then, using the first part of Theorem 2.6.2, it

Table 2.6.1 The infimum of $\pi_{12}(\boldsymbol{\mu})$ for $\boldsymbol{\mu}\notin H_1$ with $\alpha = 0.05$

k	CRITICAL VALUE C	$\frac{1}{2}P[\chi_1^2 \geqslant C]$
3	4.578	0.01620
4	6.175	0.00648
5	7.665	0.00281
6	9.095	0.00128
7	10.485	0.00060

follows that $\pi_{12}(\boldsymbol{\mu} + \delta P_{\mathrm{w}}(\boldsymbol{\mu}|\mathscr{I}))$ is nonincreasing in δ. On the other hand, using Theorem 2.6.4, it follows that for $0<\gamma<1$ and $\boldsymbol{v}\in R^k$, $\pi_{12}(P_{\mathrm{w}}(\boldsymbol{v}|H_1)) \leqslant \pi_{12}(\boldsymbol{v} - \gamma P_{\mathrm{w}}(\boldsymbol{v}|\mathscr{I}^{*\mathrm{w}})) \leqslant \pi_{12}(\boldsymbol{v})$. Applying the second observation with $\boldsymbol{v} = \boldsymbol{\mu} + \delta P_{\mathrm{w}}(\boldsymbol{\mu}|\mathscr{I})$, it follows from the first observation that $\inf_{\boldsymbol{\mu}\in E_2}\pi_{12}(\boldsymbol{\mu})$ can be obtained by taking the infimum of $\lim_{\delta\to\infty}\pi_{12}(\delta\boldsymbol{\mu})$ over all $\boldsymbol{\mu}$ in the face of H_1 corresponding to E_2 (that is $\mu_1 < \mu_2 = \mu_3$). However, for such a $\boldsymbol{\mu}$, as $\gamma\to\infty$ the probability that $\bar{\mathbf{Y}}_1 < \bar{\mathbf{Y}}_2 \wedge \bar{\mathbf{Y}}_3$ converges to one so that

$$\lim_{\gamma\to\infty} \pi_{12}(\gamma\boldsymbol{\mu}) = P[\|\bar{\mathbf{Y}} - P_{\mathrm{w}}(\bar{\mathbf{Y}}|\mathscr{I}_2)\|_{\mathrm{w}}^2 > C]$$

where $\mathscr{I}_2 = \{\mathbf{y}\in R^3 : y_2 \leqslant y_3\}$ and the probability on the right-hand side is computed under the assumption that $\mu_2 = \mu_3$. It then follows from Theorem 2.3.1 that $\inf_{\boldsymbol{\mu}\in E_2}\pi_{12}(\boldsymbol{\mu}) = \frac{1}{2}P[\chi_1^2 > C]$ where χ_1^2 denotes a standard chi-square random variable with one degree of freedom. A similar result holds for E_3.

For arbitrary k the infimum over the jth member of the partition of the complement of H_1 is equal to $\lim_{\gamma\to\infty}\pi_{12}(\gamma\boldsymbol{\mu})$ where $\boldsymbol{\mu}$ is an element of the face of H_1 corresponding to the jth member of the partition. This limit is equal to $P[\|\bar{\mathbf{Y}} - P_{\mathrm{w}}(\bar{\mathbf{Y}}|\mathscr{I}_j)\|_{\mathrm{w}}^2 > C]$ where $\mathscr{I}_j$ imposes the order restriction only between adjacent pairs which are equal in the face of H_1. For example, if $k = 4$ and the face is $\{\mathbf{y}: y_1 < y_2 = y_3 = y_4\}$ then $\mathscr{I}_j = \{\mathbf{y}: y_2 \leqslant y_3 \leqslant y_4\}$. Each of these limits is a mixture of chi-square variables from Theorem 2.3.1. The smallest of these limits corresponds to a $\mathscr{I}_j$ which imposes the fewest number of restrictions (i.e. only one). Thus, for any k, $\inf_{\boldsymbol{\mu}\notin H_1}\pi_{12}(\boldsymbol{\mu}) = \frac{1}{2}P[\chi_1^2 > C]$, where C is the critical value of the test.

This infimum is not zero for any k but does get small fairly rapidly as k grows. The value of $\inf_{\boldsymbol{\mu}\notin H_1}\pi_{12}(\boldsymbol{\mu})$ for $\alpha = 0.05$ are given in Table 2.6.1 for $k = 3, 4, \ldots, 7$. Thus, even though $\inf_{\boldsymbol{\mu}\notin H_1}\pi_{12}(\boldsymbol{\mu}) > 0$, the amount of bias in the test based upon $\bar{\chi}_{12}^2$ is substantial even for moderate values of k.

Optimality of the LRTs

The power calculations reported in Section 2.5 demonstrate that the use of information involving order restrictions often provides a substantial improvement over the omnibus procedures. One might ask, however, if these likelihood ratio tests are, in any sense, the best test for use with ordered hypotheses. Sackrowitz and Strawderman (1974) discuss the inadmissibility of the MLE for simple ordered normal means and binomial parameters. Based on their work, one would anticipate that the LRTs are not optimal either. However, the power functions of the LRT and its competitors are compared in Chapter 4. In the absence of information beyond the ordering information considered in this chapter, the LRTs seem to be the preferred tests.

2.7 POLYHEDRAL AND OTHER CONES

The order restrictions considered thus far in this chapter are of the form $\boldsymbol{\mu}\in\mathscr{I}$ where $\mathscr{I}$ is the cone of vectors isotonic with respect to a quasi-order on $X=\{1,2,\ldots,k\}$. The results obtained in this setting have been generalized to polyhedral cones by Raubertas, Lee, and Nordheim (1986).

A *polyhedral cone* in R^k is the set of all points that satisfy a finite set of homogeneous linear inequalities. For $\mathbf{A}$, a $k\times m$ matrix, the cone determined by $\mathbf{A}$ is

$$\mathscr{K}(\mathbf{A})=\{\mathbf{y}\in R^k:\mathbf{A}'\mathbf{y}\leqslant\mathbf{0}\}. \tag{2.7.1}$$

A column of $\mathbf{A}$ is *redundant* if the cone determined by $\mathbf{A}$ is the same as that determined by $\mathbf{A}$ with the column deleted. Suppose that $\mathbf{A}$ has no redundant columns. The cone of vectors isotonic with respect to a quasi-order, the orthant, $\{\mathbf{y}\in R^k: y_i\geqslant 0,\ i=1,2,\ldots,k\}$, the line, $\{\mathbf{y}\in R^k: y_1=y_2=\cdots=y_k\}$, and in fact all linear subspaces are polyhedral cones. Chapter 2 of Stoer and Witzgall (1970) contains a thorough discussion of polyhedral cones.

The dimension of a polyhedral cone $\mathscr{K}$, which will be denoted by $\dim(\mathscr{K})$, is the dimension of the linear subspace spanned by $\mathscr{K}$. The *lineality space* of $\mathscr{K}$, $\mathrm{ls}(\mathscr{K})$, is the linear subspace of highest dimension contained in $\mathscr{K}$. For $\phi\subset J\subset\{1,2,\ldots,m\}$ and $J^*=\{1,2,\ldots,m\}-J$, $\mathbf{A}_J(\mathbf{A}_{J*})$ is the matrix consisting of columns of $\mathbf{A}$ indexed by elements in $J(J^*)$. The set of points

$$\mathscr{K}_J=\{\mathbf{y}\in R^k:\mathbf{A}_J'\mathbf{Y}=\mathbf{0}\text{ and }\mathbf{A}_{J*}'\mathbf{y}\leqslant\mathbf{0}\} \tag{2.7.2}$$

is called a *face* of $\mathscr{K}=\mathscr{K}(\mathbf{A})$. If $J=\phi$ then $\mathscr{K}_J=\mathscr{K}(\mathbf{A})$ and if $J^*=\phi$ then $\mathscr{K}_J=\mathrm{ls}(\mathscr{K}(\mathbf{A}))$.

The jth column of $\mathbf{A}$ corresponds to a homogeneous linear inequality constraint. At a point $\mathbf{y}\in\mathscr{K}(\mathbf{A})$, the jth constraint is said to be *active* (*inactive*) provided $\mathbf{A}_{\{j\}}'\mathbf{y}=\mathbf{0}(\mathbf{A}_{\{j\}}'\mathbf{y}<\mathbf{0})$. The *relative interior* of the face $\mathscr{K}_J$, which is denoted by $\mathscr{K}_J^I$, is the set of all points at which the jth constraint is inactive for each $j\in J^*$, that is

$$\mathscr{K}_J^I=\{\mathbf{y}\in R^k:\mathbf{A}_J'\mathbf{y}=\mathbf{0}\text{ and }\mathbf{A}_{J*}'\mathbf{y}<\mathbf{0}\} \tag{2.7.3}$$

$(\mathscr{K}_\phi^I=\mathscr{K}(\mathbf{A})^I)$. Clearly, the $\mathscr{K}_J^I$ partition $\mathscr{K}(\mathbf{A})$ into disjoint regions, although some of the $\mathscr{K}_J^I$ may be empty.

Because a polyhedral cone, $\mathscr{K}$, is a closed, convex subset of R^k, $P(\cdot|\mathscr{K})$, the projection onto $\mathscr{K}$ with respect to the usual Euclidean norm, $\|\cdot\|$, exists and is unique. For those J with $\mathscr{K}_J^I \neq \phi$, let D_J be the set of points, $\mathbf{y}$, for which $P(\mathbf{y}|\mathscr{K}) \in \mathscr{K}_J^I$ i.e. the preimage of $\mathscr{K}_J^I$.

If $k = 3$ and $\mathscr{K}$ is the cone corresponding to the usual simple order, then

$$\mathbf{A} = \begin{bmatrix} -1 & 0 \\ 1 & -1 \\ 0 & 1 \end{bmatrix},$$

$D_\phi = \mathscr{K}_\phi^I = \{\mathbf{y}: y_1 < y_2 < y_3\}$, $\mathscr{K}_{\{1,2\}}^I = \{\mathbf{y}: y_1 = y_2 = y_3\}$, $D_{\{1,2\}} = \mathscr{K}^*$, $\mathscr{K}_{\{1\}}^I = \{\mathbf{y}: y_1 = y_2 < y_3\}$, $D_{\{1\}} = \{\mathbf{y}: y_1 \geqslant y_2, (y_1 + y_2)/2 < y_3\}$, $\mathscr{K}_{\{2\}}^I = \{\mathbf{y}: y_1 < y_2 = y_3\}$, and $D_{\{2\}} = \{\mathbf{y}: y_2 \geqslant y_3 : (y_2 + y_3)/2 > y_1\}$. Figure 2.7.1 gives a pictorial representation of such a cone and its decomposition into the relative interiors of its faces and their preimages.

Let $\mathbf{Y}$ be a k-dimensional random vector which has a multivariate normal distribution with mean vector, $\boldsymbol{\mu}$, and covariance matrix $\sigma^2\boldsymbol{\Sigma}$, that is $\mathbf{Y} \sim \mathscr{N}_k(\boldsymbol{\mu}, \sigma^2\boldsymbol{\Sigma})$. In this discussion, we assume that $\boldsymbol{\Sigma}$ is nonsingular and completely known. Let $\mathscr{K}_0$ and $\mathscr{K}_1$ be nonempty, polyhedral cones with $\mathscr{K}_0 \subset \mathscr{K}_1$ and $\mathscr{K}_1 - \mathscr{K}_0 \neq \phi$. We consider testing $\boldsymbol{\mu} \in \mathscr{K}_0$ with the alternative

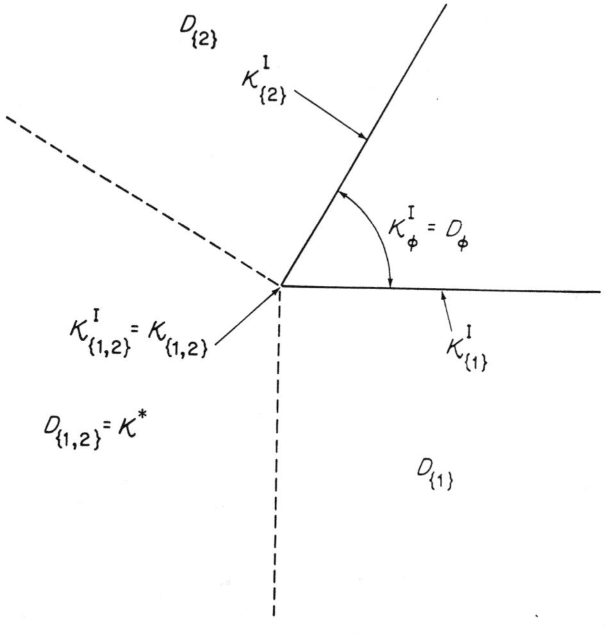

Figure 2.7.1 A cone $\mathscr{K}$, its faces, $\mathscr{K}_J$, and the preimages of faces D_J

restricted by $\boldsymbol{\mu} \in \mathscr{K}_1$. The case in which σ^2 is known exhibits the relationship between these results and those of earlier sections in this chapter. Raubertas, Lee, and Nordheim (1986) also considers the case in which σ^2 is unknown. There is a nonsingular matrix $\mathbf{D}$ for which $\sigma^2\mathbf{D}\boldsymbol{\Sigma}\mathbf{D}' = \mathbf{I}$, $\mathbf{V} = \mathbf{D}\mathbf{Y} \sim \mathscr{N}_k(\mathbf{D}\boldsymbol{\mu}, \mathbf{I})$ and $\boldsymbol{\mu} \in \mathscr{K}(\mathbf{A})$ if and only if $\mathbf{D}\boldsymbol{\mu} \in \mathscr{K}(\mathbf{B})$ where $\mathbf{B} = (\mathbf{D}^{-1})'\mathbf{A}$. Thus without loss of generality we may assume $\sigma^2\boldsymbol{\Sigma} = \mathbf{I}$.

If λ is the likelihood ratio statistic for this test, then $\boldsymbol{\mu} \in \mathscr{K}_0$ is rejected for large values of

$$T = -2\ln\lambda = \|\mathbf{Y} - P(\mathbf{Y}|\mathscr{K}_0)\|^2 - \|\mathbf{Y} - P(\mathbf{Y}|\mathscr{K}_1)\|^2. \tag{2.7.4}$$

The exact null distribution of T has been obtained for the following two cases: $\mathscr{K}_0$ is a linear subspace and $\mathscr{K}_1 = R^k$. These two cases generalize the $\bar{\chi}^2_{01}$ and $\bar{\chi}^2_{12}$ tests of this chapter. With $L \subset \mathscr{K} \subset R^k$, L a linear subspace and $\mathscr{K}$ a polyhedral cone, the LRT of $\boldsymbol{\mu} \in L$ versus $\boldsymbol{\mu} \in \mathscr{K} - L$ rejects $\boldsymbol{\mu} \in L$ for large values of

$$T_{01} = \|\mathbf{Y} - P(\mathbf{Y}|L)\|^2 - \|\mathbf{Y} - P(\mathbf{Y}|\mathscr{K})\|^2 = \|P(\mathbf{Y}|L) - P(\mathbf{Y}|\mathscr{K})\|^2$$

where the latter equality follows from Theorem 8.2.7 and by noting that $P(\mathbf{Y}|L)$, $-P(\mathbf{Y}|L) \in L \subset \mathscr{K}$. Also, the LRT of $\boldsymbol{\mu} \in \mathscr{K}$ versus $\boldsymbol{\mu} \in R^k - \mathscr{K}$ rejects $\boldsymbol{\mu} \in \mathscr{K}$ for large values of

$$T_{12} = \|\mathbf{Y} - P(\mathbf{Y}|\mathscr{K})\|^2.$$

Using cone orders like those desired in Section 2.6 (cf. Theorem 2.6.2), Raubertas, Lee, and Nordheim (1986) establish the monotonicity of the power functions of such tests and show that $\operatorname{ls}(\mathscr{K})$ is least favorable within $\mathscr{K}$ for T_{12}.

The mixing coefficients of the resulting $\bar{\chi}^2$ distributions generalize as follows: let $\mathbf{Y} \sim \mathscr{N}_k(\mathbf{0}, \mathbf{I})$ and for $\dim(\operatorname{ls}(\mathscr{K})) \leqslant l \leqslant \dim(\mathscr{K})$, set

$$p_l = \sum_{\{J:\dim(D_J) - l\}} P[\mathbf{Y} \in D_J]. \tag{2.7.5}$$

As in the proof of Theorem 2.3.1, they obtain the conditional distribution of T_{01} and T_{12} given the face (actually the relative interior of the face) in which $P(\mathbf{Y}|\mathscr{K})$ lies. Summing over the faces yields

$$P[T_{01} \geqslant C_1, T_{12} \geqslant C_2] = \sum_{l=i}^{h} p_l P[\chi^2_{l-m} \geqslant C_1] P[\chi^2_{k-l} \geqslant C_2] \tag{2.7.6}$$

where $i = \dim(\operatorname{ls}(\mathscr{K}))$, $m = \dim(L)$, and $h = \dim(\mathscr{K})$. If $\mathscr{K}$ is the cone corresponding to a partial order on $X = \{1, 2, \ldots, k\}$ and $L = \{\mathbf{y}: y_1 = y_2 = \cdots = y_k\}$ then $m = 1$ and $h = k$. However, $i = \dim(\operatorname{ls}(\mathscr{K}))$ may be greater than 1, but the corresponding $P(l, k; \mathbf{w}) = 0$ for $l < i$. Thus (2.7.6) generalizes (2.3.2). To apply these results one needs to compute $P(\mathbf{Y}|\mathscr{K})$ and the p_l. Dykstra (1983) and Wilhelmsen (1976) give algorithms for determining the projection, but except for a few partial orders little is known about the p_l.

Pincus (1975) considers testing normal means with the alternative restricted by a circular cone. With $\mathbf{Y} \sim \mathscr{N}_k(\boldsymbol{\mu}, \sigma^2\mathbf{I})$, $\mathbf{c} \neq \mathbf{0}$, the center of the cone, and ψ_0

an angle in the first quadrant, consider testing $\boldsymbol{\mu}=0$ versus $\boldsymbol{\mu}\in\mathscr{C}-\{\mathbf{0}\}$, where $\mathscr{C}=\{\mathbf{y}\in R^k:\langle\mathbf{c},\mathbf{y}\rangle/(\|\mathbf{c}\|\cdot\|\mathbf{y}\|)\geqslant\cos\psi_0\}$. Without loss of generality, one may assume that $\mathbf{c}=(1,0,\ldots,0)$. In the case in which σ^2 is known, the LRT rejects $\boldsymbol{\mu}=\mathbf{0}$ for large values of

$$T=P(\mathbf{Y}|\mathscr{K})'P(\mathbf{Y}|\mathscr{K})/\sigma^2$$
$$=\begin{cases}\dfrac{\|\mathbf{Y}\|^2}{\sigma^2} & \text{if } Y_1/\|\mathbf{Y}\|\geqslant\cos\psi_0,\\ [Y_1\cos\psi_0+(Y_2^2+\ldots+Y_k^2)^{1/2}\sin\psi_0]^2 & \text{if} \cos\psi_0>Y_1/\|\mathbf{Y}\|\geqslant-\sin\psi_0,\\ 0 & \text{if } Y_1/\|\mathbf{Y}\|<-\sin\psi_0.\end{cases} \tag{2.7.7}$$

While Pincus (1975) considers the case of unknown σ^2, similar arguments show that if $\boldsymbol{\mu}=\mathbf{0}$ then

$$P[T\geqslant t]=\sum_{l=0}^{k}q_lP[\chi_l^2\geqslant t],$$

with expressions for the q_l, which are analogues of the level probabilities, given in that reference. In the case $k=3$, it is clear that the circular cone is the limit of the polyhedral cones and it would be instructive to relate q_l to the p_l for the polyhedral cones. Shapiro (1985, 1988) discusses $\bar{\chi}^2$ distributions for arbitrary cones.

2.8 COMPLEMENTS

Several researchers began to work independently at about the same time on the problem of testing order restrictions among normal means. The first paper published on the $\bar{\chi}_{01}^2$ test was by Bartholomew (1959a), but Chacko independently obtained some of the same results although his work did not appear until 1963. Eeden (1958) considered the situation in which the null hypothesis imposed a simply ordered trend on the means, $\mu_1\leqslant\mu_2\leqslant\cdots\leqslant\mu_k$. Kudô's (1963) test for the mean of a multivariate normal distribution with a one-sided alternative is further independent work on a very closely related problem (see Section 4.6).

Apart from the two-sample normal means problem, one of the earliest discoveries of a chi-bar-squared distribution was by Chernoff (1954). He found that the asymptotic distribution of the LR statistic for a parameter point on the boundary of the null hypothesis, which is defined by linear constraints, is a mixture of chi-squares.

One test of homogeneity with a simply ordered alternative studied by Bartholomew (1959a) was based on a statistic $\bar{F}$. Bartholomew (1961b) found that its power is considerably less than the LRT based on $\bar{E}_{01}^2$ which is described in this chapter. With l the number of level sets in $\boldsymbol{\mu}^*$ and $\bar{F}$ the usual mean square ratio for the one-way analysis of variance computed after the pooling process (i.e.

after the PAVA or MLSA has been applied),

$$\bar{E}_{01}^2 = \frac{[(l-1)/(N-k)]\bar{F}}{1+[(l-1)/(N-k)]\bar{F}} \tag{2.8.1}$$

and $S_{01} = v\bar{E}_{01}^2/(1-\bar{E}_{01}^2) = [(l-1)/(N-k)]\bar{F}$. However, because l is random, critical regions of the form $\bar{E}_{01}^2 \geqslant C_1$ and $\bar{F} \geqslant C_2$ are not equivalent. On the other hand, one can use the fact that

$$B_{a/2,b/2} = \left(1 + \frac{b}{a}F_{ba}\right)^{-1}$$

to write

$$P[\bar{E}_{01}^2 \geqslant C] = \sum_{l=1}^{k} P(l,k;\mathbf{w})P\left[F_{l-1,N-l} \geqslant \frac{N-l}{l-1}\left(\frac{C}{1-C}\right)\right] \tag{2.8.2}$$

and

$$P(\bar{E}_{12}^2 \geqslant C] = \sum_{l=1}^{k} P(l,k;\mathbf{w})P\left[F_{k-l,N-k} \geqslant \frac{N-k}{k-l}\left(\frac{C}{1-C}\right)\right]. \tag{2.8.3}$$

There have been several attempts to obtain the null distributions of $\bar{\chi}_{01}^2$ and $\bar{E}_{01}^2$. Theorem 2.3.1 is easily proved for $k=2$ since $|\bar{Y}_1 - \bar{Y}_2|$ is distributed independently of $\operatorname{sgn}(\bar{Y}_1 - \bar{Y}_2)$. Bartholomew's original extension of this argument for the case of a simple order aimed to show that the distribution of $\bar{\chi}_{01}^2$, conditional on the restraints imposed on the $\bar{Y}_i$ by the averaging process, is χ^2. The restraints were divided into two types. The first type is within-block inequalities and the second is among-block inequalities. The reader will recognize that this was the approach employed here.

An alternative method of proof was given in Bartholomew (1961a) and later used by Shorack (1967). This proof, which is inductive, is based on the number of essential restrictions defining the alternative hypothesis. For the usual simple order, there are $k-1$ essential restrictions, $\mu_i \leqslant \mu_{i+1}$; $i = 1, 2, \ldots, k-1$. This argument uses the fact that once two means are pooled, they remain pooled. This is true for a simple order but not for all partial orders. The proof given here for arbitrary quasi-orders is modeled after that given in Barlow *et al.* (1972).

Shorack (1967) showed that Theorem 2.3.1 remains valid if the $\bar{Y}_i$ have covariances $\delta_{ij}w_j^{-1} - (\sum_{h=1}^{k} w_h)^{-1}$; $i, j = 1, 2, \ldots, k$ with δ_{ij} the Kronecker delta function. Of course, this is the same as the covariance structure of $\bar{Y}_j - \hat{\mu}$, and additive constants may be taken in or out of the projection operator, $P_{\mathbf{w}}(\cdot|\mathscr{I})$. This point will be met again in the discussion of nonparametric tests in Section 4.5 and in tests of trends in multinomial parameters (see Chapter 5).

The equal-weights level probabilities for a simple order, $P_S(l,k)$, arise in many contexts. They are distribution free and their study involves combinatorial ideas. The first derivation of (2.4.11) is due to Sparre Andersen (1954). Bartholomew (1959a) conjectured the following recurrence relation:

$$P_S(l,k) = \sum_{j=l-1}^{k-1} \frac{P_S(1,k-j)P_S(l-1,j)}{l}.$$

Miles (1959) established this result and Barton and Mallows (1961) gave a simpler proof, which is the one given here. Miles (1959) also showed that the average number of level sets for a simple order, under homogeneity, is of the order $\log k$ and that the average size for a level set in this case is of the order $k/\log k$.

Chacko (1963) also derived some of the theory for $P_S(l, k)$. He showed that $P_S(1, k) = 1/k$ and appealed to Sparre Andersen's work to obtain the other level probabilities. He also noted that $\sum_{l=1}^{k}(-1)^l P_S(l, k) = 0$. It has been conjectured that this is true for any quasi-order and any weight set. (See the comment at the end of this section and the discussion in Shapiro, 1987a.) Chapter 3 is devoted to a study of the level probabilities and approximations to the distributions of the $\bar{\chi}^2$ and $\bar{E}^2$.

Because of the complexity of the null distributions of the LRTs presented in this chapter, closely related conditional tests have been studied. In such tests, one chooses a critical value, depending on the number of level sets, so that conditional on the observed number of level sets the LRT statistic exceeds the critical value with probability α under the null hypothesis. Bartholomew (1961a) studied the conditional test analogue of $\bar{\chi}_{01}^2$ for a simple order and Wollan and Dykstra (1986) considered the analogue to $\bar{\chi}_{12}^2$.

Bartholomew (1959b) studied the convergence, as $k \to \infty$, of the null distribution of $\bar{\chi}_{01}^2$ with a simple order and equal weights. He computed its characteristic function, then the cumulant generating function and obtained an asymptotic expression for its cumulants. (The characteristic function, cumulant generating function, and the first two cumulants are given in Chapter 3; see (3.2.2).) The limiting distribution is shown to be normal by examining the limiting cumulants. However, the rate of approach to the limiting distribution is very slow; the cumulants depend on k through $\log k$. Hence, the asymptotic distribution is of little practical value. A two-moment chi-square approximation is presented in Section 3.2.

Pincus (1975) considers the LRT of $H_0\!:\boldsymbol{\mu} = \mathbf{0}$ versus $H_1 - H_0$: with $H_1\!:\boldsymbol{\mu} \in \mathscr{C}$ where $\mathscr{C}$ is a circular cone. An $\bar{E}^2$ distribution arises in this setting, too.

Perlman (1969) considers testing hypotheses with the null and/or alternative of the form $\boldsymbol{\mu} \in \mathscr{K}$ with $\mathscr{K}$ a closed, convex cone. He considers observations from a multivariate normal distribution with covariance matrix of the form (1) $\sigma^2\boldsymbol{\Sigma}_0$ with $\boldsymbol{\Sigma}_0$ a known positive definite matrix or (2) an unknown positive definite matrix. These results are discussed in Section 4.6. If the null hypothesis is $\boldsymbol{\mu} \in \mathscr{I}$, with $\mathscr{I}$ the cone of vectors isotonic with respect to a quasi-order, and the alternative is unrestricted, then this is the testing situation, H_1 versus $H_2 - H_1$, discussed in this chapter. Robertson and Wegman (1978) studied this problem further, obtaining the distributions of $\bar{\chi}_{12}^2$ and $\bar{E}_{12}^2$ under H_0 given in Theorem 2.3.1. They also extended the results of this chapter to exponential families (see Section 4.1).

If the usual assumptions about the variances in the one-way layout are questioned, then a two-stage procedure could be employed. Marcus (1980a, 1980b) discusses such procedures. Kudô, Sasabuchi, and Choi (1981) show that because of the ordering constraints it is possible, under mild assumptions, to test homogeneity of normal means without an independent estimate of the unknown variance.

If the $\bar{\chi}^2_{01}$ and $\bar{E}^2_{01}$ test rejects H_0 then one may wish to carry out a multiple comparison procedure to determine which means are different or which simultaneous confidence bounds may be of interest. These ideas are developed in Marcus and Peritz (1976), Spjøtvoll (1977), and Marcus (1978a, 1978b, 1982). Interesting applications are also discussed by Korn (1982) and Schoenfeld (1986). These ideas are discussed in Section 9.5.

Because of the complexity of the distribution of the $\bar{\chi}^2_{01}$ and $\bar{E}^2_{01}$ statistics, researchers have proposed tests based on contrast statistics. This approach is discussed in Section 4.2. Proctor (1971) gives an application of testing ordered hypotheses in processing survey data. He shows how the LRT and the test based on a contrast statistic would be implemented. In Section 4.3, tests based on the maximum of several contrast statistics are also studied.

A discussion of some of the basic nonparametric tests for ordered hypotheses, as well as references to other works, are given in Section 4.5. Shorack (1967) extends the results given in this chapter to other experimental designs and shows how they can be used with ranks to provide distribution-free tests.

In this chapter, tests of order hypotheses based on k independent random samples have been considered. Kudô (1963) considered tests based on a random sample from a multivariate normal distribution. His and subsequent work is discussed in Section 4.6.

Raubertas, Lee, and Nordheim (1986) generalize Bartholomew's tests to allow hypotheses involving homogeneous linear inequality restrictions. This framework includes the hypotheses of monotonicity, nonnegativity, and convexity. These constraints correspond to a polyhedral cone. The null distributions of their test statistics are mixtures of chi-square or beta distributions (cf. Section 2.7). Shapiro (1985) obtained these $\bar{\chi}^2$ and $\bar{E}^2$ distributions in a study of the distribution of a minimum discrepancy statistic (see also Shapiro, 1988).

It has recently been called to our attention that Jim Lawrence and Jon Spingarn have given a geometric argument to show that the even and the odd level probabilities each sum to one-half.

CHAPTER 3

Approximations to the $\bar{\chi}^2$ and $\bar{E}^2$ Distributions

The distributions of $\bar{\chi}_{01}^2$, $\bar{\chi}_{12}^2$, $\bar{E}_{01}^2$, and $\bar{E}_{12}^2$, under H_0, are determined by the level probabilities, $P(l, k; \mathbf{w})$. Even for a simple order with moderate k ($k > 5$) and unequal weights, good closed form expressions for these level probabilities have not been found. Thus approximations are of interest. Probably the most natural question is, how much error is introduced by using the equal-weights level probabilities when the weights are not equal? Siskind (1976) and Grove (1980) observed that the $P_S(l, k; \mathbf{w})$ are relatively insensitive to changes in the weights and conjectured that for many situations the equal-weights distributions should provide reasonable approximations. In Section 3.1, the equal-weights approximation is studied for the simple order and the simple tree ordering. The results obtained there confirm their conjecture for these orderings and suggest that similar conclusions would hold for other partial orders.

It should be noted that for large k, even if the level probabilities are known, the computation of p values for these tests may be tedious. Based on the work of Bartholomew (1959a, 1959b), two-moment chi-squared and beta approximations are developed for this case in Section 3.2. While the techniques developed could be used for arbitrary $\bar{\chi}^2$ and $\bar{E}^2$ distributions, the constants needed to implement the procedures have only been tabled for a simple order and a simple tree.

More complex approximations have been developed for the null distributions of the $\bar{\chi}^2$ and $\bar{E}^2$ statistics (cf. Sections 3.3 through 3.5). However, the use of these approximations is presently limited to the case of a simple order or a simple tree ordering.

Approximations for the situation in which one of the weights, w_i, is larger than the other weights and the other weights are (nearly) equal are discussed in Section 3.3. Approximations for arbitrary weight sets based on the pattern of large and small weights are considered for the simple order and the simple tree ordering in Sections 3.4 and 3.5, respectively. An approximation due to Siskind for the simple order is also mentioned in Section 3.4. The amount of variation in the null $\bar{\chi}^2$ and $\bar{E}^2$ distributions, as functions of the weight vector, is determined by obtaining sharp upper and lower bounds for the appropriate tail probabilities in terms of the quasi-order under consideration, cf. Section 3.6.

The level probabilities arise in other areas of statistical inference; for instance in constructing one-sided confidence bounds (cf. Bohrer and Francis, 1972a, Bohrer, 1975, and Bohrer and Chow, 1978 and in developing multivariate tests of one-sided hypotheses (cf. Kudô, 1963, Nüesch, 1964, Shorack, 1967, and related references in Chapter 4). It is hoped that either the approximations or the techniques used to develop them will prove useful in some of these situations.

It was observed in Chapter 2 that the distributions of the $\bar{\chi}^2$ and $\bar{E}^2$ test statistics, under the alternative hypotheses, are quite complicated even for equal weights and small k. There has been a limited amount of research concerning approximations for these non-null distributions and such results are discussed in Section 3.7 for a simple order and a simple tree ordering.

Because the equal-weights approximation is suitable for most practical applications, the reader may wish to omit this chapter in a first reading.

3.1 THE EQUAL-WEIGHTS DISTRIBUTION AS AN APPROXIMATION TO THE NULL $\bar{\chi}^2$ AND $\bar{E}^2$ DISTRIBUTIONS: THE SIMPLE AND SIMPLE TREE ORDERINGS

The robustness of the $P(l, k; \mathbf{w})$ will be studied by examining the changes in the tail probabilities of the null distribution of $\bar{\chi}^2_{01}$ resulting from changes in the weights, $\mathbf{w}$. In particular, for $\lesssim$ the usual simple order or the simple tree and e_α, the equal-weights critical value with a significance level of α, the value of $P_{\mathbf{w}}[\bar{\chi}^2_{01} \geqslant e_\alpha]$ under the null hypothesis will be studied as the weights vary. The subscript $\mathbf{w}$ is used to emphasize that this probability depends on the weight vector $\mathbf{w}$.

The case of a simple order is examined first. The case $k = 5$ will be studied in detail because exact probabilities can be computed for arbitrary weights in this case. In Section 3.6, sharp upper and lower bounds for the null distribution of $\bar{\chi}^2_{01}$ and $\bar{E}^2_{01}$ are obtained. For a simple order with $k = 5$, the bounds are approached by taking weight sets of the form $(\varepsilon, \sqrt{\varepsilon}, 1, \sqrt{\varepsilon}, \varepsilon)$ and $(1, \varepsilon, \varepsilon, \varepsilon, 1)$, and letting $\varepsilon \to 0$. Table 3.1.1 gives the values of $P_{\mathbf{w}}[\bar{\chi}^2_{01} \geqslant e_{0.05}]$ for such weights sets and various values of ε. The ratio, $R = \varepsilon^{-1}$, of the largest weight to the smallest is also given. Based on these values, one might conjecture that for $k = 5$ and $R \leqslant 1.5$, the true significance level lies in $[0.045, 0.053]$ or for $k = 5$ and $R \leqslant 2$, the true significance level lies in $[0.042, 0.056]$. A further study of the true significance levels corresponding to an equal-weights 0.05 significance level was conducted using random weights. Five uniform $(0, 1)$ random variates, $U_1, U_2, \ldots, U_5$, were generated and with weights $w_i = U_i$ the probability $P_{\mathbf{w}}[\bar{\chi}^2_{01} \geqslant e_{0.05}]$ was computed. This was repeated 10 000 times and the frequency distribution for these 10 000 probabilities is given in Table 3.1.2. The table also gives, for each interval, the minimum value of R for the weight sets which produced significance levels in that interval. It should be noted that every weight set with significance level

Table 3.1.1 The probability $P_{\mathbf{w}}[\bar{\chi}_{01}^2 \geqslant e_{0.05}]$ for $k = 5$ and various weights

ε	R	$\mathbf{w} = (1, \varepsilon, \varepsilon, \varepsilon, 1)$ $P_{\mathbf{w}}[\bar{\chi}_{01}^2 \geqslant e_{0.05}]$	$\mathbf{w} = (\varepsilon, \sqrt{\varepsilon}, 1, \sqrt{\varepsilon}, \varepsilon)$ $P_{\mathbf{w}}[\bar{\chi}_{01}^2 \geqslant e_{0.05}]$
1.0000	1.0	0.0500	.0500
0.6667	1.5	0.0453	0.0533
0.5000	2.0	0.0420	0.0557
0.3333	3.0	0.0375	0.0590
0.2000	5.0	0.0324	0.0629
0.1000	10.0	0.0267	0.0679
0.0000	∞	0.0130	0.0960

Table 3.1.2 Frequency distribution for $P_{\mathbf{w}}[\bar{\chi}_{01}^2 \geqslant e_{0.05}]$ with $k = 5$ and 10 000 randomly generated weights

INTERVAL	FREQUENCY	MINIMUM RATIO	INTERVAL	FREQUENCY	MINIMUM RATIO
[0.020, 0.024]	4	20.8	[0.052, 0.056]	1561	1.4
[0.024, 0.028]	22	18.3	[0.056, 0.060]	1116	2.2
[0.028, 0.032]	73	8.0	[0.060, 0.064]	670	3.4
[0.032, 0.036]	284	5.2	[0.064, 0.068]	357	5.0
[0.036, 0.040]	719	3.4	[0.068, 0.072]	145	6.9
[0.040, 0.044]	1319	2.0	[0.072, 0.076]	54	14.9
[0.044, 0.048]	1781	1.4	[0.076, 0.080]	17	34.5
[0.048, 0.052]	1876	1.1	[0.080, 0.084]	2	812.0

outside of [0.04, 0.06] had $R \geqslant 3.4$ and each weight set with significance level outside of [0.048, 0.052] had $R \geqslant 1.4$, leading one to believe that the equal-weights critical values or p values provide reasonable approximations for $k = 5$ if $R \leqslant 1.4$.

To obtain an indication of the quality of this approximation for other α and larger k, several particular weight sets have been considered. For $k > 5$, the $P_S(l, k; \mathbf{w})$ were estimated by Monte Carlo techniques. Table 3.1.3 gives $P_M(l, k; \mathbf{w})$, the Monte Carlo estimates based on 4000 iterations, and $P_S(l, k)$, the equal-weights level probabilities, for the three cases $\mathbf{w}^{(1)} = (147, 175, 150, 191, 188, 179, 200, 220, 208, 230)$, $\mathbf{w}^{(2)} = (155, 178, 215, 93, 128, 114, 172, 158, 197, 201, 95, 183)$, and $\mathbf{w}^{(3)} = (55, 20, 35, 45, 40, 25, 55, 30, 60, 45, 35, 25, 45, 35, 30)$. The agreement is reasonable, but because slight changes in the $P(l, k; \mathbf{w})$ have even less effect on the tail probabilities the agreement in significance levels is better. Table 3.1.4 gives the estimated true significance level for these three cases if e_α is used with $\alpha = 0.1$, 0.05, 0.01. It seems that for most practical purposes, even for moderately large

Table 3.1.3 Comparison of the Monte Carlo estimates with the equal-weights level probabilities for three cases

	CASE 1		CASE 2		CASE 3	
l	$P_M(l, 10; \mathbf{w})$	$P_S(l, 10)$	$P_M(l, 12; \mathbf{w})$	$P_S(l, 12)$	$P_M(l, 15; \mathbf{w})$	$P_S(l, 15)$
1	0.09425	0.10000	0.08600	0.08333	0.07175	0.06667
2	0.28975	0.28292	0.26525	0.25166	0.21800	0.21677
3	0.32600	0.32316	0.32425	0.31507	0.30000	0.29989
4	0.19850	0.19943	0.20425	0.21974	0.23950	0.23782
5	0.07300	0.07422	0.09125	0.09602	0.11900	0.12214
6	0.01600	0.01744	0.02250	0.02785	0.03875	0.04333
7	0.00225	0.00260	0.00575	0.00551	0.01175	0.01102
8	0.00025	0.00024	0.00075	0.00075	0.00125	0.00205
9	0.00000	0.00001	0.00000	0.00000	0.00000	0.00028
10[a]	0.00000	0.00000	0.00000	0.00000	0.00000	0.00003

[a]For $l > 10$, these values are zero to five places.

Table 3.14 Estimated significance levels corresponding to the equal-weights critical values

α	CASE 1	CASE 2	CASE 3
0.10	0.0991	0.0954	0.0977
0.05	0.0494	0.0474	0.0486
0.01	0.0098	0.0094	0.0096

values of k, the equal-weights approximation would be adequate for the simply ordered case if the variation in the weights is moderate. Approximations which are based on the pattern of large and small weights are discussed in Sections 3.4 and 3.5.

We now consider the quality of the equal-weights approximation for the simple tree ordering. It will be observed later in this section that for fixed k, as $R \to \infty$, the distribution of $\bar{\chi}^2_{01}$ with weights of the form $\mathbf{w} = (1, R, R, \ldots, R)$ and $\mathbf{w} = (R, 1, 1, \ldots, 1)$ approaches the upper and lower bounds for $\bar{\chi}^2_{01}$ with the tree ordering. Both of these weight sets can be written in the form $(a, 1, 1, \ldots, 1)$ with $a = 1/R$ or R and, for $l = 2, 3, \ldots, k$, $P(l, k; (a, 1, 1, \ldots, 1)) = \binom{k-1}{l-1} P_T(1, k-l+1; (a, 1, 1, \ldots, 1))\ P_T(l, l; (a+k-l, 1, 1, \ldots, 1))$. Thus, the level probabilities can be constructed recursively in this case using numerical integration on (2.4.12). With e_α denoting the α-level equal-weights critical value, Table 3.1.5 gives the value of $P_{\mathbf{w}}[\bar{\chi}^2_{01} \geq e_\alpha]$ under H_0 for various values of R and $k = 5, 8$. These results suggest that for $k = 5$, the true significance level is between 0.046 and 0.054 if R, the ratio

Table 3.1.5 The probability $P_{\mathbf{w}}[\bar{\chi}^2_{01} \geqslant e_{0.05}]$ for $k = 5$, 8 and various weights

		$k=5$		$k=8$	
ε	R	$\mathbf{w} = (\varepsilon, 1, 1, 1, 1)$	$\mathbf{w} = (1, \varepsilon, \varepsilon, \varepsilon, \varepsilon)$	$\mathbf{w} = (\varepsilon, 1, \ldots, 1)$	$\mathbf{w} = (1, \varepsilon, \ldots, \varepsilon)$
1.0000	1.0	0.0500	0.0500	0.0500	0.0500
0.6667	1.5	0.0539	0.0462	0.0533	0.0446
0.5000	2.0	0.0566	0.0436	0.0590	0.0408
0.3333	3.0	0.0601	0.0403	0.0636	0.0360
0.2000	5.0	0.0641	0.0370	0.0687	0.0307
0.1000	10.0	0.0684	0.0337	0.0740	0.0255
0.0000	∞	0.0794	0.0296	0.0863	0.0185

Table 3.1.6 Frequency distribution for $P_{\mathbf{w}}[\bar{\chi}^2_{01} \geqslant e_{0.05}]$ with $k = 5$ and 5000 randomly generated weights

INTERVAL	FREQUENCY	MINIMUM RATIO	INTERVAL	FREQUENCY	MINIMUM RATIO
[0.032, 0.036]	47	8.87	[0.056, 0.060]	397	2.06
[0.036, 0.040]	397	4.85	[0.060, 0.064]	242	3.34
[0.040, 0.044]	957	2.29	[0.064, 0.068]	180	5.87
[0.044, 0.048]	1196	1.39	[0.068, 0.072]	110	12.11
[0.048, 0.052]	868	1.14	[0.072, 0.076]	55	26.77
[0.052, 0.056]	543	1.39	[0.076, 0.080]	8	123.94

of the largest weight to the smallest, is less than 1.5 and between 0.040 and 0.060 if $R \leqslant 3.0$. Notice that the corresponding intervals for $k = 8$ are [0.045, 0.055] and [0.036, 0.064], respectively. While the quality of this approximation does decrease with k, it seems that, for small R, i.e. in the range of 1 to 2, the differences are not great. With $k = 5$ and 5000 randomly generated weight sets, as in the simply ordered case, the values of $P_{\mathbf{w}}[\bar{\chi}^2_{01} \geqslant e_{0.05}]$ were computed. The frequency distribution as well as the values of the minimum R for those weight sets are given in Table 3.1.6. Based on these results for $k = 5$, one might conjecture that if $R < 1.39$, then the true significance level is in [0.048, 0.052], and if $R < 3.34$, then it is in [0.040, 0.060]. As with the simple order, it seems that for many practical situations the equal-weights approximation should be adequate.

3.2 APPROXIMATIONS FOR THE NULL $\bar{\chi}^2$ AND $\bar{E}^2$ DISTRIBUTIONS: EQUAL WEIGHTS

For large k, the formulas in (2.3.4) may be quite tedious to use for computing p values even though the $P(l, k; \mathbf{w})$ are known. For this reason Bartholomew (1961b), based on a study of the shape coefficients of the null distribution of $\bar{\chi}^2_{01}$, proposed a two-moment gamma approximation for the case of equal weights and a simple order. This approximation can be improved by approximating the continuous part, i.e. the distribution of $\bar{\chi}^2_{01}$ given $\bar{\chi}^2_{01} > 0$, and extended to $\bar{\chi}^2_{12}$ as well as other orders. In this section such approximations are discussed for the simple order as well as the simple tree ordering. Related approximations for the null $\bar{E}^2$ distributions are also presented.

Approximations to the null $\bar{\chi}^2$ distributions: equal weights, simple, and simple tree orders

The approximations will be expressed in terms of $\bar{G}_b$, the tail function of a gamma distribution with parameters b and 1, respectively. A scale parameter, ρ, will be introduced to allow for a two-moment approximation. The following result is proved using straightforward calculations and the moments of the gamma distribution.

Remark *Let Y be a nonnegative random variable with finite second moment and a continuous distribution function, except possibly at the origin, and let κ_1 and κ_2 denote the first two cumulants (mean and variance) of the distribution of Y given that $Y > 0$. If Y_A is a nonnegative random variable which, for $y > 0$, satisfies*

$$P[Y_A \geqslant y] = P[Y > 0]\bar{G}_b\left(\frac{y}{\rho}\right) \tag{3.2.1}$$

with $\rho = \kappa_2/\kappa_1$ and $b = \kappa_1/\rho$, then the first two moments of Y and Y_A agree.

Because Y_A and Y have the same first two moments, one might use (3.2.1) as an approximation to $P[Y > y]$ for $y > 0$. To implement the approximation, the cumulants of the null distribution of $\bar{\chi}^2_{01}$ with equal weights and a simple order are obtained first. The corresponding characteristic function is given by

$$\phi_{01}(t) = \sum_{l=1}^{k} P_S(l, k)(1 - 2it)^{-(l-1)/2} = \frac{(z+1)(z+2)\cdots(z+k-1)}{k!}$$

with $z = (1 - 2it)^{-1/2}$. Thus, the cumulant generating function is

$$\psi_{01}(t) = \sum_{j=1}^{k-1} \log(j + z) - \log k!$$

and taking derivatives yields the following expressions for the first two cumulants:

$$\kappa_1^*(k, 1) = \sum_{j=2}^{k} j^{-1} \qquad \text{and} \qquad \kappa_2^*(k, 1) = \sum_{j=2}^{k} (3j^{-1} - j^{-2}). \tag{3.2.2}$$

(Bartholomew, 1961a, gives expressions for the third and fourth cumulants also.)

For computational purposes, sums of the form $2^{-r} + 3^{-r} + \cdots + k^{-r}$ can be obtained from tables of the polygamma functions. Next, the cumulants of the conditional distribution are needed. If κ_1^* and κ_2^* are the first two cumulants of Y and $p = P[Y > 0]$, then the cumulants of the distribution of Y given that $Y > 0$ are

$$\kappa_1 = \frac{\kappa_1^*}{p} \quad \text{and} \quad \kappa_2 = \frac{\kappa_2^*}{p} - (1-p)\frac{(\kappa_1^*)^2}{p^2}. \tag{3.2.3}$$

To use (3.2.1) one first computes p, which in the case of equal weights and the simple order is $1 - P[M = 1] = 1 - 1/k$. The quantities κ_1 and κ_2 are computed using (3.2.2) and (3.2.3) and then ρ and b are found based on κ_1 and κ_2. Table A.14 of the Appendix gives the values of ρ and b for $5 \leqslant k \leqslant 40$.

To give an indication of the accuracy of this approximation, the probability $P[\bar{\chi}_{01}^2 \geqslant t]$ with a simple order, equal weights, and a common mean was computed using (2.3.4) and approximated using the two-moment gamma approximation with the correction for discontinuity at zero. Table 3.2.1 gives these values for $k = 5, 10, 20$ and for values of t which yield significance levels near the common values of 0.2, 0.1, 0.05, 0.001, and 0.005. Clearly, this approximation is adequate for most purposes. The percentage error seems to increase in the far right tail. Singh and Wright (1986a) discuss a four-moment approximation that performs better in this region.

In considering the distribution, under H_0, of $\bar{\chi}_{12}^2$ for equal weights and a simple order, it should first be noted that because $P[\bar{\chi}_{12}^2 = 0] = 1/k!$, even for moderate k, it is not necessary to correct for the discrete part of $\bar{\chi}_{12}^2$. Next, the cumulants are needed. Under H_0 with equal weights and a simple order, the characteristic function of $\bar{\chi}_{12}^2$ is

$$\phi_{12}(t) = \sum_{l=1}^{k} P_S(l,k)(1-2it)^{-(k-l)/2} = \frac{P_k(u)}{u^k}$$
$$= \frac{(u+1)(u+2)\cdots(u+k-1)}{u^{k-1}k!}$$

Table 3.2.1 Exact and approximate values for $P[\bar{\chi}_{01}^2 \geqslant t]$ under H_0

$k = 5$			$k = 10$			$k = 20$		
EXACT	APPROX.	ERROR (%)	EXACT	APPROX.	ERROR (%)	EXACT	APPROX.	ERROR (%)
0.2267	0.2221	2.0	0.2297	0.2219	3.4	0.2290	0.2196	4.1
0.0836	0.0815	2.5	0.0981	0.0949	3.3	0.1077	0.1039	3.5
0.0512	0.0503	1.7	0.0413	0.0411	0.4	0.0492	0.0490	0.3
0.0119	0.0123	3.6	0.0110	0.0119	8.0	0.0097	0.0109	12.7
0.0045	0.0049	9.1	0.0045	0.0052	16.1	0.0042	0.0051	22.4

with $u = (1 - 2it)^{1/2}$. Thus, the first two cumulants are

$$\kappa_1(k, 2) = k - 1 - \sum_{j=2}^{k} j^{-1} \quad \text{and} \quad \kappa_2(k, 2) = 2(k - 1) - \sum_{j=2}^{k} (j^{-1} + j^{-2}). \quad (3.2.4)$$

(Expressions for the third and fourth cumulants are given in Singh and Wright, 1986a.) Using these cumulants ρ and b can be computed, and p values for $\bar{\chi}_{12}^2$ can be approximated using (3.2.1) with the multiplicative factor on the right-hand side set equal to one. Table A.14 in the Appendix gives the needed values of ρ and b for $5 \leqslant k \leqslant 40$. The results of a study of the accuracy of this approximation are summarized in Singh and Wright (1986a). The percentage errors tend to be smaller for $\bar{\chi}_{12}^2$ than for $\bar{\chi}_{01}^2$ and the difference is most noticeable for larger k and in the far right tail.

The simple tree case is handled similarly except that closed form expressions for the characteristic function are not available in this case. However, the rth moment of the null distribution of $\bar{\chi}_{01}^2$ can be expressed as

$$\sum_{l=1}^{k} P_T(l, k)(l - 1)(l + 1) \cdots (l + 2r - 3),$$

and the rth moment of the distribution of $\bar{\chi}_{12}^2$ under H_0 can be expressed as

$$\sum_{l=1}^{k} P_T(l, k)(k - l)(k - l + 2) \cdots (k - l + 2r - 2).$$

From these expressions, one can compute the first two cumulants of $\bar{\chi}_{01}^2$, which are denoted by $\tau_i(k, 1)$ with $i = 1, 2$, and the first two cumulants of $\bar{\chi}_{12}^2$, which are denoted by $\tau_i(k, 2)$ with $i = 1, 2$. Under H_0 with equal weights and the simple tree ordering,

$$P[\bar{\chi}_{01}^2 = 0] = P_T(1, k) \qquad \text{and} \qquad P[\bar{\chi}_{12}^2 = 0] = P_T(k, k) = \frac{1}{k}.$$

One can see from Table A.11 that even for moderate k, $P_T(1, k)$ is small and so it is not necessary to use the correction for the discontinuity at zero when approximating p values of $\bar{\chi}_{01}^2$ for the simple tree. The values of ρ and b needed to use (3.2.1) to approximate the distribution of $\bar{\chi}_{01}^2$ and $\bar{\chi}_{12}^2$ in this case are given in Table A.15 in the Appendix.

Approximations to the null $\bar{E}^2$ distributions: equal weights, simple, and simple tree orderings

Sasabuchi and Kulatunga (1985) considered two- and three-moment approximations to the null distribution of $\bar{E}_{01}^2$ for the simple order with equal weights. They found it more convenient to work with raw moments rather than the central moments. Expressing the moments of a beta distribution in terms of its parameters and solving, one obtains the following result.

Remark *Let Y be a random variable with continuous distribution function, except possibly at the orgin, let $0 \leqslant Y \leqslant 1$, and let $a = E(Y \mid Y > 0)$ and $b = E(Y^2 \mid Y > 0)$. If Y_A is a nonnegative random variable which, for $y > 0$, satisfies*

$$P[Y_A \geqslant y] = P[Y > 0]\bar{H}_{cd}(y) \tag{3.2.5}$$

with $c = a(a-b)/(b-a^2)$, $d = (1-a)(a-b)/(b-a^2)$, and $\bar{H}_{cd}$ the tail function of a beta distribution with parameters c and d, then the first two moments of Y and Y_A agree.

To apply (3.2.5) to $\bar{E}^2_{01}$, its first two conditional moments under H_0 are needed. However, for any set of level probabilities, under H_0, the rth moment of $\bar{E}^2_{01}$ is given by

$$\sum_{l=1}^{k} P(l,k;\mathbf{w}) \frac{(l-1)(l+1)\cdots(l+2r-3)}{(N-1)(N+1)\cdots(N+2r-3)}$$

with N the total sample size. Hence, the rth moment of $\bar{E}^2_{01}$ is the rth moment of $\bar{\chi}^2_{01}$ divided by $(N-1)(N+1)\cdots(N+2r-3)$. In particular, for the simple order with equal weights, the first two conditional moments of $\bar{E}^2_{01}$ under H_0 are

$$\frac{\kappa_1(k,1)}{(N-1)p} \quad \text{and} \quad \frac{\kappa_2(k,1)+[\kappa_1(k,1)]^2}{(N-1)(N+1)p} \tag{3.2.6}$$

with $p = P[\bar{E}^2_{01} > 0] = 1 - 1/k$. For the simple tree, replace $\kappa_1(k,1)$, $\kappa_2(k,1)$, and p by $\tau_1(k,1)$, $\tau_2(k,1)$, and $1 - P_T(1,k)$ with $P_T(1,k)$ the equal weights probability of one level set for the simple tree. Therefore, for the simple order or the simple tree with equal weights the moments needed to implement (3.2.5) can be obtained from the cumulants given in Tables A.16 and A.17. Sasabuchi and Kulatunga (1985) and Singh and Wright (1986b) investigated the accuracy of these approximations and found them to be more than adequate for moderate k and α.

In considering the distribution of $\bar{E}^2_{12}$, it should be noted that the denominators of the moments of the beta variables involved in these distributions depend on l and hence a different approach is needed. Recall that $\bar{E}^2_{12} \sim \bar{\chi}^2_{12}/(\bar{\chi}^2_{12} + Q)$, $Q \sim \chi^2_{N-k}$, and Q and $\bar{\chi}^2_{12}$ are independent. Therefore, if h is the chi-squared density with $N-k$ degrees of freedom, then using (3.2.1) under H_0 with $c > 0$,

$$\begin{aligned} P[\bar{E}^2_{12} \geqslant c] &= P\left[\frac{\bar{\chi}^2_{12}}{Q} \geqslant \frac{c}{1-c}\right] = \int_0^\infty P\left[\bar{\chi}^2_{12} \geqslant \frac{cu}{1-c}\right] h(u)\,du \\ &= P[\bar{\chi}^2_{12} > 0] \int_0^\infty \bar{G}_b\left(\frac{cu}{(1-c)\rho}\right) h(u)\,du \\ &= P[\bar{\chi}^2_{12} > 0] \bar{H}_{b,(N-k)/2}\left(\frac{2c}{\rho + (2-\rho)c}\right). \end{aligned} \tag{3.2.7}$$

Therefore, to approximate an equal-weights p value for $\bar{E}_{12}^2$ for a simple order (simple tree), determine ρ and b from the appropriate table, compute $P[\bar{\chi}_{12}^2 > 0] = 1 - P_S(k, k)(P[\bar{\chi}_{12}^2 > 0] = 1 - P_T(k, k))$ and use (3.2.7). Singh and Wright (1986b) found the percentage errors for these approximations to be even smaller than those for $\bar{E}_{01}^2$.

3.3 INCREASED PRECISION IN ONE OF THE SAMPLES

If one wishes to compare increasing dosage levels of a drug with a zero dose control and believes that the mean response is nondecreasing over the dosage levels studied, then the simple order $\mu_1 \leqslant \mu_2 \leqslant \cdots \leqslant \mu_k$ is of interest with μ_1 the control mean. In such situations, it is not uncommon to have larger sample sizes on the control. Chase (1974) developed an approximation for the case in which $w_1 = n_1/\sigma_1^2$, the precision of $\bar{Y}_1$, is larger than the other weights, which are assumed (nearly) equal. Using the techniques developed by Chase approximations are developed for the situation in which one of the weights is appreciably larger than the others for the simple order as well as the simple tree order.

Increased precision in one of the samples: simply ordered case

Consider the simple order with $w_1 > w_2 = w_3 = \cdots = w_k$. For both cases, that of known or unknown variances, Chase proposed interpolating between the equal-weights case and that obtained by letting $w_1 \to \infty$. For $l = 1, 2, \ldots, k$, let

$$Q(l, k) = \lim_{w_1 \to \infty} P_S(l, k; \mathbf{w}) \tag{3.3.1}$$

provided the limit exists.

Theorem 3.3.1 *Let $w_2 = w_3 = \cdots = w_k$. For each k and $1 \leqslant l \leqslant k$, the limit of $P_S(l, k; \mathbf{w})$ as $w_1 \to \infty$ exists and*

$$Q(l, k) = \sum_{v=0}^{k-l} Q(1, v+1) \frac{P_S(l-1, k-1-v)}{2^{l-1}}, \qquad l = 2, 3, \ldots, k, \tag{3.3.2}$$

with $Q(1, 1) = 1$ and $P_S(l-1, k-1-v)$ the equal weights probability for a simple order.

Proof: First it is shown that if $Y_1, Y_2, \ldots Y_l$ are independent normal variables with mean zero, then as σ_1^2, the variance of Y_1, approaches zero, $P[Y_1 < Y_2 < \cdots < Y_l] \to P[Y_2 < \cdots < Y_l]/2^{l-1}$. As $\sigma_1^2 \to 0$, $P[Y_1 < Y_2 < \cdots < Y_l] \to P[0 < Y_2 < \cdots < Y_l] = P([Y_i > 0, i = 2, \ldots, l] \cap [Y_2 < \cdots < Y_l])$. The claim follows from the independence of the signs and the magnitudes of the normal variables. This result yields $Q(k, k) = P_S(k-1, k-1)/2^{k-1}$ for $k \geqslant 2$, that is (3.3.2) holds with $l = k$.

The existence of $Q(l,k)$, for $1 \leqslant l < k$, is established inductively. They clearly exist for $k = 2$. For $2 \leqslant l \leqslant k$, $P_S(l,k;\mathbf{w})$ is computed using the recursion formula (2.4.4). An element of $\mathscr{L}_{lk}$ ($\mathscr{L}_{lk}$ is defined before Theorem 2.4.1) can be written as $\{1,2,\ldots,v+1\}$, $B_2,\ldots,B_l$ with $0 \leqslant v \leqslant k-l$ and $B_2,\ldots,B_l$ a partition of $\{v+2,\ldots,k\}$ into nonempty level sets. Fix $v \in \{0,1,\ldots,k-l\}$ and note that by the induction hypothesis

$$\lim_{w_1 \to \infty} P_S(l,l;W_{B_1},W_{B_2},\ldots,W_{B_l}) \prod_{i=1}^{l} P_S(1,C_{B_i};\mathbf{w}(B_i))$$

$$= 2^{-l+1}Q(1,C_{B_1})P(\mathrm{Av}(B_2) < \cdots < \mathrm{Av}(B_l)) \prod_{i=2}^{l} P_S(1,C_{B_i};\mathbf{w}(B_i)).$$

Summing the right-hand side over $B_2,\ldots,B_l$ yields

$$\begin{aligned} &2^{-l+1}Q(1,v+1)P_S(l-1,k-v-1;(w_{v+2},w_{v+3},\ldots,w_k)) \\ &\quad = 2^{-l+1}Q(1,v+1)P_S(l-1,k-v-1). \end{aligned}$$

Finally summing over v yields (3.3.2). Thus, $Q(l,k)$ exists for $2 \leqslant l \leqslant k$ and satisfies (3.3.2). Of course, $\lim_{w_1 \to \infty} P_S(1,k;\mathbf{w})$ exists because $P_S(l,k;\mathbf{w})$ is a probability distribution for each $\mathbf{w}$. □

In the next result the generating function for $\{Q(l,k)\}$, $Q_k(s) = \sum_{l=1}^{k} Q(l,k)s^l$, is obtained and it is used in the corollary to obtain a recurrence relationship for the $Q(l,k)$. For real numbers a and b, with b a positive integer, $\binom{a}{b}$ is defined to be $a(a-1)(a-2)\cdots(a-b+1)/b!$ and $\binom{a}{0} = 1$.

Theorem 3.3.2 *The generating function for the $Q(l,k)$ is given by*

$$Q_k(s) = s\binom{\tfrac{1}{2}s + k - \tfrac{3}{2}}{k-1}. \tag{3.3.3}$$

Proof: Using (3.3.2), the fact that $Q(l,k)$ is a probability distribution for each k, and interchanging the order of summation,

$$\begin{aligned} Q(1,k) &= 1 - \sum_{v=0}^{k-2} Q(1,v+1) \sum_{l=2}^{k-v} P_S(l-1,k-1-v)2^{-l+1} \\ &= 1 - \sum_{v=0}^{k-2} Q(1,v+1)P_{k-v-1}(\tfrac{1}{2}). \end{aligned}$$

Recalling that $P_k(s) = s(s+1)\cdots(s+k-1)/k! = \binom{s+k-1}{k}$ and applying (12.4) of Feller (1968, Chapter 2), this becomes

$$1 - \sum_{v=0}^{k-2} Q(1,v+1)(-1)^{k-v-1}\binom{-\tfrac{1}{2}}{k-v-1}.$$

Consider the generating function of $Q(1,k)$, $\psi(s)=\sum_{k=1}^{\infty}Q(1,k)s^k$. Using the expression above for $Q(1,k)$ and interchanging the order of summation, one can see that

$$\psi(s)=s(1-s)^{-1}-\sum_{v=0}^{\infty}Q(1,v+1)s^{v+1}\sum_{k=v+2}^{\infty}\binom{-\frac{1}{2}}{k-v-1}(-s)^{k-v-1}$$

$$=s(1-s)^{-1}-\psi(s)[(1-s)^{-1/2}-1].$$

Thus, $\psi(s)=s(1-s)^{-1/2}$ and so

$$Q(1,k)=\binom{-\frac{1}{2}}{k-1}(-1)^{k-1}=\binom{k-\frac{3}{2}}{k-1}.$$

Using this formula for $Q(1,k)$ and (3.3.2),

$$Q_k(s)=s\binom{k-\frac{3}{2}}{k-1}+s\sum_{l=2}^{k}\sum_{v=0}^{k-l}\binom{v-\frac{1}{2}}{v}P_S(l-1,k-v-1)\left(\frac{s}{2}\right)^{l-1}$$

$$=s\sum_{v=0}^{k-1}\binom{v-\frac{1}{2}}{v}\binom{\frac{1}{2}s+k-v-2}{k-v-1}.$$

The identity (12.16) of Feller (1968, Chapter 2) gives the desired expression for $Q_k(s)$. □

Corollary *For* $k=2,3,\ldots,$

$$Q(1,k)=\frac{\frac{1}{2}(2k-3)Q(1,k-1)}{k-1}=\frac{(2k-2)!}{2^{2k-2}[(k-1)!]^2}$$

$$Q(k,k)=\frac{Q(k-1,k-1)}{2k-2}=[2^{k-1}(k-1)!]^{-1}$$

$$Q(l,k)=\frac{1}{2k-2}Q(l-1,k-1)+\frac{2k-3}{2k-2}Q(l,k-1),\qquad \textit{for } l=2,3,\ldots,k-1. \tag{3.3.4}$$

Proof: The desired results follow from the identity

$$Q_k(s)=\frac{\frac{1}{2}s+k-\frac{3}{2}}{k-1}s\binom{\frac{1}{2}s+k-\frac{5}{2}}{k-2}=\frac{s+2k-3}{2k-2}Q_{k-1}(s)$$

$$=\frac{s(2k-3)Q(1,k-1)}{2k-2}+\sum_{l=2}^{k-1}s^l\frac{Q(l-1,k-1)+(2k-3)Q(l,k-1)}{2k-2}$$

$$+\frac{s^kQ(k-1,k-1)}{2k-2}.\ \square$$

The $Q(l,k)$, which are readily obtained from this recurrence relationship, are given in Table A.12 in the Appendix for $k\leqslant 20$. Chase considered interpolating between the critical values based on the equal weights $P_S(l, k)$ and the $Q(l, k)$, and he found that interpolation on $w^{-1/2}$ with $w = w_1/w_2$ worked quite well. In particular, if e_α is the α-level critical value for the equal-weights distribution, i.e. the distribution determined by the $P_S(l, k)$, and $c_\alpha^{(\infty)}$ is the α-level critical value for the limiting distribution, i.e. the distribution determined by the $Q(l, k)$, then the approximate α-level critical value is given by

$$c_\alpha^{(\infty)} + w^{-1/2}(e_\alpha - c_\alpha^{(\infty)}). \tag{3.3.5}$$

Robertson and Wright (1982b) found that the interpolation scheme also yields a reasonable approximation for p values. (The reader may wish to consult those references for specific numerical comparisons.) The tests of H_0 versus H_1–H_0 and H_1 versus H_2–H_1, with variances known or unknown, involve the same $P_S(l, k; \mathbf{w})$, and so this approach can be used for each of these tests. Tables A.6 and A.7 in the Appendix give the α-level critical value for $S_{01}(v) = v\bar{E}_{01}^2(v)/[1 - \bar{E}_{01}^2(v)]$ and $S_{12}(v) = v\bar{E}_{12}^2(v)/[1 - \bar{E}_{12}^2(v)]$ for $\alpha = 0.10$, 0.05, 0.01, and various values of v. For $w = 1$ the values given are the equal-weights critical values and for $w = \infty$ they are the limiting values, i.e. those corresponding to $Q(l, k)$. Thus, to obtain an approximate critical value interpolate between the values in Table A.6 for testing H_0 versus H_1–H_0 and between the values in Tables A.7 for testing H_1 versus H_2–H_1. If a p value is desired, then interpolate between the p values computed using $P_S(l, k)$ and $Q(l, k)$. The equal-weights p value could be approximated using the two-moment gamma or beta approximations discussed earlier in this section. It would be convenient to have two-moment approximations available for the p values based on $Q(l, k)$. Using the probability generating function of the $Q(l, k)$ given in (3.3.3), the characteristic function of $\bar{\chi}_{01}^2$ with these level probabilities is

$$\sum_{l=1}^{k} Q(l, k)(1 - 2it)^{-(l-1)/2} = \begin{bmatrix} \frac{1}{2}z + k - \frac{3}{2} \\ k - 1 \end{bmatrix} \text{with } z = (1 - 2it)^{-1/2}. \tag{3.3.6}$$

Taking logarithms and differentiating, the first two cumulants are found to be

$$\kappa_1'(k, 1) = \frac{1}{2}\sum_{j=1}^{k-1} j^{-1} \quad \text{and} \quad \kappa_2'(k, 1) = \frac{1}{4}\sum_{j=1}^{k-1} (6j^{-1} - j^{-2}). \tag{3.3.7}$$

Under H_0 with level probabilities $Q(l, k)$, the characteristic function for $\bar{\chi}_{12}^2$ is

$$\sum_{l=1}^{k} Q(l, k)(1 - 2it)^{-(k-l)/2} = \begin{bmatrix} \frac{1}{2}u + k - \frac{3}{2} \\ k - 1 \end{bmatrix} \Big/ u^{k-1} \text{ with } u = (1 - 2it)^{1/2}. \tag{3.3.8}$$

Taking logarithms and differentiating, the first two cumulants are found to be

$$\kappa_1'(k, 2) = k - 1 - \frac{1}{2}\sum_{j=1}^{k-1} j^{-1} \text{ and } \kappa_2'(k, 2) = 2(k-1) - \frac{1}{4}\sum_{j=1}^{k-1} \{2j^{-1} + j^{-2}\}. \tag{3.3.9}$$

For $i = 1, 2$ and $2 \leqslant k \leqslant 40$, Table A.18 contains the values of $\kappa_i'(k, 1)$ and $\kappa_i'(k, 2)$, respectively. Replacing $\kappa_i(k, 1)$ and $\kappa_i(k, 2)$ by $\kappa_i'(k, 1)$ and $\kappa_i'(k, 2)$ in the two-moment approximation for the null distributions of $\bar{\chi}^2_{01}$, $\bar{\chi}^2_{12}$, $\bar{E}^2_{01}$, and $\bar{E}^2_{12}$ for the simple order and equal weights, one obtains two-moment approximation for the simply ordered case with level probabilities $Q(l, k)$.

Because of the insensitivity of the $P_S(l, k; \mathbf{w})$ to slight variations in the weights, one would conjecture that if $w_2, w_3, \ldots, w_k$ are nearly the same, then Chase's approximation could be applied with $w = (k-1)w_1/\sum_{i=2}^k w_i$ provided w_1 is larger than the average of $w_2, \ldots, w_k$. Clearly,

$$-\{\mathbf{x} \in R^k: x_1 \leqslant x_2 \leqslant \cdots \leqslant x_k\} = \{\mathbf{x} \in R^k: x_1 \geqslant x_2 \geqslant \cdots \geqslant x_k\}.$$

It was noted earlier that $P_{\mathbf{w}}(\mathbf{x} \mid -\mathscr{K}) = -P_{\mathbf{w}}(-\mathbf{x} \mid \mathscr{K})$ and so the level probabilities for the reverse of a quasi-order are the same as for the quasi-order. Therefore,

$$Q(l, k) = \lim_{w_k \to \infty} P_S(l, k; (w_1, w_2, \ldots, w_k))$$

if $w_1 = w_2 = \cdots = w_{k-1}$. Hence, the results discussed above can be applied to the case in which $w_1 = w_2 = \cdots = w_{k-1} < w_k$. For $i = 1, 2, \ldots, k$, let all of the weights be equal except for w_i and define

$$Q_i(l, k) = \lim_{w_i \to \infty} P_S(l, k; (w_1, w_2, \ldots, w_k)), \qquad 1 \leqslant l \leqslant k$$

provided this limit exists. It has just been shown that $Q_k(l, k) = Q_1(l, k)$, which by definition is the same as $Q(l, k)$. Consider $1 < i < k$ with $w_1 = \cdots = w_{i-1} = w_{i+1} = \cdots w_k$ and recall that one may assume $\mu_1 = \mu_2 = \cdots = \mu_k = 0$ in computing the level probabilities. Intuitively, as $w_i \to \infty$ (or $\sigma_i^2/n_i \to 0$), the variable with variance w_i^{-1} converges to its mean and because the weight on this variable is infinite, in the limit, it is unchanged in the pooling process. Hence, in the limit, the restricted MLE is $(\mu_1^*, \ldots, \mu_{i-1}^*, 0, \mu_{i+1}^*, \ldots, \mu_k^*)$ with $(\mu_1^*, \ldots, \mu_{i-1}^*)$ $((\mu_{i+1}^*, \ldots, \mu_k^*))$ the projection onto $\{(x_1, \ldots, x_{i-1}): x_1 \leqslant x_2 \leqslant \cdots \leqslant x_{i-1} \leqslant 0\}$ $(\{(x_{i+1}, \ldots, x_k): 0 \leqslant x_{i+1} \leqslant \cdots \leqslant x_k\})$. Thus, in the limit, the number, M, of distinct values in the restricted MLE is M_1, the number of distinct values in $(\mu_1^*, \mu_2^*, \ldots, \mu_{i-1}^*, 0)$, plus M_2, the number of distinct values in $(0, \mu_{i+1}^*, \ldots, \mu_k^*)$, minus one. Again it is intuitively clear that $P[M_1 = l] = \lim_{w_i \to \infty} P_S(l, i; (w_1, w_2, \ldots, w_i)) = Q(l, i)$ and $P[M_2, l] = \lim_{w_i \to \infty} P_S(l, k-i+1; (w_i, w_{i+1}, \ldots, w_k)) = Q(l, k-i+1)$. Because M_1 and M_2 are independent, $Q_i(l, k)$ is the term of the convolution of $\{Q(v, i)\}$ and $\{Q(v, k-i+1)\}$ with $v = l + 1$. Therefore the PGF of $\{Q_i(l, k)\}$ is given by

$$Q^{k,i}(s) = \sum_{l=1}^{k} Q_i(l, k)s^l = \frac{Q_i(s)Q_{k-i+1}(s)}{s} = s\begin{bmatrix} \frac{1}{2}s - \frac{3}{2} + i \\ i-1 \end{bmatrix}\begin{bmatrix} \frac{1}{2}s - \frac{1}{2} + k - i \\ k-i \end{bmatrix}.$$

Using the technique employed by Chase, a rigorous proof of this result is possible, but the basic ideas are contained in the sketch above.

Increased precision on the control: the simple tree ordering

In comparing several treatments with a control under the assumption that the treatments are at least as effective as the control, one is led to the simple tree ordering: $\mu_0 \leqslant \mu_i$ for all $i = 1, 2, \ldots, k$. In such situations, it is common to have appreciably larger sample sizes on the control. Chase's idea of interpolating between the equal-weights and limiting p values or critical values provide useful approximations in this case also. For $l = 1, 2, \ldots, k+1$, let

$$R(l, k+1) = \lim_{w_0 \to \infty} P_T(l, k+1; (w_0, w_1, \ldots, w_k))$$

provided this limit exists. The reader is reminded that the level probabilities, $P_T(l, k+1; \mathbf{w})$, are computed under the assumption that $\mu_0 = \mu_1 = \cdots = \mu_k$. (Unlike the simply ordered case, it is not necessary to assume that all of the weights, except the one which is approaching infinity, are equal.) Intuitively, as $w_0 \to \infty$, the observation on the control becomes degenerate at the control mean. In the pooling process, if a treatment observation is amalgamated with the control, in the limit, because the control has infinite weight, the pooled value is equal to the control value. Hence, $R(l, k+1)$ is the probability that exactly $l-1$ of the treatment observations exceed the control with all of the treatment means equal to the control mean. This a heuristic justification of the next result. A more rigorous proof could be given, but it provides no new insights.

Theorem 3.3.3 *If $w_i \in (0, \infty)$ is fixed for $i = 1, 2, \ldots, k$, then $\lim_{w_0 \to \infty} P_T(l, k+1; \mathbf{w})$ exists and*

$$R(l, k+1) = \lim_{w_0 \to \infty} P_T(l, k+1; \mathbf{w}) = \binom{k}{l-1}\left(\frac{1}{2}\right)^k, \qquad l = 1, 2, \ldots, k+1. \qquad (3.3.10)$$

If one desires an approximate p value for $\bar{\chi}^2_{01}$, $\bar{\chi}^2_{12}$, $\bar{E}^2_{01}$, or $\bar{E}^2_{12}$ for the simple tree ordering with weights $w_0 > w_1 = w_2 = \cdots = w_k$, then one interpolates between the p values given by (2.3.4) with $P_T(l, k+1; \mathbf{w})$ replaced by the equal-weights level probabilities, $P_T(l, k+1)$, given in Table A.11 in the Appendix and the p values given by (2.3.4) with $P_T(l, k+1; \mathbf{w})$ replaced by $R(l, k+1)$ given in (3.3.10). Interpolation on $w^{-1/2}$ with $w = w_0/w_1$ is recommended (cf. (3.3.5)). If k is moderate or large, one may wish to approximate the equal-weights and the limiting p values using the two-moment approximation given in either (3.2.1) (or (3.2.5)). The first two cumulants (moments) of the null distributions of $\bar{\chi}^2_{01}$ and $\bar{\chi}^2_{12}$ are needed. However, for the special coefficients given in (3.3.10), $\bar{\chi}^2_{01}$ and $\bar{\chi}^2_{12}$ have the same null *distributions, which has the characteristic function*

$$2^{-k} \sum_{v=0}^{k} \binom{k}{v} (1 - 2it)^{-v/2} = 2^{-k}[1 + (1 - 2it)^{-1/2}]^k.$$

The first two cumulants are easily found to be $k/2$ and $5k/4$, respectively. Under H_0 with level probabilities, $R(l, k+1)$, $P[\bar{\chi}^2_{01} = 0] = P[\bar{\chi}^2_{12} = 0] = P[\bar{E}^2_{01} = 0] =$

$P[\bar{E}^2_{12}=0]=(\frac{1}{2})^k$. Thus, with $p=1-(\frac{1}{2})^k$, one can obtain the cumulants of the conditional distribution of $\bar{\chi}^2_{01}$ given $\bar{\chi}^2_{01}>0$, which are the same as those of $\bar{\chi}^2_{12}$ given $\bar{\chi}^2_{12}>0$, using (3.2.3) or the conditional moments of $\bar{E}^2_{01}$ given $\bar{E}^2_{01}>0$ using (3.2.6). (Recall that for $\bar{E}^2_{12}$ one uses (3.2.7) with the ρ and b computed for $\bar{\chi}^2_{12}$.)

Critical values for this same situation, i.e. the simple tree ordering with $w_0>w_1=w_2=\cdots=w_k$, can be obtained by interpolating between the equal-weights critical values and the limiting critical values. For $\alpha=0.10, 0.05$, and 0.01, $k=2,3,\ldots,10$ and various degrees of freedom, the equal-weights and limiting critical values for $S_{01}(v)=v\bar{E}^2_{01}(v)/[1-\bar{E}^2_{01}(v)]$ and $S_{12}(v)=v\bar{E}^2_{12}(v)/[1-\bar{E}^2_{12}(v)]$ are given in Tables A.8 and A.9 in the Appendix. The rows labeled $v=\infty$ give the critical values for $\bar{\chi}^2_{01}$ and $\bar{\chi}^2_{12}$. With v degrees of freedom, interpolation on $(v+k-3)^{-1}$ and $w^{-1/2}$ with $w=w_0/w_1$ is recommended.

Specific numerical comparisons are given in Robertson and Wright (1985), and it is noted there that these approximations work reasonably well if the $w_1, w_2, \ldots, w_k$ are not too different and w is set equal to $kw_0/\sum_{i=1}^k w_i$.

3.4 APPROXIMATIONS TO THE NULL $\bar{\chi}^2$ AND $\bar{E}^2$ DISTRIBUTIONS WITH ARBITRARY WEIGHTS AND THE SIMPLE ORDER

In this section approximations for the null $\bar{\chi}^2$ and $\bar{E}^2$ distributions are developed for the case of a simple order with arbitrary weights. The approximation discussed first is based on the pattern of the large and small weights. Siskind (1976) suggests an approximation which is based on a Taylor expansion about the equal-weights values and tends to be conservative. However, if there is not too much variation in the weights, for instance if the ratio of the largest to smallest weight does not exceed 1.4, then because of the robustness of the $P_S(l,k;\mathbf{w})$ to $\mathbf{w}$, the equal-weights approximation seems to be adequate (see Section 3.1). Furthermore, Chase's idea can be used if only one of the weights is significantly larger than the others.

An approximation based on patterns: simply ordered case

The approximation developed in this subsection is based on the pattern of large and small weights. Each weight is classified as either large or small and the limiting p value or critical value is obtained for the case in which each small weight is assumed to have value one and the large weights, which are assumed to be equal, grow without bound. The approximate p value or critical value is obtained by interpolating between the corresponding equal-weights and limiting values. As will be noted later in this section, this scheme also provides a reasonable approximation to the level probabilities, $P_S(l,k;\mathbf{w})$.

Consider an arbitrary k, $1\leqslant j\leqslant k$ and $1\leqslant i_1<i_2<\cdots<i_j\leqslant k$. Define a weight vector $\mathbf{w}$ with $w_{i_\alpha}=1$ for $\alpha=1,2,\ldots,j$ and $w_i=w$ for $i\in\{1,2,\ldots,k\}-\{i_1,i_2,\ldots,i_j\}$

with w a positive number. The limiting level probabilities used in the approximation are

$$\tilde{P}_S(l, k) = \lim_{w \to \infty} P_S(l, k; \mathbf{w}), \qquad 1 \leqslant l \leqslant k.$$

Let $I = k - j$ (the number of large weights) and let $A(B)$ be the number of consecutive weights on the left (right) that is equal to one, that is

$$w_1 = w_2 = \cdots = w_A = w_{k-B+1} = \cdots = w_k = 1 \qquad \text{and} \qquad w_{A+1} = w_{k-B} = w.$$

(Of course, A or B may be zero.) As $w \to \infty$, the variables with variance w^{-1} converge to their means and because the weight placed on them becomes infinite, they are left unchanged in the pooling process. Now M, the number of distinct values in the restricted MLE, is $M_1 + M_2 + M_3 - 2$, where M_1 is the number of distinct values $(\mu_1^*, \ldots, \mu_{A+1}^*)$, M_2 is the number of distinct values in $(\mu_{A+1}^*, \ldots, \mu_{k-B}^*)$, and M_3 is the number of distinct values in $(\mu_{k-B}^*, \ldots, \mu_k^*)$. Since, in the limit, μ_{A+1}^* and μ_{k-B}^* are forced to be constants, M_1, M_2, and M_3 are independent in the limit. Under H_0, $\lim_{w\to\infty} P[M_1 = l] = Q(l, A+1)$ and $\lim_{w\to\infty} P[M_3 = l] = Q(l, B+1)$. Next, $\lim_{w\to\infty} P[M_2 = l]$ is computed. Since, μ_{A+1}^* and μ_{k-B}^* converge to the same limit, one might conjecture that $M_2 \to 1$. However, this is not necessarily the case. For instance, if $k - B = A + 2$, then for each value of w, $P[M_2 = 1] = P[M_2 = 2] = \frac{1}{2}$. If $w_i = 1$ with $A + 1 < i < k - B$, then with probability approaching one, Y_i will be less than Y_{A+1} or greater than Y_{k-B}, which are approaching the common mean. (Recall that the level probabilities are determined by independent normal variables with a common mean, $Y_1, Y_2, \ldots, Y_k$, with the variance of Y_α equal to w_α^{-1}.) Hence, each such Y_i will be amalgamated with a Y_α with $w_\alpha = w$ and so in the limit the number of distinct values will have the same distribution as the number of level sets for equal weights, a simple order, and I populations; that is $\lim_{w\to\infty} P[M_2 = l] = P_S(l, I)$. Therefore, the PGF of the $\tilde{P}_S(l, k)$ is given by

$$\begin{aligned} \tilde{P}_k(s) &= s^{-2} Q_{A+1}(s) P_I(s) Q_{B+1}(s) \\ &= \begin{bmatrix} \frac{1}{2}s + A - \frac{1}{2} \\ A \end{bmatrix} \begin{bmatrix} s + I - 1 \\ I \end{bmatrix} \begin{bmatrix} \frac{1}{2}s + B - \frac{1}{2} \\ B \end{bmatrix}. \end{aligned} \tag{3.4.1}$$

One can compute the $\tilde{P}_S(l, k)$ using (3.4.1) or by taking the term with $v = l + 2$ in the convolution of $\{Q(v, A+1)\}$, $\{P_S(v, I)\}$, and $\{Q(v, B+1)\}$. Based on a numerical study, Robertson and Wright (1983) recommend calling a weight large if it exceeds $\tilde{w} = 0.65 \min\{w_1, w_2, \ldots, w_k\} + 0.35 \max\{w_1, w_2, \ldots, w_k\}$ and interpolating on λ, the $\frac{1}{3}$ power of the ratio of the average of the small weights to the average of the large weights. The limiting p value or critical value is obtained from (2.3.4) with $P(l, k; \mathbf{w})$ replaced by $\tilde{P}_S(l, k)$. The approximation is used to determine an $\alpha = 0.05$ critical value for $\bar{\chi}_{01}^2$ in the next example.

Example 3.4.1 Consider the usual simple order with $k = 5$ and $w_1 = (1.4, 5.6, 1.2, 1.0, 4.7)$. In this case, $\tilde{w} = 2.61$, the second and last weights are classified as large, the other three weights are small, and $\lambda = 0.6154$. Hence, $A = 1$, $B = 0$, $I = 2$, and

$$\tilde{P}_5(s) = \tfrac{1}{4}s^3 + \tfrac{1}{2}s^2 + \tfrac{1}{4}s.$$

Therefore, for $\bar{\chi}^2_{01}$ the 0.05 limiting critical value is 4.2306, the equal-weights critical value is 5.0491, and the approximate 0.05 critical value is 4.7343. The exact $P_S(l, k; \mathbf{w})$ can be computed and the true significance level for $\bar{\chi}^2_{01}$ corresponding to the approximate 0.05 critical value, 4.7343, is 0.0520 and the true significance level corresponding to the equal-weights 0.05 critical value, 5.0491, is 0.0445.

Using the 10 000 pseudo-random weight sets, $\mathbf{w} = (U_1, U_2, \ldots, U_5)$, which were generated to assess the performance of the equal-weights approximation in the simply ordered case, the behavior of this approximation was studied and it was compared with the equal-weights approximation. Recall that these weight sets were obtained by generating 50 000 uniform random variables on the interval (0, 1). With the approximate 0.05 critical value denoted by $a_{0.05}$, $P_{\mathbf{w}}[\bar{\chi}^2_{01} \geqslant a_{0.05}]$ was computed for these weight sets. These values are given in Table 3.4.1. The corresponding values for the equal-weights approximation are given in parentheses. For the approximation based on patterns, 98 percent of the values of $P_{\mathbf{w}}[\bar{\chi}^2_{01} \geqslant a_{0.05}]$ are in [0.04, 0.06] and for the equal-weights approximation 74 percent of these values are in [0.04, 0.06]. The interval [0.048, 0.052] contains 46 percent of the values of $P_{\mathbf{w}}[\bar{\chi}^2_{01} \geqslant a_{0.05}]$ but only 19 percent of $P_{\mathbf{w}}[\bar{\chi}^2_{01} \geqslant e_{0.05}]$. For $k = 5$, all of the weight sets with $P_{\mathbf{w}}[\bar{\chi}^2_{01} \geqslant a_{0.05}]$ outside of [0.04, 0.06] had the ratio of the largest weight to the smallest of at least 4.7. (The corresponding value for the equal-weights approximation is 3.4.) In order to assess the performance of the approximation for larger k, 1000 weight sets were generated with $k = 10$, and $a_{0.05}$ was computed for each weight set. Since the true value

Table 3.4.1 Frequency distribution for $P_{\mathbf{w}}[\bar{\chi}^2_{01} \geqslant a_{0.05}]$ with 10 000 randomly generated weight sets, $k = 5$ and $a_{0.05}$ computed using the approximation

INTERVAL	FREQUENCY[a]	MINIMUM[a] RATIO	INTERVAL	FREQUENCY[a]	MINIMUM[a] RATIO
[0.020, 0.024]	0 (4)	— (20.8)	[0.052, 0.056]	2549 (1561)	1.5 (1.4)
[0.024, 0.028]	0 (22)	— (18.3)	[0.056, 0.060]	492 (1116)	3.3 (2.2)
[0.028, 0.032]	1 (73)	79.3 (8.0)	[0.060, 0.064]	109 (670)	5.4 (3.4)
[0.032, 0.036]	18 (284)	18.9 (5.2)	[0.064, 0.068]	28 (357)	21.2 (5.0)
[0.036, 0.040]	86 (719)	4.7 (3.4)	[0.068, 0.072]	3 (145)	122.5 (6.9)
[0.040, 0.044]	406 (1319)	3.5 (2.0)	[0.072, 0.076]	1 (54)	1238.1 (14.9)
[0.044, 0.048]	1730 (1718)	1.5 (1.4)	[0.076, 0.080]	0 (17)	— (34.5)
[0.048, 0.052]	4577 (1876)	1.1 (1.1)	[0.080, 0.084]	0 (2)	— (812.0)

[a] The values in parentheses are for $P_{\mathbf{w}}[\bar{\chi}^2_{01} \geqslant e_{0.05}]$.

$P[\bar{\chi}_{01}^2 \geq a_{0.05}]$ cannot be calculated for an arbitrary weight set, this value was estimated by Monte Carlo techniques with 4000 iterations. The proportion of the estimates of $P_{\mathbf{w}}[\bar{\chi}_{01}^2 \geq a_{0.05}]$ in [0.04, 0.06] was 98 percent, the same as for $k = 5$, and the proportion in [0.048, 0.052] fell from 46 to 37 percent. It seems that even with $k = 10$ the approximation would be adequate for many practical purposes, and in general it outperforms the equal-weights approximation.

If one wishes to approximate the level probabilities for arbitrary weights with the usual simple order,

$$P_a(l, k; \mathbf{w}) = \tilde{P}_S(l, k) + \lambda[P_S(l, k) - \tilde{P}_S(l, k)]$$

could be used. To assess the accuracy of this approximation, three cases are considered. In Section 3.1, the Monte Carlo estimate of $P_S(l, k; \mathbf{w})$ based on 4000 iterations were reported for three weight vectors. For case 1, $k = 10$ and $\mathbf{w} = (147, 175, 150, 191, 188, 179, 200, 220, 208, 330)$; for case 2, $k = 12$ and $\mathbf{w} = (155, 178, 215, 93, 128, 114, 172, 158, 197, 207, 95, 183)$; and for case 3, $k = 15$ and $\mathbf{w} = (55, 20, 35, 45, 40, 25, 55, 30, 60, 45, 35, 25, 45, 35, 30)$. With the Monte Carlo estimates denoted by $P_M(l, k; \mathbf{w})$, these values are reported in Table 3.4.2. In case 1, the maximum absolute difference of $|P_a(l, 10; \mathbf{w}) - P_M(l, 10; \mathbf{w})|$ is 0.0153 and occurs when $l = 2$. The equal-weights $P_S(l, k)$ also agree closely with the $P_M(l, 10; \mathbf{w})$ in this case and, in fact, they fit better than the approximation studied in this section. However, both are clearly adequate and, in general, this approximation outperforms the equal-weights approximation. In case 2, the maximum absolute difference in the approximate and Monte Carlo values is 0.0075. In case 3, the corresponding value is 0.0079. In cases 2 and 3, the approximation based on

Table 3.4.2 Comparison of $P_a(l, k; \mathbf{w})$ with $P_M(l, k; \mathbf{w})$, the Monte Carlo estimates of $P(l, k; \mathbf{w})$, for three cases

	CASE 1		CASE 2		CASE 3	
l	$P_M(l, 10; \mathbf{w})$	$P_a(l, 10; \mathbf{w})$	$P_M(l, 12; \mathbf{w})$	$P_a(l, 12; \mathbf{w})$	$P_M(l, 15; \mathbf{w})$	$P_a(l, 15; \mathbf{w})$
1	0.09425	0.09555	0.08600	0.09011	0.07175	0.06388
2	0.28975	0.27445	0.26525	0.26344	0.21800	0.21254
3	0.32600	0.32092	0.32425	0.31680	0.30000	0.30041
4	0.19850	0.20498	0.20425	0.21131	0.23950	0.24174
5	0.07300	0.08018	0.09125	0.08831	0.11900	0.12459
6	0.01600	0.02021	0.02250	0.02462	0.03875	0.04375
7	0.00225	0.00333	0.00575	0.00472	0.01175	0.01085
8	0.00025	0.00035	0.00075	0.00063	0.00125	0.00195
9	0.00000	0.00002	0.00000	0.00006	0.00000	0.00026
10[a]	0.00000	0.00000	0.00000	0.00000	0.00000	0.00002

[a]For $l > 10$, these values are zero to five places.

patterns in the weight set is more accurate than the equal-weights approximation.

For moderate k this approximation can be tedious to use. A Fortran program for implementing this approximation is given in Pillers, Robertson, and Wright (1984). The approximation for p values interpolates between the equal-weights p values and the limiting p value. A two-moment approximation for the equal-weights p value was discussed in Section 3.2. The same approach could be used to provide a two-moment approximation for the limiting p value. However, to obtain such an approximation for $\bar{\chi}^2_{01}$ or $\bar{E}^2_{01}$, one needs $\tilde{P}_S(1,k)$ and the cumulants of the $\bar{\chi}^2_{01}$ distribution determined by the limiting mixing coefficients $\tilde{P}_S(l,k)$, and for $\bar{\chi}^2_{12}$ or $\bar{E}^2_{12}$, the probability, $\tilde{P}_S(k,k)$, and the cumulants of the $\bar{\chi}^2_{12}$ distribution determined by the coefficients $P_S(l,k)$ are needed. First observe that

$$\tilde{P}_S(1,k) = Q(1, A+1)P_S(1,I)Q(1,B+1) = \frac{(2A)!(2B)!}{4^{A+B}(A!B!)I^2}$$

and

$$\tilde{P}_S(k,k) = Q(A+1, A+1)P_S(I,I)Q(B+1, B+1) = (2^{A+B}A!B!I!)^{-1}$$

provided $A + B + I = k$ and $\tilde{P}_S(k,k) = 0$ if $A + B + I < k$. The cumulant generating function for the $\bar{\chi}^2_{01}$ distribution determined by $\tilde{P}_S(l,k)$ is

$$\begin{aligned}\log\left[\sum_{l=1}^{k} \tilde{P}(l,k)(1-2it)^{-(l-1)/2}\right] &= \log[z^{-1}\tilde{P}_k(z)] \\ &= \log[z^{-1}Q_{A+1}(z)] + \log[z^{-1}P_I(z)] \\ &\quad + \log[z^{-1}Q_{B+1}(z)]\end{aligned}$$

with $z = (1-2it)^{-1/2}$. Therefore, the jth cumulant is given by

$$\kappa'_j(A+1,1) + \kappa^*_j(I,1) + \kappa'_j(B+1,1) \tag{3.4.2}$$

where expressions for $\kappa'_j(k,1)$ are given in (3.3.7) and numerical values are given in Table A.18 for $2 \leqslant k \leqslant 40$ with $j = 1, 2$. The values of $\kappa^*_i(I,1)$ may be obtained from (3.2.2) or Table A.16. For $\bar{\chi}^2_{01}$ or $\bar{E}^2_{01}$, set $p = 1 - \tilde{P}_S(1,k)$. To approximate the limiting p value for $\bar{\chi}^2_{01}$, obtain the cumulants of $\bar{\chi}^2_{01}$ given $\bar{\chi}^2_{01} > 0$ using (3.2.3) with κ^*_1 and κ^*_2 replaced by the first two cumulants in (3.4.2), compute ρ and b (their definitions follow (3.2.1)) and apply (3.2.1). To approximate the limiting p value for $\bar{E}^2_{01}$, obtain the moments of $\bar{E}^2_{01}$ given $\bar{E}^2_{01} > 0$ using (3.2.6), compute c and d (their definitions follow (3.2.5)), and apply (3.2.5).

The cumulant generating function for the $\bar{\chi}^2_{12}$ distribution determined by $\tilde{P}_S(l,k)$ is

$$\begin{aligned}&\log\left[\sum_{l=1}^{k} \tilde{P}_S(l,k)(1-2it)^{-(k-l)/2}\right] = \log[u^{-k}\tilde{P}_k(u)] \\ &= -(k - A - B - I - 2)\log u \\ &\quad + \log\left[\frac{Q_{A+1}(u)}{u^{A+1}}\right] + \log\left[\frac{P_I(u)}{u^I}\right] + \log\left[\frac{Q_{B+1}(u)}{u^{B+1}}\right]\end{aligned}$$

where $u = (1 - 2it)^{1/2}$, Q_{A+1} is given by (3.3.3) and P_I is given by (2.4.10). Therefore, the jth cumulant of this $\bar{\chi}^2_{12}$ distribution is

$$2^{j-1}(j-1)!(k-A-B-I-2) + \kappa_j'(A+1,2) + \kappa_j^*(I,2) + \kappa_j'(B+1,2)$$

with $\kappa_j^*(I,2)$ taken either from (3.2.4) or Table A.18 and $\kappa_j'(A+1,2)$ and $\kappa_j'(B+1,2)$ taken either from (3.3.9) or Table A.16. Using these cumulants and $p = 1 - \tilde{P}_S(k,k)$, the cumulants of $\bar{\chi}^2_{12}$ given $\bar{\chi}^2_{12} > 0$ can be computed using (3.2.3) and then ρ and b can be computed (their definition follows (3.2.1)). Approximate p values for $\bar{\chi}^2_{01}$ are given by (3.2.1) and approximate p values for $\bar{E}^2_{12}$ are given by (3.2.7). While the accuracy of these two-moment approximations has not been studied in these cases, it would seem that their behavior would not differ markedly from that in the equal-weights case. The results in Section 3.2 show that these approximations are quite acceptable in the case of equal weights.

Siskind's approximation: simply ordered case

The mixing coefficients, $P(l,k;\mathbf{w})$, depend upon the weights through a matrix, R, of correlations. In the simply ordered case, this matrix, which is tridiagonal, is defined by (2.4.2). If the weights are equal, then the elements on the two off-diagonals are $-\frac{1}{2}$. Siskind's (1976) approximation is based on a Taylor expansion of the level probabilities in terms of the correlations about $-\frac{1}{2}$. For $k \leqslant 8$, Siskind tables the first-order partial derivatives of the $P_S(l,k;\mathbf{w})$ with respect to the correlations at the equal-weights values.

Siskind noted that

1. his approximation tends to overstate the true significance level, i.e. it tends to be conservative (see the discussion in Robertson and Wright, 1983),
2. the configuration (pattern) of the weight set seems to have more effect on the accuracy than the configuration in the correlation coefficients, and
3. the approximation is very conservative for U-shaped weights. (The approximation based on pattern in the weight sets does quite well in this case; cf. Robertson and Wright, 1983.)

3.5 APPROXIMATIONS BASED ON PATTERNS FOR THE NULL $\bar{\chi}^2$ AND $\bar{E}^2$ DISTRIBUTIONS WITH ARBITRARY WEIGHTS AND THE SIMPLE TREE ORDERING

In this section an approximation based on the pattern of large and small weights is developed for the null $\bar{\chi}^2$ and $\bar{E}^2$ distributions for the simple tree ordering with arbitrary weights. If there is not too much variation in the weights, then because of its ease of implementation, one may wish to use the equal-weights approximation which was discussed in Section 3.1. If the weight on the control population, that is w_0, is appreciably larger than the other weights which are nearly equal, then the approximation based on the results in Theorem 3.3.3 could be used.

The outline followed in developing the approximation of this subsection is the same as in the development of an approximation based on patterns for the simply ordered case. Weights are classified as large or small, the limiting p value is obtained for the situation in which the small weights are equal and remain fixed while the large weights, which are equal, grow without bound, and the approximate p value is obtained by interpolating between the equal-weights and limiting p values. Approximate critical values are obtained similarly. Consider the simple tree ordering which imposes the restrictions $\mu_0 \leqslant \mu_i$ for all $i = 1, 2, \ldots, k$. The cases in which the control weight, w_0, is large or small are considered separately. Because the ordering is symmetric in the treatment means, that is $\mu_1, \mu_2, \ldots, \mu_k$, in both of these limiting cases, one only needs to know the number of $w_1, w_2, \ldots, w_k$ which are small.

For the case in which w_0 is large, let $t \in \{1, 2, \ldots, k\}$ be the number of the treatment weights, $w_1, w_2, \ldots, w_k$, which are small and define

$$\tilde{P}_T(l, k+1) = \lim_{w \to \infty} P_T(l, k+1; \mathbf{w})$$

where $w_0 = w_1 = \cdots = w_{k-t} = w$ and $w_{k-t+1} = \cdots = w_k = 1$, provided this limit exists. Let $Y_0, Y_1, \ldots, Y_k$ be independent normal variables with zero means and the variance of Y_i equal to w_i^{-1}. As $w \to \infty$, $Y_0, Y_1, \ldots, Y_{k-t}$ become degenerate at zero. Intuitively, with $k - t + 1 \leqslant j \leqslant k$, Y_j will be amalgamated with the control observation if $Y_j < 0$ and $\{j\}$ will be a level set if $Y_j > 0$. Hence, in the limit, the number of distinct values in $(\mu_0^*, \mu_1^*, \ldots, \mu_k^*)$ is the number of distinct values in the projection of $(Y_0, Y_1, \ldots, Y_{k-t})$ plus the number of $Y_j > 0$ with $k - t + 1 \leqslant j \leqslant k$. Thus $\tilde{P}_T(l, k+1)$ is the convolution of $\{P_T(l, k-t+1)\}$ and $\{B(l, t)\}$ where $B(l, t) = \binom{t}{l} 2^{-t}$. (Recall that the equal-weights level probabilities for the simple tree ordering, $P_T(l, k+1)$, are given in Table A.11 in the Appendix.) The limiting p values or critical values in this case can be determined from (2.3.4) with $P(l, k; \mathbf{w})$ replaced by $\tilde{P}_T(l, k+1)$.

For the case in which the control weight, w_0, is small, and t other weights are small with $0 \leqslant t < k$, define a weight vector with $w_0 = w_1 = \cdots = w_t = 1$, $w_{t+1} = \cdots = w_k = w$, and

$$\tilde{P}'_T(l, k+1) = \lim_{w \to \infty} P_T(l, k+1; \mathbf{w}), \qquad 1 \leqslant l \leqslant k+1.$$

The case in which there is only one large weight will be considered first. For $1 \leqslant l \leqslant k$, define

$$S(l, k+1) = \lim_{w \to \infty} P_T(l, k+1; (1, 1, \ldots, w)), \qquad 1 \leqslant l \leqslant k+1.$$

As $w \to \infty$, the observation on the last treatment becomes degenerate at the common mean, which may be assumed to be zero, which forces $\mu_0^* \leqslant 0$ in the limit because the weight on the last treatment is tending to infinity. Hence,

$$S(k+1, k+1) = \int_{-\infty}^{0} \{1 - \Phi(x)\}^{k-1} \phi(x)\,dx = \frac{\{1 - (\tfrac{1}{2})^k\}}{k} \tag{3.5.1}$$

and, of course, $S(1, k+1) = 1 - \sum_{l=2}^{k+1} S(l, k+1)$. Consider $2 \leqslant l \leqslant k$. Recall that if the projection onto the cone determined by the simple tree ordering (see Example 1.3.2) has l level sets they are of the form $\{0\} \cup A$, with A a subset of $\{1, 2, \ldots, k\}$ with cardinality $k - l + 1$ and $l - 1$ singletons. Applying the recursion formula, (2.4.4), and the fact that $P_T(l, k+1; \mathbf{w})$ is invariant under permutations of the treatment weights, $w_1, w_2, \ldots, w_k$,

$$S(l, k+1) = \lim_{w \to \infty} \binom{k-1}{k-l} P_T(l, l; (w+k-l+1, 1, \ldots, 1)) P_T(1, k-l+2; (1, \ldots, 1, w))$$

$$+ \lim_{w \to \infty} \binom{k-1}{k-l+1} P_T(l, l; (k-l+2, 1, \ldots, 1, w)) P_T(1, k-l+2).$$

As $w \to \infty$,

$$P_T(l, l; (w+k-l+1, 1, \ldots, 1)) \to B(l-1, l-1) = 2^{-l+1}$$

and

$$P_T(l, l; (k-l+2, 1, \ldots, 1, w)) \to \int_{-\infty}^{0} [1 - \Phi\{(k-l+2)^{-1/2} x\}]^{l-2} \phi(x)\,dx$$

$$= \int_{0}^{\infty} [\Phi\{(k-l+2)^{-1/2} x\}]^{l-2} \phi(x)\,dx.$$

This establishes the following result.

Theorem 3.5.1 *The mixing coefficients, $S(l, k+1)$, are determined by*

$$S(k+1, k+1) = \frac{\{1 - (\tfrac{1}{2})^k\}}{k}$$

$$S(1, k+1) = 1 - \sum_{l=2}^{k+1} S(l, k+1)$$

and

$$S(l, k+1) = \binom{k-1}{k-l} \frac{S(1, k-l+2)}{2^{l-1}}$$

$$+\binom{k-1}{k-l+1}P_T(1,k-l+2)$$

$$\times\int_0^\infty [\Phi\{(k-l+2)^{-1/2}x\}]^{l-2}\phi(x)\,\mathrm{d}x \qquad \textit{for } 2\leqslant l\leqslant k.$$

(3.5.2)

Using numerical integration, the $S(l,k+1)$ can be obtained recursively. For $2\leqslant k\leqslant 19$ they are given in Table A.13 in the Appendix.

Now, the case is considered in which the control weight is small and t of the treatment weights are small with $t<k-1$. Let $Y_0, Y_1,\ldots, Y_k$ be independent normal variables with zero means and variances $w_0^{-1}=w_1^{-1}=\cdots=w_t^{-1}=1$ and $w_{t+1}^{-1}=\cdots=w_k^{-1}=w^{-1}$. Intuitively, as $w\to\infty$, $\mu_0^*\leqslant\min(Y_{t+1},\ldots,Y_k)$ since the weight on each of these variables is tending to infinity. The Y_j which are greater than $\min(Y_{t+1},\ldots,Y_k)$ with $t+1\leqslant j\leqslant k$ will not be affected by the pooling process and so the number of distinct values, in the limit, will have the same distribution as $k-t-1$ plus the limiting number of distinct values in the projection of $(Y_0, Y_1,\ldots, Y_{t+1})$ onto the simple tree cone. Hence, in this case

$$\tilde{P}'_T(l,k+1)=S(l-k+t+1,t+2).$$

Notice that $\tilde{P}'_T(l,k+1)=0$ for $l\leqslant k-t-1$.

Based on numerical studies, Wright and Tran (1985) recommend the following approximation procedure. Classify a weight as large if it exceeds $1.5\min(w_0,w_1,\ldots,w_k)$ and small if not. If all of the weights are small, use the equal-weights p value or critical value. If the first weight is large and $t(1\leqslant t\leqslant k, t=0$ is not possible) of the treatment weights are small, compute the limiting p value or critical value using (2.3.4) with $P(l,k;\mathbf{w})$ replaced by $\tilde{P}_T(l,k)$, the convolution of $\{P_T(l,k-t+1)\}$ and $\{B(l,t)\}$. If the first weight is small and $t(0\leqslant t<k)$ of the treatment weights are small, compute the limiting p value or critical value using (2.3.4) with $P(l,k;\mathbf{w})$ replaced by $\tilde{P}'_T(l,k)=S(l-k+t+1,t+2)$ with $S(\cdot,t+2)$ taken from Table A.13. Set γ equal to the square root of the ratio of the average of the small weights to the average of the large weights, and obtain the approximate p value or critical value by interpolating linearly on γ between the equal-weights and limiting values.

A comparison of this approximation with the equal-weights approximation for specific weights and randomly generated weight sets is given in Wright and Tran (1985). This approximation clearly outperforms the equal-weights approximation, and, as would be expected, its performance is much like that of the approximation based on patterns in the weight sets for the simply ordered case. The reference cited also contains a more rigorous development of the coefficients $\tilde{P}_T(l,k+1)$ and $\tilde{P}'_T(l,k+1)$.

Earlier in Section 3.2, two-moment approximations for the null $\bar{\chi}^2$ and $\bar{E}^2$

distributions for the simple tree ordering with equal weights were presented. The analogous approximations for the limiting p values used in the approximation based on patterns are discussed next. The steps involved in using the two-moment approximations for the limiting p values are given below, and the interested reader is referred to Singh and Wright (1986a) for the derivation of these results. The cases in which the control weight is small or large are outlined separately.

For the case in which the control weight is large, use the following:

1a. For $\bar{\chi}^2_{01}$ or $\bar{E}^2_{01}$, the first two cumulants of the $\bar{\chi}^2_{01}$ distribution determined by the $\tilde{P}_T(l, k+1)$ are

$$\tilde{\tau}_j(k+1, 1) = \tau_j(k-t+1, 1) + b_j(t), \qquad j = 1, 2,$$

where $\tau_j(k-t+1, 1)$, the jth cumulant of the $\bar{\chi}^2_{01}$ distribution for the simple tree ordering with equal weights, is given in Table A.17 and $b_1(t) = t/2$ and $b_2(t) = 5t/4$. Also, under H_0 with the limiting coefficients, $\tilde{P}(l, k+1)$,

$$p = 1 - P[\bar{\chi}^2_{01} = 0] = 1 - P[\bar{E}^2_{01} = 0] = 1 - \tilde{P}(1, k+1)$$
$$= 1 - 2^{-t} P_T(1, k-t+1)$$

with $P_T(1, k-t+1)$ given in Table A.11.

1b. For $\bar{\chi}^2_{12}$ and $\bar{E}^2_{12}$, the first two cumulants of the $\bar{\chi}^2_{12}$ distribution determined by the $\tilde{P}_T(l, k+1)$ are

$$\tilde{\tau}_j(k+1, 2) = \tau_j(k-t+1, 2) + b_j(t), \qquad j = 1, 2,$$

where $\tau_j(k-t+1, 1)$, the jth cumulant of the $\bar{\chi}^2_{12}$ distribution for the simple tree ordering with equal weights, is given in Table A.17 and $b_j(t)$ is given above. Also, under H_0 with the limiting coefficients, $\tilde{P}(l, k+1)$,

$$p = 1 - P[\bar{\chi}^2_{12} = 0] = 1 - P[\bar{E}^2_{12} = 0] = 1 - \tilde{P}(k+1, k+1)$$
$$= 1 - 2^{-t} P_T(k-t+1, k-t+1)$$

with $P_T(k-t+1, k-t+1)$ given in Table A.11.

2a. For $\bar{\chi}^2_{01}$ compute the conditional cumulants using (3.2.3) with κ_j^* replaced by $\tilde{\tau}_j(k+1, 1)$ and for $\bar{E}^2_{01}$ compute the conditional moments using (3.2.6) with κ_j replaced by $\tilde{\tau}_j(k+1, 1)$.

2b. For $\bar{\chi}^2_{12}$ and $\bar{E}^2_{12}$ compute the conditional cumulants using (3.2.3) with κ_j^* replaced by $\tilde{\tau}_j(k+1, 2)$.

3a. For $\bar{\chi}^2_{01}$ apply (3.2.1) with $P[Y > 0] = p$ and ρ and b computed according to the formulas given there. For $\bar{E}^2_{01}$ apply (3.2.5) with $P[Y > 0] = p$ and c and d computed according to the formulas given there.

3b. For $\bar{\chi}^2_{12}$ apply (3.2.1) with $P[Y > 0] = p$ and ρ and b computed according to the formulas given there. For $\bar{E}^2_{12}$ apply (3.2.7) with $P[\bar{\chi}^2_{12} > 0] = p$ and ρ and b computed as for (3.2.1).

For the case in which the control weight is small, use the following:

1a. For $\bar{\chi}^2_{01}$ and $\bar{E}^2_{01}$, the first two cumulants of the $\bar{\chi}^2_{01}$ distribution determined by the $\tilde{P}'_T(l, k+1)$ are

$$\tilde{\tau}'_j(k+1, 1) = \tau'_j(t+1, 1) + 2^{j-1}(j-1)!(k-t), \qquad j = 1, 2,$$

where $\tau'(t+1, 1)$ is given in Table A.19. Under H_0 with the limiting coefficients $\bar{P}'_T(l, k+1)$,

$$p = 1 - P[\bar{\chi}^2_{01} = 0] = 1 - P[\bar{E}^2_{01} = 0] = 1 - S(1, k+1)$$

provided $t = k-1$ and $p = 1$ if $t < k-1$ ($S(1, k+1)$ is given in Table A.13).

1b. For $\bar{\chi}^2_{12}$ and $\bar{E}^2_{12}$, the first two cumulants of the $\bar{\chi}^2_{12}$ distribution determined by the $\tilde{P}'_T(l, k+1)$ are

$$\tilde{\tau}'_j(k+1, 2) = \tau'_j(t+1, 2)$$

which are given in Table A.19. Under H_0 with the limiting coefficients $\bar{P}'_T(l, k+1)$,

$$p = 1 - P[\bar{\chi}^2_{12} = 0] = 1 - P[\bar{E}^2_{12} = 0] = 1 - S(t+1, t+1)$$

with $S(t+1, t+1)$ given in Table A.13.

2a. For $\bar{\chi}^2_{01}$ compute the conditional cumulants using (3.2.3) with κ_j^* replaced by $\tilde{\tau}'_j(k+1, 1)$ and for $\bar{E}^2_{01}$ compute the conditional moments using (3.2.6) with κ_j replaced by $\tau'_j(k+1, 1)$.

2b. For $\bar{\chi}^2_{12}$ and $\bar{E}^2_{12}$ compute the conditional cumulants using (3.2.3) with κ_j^* replaced by $\tilde{\tau}'_j(k+1, 2)$.

Steps 3a and 3b are the same as in the case in which w_0 is large.

3.6 BOUNDS ON THE NULL $\bar{\chi}^2$ AND $\bar{E}^2$ DISTRIBUTIONS

For a given k, quasi-order $\lesssim$, and observed value t, it is important to know the amounts of variation in $P_{\mathbf{w}}[\bar{\chi}^2_{01} \geqslant t]$($P_{\mathbf{w}}[\bar{\chi}^2_{12} \geqslant t]$, $P_{\mathbf{w}}[\bar{E}^2_{01} \geqslant t]$, and $P_{\mathbf{w}}[\bar{E}^2_{12} \geqslant t]$) under H_0 for the possible weight vectors, $\mathbf{w}$. Hence, sharp upper and lower bounds for these tail probabilities are determined in this section. These results also provide least favorable configurations for some testing situations, e.g. in two sample tests of stochastic ordering among multinomial distributions (cf. Section 5.4).

In the last subsection of Section 2.4, the reduction of a quasi-ordered set to a partially ordered set was discussed. It was noted that the level probabilities for the quasi-order and the corresponding partial order are the same. Hence, it suffices to find bounds for the probabilities in (2.3.4) for an arbitrary partial order, $\lesssim$. Two approaches will be employed. The first involves obtaining stochastic bounds for the distribution, under H_0, of the number of level sets. The second uses techniques like those in Perlman (1969) to obtain bounds on the distance from the observation vector to the cone by considering other cones either containing, or contained in, the original cone.

The first approach rests on the following result.

Remark *If* $\mathbf{a} = (a(1), a(2), \ldots, a(k))$ *and* $\mathbf{b} = (b(1), b(2), \ldots, b(k))$ *are probability vectors, i.e. the components are nonnegative and sum to one, and if* $\sum_{l=1}^{j} a(l) \leqslant \sum_{l=1}^{j} P(l, k; \mathbf{w}) \leqslant \sum_{l=1}^{j} b(l)$ *for* $j = 1, 2, \ldots, k-1$, *then for real* t,

$$\sum_{l=1}^{k} b(l)P[\chi_{l-1}^2 \geqslant t] \leqslant P_{\mathbf{w}}[\bar{\chi}_{01}^2 \geqslant t] \leqslant \sum_{l=1}^{k} a(l)P[\chi_{l-1}^2 \geqslant t]$$

$$\sum_{l=1}^{k} b(l)P[B_{(l-1)/2,(v+k-l)/2} \geqslant t] \leqslant P_{\mathbf{w}}[\bar{E}_{01}^2(v) \geqslant t]$$

$$\leqslant \sum_{l=1}^{k} a(l)P[B_{(l-1)/2,(v+k-l)/2} \geqslant t]$$

$$\sum_{l=1}^{k} a(l)P[\chi_{k-l}^2 \geqslant t] \leqslant P_{\mathbf{w}}[\bar{\chi}_{12}^2 \geqslant t] \leqslant \sum_{l=1}^{k} b(l)P[\chi_{k-l}^2 \geqslant t]$$

and

$$\sum_{l=1}^{k} a(l)P[B_{(k-l)/2,v/2} \geqslant t] \leqslant P_{\mathbf{w}}[\bar{E}_{12}^2(v) \geqslant t] \leqslant \sum_{l=1}^{k} b(l)P[B_{(k-l)/2,v/2} \geqslant t]. \tag{3.6.1}$$

Proof: Because of the reproductive property of the family of gamma distributions with a fixed scale parameter, the family is stochastically increasing in the shape parameter, i.e. if $X_1 \sim \Gamma(\alpha_1, \beta)$, $X_2 \sim \Gamma(\alpha_2, \beta)$, and $\alpha_1 < \alpha_2$, then $X_1 \overset{\text{st}}{\leqslant} X_2$. Hence, $c(l) = P[\chi_{l-1}^2 \geqslant t]$ is nondecreasing and $c'(l) = P[\chi_{k-l}^2 \geqslant t]$ is nonincreasing. Furthermore, expressing $B_{(l-1)/2,(v+k-l)/2}$ as $(1 + X_2/X_1)^{-1}$ with $X_1 \sim \Gamma(\frac{1}{2}(l-1), 1)$, $X_2 \sim \Gamma(\frac{1}{2}(v+k-l), 1)$, and X_1 and X_2 independent, one sees that $d(l) = P[B_{(l-1)/2,(v+k-l)/2} \geqslant t]$ is nondecreasing. Similarly, $d'(l) = P[B_{(k-l)/2,v/2} \geqslant t]$ is seen to be nonincreasing.

Using Abel's method of summation,

$$\sum_{l=1}^{k} P(l, k; \mathbf{w})c(l) = \sum_{j=1}^{k-1} [c(j) - c(j+1)] \sum_{l=1}^{j} P(l, k; \mathbf{w}) + c(k). \tag{3.6.2}$$

Because $c(l)$ is nondecreasing, (3.6.2) is no smaller (larger) if $\sum_{l=1}^{j} P(l, k; \mathbf{w})$ is replaced by $\sum_{l=1}^{j} a(l) (\sum_{l=1}^{j} b(l))$. The results for $c'(l)$, $d(l)$, and $d'(l)$ follow similarly. □

This remark with the following theorem provides bounds (which will be shown to be sharp) for the simply ordered case. Let I_A denote the indicator of the set A. Throughout this section let $\mathbf{Y} = (Y_1, Y_2, \ldots, Y_k)$ be a vector of independent normal random variables with zero means and variances $w_1^{-1}, w_2^{-1}, \ldots, w_k^{-1}$, respectively.

Theorem 3.6.1 *If* $\lesssim$ *is a simple order on* $\{1,2,\ldots,k\}$, $b(l)=\frac{1}{2}I_{\{1,2\}}(l)$ *and* $a(l) = \binom{k-1}{l-1}2^{-k+1}$, *then* $\sum_{l=1}^{j} a(l) \leqslant \sum_{l=1}^{j} P_S(l,k;\mathbf{w}) \leqslant \sum_{l=1}^{j} b(l)$ *for* $j=1,2,\ldots,k$.

Proof: For the inequality involving $b(l)$, it suffices to show that $P_S(1,k;\mathbf{w}) \leqslant \frac{1}{2}$. However, using the MLSA, one sees that

$$P_S(1,k;\mathbf{w}) = P\Big[\min_{1\leqslant\alpha\leqslant k} \mathrm{Av}(\{1,2,\ldots,\alpha\}) = \mathrm{Av}(\{1,2,\ldots,k\})\Big]$$

$$\leqslant P[\mathrm{Av}(\{1,2,\ldots,k-1\}) \geqslant \mathrm{Av}(\{1,2,\ldots,k\})] = \tfrac{1}{2},$$

where, as before, $\mathrm{Av}(A) = \sum_{i\in A} w_i Y_i / \sum_{i\in A} w_i$.

The proof that $\sum_{l=1}^{j} a(l) \leqslant \sum_{l=1}^{j} P_S(l,k;\mathbf{w})$ is by induction on k. The desired result clearly holds for $k=1$ and 2. Assume that the desired inequalities hold for $k-1$ and let $M_{k-1} = M_{k-1}(Y_1, Y_2,\ldots, Y_{k-1}; w_1, w_2,\ldots, w_{k-1})$ ($M_k = M_k(Y_1, Y_2,\ldots, Y_k; w_1, w_2,\ldots, w_k)$) denote the number of level sets in the isotonic regression of $(Y_1, Y_2,\ldots, Y_{k-1})$ ($(Y_1, Y_2,\ldots, Y_k)$). Using the PAVA, the isotonic regression of $(Y_1, Y_2,\ldots, Y_k)$, which is denoted by $\mu^*_{(k)} = (\mu^*_{1,k}, \mu^*_{2,k},\ldots, \mu^*_{k,k})$, can be computed by first computing $\mu^*_{(k-1)} = (\mu^*_{1,k-1}, \mu^*_{2,k-1},\ldots, \mu^*_{k-1,k-1})$, the isotonic regression of $(Y_1, Y_2,\ldots, Y_{k-1})$, and then combining $\mu^*_{(k-1)}$ and Y_k in the appropriate way. If $\mu^*_{k-1,k-1} < Y_k$, then $M_k = M_{k-1}+1$, and if $\mu^*_{k-1,k-1} \geqslant Y_k$, then $M_k \leqslant M_{k-1}$. Hence, for $j=1,2,\ldots,k-1$,

$$P[M_k \geqslant j+1] \leqslant P[M_{k-1}=j \text{ and } Y_k > \mu^*_{k-1,k-1}] + P[M_{k-1} \geqslant j+1]. \tag{3.6.3}$$

Examining the proof of Theorem 2.4.1, one sees that

$$P[M_{k-1}=j \text{ and } Y_k > \mu^*_{k-1,k-1}]$$

$$= \sum_{B_1,B_2,\ldots,B_j \in \mathscr{L}_{jk-1}} P[\mathrm{Av}(B_1) < \cdots < \mathrm{Av}(B_j) < Y_k] \prod_{i=1}^{j} P_S(1, C_{B_i}, \mathbf{w}(B_i)).$$

Furthermore, with $U_1 = \mathrm{Av}(B_2) - \mathrm{Av}(B_1)$, $U_2 = \mathrm{Av}(B_3) - \mathrm{Av}(B_2),\ldots, U_{j-1} = \mathrm{Av}(B_j) - \mathrm{Av}(B_{j-1})$, $U_j = Y_k - \mathrm{Av}(B_j)$, $\mathrm{cov}(U_i, U_j) \leqslant 0$ for $i=1,2,\ldots,j-1$ and thus applying the Slepian inequality (cf. Tong, 1980)

$$P[\mathrm{Av}(B_2) < \cdots < \mathrm{Av}(B_j) < Y_k] \leqslant \tfrac{1}{2} P[\mathrm{Av}(B_1) < \cdots < \mathrm{Av}(B_j)]$$

and, consequently, $P[M_{k-1}=j \text{ and } Y_k > \mu^*_{k-1,k-1}] \leqslant \frac{1}{2} P[M_{k-1}=j]$. Using this inequality in (3.6.3) and applying the inductive hypothesis, it follows that

$$\sum_{l=j+1}^{k} P_S(l,k;(w_1,w_2,\ldots,w_k)) = P[M_k \geqslant j+1]$$

$$\leqslant \frac{P[M_{k-1} \geqslant j] + P[M_{k-1} \geqslant j+1]}{2}$$

$$\leqslant 2^{-k+1}\left\{\sum_{l=j}^{k-1}\binom{k-2}{l-1}+\sum_{l=j+1}^{k-1}\binom{k-2}{l-1}\right\}$$

$$=2^{-k+1}\left\{\sum_{l=j}^{k-1}\left[\binom{k-2}{l-1}+\binom{k-2}{l}\right]\right.$$

$$\left.+\binom{k-2}{k-2}\right\}=2^{-k+1}\sum_{l=j+1}^{k}\binom{k-1}{l-1},$$

which is the desired conclusion. □

The probabilities, $a(l)$, are binomial probabilities with $k-1$ trials and the probability of success equal to $\frac{1}{2}$. However, $a(l)$ is the probability of $l-1$ successes, and so they will be referred to as shifted binomial probabilities. Because convolutions of shifted binomial probabilities are of interest, for δ an integer with $1\leqslant\delta\leqslant k$ set

$$A_\delta(l,k)=\binom{k-\delta}{l-\delta}2^{-k+\delta}\qquad\text{for } l=1,2,\ldots,k. \tag{3.6.4}$$

(Of course, $A_\delta(l,k)=0$ for $l<\delta$ and $a(l)=A_1(l,k)$.)

Theorem 3.6.1 and a decomposition result for an arbitrary partially ordered set provides one set of bounds. Some definitions are needed first.

Definition 3.6.1 Let $(X,\lesssim)$ be a partially ordered set. A subset of X is a *chain* (*antichain*) if each pair of distinct elements in the subset is comparable (noncomparable). The *breadth*, b, of X is the maximal cardinality of any antichain in X, that is $b=\max\{\text{card}(A)\text{: } A \text{ is an antichain in } X\}$.

The following result is given in Crawley and Dilworth, 1973, p. 3.

Theorem 3.6.2 *If $(X,\lesssim)$ is a partially ordered set with breadth b, then X can be written as the disjoint union of b sets which are chains with respect to $\lesssim$.*

Theorem 3.6.3 *Let $\lesssim$ be a partial order on $X=\{1,2,\ldots,k\}$ with breadth b. For $\mathbf{w}>\mathbf{0}$ and real t, under H_0,*

$$P_{\mathbf{w}}[\bar{\chi}^2_{01}\geqslant t]\leqslant\sum_{l=1}^{k}A_b(l,k)P[\chi^2_{l-1}\geqslant t]$$

$$P_{\mathbf{w}}[\bar{E}^2_{01}(v)\geqslant t]\leqslant\sum_{l=1}^{k}A_b(l,k)P[B_{(l-1)/2,(v+k-l)/2}\geqslant t],$$

$$P_{\mathbf{w}}[\bar{\chi}^2_{12}\geqslant t]\geqslant\sum_{l=1}^{k}A_b(l,k)P[\chi^2_{k-l}\geqslant t] \tag{3.6.5}$$

and

$$P_{\mathbf{w}}[\bar{E}^2_{12}(v)\geqslant t]\geqslant\sum_{l=1}^{k}A_b(l,k)P[B_{(k-l)/2,v/2}\geqslant t].$$

Proof: Let $\hat{\mu} = \sum_{i=1}^{k} w_i Y_i / \sum_{i=1}^{k} w_i$. A quasi-order, $\lesssim'$, on X is less restrictive than $\lesssim$ provided $i \lesssim' j$ implies $i \lesssim j$. If H_1' corresponds to a less restrictive quasi-order on $\{1, 2, \ldots, k\}$, that is H_1' is the collection of vectors isotonic with respect to $\lesssim'$, then $H_1' \supset H_1$, and $\| \mathbf{Y} - P_w(\mathbf{Y}|H_1) \|_w^2 \geqslant \| \mathbf{Y} - P_w(\mathbf{Y}|H_1') \|_w^2$. Furthermore, since $\| \mathbf{Y} - \hat{\mu} \|_w^2 = \| P_w(\mathbf{Y}|H) - \hat{\mu} \|_w^2 + \| \mathbf{Y} - P_w(\mathbf{Y}|H) \|_w^2$ for H the cone of functions isotonic with respect to any quasi-order, $\| P_w(\mathbf{Y}|H_1) - \hat{\mu} \|_w^2 \leqslant \| P_w(\mathbf{Y}|H_1') - \hat{\mu} \|_w^2$. Therefore, $\bar{\chi}_{01}^2 (\bar{\chi}_{12}^2)$ is not decreased (increased) if $\lesssim$ is replaced by $\lesssim'$. Furthermore, because

$$\bar{E}_{01}^2(v) = \left(1 + \frac{\bar{\chi}_{12}^2 + Q}{\bar{\chi}_{01}^2} \right)^{-1} \quad \text{and} \quad \bar{E}_{12}^2(v) = \left(1 + \frac{Q}{\bar{\chi}_{12}^2} \right)^{-1},$$

for a fixed value of Q, $\bar{E}_{01}^2(v) (\bar{E}_{12}^2(v))$ is not decreased (increased) if $\lesssim$ is replaced by $\lesssim'$. Hence, it suffices to establish (3.6.5) for a quasi-order, $\lesssim'$, which is less restrictive than $\lesssim$.

Applying Theorem 3.6.2, there are non-empty, disjoint subsets of X, $X_1, X_2, \ldots, X_b$, which are chains with respect to $\lesssim$ and $\bigcup_{j=1}^{b} X_j = X$. Define $\lesssim'$ by $i \lesssim' j$ provided $i, j \in X_\alpha$ for some α and $i \lesssim j$. Clearly, $\lesssim'$ is less restrictive than $\lesssim$. Expressions for tail probabilities of $\bar{\chi}_{01}^2$, $\bar{\chi}_{12}^2$, $\bar{E}_{01}^2(v)$, and $\bar{E}_{12}^2(v)$ under H_0 are given in (2.3.4). Denoting the level probabilities corresponding to $\lesssim'$ by $P'(l, k; \mathbf{w})$, because of the remark before Theorem 2.6.1, one only needs to show that

$$\sum_{l=1}^{j} P'(l, k; \mathbf{w}) \geqslant \sum_{l=1}^{j} A_b(l, k), \qquad j = 1, 2, \ldots, k-1.$$

Let M denote the number of level sets in the projection of $(Y_1, Y_2, \ldots, Y_k)$ onto H_1', the cone of functions isotonic with respect to $\lesssim'$, and for $\alpha = 1, 2, \ldots, b$, let M_α denote the number of level sets in the projection of $\{Y_j : j \in X_\alpha\}$ onto the cone of functions isotonic with respect to the restriction of $\lesssim'$ to X_α. Because the elements in X_α and X_β are noncomparable if $\alpha \neq \beta$, with probability one, $M = M_1 + M_2 + \cdots + M_b$. For $\alpha = 1, 2, \ldots, b$, the restriction of $\lesssim$ to X_α is a total order. Theorem 3.6.1 states that $M_\alpha \overset{\text{st}}{\lesssim} N_\alpha + 1$ where N_α is a binomial random variable with parameters card$(X_\alpha) - 1$ and $\frac{1}{2}$. The random variables $M_1, M_2, \ldots, M_b$ are independent and so $M_1 + M_2 + \cdots + M_b$ is bounded above stochastically by a binomial variable with parameters $k - b$ and $\frac{1}{2}b$, which is the desired conclusion. □

The bounds given in (3.6.5) can be shown to be sharp by proving that there exists a sequence of weight vectors $\mathbf{w}(n) = (w_1(n), w_2(n), \ldots, w_k(n))$ for which $P(l, k; \mathbf{w}(n)) \to A_b(l, k)$ for $l = 1, 2, \ldots, k$.

Definition 3.6.2 An element j in the partially ordered set $(X, \lesssim)$ is a *maximal* (minimal) element if there does not exist a $j' \in X$ with $j \lesssim j'$ ($j' \lesssim j$). Furthermore, j is said to be an *exterior* element if it is a maximal or minimal element; otherwise, it is called an *interior* element. An element which is both maximal and minimal is called *isolated*.

The first step in establishing the sharpness of the bounds in Theorem 3.6.3 is the following result.

Theorem 3.6.4 *Let $Y_1, Y_2, \ldots, Y_k$ be an independent normal variable with zero means and variances $w_1^{-1}, w_2^{-1}, \ldots, w_k^{-1}$ and let $M(\mathbf{Y};\mathbf{w})(M(\mathbf{Y}^{(j)};\mathbf{w}^{(j)}))$ denote the number of level sets in the projection of $\mathbf{Y}((Y_1, \ldots, Y_{j-1}, Y_{j+1}, \ldots, Y_k))$ onto the cone determined by $\lesssim$ ($\lesssim$ restricted to $X - \{j\}$) with weight vector $\mathbf{w}$ $((w_1, \ldots, w_{j-1}, w_{j+1}, \ldots, w_k))$. If j is an exterior element but not isolated, then as $w_j \to 0$,*

$$M(\mathbf{Y};\mathbf{w}) \xrightarrow{D} M(\mathbf{Y}^{(j)};\mathbf{w}^{(j)}) + I_{[Y_j<0]}.$$

Proof: By relabeling if necessary assume $j = k$. Suppose k is a minimal element. The proof for a maximal element is similar. By enlarging the probability space if necessary, one can obtain Z_k, a standard normal variable independent of $Y_1, Y_2, \ldots, Y_{k-1}$, and, of course, $M(\mathbf{Y};\mathbf{w}) \sim M((\mathbf{Y}^{(k-1)}, Z_k/\sqrt{w_k});\mathbf{w})$. For a fixed element in the underlying probability space, consider the cases $Z_k < 0$ and $Z_k > 0$. If $Z_k < 0$, then for sufficiently small w_k, $Z_k/\sqrt{w_k} < \min(Y_1, \ldots, Y_{k-1})$. Using the MLSA and the fact that $\{k\}$ is a lower layer, one sees that the first level set of the projection of $(\mathbf{Y}^{(k-1)}, Z_k/\sqrt{w_k})$ is $\{k\}$ and the others are the level sets of the projection of $\mathbf{Y}^{(k-1)}$ with the weight vector $\mathbf{w}^{(k-1)}$. Hence, for sufficiently small w_k,

$$M((\mathbf{Y}^{(k-1)}, Z_k/\sqrt{w_k});\mathbf{w}) = M(\mathbf{Y}^{(k-1)};\mathbf{w}^{(k-1)}) + 1.$$

Next, consider the case $Z_k > 0$. Let $B_1, \ldots, B_m$ denote the level sets for the projection of $\mathbf{Y}^{(k-1)}$ and let B_i be the first level set containing an element comparable with k (which exists by hypothesis). For w_k sufficiently small, $Z_k/\sqrt{w_k} > \max(Y_1, \ldots, Y_{k-1})$. Using the MLSA to compute the projection of $(\mathbf{Y}^{(k-1)}, Z_k/\sqrt{w_k})$, one sees that the first $i-1$ level sets are $B_1, B_2, \ldots, B_{i-1}$. Neglecting the set with probability zero on which $\text{Av}(A) = \text{Av}(B)$ for some A and B distinct subsets of X and noting that $\text{Av}(A \cup \{k\}) \to \text{Av}(A)$ as $w_k \to 0$ for $\phi \neq A$ $X - \{k\}$, one sees that for sufficiently small w_k, $B_i \cup \{k\}$ is the next level set. Hence, for $Z_k > 0$ and such w_k, $M((\mathbf{Y}^{(k-1)}, Z_k/\sqrt{w_k});\mathbf{w}) = M(\mathbf{Y}^{(k-1)};\mathbf{w}^{(k-1)})$. □

If $(X, \lesssim)$ has breadth b, then there exist b elements $i_1, i_2, \ldots, i_b$ which are noncomparable elements in X. Furthermore, because b is the breadth of X, every other element in X is comparable with at least one of $i_1, i_2, \ldots, i_b$. Applying Theorem 3.6.4 repeatedly to the elements of $X - \{i_1, i_2, \ldots, i_b\}$, one can show that $P(l, k; \mathbf{w})$ can be made arbitrarily close to

$$P[M((Y_{i_1}, \ldots, Y_{i_b}); (w_{i_1}, \ldots, w_{i_b})) + \sum_{j \notin \{i_1, \ldots, i_b\}} I_{[Y_j<0]} = l] \tag{3.6.6}$$

(details are given in Wright, 1986). Noting that $M((Y_{i_1},\ldots,Y_{i_b});(w_{i_1},\ldots,w_{i_b}))=b$ a.s., one sees that (3.6.6) is $P[B=l-b]$ with B a binomial variable with parameters $k-b$ and $\frac{1}{2}$. This shows that the bounds in (3.6.5) are sharp.

The bounds complementary to those given in (3.6.5) (i.e. lower bounds for the null distributions of $\bar{\chi}^2_{01}$ and $\bar{E}^2_{01}(v)$ and upper bounds for the null distribution of $\bar{\chi}^2_{12}$ and $\bar{E}^2_{12}(v)$) have not been established for general partial orders. However, the results discussed next apply to all of the partial orders discussed thus far, i.e. the simple order, the simple tree, the loop, the matrix order, and the unimodal order. The approach for obtaining the bounds given next is to reduce the problem to that involving a simple tree or the reverse of a simple tree ordering. Hence, the simple tree is treated first.

Lemma *If the ordering imposed by $\lesssim$ is $1\leqslant i$ for $i=2,3,\ldots,k$, then under H_0, $\sum_{l=1}^{j}P_T(l,k;\mathbf{w})\leqslant\sum_{l=1}^{j}A(l,k)$ for $j=1,2,\ldots,k$ where $A(l,k)=\binom{k-1}{l-1}2^{-k+1}$ for $l=1,2,\ldots,k$.* (Note: $A(l,k)=A_1(l,k)$.)

Before the proof of the lemma is given, note that the mixing coefficients which provide the lower bound for $\bar{\chi}^2_{01}$ and $\bar{E}^2_{01}(v)$ for the simple tree ordering also provide the upper bound for $\bar{\chi}^2_{01}$ and $\bar{E}^2_{01}(v)$ for the simple ordering. These two orderings are, in a sense, at the opposite ends of the spectrum. The simple ordering has breadth $b=1$ and a partial order which is not simple has breadth $b>1$. On the other hand, the simple tree on $X=\{1,2,\ldots,k\}$ has breadth $k-1$ which is the largest breadth on X possible for an indecomposable partial order. (Recall that $(X,\lesssim)$ is decomposable if $X=A\cup B$ with A and B nonempty, disjoint subsets with i and j noncomparable for each $i\in A$ and $j\in B$.) In Theorem 3.3.3, it was shown that as the control weight approaches infinity, $P_T(l,k;\mathbf{w})\to A(l,k)$, which shows that the bounds given in the lemma are sharp.

Proof of the Lemma: The proof is by induction. Clearly, $P(1,2;\mathbf{w})=P(2,2;\mathbf{w})=\frac{1}{2}$. Therefore, consider $k>2$ and define M_k, $B_{1,k}$, and $\mu^*_{1,k}$ to be the number of level sets, the first level set, and the value of $\boldsymbol{\mu}^*$ on the first level set in the restricted estimates based on $Y_1,\ldots,Y_k$ with weights $w_1,\ldots,w_k$. Let M_{k+1}, $B_{1,k+1}$, and $\mu^*_{1,k+1}$ be the corresponding values determined by the estimates based on $Y_1,\ldots,Y_{k+1}$ with weights $w_1,\ldots,w_{k+1}$.

First it will be shown that $M_{k+1}\geqslant M_k+I_{(\mu^*_{1,k},\infty)}(Y_{k+1})$. Because the estimates maximize the likelihood function subject to H_1, $\mu^*_{k+1,k+1}=Y_{k+1}$ and $M_{k+1}=M_k+1$ if $Y_{k+1}>\mu^*_{1,k}$. If $Y_{k+1}\leqslant\mu^*_{1,k}$, then it will be shown that $B_{1,k+1}=B_{1,k}\cup\{k+1\}$, for if not, there is an $i\in B_{1,k+1}-(B_{1,k}\cup\{k+1\})$ with $i\neq 1$ and $Y_{k+1}\leqslant\mathrm{Av}(B_{1,k})<Y_i$, which imply $\mathrm{Av}(B_{1,k}\cup\{k+1\})<Y_i$. However, by the minimum lower sets algorithm, $\mathrm{Av}(B_{1,k+1})\leqslant\mathrm{Av}(B_{1,k}\cup\{k+1\})$; hence, $\mathrm{Av}(B_{1,k+1}-\{i\})<\mathrm{Av}(B_{1,k+1})$ which contradicts the choice of $B_{1,k+1}$ according to this algorithm. Therefore, $P[M_{k+1} \geqslant l+1] \geqslant P[M_k = l, Y_{k+1} > \mu^*_{1,k}] + P[M_k \geqslant l+1]$, but the proof given for (2.4.4) shows that

$$P[M_k = l, Y_{k+1} > \mu^*_{1,k}] \\ = \sum P[\mathrm{Av}(B_{1,k}) < Y_\gamma; \gamma \in \{1,\ldots,k\} - B_{1,k}, Y_{k+1} > \mathrm{Av}(B_{1,k})]P(B_{1,k}),$$

where the summation is over all $B_{1,k} \subset \{1,\ldots,k\}$ which contain $\{1\}$ and have cardinality $k-l+1$, and $P(B_{1,k})$ is the probability that $\mathrm{Av}(B) \geqslant \mathrm{Av}(B_{1,k})$ for all B which are proper subsets of $B_{1,k}$ and contain $\{1\}$. But, because Av $(B_{1,k}) \leqslant \bar{Y}$, this sum is bounded below by

$$\tfrac{1}{2}\sum P[\mathrm{Av}(B_{1,k}) < Y_\gamma; \gamma \in \{1,\ldots,k\} - B_{1,k}]P(B_{1,k}),$$

which according to (2.4.4) is $\frac{1}{2}P(l,k;\mathbf{w})$. From the induction hypothesis, $P[M_k \geqslant l+1]$ is bounded below by

$$\tfrac{1}{2}P[M_k \geqslant l] + \tfrac{1}{2}P[M_k \geqslant l+1] \geqslant 2^{-k}\sum_{\alpha=l+1}^{k+1}\binom{k}{\alpha-1} = \sum_{\alpha=l+1}^{k+1} A(\alpha,k+1). \quad \square$$

Theorem 3.6.5 *Let $(X, \lesssim)$ be a partially ordered set which has at most one maximal or at most one minimal element and let e denote the number of exterior elements in $(X, \lesssim)$. For $\mathbf{w} > \mathbf{0}$, $\boldsymbol{\mu} \in H_0$ and real t,*

$$P_{\mathbf{w}}[\bar{\chi}^2_{01} \geqslant t] \geqslant \sum_{l=1}^{e} A(l,e)P[\chi^2_{l-1} \geqslant t]$$

$$P_{\mathbf{w}}[\bar{E}^2_{01}(\nu) \geqslant t] \geqslant \sum_{l=1}^{e} A(l,e)P[B_{(l-1)/2,(\nu+k-l)/2} \geqslant t]$$

$$P_{\mathbf{w}}[\bar{\chi}^2_{12} \geqslant t] \leqslant \sum_{l=1}^{e} A(l,e)P[\chi^2_{k-l} \geqslant t]$$

and

$$P_{\mathbf{w}}[\bar{E}^2_{12}(\nu) \geqslant t] \leqslant \sum_{l=1}^{e} A(l,e)P[B_{(k-l)/2,\nu/2} \geqslant t]. \tag{3.6.7}$$

Proof: The proof for the case in which X has at most one minimal element is given. The proof for the other case is similar. Because X is finite, it has exactly one minimal element. Suppose that the minimal element is labeled 1 and the maximal elements are $2, 3, \ldots, e$. Let $\lesssim'$ be the quasi-order which requires $1 \approx e+1 \approx e+2 \approx \cdots \approx k \lesssim' j$ for $j = 2, 3, \ldots, e$. Because $\lesssim$ is less restrictive than $\lesssim'$, replacing $\lesssim$ by $\lesssim'$ provides a lower (upper) bound for $\bar{\chi}^2_{01}$ and $\bar{E}^2_{01}(\nu)$($\bar{\chi}^2_{12}$ and $\bar{E}^2_{12}(\nu)$). At the end of Section 2.4, it was noted that the mixing coefficients for the quasi-order are the same as those for the partial order to which it reduces. However, the reduction of $\lesssim'$ is a simple tree ordering on $\{1, 2, \ldots, e\}$ and by the lemma, the appropriate bounds for this simple tree are provided by the mixing coefficients $A(l,e)$. $\square$

The bounds in Theorem 3.6.5 can be extended to decomposable partial orders.

If X is the disjoint union of the nonempty sets $X_1, X_2, \ldots, X_d$ and $i \in X_\alpha$ and $j \in X_\beta$ with $\alpha \neq \beta$ imply that i and j are noncomparable, then e, the number of exterior elements in $(X, \lesssim)$, is $e_1 + e_2 + \cdots + e_d$, where e_α is the number of exterior elements in X_α with $\lesssim$ restricted to X_α. If each X_α with $\lesssim$ restricted to X_α satisfies the hypotheses of Theorem 3.6.5, then under H_0, M_α, the number of level sets in the projection onto the cone of functions isotonic with respect to the restriction of $\lesssim$ to X_α is stochastically bounded above by $N_\alpha + 1$ with $N_\alpha \sim B(\text{card}(X_\alpha) - 1, \frac{1}{2})$ for $\alpha = 1, 2, \ldots, d$. Because $M_1, M_2, \ldots, M_d$ are independent, $M_1 + M_2 + \cdots + M_d$ is stochastically bounded above by $N + d$ with $N \sim B(k - d, \frac{1}{2})$. However, $P[N + d = l] = \binom{k-d}{l-d} 2^{-k+d}$ and so one obtains the bounds in (3.6.7) with $A(l, e)$ replaced by $A_d(l, e)$, which is defined in (3.6.4), for such a partial order.

The proof of the sharpness of these bounds rests on the following result.

Lemma *If j is an interior element in $(X, \lesssim)$, then*

$$\lim_{w_j \to 0} P(l, k; \mathbf{w}) = P(l, k-1; \mathbf{w}^{(j)})$$

where $P(\cdot, k-1; \mathbf{w}^{(j)})$ are the level probabilities for the restriction of $\lesssim$ to $X - \{j\}$ with the weight vector $\mathbf{w}^{(j)} = (w_1, \ldots, w_{j-1}, w_{j+1}, \ldots, w_k)$.

A rigorous proof of the lemma is given in Wright (1986). However, the result is intuitively quite simple. Let $Y_1, Y_2, \ldots, Y_k$ be independent normal variables with zero means and variances $w_1^{-1}, w_2^{-1}, \ldots, w_k^{-1}$, respectively. By enlarging the probability space, one can obtain a standard normal variable Z_j which is independent of the Y_i, and in computing the $P(l, k; \mathbf{w})$, one can replace Y_j by $Z_j/\sqrt{w_j}$. Because j is an interior element there are elements smaller and larger than j. Consider a fixed point in the underlying probability space. If $Z_j > 0 (Z_j < 0)$, then $Z_j/\sqrt{w_j} \to \infty (-\infty)$ as $w_j \to 0$ and thus $Z_j/\sqrt{w_j}$ must be pooled with other observations. However, in the pooling process $w_j Z_j/\sqrt{w_j} = \sqrt{w_j} Z_j$ can be made arbitrarily small and its contribution in the pooling process is negligible. Hence, the level sets are, for sufficiently small w_j, the same as those for $\lesssim$ restricted to $X - \{j\}$ with the weight vector $w^{(j)}$.

To show that these bounds are sharp, one only needs to show that the results for an indecomposable partial order are sharp. If $\lesssim$ has at most one maximal or at most one minimal element (notice that this implies indecomposability), then by repeatedly applying the last lemma, the $P(l, k; \mathbf{w})$ can be made arbitrarily close to the level probabilities of a simple tree on the exterior elements with arbitrary weights. Next, by making the control weight arbitrarily large the level probabilities can be made close to the $A(l, e)$. This provides an intuitive justification for the sharpness of bounds given in Theorem 3.6.5 and its extension to decomposable partial orders. A rigorous proof is given in Wright (1986). The following result is proved there.

Theorem 3.6.6 *If $(X, \lesssim)$ is indecomposable with e exterior elements, then there exists a sequence of positive weight vectors* $\mathbf{w}(n)$ *for which*

$$\lim_{n\to\infty} P(l, k; \mathbf{w}(n)) = A(l, e) \qquad \text{for } l = 1, 2, \ldots, k.$$

This would lead one to conjecture that the bounds given in (3.6.7) hold under the more general hypotheses of Theorem 3.6.6. However, this question is open at this time.

For purposes of application it is useful to examine these bounds for particular partial orders. For the simple order under H_0,

$$\tfrac{1}{2}P[\chi_1^2 \geqslant t] \leqslant P[\bar{\chi}_{01}^2 \geqslant t] \leqslant 2^{-k+1} \sum_{l=1}^{k} \binom{k-1}{l-1} P[\chi_{l-1}^2 \geqslant t],$$

$$2^{-k+1} \sum_{l=1}^{k} \binom{k-1}{l-1} P[\chi_{k-l}^2 \geqslant t] \leqslant P[\bar{\chi}_{12}^2 \geqslant t] \leqslant \tfrac{1}{2}(P[\chi_{k-2}^2 \geqslant t] + P[\chi_{k-1}^2 \geqslant t])$$

and the results for $\bar{E}_{01}^2(\nu)$ and $\bar{E}_{12}^2(\nu)$ are obtained by replacing the chi-squared tail probabilities by the appropriate beta probabilities (see (2.3.4)). If X has at most one maximal or at most one minimal element, e is the number of exterior elements, and b is the breadth, then under homogeneity,

$$2^{-e+1} \sum_{l=1}^{e} \binom{e-1}{l-1} P[\chi_{l-1}^2 \geqslant t] \leqslant P[\bar{\chi}_{01}^2 \geqslant t] \leqslant 2^{-k+b} \sum_{l=b}^{k} \binom{k-b}{l-b} P[\chi_{l-1}^2 \geqslant t]$$

$$2^{-k+b} \sum_{l=b}^{k} \binom{k-b}{l-b} P[\chi_{k-l}^2 \geqslant t] \leqslant P[\bar{\chi}_{12}^2 \geqslant t] \leqslant 2^{-e+1} \sum_{l=1}^{e} \binom{e-1}{l-1} P[\chi_{k-l}^2 \geqslant t]$$

and the results for $\bar{E}_{01}^2(\nu)$ and $\bar{E}_{12}^2(\nu)$ are obtained by replacing the chi-squared probabilities by the approximate beta probabilities. For the simple tree, $e = k$ and $b = k - 1$; for the unimodal order, $1 \lesssim 2 \lesssim \cdots \lesssim h \gtrsim h+1 \gtrsim \cdots \gtrsim k$ with $1 < h < k$, $e = 3$, and $b = 2$; and for the matrix order on $X = \{(i, j): 1 \leqslant i \leqslant R, 1 \leqslant j \leqslant C\}$, $e = 2$ and $b = \min(R, C)$.

It is instructive to obtain some indication of the amount of variation in the upper and lower bounds. Table 3.6.1 gives the values of the upper and lower bounds for $\bar{\chi}_{01}^2$ at $e_{0.05}(k)$, the equal-weights critical values, for the simple order with $4 \leqslant k \leqslant 10$. Table 3.6.2 contains the corresponding values for the simple tree (in this case, k denotes the total number of populations).

There is considerable variation in these bounds, particularly for the simply ordered case. However, it was noted earlier that though there are sequences of weights for which the tail probabilities, under H_0, converge to the bounds, this convergence seems to take place rather slowly. For many practical purposes the equal-weights approximation should be adequate. However, examining the proofs for sharpness of these bounds, one sees that if the weights on the interior

Table 3.6.1 Values of the lower and upper bound for $\bar{\chi}^2_{01}$ for the simple order at $e_{0.05}(k)$ with $4 \leqslant k \leqslant 10$

	$k=4$	$k=5$	$k=6$	$k=7$	$k=8$	$k=9$	$k=10$
Lower bound	0.0167	0.0123	0.0097	0.0080	0.0068	0.0059	0.0052
Upper bound	0.0777	0.0959	0.1168	0.1403	0.1662	0.1804	0.2244

Table 3.6.2 Values of the lower and upper bound for $\bar{\chi}^2_{01}$ for the simple tree at $e_{0.05}(k)$ with $4 \leqslant k \leqslant 10$

	$k=4$	$k=5$	$k=6$	$k=7$	$k=8$	$k=9$	$k=10$
Lower bound	0.0416	0.0350	0.0296	0.0252	0.0215	0.0185	0.0159
Upper bound	0.0669	0.00746	0.0794	0.0827	0.0850	0.0863	0.0881

elements of X are small in comparison to the weights on the exterior elements, then one might approximate the null $\bar{\chi}^2$ or $\bar{E}^2$ distributions by those determined by the coefficients $A(l, e)$. On the other hand, if the weights on the exterior elements are small, then the distributions determined by $A_b(l, k)$ should provide a better approximation than the equal-weights distribution. It would be helpful to obtain bounds for the $\bar{\chi}^2$ and $\bar{E}^2$ distributions corresponding to an arbitrary polyhedral cone or arbitrary convex cone.

3.7 APPROXIMATIONS FOR THE POWER FUNCTIONS

In Section 2.5 some exact expressions for the power functions of the $\bar{\chi}^2$ and $\bar{E}^2$ tests were obtained for the simple order ($\mu_1 \leqslant \mu_2 \leqslant \cdots \leqslant \mu_k$) and the simple tree ($\mu_1 \leqslant [\mu_2, \mu_3, \ldots, \mu_k]$) and small k. The power functions of the $\bar{\chi}^2$ test with $k = 3$ involve integrals which must be evaluated numerically, and for $k = 4$ ($k = 3$ with the $\bar{E}^2$ tests) computation of the powers of the $\bar{\chi}^2$ test require numerical integration of a two-dimensional integral. It is clear that for $k = 5$ and the $\bar{\chi}^2$ tests (or $k = 4$ and the $\bar{E}^2$ tests) the powers involve three-dimensional integrals. This, of course, makes it difficult to design experiments so that the LRTs have a desired power at a specified alternative. Thus approximations to the power functions of these tests are of interest. Typically in designing experiments to achieve a specified power, larger powers are of interest. In this section, attention is focused on approximations which are reasonably accurate when the power is at least 0.5. However, their accuracy will also be studied when the true power is smaller.

It has been conjectured that $\bar{\chi}^2_{01}$ with $H_1: \mu_1 \leqslant \mu_2 \leqslant \cdots \leqslant \mu_k$, equal weights, $w_1 = w_2 = \cdots = w_k = w$, and a fixed $\Delta = \{\sum_{i=1}^k (\mu_i - \bar{\mu})^2\}^{1/2} > 0$ has its minimum power over H_1 on the set, $\mu_1 = \mu_2 = \cdots = \mu_{k-1} < \mu_k$. It is believed that the minimum power for $\bar{\chi}^2_{01}$ with the simple tree ordering also occurs on this ray (for a discussion of these conjectures see Section 2.5). Furthermore, for the simple tree the maximum power for a fixed $\Delta > 0$ is believed to occur at the alternative with $\mu_1 < \mu_2 = \cdots = \mu_k$. Approximations will be developed for slippage alternatives. They are extensions of Bartholomew's (1961a) two-moment approximation for the minimum powers.

A two-moment approximation to the power of $\bar{\chi}^2_{01}$ for the simple order

Let $\bar{Y}_1, \bar{Y}_2, \ldots, \bar{Y}_k$ be the means of independent samples with $\bar{Y}_i \sim \mathcal{N}(\mu_i, w_i^{-1})$ and $w_i = n_i/\sigma_i^2$ for $i = 1, 2, \ldots, k$. Let $\boldsymbol{\mu}^{(r)} = (\mu_1^{(r)}, \mu_2^{(r)}, \ldots, \mu_k^{(r)})$ with $\mu_1^{(r)} = \mu_2^{(r)} = \cdots = \mu_r^{(r)} < \mu_{r+1}^{(r)} = \cdots = \mu_k^{(r)}$. To compute the restricted MLE, $\boldsymbol{\mu}^* = (\mu_1^*, \mu_2^*, \ldots, \mu_k^*)$, for the simple order, one may apply the PAVA to the first r sample means, apply the PAVA to the last $k - r$ sample means, and then pool between these two groups if necessary. If the difference $\mu_{r+1}^{(r)} - \mu_r^{(r)}$ is large enough for the power to be in the range under consideration, then the probability that the estimates in the first group, i.e. for the first r populations, are smaller than those in the second group will be relatively large. Of course, when this event occurs the PAVA does not pool between these groups. The proposed approximation uses the power of the LRT of H_0 versus H_1'–H_0 where $H_1': \mu_1 \leqslant \mu_2 \leqslant \cdots \leqslant \mu_r$, $\mu_{r+1} \leqslant \cdots \leqslant \mu_k$. Notice that H_1' ignores the restriction $\mu_r \leqslant \mu_{r+1}$ in H_1. The resulting ordering is decomposable and so the MLEs subject to H_1' can be computed separately for the two index sets $\{1, 2, \ldots, r\}$ and $\{r+1, r+2, \ldots, k\}$. Let $\bar{\boldsymbol{\mu}} = (\bar{\mu}_1, \bar{\mu}_2, \ldots, \bar{\mu}_k)$ denote the MLE of $\boldsymbol{\mu}$ subject to H_1' and let

$$\hat{\mu}' = \frac{\sum_{i=1}^r w_i \bar{Y}_i}{\sum_{i=1}^r w_i} \quad \text{and} \quad \hat{\mu}'' = \frac{\sum_{i=r+1}^k w_i \bar{Y}_i}{\sum_{i=r+1}^k w_i}.$$

In this discussion, it is helpful to index the test statistics by the number of populations, and so with this notation

$$\begin{aligned}
\bar{\chi}^2_{01}(k) &\doteq \sum_{i=1}^k w_i(\bar{\mu}_i - \hat{\mu})^2 \\
&= \sum_{i=1}^r w_i(\bar{\mu}_i - \hat{\mu}')^2 + (\hat{\mu}' - \hat{\mu})^2 \sum_{i=1}^r w_i \\
&\quad + 2(\hat{\mu}' - \hat{\mu}) \sum_{i=1}^r w_i(\bar{\mu}_i - \hat{\mu}') + \sum_{i=r+1}^k w_i(\bar{\mu}_i - \hat{\mu}'')^2 \\
&\quad + (\hat{\mu}'' - \hat{\mu})^2 \sum_{i=r+1}^k w_i + 2(\hat{\mu}'' - \hat{\mu}) \sum_{i=r+1}^k w_i(\bar{\mu}_i - \hat{\mu}''). \qquad (3.7.1)
\end{aligned}$$

Because $(\bar{\mu}_1, \bar{\mu}_2, \ldots, \bar{\mu}_r)$ is the projection of $(\bar{Y}_1, \bar{Y}_2, \ldots, \bar{Y}_r)$ onto the cone of functions isotonic with respect to the usual ordering on $(1, 2, \ldots, r)$, one can apply (1.3.9) to obtain $\sum_{i=1}^r w_i(\bar{\mu}_i - \hat{\mu}') = 0$. Similarly, $\sum_{i=r+1}^k w_i(\bar{\mu}_i - \hat{\mu}'')$ is shown to be zero. Furthermore, by straightforward algebra

$$(\hat{\mu}' - \hat{\mu})^2 \sum_{i=1}^{r} w_i + (\hat{\mu}'' - \hat{\mu})^2 \sum_{i=r+1}^{k} w_i = \frac{(\sum_{i=1}^{r} w_i)(\sum_{i=r+1}^{k} w_i)(\hat{\mu}' - \hat{\mu}'')^2}{\sum_{i=1}^{k} w_i}.$$

This variable, which will be denoted by $\chi_1^2(\Delta^2)$, has a noncentral chi-squared distribution with one degree of freedom and noncentrality parameter

$$\frac{(\sum_{i=1}^{r} w_i)(\sum_{i=r+1}^{k} w_i)(\mu_k^{(r)} - \mu_1^{(r)})^2}{\sum_{i=1}^{k} w_i} = \Delta^2.$$

Setting $V_1 = \sum_{i=1}^{r} w_i(\bar{\mu}_i - \hat{\mu}')^2$ and $V_2 = \sum_{i=r+1}^{k} w_i(\bar{\mu}_i - \hat{\mu}'')^2$, one obtains the approximation

$$\bar{\chi}_{01}^2(k) \doteq V_1 + V_2 + \chi_1^2(\Delta^2). \tag{3.7.2}$$

Because $(\bar{Y}_1 - \hat{\mu}', \bar{Y}_2 - \hat{\mu}', \ldots, \bar{Y}_r - \hat{\mu}'), \hat{\mu}', (\bar{Y}_{r+1} - \hat{\mu}'', \bar{Y}_{r+2} - \hat{\mu}'', \ldots, \bar{Y}_k - \hat{\mu}'')$, and $\hat{\mu}''$ are independent, so are V_1, V_2, and $\chi_1^2(\Delta^2)$. Furthermore, because $\mu_1^{(r)} = \mu_2^{(r)} = \cdots = \mu_r^{(r)}$ and $\mu_{r+1}^{(r)} = \mu_{r+2}^{(r)} = \cdots = \mu_k^{(r)}$, the distributions of V_1 and V_2 are given by Theorem 2.3.1. Bartholomew's approximation is a two-moment chi-square fit to the right-hand side of (3.7.2) with $\chi_1^2(\Delta^2)$ replaced by $\chi_1^2(\Delta^2)\, I_{[\hat{\mu}' < \hat{\mu}'']}$ (where I_A denotes the indicator of the event A). Singh and Wright (1987) study this modification numerically. In their study, for the powers greater than 0.75 these two approximations (Bartholomew's and the two-moment chi-squared fit to the right-hand side of (3.7.2)) differ by at most 0.0001; for powers greater than 0.40, they differ at most by 0.0011; and for powers around 0.20 the difference was about 0.008. Because the moments of $\chi_1^2(\Delta^2)$ are straightforward, the approximation in (3.7.2) will be studied. The jth cumulant of $\chi_v^2(\lambda)$, i.e. a chi-squared variable with v degrees of freedom and noncentrality parameter λ, is

$$\kappa_j(v, \lambda) = 2^{j-1}(j-1)!(v + j\lambda). \tag{3.7.3}$$

Let $\mathbf{w}_{(r)} = (w_1, w_2, \ldots, w_r)$, $\mathbf{w}^{(k-r)} = (w_{r+1}, w_{r+2}, \ldots, w_k)$, $\kappa_j^*(r, \mathbf{w}_{(r)})$ denote the jth cumulant of V_1 and $\kappa_j^*(k-r, \mathbf{w}^{(k-r)})$ denote the jth cumulant of V_2. The cumulants of the right-hand side of (3.7.2) are

$$\kappa_j^{(a)} = \kappa_j^*(r, \mathbf{w}_{(r)}) + \kappa_j^*(k-r, \mathbf{w}^{(k-r)}) + 2^{j-1}(j-1)!(1 + j\Delta^2), \tag{3.7.4}$$

and applying (3.2.1), the two-moment chi-squared approximation to the power of $\bar{\chi}_{01}^2(k)$ with critical value C at the alternative $\boldsymbol{\mu}^{(r)}$ is given by

$$\pi' = \pi'(\boldsymbol{\mu}^{(r)}) = P[\chi_{2b}^2 \geqslant 2C/\rho] \qquad \text{with } \rho = \kappa_2^{(a)}/\kappa_1^{(a)} \quad \text{and} \quad b = \kappa_1^{(a)}/\rho. \tag{3.7.5}$$

For powers of at least 0.5, $P[\bar{\chi}_{01}^2(k) = 0]$ is small and hence it is not necessary to correct for the discrete part of $\bar{\chi}_{01}^2(k)$ in this two-moment chi-square fit. The accuracy of the approximation, π', is discussed later.

If the weights are equal, that is $w_1 = w_2 = \cdots = w_k$, then $\kappa_j^*(r, \mathbf{w}_{(r)})$ and $\kappa_j^*(k-r, \mathbf{w}^{(k-r)})$ are given by (3.3.2) or may be found in Table A.16 in the Appendix. If the weights are not equal, then the cumulants of V_1 and V_2 must be computed or approximated. The computation of $\kappa_j^*(r, \mathbf{w}_{(r)})$ is discussed; that of $\kappa_j^*(k-r, \mathbf{w}^{(k-r)})$ is identical. If $r \leqslant 4$ then the $P_S(l, r; \mathbf{w}_{(r)})$ can be computed using the formulas in Section 2.4. Using Theorem 2.3.1, the jth moment of $\bar{\chi}_{01}^2(r)$ is

$$m_j = \sum_{l=1}^{r} P_S(l, r; \mathbf{w}_{(r)})(l-1)(l+1)\cdots(l+2j-3), \tag{3.7.6}$$

so that $\kappa_1^*(r, \mathbf{w}_{(r)}) = m_1$ and $\kappa_2^*(r, \mathbf{w}_{(r)}) = m_2 - m_1^2$. If $r \geqslant 5$, the approximation to the distribution of $\bar{\chi}_{01}^2(r)$ based on the pattern of large and small weights is recommended. The cumulants of this approximation to the distribution of $\bar{\chi}_{01}^2(r)$ are given in (3.4.2). In Section 3.1, it was noted that the null distribution of $\bar{\chi}_{01}^2$ is reasonably robust to moderate variation in the weights, and hence if the weights are nearly equal, the equal-weights probabilities should provide a reasonable approximation.

An approximation based on mixtures: the power of $\bar{\chi}_{01}^2$ for the simple order

For most practical purposes the two-moment approximation discussed in the last subsection would be adequate. Singh and Wright (1987) give a four-moment approximation which seems to be more accurate for powers of 0.8 or larger. Furthermore, one can compute tail probabilities for the right-hand side of (3.7.2). While this requires more computations, it does provide a more accurate approximation for large powers and it is a technique which can be used to provide an approximation to the powers of $\bar{E}_{01}^2$.

Let $L_1(L_2)$ denote the number of distinct values in $\bar{\mu}_1, \bar{\mu}_2, \ldots, \bar{\mu}_r(\bar{\mu}_{r+1}, \bar{\mu}_{r+2}, \ldots, \bar{\mu}_k)$. Conditioning on $L_1 = l_1$ and $L_2 = l_2$, the proof of Theorem 2.3.1 shows that

$$P[V_1 + V_2 + \chi_1^2(\Delta^2) \geqslant C]$$

$$= \sum_{l_1=1}^{r} \sum_{l_2=1}^{k-r} P_S(l_1, r; \mathbf{w}_{(r)}) P_S(l_2, k-r; \mathbf{w}^{(k-r)}) P[\chi_{l_1+l_2-1}^2(\Delta^2) \geqslant C]$$

where $\chi_\nu^2(\lambda)$ is a chi-squared variable with ν degrees of freedom and noncentrality parameter λ. Collecting together those terms for which $l_1 + l_2$ is constant, this last expression can be written as

$$\sum_{l=2}^{k} Q(l, k; \mathbf{w}) P[\chi_{l-1}^2(\Delta^2) \geqslant C] \tag{3.7.7}$$

where $\{Q(l, k; \mathbf{w})\}$ is the convolution of $\{P_S(l, r; \mathbf{w}_{(r)})\}$ with $\{P_S(l, k-r; \mathbf{w}^{(k-r)})\}$. It should be noted that the approximation in (3.7.7) is a mixture of noncentral chi-square distributions and that $Q(1, k; \mathbf{w}) = 0$. To use (3.7.7), which will be denoted by $\pi'' = \pi''(\boldsymbol{\mu}^{(r)})$, one must first find the level probabilities, $P_S(l, r; \mathbf{w}_{(r)})$ and

$P_S(l, k-r; \mathbf{w}^{(k-r)})$. If the weights are equal they can be obtained from Table A.10 in the Appendix. For unequal weights they can be computed using the formulas in Section 2.4 or approximated using the techniques in Section 3.4.

The accuracies of the two approximations

The accuracies of the two approximations, π' and π'', are studied in the case of equal weights. Since the power function is invariant under a shift by a constant vector, the power only depends on r and $\mu_k^{(r)} - \mu_1^{(r)}$ for these slippage alternatives. If one applies the PAVA with equal weights to $-\bar{Y}_k, -\bar{Y}_{k-1}, \ldots, -\bar{Y}_1$, then it is not difficult to see that the resulting estimates are $-\mu_k^*, -\mu_{k-1}^*, \ldots, -\mu_1^*$, where $\mu_1^*, \mu_2^*, \ldots, \mu_k^*$ are the estimates obtained by applying the PAVA to $\bar{Y}_1, \bar{Y}_2, \ldots, \bar{Y}_k$ with equal weights. Thus, the power function at $(-\mu_k, -\mu_{k-1}, \ldots, -\mu_1)$ is the same as at $(\mu_1, \mu_2, \ldots, \mu_k)$. Therefore, for the simple order, it is sufficient to study the powers at $\boldsymbol{\mu}^{(r)}$ with $r \leqslant [k/2]$ where [ ] denotes the greatest integer function. For the case of equal weights and the usual simple order, Table 3.7.1 gives some exact powers of $\bar{\chi}^2_{01}(k)$ at the slippage alternatives, which are denoted by $\pi = \pi(\boldsymbol{\mu}^{(r)})$. The exact powers, π', π'' and the error percentages are given for $k = 3, 4$, and 10, slippage alternatives $\boldsymbol{\mu}^{(r)}$ with $r = 1, 2, \ldots, [k/2]$, and $\Delta = 1, 2, 3, 4$. For $k = 10$ the powers are estimated using Monte Carlo techniques based on 10 000 iterations. The value π is the exact power for the simple order, π'' is the exact power for the order $\mu_1 \leqslant \cdots \leqslant \mu_r, \mu_{r+1} \leqslant \cdots \leqslant \mu_k$, and π' is the two-moment chi-squared approximation to π''.

The two-moment approximation, which is simpler to use, would seem to be satisfactory for most practical purposes, and for the smaller values of power, i.e. around 0.2, it is more accurate than the approximation based on mixtures of noncentral chi-squared variables. For values of power greater than 0.8 the second approximation, π'', is more accurate. For such values of power and $k = 3$ and 4 the largest value of $|\pi - \pi'|$ is 0.0005 and for $k = 10$ the largest of these differences is 0.008. It is instructive to note that for $k = 4$ and 10, the differences as r varies do not seem to be significant, indicating that the approximation performs well for the full range of these slippage alternatives. However, for $k = 10$ and small values of power the percentage errors tend to increase with r, but for powers 0.5 or larger the differences for increasing r are not noticeable.

Approximations to the power of $\bar{E}^2_{01}$ for the simple order

Suppose that in addition to the vector of sample means $\bar{\mathbf{Y}}$ one has an independent estimate of the common variance $\sigma_1^2 = \sigma_2^2 = \cdots = \sigma_k^2 = \sigma^2$, say S^2, with $Q(\nu) = \nu S^2/\sigma^2 \sim \chi^2_\nu$ and ν a positive integer. (The case in which the variances are known up to a multiplicative constant is treated similarly.) Unfortunately, it seems difficult to approximate the cumulants of $\bar{E}^2_{01}(k)$ under the alternative

Table 3.7.1 Power of $\bar{\chi}^2_{01}(k)$ for the simple order, $\alpha = 0.05$, and slippage alternatives with $\mu_1 = \mu_2 = \cdots = \mu_r < \mu_{r+1} = \cdots = \mu_k$

	$\Delta = 1$	$\Delta = 2$	$\Delta = 3$	$\Delta = 4$
		$k = 3, r = 1$		
Exact, π	0.2208	0.5691	0.8738	0.9833
Two-moment, π'	0.2191 (0.8%)	0.5617 (1.3%)	0.8922 (2.1%)	0.9924 (0.9%)
Mixture, π''	0.2289 (3.7%)	0.5702 (0.2%)	0.8739 (0.0%)	0.9833 (0.0%)
		$k = 4, r = 1$		
Exact, π	0.2023	0.5309	0.8486	0.9773
Two-moment, π'	0.2026 (0.1%)	0.5208 (1.9%)	0.8634 (1.7%)	0.9877 (1.1%)
Mixture, π''	0.2118 (4.7%)	0.5325 (0.3%)	0.8488 (0.0%)	0.9774 (0.0%)
		$k = 4, r = 2$		
Exact, π	0.2121	0.5451	0.8558	0.9788
Two-moment, π'	0.2200 (3.7%)	0.5381 (1.3%)	0.8712 (1.8%)	0.9886 (1.0%)
Mixture, π''	0.2292 (8.1%)	0.5490 (0.7%)	0.8563 (0.1%)	0.9788 (0.0%)
		$k = 10, r = 1$		
Estimated, π_M	0.1644	0.4465	0.7838	0.9571
Two-moment, π'	0.1679 (2.1%)	0.4287 (4.0%)	0.7810 (0.4%)	0.9676 (1.1%)
Mixture, π''	0.1748 (6.3%)	0.4432 (0.7%)	0.7761 (1.0%)	0.9554 (0.2%)
		$k = 10, r = 2$		
Estimated, π_M	0.1807	0.4794	0.8040	0.9615
Two-moment, π'	0.1980 (9.6%)	0.4640 (3.2%)	0.8023 (0.2%)	0.9720 (1.1%)
Mixture, π''	0.2057 (13.8%)	0.4771 (0.5%)	0.7960 (1.0%)	0.9606 (0.1%)
		$k = 10, r = 3$		
Estimated, π_M	0.1889	0.4900	0.8105	0.9656
Two-moment, π'	0.2152 (13.9%)	0.4828 (1.5%)	0.8130 (0.3%)	0.9741 (0.9%)
Mixture, π''	0.2233 (18.2%)	0.4951 (1.0%)	0.8060 (0.6%)	0.9631 (0.3%)
		$k = 10, r = 4$		
Estimated, π_M	0.1921	0.4935	0.8117	0.9650
Two-moment, π'	0.2243 (16.8%)	0.4925 (0.2%)	0.8183 (0.8%)	0.9751 (1.0%)
Mixture, π''	0.2327 (21.1%)	0.5043 (2.2%)	0.8110 (0.1%)	0.9643 (0.1%)
		$k = 10, r = 5$		
Estimated, π_M	0.1923	0.4950	0.8102	0.9665
Two-moment, π'	0.2272 (18.1%)	0.4954 (0.1%)	0.8200 (1.2%)	0.9754 (0.9%)
Mixture, π''	0.2356 (22.5%)	0.5072 (2.5%)	0.8125 (0.3%)	0.9647 (0.2%)

hypothesis. Hence, the techniques which led to the simple two-moment chi-squared approximation to the power of $\bar{\chi}_{01}^2$ cannot be used in this case. However, the approach using mixtures of noncentral distributions can be employed.

Recall that $\bar{E}_{01}^2(k) \geqslant C$ is equivalent to (see (2.2.12))

$$S'_{01} = \frac{\bar{\chi}_{01}^2(k)}{\bar{\chi}_{12}^2(k) + Q(v)} \geqslant \frac{C}{1-C}.$$

Hence, an approximation to $\bar{\chi}_{12}^2(k)$ is needed for the slippage alternatives. The rationale for the approximation to $\bar{\chi}_{01}^2(k)$ is to replace H_1 by H'_1. Replacing $\boldsymbol{\mu}^*$ by $\bar{\boldsymbol{\mu}}$, the restricted MLE under H'_1, the statistic

$$\bar{\chi}_{12}^2(k) \doteqdot W_1 + W_2, \tag{3.7.8}$$

where $W_1 = \sum_{i=1}^{r} w_i(\bar{Y}_i - \bar{\mu}_i)^2$, $W_2 = \sum_{i=r+1}^{k} w_i(\bar{Y}_i - \bar{\mu}_i)^2$, and W_1 and W_2 are independent. Because $(\bar{\mu}_1, \bar{\mu}_2, \ldots, \bar{\mu}_r)$ is the projection of $(\bar{Y}_1, \bar{Y}_2, \ldots, \bar{Y}_r)$ onto the cone of functions isotonic with respect to the usual simple ordering on $\{1, 2, \ldots, r\}$ and $\mu_1 = \mu_2 = \cdots = \mu_r$, the distribution of W_1 is given by Theorem 2.3.1. A similar statement applies to W_2. The independence of $(\bar{Y}_1 - \hat{\mu}', \bar{Y}_2 - \hat{\mu}', \ldots, \bar{Y}_r - \hat{\mu}')$, $\hat{\mu}'$, $(\bar{Y}_{r+1} - \hat{\mu}'', \bar{Y}_{r+2} - \hat{\mu}'', \ldots, \bar{Y}_k - \hat{\mu}'')$, $\hat{\mu}''$, and $Q(v)$ implies that (V_1, W_1), (V_2, W_2), $\bar{\chi}_1^2(\Delta^2)$, and $Q(v)$ are independent. Conditioning on $L_1 = l_1$ and $L_2 = l_2$, with L_1 and L_2 defined as in the last subsection, $V_1 \sim \chi_{l_1-1}^2$, $W_1 \sim \chi_{r-l_1}^2$, $V_2 \sim \chi_{l_2-1}^2$, $W_2 \sim \chi_{k-r-l_2}^2$, V_1 and W_1 are independent, and V_2 and W_2 are independent. (See the proof of Theorem 2.3.1.) Therefore, the power of $\bar{E}_{01}^2$ at $\boldsymbol{\mu}^{(r)}$, which will, in this subsection, be denoted by $\pi(\boldsymbol{\mu}^{(r)})$ or π when $\boldsymbol{\mu}^{(r)}$ is understood, is approximated by

$$\sum_{l_1=1}^{r} \sum_{l_2=1}^{k-r} P_S(l_1, r; \mathbf{w}_{(r)}) P_S(l_2, k-r; \mathbf{w}^{(k-r)}) P\left[\frac{\chi_{l_1+l_2-1}^2(\Delta^2)}{\chi_{v+k-l_1-l_2}^2} \geqslant \frac{C}{1-C}\right],$$

where for each l_1 and l_2 the noncentral variable, $\chi_{l_1+l_2-1}^2(\Delta^2)$, is independent of $\chi_{v+k-l_1-l_2}^2$. Again letting $Q(l, k; \mathbf{w})$ denote the convolution of $\{P_S(l, r; \mathbf{w}_{(r)})\}$ with $\{P_S(l, k-r; \mathbf{w}^{(k-r)})\}$, the approximate power for this slippage alternative is

$$\pi' = \pi'(\boldsymbol{\mu}^{(r)}) = \sum_{l=2}^{k} Q(l, k; \mathbf{w}) P\left[F_{l-1, v+k-l}(\Delta^2) \geqslant \frac{v+k-l}{l-1}\frac{C}{1-C}\right], \tag{3.7.9}$$

where C is the critical value for $\bar{E}_{01}^2(k)$ and $F_{a,b}(\Delta^2)$ is a noncentral F variable with its numerator a noncentral chi-squared variable with a degrees of freedom and noncentrality parameter Δ^2 and its denominator a central chi-squared variable with b degrees of freedom. If one wishes to avoid the use of noncentral F probabilities, the variable $F_{l-1, v+k-l}(\Delta^2)$ could be approximated by a constant times a central F variable. Details are given in Singh and Wright (1987).

Because the value of r seemed to have little effect on the accuracy of the approximation to the power of $\bar{\chi}_{01}^2(k)$ and because several values of v need to be considered, the accuracy of this approximation is only studied for $r = 1$ or, equivalently, $r = k - 1$. For $k = 3, 4, 10$, $\nu = 10, 20, 40$, and $\Delta = 1, 2, 3, 4$ the value of π and the approximation π' was computed for the $\alpha = 0.05$ test. For $k = 4$ and 10 these values were estimated by Monte Carlo techniques with 10 000 iterations. The estimates are denoted by π_M. Table 3.7.2 contains these values as well as error percentages.

Table 3.7.2 Powers and approximations for $\bar{E}^2_{01}(k)$, the simple order, $\alpha = 0.05$, and alternatives $\mu_1 < \mu_2 = \mu_3 = \cdots = \mu_k$

	$\Delta = 1$	$\Delta = 2$	$\Delta = 3$	$\Delta = 4$
		$k = 3$, $\nu = 10$		
Exact, π	0.2003	0.5034	0.8102	0.9605
Mixture, π'	0.2099 (4.8%)	0.5042 (0.2%)	0.8099 (0.0%)	0.9604 (0.0%)
		$k = 3$, $\nu = 20$		
Exact, π	0.2090	0.5339	0.8425	0.9734
Mixture, π'	0.2186 (4.5%)	0.5360 (0.4%)	0.8432 (0.1%)	0.9737 (0.0%)
		$k = 3$, $\nu = 40$		
Exact, π	0.2141	0.5500	0.8568	0.9767
Mixture, π'	0.2235 (4.4%)	0.5528 (0.5%)	0.8589 (0.2%)	0.9789 (0.2%)
		$k = 4$, $\nu = 10$		
Estimated, π_M	0.1821	0.4597	0.7729	0.9464
Mixture, π'	0.1943 (6.7%)	0.4621 (0.5%)	0.7707 (0.3%)	0.9448 (0.2%)
		$k = 4$, $\nu = 20$		
Estimated, π_M	0.1938	0.4907	0.8088	0.9656
Mixture, π'	0.2019 (4.2%)	0.4944 (0.8%)	0.8096 (0.1%)	0.9631 (0.3%)
		$k = 4$, $\nu = 40$		
Estimated, π_M	0.1974	0.5095	0.8262	0.9731
Mixture, π'	0.2064 (4.6%)	0.5125 (0.6%)	0.8292 (0.4%)	0.9708 (0.2%)
		$k = 10$, $\nu = 10$		
Estimated, π_M	0.1458	0.3722	0.6842	0.9027
Mixture, π'	0.1628 (11.7%)	0.3795 (2.0%)	0.6843 (0.0%)	0.9043 (0.2%)
		$k = 10$, $\nu = 20$		
Estimated, π_M	0.1528	0.3965	0.7197	0.9266
Mixture, π'	0.1667 (9.1%)	0.4022 (1.4%)	0.7200 (0.0%)	0.9270 (0.0%)
		$k = 10$, $\nu = 40$		
Estimated, π_M	0.1555	0.4073	0.7480	0.9382
Mixture, π'	0.1700 (9.3%)	0.4195 (3.0%)	0.7447 (0.4%)	0.9406 (0.0%)

For larger values of the power, the error percentages seem to increase slightly with v. For the cases considered with the power at least 0.7, the error percentages are at most 0.4 percent. Because $\bar{E}_{01}^2(k)$ is asymptotically equivalent to $\bar{\chi}_{01}^2(k)$ and the mixture approximation performs very well there, it seems reasonable to assume that the performance of π' over the range of $v \geqslant 10$ (this is the smallest value that has been checked) is quite satisfactory. For values of power around 0.4 and 0.5 the error percentages are at most 3.0 percent. For larger k, that is $k = 10$, the error percentages were considerably larger for smaller values of power, say under 0.2.

Approximations to the power of $\bar{\chi}_{01}^2$ for the simple tree ordering

In this subsection slippage alternatives, $\mu_1^{(r)} = \mu_2^{(r)} = \cdots = \mu_r^{(r)} < \mu_{r+1}^{(r)} = \cdots = \mu_k^{(r)}$, are considered. As was noted earlier, for the tree ordering with equal weights the maximum power of $\bar{\chi}_{01}^2$ for a fixed $\Delta > 0$ is believed to occur at alternatives of the form $\mu_1 < \mu_2 = \cdots = \mu_k$. Because the LRTs are invariant under permutations of the second, third,..., and the k population, one slippage alternative represents a wider class of alternatives than for the simple order.

As was the case with the simple order, an approximation to the power of the LRTs is obtained by ignoring the restrictions between the first r means and the last $k - r$ means, that is H_1 is replaced by H_1': $\mu_1 \leqslant \mu_i$ for $i = 2, 3, \ldots, r$. The MLE of $\boldsymbol{\mu}$ subject to H_1', which is denoted by $\bar{\boldsymbol{\mu}} = (\bar{\mu}_1, \bar{\mu}_2, \ldots, \mu_k)$, satisfies $\bar{\mu}_i = \bar{Y}_i$ for $i = r+1, r+2, \ldots, k$ and $(\bar{\mu}_1, \bar{\mu}_2, \ldots, \bar{\mu}_r)$ is the MLE of $(\mu_1, \mu_2, \ldots, \mu_r)$ subject to the tree ordering on $\{1, 2, \ldots, r\}$. An argument like the one given for (3.7.2) shows that this approximation yields

$$\bar{\chi}_{01}^2(k) \doteq \bar{\chi}_{01}^2(r) + \chi_{k-r}^2(\Delta^2), \tag{3.7.10}$$

where $\bar{\chi}_{01}^2(r) = \sum_{i=1}^r w_i(\bar{\mu}_i - \hat{\mu}')^2$, $\hat{\mu}' = \sum_{i=1}^r w_i \bar{Y}_i / \sum_{i=1}^r w_i$, $\chi_{k-r}^2(\Delta^2)$ is a noncentral chi-square variable with $k - r$ degrees of freedom and noncentrality parameter Δ^2, and $\chi_{01}^2(r)$ and $\chi_{k-r}^2(\Delta^2)$ are independent. It should be noted that if $r = 1$, $\bar{\chi}_{01}^2(r) = 0$ and $\bar{\chi}_{01}^2(k)$ is approximated by the noncentral variable $\chi_{k-r}^2(\Delta^2)$. Following the same rationale, for the simple tree and a slippage alternative,

$$\bar{\chi}_{12}^2(k) \doteq \bar{\chi}_{12}^2(r), \tag{3.7.11}$$

where $\bar{\chi}_{12}^2(r) = \sum_{i=1}^r w_i(\bar{Y}_i - \bar{\mu}_i)^2$.

To obtain the two-moment chi-squared approximation for the power of $\bar{\chi}_{01}^2(k)$, the cumulants of the right-hand side of (3.7.10) are needed. With the jth cumulant of $\bar{\chi}_{01}^2(r)$ denoted by $\tau_j(r, \mathbf{w}_{(r)})$, the desired cumulants are (the cumulants of $\chi_{k-r}^2(\Delta^2)$ are given in (3.7.3))

$$\tau_j^{(a)} = \tau_j(r, \mathbf{w}_{(r)}) + 2^{j-1}(j-1)![k - r + j\Delta^2]. \tag{3.7.12}$$

If $r = 1$, let us agree that $\tau_j(r, \mathbf{w}_{(r)}) = 0$. The two-moment approximation to the power of $\bar{\chi}^2_{01}(k)$ with the critical value C is given by

$$\pi' = \pi'(\boldsymbol{\mu}^{(r)}) = P[\chi^2_{2b} \geqslant 2C/\rho] \qquad \text{with } \rho = \tau_2^{(a)}/\tau_1^{(a)} \quad \text{and} \quad b = \tau_1^{(a)}/\rho. \quad (3.7.13)$$

If the weights, $w_1, w_2, \ldots, w_r$, are equal, then the cumulants, $\tau_j(r, \mathbf{w}_{(r)})$, can be taken from Table A.17 in the Appendix. For unequal weights, the $P_T(l, r; \mathbf{w}_{(r)})$ can be computed using the formulas in Section 2.4, the moments of $\bar{\chi}^2_{01}(r)$ computed via (3.7.6), and then the cumulants obtained from the moments. Another possibility for unequal weights is to use the approximation to $\bar{\chi}^2_{01}(r)$ given in Section 3.5. Expressions for the approximate cumulants are given there also.

Conditioning on the number of distinct values in $(\bar{\mu}_1, \bar{\mu}_2, \ldots, \bar{\mu}_r)$ and the value of $\chi^2_{k-r}(\Delta^2)$ in (3.7.10) and applying Theorem 2.3.1, one obtains the following approximation to the power of $\bar{\chi}^2_{01}(k)$:

$$\pi'' = \pi''(\boldsymbol{\mu}^{(r)}) = \sum_{l=1}^{r} P_T(l, r; \mathbf{w}_{(r)}) P[\chi^2_{k-r+l-1}(\Delta^2) \geqslant C]. \qquad (3.7.14)$$

If $r = 1$, then $P_T(1, 1; w_1) = 1$ and $\pi'' = P[\chi^2_{k-1}(\Delta^2) \geqslant C]$.

If the weights, $w_1, w_2, \ldots, w_r$, are equal, then $P_T(l, r; \mathbf{w}_{(r)})$ can be taken from Table A.11 in the Appendix. For unequal weights they can be computed using the results in Section 2.4 or approximated using the work of Section 3.5. Recall that in Section 3.1, the level probabilities were found to be robust to moderate changes in the weights. Hence, the equal-weights $P_T(l, r)$ could be in (3.7.14) if the weights are not too different.

The accuracies of these two approximations were studied in Singh and Wright (1987) for k ranging from 3 to 10 with r at the extremes, $r = 1$ and $r = k - 1$. For $r = k - 1$ and powers of 0.8 or larger the percentage error of the two-moment approximation, π', was at most 1.9 percent and for the mixture approximation, π'', they were at most 0.5 percent. For powers of this same size and $r = 1$, the error percentages were about the same size. Hence, the approximations seem to perform well over the full range of slippage alternatives. The mixture approximation was more accurate except in two of the cases studied. In these two cases the powers were about 0.2 and $r = 1$. However, except for small powers, less than 0.2, with k large and r small, either approximation was judged adequate for most practical purposes (see Singh and Wright, 1987, for details).

An approximation to the power $\bar{E}^2_{01}$ for the simple tree ordering

If, as for the simple order, one defines S'_{01} to be $\bar{E}^2_{01}/(1 - \bar{E}^2_{01})$, then applying (3.7.10) and (3.7.11) with a slippage alternative,

$$S'_{01} \doteq \frac{\bar{\chi}^2_{01}(r) + \chi^2_{k-r}(\Delta^2)}{\bar{\chi}^2_{12}(r) + Q(v)}, \qquad (3.7.15)$$

where $Q(v)$ is a chi-squared variable which is independent of $(\bar{\chi}^2_{01}(r), \bar{\chi}^2_{12}(r), \chi^2_{k-r}(\Delta^2))$. Conditioning on the number of distinct values in $(\bar{\mu}_1, \bar{\mu}_2, \ldots, \bar{\mu}_r)$, $Q(v)$, and $\chi^2_{k-r}(\Delta^2)$ and applying Theorem 2.3.1, one obtains the following approximation to the power of $\bar{E}^2_{01}(k)$:

$$\pi' = \pi'(\boldsymbol{\mu}^{(r)}) = \sum_{l=1}^{r} P_T(l, r; \mathbf{w}_{(r)}) P\left[F_{k-r+l-1, v+r-l}(\Delta^2) \geqslant \frac{v+r-l}{k-r+l-1} \frac{C}{1-C} \right], \tag{3.7.16}$$

where C is the critical value of $\bar{E}^2_{01}(k)$ and F is as in (3.7.9). Again this approximation is quite satisfactory except for larger k, $r = 1$, and values of power less than 0.2 (cf. Singh and Wright, 1986c).

Example 3.7.1 Dunnett (1955) considered the problem of determining if one of five treatment means is larger than the control mean when it is believed that they are no smaller than the control mean. He considered the alternative $\boldsymbol{\mu}/\sigma = (-1, -1, -1, -1, -1, 5)/\sqrt{30}$. Mukerjee, Robertson, and Wright (1987) found that a common sample size of $n = 16$ is needed for Dunnett's test with $\alpha = 0.05$ to insure that the power is at least 0.8. They also found that the corresponding sample size for an orthogonal contrast test, which is discussed in Chapter 4, is $n = 11$. (Unlike Dunnett's test and the LRT, the orthogonal contrast test does not have its minimum power on the ray being considered.) The approximation given in this section enables one to determine approximately the corresponding sample size for the LRT.

A trial and error approach is used; however, one would expect the LRT to require a smaller sample size than Dunnett's test (see Robertson and Wright, 1985, and Section 4.3). If $n = 13$, then $v = 72$, and interpolating in Table A.8 in the Appendix, the $\alpha = 0.05$ critical value for $\bar{E}^2_{01}$ is 0.1155. Applying (3.7.16), the approximate power is $\pi' = 0.814$. Since the approximate power at $n = 12$ is less than 0.8, it appears that $n = 13$ is the sample size needed.

3.8 COMPLEMENTS

Because of the complexity of the distributions of the $\bar{\chi}^2$ and $\bar{E}^2$ test statistics, approximations to these distributions are of considerable interest. Even if the level probabilities are known, the p value corresponding to an observed value of one of these test statistics may be tedious to compute. For the case of equal sample sizes, Bartholomew (1959a, 1959b) proposed a two-moment chi-square approximation for the null distribution of $\bar{\chi}^2_{01}$, and Sasabuchi and Kulatunga (1985) developed two- and three-moment beta approximations to the null distribution of $\bar{E}^2_{01}$. This work was extended in Singh and Wright (1986a, 1986b) to apply to $\bar{\chi}^2_{12}$ and $\bar{E}^2_{12}$, and the coefficients needed to implement these approximations for a simple order and a simple tree ordering are given there (see also Tables A.14 and A.15 in the Appendix).

Siskind (1976) and Grove (1980) conjectured that the equal-weights null distributions of these test statistics should provide reasonable approximations for the case of unequal sample sizes if the sample sizes are not too different. This conjecture was established in Robertson and Wright (1983) and Wright and Tran (1985) for the simple ordering and the simple tree ordering.

Chase (1974), in a fundamental paper on the study of these approximations, considered the simply ordered case with increased precision in the sample from the population with the smallest mean. He proposed interpolating between the equal-weights critical values (p values) and those of the limiting distribution obtained by letting the large precision approach infinity. Robertson and Wright (1982b) obtained a recursion formula for the limiting level probabilities and generalized Chase's work to allow for increased precision in any of the samples. Chase's ideas were also used to obtain an approximation based on the pattern of large and small weights (cf. Robertson and Wright, 1983). Similar results for the simple tree ordering are given in Robertson and Wright (1985) and Wright and Tran (1985).

Considering the simply ordered case, Siskind (1976) approximated the null distribution of $\bar{\chi}^2_{01}$ using a truncated Taylor's series expansion about $\rho_{i,i+1} = -\frac{1}{2}$, which is the equal-weights values of $\rho_{i,i+1}$. This approximation has the advantage of being conservative, but it has only been developed for $\bar{\chi}^2_{01}$ with a simple order and eight or fewer populations.

Bartholomew (1961a) proposed a two-moment approximation for the power of $\bar{\chi}^2_{01}$ for a simple order and a simple tree ordering. He approximated the power at the edges of these cones which are believed to produce the minimum power over the cone at a fixed distance from H_0. The approximation is extended to $\bar{E}^2_{01}$ and studied further in Singh and Wright (1987).

CHAPTER 4

Tests of Ordered Hypotheses: Generalizations of the Likelihood Ratio Tests and Other Procedures

In this chapter the k-sample LRTs of Chapter 2 are extended to exponential families and the results obtained are illustrated in several of the more common exponential families. Contrast tests and multiple contrast tests are considered as competitors to the LRT and some distribution-free tests of ordered hypotheses are described. In Section 4.6, the LRTs of Chapter 2 are extended to provide tests of hypotheses in a multivariate setting.

4.1 TESTS OF ORDERED HYPOTHESES IN EXPONENTIAL FAMILIES

Suppose ν is a σ-finite measure on the Borel subsets of the real line and consider a regular exponential family of distributions defined by the probability densities (with respect to ν) of the form

$$f(y;\theta,\tau)=\exp\left[p_1(\theta)p_2(\tau)K(y;\tau)+S(y;\tau)+q(\theta,\tau)\right] \tag{4.1.1}$$

for $y\in A$, $-\infty\leqslant\underline{\theta}<\theta<\bar{\theta}\leqslant\infty$ and $\tau\in T$. Further suppose that the family satisfies the regularity conditions (1.5.7) through (1.5.9). (In Chapter 1, it was seen that many of the commonly studied exponential families satisfy these conditions.) Let $X=\{1,2,\ldots,k\}$, let $\lesssim$ be a quasi-order on X and for $i\in X$ and $j=1,2,\ldots,n_i$, let Y_{ij} be independent with the density of Y_{ij} given by $f(y;\theta_i,\tau_i)$. Assuming the τ_i are known, let $H_0:\theta_1=\theta_2=\cdots=\theta_k$; $H_1:\boldsymbol{\theta}=(\theta_1,\theta_2,\ldots,\theta_k)$ is isotonic with respect to $\lesssim$; and $H_2:\underline{\theta}<\theta_i<\bar{\theta}$ for $i\in X$. The LRTs of H_0 versus H_1-H_0 and of H_1 versus H_2-H_1 are studied. Set $\mathbf{w}=(w_1,w_2,\ldots,w_k)$ with $w_i=n_ip_2(\tau_i)$ for $i\in X$. In Section 1.5, the unrestricted MLE of θ_i was observed to be

$$\hat{\theta}_i=\sum_{j=1}^{n_i}\frac{K(Y_{ij};\tau_i)}{n_i},$$

the MLE of the common value of the θ_i under H_0 is

$$\hat{\theta}_0 = \frac{\sum_{i=1}^k w_i \theta_i}{\sum_{i=1}^k w_i},$$

and, as given in Theorem 1.5.2, the MLE of $\boldsymbol{\theta}$ subject to $\boldsymbol{\theta} \in H_1$ is

$$\hat{\boldsymbol{\theta}}^* = P_{\mathbf{w}}(\hat{\boldsymbol{\theta}} | \mathscr{I})$$

where $\hat{\boldsymbol{\theta}} = (\hat{\theta}_1 \hat{\theta}_2, \ldots, \hat{\theta}_k)$ and $\mathscr{I}$ is the cone of functions defined on X which are isotonic with respect to $\lesssim$.

If Λ_{01} is the likelihood ratio statistic for testing H_0 versus $H_1 - H_0$ and $T_{01} = -2 \ln \Lambda_{01}$, then

$$T_{01} = 2 \sum_{i=1}^k w_i \hat{\theta}_i [p_1(\hat{\theta}_i^*) - p_1(\hat{\theta}_0)] + 2 \sum_{i=1}^k n_i [q(\hat{\theta}_i^*, \tau_i) - q(\hat{\theta}_0, \tau_i)], \tag{4.1.2}$$

and if Λ_{12} is the likelihood ratio statistic for testing H_1 versus $H_2 - H_1$ and $T_{12} = -2 \ln \Lambda_{12}$, then

$$T_{12} = 2 \sum_{i=1}^k w_i \hat{\theta}_i [p_1(\hat{\theta}_i) - p_1(\hat{\theta}_i^*)] + 2 \sum_{i=1}^k n_i [q(\hat{\theta}_i, \tau_i) - q(\hat{\theta}_i^*, \tau_i)]. \tag{4.1.3}$$

The next result provides the large sample null distributions for T_{01} and T_{12}. With $N = \sum_i^k n_i$, it will be assumed that

$$N \to \infty \quad \text{with } n_i/N \to a_i \in (0, 1) \quad \text{for } i \in X. \tag{4.1.4}$$

The notation $P\xi$ is used to indicate that the probability is computed with $\boldsymbol{\theta} = \xi$.

Theorem 4.1.1 *Let* $\{Y_{ij}\}_{j=1}^{n_i}$ *for* $i = 1, 2, \ldots, k$ *be independent samples; let* $f(y; \theta_i, \tau_i)$ *be the density of* Y_{ij} *with* f *given by (4.1.1); let the exponential family satisfy the regularity conditions (1.5.7) through (1.5.9); and let the sample sizes satisfy (4.1.4). If* $\theta_1 = \theta_2 = \cdots = \theta_k$, *then, for all real* C,

$$\lim_{N \to \infty} P_{\boldsymbol{\theta}}[T_{01} \geqslant C] = \sum_{l=1}^k P(l, k; \mathbf{w}') P[\chi_{l-1}^2 \geqslant C] \tag{4.1.5}$$

with $\mathbf{w}' = (w_1', w_2', \ldots, w_k')$ *and* $w_i' = a_i p_2(\tau_i)$ *for* $i \in X$. *If* $\boldsymbol{\theta}$ *is isotonic with respect to* $\lesssim$, *then, for all* $\xi \in H_0$,

$$\lim_{N \to \infty} P_{\boldsymbol{\theta}}[T_{12} \geqslant C] \leqslant \lim_{N \to \infty} P\xi[T_{12} \geqslant C) \quad \textit{for all real } C.$$

Moreover, for all $\xi \in H_0$ *and real* C,

$$\lim_{N \to \infty} P\xi[T_{12} \geqslant C] = \sum_{l=1}^k P(l, k; \mathbf{w}') P[\chi_{k-l}^2 \geqslant C]. \tag{4.1.6}$$

The coefficients $P(l, k; \mathbf{w}')$ *are the level probabilities associated with the normal means problem discussed in Chapters 2 and 3.*

Proof: The asymptotic distribution of T_{01} is considered first. Suppose $\boldsymbol{\theta} \in H_0$. Expanding p_1 and $q(\cdot, \tau_1)$ about $\hat{\theta}_i$ using Taylor's theorem with a second-degree remainder term and recalling that $q'(\theta, \tau) = -\theta p_1'(\theta) p_2(\tau)$, one obtains

$$T_{01} = \sum_{i=1}^{k} n_i p_2(\tau_i) \hat{\theta}_i [(\hat{\theta}_i^* - \hat{\theta}_i)^2 p_1''(\alpha_i) - (\hat{\theta}_0 - \hat{\theta}_i)^2 p_1''(\beta_i)]$$

$$+ \sum_{i=1}^{k} n_i [(\hat{\theta}_i^* - \hat{\theta}_i)^2 q''(\gamma_i, \tau_i) - (\hat{\theta}_0 - \hat{\theta}_i)^2 q''(\delta_i, \tau_i)]$$

where α_i and γ_i are between $\hat{\theta}_i^*$ and $\hat{\theta}_i$ and β_i and δ_i are between $\hat{\theta}_0$ and $\hat{\theta}_i$. It was observed in Section 1.5 that $E[K(Y_{ij}; \tau_i)] = \theta_i$ and $\mathrm{Var}\ [K(Y_{ij}; \tau_i)] = [p_1'(\theta_i) p_2(\tau_i)]^{-1}$ and thus by the law of large numbers $\hat{\theta}_i \xrightarrow{\text{a.s.}} \theta_i$ for $i = 1, 2, \ldots, k$. Because $\boldsymbol{\theta} \in H_0$, it also follows that $\hat{\theta}_0 \xrightarrow{\text{a.s.}} \theta_1 = \theta_2 = \cdots = \theta_k$. If $\boldsymbol{\theta} \in H_1$, then by the norm reducing property (see the second Corollary to Theorem 1.6.1), $\max_{1 \leqslant i \leqslant k} |\hat{\theta}_i^* - \theta_i| \leqslant \max_{1 \leqslant i \leqslant k} |\hat{\theta}_i - \theta_i| \xrightarrow{\text{a.s.}} 0$. Because $\boldsymbol{\theta}$ is assumed to be in H_0, $\alpha_i \xrightarrow{\text{a.s.}} \theta_i$, $\beta_i \xrightarrow{\text{a.s.}} \theta_i$, $\gamma_i \xrightarrow{\text{a.s.}} \theta_i$ and $\delta_i \xrightarrow{\text{a.s.}} \theta_i$. Furthermore, by the central limit theorem,

$$\begin{aligned}
&[N p_1'(\boldsymbol{\theta})]^{1/2} (\hat{\boldsymbol{\theta}} - \boldsymbol{\theta}) \\
&= \sqrt{N} \{ p_1'(\theta_1)^{1/2} (\hat{\theta}_1 - \theta_1), \ldots, (p_1'(\theta_k))^{1/2} (\hat{\theta}_k - \theta_k) \} \xrightarrow{D} \mathbf{U} \\
&= (U_1, U_2, \ldots, U_k)
\end{aligned}$$

where the U_i are independent normal variables with zero means and $\mathrm{Var}(U_i) = 1/w_i'$ for $i \in X$. Let $\mathbf{w}'/N = (w_1'/N, w_2'/N, \ldots, w_k'/N)$. Because $\theta_1 = \theta_2 = \cdots = \theta_k$, $p_1'(\theta_1) = p_1'(\theta_2) = \cdots = p_1'(\theta_k) > 0$ and dividing the weights by N does not affect the projection,

$$[N p_1'(\boldsymbol{\theta})]^{1/2} (\hat{\boldsymbol{\theta}}^* - \boldsymbol{\theta}) = P_{\mathbf{w}/N} \{ [N p_1'(\boldsymbol{\theta})]^{1/2} (\hat{\boldsymbol{\theta}} - \boldsymbol{\theta}) | \mathscr{I} \}.$$

Applying the continuity of the projection with respect to its argument and weights and Theorem 4.4 of Billingsley (1968), the last expression converges in distribution to $\mathbf{U}^* = P_{\mathbf{w}}(\mathbf{U} | \mathscr{I})$. Applying Theorem 4.4 of Billingsley (1968) again, under the assumption $\boldsymbol{\theta} \in H_0$,

$$T_{01} \xrightarrow{D} \sum_{i=1}^{k} \frac{a_i}{p_1'(\theta_i)} [\theta_i p_1''(\theta_i) p_2(\tau_i) + q''(\theta_i, \tau_i)] [(U_i^* - U_i)^2 - (U_i - \bar{U})^2]$$

where $\bar{U} = \sum_{i=1}^{k} w_i' U_i / \sum_{i=1}^{k} w_i'$. However, $q''(\theta_i, \tau_i) = -\theta_i p_1''(\theta_i) p_2(\tau_1) - p_1'(\theta_i) \cdot p_2(\tau_i)$ and hence

$$T_{01} \xrightarrow{D} \sum_{i=1}^{k} w_i' [(U_i - \bar{U})^2 - (U_i^* - U_i)^2] = \sum_{i=1}^{k} w_i' (U_i^* - \bar{U})^2.$$

Applying Theorem 2.3.1 establishes the conclusion (4.1.5).

The asymptotic distribution of T_{12} is considered next. Suppose $\boldsymbol{\theta} \in H_1$. Expanding p_1 and $q(\cdot, \tau_i)$ about $\hat{\theta}_i$ with a second-degree remainder term and using

$q'(\theta,\tau) = -\theta p_1'(\theta)p_2(\tau)$, one obtains

$$T_{12} = -\sum_{i=1}^{k} n_i[p_2(\tau_i)\hat{\theta}_i p_1''(\alpha_i) + q''(\gamma_i,\tau_i)](\hat{\theta}_i^* - \hat{\theta}_i)^2$$

where α_i and γ_i are between $\hat{\theta}_i^*$ and $\hat{\theta}_i$. Because $\boldsymbol{\theta}\in H_1, \alpha_i \xrightarrow{\text{a.s.}} \theta_i$ and $\gamma_i \xrightarrow{\text{a.s.}} \theta_i$. Define a new quasi-order $\lesssim_\theta$ induced by $\lesssim$ and $\boldsymbol{\theta}$ on X which requires $i \lesssim_\theta j$ if and only if $i \lesssim j$ and $\theta_i = \theta_j$. Let $\mathscr{I}_\theta$ be the collection of functions on X which are isotonic with respect to $\lesssim_\theta$. Note that $\mathscr{I} \subset \mathscr{I}_\theta$. Let $\eta_1 < \eta_2 < \cdots < \eta_h$ be the distinct values among $\theta_1, \theta_2, \ldots, \theta_k$ and set $S_i = \{j:\theta_j = \eta_i\}$ for $i = 1, 2, \ldots, h$. Since $\hat{\theta}_i \xrightarrow{\text{a.s.}} \theta_i$ for each i, for almost all ω in the underlying probability space and sufficiently large N,

$$\max_{j\in S_1}\hat{\theta}_j < \min_{j\in S_2}\hat{\theta}_j \leqslant \max_{j\in S_2}\hat{\theta}_j < \cdots < \min_{j\in S_h}\hat{\theta}_j. \tag{4.1.7}$$

If $\bar{\boldsymbol{\theta}} = P_{\mathbf{w}'}(\hat{\boldsymbol{\theta}}|\mathscr{I}_\theta)$, then applying Theorem 1.3.4, (4.1.7) holds with $\hat{\theta}_j$ replaced by $\bar{\theta}_j$. Thus, $\bar{\boldsymbol{\theta}}\in\mathscr{I}$ and for almost all ω and sufficiently large N, $\bar{\boldsymbol{\theta}} = \hat{\boldsymbol{\theta}}^*$. However, because $\bar{\boldsymbol{\theta}}$ can be computed separately on the S_i, $\boldsymbol{\theta}$ is constant on the S_i and $p_1'(\boldsymbol{\theta})$ is positive and constant on the S_i, it follows that

$$[Np_1'(\boldsymbol{\theta})]^{1/2}(P_{\mathbf{w}'}(\hat{\boldsymbol{\theta}}|\mathscr{I}_\theta) - \boldsymbol{\theta}) = P_{\mathbf{w}'/N}\{[Np_1'(\boldsymbol{\theta})]^{1/2}(\hat{\boldsymbol{\theta}} - \boldsymbol{\theta})|\mathscr{I}_\theta\}.$$

Therefore, with $\mathbf{U}$ defined as in the first part of the proof, one may again apply Theorem 4.4 of Billingsley (1968) to obtain

$$\begin{aligned} T_{12} &\xrightarrow{D} -\sum_{i=1}^{k}\frac{a_i}{p_1'(\theta_i)}[\theta_i p_1''(\theta_i)p_2(\tau_i) + q''(\theta_i,\tau_i)][P_{\mathbf{w}'}(\mathbf{U}|\mathscr{I}_\theta)_i - U_i]^2 \\ &= \sum_{i=1}^{k} w_i'(P_{\mathbf{w}'}(\mathbf{U}|\mathscr{I}_\theta)_i - U_i)^2. \end{aligned} \tag{4.1.8}$$

The weights, w_i', and the variables, U_i, do not depend on $\boldsymbol{\theta}$ and thus the right-hand side of (4.1.8) is maximized by choosing $\boldsymbol{\theta}$ so that $\mathscr{I}_\theta$ is smallest. Because $\mathscr{I} \subset \mathscr{I}_\theta$ for all $\boldsymbol{\theta}$ and $\mathscr{I} = \mathscr{I}_\theta$ if $\boldsymbol{\theta}$ is constant, choose $\theta_i' = \sum_{j=1}^{k}\theta_j/k$ for all i. Thus $\mathbf{H}_0$ is asymptotically least favorable and (4.1.6) follows from (4.1.8) by taking $\boldsymbol{\theta}$ to be a constant vector and applying Theorem 2.3.1. □

The application of Theorem 4.1.1 is illustrated for several exponential families.

LRTs for trends among binomial proportions

Let $P[Y_{ij} = 1] = \theta_i$ and $P[Y_{ij} = 0] = 1 - \theta_i$ for $j = 1, 2, \ldots, n_i$ and $i = 1, 2, \ldots, k$ and let $\lesssim$ be a quasi-order on $X = \{1, 2, \ldots, k\}$. As was noted in Example 1.5.1, this exponential family satisfies the regularity conditions assumed in Theorem

4.1.1 and it was observed that $w_i = n_i$ and $\hat{\theta}_i = \sum_{j=1}^{n_i} Y_{ij}/n_i$. The LRT statistics are

$$T_{01} = 2 \sum_{i=1}^{k} \left[n_i \hat{\theta}_i \ln\left(\frac{\hat{\theta}_i^*}{\hat{\theta}_0}\right) + n_i(1-\hat{\theta}_i) \ln\left(\frac{1-\hat{\theta}_i^*}{1-\hat{\theta}_0}\right)\right]$$

and (4.1.9)

$$T_{12} = 2 \sum_{i=1}^{k} \left[n_i \hat{\theta}_i \ln\left(\frac{\hat{\theta}_i}{\hat{\theta}_i^*}\right) + n_i(1-\hat{\theta}_i) \ln\left(\frac{1-\hat{\theta}_i}{1-\hat{\theta}_i^*}\right)\right],$$

and their large sample p values can be computed from (4.1.5) and (4.1.6), respectively. One would expect that the rate of convergence of the null distributions of these statistics to their limits would be like that of a binomial distribution with a fixed proportion to the appropriate normal distribution. For instance, if $n_i\hat{\theta}_i$ and $n_i(1-\hat{\theta}_i)$ are at least 5 for each i, then these approximations to the p values should be adequate for practical purposes.

Other tests for trends among binomial parameters

As was noted above, the proportions $\hat{\theta}_i$ are approximately normal by the central limit theorem, but their variances depend on the unknown θ_i and so the application of the $\bar{\chi}^2$ tests developed for normal means is not straightforward. There are two ways of dealing with this situation. The first is to transform the observations by $U_i = \sin^{-1}(\hat{\theta}_i)$ for $i = 1, 2, \ldots, k$. It is well known that U_i has an approximate $\mathcal{N}(\sin^{-1}(\theta_i), (4n_i)^{-1})$ distribution. Since the arcsin transformation is increasing, the hypotheses $H_0: \theta_1 = \theta_2 = \cdots = \theta_k$ and $H_1: \boldsymbol{\theta}$ is isotonic with respect to $\lesssim$ are unchanged if $\boldsymbol{\theta}$ is replaced by $\boldsymbol{\eta}$ with $\eta_i = E(U_i)$ for $i = 1, 2, \ldots, k$. Hence, approximate tests of H_0 versus $H_1 - H_0$ and of H_1 versus $H_2 - H_1$ are obtained by applying the $\bar{\chi}^2$ tests for normal means to the U_i. The second method was proposed by Bartholomew (1959a) and Shorack (1967) for testing H_0 versus $H_1 - H_0$. Under H_0, $\hat{\theta}_0(1-\hat{\theta}_0)/n_i$ is an estimate of the variance of $\hat{\theta}_i$ and thus $\hat{\theta}_i$ is approximately normal with mean θ_i and variance $\hat{\theta}_0(1-\hat{\theta}_0)/n_i$. Therefore the corresponding χ^2 statistic,

$$\sum_{i=1}^{k} \frac{n_i(\hat{\theta}_i^* - \hat{\theta}_0)^2}{\hat{\theta}_0(1-\hat{\theta}_0)},$$

has an approximate distribution give by (2.3.4) under H_0. To actually justify this procedure one would need to find the distribution of this statistic conditional on $\hat{\theta}_0$. This approach has been much debated, but is commonly accepted in the analysis of contingency tables. Shorack (1967) showed that the distribution of the $\bar{\chi}^2$ statistic was unaffected by the fact that the proportions are subject to the restraint

$$\frac{\sum_{i=1}^{k} n_i \hat{\theta}_i}{\sum_{i=1}^{k} n_i} = \hat{\theta}_0$$

(see the discussion of this point in Section 2.7).

Another class of approximate tests can be obtained by considering contrast tests. Contrast tests for H_0 versus H_1-H_0 and H_1 versus H_2-H_1 for the simple order and the simple tree are discussed in detail in Section 4.2 and the results given there will apply approximately if the normal variates are replaced by proportions.

Closely related to the contrast tests are tests based on rank correlation coefficients. Suppose $\lesssim$ is the usual simple order on X and that one is interested in testing H_0 versus H_1-H_0. With $N=\sum_{i=1}^{k} n_i$, consider the N pairs, (i, Y_{ij}) for $j=1,2,\ldots,n_i$ and $i=1,2,\ldots,k$. One could compute a rank correlation coefficient allowing for ties in both variables, and reject H_0 in favor of H_1 for large values of the correlation coefficient. Armitage (1955) proposed Kendall's tau and gave the theory necessary for its application. Stuart (1963) gave the corresponding theory for Spearman's rho. Nothing directly appears to be known about the power of these tests but insofar as they depend on normal approximations one expects the conclusions of Chapter 2 regarding the comparison of their powers with unrestricted tests to apply in general terms here. These tests will be met again in Section 4.5.

Poon (1980) conducted a Monte Carlo study of the powers of a test based on the regression of $\hat{\theta}_i$ onto fixed scores, which were thought of as dose levels, a contrast test (using the Abelson–Tukey contrast coefficients, cf. Section 4.2), and an analogue of the $\bar{\chi}^2_{01}$ test statistic. The $\bar{\chi}^2$ statistic used was the one based on the pooled estimate of the variance, $\hat{\theta}_0(1-\hat{\theta}_0)$, but it is believed that the one defined in (4.1.9) and the one based on the arcsin transformation should behave similarly. Her conclusion was that, in general, the $\bar{\chi}^2$ analogue is preferred.

Tests for trends among Poisson parameters

Suppose one observes k independent Poisson processes for possibly different lengths of time. Let the intensity of the ith process be denoted by θ_i; the length of time it is observed by T_i; the number of occurrences in the ith process during T_i units of time be N_i; and $\lesssim$ a quasi-order on $X=\{1,2,\ldots,k\}$. We consider the LRTs of $H_0:\theta_1=\theta_2=\cdots=\theta_k$ versus H_1-H_0 with $H_1:\boldsymbol{\theta}$ is isotonic with respect to $\lesssim$ and of H_1 versus H_2-H_1 with $H_2:\theta_i>0$ for $i=1,2,\ldots,k$. In Example 1.5.6, this exponential family with $\tau_i=T_i$ was shown to satisfy the regularity conditions imposed in Theorem 4.1.1 and the restricted MLE of $\boldsymbol{\theta}=(\theta_1,\theta_2,\ldots,\theta_k)$ subject to $\boldsymbol{\theta}\in H_1$ was shown to be $\hat{\boldsymbol{\theta}}^*=(\hat{\theta}_1^*,\hat{\theta}_2^*,\ldots,\hat{\theta}_k^*)$, the isotonic regression of $\hat{\boldsymbol{\theta}}=(\hat{\theta}_1,\hat{\theta}_2,\ldots,\hat{\theta}_k)$ with weight $w_i=T_i$, where $\hat{\theta}_i=N_i/T_i$. Of course, $\hat{\boldsymbol{\theta}}$ is the unrestricted MLE of $\boldsymbol{\theta}$ and $\hat{\theta}_0=\sum_{i=1}^{k}N_i/\sum_{i=1}^{k}T_i$ is the MLE of the common value of θ_i if $\boldsymbol{\theta}$ is restricted to H_0. The LRT statistics in this setting are

$$T_{01}=2\sum_{i=1}^{k}N_i(\ln\hat{\theta}_i^*-\ln\hat{\theta}_0) \quad\text{and}\quad T_{12}=2\sum_{i=1}^{k}N_i(\ln\hat{\theta}_i-\ln\hat{\theta}_i^*). \tag{4.1.10}$$

Theorem 4.1.1 gives the asymptotic null distribution of T_{01} and T_{12}.

If $N_1, N_2, \ldots, N_k$ are independent random variables with $N_i \sim P(\mu_i)$ for $i = 1, 2, \ldots, k$, then conditional on $\sum_{i=1}^k N_i = n$ the distribution of $(N_1, N_2, \ldots, N_k)$ is multinomial with n trials and probabilities $p_i = \mu_i / \sum_{j=1}^k \mu_j$ for $i = 1, 2, \ldots, k$. Magel and Wright (1984a) discuss tests for trends of the following form for a multinomial setting: p_i/T_i is isotonic with respect to a quasi-order $\lesssim$ on $X = \{1, 2, \ldots, k\}$. This is an alternate approach for obtaining the same tests given in (4.1.10), and in fact this technique provides solutions to a large number of problems involving independent Poisson observations. Chapter 5 is devoted to testing ordered restricted hypotheses in a multinomial setting. This observation also suggests that the rate of convergence of the null distribution of these LRTs to their limits should be like that of a multinomial to the appropriate multivariate normal. Thus, one would expect that if $N_i \geqslant 5$ for $i = 1, 2, \ldots, k$, these approximations should be acceptable.

Order restricted hypotheses are also of interest when studying the intensity of a nonhomogeneous Poisson process. Some tests for this setting are discussed in Section 9.5.

Tests for trends in proportions in negative binomial distributions

There are many applications of the negative binomial distribution, especially in biology and medicine. Meelis (1974) considers testing homogeneity in the proportions of k independent negative binomial random variables and references the applications given in Patil (1962) and Bennett (1964). One alternative which was considered was that of a simply ordered trend.

Suppose $Y_1, Y_2, \ldots, Y_k$ are independent, negative binomial random variables with the density of Y_i, with respect to the counting measure on $\{0, 1, 2, \ldots\}$, given by

$$f(y; \theta_i, \tau_i) = \binom{y + \tau_i - 1}{\tau_i - 1} \theta_i^y (1 - \theta_i)^{\tau_i}, \qquad y = 0, 1, \ldots, 0 < \theta < 1,$$

with $\tau_1, \tau_2, \ldots, \tau_k$ known positive integers. As in Example 1.5.2, it is convenient to reparameterize with $\theta_i = \eta_i/(1 + \eta_i)$. Because this transform is monotonic the hypotheses H_0, H_1 and H_2 are invariant under it. With $K(y, \tau) = y/\tau$ and $p_2(\tau) = \tau$ it is not difficult to show that, with this parameterization, the negative binomial density satisfies the regularity conditions imposed in Theorem 4.1.1. Under H_0, the MLE of $\eta_1 = \eta_2 = \cdots = \eta_k$ is $\hat{\eta}_0 = \sum_{i=1}^k Y_i / \sum_{i=1}^k \tau_i$, the unrestricted MLEs are $\hat{\eta}_i = Y_i/\tau_i$, and the restricted MLE is given by $\hat{\eta}^* = (\hat{\eta}_1^*, \hat{\eta}_2^*, \ldots, \hat{\eta}_k^*)$, the isotonic regression of $\hat{\boldsymbol{\eta}} = (\hat{\eta}_1, \hat{\eta}_2, \ldots, \hat{\eta}_k)$ with weights $w_i = \tau_i$ for $i = 1, 2, \ldots, k$ (cf. Theorem 1.5.2).

The LRT of H_0 versus $H_1 - H_0$ rejects for large values of

$$T_{01} = 2 \sum_{i=1}^k \left\{ y_i \log\left[\frac{\hat{\eta}_i^*(1 + \hat{\eta}_0)}{(1 + \hat{\eta}_i^*)\hat{\eta}_0}\right] + \tau_i \log\left(\frac{1 + \hat{\eta}_0}{1 + \hat{\eta}_i^*}\right) \right\}$$

and the LRT of H_1 versus $H_2 - H_1$ rejects for large values of

$$T_{12} = 2 \sum_{i=1}^{k} \left\{ y_i \log \left[\frac{\hat{\eta}_i(1+\hat{\eta}_i^*)}{(1+\hat{\eta}_i)\hat{\eta}_i^*} \right] + \tau_i \log \left(\frac{1+\hat{\eta}_i^*}{1+\hat{\eta}_i} \right) \right\}.$$

Theorem 4.1.1 gives approximate p values for these tests.

LRTs for trends among exponential means

In this subsection, the application of Theorem 4.1.1 to exponential distributions with sample sizes $n_1 = n_2 = \cdots = n_k = 1$ is discussed. For a sample size of one the exact null distribution of the test statistic for H_0 versus $H_1 - H_0$ can be computed. Hence, the accuracy of the approximation can be assessed in this case. If the approximation given by Theorem 4.1.1 is reasonable for this case we would expect it to provide useful results for larger sample sizes. The case of general sample sizes is considered in the next subsection which deals with gamma distribution.

Writing the density of an exponential distribution in the form

$$f(y;\theta) = \frac{1}{\theta} \exp\left(\frac{-y}{\theta} \right) \qquad \text{for } y > 0 \text{ and } \theta > 0,$$

one sees that it satisfies the regularity hypotheses of Theorem 4.1.1 (see Example 1.5.3). If $Y_1, Y_2, \ldots, Y_k$ are independent, exponentially distributed random variables with $E(Y_i) = \theta_i$, then the unrestricted MLE of θ_i is $\hat{\theta}_i = Y_i$. With $\lesssim$ a quasi-order on $X = \{1, 2, \ldots, k\}$, $\hat{\boldsymbol{\theta}}^* = (\hat{\theta}_1^*, \hat{\theta}_2^*, \ldots, \hat{\theta}_k^*)$, the isotonic regression of $\hat{\boldsymbol{\theta}} = (\hat{\theta}_1, \hat{\theta}_2, \ldots, \hat{\theta}_k)$ with weights $w_i = 1$, is the MLE restricted to be isotonic with respect to $\lesssim$. The MLE, $\hat{\theta}_0$, of the common value of $\boldsymbol{\theta}$, if $\boldsymbol{\theta}$ is restricted to $H_0{:}\theta_1 = \theta_2 = \cdots = \theta_k$, is the mean of the Y_i, which will be denoted by $\bar{Y}$. In this setting,

$$T_{01} = 2 \sum_{i=1}^{k} \log\left(\frac{\hat{\theta}_0}{\hat{\theta}_i^*} \right) \quad \text{and} \quad T_{12} = 2 \sum_{i=1}^{k} \log\left(\frac{\hat{\theta}_i^*}{\hat{\theta}_i} \right). \tag{4.1.11}$$

Boswell and Brunk (1969) were able to compute the exact distribution of T_{01} under H_0. They tabled the critical values of $\Lambda_{01} = \exp(-T_{01}/2)$. (It should be noted that one rejects for small values of Λ_{01}.) Some of these values are given in Table 4.1.1. These values enable one to carry out the test and they also make it possible to assess the accuracy of the approximation given in Theorem 4.1.1. Let $F(\lambda)$ be the probability under H_0 that $\Lambda_{01} \leqslant \lambda$ and let $F_{(1)}(\lambda)$ be the approximation provided by Theorem 4.1.1. The statistics T_{01} and T_{12} are related to the logarithm of the ratio of the arithmetic and geometric means. Guffey and Wright (1986) make use of existing approximations for such statistics to obtain a second approximation for this setting. Their technique will be illustrated in the simply ordered case and the resulting approximation will be compared with that given by Theorem 4.1.1. To use (1.5) of Boswell and Brunk (1969), one needs to be able

Table 4.1.1 Critical values for the likelihood ratio statistic, Λ_{01}, for $2 \leqslant k \leqslant 10$

	k								
α	2	3	4	5	6	7	8	9	10
0.005	0.020	0.010	0.004	0.004	0.002	0.002	0.001	0.001	0.001
0.010	0.040	0.020	0.011	0.009	0.006	0.005	0.004	0.004	0.003
0.020	0.076	0.040	0.025	0.020	0.016	0.014	0.010	0.010	0.009
0.050	0.175	0.103	0.070	0.057	0.046	0.039	0.035	0.030	0.026
0.100	0.390	0.214	0.153	0.123	0.101	0.089	0.077	0.070	0.064

to approximate, for $B_1, B_2, \ldots, B_l$ possible blocks in the isotonic regression of the Y_i,

$$P\left\{2 \sum_{i=1}^{l} \operatorname{card}(B_i) \ln\left[\frac{\bar{Y}}{\operatorname{Av}(B_i)}\right] \geqslant t\right\},$$

where $\operatorname{Av}(B_i) = \sum_{j \in B_i} Y_j / \operatorname{card}(B_i)$. However, if $U_i = \sum_{j \in B_i} Y_j$, $\bar{U} = \sum_{i=1}^{l} U_i / l$, and $\tilde{U} = (\prod_{i=1}^{l} U_i)^{1/l}$, then the approximation is obtained by assuming $\operatorname{card}(B_1) = \operatorname{card}(B_2) = \cdots = \operatorname{card}(B_l)$. Under that assumption,

$$2 \sum_{i=1}^{l} \operatorname{card}(B_i) \ln\left[\frac{\bar{Y}}{\operatorname{Av}(B_i)}\right] \doteq 2l \operatorname{card}(B_1) \ln\left(\frac{\bar{U}}{\tilde{U}}\right).$$

Applying the two-moment chi-squarc approximation of Bain and Engelhardt (1975), letting χ_a^2 denote a chi-square variable with a degrees of freedom, and setting $\operatorname{card}(B_1) = k/l$, one obtains

$$2 \sum_{i=1}^{l} \operatorname{card}(B_i) \ln\left[\frac{\bar{Y}}{\operatorname{Av}(B_i)}\right] \doteq \frac{1}{c(\tau, l)} \chi^2_{b(\tau,l)(l-1)}$$

where

$$\begin{aligned} \tau &= k/l \\ c(\tau, l) &= [l\phi_1(\tau) - \phi_1(l\tau)]/[l\phi_2(\tau) - \phi_2(l\tau)] \qquad (4.1.12) \\ b(\tau, l) &= [l\phi_1(\tau) - \phi_1(l\tau)]c(\tau, l)/(l-1) \end{aligned}$$

and $\phi_1(\tau)$ and $\phi_2(\tau)$ are approximated by

$$\phi_1(\tau) \doteq 3 - \frac{1}{1+\tau} + \frac{\tau}{6(1+\tau)^2} - 2\tau \ln\left(1 + \frac{1}{\tau}\right)$$

and

$$\phi_2(\tau) \doteq 1 + \frac{2+\tau}{3(1+\tau)^2} + \frac{1}{3(1+\tau)^3}.$$

Finally, for $t > 0$ and $v(\tau, l) = b(\tau, l)(l-1)$,

$$P[T_{01} \geqslant t] \doteq \sum_{l=2}^{k} P_S(l, k) P[\chi^2_{v(\tau,l)} \geqslant c(\tau, l)t]. \qquad (4.1.13)$$

Table 4.1.2 Comparison of the exact and approximate distribution of the likelihood ratio statistic, Λ_{01}, for the exponential distribution

	$k=2$			$k=5$			$k=10$		
λ	$F(\lambda)$	$F_{(1)}(\lambda)$	$F_{(2)}(\lambda)$	$F(\lambda)$	$F_{(1)}(\lambda)$	$F_{(2)}(\lambda)$	$F(\lambda)$	$F_{(1)}(\lambda)$	$F_{(2)}(\lambda)$
0.05	0.013	0.007	0.013	0.046	0.032	0.044	0.083	0.063	0.077
0.10	0.026	0.016	0.025	0.082	0.062	0.080	0.143	0.116	0.134
0.20	0.053	0.036	0.052	0.153	0.123	0.147	0.243	0.209	0.230
0.30	0.082	0.060	0.080	0.218	0.184	0.211	0.328	0.295	0.316
0.40	0.113	0.088	0.112	0.282	0.247	0.275	0.408	0.375	0.395
0.50	0.146	0.120	0.146	0.345	0.312	0.339	0.482	0.452	0.470

The corresponding approximation to $P[\Lambda_{01} \leqslant \lambda]$ is denoted by $F_{(2)}(\lambda)$. Table 4.1.2 gives the exact values, $F(\lambda)$, and the values of the two approximations for $k=2$, 5, 10 and several values of λ. The approximation based on Theorem 4.1.1 would be adequate for many practical purposes. The error in this approximation would seem to be due to the high degree of nonnormality in the exponential distribution. Consequently, one would expect better agreement between the exact and approximate results of Theorem 4.1.1 for distributions with smaller departures from normality. The second approximation given here requires more computation but also gives a greater degree of accuracy.

Theorem 4.1.1 can also be used to give approximate significance levels for T_{12}. The techniques mentioned above can be modified to provide an approximation for the distribution of T_{12} under the hypothesis, H_0, which is asymptotically least favorable within H_1. Applying (1.5) of Boswell and Brunk (1969), with B_1, $B_2, \ldots, B_l$ possible blocks for the isotonic regression of the Y_i, $\bar{V}_i = \mathrm{Av}(B_i)$, and $\tilde{V}_i = (\prod_{j \in B_i} Y_j)^{1/\mathrm{card}(B_i)}$, one needs to approximate

$$P\left[2 \sum_{i=1}^{l} \mathrm{card}(B_i) \ln\left(\frac{\bar{V}_i}{\tilde{V}_i}\right) \geqslant t\right]. \tag{4.1.14}$$

However, for $i=1,2,\ldots,l$, $\ln(\bar{V}_i/\tilde{V}_i)$ are independent and so if each term in the sum in (4.1.14) is approximated by the same constant times a chi-squared variable, then the sum can be approximated by that constant times a chi-squared variable. Assuming that $\mathrm{card}(B_1) = \cdots = \mathrm{card}(B_l) = k/l$, for positive t, (4.1.14) is approximated by

$$P[\chi^2_{b(\tau,k/l)(k-l)} \geqslant tc(\tau, k/l)]$$

where $\tau = 1$. Using this approximation in (1.5) of Boswell and Brunk (1969) yields an approximate p value for T_{12} given by

$$\sum_{l=1}^{k-1} P_S(l,k) P[\chi^2_{b(1,k/l)(k-l)} \geqslant tc(1, k/l)]. \tag{4.1.15}$$

Guffey and Wright (1986) found that this approximation was more accurate than that given by Theorem 4.1.1 in a closely related problem involving nonhomogeneous Poisson processes. Because of the similarity of the problems, this should be the case in this setting, too.

Other tests for trend among exponential means

There are several other classes of tests which could be used to test H_0 versus $H_1 - H_0$ for exponential distributions in the simply ordered case. If the exponential variables, $Y_1, Y_2, \ldots, Y_k$, have the same mean, then the joint distribution of

$$U_i = \frac{\sum_{j=1}^{i} Y_j}{\sum_{j=1}^{k} Y_j} \qquad i = 1, 2, \ldots, k-1$$

is that of the order statistics of a sample of size $k-1$ from the uniform distribution on the interval $(0, 1)$. Since $E(U_i) = i/k$ (whatever the distribution of the Y_i, providing they are nonnegative), evidence of an increasing trend in the means of the Y_i would be revealed by a tendency for the U_i to be smaller than their expectation under the null hypothesis: This suggests constructing statistics as functions of the differences $(U_i - i/k)$ which preserve their signs. One such statistic is

$$D_k^- = \min_i \left(U_i - \frac{i}{k} \right). \tag{4.1.16}$$

It is known (see, for example, Birnbaum and Tingey, 1951) that

$$\lim_{k \to \infty} P[D_k^- < D] = 1 - e^{-2D^2}$$

and tables of percentage points for finite k are given in Owen (1962). A second statistic is the arithmetic mean of the U_i. It is easy to show (see Bartholomew, 1956) that this has the following representation:

$$\bar{U} = -\left\{ \int_0^1 [F_k(u) - u] \, du - \tfrac{1}{2} \right\}$$

where $F_k(u)$ is the empirical distribution function computed from the U_i. Under the null hypothesis $\bar{U}$ has a distribution which is very close to normal, even for small k. The result required for tests of significance is

$$\bar{U} \sim \mathcal{N}\left(\frac{1}{2}, \frac{1}{12k} \right).$$

The average, or sum, of any monotonic function of the U_i will also be sensitive to a trend in the means of the Y_i. One such statistic,

$$Q = -2 \sum_{i=1}^{k-1} \log U_i,$$

has the useful property that it is distributed like χ^2 with $2(k-1)$ degrees of freedom under H_0.

Most of the results which are available on the power of the statistics based on the U_i relate to the case where the alternative hypothesis specifies a distribution for U other than the uniform. This kind of alternative is not relevant for the application here. Marcus (1976a) studied the levels and powers of the LRT and tests based on these U_i. Neither test dominates the other. For further details the reader may consult this work.

The foregoing discussion does not exhaust the possibilities for testing for trend under the exponential distribution. Before leaving the subject one further class of tests is mentioned which is easily generalized to the gamma distribution discussed in the following subsection. Suppose the sequence $\{Y_i\}$ is divided arbitrarily into two parts. Let

$$S_1 = \sum_{i=1}^{d} Y_i, \qquad S_2 = \sum_{i=d+1}^{k} Y_i,$$

where the division takes place between Y_d and Y_{d+1}. Then, under the null hypothesis, $dS_2/[(k-d)S_1]$ is distributed as an F random variable with $2d$ and $2(k-d)$ degrees of freedom. In the presence of an increasing trend the ratio will tend to be larger so large values of F would be significant. After reading the next section the reader should be able to devise other tests by breaking the sequence at more than one point.

The tests discussed above are, in a sense, one sided and it is not clear how they could be used to test H_1 versus H_2–H_1.

Tests for trend in the scale parameters of gamma distributions

The problem of testing for trend in gamma variables may arise in several different ways. For example, if instead of taking $n_i = 1$ in the previous subsection, $n_i > 1$ had been taken, then gamma variables would have arisen as follows. The sample sums

$$Y_i = \sum_{j=1}^{n_i} Y_{ij} \qquad \text{for } i = 1, 2, \ldots, k$$

are sufficient statistics for the population means θ_i, and so a test can be based on them without loss of information. However, these sums have gamma distributions. A second important example arises when testing the homogeneity of normal variances against ordered alternatives because the sample variances have gamma distributions.

Both of the above problems may be expressed formally as follows. Independent observations, $Y_1, Y_2, \ldots, Y_k$, are available; Y_i has density

$$f(y;\theta_i,\tau_i)=\frac{y^{\tau_i-1}\exp(-y/\theta_i)}{\theta_i^{\tau_i}\Gamma(\tau_i)},\qquad y>0,\quad \theta_i>0,\quad \tau_i>0,$$

and $\lesssim$ is a quasi-order on $X=\{1,2,\ldots,k\}$. For the application involving exponential variables, set $\tau_i=n_i$ for $i=1,2,\ldots,k$. For testing the equality of k normal variances, set $\tau_i=(n_i-1)/2$, $\theta_i=\sigma_i^2$, and $Y_i=(n_i-1)S_i^2/2$ where S_i^2 is the unbiased sample estimator for σ_i^2 based on n_i observations. The unrestricted MLE of θ_i is $\hat{\theta}_i=Y_i/\tau_i$ and the MLE of the common value of θ_i under H_0 is $\hat{\theta}_0=\sum_{i=1}^k Y_i/\sum_{i=1}^k \tau_i$. In Example 1.5.3, it was observed that the gamma family with known shape parameters satisfies the regularity conditions of Theorem 4.1.1 and that the restricted MLE of $\boldsymbol{\theta}=(\theta_1,\theta_2,\ldots,\theta_k)$ is $\hat{\boldsymbol{\theta}}^*$, the isotonic regression of $\hat{\boldsymbol{\theta}}=(\hat{\theta}_1,\hat{\theta}_2,\ldots,\hat{\theta}_k)$ with weights $w_i=\tau_i$ for $i=1,2,\ldots,k$. Hence, the LRT for H_0 versus H_1 rejects for large values of

$$T_{01}=2\sum_{i=1}^{k}\left[\frac{Y_i}{\hat{\theta}_i^*}-\frac{Y_i}{\hat{\theta}_0}+\tau_i\ln\left(\frac{\hat{\theta}_0}{\hat{\theta}_i^*}\right)\right],$$

which, applying Theorem 1.3.6, can be written as

$$T_{01}=2\sum_{i=1}^{k}\tau_i\ln\left(\frac{\hat{\theta}_0}{\hat{\theta}_i^*}\right). \tag{4.1.17}$$

The LRT for H_1 versus H_2–H_1 rejects for large values of

$$T_{12}=2\sum_{i=1}^{k}\left[\frac{Y_i}{\hat{\theta}_i^*}-\frac{Y_i}{\hat{\theta}_i}+\tau_i\ln\left(\frac{\hat{\theta}_i^*}{\hat{\theta}_i}\right)\right]=2\sum_{i=1}^{k}\tau_i\ln\left(\frac{\hat{\theta}_i^*}{\hat{\theta}_i}\right). \tag{4.1.18}$$

Theorem 4.1.1 provides large sample approximations for the significance levels of these tests. For $\tau_i>1$, the gamma distributions are more like normal variables than for $\tau_i=1$ and thus the approximation for such cases should be more accurate than for exponential distributions ($\tau_1=1$). However, if $\lesssim$ is the usual simple order and $\tau_1=\tau_2=\cdots=\tau_k=\tau$, then the techniques used in the last subsection can be modified to provide an approximation for the null distribution of these statistics in the gamma setting. Under H_0, with $t>0$ and $c(\cdot,\cdot)$ and $b(\cdot,\cdot)$ given by (4.1.12), $P[T_{01}\geqslant t]$ can be approximated by

$$\sum_{l=2}^{k}P_S(l,k)P[\chi^2_{b(\tau k/l,l)(l-1)}\geqslant c(\tau k/l,l)t] \tag{4.1.19}$$

and $P[T_{12}\geqslant t]$ can be approximated by

$$\sum_{l=1}^{k-1}P_S(l,k)P[\chi^2_{b(\tau,k/l)(k-l)}\geqslant tc(\tau,k/l)]. \tag{4.1.20}$$

Studying Table 1 of Bain and Engelhardt (1975), which gives some values of $c(\tau,n)$

and $b(\tau, n)$, one sees that they depend, primarily, on τ. Also, if $\tau \geqslant 10$, these values are close to one, and in that case (4.1.19) and (4.1.20) coincide with the results in Theorem 4.1.1.

Some of the *ad hoc* tests discussed in the last subsection for testing H_0 versus $H_1 - H_0$ are easily generalized to this setting. The U transformation which was used in the exponential case can be applied here also, but its use is more limited. If $\tau_1 = \tau_2 = \cdots = \tau_k = \tau$ is an integer, the variable

$$U_i = \frac{\sum_{j=1}^{i} Y_j}{\sum_{j=1}^{k} Y_j}$$

will now have the same distribution as the $i\tau$th order statistic in a sample of size $k\tau - 1$ from a uniform distribution. The statistics D_k^- (cf. equation (4.1.16)) and $\bar{U}$, the average of the U_i, will still be suitable for testing for trend but their distributions will be more difficult to determine since the joint distribution of the U_i is that of a particular subset of the order statistics from a uniform distribution instead of all of them.

The F-ratio test described for the exponential case is still available. The statistic in this case is $(\sum_{i=1}^{d} \tau_i) S_2 / [(\sum_{i=d+1}^{k} \tau_i) S_1]$ but the degrees of freedom are now $2\sum_{i=1}^{d} \tau_i$ and $2\sum_{i=d+1}^{k} \tau_i$.

A test for inequality of variances against ordered alternatives was proposed by Vincent (1961). This test was derived under the assumption that $\sigma_i^2 (i = 1, 2, \ldots, k)$ is a linear function of i. Under this assumption a reasonable statistic to use is the regression coefficient of $\hat{\sigma}_i^2$ on i. This leads to a statistic belonging to the family of contrast tests considered in the next section.

4.2 TESTS BASED ON CONTRASTS

Partly because of the difficulties involved in applying LRTs for ordered hypotheses, several researchers, including Abelson and Tukey (1963), Hogg (1965), Schaafsma and Smid (1966), and more recently Snidjers (1979), considered testing H_0 versus $H_1 - H_0$ using tests based on contrasts. In this section, attention is primarily focused on the k-sample normal means problems. (Snidjers, 1979, considers in detail the use of linear tests in exponential families.) One advantage of these tests is that the contrast statistic is normally distributed with easily computed mean and variance under both the null and alternative hypotheses. Such a contrast test is easily shown to be uniformly most powerful (UMP) for alternatives in a certain direction and consequently is very powerful is some subregion of the alternative hypothesis. However, as will be shown, the power may be quite low in other directions. An *optimal contrast test* is chosen to maximize its minimum power. While the LRT is not most powerful at any particular point, it does maintain a more uniformly reasonable power over all the alternative regions. In fact, its region of consistency is larger than that of an optimal contrast test. One explanation of this fact is given in Theorem 4.2.1 which can be interpreted as saying that the LRT is based on an 'adaptive' contrast statistic. In other words, the parameters are estimated from the data and then the contrast coefficients which are optimal for the estimated point are chosen. In our opinion, the approximations of Chapter 3 alleviate many of the difficulties involved with using the LRTs for the simple order and the simple tree, and except in cases in which the researcher has specific information as to what part of the alternative region $\boldsymbol{\mu}$ is in, the LRT is preferred over contrast tests. However, these difficulties persist for other partial orders such as the matrix order and in problems involving such orders a well-chosen contrast test may be a reasonable alternative to the LRT.

In this section, optimal contrast tests are considered for testing H_0 versus $H_1 - H_0$ and H_1 versus $H_2 - H_1$ for the simple order and the simple tree with H_0, H_1, and H_2 defined as in Chapter 2. Their powers and regions of consistency are compared with those of the LRTs.

Contrast test statistics for H_0 versus $H_1 - H_0$

Let $\bar{Y}_1, \bar{Y}_2, \ldots, \bar{Y}_k$ and S^2 be independent variables with $\bar{Y}_i \sim \mathcal{N}(\mu_i, w_i^{-1})$ where $w_i = n_i/\sigma^2$ for $i = 1, 2, \ldots, k$ and $\nu S^2/\sigma^2 \sim \chi^2_\nu$ with ν a positive integer. (The case in which the variances are assumed known up to a multiplicative constant is treated similarly.) With $\mathbf{x} \in R^k$, let $\|\mathbf{x}\|_{\mathrm{w}} = (\sum_{i=1}^k w_i x_i^2)^{1/2}$. For the case of σ^2 known, a contrast test would reject $H_0: \mu_1 = \mu_2 = \cdots = \mu_k$ in favor of $H_1: \mu$ is isotonic with respect to $\lesssim$, for large values of

$$T_{\mathbf{c}} = \frac{\sum_{i=1}^k w_i c_i \bar{Y}_i}{\|\mathbf{c}\|_{\mathrm{w}}} \qquad \text{with } \sum_{i=1}^k w_i c_i = 0 \tag{4.2.1}$$

for a suitable choice of $\mathbf{c}$. The condition $\sum_{i=1}^k w_i c_i = 0$ ensures that the distribution of $T_{\mathbf{c}}$ is constant over H_0. Under H_0, $T_{\mathbf{c}} \sim \mathcal{N}(0, 1)$ and so the α-level test based on $T_{\mathbf{c}}$ rejects for $T_{\mathbf{c}} \geqslant z_\alpha$ with z_α chosen to satisfy $\Phi(z_\alpha) = 1 - \alpha$. Optimal contrast coefficients will be obtained for the case of known variances and the resulting test Studentized for the case of unknown variances. The power of the α-level test based on $T_{\mathbf{c}}$ is given by

$$\pi_{\mathbf{c}}(\boldsymbol{\mu}) = \Phi\left(\frac{\langle \mathbf{c}, \boldsymbol{\mu} \rangle_{\mathrm{w}}}{\|\mathbf{c}\|_{\mathrm{w}}} - z_\alpha\right) \tag{4.2.2}$$

with $\langle \mathbf{c}, \boldsymbol{\mu} \rangle_{\mathrm{w}} = \sum_{i=1}^k w_i c_i \mu_i$. If σ^2 is not known, then, with $\mathbf{c}$ satisfying the last condition in (4.2.1), one would reject H_0 for the Studentized contrast test $T'_{\mathbf{c}} = \sigma T_{\mathbf{c}}/S \geqslant t_\alpha(\nu)$ (note that $\sigma T_{\mathbf{c}} = \sum_{i=1}^k n_i c_i \bar{Y}_i / (\sum_{i=1}^k n_i c_i^2)^{1/2}$ is a statistic), where $t_\alpha(\nu)$ is the $(1-\alpha)$th quantile of Student's t-distribution with ν degrees of freedom. The powers of $T'_{\mathbf{c}}$ are noncentral-t tail probabilities.

Optimal contrast coefficients for testing H_0 versus $H_1 - H_0$

As in Chapter 2, define $\bar{\mu} = \sum_{i=1}^k w_i\mu_i / \sum_{i=1}^k w_i$ and $\Delta = [\sum_{i=1}^k w_i(\mu_i - \bar{\mu})^2]^{1/2}$. For $\boldsymbol{\mu} \in H_1 - H_0$ the $\mathbf{c}$ which yields the $T_\mathbf{c}$ with maximum power at $\boldsymbol{\mu}$ maximizes $\langle \mathbf{c}, \boldsymbol{\mu} \rangle_\mathbf{w} / \| \mathbf{c} \|_\mathbf{w}$. Equivalently, since fixing $\boldsymbol{\mu}$ fixes Δ, $\mathbf{c}$ maximizes $r(\mathbf{c}, \boldsymbol{\mu})$, where

$$r(\mathbf{c}, \boldsymbol{\mu}) = \frac{\sum_{i=1}^k w_i(\mu_i - \bar{\mu})c_i}{\| \mathbf{c} \|_\mathbf{w} \Delta}. \tag{4.2.3}$$

(Recall that $\bar{c} = \sum_{i=1}^k w_i c_i / \sum_{i=1}^k w_i = 0$.) If $c_i = \mu_i - \bar{\mu}$, then $\sum_{i=1}^k w_i c_i = 0$ and $r(\mathbf{c}, \boldsymbol{\mu}) = 1$, which is the maximum. This clearly indicates that a contrast test with $\mathbf{c} \in H_1$ will have large power around the point $\mathbf{c}$. In fact, if one wishes to test $H_0' : \boldsymbol{\mu} = \boldsymbol{\mu}_0$ versus $H_1 : \boldsymbol{\mu} \in \{\boldsymbol{\mu}_0 + b\boldsymbol{\eta} : b > 0\}$ with $\sum_{i=1}^k w_i \eta_i = 0$ and rejects for large values of $T_{\boldsymbol{\eta}}$, then, using the Neyman–Pearson theorem, it is straightforward to show that $T_{\boldsymbol{\eta}}$ is UMP. However, it is the essence of the problem considered here that only the ordering of the means are known. If more is known about $\boldsymbol{\mu}$, then one may be able to develop a very powerful contrast test. In the absence of such knowledge one might use the estimate of $\boldsymbol{\mu}$, namely $\boldsymbol{\mu}^* = (\mu_1^*, \mu_2^*, \ldots, \mu_k^*)$, the isotonic regression of $\bar{\mathbf{Y}}$ with weights w_i. Because $\sum_{i=1}^k w_i \mu_i^* / \sum_{i=1}^k w_i = \sum_{i=1}^k w_i \bar{Y}_i / \sum_{i=1}^k w_i = \hat{\mu}$ (see Theorem 1.3.3),

$$\sum_{i=1}^k w_i(\mu_i^* - \hat{\mu})\bar{Y}_i = \sum_{i=1}^k w_i(\mu_i^*)^2 - \left(\sum_{i=1}^k w_i\right)\hat{\mu}^2 = \sum_{i=1}^k w_i(\mu_i^* - \hat{\mu})^2 = \bar{\chi}_{01}^2.$$

This proves the following result which gives another justification for the LRT and indicates that it would be difficult to find a test which has a power function which exceeds that of the LRT over a substantial portion of H_1.

Theorem 4.2.1 *The adaptive contrast test with coefficients* $\mathbf{c} = \boldsymbol{\mu}^* - \hat{\mu}$ *is the same as* $\bar{\chi}_{01}^2$.

For a standard contrast test the optimal contrast coefficients depend upon the quasi-order, $\lesssim$. The usual simple order is considered first with equal weights. Since this order requires $\mu_i > \mu_j$ for each pair (i, j) with $i > j$, one might consider the contrast statistic

$$T = \sum_{i=2}^k \sum_{j=1}^{i-1} (\bar{Y}_i - \bar{Y}_j) = \sum_{i=1}^k (2i - k - 1)\bar{Y}_i \qquad \text{(which needs to be normalized).} \tag{4.2.4}$$

However, it is not clear that each of the differences in the left-hand sum should be given the same weight. Armitage (1955), using regression arguments, recommends a class of contrast statistics of the form $T_\mathbf{c}$ with $\mathbf{c} = \mathbf{s} - \bar{s}$ and $s_1 < s_2 < \cdots < s_k$, and clearly the T in (4.2.4) is a member of this class. In many settings it may not be clear how $\mathbf{s}$ should be chosen. Abelson and Tukey (1963) sought to choose $\mathbf{c}$ so that it is as highly correlated with various $\boldsymbol{\mu}$ in H_1 as possible. This choice is natural in view of the way the power of T_c depends on the correlation between $\mathbf{c}$ and $\boldsymbol{\mu}$ (see equations (4.2.2) and (4.2.3)). For any given $\mathbf{c}$, r given by (4.2.3) has an upper limit of 1 and a lower limit over H_1, called $r_{\min}$. One should hope for this value to be positive and reasonably large because the rank correlation between the $\mathbf{c}$ and $\boldsymbol{\mu}$ is necessarily unity. Abelson and Tukey suggested choosing $\mathbf{c}$ so as to maximize the value of $r_{\min}$ and they called this value maximin r. It is clear from (4.2.2) that maximin r also maximizes the minimum power for given Δ. It was seen in Section 2.5 that there is a power 'band' associated with the $\bar{\chi}^2$ test. The effect of choosing $\mathbf{c}$ in the way just described can therefore be thought of as maximizing the lower limit of the corresponding power band for a test based on scores.

The details of the determination of maximin r will not be given, but the principle of the method is easily described in geometric terms. Consider a k-dimensional Euclidean space of points $\boldsymbol{\mu}$. The null hypothesis $\mu_1 = \mu_2 = \cdots = \mu_k$ is then represented by a line in this space; each inequality $\mu_i < \mu_j$ defines a region on one side of the $(k-1)$ dimensional subspace $\mu_i = \mu_j$. The intersection of the half-spaces defined by the set of inequalities is the region in which $\boldsymbol{\mu}_0 = (\mu_{01}, \mu_{02}, \ldots, \mu_{0k})'$, the true value, is constrained to lie. Let

$$\bar{\mu}_0 = \sum_{i=1}^{k} \frac{\mu_{0i}}{k};$$

then $\boldsymbol{\mu}_0$ is a point in the subspace

$$\left\{\boldsymbol{\mu}\colon \sum_{i=1}^{k} (\mu_i - \bar{\mu}_0) = 0\right\}.$$

Any $\mathbf{c}$ satisfying

$$\sum_{i=1}^{k} c_i = 0$$

can be thought of as another point in this subspace and r is then the cosine of the angle between the lines joining the point $\mu_i = \bar{\mu}_0 (i = 1, 2, \ldots, k)$ to the points $\boldsymbol{\mu}_0$ and $\mathbf{c}$. Thus if the two lines are coincident the angle is zero and $r = 1$. For any fixed point $\mathbf{c}$, r will vary as $\boldsymbol{\mu}_0$ moves within the restricted parameter space and there will be a certain $\boldsymbol{\mu}_0$ for which r is a minimum. This minimum value will vary with $\mathbf{c}$ and the problem is to find that $\mathbf{c}$ for which the minimum r is the greatest. One would expect the solution for $\mathbf{c}$ to lie in the 'middle' of the constraint region. This is, in fact, the case and Abelson and Tukey argued in general that the solution is the vector which makes equal angles with the generators of this polyhedral cone. Specifically they showed that maximin r for the simple order was obtained by taking

$$c_i^{(01)} \propto \left[(i-1)\left(1 - \frac{i-1}{k}\right)\right]^{1/2} - \left[i\left(1 - \frac{i}{k}\right)\right]^{1/2} \quad \text{for } i = 1, 2, \ldots, k. \tag{4.2.5}$$

For the simple order cone and equal weights, it is interesting to note that, except when $k = 2$ and 3, the scores given by this formula are not equally spaced (that is $c_{i+1} - c_i \neq c_i - c_{i-1}$). The original statistic T of (4.2.4) is therefore not optimum in this sense if $k > 3$, even though it would seem to be in the center of the cone. The value of maximin r, as a function of k, decreases from 1 when $k = 2$ to zero as $k \to \infty$. For large k,

$$\text{maximin } r \approx \sqrt{2}[2 + \log(k-1)]^{-1/2},$$

indicating that the approach to zero with k is extremely slow. The minimum value of the power of this contrast test is attained when all but one of the differences $\{\mu_{i+1} - \mu_i\}$ are zero. A table of optimum contrast coefficients or scores is given in Abelson and Tukey (1963).

For the simply ordered case with possibly different weights, Schaafsma and Smid (1966) obtained the most stringent somewhere most powerful (MSSMP) test. They considered a different optimality condition, but, for testing H_0 versus H_1–H_0, their criterion agrees with that of Abelson and Tukey. Any test with $\mathbf{c} \in H_1$ will be most powerful at $\mathbf{c}$. Its shortcoming, as a function of $\boldsymbol{\mu}$, is the amount its power falls below the maximum attainable. The MSSMP test chooses $\mathbf{c}$ to minimize the maximum shortcoming over H_1. Their approach is essentially the same as that of Abelson and Tukey (1963). For the simply ordered case,

$$w_i c_i^{(01)} = [W_{i-1}(W_k - W_{i-1})]^{1/2} - [W_i(W_k - W_i)]^{1/2} \qquad \text{for } i = 1, 2, \ldots, k, \tag{4.2.6}$$

where $W_0 = 0$ and $W_i = \sum_{j=1}^{i} w_j$. If the weights are equal, these coefficients agree with those obtained by Abelson and Tukey (1963).

Optimal contrast coefficients for testing H_0 versus H_1–H_0 with $\lesssim$ the simple tree ordering are considered next. In order that the same notation that was used for the simple ordering can be used in this case, let the control mean be denoted by μ_1 and the treatment means by $\mu_2, \mu_3, \ldots, \mu_k$. For the simple tree ordering, $H_1: \mu_1 \leqslant [\mu_2, \mu_3, \ldots, \mu_k]$, Abelson and Tukey (1963) show that the contrast vector which maximizes the minimum power over H_1 makes equal angles with generators of H_1. This condition gives rise to a set of equations which are most easily solved if we note that H_1 is generated by $(-1, -1, \ldots, -1)$ and $(1, 1, \ldots, 1)$ along with

$$\mathbf{e}_1 = (0, 1, 0, \ldots, 0)', \qquad \mathbf{e}_2 = (0, 0, 1, \ldots, 0)', \qquad \ldots, \qquad \text{and} \qquad \mathbf{e}_{k-1} = (0, 0, \ldots, 0, 1)'.$$

Note that $\Delta^2(\mathbf{e}_i) = \sum_{j=1}^{k} w_j(e_{ij} - \bar{e}_i)^2 = (W - w_{i+1})w_{i+1}/W$ with $W = \sum_{j=1}^{k} w_j$. Setting $\langle \mathbf{e}_i, \mathbf{c}^{(01)} \rangle_{\mathbf{w}} = \Delta(\mathbf{e}_i)$ gives $c_i \propto (w_i^{-1} - W^{-1})^{1/2}$ for $i = 2, 3, \ldots, k$, and $\sum_{j=1}^{k} w_j c_j^{(01)} = 0$ is required so that the power function will be constant over H_0. Hence, the optimal contrast coefficients for testing H_0 versus H_1–H_0 for the simple tree ordering are determined, up to a positive multiplier, by

$$c_i^{(01)} = (w_i^{-1} - W^{-1})^{1/2} \qquad \text{for } i = 2, 3, \ldots, k, \quad \text{and} \quad c_1^{(01)} = -\sum_{j=2}^{k} \frac{w_j c_j^{(01)}}{w_1}. \tag{4.2.7}$$

If the treatment sample sizes are equal, then the optimal contrast test statistic can be written as

$$T_{\mathbf{c}^{(01)}} = \frac{\sum_{j=2}^{k} \bar{Y}_j/(k-1) - (w_1/w)\bar{Y}_1}{\|\mathbf{c}^{(01)}\|_{\mathbf{w}}}$$

where $w = w_2 = \cdots = w_k$, and if all the sample sizes are equal, $n_1 = n_2 = \cdots = n_k$, then the optimal contrast vector is

$$\mathbf{c} \propto (-(k-1), 1, \ldots, 1)' \tag{4.2.8}$$

and the optimal contrast test compares $(k-1)^{-1}\sum_{j=2}^{k} \bar{Y}_j$ with Y_1. However, for arbitrary weights, the contrast statistic which compares the average treatment mean with the control mean is not optimal. The vector, $\mathbf{c} \in H_1$, which has $c_2 = c_3 = \cdots = c_k$, makes equal angles with the faces of H_1, but not, in general, the edges.

For this simple tree ordering with equal weights, the maximin $r = (k-1)^{-1}$, which is attained when

$$(\boldsymbol{\mu} - \bar{\boldsymbol{\mu}}) \propto (-1, -1, \ldots, -1, k-1)',$$

where $\bar{\boldsymbol{\mu}}$ is a vector in which each element is $\bar{\mu} = \sum_{i=1}^{k} \mu_i/k$. The lower limit of the power function of the α level test based on the optimal $\mathbf{c}$ is

$$\Phi\left(\frac{\Delta}{k-1} - z_\alpha\right). \tag{4.2.9}$$

It is clear from this expression that the lower bound for the power falls away much faster, with increasing k, than was the case for simple order. This result is typical of what may be expected when the $\boldsymbol{\mu}$ is subject only to a partial ordering, because such constraint regions are usually much wider than the simple order.

It will be shown in Section 4.4 that, except in the case in which the cone, H_1, is narrow, as is the case for a simple order and small k, the minimum power of the optimal contrast test for H_0 versus $H_1 - H_0$ is much smaller than that of the LRT. Hence, if no prior information is available concerning the location of the true mean vector, other than $\boldsymbol{\mu} \in H_1$, and if the LRT is viable, these contrast tests cannot be recommended.

Optimal contrast coefficients for testing H_1 versus $H_2 - H_1$

We first argue that, for testing H_1 as a null hypothesis, only contrast vectors which are located in the dual of H_1 should be considered. Next it is shown that the

LRT for this problem is an adaptive contrast test so that its power properties will be difficult to beat in any global sense. It is then argued that the criteria used by Abelson and Tukey and by Schaafsma and Smid for selecting an optimal contrast test are not useful in this testing problem and must somehow be modified. This modification is described before Theorem 4.2.4 and optimal contrast coefficients are derived in Theorems 4.2.4 and 4.2.5 for the cases when H_1 prescribes a simple linear order and a simple tree order. The section is closed with a discussion of monotonicity and consistency of the power functions of these contrast tests.

As in the last subsection, let $\bar{Y}_1, \bar{Y}_2, \ldots, \bar{Y}_k$ and S^2 be independent with $\bar{Y}_i \sim \mathcal{N}(\mu_i, w_i^{-1})$ where $w_i = n_i/\sigma^2$ for $i = 1, 2, \ldots, k$ and $\nu S^2/\sigma^2 \sim \chi^2_\nu$ with ν a positive integer. Let $\bar{\mu} = \sum_{i=1}^k w_i\mu_i / \sum_{i=1}^k w_i$ and $\Delta^2(\boldsymbol{\mu}) = \sum_{i=1}^k w_i(\mu_i - \bar{\mu})^2$. Optimal contrast tests for the null hypothesis $H_1 : \boldsymbol{\mu}$ is isotonic with respect to $\lesssim$ will be derived for the case in which σ^2 is known with $\lesssim$ both the usual simple order and the simple tree ordering. The resulting test statistics can be 'Studentized' using S^2 for the case in which σ^2 is unknown. Contrast tests which reject H_1 for large values of $T_{\mathbf{c}} = \sum_{i=1}^k w_i c_i \bar{Y}_i / \|\mathbf{c}\|_{\mathbf{w}}$ are considered, and the power of the test based on $T_{\mathbf{c}}$ with critical value t is

$$\pi_{\mathbf{c}}(\boldsymbol{\mu}) = \Phi\left(\frac{\langle \mathbf{c}, \boldsymbol{\mu}\rangle_{\mathbf{w}}}{\|\mathbf{c}\|_{\mathbf{w}}} - t\right).$$

Since the distribution of $T_{\mathbf{c}}$ is not the same for all $\boldsymbol{\mu} \in H_1$, the level of significance is $\sup_{\boldsymbol{\mu}\in H_1} \pi_{\mathbf{c}}(\boldsymbol{\mu})$. If there is a $\boldsymbol{\mu} \in H_1$ with $\langle \mathbf{c}, \boldsymbol{\mu}\rangle_{\mathbf{w}} > 0$, then using the fact that H_1 is a cone, the significance level is easily seen to be 1. Thus, it is assumed that $\langle \boldsymbol{\mu}, \mathbf{c}\rangle_{\mathbf{w}} \leqslant 0$ for all $\boldsymbol{\mu} \in H_1$, or equivalently $\mathbf{c} \in H_1^{*\mathbf{w}}$. For such $\mathbf{c}$, the level of significance is $\sup_{\boldsymbol{\mu}\in H_1} \Phi(\langle \mathbf{c}, \boldsymbol{\mu}\rangle_{\mathbf{w}}/\|\mathbf{c}\|_{\mathbf{w}} - t) = \Phi(-t)$. Thus if z_α satisfies $\Phi(z_\alpha) = 1 - \alpha$, then $t = z_\alpha$ gives a test of size α.

The optimal contrast coefficients for testing H_1 against a fixed point $\boldsymbol{\mu} \notin H_1$ are determined next. Fix $\boldsymbol{\mu} \notin H_1$ and consider the contrast test which maximizes the power at $\boldsymbol{\mu}$, that is $\mathbf{c}$ maximizes $\langle \mathbf{c}, \boldsymbol{\mu}\rangle_{\mathbf{w}}/\|\mathbf{c}\|_{\mathbf{w}}$ over all $\mathbf{c} \in H_1^{*\mathbf{w}} - \{\mathbf{0}\}$. Since $\boldsymbol{\mu} \notin H_1$, there exists a pair $i \lesssim j$ with $\mu_i > \mu_j$. Consider $\mathbf{c}$ with $c_l = 0$ for $l \neq i, j$ and $w_i c_i = w_j c_j = 1$, then $r(\mathbf{c}, \boldsymbol{\mu}) > 0$ where $r(\mathbf{c}, \boldsymbol{\mu})$ is defined in (4.2.3). With the agreement that $\langle \mathbf{0}, \boldsymbol{\mu}\rangle_{\mathbf{w}}/\|\mathbf{0}\|_{\mathbf{w}} = 0$, the maximization problem is unchanged if $H_1^{*\mathbf{w}} - \{\mathbf{0}\}$ is replaced by $H_1^{*\mathbf{w}}$. Since $\boldsymbol{\mu}$ is fixed and $\bar{c} = 0$ for $\mathbf{c} \in H_1^{*\mathbf{w}}$, the above is equivalent to (recall, $r(\mathbf{0}, \boldsymbol{\mu}) = 0$)

$$\text{maximize } r(\mathbf{c}, \boldsymbol{\mu}) = \sum_{i=1}^{k} \frac{w_i(c_i - \bar{c})(\mu_i - \bar{\mu})}{\|\mathbf{c} - \bar{c}\|_{\mathbf{w}} \|\boldsymbol{\mu} - \bar{\mu}\|_{\mathbf{w}}} \quad \text{with } \mathbf{c} \in H_1^{*\mathbf{w}}. \tag{4.2.10}$$

Because $H_1^{*\mathbf{w}}$ is a closed, convex cone, it is not difficult to show that $H_1^{*\mathbf{w}} \oplus H_0$ is a closed, convex cone. (Recall $A \oplus B = \{a + b : a \in A, b \in B\}$.) To show closure, let $\boldsymbol{\mu}_n = \boldsymbol{\mu}_n' + \boldsymbol{\mu}_n'' \to \boldsymbol{\mu}$ with $\mu_n' \in H_1^{*\mathbf{w}}$ and $\boldsymbol{\mu}_n'' \in H_0$ and note that with $\bar{\mu}_n$ the common value of $\boldsymbol{\mu}_n''$, $\bar{\mu}_n \to \bar{\mu}$. Furthermore, $\mathbf{c}$ solves maximize $r(\mathbf{c}, \boldsymbol{\mu})$ over $\mathbf{c} \in H_1^{*\mathbf{w}} \oplus H_0$ if and only if $\mathbf{c} - \bar{c}$ solves (4.2.10). Applying (b) of Corollary E of

Theorem 8.2.7, $P_w(\boldsymbol{\mu}|H_1^{*w} \oplus H_0)$ maximizes $r(\mathbf{c}, \boldsymbol{\mu})$ for $\mathbf{c} \in H_1^{*w} \oplus H_0$. Using (1.5.5) and (1.5.6) it is easily shown that $P_w(\boldsymbol{\mu}|H_1^{*w} \oplus H_0) = P_w(\boldsymbol{\mu}|H_1^{*w}) + \bar{\mu}$. Since $\sum_{i=1}^k w_i(P_w(\boldsymbol{\mu}|H_1^{*w})_i + \bar{\mu})/\sum_{i=1}^k w_i = \bar{\mu}$, $P_w(\boldsymbol{\mu}|H_1^{*w})$ solves (4.2.10). The power function of the resulting tests is $\Phi(\langle P_w(\boldsymbol{\mu}|H_1^{*w})_w, \boldsymbol{\mu}\rangle / \| P_w(\boldsymbol{\mu}|H_1^{*w})\|_w - z_\alpha)$, which by (1.3.7) can be written as $\Phi(\|P_w(\boldsymbol{\mu}|H_1^{*w} \oplus H_0)\|_w - z_\alpha)$. The following result has been proved.

Theorem 4.2.2 *Let $\boldsymbol{\mu} \notin H_1$. The contrast test with maximum power at $\boldsymbol{\mu}$ is determined by $\mathbf{c} = P_w(\boldsymbol{\mu}|H_1^{*w})$. The power function is $\pi_c(\boldsymbol{\mu}) = \Phi(\|P_w(\boldsymbol{\mu}|H_1^{*w})\|_w - z_\alpha)$.*

Since the optimum $\mathbf{c}$ depends on the unknown $\boldsymbol{\mu}$, one could estimate $\mathbf{c}$ using $P_w(\bar{\mathbf{Y}}|H_1^{*w}) = \bar{\mathbf{Y}} - P_w(\bar{\mathbf{Y}}|H_1)$. However, $\sum_{i=1}^k w_i(\bar{Y}_i - P_w(\bar{\mathbf{Y}}|H_1)_i)\bar{Y}_i = \|\bar{\mathbf{Y}} - P_w(\bar{\mathbf{Y}}|H_1)\|_w^2 = \bar{\chi}_{12}^2$ (cf. equation (2.2.5)), which proves the following result.

Theorem 4.2.3 *The adaptive contrast test for testing H_1 versus $H_2 - H_1$ which has coefficient vector $\mathbf{c} = P_w(\bar{\mathbf{Y}}|H_1^{*w}) = \bar{\mathbf{Y}} - P_w(\bar{\mathbf{Y}}|H_1)$ is $\bar{\chi}_{12}^2$, i.e. for any quasi-order, $\lesssim$, $\bar{\chi}_{12}^2$ is an adaptive contrast test.*

First, optimal contrast coefficients are sought for the usual simple order. Next, the simple tree ordering is treated. Considering the criterion used by Abelson and Tukey (1963), with $\delta > 0$ fixed, one wants contrast coefficients which maximize the minimum power over points at a distance δ from the null hypothesis, H_1. Hence, $\mathbf{c}$ solves

$$\sup_{\mathbf{c} \in H_1^{*w} - \{\mathbf{0}\}} \inf_{\boldsymbol{\mu}: \|\boldsymbol{\mu} - P_w(\boldsymbol{\mu}|H_1)\|_w = \delta} \Phi\left(\frac{\langle \mathbf{c}, \boldsymbol{\mu}\rangle_w}{\|\mathbf{c}\|_w} - z_\alpha\right).$$

However, it will be shown that for $\mathbf{c} \in H_1^{*w} - \{\mathbf{0}\}$ and $\delta > 0$,

$$\inf_{\boldsymbol{\mu}: \|\boldsymbol{\mu} - P_w(\boldsymbol{\mu}|H_1)\|_w = \delta} \pi_c(\boldsymbol{\mu}) = 0$$

so that this criterion is not useful.

Lemma *If $\mathbf{c} \in H_1^{*w} - \{\mathbf{0}\}$, $\delta > 0$ and $k > 2$, then there exists a $\boldsymbol{\mu} \notin H_1$ with $\|\boldsymbol{\mu} - P_w(\boldsymbol{\mu}|H_1)\|_w = \delta$ and $\langle \mathbf{c}, P_w(\boldsymbol{\mu}|H_1)\rangle_w < 0$.*

Proof: Let $\mathbf{v}_1 = (w_1^{-1}, -w_2^{-1}, 0, \ldots, 0)$, $\mathbf{v}_2 = (0, w_2^{-1}, -w_3^{-1}, 0, \ldots, 0), \ldots, \mathbf{v}_{k-1} = (0, \ldots, 0, w_{k-1}^{-1}, -w_k^{-1})$. It is easy to show that $H_1^{*w} = \{a_1\mathbf{v}_1 + \cdots + a_{k-1}\mathbf{v}_{k-1} : a_i \geq 0\}$. Furthermore, $\langle \mathbf{v}_i, P_w(\boldsymbol{\mu}|H_1)\rangle_w \leq 0$ for each i and $\boldsymbol{\mu}$. Let $\mathbf{c} = a_1\mathbf{v}_1 + \cdots + a_{k-1}\mathbf{v}_{k-1}$. If $a_j > 0$, $1 \leq j < k-1$, then let $\boldsymbol{\mu}$ be $(1, 2, \ldots, k)$ with the $(j+1)$th and the $(j+2)$th coordinates interchanged. So, $P_w(\boldsymbol{\mu}|H_1)_i = i$ for $i \neq j+1$, $j+2$ and $P_w(\boldsymbol{\mu}|H_1)_i = [(j+2)w_{j+1} + (j+1)w_{j+2}]/(w_{j+1} + w_{j+2})$ for $i = j+1$, $j+2$. Thus, $a_j\langle \mathbf{v}_j, P_w(\boldsymbol{\mu}|H_1)\rangle_w = a_j(j - P_w(\boldsymbol{\mu}|H_1)_{j+1}) < 0$ and $\langle \mathbf{c}, P_w(\boldsymbol{\mu}|H_1)\rangle_w < 0$. If $a_1 = \cdots = a_{k-2} = 0$ and

$a_{k-1} > 0$ (recall, $\mathbf{c} \neq \mathbf{0}$), then let $\boldsymbol{\mu} = (2, 1, \ldots, k)$. It is easy to show that $a_{k-1}\langle \mathbf{v}_{k-1}, P_{\mathbf{w}}(\boldsymbol{\mu}|H_1)\rangle_{\mathbf{w}} < 0$. Thus, in either case, one can find $\boldsymbol{\mu} \notin H_1$ with $\langle \mathbf{c}, P_{\mathbf{w}}(\boldsymbol{\mu}|H_1)\rangle_{\mathbf{w}} < 0$. Multiplying by the appropriate positive constant, we obtain $\boldsymbol{\mu} \notin H_1$ with $\langle \mathbf{c}, P_{\mathbf{w}}(\boldsymbol{\mu}|H_1)\rangle_{\mathbf{w}} < 0$ and $\|\boldsymbol{\mu} - P_{\mathbf{w}}(\boldsymbol{\mu}|H_1)\|_{\mathbf{w}} = \delta$. □

For $a \geqslant -1$, set $\boldsymbol{\mu}_a = \boldsymbol{\mu} + aP_{\mathbf{w}}(\boldsymbol{\mu}|H_1)$. It was noted in Section 2.6 (see also Theorem 8.2.7) that $\|\boldsymbol{\mu}_a - P_{\mathbf{w}}(\boldsymbol{\mu}_a|H_1)\|_{\mathbf{w}} = \|\boldsymbol{\mu} - P_{\mathbf{w}}(\boldsymbol{\mu}|H_1)\|_{\mathbf{w}} = \delta > 0$. Thus, $\boldsymbol{\mu}_a \notin H_1$, the distance from $\boldsymbol{\mu}_a$ to H_1 is δ and

$$\lim_{a \to \infty} \langle \mathbf{c}, \boldsymbol{\mu}_a \rangle_{\mathbf{w}} = \langle \mathbf{c}, \boldsymbol{\mu} \rangle_{\mathbf{w}} + \lim_{a \to \infty} a \langle \mathbf{c}, P_{\mathbf{w}}(\boldsymbol{\mu}|H_1)\rangle_{\mathbf{w}} = -\infty.$$

Therefore, for each $\mathbf{c} \in H_1^{*\mathbf{w}} - \{\mathbf{0}\}$ and $\delta > 0$,

$$\inf_{\boldsymbol{\mu}:\, \|\boldsymbol{\mu} - P_{\mathbf{w}}(\boldsymbol{\mu}|H_1)\|_{\mathbf{w}} = \delta} \pi_{\mathbf{c}}(\boldsymbol{\mu}) = 0.$$

Other criteria must be considered.

Following Schaafsma and Smid (1966), the contrast coefficients which minimize the maximum 'shortcoming' among all contrast tests are considered. Recall that for a given $\boldsymbol{\mu} \notin H_1$, the contrast test with maximum power at $\boldsymbol{\mu}$ is obtained by taking $\mathbf{c} = \boldsymbol{\mu} - P_{\mathbf{w}}(\boldsymbol{\mu}|H_1)$ and has power $\Phi(\|\boldsymbol{\mu} - P_{\mathbf{w}}(\boldsymbol{\mu}|H_1)\|_{\mathbf{w}} - z_\alpha)$. So, for any contrast test its shortcoming at $\boldsymbol{\mu}$ is

$$\Phi(\|\boldsymbol{\mu} - P_{\mathbf{w}}(\boldsymbol{\mu}|H_1)\|_{\mathbf{w}} - z_\alpha) - \Phi\left(\frac{\langle \mathbf{c}, \boldsymbol{\mu}\rangle_{\mathbf{w}}}{\|\mathbf{c}\|_{\mathbf{w}}} - z_\alpha\right).$$

If there is no constraint on $\boldsymbol{\mu}$ other than $\boldsymbol{\mu} \notin H_1$, one sees from the preceding analysis that the supremum is at least as large as $\Phi(\delta - z_\alpha)$ for each $\delta > 0$, and so the maximum shortcoming over all $\boldsymbol{\mu} \notin H_1$ is 1. Even if $\boldsymbol{\mu}$ is constrained so that $\|\boldsymbol{\mu} - P_{\mathbf{w}}(\boldsymbol{\mu}|H_1)\|_{\mathbf{w}} = \delta > 0$, the maximum shortcoming is $\Phi(\delta - z_\alpha)$ which does not depend on $\mathbf{c}$. Neither of these criteria are helpful in choosing a good contrast test.

The vector, $\boldsymbol{\mu} + aP_{\mathbf{w}}(\boldsymbol{\mu}|H_1)$, used in the lemma above remains at a fixed distance from H_1, but it is moving away from H_0 as a increases (see Figure 4.2.1). Therefore, consider the contrast test which maximizes the minimum power over all $\boldsymbol{\mu} \notin H_1$ with $\Delta(\boldsymbol{\mu}) = \|\boldsymbol{\mu} - \bar{\mu}\|_{\mathbf{w}} = \delta > 0$. Let $a_i = (w_i^{-1} + w_{i+1}^{-1})^{1/2}$ for $i = 1, 2, \ldots, k-1$, $b_1 = 0$, $b_i = \sum_{j=1}^{i-1} a_j$ for $i = 2, \ldots, k$, and let $\mathbf{c}^{(12)} = \bar{b} - \mathbf{b}$.

Theorem 4.2.4 *Let $\delta > 0$. The contrast test which has coefficients $\mathbf{c}^{(12)}$ and rejects for large values of $T_{\mathbf{c}^{(12)}}$ maximizes the minimum power over all $\boldsymbol{\mu} \notin H_1$ with $\Delta(\boldsymbol{\mu}) = \delta$. Furthermore, such contrast coefficients are unique up to a positive multiplier.*

Before the proof of the theorem is given, the following lemma is established.

Lemma *If $\boldsymbol{\mu}, \mathbf{v} \in H_1$, then $\langle \boldsymbol{\mu} - \bar{\mu}, \mathbf{v} - \bar{v}\rangle_{\mathbf{w}} \geqslant 0$.*

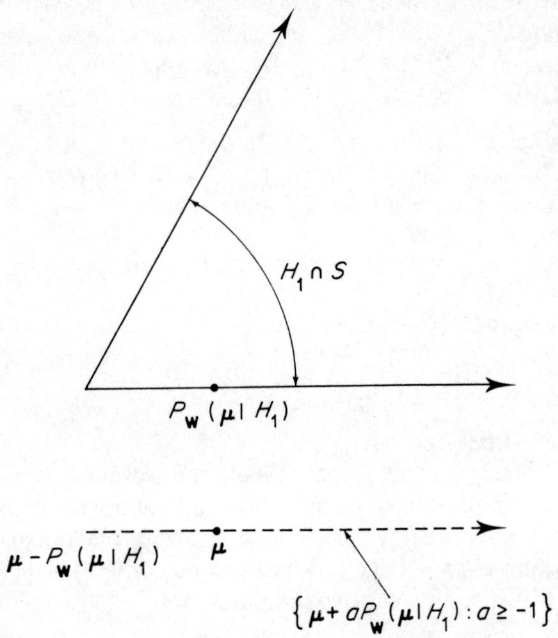

Figure 4.2.1 Representation of $H_1 \cap S$, $\boldsymbol{\mu}$ and $\{\boldsymbol{\mu} + aP_w(\boldsymbol{\mu}|H_1): a \geqslant -1\}$ for H_1, the simply ordered cone, and $S = \{\mathbf{y}: \sum_{i=1}^k w_i y_i = 0\}$

Proof: $\boldsymbol{\mu} - \bar{\mu} \in H_1$ and $\boldsymbol{\nu} - \bar{\nu} \in -H_1^{*w}$ (see equation (1.7.6)) and the conclusion is immediate. □

Proof of Theorem 4.2.4: We wish to find $\mathbf{c}^{(12)}$ which yields

$$\sup_{\mathbf{c} \in H_1^{*w} - \{0\}} \ \inf_{\mu \notin H_1, \Delta(\mu) = \delta} \Phi\left(\frac{\langle \mathbf{c}, \boldsymbol{\mu} \rangle_w}{\|\mathbf{c}\|_w} - z_\alpha\right),$$

or since $\bar{c} = 0$, equivalently,

$$\sup_{\mathbf{c} \in H_1^{*w} - \{0\}} \ \inf_{\mu \notin H_1} r(\mathbf{c}, \boldsymbol{\mu}). \tag{4.2.11}$$

If $-\mathbf{c} \notin H_1$, then consider $\boldsymbol{\mu} = -\mathbf{c}$. Since $r(\mathbf{c}, -\mathbf{c}) = -1$, we may omit such $\mathbf{c}$ from the supremum. Because $H_1^{*w} \cap (-H_1) = (-H_1) \cap \{\boldsymbol{\mu}: \bar{\mu} = 0\}$, (4.2.11) is equivalent to

$$\sup_{-\mathbf{c} \in H_1^{*w} - \{0\}, \bar{c} = 0} \ \inf_{\mu \notin H_1} r(\mathbf{c}, \boldsymbol{\mu}) = - \inf_{\mathbf{d} \in H_1 - \{0\}, \bar{d} = 0} \ \sup_{\mu \notin H_1} r(\mathbf{d}, \boldsymbol{\mu}). \tag{4.2.12}$$

We will solve for $\mathbf{d}$ and remember that $\mathbf{c}^{(12)} = -\mathbf{d}$. Because of the continuity of $r(\mathbf{d}, \cdot)$, the supremum in the right-hand side could also be taken over $\boldsymbol{\mu} \notin H_1^0 \cup H_0$. ($A^0$ is the interior of A.) However, if $\boldsymbol{\mu} \in \partial H_1$, then applying the lemma above,

$r(\mathbf{d},\boldsymbol{\mu}) \geqslant 0$ and hence the supremum could be restricted to $\boldsymbol{\mu}\notin H_1^0 \cup (H_0 \oplus H_1^{*w})$ (for $r(\mathbf{d},\boldsymbol{\mu}) \leqslant 0$ for $\boldsymbol{\mu}\in H_1^{*w}$). Therefore, we seek $\mathbf{d}$ which solves

$$\inf_{\mathbf{d}\in H_1-\{\mathbf{0}\},\bar{d}=0} \ \sup_{\boldsymbol{\mu}\notin H_1\cup(H_0\oplus H_1^{*w})} r(\mathbf{d},\boldsymbol{\mu}).$$

Furthermore, if $\boldsymbol{\mu}\notin H_1^{*w}\oplus H_0$ then $P_w(\boldsymbol{\mu}|H_1)\notin H_0$ and so $\Delta(P_w(\boldsymbol{\mu}|H_1))>0$. Applying (2.2.7) and the fact that $\sum_{i=1}^k w_i P_w(\boldsymbol{\mu}|H_1)_i = \sum_{i=1}^k w_i\mu_i$, it follows that $0<\Delta(P_w(\boldsymbol{\mu}|H_1)) = \|P_w(\boldsymbol{\mu}|H_1) - \bar{\mu}\|_w \leqslant \|\boldsymbol{\mu}-\bar{\mu}\|_w = \Delta(\boldsymbol{\mu})$. For fixed $\mathbf{d}\in H_1-\{\mathbf{0}\}$ with $\bar{d}=0$, $r(\mathbf{d},\boldsymbol{\mu}) = (\|\mathbf{d}\|_w\|\boldsymbol{\mu}-\bar{\mu}\|_w)^{-1}\langle\mathbf{d},\boldsymbol{\mu}\rangle_w \leqslant (\|\mathbf{d}\|_w \|\boldsymbol{\mu}-\bar{\mu}\|_w)^{-1}\langle\mathbf{d},P_w(\boldsymbol{\mu}|H_1)\rangle_w$, which is nonnegative by the last lemma. Thus, $r(\mathbf{d},\boldsymbol{\mu}) \leqslant r(\mathbf{d},P_w(\boldsymbol{\mu}|H_1))$ for $\boldsymbol{\mu}\notin H_1^{*w}\oplus H_0$. Therefore, $\mathbf{d}$ solves

$$\inf_{\mathbf{d}\in H_1-\{\mathbf{0}\},\bar{d}=0} \ \sup_{\boldsymbol{\mu}\notin H_1\cup(H_1^{*w}\oplus H_0)} r(\mathbf{d},P_w(\boldsymbol{\mu}|H_1)) = \inf_{\mathbf{d}\in H_1-\{\mathbf{0}\},\bar{d}=0} \ \sup_{\boldsymbol{\mu}\in\partial H_1-H_0} r(\mathbf{d},\boldsymbol{\mu}).$$

The boundary of H_1 is the union of

$$A_1 = \{\mathbf{y}\in R^k: y_1 = y_2 \leqslant y_3 \leqslant\cdots\leqslant y_k\},$$
$$A_2 = \{\mathbf{y}\in R^k: y_1 \leqslant y_2 = y_3 \leqslant\cdots y_k\},\ldots,$$
$$A_{k-1} = \{\mathbf{y}\in R^k: y_1 \leqslant\cdots\leqslant y_{k-1} = y_k\}.$$

Because of the convention, $r(\mathbf{d},\mathbf{0})=0$, we seek $\mathbf{d}$ that solves

$$\inf_{\mathbf{d}\in H_1-\{\mathbf{0}\},\bar{d}=0} \ \max_{1\leqslant i\leqslant k-1} \ \max_{\boldsymbol{\mu}\in A_i} r(\mathbf{d},\boldsymbol{\mu}). \tag{4.2.13}$$

Each A_i is a closed, convex cone in R^k containing the constant functions and $r(\mathbf{d},\boldsymbol{\mu})\geqslant 0$ for any $\boldsymbol{\mu}\in A_i$. So, by Corollary E of Theorem 8.2.7, $\max_{\boldsymbol{\mu}\in A_i} r(\mathbf{d},\boldsymbol{\mu}) = r(\mathbf{d},P_w(\mathbf{d}|A_i))$. It is easy to show that $\mathbf{d}^* = (d_1^*,\ldots,d_k^*)$, with $d_j^* = d_j$ for $j\neq i$, $i+1$ and $d_j^* = (w_i d_i + w_{i+1}d_{i+1})/(w_i+w_{i+1})$ for $j=i$, $i+1$, is the point in A_i closest to $\mathbf{d}\in H_1$, that is $\mathbf{d}^* = P_w(\mathbf{d}|A_i)$. Also $r(\mathbf{d},P_w(\mathbf{d}|A_i)) = \|P_w(\mathbf{d}|A_i)\|_w/\|\mathbf{d}\|_w$. Hence $\mathbf{d}$ solves (4.2.13) if and only if $\mathbf{d}/\|\mathbf{d}\|_w$ solves

$$\min_{\mathbf{d}\in H_1,\|\mathbf{d}\|_w=1,\bar{d}=0} \ \max_{1\leqslant i\leqslant k-1} \|P_w(\mathbf{d}|A_i)\|_w.$$

However, $\|\mathbf{d}\|_w - \|P_w(\mathbf{d}|A_i)\|_w = w_i w_{i+1}(d_{i+1}-d_i)^2/(w_i+w_{i+1}) = (d_{i+1}-d_i)^2/a_i^2$, with a_i defined as above. Therefore, $\mathbf{d}$ solves

$$\max_{\mathbf{d}\in H_1,\|\mathbf{d}\|_w=1,\bar{d}=0} \ \min_{1\leqslant i\leqslant k-1} \frac{(d_{i+1}-d_i)^2}{a_i^2}. \tag{4.2.14}$$

With $\mathbf{b}$ defined as in the paragraph before the statement of the theorem, let $\mathbf{d}_\gamma = \gamma(\mathbf{b}-\bar{b})$. Note that $\mathbf{d}_\gamma\in H_1$, $\bar{\mathbf{d}}_\gamma = 0$ and $\|\mathbf{d}_\gamma\|_w = \gamma\|\mathbf{d}_1\|_w > 0$ for all $\gamma>0$. Therefore, choose γ so that $\|\mathbf{d}_\gamma\|_w = 1$.

We now show that the $\mathbf{d}_\gamma$ chosen above is the unique solution to (4.2.1.4), which implies that $-\mathbf{d}_\gamma$ is the unique, up to a positive multiplier, set of contrast coefficients which is being sought. Note that if $\mathbf{d}_\gamma = (\mathbf{d}_{\gamma,1},\mathbf{d}_{\gamma,2},\ldots,d_{\gamma,k})$, then

$(d_{\gamma,i+1} - d_{\gamma,i})^2/a_i^2 = \gamma^2$ for $i = 1, 2, \ldots, k-1$. Suppose $\mathbf{z} \in H_1$ with $\bar{z} = 0$, $\|\mathbf{z}\|_{\mathbf{w}} = 1$, and $\min_{1 \leqslant i \leqslant k-1} (z_{i+1} - z_i)^2/a_i^2 \geqslant \gamma^2$. Then $(z_{i+1} - z_i) \geqslant (d_{\gamma,i+1} - d_{\gamma,i})$ or $z_{i+1} - d_{\gamma,i+1} > z_i - d_{\gamma,i}$ for $i = 1, 2, \ldots, k-1$. Hence, $\mathbf{z} - \mathbf{d}_\gamma \in H_1$ and, applying the lemma,

$$1 = \|\mathbf{z}\|_{\mathbf{w}}^2 = \|\mathbf{d}_\gamma\|_{\mathbf{w}}^2 + \|\mathbf{z} - \mathbf{d}_\gamma\|_{\mathbf{w}}^2 + 2\langle \mathbf{d}_\gamma, \mathbf{z} - \mathbf{d}_\gamma \rangle_{\mathbf{w}} \geqslant 1 + \|\mathbf{z} - \mathbf{d}_\gamma\|_{\mathbf{w}}^2.$$

Therefore, $\|\mathbf{z} - \mathbf{d}_\gamma\|_{\mathbf{w}}^2 = 0$ or $\mathbf{z} = \mathbf{d}_\gamma$. □

If the weights are equal, $w_1 = w_2 = \cdots = w_k$, then the a_i, which are defined before the statement of Theorem 4.2.4, are equal and the contrast vector, $\mathbf{c}^{(12)}$, has equal increments, that is $c_{i+1}^{(12)} - c_i^{(12)}$ is constant. For equal weights, the negative of $\mathbf{c}^{(12)}$ is in H_1 and in fact is a center of H_1 in the sense that it makes equal angles with the faces of H_1. It, in fact, might be used to test H_0 versus H_1. It is the original T-statistic suggested in (4.2.4). On the other hand, for the simple order and equal weights, $\mathbf{c}^{(01)}$ makes equal angles with the edges and is another center of H_1.

Optimal contrast coefficients for the simple tree ordering are discussed next. Using the same notation as in the simply ordered case, label the control mean μ_1 and the treatment means $\mu_2, \mu_3, \ldots, \mu_k$. With $w_i = n_i/\sigma^2$ for $i = 1, 2, \ldots, k$, set $a_i = (w_1^{-1} + w_{i+1}^{-1})^{1/2}$ for $i = 1, 2, \ldots, k-1$, $b_1 = 0$ and $b_i = \sum_{j=1}^{i-1} a_j$ for $i = 2, 3, \ldots, k$. As in the simply ordered case,

$$\mathbf{c}^{(12)} = \bar{b} - \mathbf{b}.$$

For the simple tree ordering, $\mu_1 \leqslant [\mu_2, \mu_3, \ldots, \mu_k]$, one can, using techniques like those used to establish (1.7.6), show that

$$H_1^{*\mathbf{w}} = \{\mathbf{y} \in R^k : \textstyle\sum_{i=1}^k w_i y_i = 0 \text{ and } y_i \leqslant 0;\ i = 2, 3, \ldots, k\}.$$

While $\mathbf{c}^{(12)}$ is not always in $H_1^{*\mathbf{w}}$, it is in this set in the special case in which the treatment sample sizes are equal, that is $n_2 = n_3 = \cdots = n_k$. Mukerjee, Robertson, and Wright (1986a) show that, as in the simply ordered case, the criteria used by Abelson and Tukey and Schaafsma and Smid for testing H_0 against $H_1 - H_0$ needs to be modified for testing H_1 against $H_2 - H_1$. Applying the same criterion as in the case of a simple order, they prove the following result.

Theorem 4.2.5 *Let $\mathbf{c}^{(12)} \in H_1^{*\mathbf{w}}$ and $\delta > 0$. The contrast test which rejects H_1 for large values of $T_{\mathbf{c}^{(12)}}$ maximizes, over all contrast tests, the minimum of the power over all $\boldsymbol{\mu} \notin H_1$ with $\Delta(\boldsymbol{\mu}) = \delta$. Furthermore, the contrast coefficients are unique up to a positive multiplier.*

The proof which is similar to that given for Theorem 4.2.4 is omitted.

If H_1 is a half-space, for instance $\{\boldsymbol{\mu} : \mu_1 \leqslant \mu_2\}$, then a contrast provides a good test of H_1 versus $H_2 - H_1$. However, typically H_1 is considerably narrower than a half-space (notice that a smaller H_1 corresponds to a larger alternative region) and even the optimal contrast test is not very efficient in such cases.

Properties of contrast tests

Properties of the power function of a contrast test depend only on the contrast coefficients and not the hypotheses or the quasi-order of interest. Considering the case in which σ^2 is known, suppose that a contrast test rejects for large values of $T_{\mathbf{c}}$, so that the power function is given by $\pi_{\mathbf{c}}(\boldsymbol{\mu})$ in (4.2.2).

Theorem 4.2.6 *Let $\boldsymbol{\mu}$, $\boldsymbol{\nu} \in R^k$. If $\langle \mathbf{c}, \boldsymbol{\nu} \rangle_{\mathbf{w}} = 0$, then $\pi_{\mathbf{c}}(\boldsymbol{\mu} + \gamma\boldsymbol{\nu}) = \pi_{\mathbf{c}}(\boldsymbol{\mu})$ for all $\gamma \in (-\infty, \infty)$. If $\langle \mathbf{c}, \boldsymbol{\nu} \rangle_{\mathbf{w}} > 0 (\langle \mathbf{c}, \boldsymbol{\nu} \rangle_{\mathbf{w}} < 0)$, than $\pi_{\mathbf{c}}(\boldsymbol{\mu} + \gamma\boldsymbol{\nu})$ is increasing (decreasing) in γ with $\lim_{\gamma \to \infty}[\pi_{\mathbf{c}}(\boldsymbol{\mu} + \gamma\boldsymbol{\nu})] = 1(0)$ and $\lim_{\gamma \to -\infty}[\pi_{\mathbf{c}}(\boldsymbol{\mu} + \gamma\boldsymbol{\nu})] = 0(1)$.*

Proof: The result follows immediately from (4.2.2) since $\langle \mathbf{c}, \boldsymbol{\mu} + \gamma\boldsymbol{\nu} \rangle_{\mathbf{w}} = \langle \mathbf{c}, \boldsymbol{\mu} \rangle_{\mathbf{w}} + \gamma \langle \mathbf{c}, \boldsymbol{\nu} \rangle_{\mathbf{w}}$. □

In the next result the regions of consistency are determined for such contrast tests.

Theorem 4.2.7 *Let $\boldsymbol{\mu}$, $\boldsymbol{\gamma} \in R^k$ with $\gamma_i > 0$ for $i = 1, 2, \ldots, k$. For all n let $\mathbf{w}_n = n\boldsymbol{\gamma}$ and fix the level of the contrast test at $\alpha \in (0, 1)$. If $\langle \mathbf{c}, \boldsymbol{\mu} \rangle_{\boldsymbol{\gamma}} > 0$ $(\langle \mathbf{c}, \boldsymbol{\mu} \rangle_{\boldsymbol{\gamma}} < 0)$, then $\pi_{\mathbf{c}}(\boldsymbol{\mu}) \to 1 (0)$ as $n \to \infty$. If $\langle \mathbf{c}, \boldsymbol{\mu} \rangle_{\boldsymbol{\gamma}} = 0$, then $\pi_{\mathbf{c}}(\boldsymbol{\mu}) = \alpha$ for all n.*

Proof: Since $\langle \mathbf{c}, \boldsymbol{\mu} \rangle_{\mathbf{w}_n} / \| \mathbf{c} \|_{\mathbf{w}_n} = n^{1/2} \langle \mathbf{c}, \boldsymbol{\mu} \rangle_{\boldsymbol{\gamma}} / \| \mathbf{c} \|_{\boldsymbol{\gamma}}$, the desired conclusion follows from (4.2.2). □

It is instructive to compare the regions of consistency for $\bar{\chi}^2_{01}$ and $T_{\mathbf{c}^{(01)}}$ in testing H_0 versus $H_1 - H_0$ with $\lesssim$ the usual simple order. It is first shown that $\mathbf{c}^{(01)} \in H_1^0$. Let $x_i = W_i/W_k$ for $i = 0, 1, 2, \ldots, k$ and note that (cf. equation (4.2.6)) $c_i^{(01)} \propto [g(x_{i-1}) - g(x_i)]/(x_i - x_{i-1})$ where $g(x) = [x(1-x)]^{1/2}$. The desired conclusion follows from the strict concavity of g on $[0, 1]$. Applying Theorem 4.2.7 the contrast test is consistent for $\boldsymbol{\mu} \in A^+(\mathbf{c}^{(01)}) = \{\boldsymbol{\mu} : \langle \mathbf{c}^{(01)}, \boldsymbol{\mu} \rangle_{\mathbf{w}} > 0\}$, and from Theorem 2.6.1, $\bar{\chi}^2_{01}$ is consistent for $\boldsymbol{\mu} \notin H_1^{*\mathbf{w}} \oplus H_0$. If $\boldsymbol{\mu} = \boldsymbol{\mu}' + \boldsymbol{\mu}''$ with $\boldsymbol{\mu}' \in H_1^{*\mathbf{w}}$ and $\boldsymbol{\mu}'' \in H_0$, then $\langle \mathbf{c}^{(01)}, \boldsymbol{\mu} \rangle_{\mathbf{w}} = \langle \mathbf{c}^{(01)}, \boldsymbol{\mu}' \rangle_{\mathbf{w}} \leqslant 0$. So, $A^+(\mathbf{c}^{(01)}) \subset (H_1^{*\mathbf{w}} \oplus H_0)^c$. Examining Figure 2.5.2 gives one an idea of the size of $(H_1^{*\mathbf{w}} \oplus H_0)^c - A^+(\mathbf{c}^{(01)})$ for $k = 3$ and equal weights.

In testing H_1 versus H_2, $T_{\mathbf{c}^{(12)}}$ is consistent for $\boldsymbol{\mu} \in A^+(\mathbf{c}^{(12)})$ and $\bar{\chi}^2_{12}$ is consistent for $\boldsymbol{\mu} \notin H_1$. Since $-\mathbf{c}^{(12)} \in H_1$ and $\bar{c}^{(12)} = 0$, $\mathbf{c}^{(12)} \in H_1^{*\mathbf{w}}$. If $\boldsymbol{\mu} \in H_1$, then $\langle \mathbf{c}^{(12)}, \boldsymbol{\mu} \rangle_{\mathbf{w}} \leqslant 0$ and $\boldsymbol{\mu} \in [A^+(\mathbf{c}^{(12)})]^c$. Hence, $A^+(\mathbf{c}^{(12)}) \subset H_1^c$. Again one can obtain an idea of the size of $H_1^c - A^+(\mathbf{c}^{(12)})$ for $k = 3$ and equal weights from Figure 2.5.2.

4.3 MULTIPLE CONTRAST TESTS

While the contrast tests studied in the last section are very simple to use, their power characteristics are such that they cannot be recommended in general as competitors to the LRT. (The reader may wish to look at the results of the power study which are given in Section 4.4.) On the other hand, it can be shown that the LRT statistic may be expressed as the maximum of an infinite number of contrast statistics. In this section the possibility of using the maximum of a finite number of contrast statistics is considered for testing H_0 versus H_1-H_0 and H_1 versus H_2-H_1 for some special partial orders (H_0, H_1, H_2 are defined as in the last section). While some of the *ad hoc* tests in the literature are such multiple contrast tests (MCTs), they do not seem to have been developed from this point of view. The question of which MCTs are optimal is an important and challenging open problem in order restricted inference.

As in the last section, suppose $\bar{\mathbf{Y}} = (\bar{Y}_1, \bar{Y}_2, \ldots, \bar{Y}_k)$ and S^2 are stochastically independent, $\bar{Y}_i \sim \mathcal{N}(\mu_i, \sigma^2/n_i)$, $i = 1, 2, \ldots, k$, and $\nu S^2/\sigma^2 \sim \chi^2_\nu$ with ν a positive integer. The case in which σ^2 is known is discussed first. MCTs with k_0 contrasts are considered. Specifically, let $w_i = n_i/\sigma^2$ for $i = 1, 2, \ldots, k$, let $\mathbf{c}_1, \mathbf{c}_2, \ldots, \mathbf{c}_{k_0}$ be k-dimensional vectors, let $\langle \mathbf{x}, \mathbf{y} \rangle_{\mathbf{w}} = \sum_{i=1}^k w_i x_i y_i$, let $\| \mathbf{x} \|_{\mathbf{w}}^2 = \sum_{i=1}^k w_i x_i^2$, and let

$$T_l = \frac{\sum_{i=1}^k w_i c_{li} \bar{Y}_i}{\| \mathbf{c}_l \|_{\mathbf{w}}} \qquad \text{for } l = 1, 2, \ldots, k_0. \tag{4.3.1}$$

The MCT statistic based on $T_1, T_2, \ldots, T_{k_0}$ is

$$T = \max_{1 \leqslant l \leqslant k_0} T_l. \tag{4.3.2}$$

The distribution of T, which depends on the correlation coefficients

$$\rho(T_l, T_{l'}) = \frac{\sum_{i=1}^k w_i c_{li} c_{l'i}}{\| \mathbf{c}_l \|_{\mathbf{w}} \| \mathbf{c}_{l'} \|_{\mathbf{w}}} = \frac{\langle \mathbf{c}_l, \mathbf{c}_{l'} \rangle_{\mathbf{w}}}{\| \mathbf{c}_l \|_{\mathbf{w}} \| \mathbf{c}_{l'} \|_{\mathbf{w}}} \tag{4.3.3}$$

may be rather complex. There are two ways to deal with this difficulty in computing significance levels: Bonferroni's inequality can be used or the $\mathbf{c}_l$ can be chosen so that the correlations $\rho(T_l, T_{l'})$ are the same for $1 \leqslant l \neq l' \leqslant k_0$. As a special case of the latter, orthogonal multiple contrast tests (OCTs) will be considered, i.e. the $\mathbf{c}_l$ will be chosen so that (4.3.3) is zero for $1 \leqslant l \neq l' \leqslant k_0$.

For MCTs and for testing H_0 versus H_1, each $\mathbf{c}_l$ should be in $H_0^{\perp}$, for if $\langle \mathbf{c}_l, \boldsymbol{\mu} \rangle_{\mathbf{w}} \neq 0$ for some $\boldsymbol{\mu} \in H_0$ then there is a $\boldsymbol{\mu}' \in H_0$ (either $\boldsymbol{\mu}'$ or $-\boldsymbol{\mu}'$) with $\langle \mathbf{c}_l, \boldsymbol{\mu}' \rangle_{\mathbf{w}} > 0$, and as $\gamma \to \infty$, $T_l \xrightarrow{P} \infty$ at the points $\gamma \boldsymbol{\mu}' \in H_0$. Therefore, each $\mathbf{c}_l$ is chosen so that

$$\sum_{i=1}^k w_i c_{li} = 0.$$

The arguments given in Section 4.3 show that for testing H_1 versus H_2-H_1, each $\mathbf{c}_l$ needs to be in $H_1^{*\mathbf{w}}$, and we make this assumption in these testing situations.

For the case in which σ^2 is unknown, each $T_l/(S/\sigma)$ is a statistic and has a Student's t-distribution, possibly noncentral, and hence the test statistic $T/(S/\sigma)$ could be used in this case.

Ad hoc MCTs for the simple order

Williams (1971, 1972), in the simply ordered case, that is $\mu_1 \leqslant \mu_2 \leqslant \cdots \leqslant \mu_k$, considers the situation in which the μ_i are mean responses to increasing levels of a drug with μ_1 the mean response to a zero dose control. Denoting the sample sizes by $n_1, n_2, \ldots, n_k$, he considers the sampling situation in which $n_1 \geqslant n_2 = n_3 = \cdots = n_k = n$ and proposes rejecting $H_0: \mu_1 = \mu_2 = \cdots = \mu_k$ in favor of $H_1: \mu_1 \leqslant \mu_2 \leqslant \cdots \leqslant \mu_k$ for large values of

$$\bar{t} = \frac{\mu_k^* - \bar{Y}_1}{S((1/n_1) + (1/n))^{1/2}}$$

where μ_k^* is the MLE of μ_k subject to $\mu_2 \leqslant \mu_3 \leqslant \cdots \leqslant \mu_k$. Williams gives approximate $\alpha = 0.05$ and 0.01 critical values for this test. (He omitted the control mean from this restriction in order to simplify the distribution theory.) Using the upper sets algorithm (cf. Section 1.4),

$$\mu_k^* = \max_{2 \leqslant i \leqslant k} \frac{\sum_{j=i}^k n_j \bar{Y}_j}{\sum_{j=i}^k n_j}$$

and thus $S\bar{t}/\sigma$ is the maximum of $k-1$ contrasts. However, the contrast vectors do not have the same length even if $n_1 = n_2 = \cdots = n_k$. In the case in which all the sample sizes are equal, one contrast vector is $\mathbf{c}_1 = \sigma(-1, 0, 0, \ldots, 0, 1)/\sqrt{2n}$ which has $\|\mathbf{c}_1\|_w = 1$ and another is $\mathbf{c}_{k-1} = \sigma(-(k-1), 1, \ldots, 1)/[(k-1)\sqrt{2n}]$ which has $\|\mathbf{c}_{k-1}\|_w = \{k/[2(k-1)]\}^{1/2}$.

Chase (1974) and Williams (1971) compare the powers of the $\bar{t}$-test with those of $\bar{\chi}_{01}^2$ and $\bar{E}_{01}^2$ and conclude that except for $k=3$ (for $k=2$ the tests based on $\bar{t}$ and $\bar{E}_{01}^2$ are equivalent) the LRT is preferred.

Eeden (1958) proposed rejecting $H_1: \mu_1 \leqslant \mu_2 \leqslant \cdots \leqslant \mu_k$ in favor of $\sim H_1$ for large values of

$$\max_{1 \leqslant l \leqslant k-1} \frac{\bar{Y}_l - \bar{Y}_{l+1}}{S((1/n_l) + (1/n_{l+1}))^{1/2}}, \tag{4.3.4}$$

which is a 'Studentization' of a MCT of the form (4.3.2). Applying the Bonferroni inequality, a conservative critical value for (4.3.4) is $t_{\alpha/(k-1)}(\nu)$, the $100[1-\alpha/(k-1)]$ percentile for the central Student t-distribution with ν degrees of freedom. In the power study conducted by Robertson and Wegman (1978), the LRT was seen to outperform this MCT except for very special slippage alternatives.

Because of the results that are given in the next subsection, one would conjecture that there are MCTs which should perform much like the LRTs for testing H_0 versus $H_1 - H_0$ and H_1 versus $H_2 - H_1$ in the simply ordered case. This would seem to be an important and fruitful area for study.

MCTs for H_0 versus $H_1 - H_0$: The simple tree

With $\bar{\mathbf{Y}}$ and S^2 defined as at the beginning of this section, MCTs for $H_0: \mu_1 = \mu_2 = \cdots = \mu_k$ versus $H_1 - H_0$ with $H_1: \mu_1 \leqslant [\mu_2, \mu_3, \ldots, \mu_k]$ are considered. The simple tree cone is easier to work with than the simply ordered cone because of the symmetry among the treatments provided $n_2 = n_3 = \cdots = n_k = n$ (n_1 need not be equal to n). Hence, attention will be focused on this case. However, if $n_2, n_3, \ldots, n_k$ are not too different, then the distributional results given below should provide reasonable approximations with n replaced by $\bar{n} = (n_2 + \cdots + n_k)/(k-1)$. In order to keep the distribution theory simple, we restrict attention to $\mathbf{c}_l$ for which $\rho(T_l, T_{l'})$ is the same for $1 \leqslant l \neq l' \leqslant k_0$. It was noted earlier in this subsection that $\mathbf{c}_l$ should be required to be in $H_0^{\perp} = \{\mathbf{x} \in R^k : \sum_{i=1}^k w_i x_i = 0\}$. Because large correlations between the $\mathbf{c}_l$ and all $\boldsymbol{\mu} \in H_1$ are desired, the $\mathbf{c}_l$ will be chosen from H_1. Thus, if $H_1' = H_1 \cap H_0^{\perp}$, then $\mathbf{c}_l$ needs to be in H_1' for $l = 1, 2, \ldots, k_0$. Because H_1' has dimension $k-1$, k_0 will be chosen to be $k-1$. As in the previous section, $w_i = n_i/\sigma^2$ for $i = 1, 2, \ldots, k$. The ratio $v = w_1/w_2$ occurs in that which follows.

The cone H_1' is generated by convex combinations of nonnegative multiples of the $k-1$ generators

$$\mathbf{e}_i = (-1, -1, \ldots, -1, v + k - 2, -1, -1, \ldots, -1) \qquad \text{for } i = 1, 2, \ldots, k-1 \tag{4.3.5}$$

with $e_{i,i+1} = v + k - 2$. The angle between any two generators has cosine

$$\frac{\langle \mathbf{e}_i, \mathbf{e}_j \rangle_w}{\| \mathbf{e}_i \|_w \| \mathbf{e}_j \|_w} = -\frac{1}{v + k - 2}$$

and because of symmetry, the cone, H_1', has a unique center

$$\mathbf{c} = \sum_{i=1}^{k-1} \mathbf{e}_i = (-k + 1, v, v, \ldots, v). \tag{4.3.6}$$

Clearly, $\mathbf{c} \in H_1'$ and makes equal angles with all the generators of H_1'. This suggests a family of MCTs which have contrast vectors

$$\mathbf{c}_l(r) = r\mathbf{c} + (1 - r)\mathbf{e}_l \qquad \text{for } l = 1, 2, \ldots, k-1 \text{ and } 0 \leqslant r \leqslant 1. \tag{4.3.7}$$

The corresponding $T_l(r) = \langle \mathbf{c}_l(r), \bar{\mathbf{Y}} \rangle_w / \| \mathbf{c}_l(r) \|_w = \sum_{i=1}^k w_i c_{li}(r) \bar{Y}_i / \| \mathbf{c}_l(r) \|_w$ and the MCT statistics are $T(r) = \max_{1 \leqslant l \leqslant k-1} T_l(r)$ for $0 \leqslant r \leqslant 1$. Each $\mathbf{c}_l(r) \in H_1'$ and because each $\mathbf{e}_l$ has the same length (that is $\| \mathbf{e}_l \|_w^2 = w_2(v + k - 1)(v + k - 2)$) and $\mathbf{e}_l$ makes the same angle with $\mathbf{c}$, $\| \mathbf{c}_l(r) \|_w$ is the same for each l. Furthermore, for $1 \leqslant l \neq l' \leqslant k-1$, $\langle \mathbf{c}_l(r), \mathbf{c}_{l'}(r) \rangle_w = r^2 \| \mathbf{c} \|_w^2 + 2r(1-r)\langle \mathbf{c}, \mathbf{e}_1 \rangle_w + (1-r)^2 \langle \mathbf{e}_1, \mathbf{e}_2 \rangle_w$ which implies that $\rho(T_l(r), T_{l'}(r)) = \langle \mathbf{c}_l(r), \mathbf{c}_{l'}(r) \rangle_w / (\| \mathbf{c}_l(r) \|_w \| \mathbf{c}_{l'}(r) \|_w)$ is the same for all $1 \leqslant l \neq l' \leqslant k-1$.

If $r = 1$, then $T_l(r)$ is the same for each l and in fact is the Abelson–Tukey–Schaafsma–Smid single contrast statistic (cf. equation (4.2.7) and recall that $n_2 = n_3 = \cdots = n_k$ is assumed in this subsection). If it is believed that the treatment effects are reasonably homogeneous, this member of the family could be used since it is quite powerful at the center of the cone H_1'. At the other extreme, if $r = 0$ then the MCT statistic is the maximum of the contrasts of the data vector with the generators of H_1'. This test has good power if one of the treatments dominates all the others and the other treatments are fairly homogeneous, but it seems that a researcher would rarely have this kind of information. If $r=(v+1)^{-1}$, then $T_l(r)=(\bar{Y}_{l+1}-\bar{Y}_1)/[\sigma((1/n_1)+(1/n_{l+1}))^{1/2}]$ for $l=1,2\ldots,$ $k-1$. Hence, $T(r)$ is the version of Dunnett's test (cf. Dunnett, 1955) which is appropriate if σ^2 is known. Thus, the family of MCTs $\{T(r):0\leqslant r\leqslant 1\}$ is flexible and contains two of the best-known tests for testing H_0 versus H_1-H_0.

Under $H_0, \sigma T_l(r)/S$ has a Student t-distribution with v degrees of freedom for $l=1,2,\ldots,k-1$ and $\rho=\rho(T_l(r),T_{l'}(r))$ is constant for $1\leqslant l\neq l'\leqslant k-1$. The joint distribution of $\{\sigma T_l(r)/S\}$ is a multivariate t-distribution with common denominator. Johnson and Kotz (1976, p. 134) discuss this distribution and give references to available tables of critical values for tests based on the maximum of the components of such a random vector. The α-level critical value for $\sigma T(r)/S$ will be denoted by $t_\alpha(v,k-1,\rho)$. Hence, for any tabled value of ρ these MCTs are not difficult to implement. However, the distribution under the alternative hypothesis is more complex and consequently designing experiments so that the power is a prespecified value at a particular alternative is much more difficult. If the contrasts are chosen to be orthogonal, then the distribution theory is simpler.

An orthogonal contrast test of H_0 versus H_1–H_0

The OCT is obtained by choosing r so that $\rho=0$. However, $\rho=0$ implies

$$\{(k-3)v-1\}r^2+2(v+1)r-1=0$$

and solving for r gives

$$r_0=\begin{cases}[2(v+1)]^{-1} & \text{if } (k-3)v=1\\ \dfrac{-(v+1)+[(v+1)^2+v(k-3)-1]^{1/2}}{v(k-3)-1} & \text{otherwise.}\end{cases} \qquad (4.3.8)$$

The OCT, that is $T(r_0)$, has the critical value $t_\alpha(\infty,k-1,0)=z_{\alpha/(k-1)}$ if σ^2 is known ($v=\infty$), and for σ^2 unknown, the critical value for $\sigma T(r_0)/S$, that is $t_\alpha(v,k-1,0)$, can be obtained from the tables given by Krishnaiah and Armitage (1966) for $\alpha=0.05$ or 0.10. More detailed tables are given in Krishnaiah and Armitage (1965) and Hahn (1970).

Approximate critical values can be easily obtained. If U is a chi-square random variable with v degrees of freedom, then conditioning on S^2 yields $P[\sigma T(r_0)/S\geqslant t]=1-E\{[\Phi((U/v)^{1/2}t)]^{k-1}\}$ under H_0. Applying Jensen's inequality, under H_0,

$$P[\sigma T(r_0)/S\geqslant t]=1-E\{[\Phi((U/v)^{1/2}t)]^{k-1}\}\geqslant 1-[G(t)]^{k-1}$$

with $G(t)$ the CDF of the Student t-distribution with v degrees of freedom. In particular, a conservative critical value for an α-level test is given by $G^{-1}[(1-\alpha)^{1/(k-1)}]=t_{\alpha'}(v)$ with $\alpha'=1-(1-\alpha)^{1/(k-1)}$, where $t_\alpha(v)$ is the $100(1-\alpha)$

percentile of a Student t-distribution with ν degrees of freedom. Mukerjee, Robertson, and Wright (1987) studied the error introduced by using this approximation and found that it is quite minimal for $\nu \geqslant 15$ and was not more than 10 percent in all the cases considered.

If σ^2 is known ($\nu = \infty$), then the power of the α-level OCT is given by

$$\pi_\infty(\boldsymbol{\mu}) = 1 - \prod_{l=1}^{k-1} \Phi\left(t_\alpha(\infty, k-1, 0) - \frac{\langle \mathbf{c}_l(r_0), \boldsymbol{\mu} \rangle_{\mathbf{w}}}{\| \mathbf{c}_l(r_0) \|_{\mathbf{w}}} \right). \tag{4.3.9}$$

Equation (4.3.9), with $t_\alpha(\infty, k-1, 0)$ replaced by $t_\alpha(\nu, k-1, 0)$, provides a reasonable approximation to the power if ν is large. The exact power for finite ν can be expressed as

$$\pi_\nu(\boldsymbol{\mu}) = 1 - \int_0^\infty \prod_{l=1}^{k} \Phi\left((u/\nu)^{1/2} t_\alpha(\nu, k-1, 0) - \frac{\langle \mathbf{c}_l(r_0), \boldsymbol{\mu} \rangle_{\mathbf{w}}}{\| \mathbf{c}_l(r_0) \|_{\mathbf{w}}} \right) h_\nu(u)\, du \tag{4.3.10}$$

where h_ν is the density of the chi-square distribution with ν degrees of freedom.

Sample size determination

For large degrees of freedom and equal sample sizes $n_1 = n_2 = \cdots = n_k = n$, one can use (4.3.9) to determine the value of n which gives a prespecified power at a fixed alternative $\boldsymbol{\mu}$. In Example 3.7.1, testing H_0 versus $H_1 - H_0$ is considered with $\lesssim$ the simple tree ordering. Following Dunnett (1955) the alternative $\boldsymbol{\mu} = (-1, 5, -1, -1, -1, -1)/\sqrt{30}$ ($k = 6$) is considered with $\sigma^2 = 1$ and $\alpha = 0.05$. In that example, the sample size, n, needed to obtain a power of at least 0.80 at $\boldsymbol{\mu}$ for the LRT, $\bar{E}^2_{01}$, was found to be $n = 13$.

For the OCT, the large sample critical value is $t_{0.05}(\infty, 5, 0) = \Phi^{-1}((0.95)^{1/5}) = 2.3191$ and the power is given by

$$1 - \prod_{l=1}^{5} \Phi(2.3191 - \langle \mathbf{c}_l(r_0), \boldsymbol{\mu} \rangle_{\mathbf{w}}) > 1 - \Phi(2.3191 - \langle \mathbf{c}_1(r_0), \boldsymbol{\mu} \rangle_{\mathbf{w}}). \tag{4.3.11}$$

Thus it suffices to choose n to make the right-hand side equal to 0.80.

However, $r_0 = -1 + \sqrt{6}/2$ and $\langle \mathbf{c}_1(r_0), \boldsymbol{\mu} \rangle_{\mathbf{w}} / \| \mathbf{c}_1(r_0) \|_{\mathbf{w}} = \sqrt{n}\,(9 - 2\sqrt{6}) / [5(33 - 12\sqrt{6})]^{1/2} = 0.9658\sqrt{n}$. Solving for n yields $n = 11$. It is interesting to observe that the factors neglected in (4.3.11) are near one. With $n = 11$, the left-hand side of (4.3.11) is 0.8177 while the right-hand side is 0.8139. For Dunnett's test, large ν and approximate critical value $t = t_\alpha(\infty, k-1, (1 + n_1/n)^{-1})$, the power is approximated by

$$1 - \int_{-\infty}^{\infty} \prod_{l=1}^{k} \Phi\left[t\left(1 + \frac{n}{n_1}\right)^{1/2} + \left(\frac{n}{n_1}\right)^{1/2} x - \frac{\sqrt{n}(\mu_{l+1} - \mu_1)}{\sigma} \right] \phi(x)\, dx. \tag{4.3.12}$$

Mukerjee, Robertson, and Wright (1987), using trial and error, found that $n = 16$ is required for Dunnett's test with $\alpha = 0.05$ to have power at least 0.80 at this alternative. This comparison favors the OCT because it is believed that both the LRT and Dunnett's test have their minimum power at points in H_1 with $\Delta = 1$ at this $\boldsymbol{\mu}$. However, the power at the center of H_1' is smaller for the OCT than at the edges. In fact, one can, using (4.3.9), show that the power at the center with $n = 11$ is 0.6756 and $n = 15$ is required to ensure a power of at least 0.80 at the center of H_1' for the $\alpha = 0.05$ OCT. In the next subsection, based on some power computations, it is conjectured that the MCTs either have their minimum powers at the edges or at the center of H_1'.

Comparison of MCTs

The members of the class of MCTs studied here can be indexed by either r or ρ, the correlation between any two contrast vectors, and of course ρ can be thought of as the cosine of the angle between any two of the contrast vectors. With $\alpha = 0.05$, $k = 3$ and $w_1 = w_2 = w_3$, the powers of some of the MCTs were computed for the case $\nu = \infty$. In Section 2.5, the cone H_1' was parameterized by Δ and β with $0° \leqslant \beta \leqslant 120°$. Using this parameterization, $\Delta = 2$, and $0° \leqslant \beta \leqslant 120°$, these powers are shown in Figure 4.3.1. For $k = 3$ and $w_1 = w_2 = w_3$, it can be shown that $-0.5 \leqslant \rho \leqslant 1$ and consequently the angle between two contrast vectors, θ, varies from 0 to 120°. The MCTs studied have $60° \leqslant \theta \leqslant 120°$. For smaller angles the power function has such small values at $\beta = 0$ and 120° that they were not included in the study. The MCT with $\theta = 60°$ is Dunnett's test for σ^2 known; with $\theta = 90°$, one has the OCT; and the test statistic corresponding to $\theta = 120°$ is the maximum contrast of the edges with the data vector.

For $\theta = 60$, 70, and 80°, the powers are low at the edges of H_1', but grow rapidly to high values near the center of H_1'. For $\theta = 90$ and 100° the powers are fairly high at the edges, grow as β moves toward 60°, and then drop again near the center of H_1'. For $\theta = 110$ and 120° the powers are high near the edges and drop rapidly at the center of H_1'. Based on these calculations, one might conjecture that a MCT has its minimum power over points in H_1 at distance Δ from H_0 at either the center or the edges of H_1 (of course, the power is the same at all the edges).

If one has information concerning the treatment means that indicates they are homogeneous, then the Abelson–Tukey–Schaafsma–Smid contrast should be considered. If some protection against the possibility of nonhomogeneous μ_i is desired then Dunnett's test is a viable candidate. It has the following advantage: that it can be used with the Bonferroni inequality if the sample sizes are quite different (see the discussion in Mukerjee, Robertson, and Wright, 1987). Also, confidence intervals for the differences $\mu_i - \mu_1$ can be based on the individual contrasts. If the treatment means are believed to be heterogeneous with one mean (or a few means) dominating the others, the researcher may wish to use the MCT based on the edges of H_1'. Lacking such information, in this family of MCTs, the OCT is

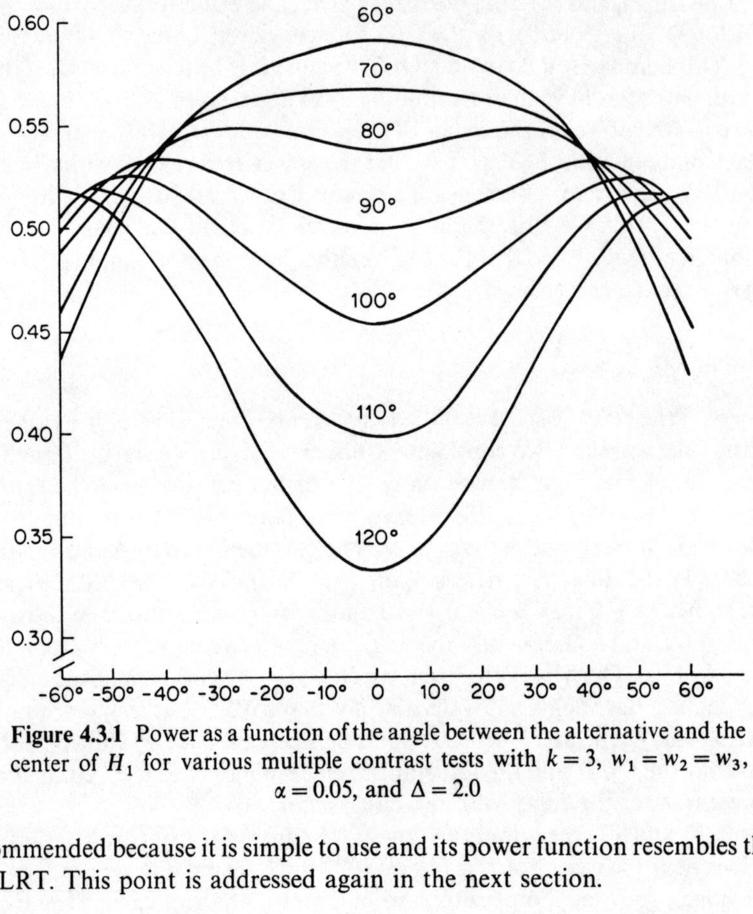

Figure 4.3.1 Power as a function of the angle between the alternative and the center of H_1 for various multiple contrast tests with $k=3$, $w_1=w_2=w_3$, $\alpha=0.05$, and $\Delta=2.0$

recommended because it is simple to use and its power function resembles that of the LRT. This point is addressed again in the next section.

MCTs for H_1 versus H_2-H_1: The simple tree

It was noted in Chapter 2 that if the assumption $\boldsymbol{\mu}\in H_1$ were in question, one may wish to test H_1 versus H_2-H_1. With $\lesssim$ the simple tree ordering (that is $\mu_1 \leqslant [\mu_2, \mu_3, \ldots, \mu_k]$) MCTs for H_1 versus H_2-H_1 will be considered for the case in which $n_2=n_3=\cdots=n_k=n$. Such a test would also be useful in any situation in which one wishes to determine if a collection of means are all larger (smaller) than a given mean. The case in which σ^2 is known will be considered first. The MCT rejects H_1 for large values of $T'=\max_{1\leqslant l\leqslant k-1} T_l$ with T_l defined as in (4.3.1) and each $\mathbf{c}_l \in H_1^{*\mathbf{w}}$ (so that the size of the test will be less than one). Recall that $w_i=n_i/\sigma^2, i=1,2,\ldots,k$. In (4.2.15) the dual of H_1 was characterized as follows: $H_1^{*\mathbf{w}}=\{\mathbf{y}\in R^k : \sum_{i=1}^k w_i y_i=0 \text{ and } y_i>0;\ i=2,3,\ldots,k\}$. Because the OCT was effective in testing H_0 versus H_1-H_0, one might search for an OCT in this setting also. However, it is not difficult to show that if two nonzero vectors are orthogonal, then they both cannot be in H_1^{*w}; H_1^{*w} is narrow since H_1 is wide (see Mukerjee, Robertson, and Wright, 1986b). In the last subsection, we observed that if the common angle between each pair of contrast vectors is too small, then the power will be large at the center of the contrast vectors and will drop off rapidly. For this reason, Mukerjee, Robertson, and Wright (1986a) recommend the MCT with contrast vectors which are the generators of H_1^{*w}. The generators of H_1^{*w} satisfy $c_{l1} = -w_i^{-1}$, $c_{l,l+1} = w_{l+1}^{-1}$, and $c_{l,j} = 0$ for $j \in \{2, 3, \ldots, k\} - \{l+1\}$, which gives rise to the test statistic

$$T' = \max_{1 \leqslant l \leqslant k-1} \frac{\bar{Y}_1 - \bar{Y}_{l+1}}{\sigma(1/n_1 + 1/n_{l+1})^{1/2}}. \tag{4.3.13}$$

For the case of unknown σ^2, one would use $\sigma T'/S$, which is Dunnett's test statistic with $\bar{Y}_1$ and $\bar{Y}_{l+1}$ interchanged. Clearly, H_0 is least favorable within H_1 and the critical values for $\sigma T'/S$ are $t_\alpha(v, k-1, (1 + n_1/n)^{-1})$, which are the same as those for Dunnett's test. For $v = \infty$ and $t = t_\alpha(v, k-1, (1 + n_1/n)^{-1})$ the power of T' is given by (4.3.12) with μ_1 and μ_{l+1} interchanged. Of course, this also provides an approximation to the power for large v.

The power of this MCT, the LRT, and the optimal single contrast test are compared in the next section.

4.4 COMPARISONS OF THE POWERS OF THE LRT, MCT, AND SINGLE CONTRAST TESTS

In Section 2.5 the powers of the restricted LRTs were compared to those of the unrestricted LRTs to give some indication of the increase in power obtained by using the restricted LRTs when the assumed ordering holds. In this section the restricted LRTs are compared with their competitors to provide some guidance as to which restricted test should be used. Attention is focused on the simple ordering and the simple tree ordering. To simplify the comparisons the weights will be assumed to be equal, $w_1 = w_2 = \cdots = w_k$, and σ^2 will be assumed known.

Powers of tests of H_0 versus H_1–H_0: the simply ordered case

In Section 4.3, an *ad hoc* multiple contrast test for H_0 versus H_1–H_0 in the simply ordered case was discussed. This MCT is not considered further in this section, because the work of Williams (1971) and Chase (1974) showed that for $k > 3$ the LRT is preferred. A conditional $\bar{\chi}^2_{01}$ test has been proposed. To be specific, if L, the number of level sets in $P_w(\bar{\mathbf{Y}}|\mathscr{I})$, is greater than one, then the conditional test rejects H_0 if $\bar{\chi}^2_{01} \geqslant \chi^2_{1-\alpha', l-1}$, where $L = l$, $\alpha' = \alpha/[1 - P(1, k; \mathbf{w})]$ and $\chi^2_{1-\alpha, v}$ is the $100(1-\alpha)$ percentile of the χ^2 distribution with v degrees of freedom. Bartholomew (1961a) studied the power function of this conditional test. It is not clear where its minimum over the cone H_1 occurs. For $k = 3$ and small Δ the power of the conditional test is larger at the center than the edges, but for larger Δ

Table 4.4.1 Comparison of the powers[a] of $\bar{\chi}^2_{01}$ and T, the optimum contrast test for the simple order with equal weights and $\alpha = 0.05$

		Δ			
k	STATISTIC	1	2	3	4
3	T(max)	0.260	0.639	0.912	0.991
	$\bar{\chi}^2_{01}$(max)	0.244	0.605	0.892	0.987
	$\bar{\chi}^2_{01}$(min)	0.221	0.569	0.872	0.983
	T(min)	0.218	0.535	0.830	0.966
4	T(max)	0.260	0.639	0.912	0.991
	$\bar{\chi}^2_{01}$(max)	0.238	0.594	0.885	0.985
	$\bar{\chi}^2_{01}$(min)	0.202	0.531	0.849	0.977
	T(min)	0.201	0.488	0.781	0.943
12[a]	T(max)	0.260	0.639	0.912	0.991
	$\bar{\chi}^2_{01}$(max)	0.234	0.572	0.868	0.978
	$\bar{\chi}^2_{01}$(min)	0.162	0.429	0.762	0.949
	T(min)	0.166	0.385	0.649	0.855

[a]For $k = 12$, the powers of $\bar{\chi}^2_{01}$ were estimated by Monte Carlo techniques with 10000 iterations. For $k = 12$ and $\Delta = 1$, $\bar{\chi}^2_{01}$ (min) should be greater than T(min), but because of Monte Carlo error, the estimate is not.

the reverse is true (cf. Table 8 of Bartholomew, 1961a). Furthermore, its power seems to be noticeably less than that of $\bar{\chi}^2_{01}$. Because this test would be of most interest in situations in which the level probabilities cannot be computed, further power comparisons in such cases would be helpful.

Table 4.4.1 contains the powers of T, the Abelson–Tukey–Schaafsma–Smid optimal contrast test, and $\bar{\chi}^2_{01}$ for the simple order, equal-weights, $\alpha = 0.05$, $k = 3, 4, 12$, and $\Delta = 1, 2, 3, 4$ (recall that $\Delta^2 = \sum_{i=1}^k w_i(\mu_i - \bar{\mu})^2$ and $\bar{\mu} = \sum_{i=1}^k w_i\mu_i / \sum_{i=1}^k w_i$). The value labeled maximum is the power at 'equally spaced' alternatives and the labeled minimum is for the points with $\mu_1 = \mu_2 = \cdots = \mu_{k-1} < \mu_k$. (The power is known to be largest and smallest at these points for $k = 3$, and this is conjectured to be so for larger k.)

From equation (4.2.2) one sees that the power of a contrast test is $\Phi(r\Delta - z_\alpha)$. The maximum power is obtained by setting $r = 1$ and the minimum by setting $r =$ maximin r. For $k = 3, 4$, there is little difference in the power bands for T and $\bar{\chi}^2_{01}$, and hence T, because of its ease of implementation, might be preferred. However, for large k, that is $k = 12$, the minimum power of $\bar{\chi}^2_{01}$ may be 10 percent or more larger than that of T. The maximum power of T is larger, but if one does not know that the mean vector is near the center of the cone then the $\bar{\chi}^2$ test would be preferred because of its more uniform power function.

Powers of H_0 versus H_1-H_0: the simple tree ordering

As in the simply ordered case, little is known about the power of the conditional test for the simple tree ordering. Bartholomew (1961a) gives some results for three populations. Further work along these lines could be helpful in choosing a test, particularly for larger k and unequal weights.

For the simple tree ordering with k populations, the optimum contrast test T, Dunnett's test D, the orthogonal contrast test OCT, and the LRT, $\bar{\chi}^2_{01}$, will be compared. Again the variance is assumed known, the weights will be assumed to be equal, and $\alpha = 0.05$. The power bands for these tests are given in Table 4.4.2. The maximum and minimum powers for T, $\bar{\chi}^2_{01}$, and D are conjectured to occur, respectively, at the center of the cone and at the edges with all treatments but one equal to the control. For $k = 5$, the power of $\bar{\chi}^2_{01}$ is estimated by Monte Carlo techniques. For $k = 3$, the cone H_1 can be parameterized by Δ and β with $0° \leqslant \beta \leqslant 120°$; for $k = 4$ by Δ, β, and θ; and for $k = 5$ by Δ, β, θ, and ϕ. For $k = 3$ and 4 and fixed Δ, the power of the OCT was computed for those angles which are multiples of one degree and are consistent with the cone restriction; i.e. the maximum and minimum powers were approximated by a 1° search. For $k = 5$, a 5° search was employed.

For the tree ordering, $\mu_1 \leqslant [\mu_2, \ldots, \mu_k]$, the minimum power of T is quite low even for small k. Therefore, unless something more is known about the alternative, T cannot be recommended for this ordering. The cone for the simple tree ordering is wider than for the simple order. Similar results would be expected for any 'wide' cone.

To summarize the results concerning contrast tests for testing H_0 versus H_1-H_0, note that the contrast test is easy to compute and has a straightforward distribution under the null and alternative hypotheses. If the cone is narrow, as in the simply ordered case with small k, its power band over H_1 is much like that of the LRT. However, for larger k and the simple order or for any wide cone its minimum power is much lower than that of the LRT. If the assumption $\boldsymbol{\mu} \in H_1$ is violated one would still like the test to be consistent. We have seen that the region of consistency of the LRT is substantially larger than that of the contrast test. In conclusion, if one cannot restrict the alternative more than $\boldsymbol{\mu} \in H_1$, the LRT is preferred over a contrast test except in the special case of a narrow cone.

The MCTs behave much better than a single contrast, and either Dunnett's test or the OCT might be acceptable in the case of a simple tree ordering. Dunnett's test has the advantage that it can be used with the Bonferroni inequality for any sample size and multiple comparison procedures exist which use the same

Table 4.4.2 Comparison of the powers of $\bar{\chi}^2_{01}$, the optimum contrast test T, Dunnett's test D, and the orthogonal contrast test OCT, for the simple tree ordering, $\mu_1 \leqslant [\mu_2, \ldots, \mu_k]$, equal weights, and $\alpha = 0.05$

		Δ			
k	STATISTIC	1	2	3	4
3	T(max)	0.260	0.639	0.912	0.991
	D(max)	0.238	0.590	0.881	0.984
	OCT(max)	0.201	0.533	0.856	0.980
	$\bar{\chi}^2_{01}$(max)	0.210	0.543	0.856	0.979
	$\bar{\chi}^2_{01}$(min)	0.173	0.495	0.831	0.974
	OCT(min)	0.173	0.494	0.812	0.963
	D(min)	0.159	0.431	0.753	0.939
	T(min)	0.126	0.260	0.442	0.639
4	T(max)	0.260	0.639	0.912	0.991
	D(max)	0.231	0.574	0.869	0.981
	OCT(max)	0.173	0.475	0.818	0.971
	$\bar{\chi}^2_{01}$(max)	0.187	0.489	0.813	0.967
	$\bar{\chi}^2_{01}$(min)	0.140	0.416	0.767	0.955
	OCT(min)	0.142	0.387	0.700	0.915
	D(min)	0.126	0.342	0.653	0.886
	T(min)	0.092	0.159	0.254	0.370
5	T(max)	0.260	0.639	0.912	0.991
	D(max)	0.227	0.566	0.862	0.979
	OCT(max)	0.159	0.438	0.791	0.963
	$\bar{\chi}^2_{01}$(max)	0.169	0.449	0.778	0.955
	$\bar{\chi}^2_{01}$(min)	0.122	0.357	0.711	0.934
	OCT(min)	0.126	0.329	0.608	0.848
	D(min)	0.109	0.291	0.586	0.842
	T(min)	0.082	0.126	0.185	0.259

contrasts that were computed in computing the test statistic. The OCT has the advantage that its minimum power is larger than Dunnett's test. In Section 4.3, we saw that, with the OCT, it is easier to design experiments with power approximately a specified value at a given alternative.

The LRT tended to have the largest minimum power. (At points with powers less than 0.2 the OCT had minimum power equal to, or slightly larger than, that of the LRT, but this was not true for moderate or large powers.) With the approximations given in Chapter 3, it is not difficult to approximate the null distribution or the power of the LRT at slippage alternatives for a simple order or a simple tree ordering. For both of these testing situations, the LRT would seem to be the preferred test.

Tests of H_1 versus H_2-H_1

One can see from the discussion at the end of Section 4.2 that the optimal contrast tests for H_1 versus H_2-H_1 are not consistent for all $\boldsymbol{\mu} \notin H_1$ for the simply ordered case or the simple tree ordering. This is typical of all partial orderings. However, it was shown in Theorem 2.6.1 that $\bar{\chi}^2_{12}$ and $\bar{E}^2_{12}$ are consistent for all $\boldsymbol{\mu} \notin H_1$ for any quasi-order. Without further information as to where $\boldsymbol{\mu}$ lies in H_1^c, a contrast test should not be used.

In Section 4.3 in the case of a simple tree ordering, a MCT for H_1 versus H_2-H_1 was discussed. It is the dual to Dunnett's test. It was also shown that a reasonable OCT does not exist in this setting. Mukerjee, Robertson, and Wright (1986a) study the power functions of the contrast test, the MCT, and the LRT for the case of known variance in this setting. They noted that the LRT is more powerful in the 'middle' of H_1^c than the MCT, but the MCT has a more uniform power function than the LRT. (For two or three treatments the two power functions are quite similar.) The loss in power at the middle is more than offset at the edges of the dual H_1^{*w}. The MCT has the advantage that, for any k, its large sample power at any point can be computed with a single numerical integration. If slippage alternatives are of interest, presumably the techniques discussed in Section 3.7 could be used to approximate the power of $\bar{\chi}^2_{12}$. However, the accuracy of such an approximation has not been studied. For the special case of testing H_1 versus $H_2-\mathrm{H}_1$ with a simple tree ordering the MCT has more uniform power characteristics and is simpler to use.

All three of the tests studied in this setting are biased. This is typical of all partial orderings. Wollan and Dykstra (1986) consider the use of a conditional test of H_1 versus H_2-H_1 and found that this conditional test is less biased than the LRT. A comparison of the power functions of the LRT and conditional tests would be informative. Such a study would give some idea as to the amount of power lost in the middle of H_1^c, as well as the gain in power near the boundary of H_1, if the conditional test were used.

4.5 DISTRIBUTION-FREE TESTS FOR ORDERED ALTERNATIVES

In Chapter 2 the LRT for homogeneity of normal means with the alternative restricted by an isotonic trend and the LRT for the trend versus all alternatives were studied. If the populations under consideration are not normal, then the level of the $\bar{\chi}^2$ and $\bar{E}^2$ tests may not be correct and the tests may be quite inefficient. If the populations are members of one of the exponential families discussed in Section 4.1 then the level will be asymptotically correct, but of course even in this case other tests may outperform these normal theory tests. There is a considerable body of literature on distribution-free order restricted tests. However, most of the work involves testing homogeneity with an ordered alternative in a one-way layout or a randomized block design. Some of the basic work in this area is described below and references to modifications and generalizations are mentioned.

The one-way layout: linear rank tests

Consider k continuous distributions with CDFs $F_1, F_2, \ldots, F_k$ and let $F_i \overset{\text{st}}{\leqslant} F_j (F_i = F_j)$ mean $F_i(x) \geqslant F_j(x) (F_i(x) = F_j(x))$ for all $x \in R$. The hypothesis of homogeneity can be expressed as $H_0: F_1 = F_2 = \cdots = F_k$ and the hypothesis of an isotonic trend is $H_1: F_i \overset{\text{st}}{\leqslant} F_{i'}$, for all $i \lesssim i'$ with $\lesssim$ a quasi-order on $X = \{1, 2, \ldots, k\}$. Much of the work in the literature concerns the usual simple order, that is $1 \lesssim 2 \lesssim \cdots \lesssim k$. Thus, unless mentioned otherwise $\lesssim$ is this simple order. Suppose independent random samples, $Y_{ij} \sim F_i, j = 1, 2, \ldots, n_i$ and $i = 1, 2, \ldots, k$, are available and $N = \sum_{i=1}^{k} n_i$.

The statistic proposed independently by Terpstra (1952) and Jonckheere (1954a) is

$$J = \sum_{1 \leqslant a < b \leqslant k} M_{ab} \tag{4.5.1}$$

with M_{ab} the Mann–Whitney statistic for comparing the ath and bth samples, that is $M_{ab} = \sum_{s=1}^{n_a} \sum_{t=1}^{n_b} I_{[Y_{as} < Y_{bt}]}$ where I_A is the indicator of A. An increasing trend is evidenced by larger J and so H_0 is rejected for large values of J. Terpstra (1955, 1956) found that the set of alternatives for which this test is consistent depends upon the ratio of the sample sizes and that if M_{ab} is divided by $n_a n_b$ then this is no longer the case. Because the statistic, J, given in (4.5.1) has a simpler distribution theory, it will be considered here.

Selected critical values for J are given in Jonckheere (1954a) and Odeh (1971). Because the asymptotic null distribution of the Mann–Whitney statistic is normal, this is also true for J. Under H_0, its mean and variance are, with $N = \sum_{i=1}^{k} n_i$, given by

$$E(J) = \frac{N^2 - \sum_{i=1}^{k} n_i^2}{4} \quad \text{and} \quad \operatorname{Var}(J) = \frac{N^2(2N+3) - \sum_{i=1}^{k} n_i^2(2n_i + 3)}{72} \tag{4.5.2}$$

(cf. Randles and Wolfe, 1979). The power of the Jonckheere–Terpstra test is studied by Odeh (1972), Nelson and Toothaker (1975), and Potter and Sturm (1981). In the latter reference, it is shown that in the normal means problem the power of this test is bounded away from one as the mean vector moves away from H_0 along certain rays in H_1. In particular, if $\boldsymbol{\mu} = \gamma(-2, 1, 1)$, and $n_1 = n_2 = n_3 = 3$, then for all positive γ, the power of the test based on J is bounded above by 0.731.

There are several other ways to view the Jonckheere–Terpstra test statistic. Recall that Kendall's correlation coefficient for $(X_1, Y_1), (X_2, Y_2), \ldots, (X_n, Y_n)$, with possible ties, can be expressed as a nondecreasing function of

$$K = \sum_{i=1}^{n-1} \sum_{j=i+1}^{n} \zeta(X_i, X_j, Y_i, Y_j) \text{ with } \zeta(a, b, c, d) = \operatorname{sgn}((a-b)(c-d)) \quad (4.5.3)$$

where $\operatorname{sgn}(0) = 0$. If one applies K to the N pairs (i, Y_{ij}), $j = 1, 2, \ldots, n_i$ and $i = 1, 2, \ldots, k$, then for two pairs with the same first element $\zeta = 0$. Hence, K applied to these N pairs is $J - J'$, where

$$J' = \sum_{1 \leqslant a < b \leqslant k} M'_{ab} \qquad \text{and} \qquad M'_{ab} = \sum_{j=1}^{n_a} \sum_{j'=1}^{n_b} I_{[Y_{aj} > Y_{bj'}]}$$

and with probability one, $J + J' = [N^2 - \sum_{i=1}^{k} n_i^2]/2$. Therefore,

$$J = \frac{K + (N^2 - \sum_{i=1}^{k} n_i^2)/2}{2},$$

and consequently tests based on J and K are equivalent.

Assuming there are no ties in the sample obtained by pooling the ath and bth samples, the rank of Y^*, whether Y^* is in the ath or bth sample, in this pooled sample is $\sum_{j=1}^{n_a} I_{[Y_{aj} < Y^*]} + \sum_{j=1}^{n_b} I_{[Y_{bj} < Y^*]} + 1$. Consequently, if $\bar{R}_a^{(ab)}(\bar{R}_b^{(ab)})$ is the average rank of the elements in the ath (bth) sample when the two samples are pooled, then $\bar{R}_b^{(ab)} - \bar{R}_a^{(ab)} = M_{ab}/n_b - M'_{ab}/n_a + (n_b - n_a)/2$. However, $M'_{ab} = n_a n_b - M_{ab}$ and thus

$$\frac{n_a n_b}{n_a + n_b}(\bar{R}_b^{(ab)} - \bar{R}_a^{(ab)}) = M_{ab} - \tfrac{1}{2} n_a n_b = \frac{M_{ab} - M'_{ab}}{2}. \quad (4.5.4)$$

Therefore, summing (4.5.4) over $1 \leqslant a < b \leqslant k$ yields a test statistic which produces a test equivalent to that based on J. If the sample sizes are equal, then the test based on J is a sum over $a < b$ of the difference in average ranks with each pair given the same weight. In the study of contrast tests given in Section 4.2, this weighting was found not to be optimal in the normal means problem.

There are several generalizations of J that could be considered. Puri (1965) studied the statistic which replaces M_{ab} in (4.5.1) by a Chernoff–Savage statistic (cf. Chernoff and Savage, 1958). Because this again assigns the same weight to each pair, $a < b$, Tryon and Hettmansperger (1973) proposed the family of test statistics

$$H_c = \sum_{1 \leqslant a < b \leqslant k} c_{ab} H_{ab} \quad (4.5.5)$$

where $c_{ab} \geqslant 0$ and H_{ab} is a Chernoff–Savage statistic. Let $m_{ab} = n_a + n_b$ and let $R_j^{(ab)}$ be the rank of Y_{bj} in the sample of m_{ab} items obtained by combining the ath and bth samples. For each m_{ab} the scores, $\{a_{m_{ab}}(i)\}_{i=1}^{m_{ab}}$, satisfy

$a_{m_{ab}}(1) \leqslant a_{m_{ab}}(2) \leqslant \cdots \leqslant a_{m_{ab}}(m_{ab})$ and

$$H_{ab} = \sum_{j=1}^{n_b} a_{m_{ab}}(R_j^{(ab)}).$$

Typically, the scores correspond to a score generating function ϕ:[0, 1]$\rightarrow R$ where, with [·] the greatest integer function,

$$\int_0^1 \{a_\gamma(1+[u\gamma]) - \phi(u)\}^2 \, du \rightarrow 0 \quad \text{as} \quad \gamma \rightarrow \infty.$$

Nondecreasing scores, such as these considered here, correspond to a nondecreasing ϕ. Randles and Wolfe (1979, Chap. 9) discuss some useful score functions. Tryon and Hettmansperger (1973) show that if each of the two-sample statistics are of the same type, then for a given c there is a vector $\mathbf{c}' = (c'_1, c'_2, \ldots, c'_{k-1})$ for which $H'_{\mathbf{c}'} = \sum_{a=1}^{k-1} c'_a H_{a,a+1}$ is asymptotically equivalent to $H_{\mathbf{c}}$, and the statistics $H_{\mathbf{c}}$ and $H'_{\mathbf{c}'}$ have the same Pitman efficacy. To investigate asymptotically optimal choices for $\mathbf{c}'$, they consider the location parameter problem with $F_i(x) = F(x - \theta_i)$ for $i = 1, 2, \ldots, k$. For $\boldsymbol{\theta} = (\theta_1, \theta_2, \ldots, \theta_k)$ with $\theta_1 \leqslant \theta_2 \leqslant \cdots \leqslant \theta_k$, the spacings are defined by $\delta_i = (\theta_{i+1} - \theta_i)/(\theta_k - \theta_1)$, $i = 1, 2, \ldots, k-1$. For $\boldsymbol{\delta} = (\delta_1, \delta_2, \ldots, \delta_{k-1})$ fixed with nonnegative entries, equal sample sizes and a fixed type of Chernoff–Savage statistic, they determine the member of their class with the maximal Pitman efficacy. For equal spacings $\delta_1 = \delta_2 = \cdots = \delta_{k-1}$, and equal sample sizes the optimal coefficients in (4.5.5) are constant, that is $c_{ab} = c$. The asymptotic null distribution of $H_{\mathbf{c}}$ is normal with mean zero. Tryon and Hettmansperger (1973) give the variance of H_{ab} and the covariance of any pair H_{ab} and H_{cd} (cf. their equation (5)). Thus the variance of the limiting distribution can be computed, but the computation may be tedious.

Expressing J as a sum of the difference of average ranks as in (4.5.4) shows its relationship to the contrast tests defined in (4.2.4). The weighted versions proposed by Tryon and Hettmansperger (1973) (cf. equation (4.5.5)) parallel the general contrast statistics given by (4.2.1). Tests which more nearly resemble the $\bar{\chi}^2$ tests of Chapter 2 are discussed next, but first a couple of additional generalizations of the Jonckheere–Terpstra test are mentioned. Randles and Wolfe (1979, p. 397) discuss a version of J for testing a simple tree ordering. This test statistic is the sum of the one-sided Mann–Whitney statistics for comparing each treatment with the control. Mack and Wolfe (1981) use this same approach for the unimodal ordering. Shirahata (1980) found the asymptotically optimal $H_{\mathbf{c}}$ for the simple order, the simple tree, and the simple loop. He actually considered joint rankings among all the N observations, but he established the asymptotic equivalence of tests based on pairwise and joint rankings.

Let R_{ij} be the rank of Y_{ij} when all k samples are pooled and let $\bar{R}_i = \sum_{j=1}^{n_i} R_{ij}/n_i$ be the average rank of the ith sample. A direct parallel to the contrast statistics of

Section 4.2 is given by

$$T_c(\mathbf{R}) = \sum_{i=1}^{k} n_i c_i \bar{R}_i \qquad \text{with } \sum_{i=1}^{k} n_i c_i = 0.$$

Recall that Spearman's rank correlation coefficient is Pearson's correlation applied to the ranks of the N pairs (i, Y_{ij}). Because there are ties in the first position, the rank of each i is $s_i = \sum_{a=0}^{i-1} n_a + (n_i + 1)/2$, with $n_0 = 0$. With $c_i = s_i - (N+1)/2$, $T_c(\mathbf{R})$ is proportional to Spearman's rank correlation coefficient. The statistic $T_c(\mathbf{R})$ can easily be generalized to include scores. With scores, $a_N(1) \leqslant a_N(2) \leqslant \cdots \leqslant a_N(N)$ and $\bar{S}_i = \sum_{j=1}^{n_i} a_N(R_{ij})/n_i$,

$$T_c(\mathbf{S}) = \sum_{i=1}^{k} n_i c_i \bar{S}_i \qquad \text{with } \sum_{i=1}^{k} n_i c_i = 0$$

has, under H_0, an asymptotic normal distribution with mean zero and variance $\sigma_{c,\phi}^2 = \sum_{i=1}^{N} c_i^2 \int_0^1 [\phi(u) - \bar{\phi}]^2\, du$ where $\bar{\phi} = \int_0^1 \phi(u)\, du$ (cf. Theorem 1, p. 160 of Hájek and Sidák, 1967, for the necessary regularity conditions). For suitable $\mathbf{c}$, one rejects H_0 for large values of $T_c(\mathbf{S})$.

One-way layout: $\bar{\chi}^2$ and related tests

Chacko (1963) proposed a distribution-free analogue of the $\bar{\chi}_{01}^2$ test of Chapter 2. Chacko's work, which dealt with equal sample sizes, was extended by Shorack (1967). As in the last subsection, let R_{ij} be the rank of Y_{ij} among the N observations in the k samples; let $\bar{R}_i = \sum_{j=1}^{n_i} R_{ij}/n_i$ be the average rank in the ith sample; let $\lesssim$ be a quasi-order on $X = \{1, 2, \ldots, k\}$; and let $\bar{\mathbf{R}}^* = (\bar{R}_1^*, \bar{R}_2^*, \ldots, \bar{R}_k^*)$ be the isotonic regression of $\bar{\mathbf{R}} = (\bar{R}_1, \bar{R}_2, \ldots, \bar{R}_k)$ with weight vector $\mathbf{w} = (w_1, w_2, \ldots, w_k)$, where $w_i = 12n_i/[N(N+1)]$. (Of course for the purpose of computing $\bar{\mathbf{R}}^*$ one may use $w_i = n_i$). Computing the $\bar{\chi}_{01}^2$ statistic of Chapter 2 on $\bar{\mathbf{R}}$ yields

$$\bar{\chi}_{01}^2(\mathbf{R}) = \frac{12}{N(N+1)} \sum_{i=1}^{k} n_i \left(\bar{R}_i^* - \frac{N+1}{2} \right)^2 \tag{4.5.6}$$

and $\sigma_R^2 = \sum_{i=1}^{k} \sum_{j=1}^{n_i} [R_{ij} - (N+1)/2]^2/(N-1) = N(N+1)/12$. The statistic, $\bar{\chi}_{01}^2(\mathbf{R})$, is a natural extension of the Kruskal–Wallis statistic for ordered alternatives. The form of the statistic $\bar{\chi}_{01}^2(\mathbf{R})$ suggests a distribution-free analogue to $\bar{\chi}_{12}^2$, which is given by

$$\bar{\chi}_{12}^2(\mathbf{R}) = \frac{12}{N(N+1)} \sum_{i=1}^{k} n_i (\bar{R}_i - \bar{R}_i^*)^2. \tag{4.5.7}$$

Of course, one rejects H_1 for large values of $\bar{\chi}_{12}^2(\mathbf{R})$. Clearly, one of the appealing features of $\bar{\chi}_{01}^2(\mathbf{R})$ and $\bar{\chi}_{12}^2(\mathbf{R})$ is that they are defined for any quasi-order, $\lesssim$.

Theorem 4.5.1 *If $F_1 = F_2 = \cdots = F_k$ and if $N \to \infty$ with $n_i/N \to \gamma_i$ and $\gamma_i \in (0, 1)$*

for $i = 1, 2, \ldots, k$, *then* $\bar{\chi}^2_{01}(\mathbf{R}) \xrightarrow{D} \| P_\gamma(\mathbf{Y}|\mathscr{I}) - \hat{\mu} \|^2_\gamma$ *and* $\bar{\chi}^2_{12}(\mathbf{R}) \xrightarrow{D} \| \mathbf{Y} - P_\gamma(\mathbf{Y}|\mathscr{I}) \|^2_\gamma$, *where* $Y_1, Y_2, \ldots, Y_k$ *are independent with* $Y_i \sim \mathcal{N}(0, \gamma_i^{-1})$, $\hat{\mu} = \sum_{i=1}^k \gamma_i Y_i / \sum_{i=1}^k \gamma_i$, *and* $\mathscr{I}$ *is the collection of vectors isotonic with respect to* $\lesssim$.

Before the proof of Theorem 4.5.1 is given it should be observed that the distribution of $\| P_\gamma(\mathbf{Y}|\mathscr{I}) - \hat{\mu} \|^2_\gamma$ and $\| \mathbf{Y} - P_\gamma(\mathbf{Y}|\mathscr{I}) \|^2_\gamma$ are given in Theorem 2.3.1 and are the null distributions of $\bar{\chi}^2_{01}$ and $\bar{\chi}^2_{12}$ with quasi-order $\lesssim$ and weight vector γ. Parsons (1979a) considers a more general version of $\bar{\chi}^2_{12}(\mathbf{R})$ and gives some conditions which ensure that H_0 is asymptotically least favorable in H_1. However, these conditions seem to be too restrictive. Further research on the least favorable status of H_0 in H_1 is needed.

Proof of Theorem 4.5.1: Now

$$\bar{\chi}^2_{01}(\mathbf{R}) = \frac{N}{N+1} \sum_{i=1}^k \frac{n_i}{N} \left\{ P_{\mathbf{n}/N} \left[\left(\frac{12}{N} \right)^{1/2} \left(\mathbf{R} - \frac{N+1}{2} \right) \middle| \mathscr{I} \right]_i \right\}^2$$

and

$$\bar{\chi}^2_{12}(\mathbf{R}) = \frac{N}{N+1} \sum_{i=1}^k \frac{n_i}{N} \left\{ \left(\frac{12}{N} \right)^{1/2} \left(R_i - \frac{N+1}{2} \right) - P_{\mathbf{n}/N} \left[\left(\frac{12}{N} \right)^{1/2} \left(\mathbf{R} - \frac{N+1}{2} \right) \middle| \mathscr{I} \right]_i \right\}^2.$$

Under the hypotheses of the theorem, Kruskal (1952) showed that $12/N)^{1/2}$ $[\mathbf{R} - (N+1)/2]$ has multivariate normal limiting distribution with mean zero and covariance matrix $\sigma_{ij} = (\delta_{ij}/\gamma_i - 1)$ with δ_{ij} the Kronecker delta function. However, $\mathbf{Y} - \hat{\mu}$ has the same multivariate normal distribution. By the continuity of the projection with respect to its weights and argument,

$$\bar{\chi}^2_{01}(\mathbf{R}) \xrightarrow{D} \sum_{i=1}^k \gamma_i [P_\gamma(\mathbf{Y} - \hat{\mu}|\mathscr{I})_i]^2 = \| P_\gamma(\mathbf{Y}|\mathscr{I}) - \hat{\mu} \|^2_\gamma$$

and

$$\bar{\chi}^2_{12}(\mathbf{R}) \xrightarrow{D} \sum_{i=1}^k \gamma_i [\mathbf{Y} - \hat{\mu} - P_\gamma(\mathbf{Y} - \hat{\mu}|\mathscr{I})_i]^2 = \| \mathbf{Y} - P_\gamma(\mathbf{Y}|\mathscr{I}) \|^2_\gamma. \qquad \square$$

Parsons (1979b, 1981), using techniques developed by Boswell and Brunk (1969), derived the exact null distribution for $\bar{\chi}^2_{01}(\mathbf{R})$ and gave tables of some critical values for $k = 3$ and the usual simple order. Shiraishi (1982) generalized Chacko's test to allow for scores other than the Wilcoxon scores. With scores $a_N(1) \leqslant a_N(2) \leqslant \cdots \leqslant a_N(N)$ satisfying the regularity conditions imposed by Chernoff and Savage (1958), set $\bar{S}_i = \sum_{j=1}^{n_i} a_N(R_{ij})/n_i$, $\bar{\mathbf{S}}^* = P_{\mathbf{n}/N}(\bar{\mathbf{S}}|\mathscr{I})$, $\sigma^2_{a_N} = \sum_{i=1}^N [a_N(i) - a_N]^2/(N-1)$ with $\bar{a}_N = \sum_{i=1}^N a_N(i)/N$, and

$$\bar{\chi}^2_{01}(\mathbf{S}) = \sum_{i=1}^k \frac{n_i[\bar{S}^*_i - \bar{a}_N]^2}{\sigma^2_{a_N}} \quad \text{and} \quad \bar{\chi}^2_{12}(\mathbf{S}) = \sum_{i=1}^k \frac{n_i[\bar{S}_i - \bar{S}^*_i]^2}{\sigma^2_{a_N}}. \qquad (4.5.8)$$

Using the results of Hájek and Sidák (1967, p. 171) rather than those of Kruskal in the proof of Theorem 4.5.1, the same asymptotic results given for $\bar{\chi}^2_{01}(\mathbf{R})$ and $\bar{\chi}^2_{12}(\mathbf{R})$ can be established for $\bar{\chi}^2_{01}(\mathrm{S})$ and $\bar{\chi}^2_{12}(\mathrm{S})$. Also, Shiraishi obtained the exact distribution for $\bar{\chi}^2_{01}(\mathrm{S})$ for equal sample sizes, $n_1 = n_2 = \cdots = n_k = n$, and he tabled critical values for $3 \leqslant k \leqslant 6$ and $kn \leqslant 15$ with the Fisher–Yates normal scores. Presumably one could, using similar techniques, obtain the small sample distribution of $\bar{\chi}^2_{12}(\mathrm{S})$ (which would include $\bar{\chi}^2_{12}(\mathbf{R})$ as a special case), but this has not been done.

Another test for H_0 versus H_1–H_0, which is implicit in the work of Sparre Andersen (1954), has been proposed by Brunk (1960). It rejects H_0 for large values of L, the number of level sets in $P_w(\bar{\mathbf{Y}}|\mathscr{I})$ with $w_i = n_i$. If $\lesssim$ is a simple order and $n_1 = n_2 = \cdots = n_k$, then, as was noted in Section 2.3, the distribution of L does not depend on F provided F is continuous. The authors know of no work comparing the power of the L-test and $\bar{\chi}^2_{01}(\mathrm{S})$.

One-way layout: comparison of the linear rank tests and $\bar{\chi}^2_{01}(\mathrm{S})$

Magel (1983) conducted a Monte Carlo study of the power functions of the Kruskal–Wallis, the Jonckheere–Terpstra, and the Chacko–Shorack tests for testing homogeneity of location parameters against a simply ordered trend. With $\alpha = 0.05, 0.01$, $k = 3, 4, 5$, and six observations from each population, the power functions were estimated for normal, uniform, and Cauchy populations. For all parameter vectors satisfying H_1 and each of the distributions considered, the Kruskal–Wallis test has smaller power, and for those points located at a reasonable distance from H_0 it was substantially smaller. For points in the center of H_1, that is $\theta_i - \theta_{i-1} = \theta_{i+1} - \theta_i$, $i = 2, 3, \ldots, k-1$, the Jonckheere–Terpstra test was the most powerful of the three, but for points with all parameters equal but one, the Chacko–Shorack test was the most powerful. For example, with normal populations, $\sigma^2 = 1, k = 3, \boldsymbol{\theta} = (0, 1, 2)$, and $\alpha = 0.05$, the estimated powers for the Kruskal–Wallis, the Jonckheere–Terpstra, and the Chacko–Shorack tests were 0.748, 0.938, 0.912, respectively, and for $\boldsymbol{\theta} = (0, 2, 2)$, they were 0.858, 0.887, 0.955. For normal distributions, the same values of α and σ^2 and $k = 5$, the estimated values of the power functions at $(0, 0.5, 1.0, 1.5, 2.0)$ were 0.232, 0.612, and 0.504, but at $(0, 2, 2, 2, 2)$ they were 0.235, 0.365, and 0.456.

The conclusions are like those reached in Section 4.2 where the contrast tests were compared with the LRTs. If one believes the distributions are nearly 'equally spaced', then the Jonckheere–Terpstra test, or Puri's modification of it using scores, should be used. Presumably the optimal linear rank tests studied by Shirahata (1980) would provide some protection against the violation of equal spacings. However, if nothing is known about the populations other than that they satisfy H_1, then the Chacko–Shorack test (or its analogue using scores) should be used because its power function is more uniform over H_1. Magel and Magel (1986) conducted a similar study for the simple tree ordering and the conclusions are like those in the normal theory case (cf. Section 4.2).

Bartholomew (1961b), Puri (1965), and Berenson (1982a) conducted power studies of the normal theory tests ($\bar{\chi}^2_{01}$ and $\bar{E}^2_{01}$) and some of the nonparametric tests described above. Their work further corroborates these conclusions. These references are interesting because they give an indication of the power gained by using nonparametric procedures for nonnormal distributions.

One-way layout: asymptotic relative efficiencies

Let F be an absolutely continuous distribution function with density f and let $F_i(y) = F(y - \theta_i/\sqrt{N})$. The asymptotic powers of $\bar{\chi}^2_{01}$ and $\bar{\chi}^2_{01}(\mathbf{R})$ as tests of $H_0 : \theta_1 = \theta_2 = \cdots = \theta_k$ versus $H_1 - H_0$ with $H_1 : \theta_1 \leqslant \theta_2 \leqslant \cdots \leqslant \theta_k$ will be compared. Suppose that $n_i/N \to \gamma_i \in (0, 1)$ for $i = 1, 2, \ldots, k$. Under these assumptions on F_i and n_i, Andrews (1954) showed that the mean ranks, $\bar{R}_i$, $i = 1, 2, \ldots, k$, have an asymptotic multivariate normal distribution; the limiting covariance matrix of $(12/N)^{1/2}\bar{\mathbf{R}}$ is as in the null case (see the proof of Theorem 4.5.1); and the means are given by

$$\begin{aligned} E(\bar{R}_i) = E(R_{ij}) &= \sum_{a=1}^{k} \sum_{l=1}^{n_a} P[Y_{al} < Y_{ij}] + 1 \\ &= \sum_{a=1}^{k} \sum_{l=1}^{n_a} \int_{-\infty}^{\infty} [1 - F_i(x)]\, \mathrm{d}F_a(x) + \tfrac{1}{2} \\ &= \sum_{a=1}^{k} n_a \int_{-\infty}^{\infty} [1 - F(x - \theta_i/\sqrt{N})] f(x - \theta_a/\sqrt{N})\, \mathrm{d}x + \tfrac{1}{2}. \end{aligned}$$

Expanding F and f about $\theta_i = 0$ and $\theta_a = 0$, respectively, and assuming that the appropriate integrals of $f'(x)$ are finite,

$$\begin{aligned} E(\bar{R}_i) &= \frac{1}{2} + \sum_{a=1}^{k} n_a \left\{ \int_{-\infty}^{\infty} [1 - F(x)] f(x)\, \mathrm{d}x + \frac{\theta_i}{\sqrt{N}} \int_{-\infty}^{\infty} f^2(x)\, \mathrm{d}x \right. \\ &\quad \left. - \frac{\theta_a}{\sqrt{N}} \int_{-\infty}^{\infty} f'(x)[1 - F(x)]\, \mathrm{d}x + O(N^{-1}) \right\} \\ &= \frac{N+1}{2} + \sqrt{N}(\theta_i - \bar{\theta}) \int_{-\infty}^{\infty} f^2(x)\, \mathrm{d}x + O(1) \qquad (4.5.9) \end{aligned}$$

where $\bar{\theta} = \sum_{i=1}^{k} \gamma_i \theta_i$. Therefore

$$E\left[\left(\frac{12}{N}\right)^{1/2}\left(\bar{R}_i - \frac{N+1}{2}\right)\right] = \sqrt{12}(\theta_i - \bar{\theta}) \int_{-\infty}^{\infty} f^2(x)\,\mathrm{d}x + O(N^{-1/2}) \quad (4.5.10)$$

In the normal theory case, the corresponding quantity, $\sqrt{N}(\bar{Y}_i - \hat{\mu})/\sigma$, has mean $(\theta_i - \bar{\theta})/\sigma$. (Recall that the means are $\theta_i/\sqrt{N}$.) It follows by comparing this mean with that in (4.5.10) that the power of $\bar{\chi}_{01}^2$ with location parameters $12\sigma\theta_i \int_{-\infty}^{\infty} f^2(x)\,\mathrm{d}x/\sqrt{N}$ is the same as $\bar{\chi}_{01}^2(\mathbf{R})$ with parameters $\theta_i/\sqrt{N}$. Therefore, if N_R is the sample size needed so that $\bar{\chi}_{01}^2(\mathbf{R})$ will have the same asymptotic power as $\bar{\chi}_{01}^2$ with the sample size N, then

$$\frac{N}{N_R} = 12\sigma^2\left[\int_{-\infty}^{\infty} f^2(x)\,\mathrm{d}x\right]^2. \quad (4.5.11)$$

The asymptotic relative efficiency (ARE) given in (4.5.11) is that which arises in the one- and two-sample location problems when comparing the Student t-tests with those based on rank statistics with Wilcoxon scores. Shiraishi (1982) obtains the ARE of $\bar{\chi}_{01}^2(\mathbf{S})$ when compared to $\bar{\chi}_{01}^2$ for score functions satisfying the usual regularity conditions. Again, these AREs are like those for the one- and two-sample location testing problems.

Next, the ARE of the test based on $T_{\mathbf{c}}(\mathbf{R}) = \sum_{i=1}^{k} n_i c_i \bar{R}_i$ with respect to its normal theory counterpart, i.e. the one based on $T_{\mathbf{c}} = \sum_{i=1}^{k} w_i c_i \bar{Y}_i / \|\mathbf{c}\|_{\mathbf{w}}$, is considered with the same vector $\mathbf{c}$ used in both tests. The two tests reject for large values of (recall that $\sum_{i=1}^{k} n_i c_i = 0$)

$$\sum_{i=1}^{k} n_i c_i \left[\left(\frac{12}{N}\right)^{1/2}\left(\bar{R}_i - \frac{N+1}{2}\right)\right] \quad \text{and} \quad \sum_{i=1}^{k} n_i c_i \frac{\sqrt{N}(\bar{Y}_i - \hat{\mu})}{\sigma},$$

respectively. Under the assumptions that $F_i(y) = F(y - \theta_i/\sqrt{N})$ and $n_i/N \to \gamma_i \in (0, 1)$ for $i = 1, 2, \ldots, k$, the argument above shows that the ARE of the rank version of the contrast test is given by (4.5.11). Thus the optimum choice for the normal theory contrast vector is asymptotically optimum for its rank analogue.

Replacing $\bar{R}_b^{(ab)} - \bar{R}_a^{(ab)}$ by $\bar{Y}_b - \bar{Y}_a$ in (4.5.4) and summing over $a < b$ gives a constrast statistic, which is equivalent to the T-statistic in (4.2.4). Bartholomew (1961b) shows that the ARE of J, the Jonckheere–Terpstra test statistic, with respect to this normal theory test is also given by (4.5.11). Puri (1965) obtains the ARE of his scored version of the Jonckheere–Terpstra test compared to the same normal theory constrast test.

Aiyar, Guillier, and Albers (1979) give a detailed discussion of the AREs of the nonparametric tests discussed here. In particular, the AREs of the $T_{\mathbf{c}}(\mathbf{S})$ with respect to their normal theory contrast tests are derived. Shiraishi (1982) obtains the AREs for the tests based on $\bar{\chi}_{01}^2(\mathbf{S})$.

Randomized block design

As in the last subsection, k-treatment effects are to be compared. In the case of a randomized block design there are other effects which are present in the experiment, but they are not of direct interest. The types of tests that are available will be illustrated by considering the situation in which there is one observation per cell, i.e. per combination of treatment and block. Suppose Y_{ij} for $i = 1, 2, \ldots, r$ and $j = 1, 2, \ldots, k$ are independent continuous random variables with the CDF of Y_{ij} given by $F_j(y - b_i)$ where $b_1, b_2, \ldots, b_r$ are the block effects. If the treatment effects were believed to be simply ordered, one might wish to test

$$H_0: F_1 = F_2 = \cdots = F_k \qquad \text{versus} \qquad H_1: F_1 \overset{\text{st}}{\leqslant} F_2 \overset{\text{st}}{\leqslant} \cdots \overset{\text{st}}{\leqslant} F_k.$$

A common method of eliminating the block effects is to rank the observations within each block separately. For instance, if $r = 3$ and $k = 5$, one such possible ranking is given below:

		Treatment				
		I	II	III	IV	V
	I	2	1	4	3	5
Blocks	II	1	3	2	5	4
	III	1	2	5	4	3

Randomized block designs: tests based on correlation coefficients

In the discussion of the one-way layout, Kendall's tau and Spearman's rho were found to provide useful statistics for testing for trend. The correlation coefficients for the pairs (i, Y_{ij}) for $j = 1, 2, \ldots, n_i$ and $i = 1, 2, \ldots, k$ were used. With J_i the Jonckheere–Terpstra statistic computed on the ith block, $i = 1, 2, \ldots, r$, Jonckheere (1954b) proposed rejecting H_0 in favor of H_1 for large values of

$$J = \sum_{i=1}^{r} J_i, \tag{4.5.12}$$

which is equivalent to rejecting for large values of the sum of the r Kendall's taus. These tests can be employed with an arbitrary number of observations in a cell (i.e. arbitrary n_{ij}). Skillings (1980) gives exact critical values for small r and k. Because the J_i are independent with asymptotic normal distributions, J has an asymptotic normal distribution as k or r approach infinity, and under H_0, $E(J) = \sum_{i=1}^{r} E(J_i)$ and Var $(J) = \sum_{i=1}^{r}$ Var (J_i). The mean and variance of J_i are given in (4.5.2). Skillings and Wolfe (1977, 1978) consider statistics with J_i replaced by weighted Chernoff–Savage statistics on the ith block and allow for different weights for different blocks. With H defined as in (4.5.5), they consider statistics of the form $H = \sum_{i=1}^{r} b_i H_{c_i}$.

Page (1963) proposed a statistic which, in the case considered, produces a test equivalent to that based on the sum over the blocks of the Spearman rank correlation coefficient within a block. With R_{ij} the rank of Y_{ij} in the ith block and $T_i(\mathbf{R}) = \sum_{j=1}^{k} jR_{ij}$, Page's test rejects H_0 for large values of

$$T(\mathbf{R}) = \sum_{i=1}^{r} T_i(\mathbf{R}). \tag{4.5.13}$$

Page, as well as Odeh (1977), gives extensive tables of critical values of $T(\mathbf{R})$. Of course, $T(\mathbf{R})$ has an asymptotic normal distribution as k or r approach infinity and under H_0, $E(T(\mathbf{R})) = rk(k+1)^2/4$ and $\operatorname{Var}(T(\mathbf{R})) = rk^2(k+1)^2(k-1)/144$ (cf. Page, 1963, eq. (4)). Pirie and Hollander (1972) develop a version of Page's test using scores. With scores $a_k(1) \leqslant a_k(2) \leqslant \cdots \leqslant a_k(k)$,

$$T(\mathbf{S}) = \sum_{i=1}^{r} T_i(\mathbf{S}) \qquad \text{and} \qquad T_i(\mathbf{S}) = \sum_{j=1}^{k} ja_k(R_{ij}).$$

They give a table of critical values for the statistic based on normal scores and derive a large-sample normal approximation. Hettmansperger (1975) extends the definition of $T(\mathbf{R})$ to allow for more than one observation per cell. He also develops interval estimates and multiple comparison procedures for this setting. For a discussion of related multiple comparison procedures see Marcus (1978b).

Hollander (1967) studied the asymptotic behavior of J and $T(\mathbf{R})$. If $F_j(x) = F(x-(j-1)cr^{-1/2})$ for $j = 1, 2, \ldots, k$ where F satisifies certain regularity conditions, then the ARE of $T(\mathbf{R})$ with respect to J as $r \to \infty$ is

$$\frac{k(2k+5)}{2(k+1)^2}. \tag{4.5.14}$$

When $k = 2$ this ratio is one; it increases with k until it reaches its maximum of 1.042 at $k = 5$; and for $k \geqslant 5$, it decreases to one as $k \to \infty$. Thus, for the 'equally spaced' alternatives the test based on Spearman's rho has a slight advantage in terms of AREs. It would be instructive to study the ARE of these two tests for alternatives at the edge of the simply ordered cone.

Randomized block designs: an extension of Chacko's test

Shorack (1967) extended Chacko's test for the one-way layout to a randomized block design. This test may also be viewed as an ordered version of Friedman's (1937) test of concordance for a set of rankings. As in the previous subsection, let R_{ij} be the rank of Y_{ij} in the ith block. With $\bar{R}_{\cdot j} = \sum_{i=1}^{r} R_{ij}/r$ for $j = 1, 2, \ldots, k$, Friedman's test rejects for large values of

$$\frac{12r}{k(k+1)} \sum_{j=1}^{k} \left(\bar{R}_{\cdot j} - \frac{k+1}{2} \right)^2 = \frac{12r}{k(k+1)} \sum_{j=1}^{k} \left(\sum_{i=1}^{r} R_{ij} \right)^2 - 3r(k+1),$$

which under H_0 has an asymptotic chi-square distribution with $k-1$ degrees of freedom.

Shorack's proposal is to compute $\bar{\mathbf{R}}^*_. = P(\bar{\mathbf{R}}_.|\mathscr{I})$, the equal-weights projection of $\bar{\mathbf{R}}_. = (\bar{R}_{.1}, \bar{R}_{.2}, \ldots, \bar{R}_{.k})$ onto the cone of isotonic functions, and reject H_0 for large values of

$$\bar{\chi}^2_{01}(\mathbf{R}) = \frac{12r}{k(k+1)} \sum_{j=1}^{k} \left(\bar{R}^*_{.j} - \frac{k+1}{2} \right)^2. \tag{4.5.15}$$

In Section 2.2, the parametric analogue for this order restricted test was developed. Let $\bar{Y}_{.j} = \sum_{i=1}^{r} Y_{ij}/r$, $\bar{Y}_{..} = \sum_{j=1}^{k} Y_{.j}/k$ and $\bar{\mathbf{Y}}^*_. = (\bar{Y}^*_{.1}, \bar{Y}^*_{.2}, \ldots, \bar{Y}^*_{.k})$ be the projection, with equal weights, of $\bar{\mathbf{Y}}_. = (\bar{Y}_{.1}, \bar{Y}_{.2}, \ldots, \bar{Y}_{.k})$ onto the cone of isotonic functions. In this situation with one observation per cell, the statistic is

$$r \sum_{j=1}^{k} \frac{(\bar{Y}^*_{.j} - \bar{Y}_{..})^2}{\sigma^2} \tag{4.5.16}$$

and under H_0 this statistic has a $\bar{\chi}^2$ distribution with unit weights, i.e. the distribution given in the first equation in (2.3.4) with $w_1 = w_2 = \cdots = w_k = 1$.

If $k \to \infty$, each vector $(R_{i1}, R_{i2}, \ldots, R_{ik})$, when suitably normalized, has an asymptotic normal distribution. Because of the independence of these vectors over i, $\bar{\mathbf{R}}_.$ also has an asymptotic normal distribution. Under H_0,

$$E(\bar{R}_{.j}) = \frac{k+1}{2} \quad \text{and} \quad \operatorname{cov}(\bar{R}_{.j}, \bar{R}_{.j'}) = \frac{k+1}{12r}(k\delta_{jj'} - 1) \tag{4.5.17}$$

with $\delta_{jj'}$ the Kronecker delta function. Therefore,

$$\left[\frac{12}{k(k+1)}\right]^{1/2} \left(\bar{\mathbf{R}}_. - \frac{k+1}{2} \right) \quad \text{and} \quad \frac{\bar{\mathbf{Y}}_. - \bar{Y}_{..}}{\sigma}$$

have the same limiting distributions. Thus, as in the one-way layout, the asymptotic null distribution of (4.5.15) is the same as the distribution of (4.5.16). Using a multivariate central limit theorem, one can show that as $n \to \infty$ the distribution of (4.5.15) also approaches that of (4.5.16).

The authors know of no small sample critical values for $\bar{\chi}^2_{01}(\mathbf{R})$ in this case nor of any work which introduces scores into the definition of $\bar{\chi}^2_{01}(\mathbf{R})$.

Randomized block designs: utilizing information among the blocks

Several researchers have noted that the tests which are based on the ranks within the blocks do not make use of the between-block information available in the data. Several proposals to make use of this information have been made.

Hollander (1967) and Doksum (1967a) have devised closely related tests which are asymptotically more powerful than either Jonckheere's or Page's tests. To provide some motivation for their proposals we first observe that all of the statistics discussed thus far depend on a function of the form

$$\sum_{j=1}^{k-1} \sum_{j'=j+1}^{k} S_{jj'}$$

where $S_{jj'}$ is some measure of the difference between group j and group j'. If we score 1 if the gth member of group j' is greater than the hth member of group j and zero otherwise, and sum over g and h, we obtain J. If, instead, $S_{jj'}$ is taken as the difference in mean ranks, $T(\mathbf{R})$ is obtained. That there exists a possibility of improvement can be seen by considering J. In this case there is a contribution of 1 to the total score if $Y_{hj} < Y_{hj'}$, regardless of how big the difference is. However, 'big' differences are more 'significant' than small ones so some means of incorporating the size of the differences should improve the test. Hollander and Doksum proposed to do this by ranking these differences. They base their tests on

$$S_{jj'} = \sum_{h=1}^{r} I_{[Y_{hj} < Y_{hj'}]} R_{jj'}^{(h)}$$

where, with j and j' fixed, $R_{jj'}^{(h)}$ is the rank of $|Y_{hj} - Y_{hj'}|$ in the ranking from the least to the greatest of these r values. Hollander's statistic is then

$$Y = \sum_{j=1}^{k-1} \sum_{j'=j+1}^{k} S_{jj'}.$$

Doksum's statistic is

$$Y' = \sum_{j=1}^{k-1} \sum_{j'=j+1}^{k} (S_{j\cdot} - S_{\cdot j'}).$$

Although the scores $S_{jj'}$ are distribution free, the statistics Y and Y' are not because the correlation coefficient between any pair with a common j depends on the form of $F(x)$. In order to circumvent this difficulty the correlation coefficients can be estimated from the data using the device first proposed by Lehmann (1964). The resulting test is asymptotically distribution free. The execution of these tests is more tedious than the others which have been described and the adequacy of the asymptotic distribution for finite samples remains unexplored. The reader is referred to the original papers for further details, but in order to show the advantages of the tests, the following results are quoted on their asymptotic efficiency. Doksum showed that Y' is always asymptotically more powerful than Y for all F and k. However, the difference is negligible; values of the maximum of the ARE given by Hollander are 1.004 for F normal, 1.002 for F rectangular, and 1.008 when F is exponential. The asymptotic gain using Y (or Y') as compared with $T(R)$ can be seen from Table 4.5.1. Puri and Sen (1968) introduce a version of Hollander's test using scores.

Sen (1968) proposed a test based on ranks of 'aligned' observations. In the situation under study, let $Y_{ij}^* = Y_{ij} - \bar{Y}_{i\cdot}$ and let R_{ij} be the rank of Y_{ij}^* among the rk aligned observations. Subtracting $\bar{Y}_{i\cdot}$ is to compensate for the block effects and the aligned observations are then comparable. With $N = rk$, let $\{a_N(i)\}_{i=1}^{N}$ be

Table 4.5.1 The ARE of Y and $T(\mathbf{R})$ compared with the statistic $T=\sum_{j=1}^{k}(2j-k-1)\bar{Y}_{.j}$ when the θ's are equally spaced and normal distributions are assumed

k	2	3	4	6	10	20	50	∞
Y with T	0.955	0.963	0.968	0.974	0.980	0.984	0.987	0.989
$T(R)$ with T	0.637	0.716	0.764	0.819	0.868	0.909	0.936	0.955

scores corresponding to a score function ϕ; let

$$T_{N,j}=\sum_{i=1}^{r}\frac{a_N(R_{ij})}{r} \qquad \text{for } j=1,2,\ldots,k$$

$$\bar{E}_N=\sum_{l=1}^{N}\frac{a_N(l)}{N}$$

and

$$\sigma_0^2=\frac{1}{r(k-1)}\sum_{i=1}^{r}\sum_{j=1}^{k}[a_N(R_{ij})-\frac{1}{k}\sum_{l=1}^{k}a_N(R_{il})]^2.$$

Sen's aligned rank test rejects H_0 for large values of

$$S_N=r\sum_{j=1}^{k}\frac{(T_{N,j}-\bar{E}_N)^2}{\sigma_0^2},$$

which, under H_0 and suitable regularity conditions, has an asymptotic χ_{k-1}^2 distribution. We shall see in the next subsection that using the between-block information has improved the ARE, which for Sen's test is the same improvement as is typical for one or two sample location problems. Sen shows how this test can be extended to more than one observation per cell.

Salama and Quade (1981) propose a scheme for using interblock information based on the following idea. A block with a large apparent variability is more likely to reflect the true ordering than one with smaller apparent variability. Hence, nonnegative weights will be assigned to the r blocks based on the apparent variability in them. Let C_i be a measure of the agreement of the data in block i with the postulated ordering; let $D_i=D(Y_{i1},Y_{i2},\ldots,Y_{ir})$ with D a location-free statistic; and let Q_i be the rank of D_i among $D_1,D_2,\ldots,D_r$. They propose rejecting H_0 for large values of

$$W=\frac{\sum_{i=1}^{r}b_{Q_i}C_i}{\sum_{i=1}^{r}b_i}$$

with $0\leqslant b_1\leqslant b_2\leqslant\cdots\leqslant b_r$. They discuss the limiting null distribution of W and choices for the b_i. This is a special case of the statistic proposed by Skillings and Wolfe (1977).

Berenson (1982b) discusses the use of block weights in a weighted sum of weighted Chernoff–Savage statistics, i.e.

$$\sum_{i=1}^{r} b_i \sum_{j=1}^{k} c_j a_k(R_{ij}) \qquad \text{with} \sum_{j=1}^{k} c_j = 0$$

for the special cases $c_j = (2j-k-1)$ and $c_j = \{(j-1)[1-(j-1)/k]\}^{1/2} - [j(1-j/k)]^{1/2}$ (the Abelson–Tukey contrast coefficients).

Randomized block design: AREs

First the asymptotic power of Shorack's test and the normal theory $\bar{\chi}^2$ test (cf. equation (4.5.16)) are compared as $r \to \infty$ with k fixed. Suppose that the CDF of Y_{ij} is given by $F(y - b_i - \theta_j/\sqrt{r})$ for $i = 1, 2, \ldots, r$ and $j = 1, 2, \ldots, k$ with $\theta_1 \leqslant \theta_2 \leqslant \cdots \leqslant \theta_k$. Let the rank of Y_{ij}, among the members of the ith block, be R_{ij} and let $\bar{R}_{\cdot j}$ be the average rank of the jth treatment over the r blocks. Setting $n_a = 1$ and $N = r$ in the argument which gives (4.5.9) shows that, in this case,

$$E(\bar{R}_{\cdot j}) = E(R_{ij}) = \frac{k+1}{2} + \frac{k}{\sqrt{r}}(\theta_j - \bar{\theta}) \int_{-\infty}^{\infty} f^2(x)\,\mathrm{d}x + O\left(\frac{k}{r}\right) \qquad (4.5.18)$$

where $\bar{\theta} = \sum_{j=1}^{k} \theta_j/k$ (i.e. the weights $v_a = 1/k$). Asymptotically, as $r \to \infty$, the covariance matrix of $(\bar{R}_{\cdot 1}, \bar{R}_{\cdot 2}, \ldots, \bar{R}_{\cdot k})$ is the same as under H_0, namely

$$\operatorname{cov}(\bar{R}_{\cdot j}, \bar{R}_{\cdot l}) = \frac{k+1}{12r}(\delta_{jl}k - 1) \qquad \text{for } j, l = 1, 2, \ldots, k,$$

and the limiting distribution is multivariate normal.

Next, the distribution of $\sqrt{r}(\bar{Y}_{\cdot 1} - \bar{Y}_{\cdot\cdot}, \bar{Y}_{\cdot 2} - \bar{Y}_{\cdot\cdot}, \ldots, \bar{Y}_{\cdot k} - \bar{Y}_{\cdot\cdot})/\sigma$ is determined. It is multivariate normal with mean and covariance

$$E\left[\frac{\sqrt{r}}{\sigma}(\bar{Y}_{\cdot j} - \bar{Y}_{\cdot\cdot})\right] = \frac{1}{\sigma}(\theta_j - \bar{\theta})$$

and (4.5.19)

$$\operatorname{cov}\left[\frac{\sqrt{r}}{\sigma}(\bar{Y}_{\cdot j} - \bar{Y}_{\cdot\cdot}), \frac{\sqrt{r}}{\sigma}(\bar{Y}_{\cdot l} - \bar{Y}_{\cdot\cdot})\right] = \frac{1}{k}(\delta_{jl}k - 1).$$

Consequently, $\bar{\chi}_{01}^2(\mathbf{R})$ defined in (4.5.15) and its normal theory counterpart given in (4.5.16) are the same function of the vectors

$$\left[\frac{12r}{k(k+1)}\right]^{1/2}\left(\bar{\mathbf{R}}_{\cdot} - \frac{k+1}{2}\right) \qquad \text{and} \qquad \frac{\sqrt{r}}{\sigma}(\bar{\mathbf{Y}}_{\cdot} - \bar{Y}_{\cdot\cdot})$$

which have asymptotic multivariate normal distributions with the same covariance structure. The limiting means of the jth components are

$$\left(\frac{12k}{k+1}\right)(\theta_j - \bar{\theta}) \int_{-\infty}^{\infty} f^2(x)\,dx \qquad \text{and} \qquad \frac{1}{\sigma}(\theta_j - \bar{\theta}).$$

Thus the ARE of Shorack's test to the $\bar{\chi}^2_{01}$ for a randomized block design is

$$\frac{12k\sigma^2}{k+1}\left[\int_{-\infty}^{\infty} f^2(x)\,dx\right]^2. \tag{4.5.20}$$

It should be noted that this is like the ARE for the one-way layout except there is a factor of $k/(k+1)$ which is less than one.

Using similar techniques, Hollander (1967) found that the ARE of Page's test with respect to the contrast test based on

$$T_c = \frac{\sum_{j=1}^{k} c_j \bar{Y}_{ij}}{\sum_{j=1}^{k} c_j^2} \qquad \text{with} \sum_{j=1}^{k} c_j = 0$$

is

$$\frac{12k\sigma^2}{k+1}\left[\int_{-\infty}^{\infty} f^2(x)\,dx\right]^2 \left[\frac{\sum_{j=1}^{k}(2j-k-1)\theta_j}{\sum_{j=1}^{k} c_j\theta_j}\right]^2, \tag{4.5.21}$$

provided the c_j's are normalized so that $\sum_{j=1}^{k} c_j^2 = \sum_{j=1}^{k}(2j-k-1)^2 = k(k^2-1)/3$. Page's test uses equally spaced c_j and in that case (4.5.20) and (4.5.21) agree. Combining (4.5.21) with (4.5.14), one may compute the ARE of Jonckheere's test with respect to T_c. Skillings and Wolfe (1978) and Hettmansperger (1975) discuss the AREs of their modification of Jonckheere's and Page's tests, respectively.

Hollander (1967) studied the AREs of his test and Doksum's test with respect to the contrast test T with equally spaced contrast coefficients. These values are more like those for a one- and two-sample location problem. For equally spaced alternatives and normal distributions the AREs of Hollander's test with respect to T and of Page's $T(\mathbf{R})$ with respect to T are given in Table 4.5.1. It is clear that Hollander's test (and Doksum's test) have benefited in terms of AREs from using interblock information. Puri and Sen (1968) give the ARE of their scored version of Hollander's test.

Pirie (1974) studied AREs for some of the tests for randomized blocks as $k \to \infty$ with r fixed. These AREs are larger for the tests using only within-block information than for those using among-block information. This is the opposite of what was observed as $r \to \infty$ with k fixed.

Sen (1968) studied the ARE of the aligned rank tests he proposed and found that they are like those for the one- and two-sample location problems. Thus, it seems that aligning over the blocks is also an effective way to make use of the information in the data between the blocks.

Bernson (1982b) has conducted a large-scale study of the small sample power characteristics of the parametric and nonparametric tests for a randomized block design. The conclusions are too detailed to report here, but they certainly provide some guidance for selecting a test in this situation.

Other distribution-free tests

Chinchilli and Sen (1981a, 1981b) use the union-intersection principle to develop rank tests for ordered hypotheses. Sen (1982) proposed combining the union-intersection principle with the idea of locally most powerful rank tests, and Boyd and Sen (1983, 1984, 1986) discuss the use of such tests for ordered alternatives in linear models, randomized block designs, and analysis of covariance applications, respectively.

Boyett and Schuster (1977) discuss the use of rank tests for some problems in multivariate analysis with medical applications.

The discussion in this section has been directed primarily toward orderings in location parameters. There are several papers dealing with nonparametric tests to detect orderings in scale parameters. See Govindarajulu and Haller (1977), Govindarajulu and Gupta (1978), Rao (1982), Govindarajulu and Mansouri–Ghiassi (1986), and Kochar and Gupta (1986).

4.6 ONE-SIDED TESTS IN MULTIVARIATE ANALYSIS

Consider again the problem discussed in Chapter 2 of testing the homogeneity of k normal means against an order restricted alternative. Specifically, suppose $\bar{Y}_1$, $\bar{Y}_2, \ldots, \bar{Y}_k$ are the means of independent random samples of sizes $n_1, n_2, \ldots, n_k$ from normal populations with means $\mu_1, \mu_2, \ldots, \mu_k$ and known variance $\sigma_1^2, \sigma_2^2, \ldots, \sigma_k^2$, and consider testing the null hypothesis

$$\mu_1 = \mu_2 = \cdots = \mu_k \tag{4.6.1}$$

against the alternative hypothesis that

$$\mu_1 \leqslant \mu_2 \leqslant \cdots \leqslant \mu_k. \tag{4.6.2}$$

If one makes the transformations

$$Z_i = \bar{Y}_{i+1} - \bar{Y}_i \quad \text{and} \quad \theta_i = \mu_{i+1} - \mu_i \quad \text{for } i = 1, 2, \ldots, k-1, \tag{4.6.3}$$

then (4.6.1) and (4.6.2) are equivalent to

$$H_0\colon \theta_i = 0; \quad i = 1, 2, \ldots, k-1 \quad \text{and} \quad H_1\colon \theta_i \geqslant 0; \quad i = 1, 2, \ldots, k-1, \tag{4.6.4}$$

respectively. A test of H_0 versus $H_1 - H_0$ can be based upon the random vector $\mathbf{Z} = (Z_1, Z_2, \ldots, Z_{k-1})'$, which is sufficient for $\boldsymbol{\theta} = (\theta_1, \theta_2, \ldots, \theta_{k-1})'$. The distribution of $\mathbf{Z}$ is multivariate normal with mean vector $\boldsymbol{\theta}$ and variance-covariance matrix $\boldsymbol{\Sigma}$ with elements

$$\sigma_{i,i}^2 = w_{i+1}^{-1} + w_i^{-1} = \frac{\sigma_{i+1}^2}{n_{i+1}} + \frac{\sigma_i^2}{n_i} \qquad \text{for } i = 1, 2, \ldots, k-1$$

$$\sigma_{i,i+1}^2 = -w_{i+1}^{-1} = -\frac{\sigma_{i+1}^2}{n_{i+1}} \qquad \text{for } i = 1, 2, \ldots, k-2$$

and

$$\sigma_{i,j}^2 = 0 \qquad \text{for } |i-j| \geqslant 2.$$

Thus, the testing problems discussed in Chapter 2 are related to testing problems involving constraints on a collection of means from a multivariate normal distribution. The relationship between these two types of problems goes deeper as $\bar{\chi}^2$ and $\bar{E}^2$ types of distributions also arise under very general settings in the latter problems (cf. Theorems 4.6.1 and 4.6.2).

For the hypothesis H_1 given in (4.6.4) the constraints require the parameter vector to be in the positive orthant. Partial order restrictions on means may lead to more, or fewer, constraints than the dimension of the parameter space for $\boldsymbol{\theta}$ and thus do not, in general, transform into problems requiring $\boldsymbol{\theta}$ to be in an orthant.

If there are fewer contraints than the dimension of the parameter space of means, the new transformed mean vector can be partitioned as $\boldsymbol{\theta}' = (\boldsymbol{\theta}_1', \boldsymbol{\theta}_2')$ and the hypotheses expressed as

$$H_0 : \boldsymbol{\theta}_1' = 0 \qquad \text{versus} \qquad H_1 : \boldsymbol{\theta}_1' \geqslant 0$$

where $\boldsymbol{\theta}$ is the vector of means of the new multivariate normal random vector $\mathbf{Z} = (Z_1, Z_2, \ldots, Z_k)'$. However, if the covariate structure is totally known, the likelihood ratio test for the entire vector $\mathbf{Z}$ will be identical to the likelihood ratio test based on the reduced vector $\tilde{\mathbf{Z}}_1$ which contains only those coordinates corresponding to $\boldsymbol{\theta}_1$. Thus if the covariate structure is known and the number of linear constraints is no more than the dimension of the parameter space, we can assume that the alternative region is the entire positive orthant.

It may happen that there are more constraints than the dimension of the space. For example, if $k = 4$ and H_1 is the loop partial order, $\mu_1 \leqslant [\mu_2, \mu_3] \leqslant \mu_4$, then the restrictions on $\boldsymbol{\theta}$ corresponding to the transformation in (4.6.3) are $\theta_1 \geqslant 0$, $\theta_1 + \theta_2 \geqslant 0$, $\theta_2 + \theta_3 \geqslant 0$, and $\theta_3 \geqslant 0$. There have been two approaches in the literature to handling such partial order restrictions. One approach is illustrated in this example by transforming $(\mu_1, \mu_2, \mu_3, \mu_4)$ into $\boldsymbol{\theta} = (\mu_2 - \mu_1, \mu_3 - \mu_1, \mu_4 - \mu_2, \mu_4 - \mu_3)'$ and basing a test of H_0 against the alternative that $\boldsymbol{\theta}$ lies in the positive orthant upon the random vector $(\bar{Y}_2 - \bar{Y}_1, \bar{Y}_3 - \bar{Y}_1, \bar{Y}_4 - \bar{Y}_2, \bar{Y}_4 - \bar{Y}_3)'$, which has a singular normal distribution (cf. Kudô and Choi, 1975). The second approach is to assume a nonsingular covariance structure but to allow more general types of cones in the alternative restriction, and this is the approach discussed here.

Consider a model which consists of n independent observations $\mathbf{Z}_1, \mathbf{Z}_2, \ldots, \mathbf{Z}_n$ from a p-dimensional normal distribution with mean vector $\boldsymbol{\theta}$ and nonsingular variance-covariance matrix $\boldsymbol{\Sigma}$. Consider the problem of testing the null hypothesis, H_0, that $\boldsymbol{\theta}$ is identically zero against the alternative $H_1 - H_0$ where $H_1 : \boldsymbol{\theta} \in \mathscr{K}$ and $\mathscr{K}$ is a closed, convex cone in Euclidean p-space. If $\mathscr{K}$ is a subspace then this problem is a standard topic in multivariate analysis. We assume that $\boldsymbol{\Sigma}$ is nonsingular.

There is a substantial literature on this problem. Kudô (1963) and independently Nüesch (1964, 1966) studied the case in which $\boldsymbol{\Sigma}$ is known and nonsingular and $\mathscr{K}$ is the positive orthant in R^p. Shorack (1967) considered variations of the work of Kudô and Nüesch including a testing problem in which $\boldsymbol{\Sigma} = \sigma^2 \boldsymbol{\Sigma}_0$ with σ^2 unknown and $\boldsymbol{\Sigma}_0$ a particular, known, nonsingular matrix. Perlman (1969) studied this problem and other restricted testing problems under the assumption that $\boldsymbol{\Sigma}$ is completely unknown. Kudô and Choi (1975) studied these estimation and testing problems under the assumptions that $\mathscr{K}$ is a convex, polyhedral cone and that the matrix $\boldsymbol{\Sigma}$ may be singular. Shapiro (1985) assumes that the matrix $\boldsymbol{\Sigma}$ is known and nonsingular but considers arbitrary convex cones, $\mathscr{K}$.

Projections

When Σ is known (up to a scale parameter) the MLE of $\boldsymbol{\theta}$ constrained to be in the cone $\mathscr{K}$ minimizes the quadratic form

$$(\bar{\mathbf{Z}} - \boldsymbol{\theta})' \boldsymbol{\Sigma}^{-1} (\bar{\mathbf{Z}} - \boldsymbol{\theta}) \tag{4.6.5}$$

where $\bar{\mathbf{Z}} = \sum_{i=1}^{n} \mathbf{Z}_i / n$. Central to this theory is the determination of $\boldsymbol{\theta}^*$, which is the projection of $\bar{\mathbf{Z}}$ onto $\mathscr{K}$ with distance determined by the inner product $\langle \mathbf{x}, \mathbf{y} \rangle_{\Sigma^{-1}} = \mathbf{x}' \boldsymbol{\Sigma}^{-1} \mathbf{y}$ and will be denoted by $P_{\Sigma^{-1}}(\bar{\mathbf{Z}} | \mathscr{K})$. It is not difficult to show that $P_{\Sigma^{-1}}(\mathbf{x} | \mathscr{K})$ is the unique element of $\mathscr{K}$ (recall that $\mathscr{K}$ is a closed, convex cone) which satisfies

$$(\mathbf{x} - P_{\Sigma^{-1}}(\mathbf{x} | \mathscr{K}), P_{\Sigma^{-1}}(\mathbf{x} | \mathscr{K}))_{\Sigma^{-1}} = 0 \quad \text{and} \quad (\mathbf{x} - P_{\Sigma^{-1}}(\mathbf{x} | \mathscr{K}), \mathbf{y})_{\Sigma^{-1}} \leqslant 0 \quad \text{for all } \mathbf{y} \in \mathscr{K}.$$

This is a special case of Theorem 8.2.7. However, this and similar results can be derived from those in Chapter 1 for diagonal $\boldsymbol{\Sigma}$ by noting that for $\mathbf{N}$ a nonsingular $p \times p$ matrix, $\mathbf{N}\mathscr{K}$ is a closed, convex cone and $\mathbf{N} P_{\Sigma^{-1}}(\mathbf{x} | \mathscr{K}) = P_{(\mathbf{N}^{-1})'\Sigma^{-1}\mathbf{N}^{-1}}(\mathbf{N}\mathbf{x} | \mathbf{N}\mathscr{K})$ and by choosing $\mathbf{N}$ so that $(\mathbf{N}^{-1})' \boldsymbol{\Sigma}^{-1} \mathbf{N}^{-1}$ is diagonal. If $\mathscr{K}^{*\Sigma^{-1}}$ is the dual of $\mathscr{K}$ (see Section 1.7), then this same technique can be used to show that $\mathbf{y}^{**} = P_{\Sigma^{-1}}(\mathbf{y} | \mathscr{K}^{*\Sigma^{-1}}) = \mathbf{y} - P_{\Sigma^{-1}}(\mathbf{y} | \mathscr{K})$.

Obviously if $\bar{\mathbf{Z}}$ is in $\mathscr{K}$ then $\boldsymbol{\theta}^* = \bar{\mathbf{Z}}$. Otherwise $\boldsymbol{\theta}^*$ must lie on the boundary of $\mathscr{K}$. If $\mathscr{K}$ is the positive orthant of R^p, then the boundary of $\mathscr{K}$ consists of $2^p - 1$ distinct 'faces', each of which is defined by restricting a particular subset of $\{\theta_1, \theta_2, \ldots, \theta_p\}$ to be zero and the others to be positive. If we know which of these 'faces' is the one which contains $\boldsymbol{\theta}^*$, then $\boldsymbol{\theta}^*$ is the projection, with respect to the inner product $\langle \cdot, \cdot \rangle_{\Sigma^{-1}}$ of $\bar{\mathbf{Z}}$ onto the linear subspace generated by that face. (The negative of (4.6.5) is twice the exponent of a multivariate normal p.d.f. in $\boldsymbol{\theta}$ with mean $\bar{\mathbf{Z}}$ and if some of the θ_i are fixed at 0, then this p.d.f. is unimodal in the other θ_i.) For example, if $\boldsymbol{\Sigma}^{-1}$ is diagonal and the 'face' containing $\boldsymbol{\theta}^*$ is $\{\boldsymbol{\theta}:\theta_1 = \theta_2 = \cdots = \theta_m = 0; \theta_i > 0; m < i \leqslant p\}$, then $\theta_i^* = 0$ if $i \leqslant m$ and $\theta_i^* = \bar{Z}_i$ if $i > m$. In general, $\boldsymbol{\theta}^*$ can be found by finding all 2^p such projections and choosing that one which gives the smallest value of $(\bar{\mathbf{Z}} - \boldsymbol{\theta})'\boldsymbol{\Sigma}^{-1}(\bar{\mathbf{Z}} - \boldsymbol{\theta})$. This method may be feasible for small values of p but becomes intractable for larger values of p since the computations quickly become excessive. It would be desirable to have a simple algorithm which reduces the amount of computation needed.

Wynn (1975) and Shapiro (1985) give further insight into these projections onto a polyhedral cone, $\mathscr{K}$. They show that for F, a face of $\mathscr{K}$, there corresponds the polar face, F^* of $\mathscr{K}^{*\Sigma^{-1}}$ with the linear spaces generated by F and F^* orthogonal complements to each other. If $\mathbf{S}_F$ is the symmetric, idempotent matrix giving the orthogonal projection onto the space generated by F, then $\mathbf{S}_{F^*} = \mathbf{I} - \mathbf{S}_F$. Furthermore, $P_{\Sigma^{-1}}(\mathbf{y}|\mathscr{K}) \in F$ if and only if $\mathbf{S}_F\mathbf{y} \in \mathscr{K}$ and $\mathbf{S}_{F^*}\mathbf{y} \in \mathscr{K}^{*\Sigma^{-1}}$.

Kudô describes a heuristic method which avoids the examination of many of the above projections. His method is not, strictly speaking, an algorithm as it involves intuitive judgment in the selection of subspaces onto which to project. It is therefore most suitable for use with a desk calculator. For computer applications Barlow *et al.* (1972) mention the existence of an algorithm based upon Kudô's method (Bremner, 1967). However, no proof exists that this algorithm always leads to a solution. Kudô and Choi (1975) give necessary and sufficient conditions for a given 'face' of $\mathscr{K}$ to contain the solution $\boldsymbol{\theta}^*$. Using their algorithm all 'faces' may not need to be checked, but this algorithm is not guaranteed to yield a solution in a reasonable number of steps. Wollan and Dykstra (1985) give a procedure and computer program for computing $\boldsymbol{\theta}^*$ based on an infinite iterative algorithm which is guaranteed to converge correctly. This method is related to the iterative algorithm for the matrix partial order discussed in Section 1.4. The authors know of no practical finite step general algorithm for computing $\boldsymbol{\theta}^*$ which is applicable for general covariance matrices of large dimensions and general polyhedral cones.

$\boldsymbol{\Sigma}$ is known

In this subsection, $\boldsymbol{\Sigma}$ is assumed to be known. The LRT rejects $H_0:\boldsymbol{\theta} = \mathbf{0}$ in favor of $H_1 - H_0$ with $H_1:\boldsymbol{\theta} \in \mathscr{K}$ for large values of

$$n[\bar{\mathbf{Z}}'\boldsymbol{\Sigma}^{-1}\bar{\mathbf{Z}} - (\bar{\mathbf{Z}} - \boldsymbol{\theta}^*)'\boldsymbol{\Sigma}^{-1}(\bar{\mathbf{Z}} - \boldsymbol{\theta}^*)]. \tag{4.6.6}$$

Since $\boldsymbol{\theta}^*$ is the projection of $\bar{\mathbf{Z}}$ onto a linear subspace, it follows that

$$\bar{\mathbf{Z}}'\boldsymbol{\Sigma}^{-1}\bar{\mathbf{Z}} = (\bar{\mathbf{Z}} - \boldsymbol{\theta}^*)'\boldsymbol{\Sigma}^{-1}(\bar{\mathbf{Z}} - \boldsymbol{\theta}^*) + \boldsymbol{\theta}^{*\prime}\boldsymbol{\Sigma}^{-1}\boldsymbol{\theta}^*$$

so that the test statistic may be written as

$$\bar{\chi}^2 = n\boldsymbol{\theta}^{*\prime}\boldsymbol{\Sigma}^{-1}\boldsymbol{\theta}^*. \tag{4.6.7}$$

Also, (4.6.7) can be expressed in terms of the norm associated with the inner product $\langle \mathbf{x}, \mathbf{y} \rangle_{\Sigma^{-1}} = \mathbf{x}'\boldsymbol{\Sigma}^{-1}\mathbf{y}$ as follows: $\bar{\chi}^2 = n\| \boldsymbol{\theta}^* \|^2_{\Sigma^{-1}}$, or, since $\boldsymbol{\theta}^* = \bar{\mathbf{Z}} - \boldsymbol{\theta}^{**}$, where $\boldsymbol{\theta}^{**}$ is the projection of $\bar{Z}$ onto the dual of $\mathscr{K}$,

$$\bar{\chi}^2 = n\| \bar{\mathbf{Z}} - \boldsymbol{\theta}^{**} \|_{\Sigma^{-1}}.$$

Proofs of the following result for $\mathscr{K}$, the positive orthant of R^p, may be found in Kudô (1963) or Nüesch (1964, 1966). A proof for $\mathscr{K}$ an arbitrary, closed, convex cone is given in Shapiro (1985).

Theorem 4.6.1 *If H_0 is true, if $\boldsymbol{\Sigma}$ is known and nonsingular, and if $\mathscr{K}$ is an arbitrary closed, convex cone in R^p, then for any real number t,*

$$P[\bar{\chi}^2 \geqslant t] = \sum_{j=0}^{p} Q(j, p) P[\chi_j^2 \geqslant t]$$

where the weights $Q(j, p)$, $j = 0, 1, 2, \ldots, p$, sum to one and χ_j^2 is a standard chi-squared variable with j degrees of freedom ($\chi_0^2 \equiv 0$).

The similarity between the roles played by the level probabilities discussed in Chapters 2 and 3 and the $Q(l, p)$ in Theorem 4.6.1 is striking. The weight $Q(l, p) = Q(l, p; \boldsymbol{\Sigma}^{-1}, \mathscr{K})$ depends upon both the matrix $\boldsymbol{\Sigma}^{-1}$ and the cone $\mathscr{K}$. For the pairs $(\boldsymbol{\Sigma}^{-1}, \mathscr{K})$ corresponding to some of the testing problems discussed in Chapter 2, these weights have been researched extensively and satisfactory computational algorithms have been derived. If the cone, $\mathscr{K}$, is polyhedral then $Q(l, p)$ is the sum of the probabilities that $\boldsymbol{\theta}^*$ belongs to the various l-dimensional faces of $\mathscr{K}$ (see Shapiro, 1985). Thus, for example, if $\mathscr{K}$ is an orthant and if $\boldsymbol{\Sigma}^{-1}$ is diagonal, then these $Q(l, p)$ are symmetric binomial probabilities (cf. Section 5.3). Shapiro (1985) notes that $Q(p - l, p; \boldsymbol{\Sigma}^{-1}, \mathscr{K}) = Q(l, p; \boldsymbol{\Sigma}^{-1}, \mathscr{K}^{*\Sigma^{-1}})$ where $\mathscr{K}^{*\Sigma^{-1}}$ is the Fenchel dual of $\mathscr{K}$ with respect to the inner product $\langle \mathbf{x}, \mathbf{y} \rangle_{\Sigma^{-1}}$ on R^p. Shapiro (1985) gives formulae for the $Q(l, p; \boldsymbol{\Sigma}^{-1}, \mathscr{K})$ when $p \leqslant 4$ and $\mathscr{K}$ is the positive orthant. Other special cases have been considered by Gourieroux, Holly, and Monfort (1982).

Similar methodology is available for more general testing situations. For example, suppose that H_0 restricts $\boldsymbol{\theta}$ to lie within a subspace, $\mathscr{S}$, properly contained in $\mathscr{K}$ (of course $\mathscr{S} = \{\mathbf{0}\}$ is a special case), i.e. one might test

$$H_0 : \boldsymbol{\theta} \in \mathscr{S} \qquad \text{versus} \qquad H_1 : \boldsymbol{\theta} \in \mathscr{K} - \mathscr{S}.$$

This implies that the cone $\mathscr{K}$ can be written as the direct sum of $\mathscr{S}$ and $\mathscr{K} \cap \mathscr{S}^{\perp}$, where $\mathscr{S}^{\perp}$ is the orthogonal complement of $\mathscr{S}$. Since these regions are orthogonal, the projection of $\bar{\mathbf{Z}}$ onto $\mathscr{K}$ can be written as the sum of $\hat{\boldsymbol{\theta}}$, the projection of $\mathbf{Z}$ onto $\mathscr{S}$ and $\tilde{\boldsymbol{\theta}}$, the projection of $\bar{Z}$ onto $\mathscr{K} \cap \mathscr{S}^{\perp}$. From this it follows that the likelihood ratio statistic will be of the form

$$\|\bar{\mathbf{Z}} - \hat{\boldsymbol{\theta}}\|^2_{\Sigma^{-1}} - \|\bar{\mathbf{Z}} - \boldsymbol{\theta}^*\|^2_{\Sigma^{-1}} = \|\bar{\mathbf{Z}} - \tilde{\boldsymbol{\theta}}\|^2_{\Sigma^{-1}}.$$

This statistic also has a chi-bar-square distribution, but the coefficient of $P[\chi^2_j \geqslant t]$ will now be the probability that $\boldsymbol{\theta}^*$ lies in a $(j+m)$-dimensional face of $\mathscr{K}$ where m is the dimension of $\mathscr{S}$ (assuming $\mathscr{K}$ is polyhedral).

Another variation is to define the null hypothesis to be that $\boldsymbol{\theta}$ lies within the closed, convex cone $\mathscr{K}$, i.e. test

$$H_0 : \boldsymbol{\theta} \in \mathscr{K} \qquad \text{versus} \qquad H_1 : \boldsymbol{\theta} \in R^p - \mathscr{K}.$$

In this situation, the likelihood ratio statistic is of the form

$$(\bar{\mathbf{Z}} - \boldsymbol{\theta}^*)'\boldsymbol{\Sigma}^{-1}(\bar{\mathbf{Z}} - \boldsymbol{\theta}^*) = \|\bar{\mathbf{Z}} - \boldsymbol{\theta}^*\|^2_{\Sigma^{-1}},$$

or equivalently of the form $\|\boldsymbol{\theta}^{**}\|^2_{\Sigma^{-1}}$, where $\boldsymbol{\theta}^{**}$ is the projection of $\bar{\mathbf{Z}}$ onto $\mathscr{K}^{*\Sigma^{-1}}$.

Using techniques like those employed in Chapter 2, it can be shown that the least favorable distribution occurs at $\boldsymbol{\theta} = \mathbf{0}$. At $\boldsymbol{\theta} = \mathbf{0}$, the LRT statistic has a chibar-square distribution with coefficients given by

$$Q(l, p; \boldsymbol{\Sigma}^{-1}, \mathscr{K}^{*\Sigma^{-1}}) = Q(p - l, p; \boldsymbol{\Sigma}^{-1}, \mathscr{K}).$$

Further research on the mixing coefficients, $Q(l, p)$, and on computational algorithms for $\boldsymbol{\theta}^*$ is needed before these results can receive widespread application.

$\boldsymbol{\Sigma} = \sigma^2 \boldsymbol{\Sigma}_0$ with $\boldsymbol{\Sigma}_0$ known

In many applications the covariance matrix $\boldsymbol{\Sigma}$ is unknown but is believed to be of the form $\sigma^2\boldsymbol{\Sigma}_0$ with $\boldsymbol{\Sigma}_0$ known but σ^2 unknown (cf. Shorack, 1967, and Barlow *et al.*, 1972, p. 178). In this setting, the LRT is related to the $\bar{E}^2$ test of the previous chapter. This LRT of $H_0 : \boldsymbol{\theta} = \mathbf{0}$ versus $H_1 - H_0$ with $H_1 : \boldsymbol{\theta} \in \mathscr{K}$ rejects H_0 for large values of the statistic

$$\bar{E}^2 = \frac{n\boldsymbol{\theta}^{*\prime}\boldsymbol{\Sigma}_0^{-1}\boldsymbol{\theta}^*}{\sum_{i=1}^n \mathbf{Z}_i'\boldsymbol{\Sigma}_0^{-1}\mathbf{Z}_i}$$

where $\boldsymbol{\theta}^*$ solves

$$\min_{\boldsymbol{\theta} \in \mathscr{K}} (\bar{\mathbf{Z}} - \boldsymbol{\theta})'\boldsymbol{\Sigma}_0^{-1}(\bar{\mathbf{Z}} - \boldsymbol{\theta}).$$

The null hypothesis distribution of $\bar{E}^2$ is given in the following theorem for an important special case. (See Barlow *et al.*, 1972, and Theorem 3 of Shorack, 1967.)

Theorem 4.6.2 *If H_0 is true and if $\mathscr{K}$ is the positive orthant, then for each real number t,*

$$P[\bar{E}^2 \geqslant t] = \sum_{l=0}^{p} Q(l, p) P[B_{j/2, (np-j)/2} \geqslant t]$$

where $Q(l, p)$ is defined as in the discussion after Theorem 4.6.1 and $B_{a,b}$ denotes a random variable having a beta distribution with parameters a and b ($B_{0,b} \equiv 0$).

The above problem can also be generalized to the case where H_0 specifies that $\boldsymbol{\theta}$ lies within a subspace properly contained in the closed, convex cone and H_1 requires $\boldsymbol{\theta}$ to lie within $\mathscr{K}$. In this situation, the distribution of the LRT statistic is also a mixture of beta distributions if H_0 is true, but the parameters of the beta distribution may take on different values than those given in Theorem 4.6.2.

One can also construct the LRT for testing the null hypothesis that $\boldsymbol{\theta}$ lies within a convex cone. As would be expected, the distribution of the LRT statistic under the least favorable distribution is again a mixture of beta distributions.

Other variations are possible. For example, problems need not be couched in terms of random samples, since the central idea deals with cone restrictions on the mean of a multivariate normal distribution. For instance, hypotheses testing problems involving order constraints for normal regression models can be handled in a similar fashion.

$\boldsymbol{\Sigma}$ completely unknown

Perlman (1969) has studied tests of restricted hypotheses in a very general framework which includes the problems discussed in this section. He makes no assumption regarding the matrix $\boldsymbol{\Sigma}$ other than it be positive definite. Assume that $n \geqslant p+1$ so that the estimate, $\mathbf{S} = n^{-1}\sum_{i=1}^{n}(\mathbf{Z}_i - \bar{\mathbf{Z}})(\mathbf{Z}_i - \bar{\mathbf{Z}})'$ of $\boldsymbol{\Sigma}$ is positive definite with probability one. The estimate $\boldsymbol{\theta}^*$ of $\boldsymbol{\theta}$ is the vector in $\mathscr{K}$ which minimizes $(\bar{\mathbf{Z}} - \boldsymbol{\theta})'\mathbf{S}^{-1}(\bar{\mathbf{Z}} - \boldsymbol{\theta})$ and the LRT of $H_0\!:\!\boldsymbol{\theta} = \mathbf{0}$ with alternative $H_1\!:\!\boldsymbol{\theta} \in \mathscr{K}$ rejects H_0 for large values of the test statistic

$$U = \frac{\boldsymbol{\theta}^{*\prime}\mathbf{S}^{-1}\boldsymbol{\theta}^*}{1 + (\bar{\mathbf{Z}} - \boldsymbol{\theta}^*)'\mathbf{S}^{-1}(\bar{\mathbf{Z}} - \boldsymbol{\theta}^*)}.$$

The null hypothesis distribution of U depends upon $\boldsymbol{\Sigma}$ and Perlman gives sharp upper and lower bounds for those distributions. Critical values for the test of H_0 against $H_1 - H_0$ based upon U can be obtained from these bounds.

Theorem 4.6.3 *If H_0 is true then for any real number t,*

$$\inf_{\Sigma > 0} P[U \geqslant t] = \frac{1}{2} P\left[\frac{\chi_1^2}{\chi_{n-p}^2} \geqslant t\right]$$

and if $\mathscr{K}$ contains a p-dimensional open set then

$$\sup_{\Sigma > 0} P[U \geqslant t] = \frac{1}{2} P\left[\frac{\chi_{p-1}^2}{\chi_{n-p}^2} \geqslant t\right] + \frac{1}{2} P\left[\frac{\chi_p^2}{\chi_{n-p}^2} \geqslant t\right]$$

where $\chi_1^2, \chi_{p-1}^2, \chi_p^2, \chi_{n-p}^2$ denote independent chi-square variates.

Because this null hypothesis distribution varies over such a wide range as a function of $\mathbf{\Sigma}$, the test based upon U has low power over large portions of the alternative region. In fact, as Perlman notes, Theorem 4.6.3 implies that this test is not unbiased for testing H_0 against $H_1 - H_0$ (assuming $\mathscr{K}$ contains an open set). Hotelling's T^2-test of H_0 against $\boldsymbol{\theta} \neq \mathbf{0}$, which rejects for large values of $\bar{\mathbf{Z}}'\mathbf{S}^{-1}\bar{\mathbf{Z}}$, is unbiased so that in some subregions of the alternative, U does not compare very favorably with the power of this unrestricted test. One possible explanation for this is that the bounds given in Theorem 4.6.3 do not even depend upon $\mathscr{K}$.

Perlman (1969) also discusses the problem of testing $\boldsymbol{\theta} \in \mathscr{K}_1$ against the alternative that $\boldsymbol{\theta} \in \mathscr{K}_2 - \mathscr{K}_1$, where $\mathscr{K}_1$ and $\mathscr{K}_2$ are closed, convex cones in R^p. This of course includes the problem of testing $\boldsymbol{\theta} \in \mathscr{K}$ versus $\boldsymbol{\theta} \in R^p - \mathscr{K}$.

Ordered tests for the general linear hypothesis

In Section 2.2 it was shown that the $\bar{E}^2_{01}$ test could be used in the analysis of orthogonal designs to provide a test of the equality of main effects against the alternative that they are ordered. The method used there breaks down if the design is not orthogonal. A more general approach, related to the multivariate problem of this section, removes this restriction and is more widely applicable. This approach will be described in the context of the general linear model and, as an example, we shall indicate how it may be applied to the two-way classification with unequal frequencies.

The model used is well known. A vector $\mathbf{Y}$ denotes the observations, $Y_1, Y_2, \ldots, Y_N$, and the joint distribution of the observations is given by

$$\mathbf{Y} = \mathbf{X}\boldsymbol{\theta} + \mathbf{e},$$

where $\mathbf{X}$ is a known matrix of order $N \times p$ and rank p, $\boldsymbol{\theta} = (\theta_1, \theta_2, \ldots, \theta_p)'$ is a vector of unknown parameters, and $\mathbf{e}$ is a vector of independent random variables, each having a normal distribution with mean zero and variance σ^2. On the basis of the data, $\mathbf{Y}$, it is desired to test the null hypothesis that the first $q (1 \leqslant q \leqslant p)$ of the θ_i are zero against the alternative that at least one of these parameters is nonzero and the first r of them $(1 \leqslant r \leqslant q)$ are nonnegative. It should be noted that the majority of problems likely to be encountered involving ordered hypotheses in linear models will not appear in this form when expressed in terms of the natural parametrization of the model. However, nearly all such problems, whether the model is of full rank or not, may be expressed in this form after suitable reparametrization. The only exceptions occur when the number of inequality restrictions imposed by the alternative hypothesis exceeds the number of independent constraints imposed by the null hypothesis. Then a transformation to the above form cannot be made; however, as has already been remarked, the methods of this section readily extend to such situations.

In what follows it will be found convenient to partition $\boldsymbol{\theta}$ into two components, $\boldsymbol{\theta}_1 = (\theta_1, \theta_2, \ldots, \theta_q)'$ and $\boldsymbol{\theta}_2 = (\theta_{q+1}, \theta_{q+2}, \ldots, \theta_p)'$. Thus the argument assumes that $q < p$; a simplified version of this argument shows that the same results are valid for $q = p$. The null hypothesis is now

$$H_0 : \boldsymbol{\theta}_1 = \mathbf{0},$$

and the alternative hypothesis is

$$H_1 : \boldsymbol{\theta}_1 \neq \mathbf{0},\ \theta_i \geqslant 0,\ i = 1, 2, \ldots, r.$$

Following the earlier conventions, the notation H_0 will also be used to denote the region $\boldsymbol{\theta}_1 = \mathbf{0}$ and the notation H_1 to denote the region $\theta_i \geqslant 0$, $i = 1, 2, \ldots, r$.

First suppose that σ^2 is known. The likelihood ratio test then rejects H_0 for large values of

$$\min_{\boldsymbol{\theta} \in H_0} S - \min_{\boldsymbol{\theta} \in H_1} S, \tag{4.6.8}$$

where $S = (\mathbf{Y} - \mathbf{X}\boldsymbol{\theta})'(\mathbf{Y} - \mathbf{X}\boldsymbol{\theta})$. Now it is a familiar result of least squares theory that

$$S = (\mathbf{Y} - \mathbf{X}\hat{\boldsymbol{\theta}})'(\mathbf{Y} - \mathbf{X}\hat{\boldsymbol{\theta}}) + (\hat{\boldsymbol{\theta}} - \boldsymbol{\theta})'\boldsymbol{\Gamma}(\hat{\boldsymbol{\theta}} - \boldsymbol{\theta}), \tag{4.6.9}$$

where $\boldsymbol{\Gamma} = \mathbf{X}'\mathbf{X}$ and $\hat{\boldsymbol{\theta}} = \boldsymbol{\Gamma}^{-1}\mathbf{X}'\mathbf{Y}$ is the vector of (unrestricted) least squares estimators of the parameters. Further, if $\boldsymbol{\Gamma}$ is partitioned as

$$\boldsymbol{\Gamma} = \begin{bmatrix} \boldsymbol{\Gamma}_{11} & \boldsymbol{\Gamma}_{12} \\ \boldsymbol{\Gamma}_{21} & \boldsymbol{\Gamma}_{22} \end{bmatrix},$$

where $\boldsymbol{\Gamma}_{11}$ is a $q \times q$ matrix, the second term of the right-hand side of (4.6.9) may be rewritten so that

$$\begin{aligned} S = {} & (\mathbf{Y} - \mathbf{X}\hat{\boldsymbol{\theta}})'(\mathbf{Y} - \mathbf{X}\hat{\boldsymbol{\theta}}) + (\hat{\boldsymbol{\theta}}_1 - \boldsymbol{\theta}_1)'(\boldsymbol{\Gamma}_{11} - \boldsymbol{\Gamma}_{12}\boldsymbol{\Gamma}_{22}^{-1}\boldsymbol{\Gamma}_{21})(\hat{\boldsymbol{\theta}}_1 - \boldsymbol{\theta}_1) \\ & + [\boldsymbol{\Gamma}_{22}^{-1}\boldsymbol{\Gamma}_{21}(\hat{\boldsymbol{\theta}}_1 - \boldsymbol{\theta}_1) + \hat{\boldsymbol{\theta}}_2 - \boldsymbol{\theta}_2]^{-1}\boldsymbol{\Gamma}_{22}[\boldsymbol{\Gamma}_{22}^{-1}\boldsymbol{\Gamma}_{21}(\hat{\boldsymbol{\theta}}_1 - \boldsymbol{\theta}_1) + \hat{\boldsymbol{\theta}}_2 - \boldsymbol{\theta}_2]. \end{aligned} \tag{4.6.10}$$

Of the three terms in this expression, the first does not involve $\boldsymbol{\theta}$ and the minimum of the third with respect to $\boldsymbol{\theta}_2$ is zero irrespective of the value of $\boldsymbol{\theta}_1$. Thus the test statistic (4.6.8) becomes

$$\begin{aligned} & \min_{\boldsymbol{\theta} \in H_0} (\hat{\boldsymbol{\theta}}_1 - \boldsymbol{\theta}_1)'\boldsymbol{\Lambda}_{11}^{-1}(\hat{\boldsymbol{\theta}}_1 - \boldsymbol{\theta}_1) - \min_{\boldsymbol{\theta} \in H_1} (\hat{\boldsymbol{\theta}}_1 - \boldsymbol{\theta}_1)'\boldsymbol{\Lambda}_{11}^{-1}(\hat{\boldsymbol{\theta}}_1 - \boldsymbol{\theta}_1) \\ & \quad = \hat{\boldsymbol{\theta}}_1'\boldsymbol{\Lambda}_{11}^{-1}\hat{\boldsymbol{\theta}}_1 - (\hat{\boldsymbol{\theta}}_1 - \boldsymbol{\theta}_1^*)'\boldsymbol{\Lambda}_{11}^{-1}(\hat{\boldsymbol{\theta}}_1 - \boldsymbol{\theta}_1^*) \end{aligned} \tag{4.6.11}$$

where $\boldsymbol{\Lambda}_{11} = (\boldsymbol{\Gamma}_{11} - \boldsymbol{\Gamma}_{12}\boldsymbol{\Gamma}_{22}^{-1}\boldsymbol{\Gamma}_{21})^{-1}$ and $\boldsymbol{\theta}_1^*$ is the value of $\boldsymbol{\theta}_1$ achieving the second minimization above. As $\hat{\boldsymbol{\theta}}_1$ is distributed with covariance matrix $\sigma^2\boldsymbol{\Lambda}_{11}$, comparison of (4.6.11) and (4.6.6) shows that the test procedure amounts to ignoring $\hat{\boldsymbol{\theta}}_2$ and performing a test of the type discussed in the earlier parts of this section, based on the single observation $\hat{\boldsymbol{\theta}}_1$. Thus the test rejects H_0 for large values of

$$\bar{\chi}^2 = \frac{\boldsymbol{\theta}_1^{*\prime}\boldsymbol{\Lambda}_{11}^{-1}\boldsymbol{\theta}_1^*}{\sigma^2}, \tag{4.6.12}$$

the null hypothesis distribution of this statistic being as in Theorem 4.6.1 with q substituted for p.

When σ^2 is unknown the likelihood ratio test of H_0 against H_1 rejects the null hypothesis for large values of

$$\bar{E}^2 = \frac{\boldsymbol{\theta}_1^{*\prime}\boldsymbol{\Lambda}_{11}^{-1}\boldsymbol{\theta}_1^*}{\hat{\boldsymbol{\theta}}_1'\boldsymbol{\Lambda}_{11}^{-1}\hat{\boldsymbol{\theta}}_1 + (\mathbf{Y}-\mathbf{X}\hat{\boldsymbol{\theta}})'(\mathbf{Y}-\mathbf{X}\hat{\boldsymbol{\theta}})}. \tag{4.6.13}$$

The null hypothesis distribution of this statistic is given by

$$P[\bar{E}^2 \geqslant C] = \sum_{j=1}^{q} Q(j,p)P[B_{j/2,(N-p+q-j)/2} \geqslant C] \text{ for } C>0 \tag{4.6.14}$$

and $P[\bar{E}^2 = 0] = Q(0,q)$.

The above methods will now be illustrated by applying them to the problem of testing the equality of main effects in the two-way cross-classification against the alternative of an increasing trend. The problem has already been examined in Section 2.2 under the assumption of equal cell frequencies; that restriction is now relaxed. The model is

$$Y_{tij} = \mu + \alpha_t + \beta_i + \gamma_{ti} + e_{tij} \qquad \text{for } t=1,2,\ldots,k;\ i=1,2,\ldots,m;\ j=1,2,\ldots,n_{ti} \tag{4.6.15}$$

where

$$\sum_{t=1}^{k}\alpha_t = \sum_{i=1}^{m}\beta_i = \sum_{t=1}^{k}\gamma_{ti} = \sum_{i=1}^{m}\gamma_{ti} = 0, \tag{4.6.16}$$

and the e's are uncorrelated normal random variables with zero means and the same (unknown) variance. A brief description will be given of the LRTs of the null hypothesis

$$\alpha_1 = \alpha_2 = \cdots \alpha_k$$

against the simple order alternative

$$\alpha_1 \leqslant \alpha_2 \leqslant \cdots \leqslant \alpha_k,$$

with at least one inequality strict. Suppose that we are prepared to postulate an additive model, i.e. to delete the γ's from (4.6.15). If the reparameterization given by

$$\begin{aligned} \theta_i &= \alpha_{i+1} - \alpha_i, & i &= 1,2,\ldots,k-1 \\ \theta_{k-1+i} &= \beta_{i+1} - \beta_i, & i &= 1,2,\ldots,m-1 \\ \theta_{k+m-1} &= \mu \end{aligned} \tag{4.6.17}$$

is made, the problem is exactly in the form that has been described, with $p = k + m - 1$, $q = r = k - 1$. The appropriate test statistic is of the form (4.6.13). Its null hypothesis distribution is as given by (4.6.14), with $q = k - 1$ and its second parameter of the beta distribution replaced by $(N - m - j)/2$.

Barlow *et al.* (1972, pp. 182–3) gives a further discussion of these ideas, considering models with possible interaction terms.

Other tests

The discussion given here has primarily been concerned with likelihood ratio tests. Other tests are, of course, possible. Tests of the Wald type and Lagrange multiplier tests are two possibilities for alternative tests (cf. Gourieroux, Holly, and Monfort, 1982). However, they are identical to the LRT if Σ is known and asymptotically equivalent otherwise. While the small sample properties are different for unknown Σ, none seems to dominate the others. The topic of Lagrange multiplier tests is discussed further in Chapter 6.

Schaafsma (1966), Schaafsma and Smid (1966), and Perlman (1969) discuss the possibility of using an optimal contrast test for testing with restricted alternatives in a multivariate setting. Further results along these lines can be found in Shi and Kudô (1986) and Shi (1986).

4.7 COMPLEMENTS

Likelihood ratio tests for testing homogeneity of a collection of parameters in the generality of exponential families seems to have been first considered by Boswell and Brunk (1969) although research on estimation in this context appeared earlier. Boswell and Brunk's characterization of an exponential family differs somewhat from the form hypothesized in (4.1.1) and their work focuses on the simple order restriction as an alternative. Of course, Chernoff (1954) and later Feder (1968) consider properties of LRTs and found rudimentary forms of the chi-bar-square distribution. Robertson and Wegman (1978) give the first part of Theorem 4.1.1 and this result holds for alternatives which place partial order restrictions on the parameters. As discussed in Chapter 2, for the case in which populations are normal, Bartholomew (1959a, 1959b, 1961a) found the distribution of the LRT statistic. Boswell (1966) considered testing problems for stochastic processes of the Poisson type.

Chotai (1983) considered likelihood ratio tests for testing order restrictions on the endpoints of uniform $(0, \theta_i)$ distributions. He was able to find nice expressions for the restricted maximum likelihood estimators. If Λ is the likelihood ratio statistic, he showed that $-2 \ln \Lambda$ is distributed *exactly* as a chi-bar-square distribution. Notably, his derivation of this distribution is based on the lack of memory property of the exponential distribution rather than conditional normal theory. Chotai also found some interesting relationships among the weights of the chi-bar-square distributions. His results also extend to other distributions for which the largest order statistics are sufficient statistics.

The problem of testing an order restriction as a null hypothesis was considered by Eeden (1958) who proposed a MCT as discussed in Section 4.3. Robertson and Wegman (1978) obtain the second conclusion in Theorem 4.1.1.

Contrast tests have a long history in the theory of statistics. The idea of using one of these tests to test against an order restricted alternative seems to have been first considered by Abelson and Tukey (1963) although, as mentioned at the beginning of Section 4.2, a number of researchers seem to have been thinking about this idea at about the same time.

A disadvantage of the optimal contrast test for a simple order is that the contrast coefficients lack the simplicity of the 'equally spaced' version given in (4.2.4). In an attempt to combine simplicity with near optimality, Abelson and Tukey (1963) proposed the use of 'linear-2-4' scores. These are obtained from the equally spaced scores by doubling the scores next to the end and quadrupling those at the end. Thus for $k = 11$ the contrast vector would be $(-20, -8, -3, -2, -1, 0, 1, 2, 3, 8, 20)$. For $k \leqslant 20$ these contrasts have a minimum of r which are at least 97.5 per cent of the value for the maximin r. Schaafsma and Smid (1966) and Schaafsma (1968) study optimal contrast test analogues to $\bar{\chi}^2_{01}$ for the simply ordered case with unequal sample sizes. Snidjers (1979) considers asymptotically optimal tests for order restricted hypotheses. The asymptotically optimal tests in these problems are also contrast tests.

The relationship between the LRT and contrast tests given in Theorem 4.2.1 was first pointed out in Bartholomew (1961b) and again by Hogg (1965). This last paper also raised the important question of what happens if the prior information on which an order restriction is based proves to be wrong. As has been remarked elsewhere, it is desirable that an ordered restricted test should have reasonable power if the true alternative lies outside the restricted parameter space. Order restrictions which are somewhat less restrictive than simple order restrictions and yet retain much of the same flavor are considered in Dykstra and Robertson (1983).

Multiple contrast tests also have a long history in statistics but, as mentioned at the beginning of Section 4.3, most seem to have been proposed on an *ad hoc* basis. Dunnett's test (1955) for testing against a simple tree alternative is surely the best known and most widely used. The paper by Mukerjee, Robertson, and Wright (1987) develops these multiple contrast tests in a unified context.

Chacko (1963) introduced a distribution-free version of the $\bar{\chi}^2$ test. To apply Chacko's test, one ranks the combined samples and replaces each sample mean in $\bar{\chi}^2$ by the average, over the corresponding sample, of the ranks. His work was extended by Shorack (1967). Terpstra (1952) and Jonckheere (1954a) developed a test for simply ordered alternatives. Their test statistic is the sum, over all pairs (i, j) with $i < j$, of the Mann–Whitney test statistic for comparing the ith and jth samples. This approach is easily extended to other orderings. There has been a vigorous growth of the literature concerning nonparametric tests for ordered alternatives, much of which is mentioned in Section 4.5.

The tests for the multivariate normal mean described in Section 4.6 were originally developed by Kudô (1963) and Nüesch (1964, 1966). Perlman (1969) considered testing hypotheses of the form $\boldsymbol{\mu} \in \mathscr{K}$, with $\mathscr{K}$ a closed, convex cone and an arbitrary nonsingular covariance matrix. There is substantial literature on this problem and much of this research is discussed in Section 4.6.

CHAPTER 5

Inferences about a Set of Multinomial Parameters

5.1 MAXIMUM LIKELIHOOD ESTIMATES SUBJECT TO A QUASI-ORDER RESTRICTION

As in Example 1.5.7, suppose the outcome of an experiment must be one of k mutually exclusive events having probabilities $p_1, p_2, \ldots, p_k$. Assume that n independent trials of this experiment are performed and that $\hat{p}_i$ is the relative frequency of the event having probability p_i. Suppose $\lesssim$ is a quasi-order on $\{1, 2, \ldots, k\}$. It is shown in Example 1.5.7 that the maximum likelihood estimate of the vector $\mathbf{p}$ of probabilities subject to the constraint that the estimate be isotonic with respect to this partial order is the isotonic regression, $\hat{\mathbf{p}}^*$, of the vector $\hat{\mathbf{p}}$ with weights $w_i = 1$, $i = 1, 2, \ldots, k$. It follows from (1.6.6) that

$$\max_{1 \leqslant i \leqslant k} |\hat{p}_i^* - p_i| \leqslant \max_{1 \leqslant i \leqslant k} |\hat{p}_i - p_i| \tag{5.1.1}$$

provided $\mathbf{p}$ is isotonic. This inequality together with well-known consistency properties of $\hat{p}_i$ as an estimator of p_i can be used to derive consistency properties of $\hat{\mathbf{p}}^*$. For example, the law of large numbers implies that $\hat{p}_i$ converges almost surely to p_i as $n \to \infty$ $(i = 1, 2, \ldots, k)$. Thus, $\hat{p}_i^*$ converges almost surely to p_i as $n \to \infty$ $(i = 1, 2, \ldots, k)$.

More precise results can be obtained using the law of the iterated logarithm. It follows from Kolmogorov's law of the iterated logarithm that, with probability one,

$$\limsup_{n \to \infty} \left(\frac{n}{\log \log n} \right)^{1/2} |\hat{p}_i - p_i| = [2p_i(1 - p_i)]^{1/2}$$

for $i = 1, 2, \ldots, k$. Hence, using (5.1.1),

$$\limsup_{n \to \infty} \left(\frac{n}{\log \log n} \right)^{1/2} \max_{1 \leqslant i \leqslant k} |\hat{p}_i^* - p_i| \leqslant \max_{1 \leqslant i \leqslant k} [2p_i(1 - p_i)]^{1/2} \tag{5.1.2}$$

with probability one. Choose p so that $[p(1-p)]^{1/2} = \max_{1 \leqslant i \leqslant k}[p_i(1-p_i)]^{1/2}$ and for each nonempty subset A of $\{1, 2, \ldots, k\}$ let $\mathrm{Av}(A)$ denote the average value of $\hat{p}$ for subscripts in A. Consider the upper sets $L_1(p) = \{i: p_i \geqslant p\}$ and $L_2(p) = \{i: p_i > p\}$ so that $L_1(p) - L_2(p) = \{i: p_i = p\}$. It can be shown by induction that there is a point $i_0 \in L_1(p) - L_2(p)$ and an upper layer $L_0 \subset L_1(p)$ such that $L_0 \cap [L_1(p) - L_2(p)] = \{i_0\}$. Hence, by Theorem 1.4.4, $\hat{p}^*_{i_0} \geqslant \min_{i_0 \notin L'} \mathrm{Av}(L_0 - L')$. However, by the strong law of large numbers there is a set of ω's with probability one such that for $n \geqslant N(\omega)$, $\hat{p}_i > \hat{p}_{i_0}$ for all $i \in L_2(p)$. Using the averaging property, $\hat{p}^*_{i_0} \geqslant \hat{p}_{i_0}$ for $n \geqslant N(\omega)$ and ω in this sure event. Hence

$$
\begin{aligned}
&P\left[\limsup_{n\to\infty}\left(\frac{n}{\log\log n}\right)^{1/2} \max_{1\leqslant i\leqslant k} |\hat{p}_i^* - p_i| \geqslant [2p(1-p)]^{1/2}\right] \\
&\geqslant P\left[\limsup_{n\to\infty}\left(\frac{n}{\log\log n}\right)^{1/2} \max_{1\leqslant i\leqslant k} (\hat{p}_i^* - p_i) \geqslant [2p(1-p)]^{1/2}\right] \\
&\geqslant P\left[\limsup_{n\to\infty}\left(\frac{n}{\log\log n}\right)^{1/2} (\hat{p}_{i_0}^* - p_{i_0}) \geqslant [2p(1-p)]^{1/2}\right] \\
&\geqslant P\left[\limsup_{n\to\infty}\left(\frac{n}{\log\log n}\right)^{1/2} (\hat{p}_{i_0} - p_{i_0}) \geqslant [2p(1-p)]^{1/2}\right] \\
&= 1.
\end{aligned}
$$

Combining this with (5.1.1) we obtain

$$
\limsup_{n\to\infty}\left(\frac{n}{\log\log n}\right)^{1/2} \max_{1\leqslant i\leqslant k} |\hat{p}_i^* - p_i| = \max_{1\leqslant i\leqslant k} [2p_i(1-p_i)]^{1/2} \tag{5.1.3}
$$

with probability one.

If $Y_1, Y_2, \ldots, Y_k$ are independent Poisson random variables with means $\mu_1, \mu_2, \ldots, \mu_k$, respectively, then the distribution of $\mathbf{Y} = (Y_1, Y_2, \ldots, Y_k)$ given $\sum_{i=1}^k Y_i = s$ is multinomial with s trials and probability vector $\mathbf{p}$ with $p_i = \mu_i / \sum_{j=1}^k \mu_j$, $i = 1, 2, \ldots, k$. Thus the estimates mentioned here and the tests discussed in the next section provide order restricted inferences for the analogous Poisson problems (cf. Section 4.1).

5.2 TESTS FOR ORDERED ALTERNATIVES

Throughout this section it is assumed that $\lesssim$ is a given quasi-order on $X = \{1, 2, \ldots, k\}$; that $\mathscr{I}$ and $\mathscr{A}$ are the collections of isotonic and antitonic functions on X; that $\mathbf{q}$ is a given possibility for $\mathbf{p}$; and that the necessary assumptions are made to assure that the arguments of the logarithm function are positive and denominators are not zero. Define the hypotheses $H_i: i = 0, 1, 2$ by

$$H_0: \mathbf{p} = \mathbf{q},$$
$$H_1: \mathbf{p} \text{ is isotonic},$$

and H_2 places no restriction on $\mathbf{p}$ other than $p_i \geqslant 0$ and $\sum_{i=1}^{k} p_i = 1$. It is assumed that $\mathbf{q}$ is isotonic so that $H_0 \subset H_1$. As in Chapter 2, the symbol H_i is used to denote a hypothesis and the set of functions on X which satisfy the hypothesis. Thus, for example, $H_1 = H_2 \cap \mathscr{I}$.

Likelihood ratio tests

Let $T_{01} = -2 \ln \Lambda_{01}$ where Λ_{01} is the likelihood ratio for testing H_0 against the alternative $H_1 - H_0$. Theorem 4.2.1, below, gives the asymptotic null hypothesis distribution of T_{01}.

Consider the quasi-order restriction, $\lesssim_{\mathbf{q}}$, induced by $\mathbf{q}$ and $\lesssim$ on X which requires that $i \lesssim_{\mathbf{q}} j$ only when $i \lesssim j$ and $q_i = q_j$. For example, if $k = 4$, $1 \lesssim 2 \lesssim 3 \lesssim 4$, and $\mathbf{q} = (\frac{1}{6}, \frac{1}{6}, \frac{1}{3}, \frac{1}{3})$, then the induced order restriction requires that $1 \lesssim_{\mathbf{q}} 2$ and $3 \lesssim_{\mathbf{q}} 4$. Let $\mathscr{I}_{\mathbf{q}}$ be the set of functions on X isotonic with respect to $\lesssim_{\mathbf{q}}$. Note that $\mathscr{I} \subset \mathscr{I}_{\mathbf{q}}$ regardless of the value of $\mathbf{q}$. Let $\mathbf{Z} = (Z_1, Z_2, \ldots, Z_k)$ where $Z_1, Z_2, \ldots, Z_k$ are independent standard normal variables and for $l = 1, 2, \ldots, k$, let $P_{\mathbf{q}}(l, k)$ denote the probability that $P(\mathbf{Z} | \mathscr{I}_{\mathbf{q}})$, the equal-weights projection of $\mathbf{Z}$ onto $\mathscr{I}_{\mathbf{q}}$, takes on exactly l distinct values. (Notice that the $P_{\mathbf{q}}(l, k)$ are the equal-weights level probabilities for the quasi-order $\lesssim_{\mathbf{q}}$ and that the subscript $\mathbf{q}$ relates to the quasi-order and not the weights in the appropriate projection.)

Theorem 5.2.1 *If H_0 is true, then for any real number t,*

$$\lim_{n \to \infty} P[T_{01} \geqslant t] = \sum_{l=1}^{k} P_{\mathbf{q}}(l, k) P[\chi^2_{l-1} \geqslant t]. \tag{5.2.1}$$

The following lemmas are used in the proof of Theorem 5.2.1. Let $v_1 > v_2 > \cdots > v_h$ be the distinct values among $q_1, q_2, \ldots, q_k$ and define the partition, $S_1, S_2, \ldots, S_h$, of $\{1, 2, \ldots, k\}$ by $S_i = \{j : q_j = v_i\}$, $i = 1, 2, \ldots, h$.

Lemma A *If $\mathbf{u} = (u_1, u_2, \ldots, u_k)$ and $\mathbf{v} = (v_1, v_2, \ldots, v_k)$ are functions on X and if $v_i = v_j$ whenever $q_i = q_j$, then*

$$P_{\mathbf{q}}(\mathbf{u} | \mathscr{I}_{\mathbf{q}}) = P(\mathbf{u} | \mathscr{I}_{\mathbf{q}}) \qquad \textit{and} \qquad P(\mathbf{u} - \mathbf{v} | \mathscr{I}_{\mathbf{q}}) = P(\mathbf{u} | \mathscr{I}_{\mathbf{q}}) - \mathbf{v}.$$

If, in addition, $v_i \geqslant 0$, $i = 1, 2, \ldots, k$, then

$$P(\mathbf{v} \cdot \mathbf{u} | \mathscr{I}_{\mathbf{q}}) = \mathbf{v} \cdot P(\mathbf{u} | \mathscr{I}_{\mathbf{q}}).$$

(We are using the notation $\mathbf{v} \cdot \mathbf{u}$ to denote the vector which is the coordinatewise product of $\mathbf{v}$ and $\mathbf{u}$.)

Proof: This lemma follows by a straightforward application of the following three facts: (1) the projection $P(\cdot | \mathscr{I}_{\mathbf{q}})$ may be found by independently computing its values for subscripts in the sets $S_1, S_2, \ldots, S_h$; (2) for any constant function, $\mathbf{v}$ on X, and any closed convex cone, $\mathscr{K}$, of functions on X which contains the constant functions, $P(\mathbf{u}-\mathbf{v}|\mathscr{K}) = P(\mathbf{u}|\mathscr{K}) - \mathbf{v}$; and (3) if $\mathscr{K}$ is a closed convex cone and $\mathbf{v}$ is a constant function with a nonnegative value, then $P(\mathbf{v}\cdot\mathbf{u}|\mathscr{K}) = \mathbf{v}\cdot P(\mathbf{u}|\mathscr{K})$. The latter two properties are easily verified using the characterization of $P(\cdot|\mathscr{K})$ in Theorem 1.3.2. □

The next lemma is a consequence of Theorem 1.3.4 together with the assumption that $\mathbf{q}\in\mathscr{I}$.

Lemma B *If* $\mathbf{u}$ *is any function on* X *and if*

$$\min_{j\in S_i} u_j > \max_{j\in S_{i+1}} u_j \qquad \textit{for } i = 1, 2, \ldots, h-1,$$

then

$$P(\mathbf{u}|\mathscr{I}) = P(\mathbf{u}|\mathscr{I}_{\mathbf{q}}).$$

Proof of Theorem 5.2.1: The statistic T_{01} is equal to $2\sum_{i=1}^{k} n\hat{p}_i[\ln \hat{p}_i^* - \ln q_i]$. Assuming H_0 is true, using Taylor's theorem with a second-degree remainder term and expanding $\ln \hat{p}_i^*$ and $\ln q_i$ about $\hat{p}_i$, one can write

$$T_{01} = \sum_{i=1}^{k} n\hat{p}_i[\alpha_i^{-2}(q_i - \hat{p}_i)^2 - \beta_i^{-2}(\hat{p}_i^* - \hat{p}_i)^2],$$

where α_i and β_i are random variables converging almost surely to q_i. In fact, with probability one for sufficiently large n, α_i is between q_i and $\hat{p}_i$ and β_i is between $\hat{p}_i^*$ and $\hat{p}_i$. The almost sure convergence of α_i and β_i follows from the discussion in Section 4.1. The first-order terms are zero because

$$\sum_{i=1}^{k} q_i = \sum_{i=1}^{k} \hat{p}_i = \sum_{i=1}^{k} \hat{p}_i^* = 1.$$

Now with probability one for sufficiently large n, $\hat{\mathbf{p}}$ satisfies the hypothesis imposed on $\mathbf{u}$ in Lemma B so that $\hat{\mathbf{p}}^* = P(\hat{\mathbf{p}}|\mathscr{I}) \doteq P(\hat{\mathbf{p}}|\mathscr{I}_{\mathbf{q}})$ where $\doteq$ represents equality for sufficiently large n with probability one. Thus, using Lemma A,

$$T_{01} \doteq \sum_{i=1}^{k} \hat{p}_i\{\alpha_i^{-2}[\sqrt{n}(\hat{p}_i - q_i)]^2 - \beta_i^{-2}[P(\sqrt{n}(\hat{\mathbf{p}} - \mathbf{q})|\mathscr{I}_{\mathbf{q}})_i - \sqrt{n}(\hat{p}_i - q_i)]^2\}.$$

The random vector $\sqrt{n}(\hat{\mathbf{p}} - \mathbf{q})$ converges in law to a singular normal distribution with zero mean and variance-covariance matrix given by $\mathbf{M} = (m_{ij})$ where $m_{ij} = q_i(\delta_{ij} - q_j)$ and $\delta_{ij} = 1$ if $i = j$ and $\delta_{ij} = 0$ if $i \neq j$. Now let $\mathbf{U} = (U_1, U_2, \ldots, U_k)$, where $U_1, U_2, \ldots, U_k$ are independent random variables such that U_i is normal with mean zero and variance q_i^{-1} and let $\bar{U} = \sum_{i=1}^{k} q_i U_i$. The random vector $(q_1(U_1 - \bar{U}), q_2(U_2 - \bar{U}), \ldots, q_k(U_k - \bar{U}))$ has this singular normal distribution. The projection operator $P(\cdot|\mathscr{I}_{\mathbf{q}})$ is continuous and T_{01} is a continuous function of its arguments so that by Theorem 4.4 and Corollary 1 of Theorem 5.1 of

Billingsley (1968), T_{01} converges in law to

$$\sum_{i=1}^{k} q_i^{-1}\{[q_i(U_i-\bar{U})]^2-[P(\mathbf{q}\cdot(\mathbf{U}-\bar{U})|\mathscr{I}_{\mathbf{q}})_i-q_i(U_i-\bar{U})]^2\}.$$

Using Lemma A, the fact that $\mathbf{q}$ is constant on the sets $S_1, S_2, \ldots, S_h$, and some algebra, this is equal to

$$\sum_{i=1}^{k} q_i\{(U_i-\bar{U})^2-[P(\mathbf{U}|\mathscr{I}_{\mathbf{q}})_i-U_i]^2\}.$$

However,

$$\begin{aligned}\sum_{i=1}^{k} q_i(U_i-\bar{U})^2 &= \sum_{i=1}^{k} q_i[U_i-P(\mathbf{U}|\mathscr{I}_{\mathbf{q}})_i]^2\\ &\quad+2\sum_{i=1}^{k} q_i[U_i-P(\mathbf{U}|\mathscr{I}_{\mathbf{q}})_i][P(\mathbf{U}|\mathscr{I}_{\mathbf{q}})_i-\bar{U}]\\ &\quad+\sum_{i=1}^{k} q_i[P(\mathbf{U}|\mathscr{I}_{\mathbf{q}})_i-\bar{U}]^2\end{aligned}$$

and

$$\sum_{i=1}^{k} q_i[U_i-P(\mathbf{U}|\mathscr{I}_{\mathbf{q}})_i]P(\mathbf{U}|\mathscr{I}_{\mathbf{q}})_i=\sum_{i=1}^{k} q_i[U_i-P(\mathbf{U}|\mathscr{I}_{\mathbf{q}})_i]\bar{U}=0$$

using Theorem 1.3.3 since $\mathbf{q}$ is constant on the sets $S_1, S_2, \ldots, S_h$. The distribution of $\sum_{i=1}^{k} q_i[P(\mathbf{U}|\mathscr{I}_{\mathbf{q}})_i-\bar{U}]^2$ is given by Theorem 2.3.1, so that Theorem 5.2.1 follows since the probability that $P(\mathbf{U}|\mathscr{I}_{\mathbf{q}})$ assumes exactly l values is the same as the probability that $P(\mathbf{Z}|\mathscr{I}_{\mathbf{q}})$ assumes l values. (Recall that $P(\cdot|\mathscr{I}_{\mathbf{q}})$ can be computed independently on the sets $S_1, S_2, \ldots, S_h$.) □

Note that the distribution on the right-hand side of (5.2.1) is the distribution of $\sum_{i=1}^{k}[P(\mathbf{Z}|\mathscr{I}_{\mathbf{q}})_i-\bar{Z}]^2$ and depends upon $\mathbf{q}$ only through the sets on which $\mathbf{q}$ is constant. Moreover, the probability in (5.2.1) is in a sense an increasing function of the number of distinct values in $\mathbf{q}$. Specifically, if $\mathbf{q}$ and $\mathbf{r}$ both satisfy H_1 and if $r_i=r_j$ for all i and j such that $q_i=q_j$, then $\mathscr{I}_{\mathbf{q}}\supset\mathscr{I}_{\mathbf{r}}$. Thus, $\sum_{i=1}^{k}[Z_i-P(\mathbf{Z}|\mathscr{I}_{\mathbf{q}})_i]^2\leqslant\sum_{i=1}^{k}[Z_i-P(\mathbf{Z}|\mathscr{I}_{\mathbf{r}})_i]^2$, and since

$$\sum_{i=1}^{k}[Z_i-\bar{Z}]^2=\sum_{i=1}^{k}[Z_i-P(\mathbf{Z}|\mathscr{I}_{\mathbf{q}})_i]^2+\sum_{i=1}^{k}[P(\mathbf{Z}|\mathscr{I}_{\mathbf{q}})_i-\bar{Z}]^2$$

and this equality is valid if $\mathscr{I}_{\mathbf{q}}$ is replaced by $\mathscr{I}_{\mathbf{r}}$, it can be concluded that $\sum_{i=1}^{k}[P(\mathbf{Z}|\mathscr{I}_{\mathbf{q}})_i-\bar{Z}]^2\geqslant\sum_{i=1}^{k}[P(\mathbf{Z}|\mathscr{I}_{\mathbf{r}})_i-\bar{Z}]^2$. It follows that

$$\sum_{l=1}^{k} P_{\mathbf{q}}(l,k)P[\chi_{l-1}^2\geqslant t]\geqslant\sum_{l=1}^{k} P_{\mathbf{r}}(l,k)P[\chi_{l-1}^2\geqslant t].$$

The two extreme cases are of interest. If $q_i = k^{-1}$, $i = 1, 2, \ldots, k$, then $\mathscr{I}_{\mathbf{q}} = \mathscr{I}$ and $P_{\mathbf{q}}(l, k), l = 1, 2, \ldots, k$, are the equal-weights level probabilities for the quasi-order $\lesssim$. They, of course, depend upon $\mathscr{I}$. In this case $\sum_{l=1}^{k} P_{\mathbf{q}}(l, k) P[\chi_{l-1}^2 \geqslant t]$ is the smallest possible value for any $\mathbf{q}$. On the other extreme, if H_1 imposes no equality constraints so that there exists a $\mathbf{q}$ in H_1 having k distinct values then, for this $\mathbf{q}$, $\mathscr{I}_{\mathbf{q}} = R^k$ and $P(\mathbf{Z} | \mathscr{I}_{\mathbf{q}}) = \mathbf{Z}$. In this case $P_{\mathbf{q}}(l, k) = 0$, $l = 1, 2, \ldots, k - 1$, and $P_{\mathbf{q}}(k, k) = 1$. Thus the asymptotic distribution of T_{01} is the standard chi-square distribution with $k - 1$ degrees of freedom. Note that in this latter case $\mathbf{q}$ is in the interior of $\mathscr{I}$.

Example 5.2.1 Mendel classified 556 peas which resulted from crosses of plants which issued from round yellow seeds and plants which issued from wrinkled green seeds. Let p_1 be the probability that a pea resulting from such a mating is round and yellow; p_2 be the probability that it is wrinkled and yellow; p_3 be the probability that it is round and green; and p_4 be the probability that it is wrinkled and green. The Mendelian theory of inheritance states that the probabilities should satisfy $H_0\colon p_1 = \frac{9}{16}, p_2 = \frac{3}{16}, p_3 = \frac{3}{16}, p_4 = \frac{1}{16}$. The experiments resulted in respective frequencies of 315, 101, 108, and 32. If one tests H_0 against all alternatives, the likelihood ratio statistic has a value of 0.4754 and the probability that χ_3^2 assumes a value this large or larger (i.e. the p-value) is approximately 0.9246.

Assume the simple model described in Feller (1968 Chap. 5, Sec. 5) for each of the characteristics, shape (round or wrinkled) and color, and assume that these two characteristics are independent. The probabilities p_i, $i = 1, 2, 3, 4$, can be expressed as follows:

$$p_1 = \beta_1 \beta_2, \qquad p_2 = (1 - \beta_1)\beta_2,$$
$$p_3 = \beta_1(1 - \beta_2), \qquad p_4 = (1 - \beta_1)(1 - \beta_2),$$

where

$$\beta_1 = \alpha_1^2 + 2\alpha_1(1 - \alpha_1), \qquad \beta_2 = \alpha_2^2 + 2\alpha_2(1 - \alpha_2),$$

α_1 is the gene relative frequency of round in the parent population, and α_2 is the gene relative frequency of yellow in the parent population. Now, depending on the nature of the parent population, any one of several alternatives might be of interest. The assumption that $\beta_1 \geqslant \frac{1}{2}$ and $\beta_2 \geqslant \frac{1}{2}$ is equivalent to $p_1 \geqslant p_2 \cdot p_1 \geqslant p_3$, $p_2 \geqslant p_4$, and $p_3 \geqslant p_4$ (that is $p_1 \geqslant [p_2, p_3] \geqslant p_4$). (The assumption that $\beta_1 \geqslant \frac{1}{2}$ is equivalant to assuming that $\alpha_1 \geqslant 1 - \sqrt{2}/2 \doteq 0.29$. The Mendelian theory is consistent with the assumption that $\alpha_1 = \alpha_2 = \frac{1}{2}$.) If, in addition to assuming that $\beta_1 \geqslant \frac{1}{2}$, one assumes that $\beta_1 = \beta_2$ then $p_1 \geqslant p_2 = p_3 \geqslant p_4$ would also be an appropriate alternative.

First, consider testing H_0 against $H_1 - H_0$, where

$$H_1\colon p_1 \geqslant [p_2, p_3] \geqslant p_4$$

(recall that $H_0:\mathbf{p}=(\frac{9}{16},\frac{3}{16},\frac{3}{16},\frac{1}{16})$). In this case the relative frequencies $\hat{p}_i$, $i=1,2,3,4$, satisfy H_1 so that $\hat{\mathbf{p}}^*=\hat{\mathbf{p}}$ and the value of the likelihood ratio statistic T_{01} is 0.4754. Now the null hypothesized value of $\mathbf{p}$ has equality between p_2 and p_3, but the order restriction does not relate these values. Thus $\mathscr{I}_{\mathbf{q}}=R^4$ and $P(\mathbf{Z}|\mathscr{I}_{\mathbf{q}})=\mathbf{Z}$. Since the Z_i's have absolutely continuous distributions, $P_{\mathbf{q}}(1,4)=P_{\mathbf{q}}(2,4)=P_{\mathbf{q}}(3,4)=0$ and $P_{\mathbf{q}}(4,4)=1$. Thus the p value is again $P[\chi_3^2\geqslant 0.4754]=0.9246$.

Consider testing H_0 against H_1–H_0, where

$$H_1:p_1\geqslant p_2=p_3\geqslant p_4.$$

In this case the maximum likelihood estimates subject to H_1 are 315/556, 209/1112, 209/1112, and 32/556, respectively, and the value of the likelihood ratio statistic is 0.2409. In order to compute the corresponding p-value, note that $\mathscr{I}_{\mathbf{q}}=\{\boldsymbol{\theta}:\theta_2=\theta_3\}$ so that with probability one $P(\mathbf{Z}|\mathscr{I}_{\mathbf{q}})$ has three distinct values. Thus the p value is $P[\chi_2^2\geqslant 0.2409]=0.8865$.

Finally, consider testing H_0 against H_1–H_0, where

$$H_1:p_1\geqslant p_2\geqslant p_3\geqslant p_4.$$

The value of the test statistic is 0.2409, $P_{\mathbf{q}}(1,4)=P_{\mathbf{q}}(2,4)=0$, and $P_{\mathbf{q}}(3,4)=P_{\mathbf{q}}(4,4)=\frac{1}{2}$. The p value is given by $(P[\chi_2^2\geqslant 0.2409]+P[\chi_3^2\geqslant 0.2409])/2=0.9287$.

Example 5.2.2 Frequently, in practice, a parametric model is fitted to one data set and its validity is subsequently checked on an independently chosen set of data. The traffic accident data of Froggatt (1970, p. 36) were, using a table of random numbers, randomly allocated to two groups. The allocation was done in such a way that the probability that any data point was assigned to group 1 was $\frac{9}{10}$ and to group 2 was $\frac{1}{10}$. As a result, 646 and 62 points were assigned to groups 1 and 2, respectively. The results are given in Table 5.2.1. A negative binomial distribution, i.e.

$$p_x=\binom{\alpha+x-1}{x}\frac{p^x}{(1+p)^{\alpha+x}},$$

was fitted to the group 1 data using the scoring technique presented in Bliss and Fisher (1953) with resulting values of $\alpha=4.78$ and $p=0.48$. This distribution seems to fit the data very well. As a measure of this fit, if b_x is the observed frequency and n_x is the expected frequency for this negative binomial distribution, then the value of $2\sum_x b_x\ln(b_x/n_x)$ is 3.3 and $P[\chi_{10}^2\geqslant 3.3]=0.97$.

Now consider the problem of testing this negative binomial distribution as a null hypothesis distribution for the group 2 data set. If one tests against all alternatives, then the value of the likelihood ratio statistic (that is $2\sum_x a_x\ln(a_x/m_x)$) is 7.138 and $P[\chi_{12}^2\geqslant 7.138]=0.848$. Bliss (1953, p. 177) remarks that 'The curve defined by the P's (or ϕ's) is unimodal, so that in fitting the negative

Table 5.2.1 Observations of number of traffic accidents incurred during 1952–5 by each of 708 bus drivers

NUMBER OF TRAFFIC ACCIDENTS, x, DURING THE PERIOD	OBSERVED FREQUENCY OF x ACCIDENTS	GROUP 1 FREQUENCY b_x	GROUP 2 FREQUENCY a_x	EXPECTED FREQUENCIES NEGATIVE BINOMIAL ($\alpha = 4.78$, $p = 0.48$, $N = 62$)	$\bar{a}_x = 62p_x$
0	117	106	11	9.51	11
1	157	143	14	14.75	16.5
2	158	139	19	13.83	16.5
3	115	110	5	10.14	5
4	78	75	3	6.40	4
5	44	39	5	3.65	4
6	21	18	3	1.93	3
7	7	7	0	0.96	0.5
8	6	5	1	0.46	0.5
9	1	1	0	0.21	0.5
10	3	2	1	0.10	0.5
11	1	1	0	0.04	0
12 or more	0	0	0	0.03	0
Total	708	646	62		

binomial to an observed distribution any apparent bimodality (or multimodality) is attributed to random sampling.' Now, this particular negative binomial distribution is unimodal with mode at 1, so consider testing against the alternative that $p_0 \leqslant p_1 \geqslant p_2 \geqslant \cdots \geqslant p_{11}$. Using the minimum lower sets algorithm, the maximum likelihood estimates are computed as follows. First find the maximum average value of $\hat{p}_x$ over sets of consecutive integers containing 1. For this data this set is $\{1,2\}$ and the corresponding maximum is $(p_1+p_2)/2 = 0.266$ $(0.266 \times 62 = 16.5)$. This is the value of the restricted estimate $\hat{p}_x^*$ at $x=1,2$ (that is $\hat{p}_1^* = \hat{p}_2^* = 0.266$). The remaining values may be computed independently using the pool adjacent violators algorithm on the two sets $\{0\}$ and $\{3,4,\ldots,11\}$. (For example, $\hat{p}_4 < \hat{p}_5$ is a violator, so they are both replaced by their average $(\hat{p}_4+\hat{p}_5)/2$.) The value of the likelihood ratio statistic is $2\sum_x a_x \ln(\bar{a}_x/m_x) = 5.119$. Moreover, since none of the hypothesized p_x's are equal, $P_{\mathbf{q}}(13,13)=1$ and the p-value is $P[\chi_{12}^2 \geqslant 5.119] = 0.95$.

In another example Lee (1986) considers a natural order restriction in studying the advantage of post-positions in horse racing. His work is a further study of the data given in Siegel (1956, p. 44).

Testing an order restriction as a null hypothesis

Let $T_{12} = -2\ln\Lambda_{12}$ where Λ_{12} is the likelihood ratio for testing the null hypothesis, H_1, that $\mathbf{p}$ is isotonic with respect to $\lesssim$. The power function of T_{12} is not constant as a function of $\mathbf{p}$ as $\mathbf{p}$ ranges over H_1. Thus, the significance level of the test which rejects for all values of T_{12} at least as large as t would be given by $\sup_{\mathbf{p}\in H_1} P_{\mathbf{p}}[T_{12} \geqslant t]$, where $P_{\mathbf{p}}[T_{12} \geqslant t]$ is the probability that $T_{12} \geqslant t$ computed under the assumption that $\mathbf{p}$ is the vector of parameter values. Finding this supremum is difficult for finite samples. However, in Theorem 5.2.2 we argue that the hypothesis $p_1 = p_2 = \cdots = p_k = k^{-1}$ is asymptotically least favorable for this test and in addition that under this hypothesis the asymptotic distribution of T_{12} is the distribution given in Theorem 2.3.1 for testing an analogous hypothesis about a set of normal means. Probabilities determined by $p_1 = p_2 = \cdots = p_k = 1/k$ are denoted by $P_0(\cdot)$ and the level probabilities $P_{\mathbf{p}}(l,k)$ for this $\mathbf{p}$ are denoted by $P_0(l,k)$.

Theorem 5.2.2 *If* $\mathbf{p}$ *satisfies* H_1 *then, for any real* t,

$$\lim_{n\to\infty} P_{\mathbf{p}}[T_{12} \geqslant t] = \sum_{l=1}^{k} P_{\mathbf{p}}(l,k) P[\chi_{k-l}^2 \geqslant t]. \tag{5.2.2}$$

Moreover,

$$\lim_{n\to\infty} P_{\mathbf{p}}[T_{12} \geqslant t] \leqslant \lim_{n\to\infty} P_0[T_{12} \geqslant t]$$

for any real t.

In addition to providing the value of

$$\sup_{\mathbf{p}\in H_1} \lim_{n\to\infty} P_{\mathbf{p}}[T_{12} \geqslant t],$$

Theorem 5.2.2 also gives a method for investigating the behavior of $\lim_{n\to\infty} P_{\mathbf{p}}[T_{12} \geqslant t]$ for various values of **p** satisfying H_1. For example, if there is a **p** satisfying H_1 which has k distinct values, then for this **p**, $\mathscr{I}_{\mathbf{p}} = R^k$ and $\lim_{n\to\infty} P_{\mathbf{p}}[T_{12} \geqslant t] = 0$ and $\inf_{\mathbf{p}\in H_1} \lim_{n\to\infty} P_{\mathbf{p}}[T_{12} \geqslant t] = 0$. Note that $\lim_{n\to\infty} P_{\mathbf{p}}[T_{12} \geqslant t]$ depends on **p** only through the sets on which **p** is constant. Moreover, in this case the limiting distribution encountered here is a decreasing function of the number of distinct values of **p** in the same sense that the limiting distribution of T_{01} was an increasing function of this number of distinct values.

Proof of Theorem 5.2.2: The statistic T_{12} can be written as $2\sum_{i=1}^{k} n\hat{p}_i[\ln \hat{p}_i - \ln \hat{p}_i^*]$. Expanding $\ln \hat{p}_i^*$ about $\hat{p}_i$ and proceeding as in the proof of Theorem 5.2.1,

$$T_{12} = \sum_{i=1}^{k} \frac{n\hat{p}_i}{\alpha_i^2}(\bar{p}_i - \hat{p}_i)^2 \xrightarrow{D} \sum_{i=1}^{k} p_i^{-1}[P(\mathbf{p}(\mathbf{U} - \bar{U})|\mathscr{I}_{\mathbf{p}})_i - p_i(U_i - \bar{U})]^2$$

$$= \sum_{i=1}^{k} [P(\sqrt{\mathbf{p}}\mathbf{U}|\mathscr{I}_{\mathbf{p}})_i - \sqrt{p_i}U_i]^2.$$

The first conclusion of Theorem 5.2.2 follows from Theorem 2.3.1 and the second conclusion follows from the fact that $\sum_{i=1}^{k}[P(\mathbf{Z}|\mathscr{I}_{\mathbf{p}})_i - Z_i]^2$ is the square of the distance from **Z** to $\mathscr{I}_{\mathbf{p}}$ and $\mathscr{I} \subset \mathscr{I}_{\mathbf{p}}$. □

Example 5.2.1 (continued) Suppose one is interested in determining if Mendel's data are compatible with the hypothesis that $p_1 \geqslant p_2 \geqslant p_3 \geqslant p_4$. The value of T_{12} is 0.2345 and the asymptotic p value is equal to

$$\sup_{\mathbf{p}\in H_1} \lim_{n\to\infty} P_{\mathbf{p}}[T_{12} \geqslant 0.2345] = \sum_{l=1}^{k} P_0(l,4) P[\chi_{4-l}^2 \geqslant 0.2345].$$

The values of $P_0(l,4) = P_S(l,4)$ are given in Table A.10. Thus the asymptotic p value is $(1/4)P[\chi_3^2 \geqslant 0.2345] + (11/24)P[\chi_2^2 \geqslant 0.2345] + (1/4)P[\chi_1^2 \geqslant 0.2345] = 0.8114$. Clearly, there are **p**'s satisfying H_1 which have four distinct values so that

$$\inf_{\mathbf{p}\in H_1} \lim_{n\to\infty} P_{\mathbf{p}}[T_{12} \geqslant 0.2345] = 0.$$

If the Mendelian theory of inheritance is true then the actual value of **p** is $(\frac{9}{16}, \frac{3}{16}, \frac{3}{16}, \frac{1}{16})$. The limit, $\lim_{n\to\infty} P_{\mathbf{p}}[T_{12} \geqslant t]$, is the same for any **p** such that $p_1 > p_2 = p_3 > p_4$. In computing $P(\bar{\mathbf{Y}}|\mathscr{I}_{\mathbf{p}})$ one averages Y_2 and Y_3 when $Y_2 < Y_3$. Thus $P_{\mathbf{p}}(1,4) = P_{\mathbf{p}}(2,4) = 0$ and $P_{\mathbf{p}}(3,4) = P_{\mathbf{p}}(4,4) = \frac{1}{2}$. The asymptotic p value is $(P[\chi_1^2 \geqslant 0.2345] + P[\chi_0^2 \geqslant 0.2345])/2 = 0.3217$. Asymptotic p values for all possible configurations are given in Table 5.2.2.

Table 5.2.2 Asymptotic p values for testing $p_1 \geqslant p_2 \geqslant p_3 \geqslant p_4$

CONFIGURATION	$\lim_{n\to\infty} P_{\mathbf{p}}[T_{12} \geqslant 0.2345]$
$p_1 = p_2 = p_3 = p_4$	0.8114
$p_1 = p_2 = p_3 > p_4$	0.6182
$p_1 > p_2 = p_3 = p_4$	0.6182
$p_1 = p_2 > p_3 = p_4$	0.5441
$p_1 > p_2 > p_3 = p_4$	0.3217
$p_1 > p_2 = p_3 > p_4$	0.3217
$p_1 = p_2 > p_3 > p_4$	0.3217
$p_1 > p_2 > p_3 > p_4$	0

Table 5.2.3 Asymptotic p values for testing $p_1 \geqslant p_2 = p_3 \geqslant p_4$

CONFIGURATION	$\lim_{n\to\infty} P_{\mathbf{p}}[T_{12} \geqslant 0.2345]$
$p_1 = p_2 = p_3 = p_4$	0.8363
$p_1 = p_2 = p_3 > p_4$	0.7588
$p_1 > p_2 = p_3 = p_4$	0.7588
$p_1 > p_2 = p_3 > p_4$	0.6282

Finally, consider testing

$$H_1 : p_1 \geqslant p_2 = p_3 \geqslant p_4.$$

The value of T_{12} is again 0.2345. In this case the values of $P_0(l, 4)$ have not been tabled but can be computed using the formulas in Section 2.4 for $P_S(l, 3; (1, 2, 1))$. The asymptotic p value is given by $0.3041P[\chi_3^2 \geqslant 0.2345] + 0.5P[\chi_2^2 \geqslant 0.2345] + 0.1959P[\chi_1^2 \geqslant 0.2345] = 0.8363$. Asymptotic p values for other configurations are given in Table 5.2.3.

Other tests for H_0 versus $H_1 - H_0$

One might consider using a chi-square-type statistic for testing H_0 against $H_1 - H_0$. One such test would reject H_0 for large values of the statistic

$$C_{01} = \sum_{i=1}^{k} \frac{(n\hat{p}_i^* - nq_i)^2}{nq_i} = \sum_{i=1}^{k} (\hat{p}_i^* - q_i)^2 \frac{n}{q_i}.$$

Using the ideas developed in the proof of Theorem 4.2.1 it is relatively easy to see that C_{01} has the same asymptotic distribution as that of the likelihood ratio statistic T_{01}.

Y. J. Lee (1977) considered maximin tests of H_0 against various subclasses of alternatives in $H_1 - H_0$ where H_1 imposes a simple order on $\mathbf{p}$ (that is

$H_1: p_1 \leqslant p_2 \leqslant \cdots \leqslant p_k$). For example, suppose $\boldsymbol{\delta} = (\delta_1, \delta_2, \ldots, \delta_{k-1})$ is a $(k-1)$-tuple of nonnegative constants (at least one nonzero) and let $K(\boldsymbol{\delta}) = \{\mathbf{p} \in H_1 : p_{i+1} - p_i \geqslant \delta_i, i = 1, 2, \ldots, k-1\}$. A test which maximizes the minimum power over points $\mathbf{p}$ in $K(\boldsymbol{\delta})$ rejects for large values of a contrast of the form

$$\sum_{i=1}^{k} n\hat{p}_i \log p_i'$$

where $p_{i+1}' - p_i' = \delta_i$. No power comparisons of these maximin tests with the likelihood ratio tests have been made. However, based upon the comparisons made in Chapter 4, in the analogous normal means problem one would expect that these maximin tests would be acceptable alternatives to the likelihood ratio test for a simple order and small k. They can be made very powerful if one has additional prior information about $\mathbf{p}$, such as that the differences $p_{i+1} - p_i$ are all of the same magnitude. However, our conclusion is that the likelihood ratio test is, in general, a more satisfactory procedure for testing H_0 against $H_1 - H_0$.

Chi-square-type tests for an order restriction as a null hypothesis

Several chi-square-type tests of the null hypothesis H_1, that $\mathbf{p} = (p_1, p_2, \ldots, p_k)$ is isotonic with respect to the quasi-order, $\lesssim$, on $X = \{1, 2, \ldots, k\}$, have been proposed. Two possibilities are to reject the null hypothesis for large values of

$$C_{12} = \sum_{i=1}^{k} \frac{(n\hat{p}_i^* - n\hat{p}_i)^2}{n\hat{p}_i^*} \tag{5.2.3}$$

or for large values of

$$D_{12} = \sum_{i=1}^{k} \frac{(n\hat{p}_i^* - n\hat{p}_i)^2}{n\hat{p}_i}. \tag{5.2.4}$$

Recall that in deriving the asymptotic distribution of the likelihood ratio statistic T_{12}, it was argued that T_{12} was equal to $\sum_{i=1}^{k} \hat{p}_i(n\hat{p}_i^* - n\hat{p}_i)^2/(n\alpha_i^2)$ where $\boldsymbol{\alpha}^2/\hat{\mathbf{p}}$ converges almost surely to $\mathbf{p}$ provided $\mathbf{p} \in H_1$. Both C_{12} and D_{12} have similar forms. It follows that both C_{12} and D_{12} converge in distribution to $\sum_{i=1}^{k} [P(\mathbf{Z}|\mathscr{I}_{\mathbf{p}})_i - Z_i]^2$ where $Z_1, Z_2, \ldots, Z_k$ are independent standard normal variables. Thus, the asymptotic distribution of C_{12} and D_{12} is the same as that of T_{12} given in Theorem 5.2.2. Critical values for these tests can be found using the least favorable configuration, $p_i = k^{-1}$, $i = 1, 2, \ldots, k$.

Other possibilities for chi-square-type tests are to reject H_1 for large values of a Pearson-type statistic

$$C_{12}' = \inf_{\mathbf{r} \in H_1} \sum_{i=1}^{k} \frac{(nr_i - n\hat{p}_i)^2}{nr_i} \tag{5.2.5}$$

or for large values of a Neyman-type statistic

$$C''_{12} = \inf_{\mathbf{r} \in H_1} \sum_{i=1}^{k} \frac{(nr_i - n\hat{p}_i)^2}{n\hat{p}_i}. \tag{5.2.6}$$

First consider the problem of finding the point in H_1 that solves the minimization problem in (5.2.5). Let $p'_i = [\sum_{j=1}^{k} P(\hat{\mathbf{p}}^2|\mathscr{I})_j^{1/2}]^{-1} P(\hat{\mathbf{p}}^2|\mathscr{I})_i^{1/2}$ and suppose $\mathbf{r}$ is an arbitrary point in H_1. Note that for any $\lambda \in [0, 1]$ the point $\lambda\mathbf{r} + (1 - \lambda)\mathbf{p}'$ is also in H_1. Define the function g on $[0, 1]$ by

$$\begin{aligned} g(\lambda) &= n \sum_{i=1}^{k} \frac{[\lambda r_i + (1 - \lambda)p'_i - \hat{p}_i]^2}{\lambda r_i + (1 - \lambda)p'_i} \\ &= n \sum_{i=1}^{k} \frac{\hat{p}_i^2}{\lambda r_i + (1 - \lambda)p'_i} - 1. \end{aligned}$$

The second expression follows because $\sum_{i=1}^{k} [\lambda r_i + (1 - \lambda)p'_i] = \sum_{i=1}^{k} \hat{p}_i = 1$. By taking derivatives with respect to λ, it is seen that

$$g''(\lambda) = \sum_{i=1}^{k} 2\hat{p}_i^2 [\lambda r_i + (1 - \lambda)p'_i]^{-3} (r_i - p'_i)^2 \geqslant 0$$

and hence g is convex on $[0, 1]$. Moreover,

$$g'(0) = -\sum_{i=1}^{k} \frac{\hat{p}_i^2 r_i}{p_i'^2} + \sum_{i=1}^{k} \frac{\hat{p}_i^2}{p'_i}. \tag{5.2.7}$$

With $\mathbf{s} = P(\hat{\mathbf{p}}^2|\mathscr{I})^{1/2}$, and $p'_i = s_i [\sum_{j=1}^{k} s_j]^{-1}$ the second term in the expression for $g'(0)$ can be written as

$$\sum_{i=1}^{k} \frac{\hat{p}_i^2}{s_i} \left[\sum_{j=1}^{k} s_j \right].$$

However, by Theorem 1.3.6, $\sum_{i=1}^{k} \hat{p}_i^2 / s_i = \sum_{i=1}^{k} s_i^2 / s_i$ so that the second term in (5.2.7) is equal to $[\sum_{j=1}^{k} s_j]^2$. The first term in (5.2.7) is equal to

$$-\left[\sum_{j=1}^{k} s_j \right]^2 \sum_{i=1}^{k} \frac{\hat{p}_i r_i}{s_i^2}$$

which by Theorem 1.8.1 is at least as large as

$$-\left[\sum_{j=1}^{k} s_j \right]^2 \sum_{i=1}^{k} \frac{s_i^2 r_i}{s_i^2} = -\left[\sum_{j=1}^{k} s_j \right]^2.$$

Combining these two results it follows that $g'(0) \geqslant 0$. Since g is convex on $[0, 1]$ and since $g'(0) \geqslant 0$ it follows that $g(0) \leqslant g(\lambda)$ for all $\lambda \in [0, 1]$. Applying this result for $\lambda = 0, 1$, one concludes that $n\sum_{i=1}^{k} [p'_i - \hat{p}_i]^2 / p'_i \leqslant n\sum_{i=1}^{k} [r'_i - \hat{p}_i]^2 / r_i$ and since $\mathbf{r}$ is arbitrary, $\mathbf{p}'$ solves the minimization problem in (5.2.5). It seems appropriate to call $\mathbf{p}'$ Pearson's restricted minimum chi-square estimator for $\mathbf{p}$.

Using a similar derivation the solution to the minimization problem in (5.2.6) is given by $\mathbf{p}''$ where $p_i'' = [\sum_{j=1}^k P_{1/\hat{\mathbf{p}}}(\hat{\mathbf{p}}|\mathscr{I})_j]^{-1} P_{1/\hat{\mathbf{p}}}(\hat{\mathbf{p}}|\mathscr{I})_i$. It is of interest to note that by Theorem 1.8.2, $P_{1/\hat{\mathbf{p}}}(\hat{\mathbf{p}}|\mathscr{I}) = P(1/\hat{\mathbf{p}}|\mathscr{I}^c)^{-1}$ where $\mathscr{I}^c$ is the set of antitonic functions on X. Thus, the solutions to both (5.2.5) and (5.2.6) can be found by taking the isotonic regression of a monotone function of $\hat{\mathbf{p}}$, applying the inverse function coordinatewise to the isotonic regression ($(P(\hat{\mathbf{p}}^2|\mathscr{I})^{1/2}$ for (5.2.5) and $P(1/\hat{\mathbf{p}}|\mathscr{I}^c)^{-1}$ for (5.2.6)), and normalizing to obtain a probability vector. In fact, both (5.2.5) and (5.2.6) are special cases of a minimum directed divergence problem (cf. Cressie and Read, 1984).

As with C_{12} and D_{12} both C'_{12} and C''_{12} converge in law to $\sum_{i=1}^k [P(\mathbf{Z}|\mathscr{I}_{\mathbf{p}})_i - Z_i]^2$ if $\mathbf{p}$ is a point satisfying H_1. A sketch of the proof for $C'_{12} = n\sum_{i=1}^k [(p_i' - \hat{p}_i)^2/p_i']$, where $\mathbf{p}'$ is proportional to $P(\hat{\mathbf{p}}^2|\mathscr{I})^{1/2}$, is given. This proof shows the asymptotic equivalence of a large number of statistics. Suppose a is a positive number and consider the random vector, $P(\hat{\mathbf{p}}^a|\mathscr{I})^{1/a}$. Using an argument like the one given for Theorem 5.2.2, $P(\hat{\mathbf{p}}^a|\mathscr{I}) = P(\hat{\mathbf{p}}^a|\mathscr{I}_{\mathbf{p}})$ for sufficiently large n with probability one. Using the max-min formula, Theorem 1.4.4, and the fact that $y^{1/a}$ is an increasing function of y,

$$P(\hat{\mathbf{p}}^a|\mathscr{I}_{\mathbf{p}})^{1/a} = \max\min\left[\frac{\sum_{i\in B}\hat{p}_i^a}{\text{card}(B)}\right]^{1/a} = \min\max\left[\frac{\sum_{i\in B}\hat{p}_i^a}{\text{card}(B)}\right]^{1/a} \tag{5.2.8}$$

where card(B) denotes the number of points in the block, B. Consider the collection of random variables, $g_B(\hat{\mathbf{p}}) = [\sum_{i\in B}\hat{p}_i^a/\text{card}(B)]^{1/a}$, where B ranges over the collection, $\mathscr{B}$, of blocks involved in the max-min formula. By the central limit theorem, $\hat{\mathbf{p}}$ has an asymptotic normal distribution with mean $\mathbf{p}$ and variance-covariance matrix, $\mathbf{M}$, where $\mathbf{M}$ is given in the proof of Theorem 5.2.1. Using the delta method (cf. Theorem A on page 122 of Serfling, 1980) the joint distribution of the collection of random variables $\{g_B(\hat{\mathbf{p}}): B\in\mathscr{B}\}$ is seen to be asymptotically normal with mean vector whose component corresponding to B is $g_B(\mathbf{p})$. However, in computing $P(\hat{\mathbf{p}}^a|\mathscr{I}_{\mathbf{p}})$ the smoothing is done only over subsets of X on which $\mathbf{p}$ is constant. Thus $\mathbf{p}$ is constant on all $B\in\mathscr{B}$ and $g_B(\mathbf{p})$ does not depend upon a. Moreover, the variance-covariance matrix of this asymptotic distribution is of the form $n\mathbf{DMD}'$, where $\mathbf{D}$ is a matrix whose typical entry is of the form $\partial g_B(\mathbf{x})/(\partial x_i)|_{\mathbf{x}=\mathbf{p}}$. Now,

$$\frac{\partial g_B(\mathbf{x})}{\partial x_i} = \begin{cases} 0, & \text{if } i\notin B \\ \left(\dfrac{1}{a}\right)\left[\dfrac{\sum_{j\in B}x_j^a}{\text{card}(B)}\right]^{1/a-1}[a x_i^{a-1}] & \text{if } i\in B. \end{cases}$$

Thus, if $i\in B$,

$$\left.\frac{\partial g_B(\mathbf{x})}{\partial x_i}\right|_{\mathbf{x}=\mathbf{p}} = \left[\frac{\sum_{j\in B}p_j^a}{\text{card}(B)}\right]^{(1-a)/a} p_i^{a-1} = p_i^{1-a}p_i^{a-1} = 1$$

since $\mathbf{p}$ is constant on B. This limiting variance-covariance matrix does not depend upon a. It follows from (5.2.8) that the asymptotic distribution of $P(\hat{\mathbf{p}}|\mathscr{I}_{\mathbf{p}})^{1/a}$ does not depend upon a. Thus, the asymptotic distribution of C'_{12} is the same as that of C_{12} which, as was already noted, is the distribution of $\sum_{i=1}^{k}[P(\mathbf{Z}|\mathscr{I}_{\mathbf{p}})_i - Z_i]^2$. The result for C''_{12} follows similarly.

5.3 ESTIMATION AND TESTING: THE STAR-SHAPED RESTRICTION

In this section, let H_0 denote the hypothesis: $p_i = 1/k$, $i = 1, 2, \ldots, k$, and let the hypothesis H_1 be defined by

$$H_1 : p_1 \geqslant \frac{p_1 + p_2}{2} \geqslant \cdots \geqslant \frac{p_1 + p_2 + \cdots + p_{k-1}}{k-1} \geqslant \frac{1}{k}. \tag{5.3.1}$$

The restriction imposed upon $\mathbf{p}$ by H_1 is essentially different from the restrictions considered thus far in that H_1 is not equivalent to the hypothesis that $\mathbf{p}$ is isotonic with respect to some partial order on $\{1, 2, \ldots, k\}$. Restriction (5.3.1) is encountered in reliability problems and is called a *star-shaped* restriction. A vector $\mathbf{x} = (x_1, x_2, \ldots, x_k)$ is said to be *lower star-shaped* provided

$$x_1 \geqslant \frac{x_1 + x_2}{2} \geqslant \cdots \geqslant \frac{x_1 + x_2 + \cdots + x_k}{k} \geqslant 0 \tag{5.3.2}$$

with an analogous restriction defining an upper star-shaped vector. A vector which satisfies (5.3.2) except possibly for the nonnegativity requirement is termed *decreasing on the average*.

The term *star-shaped* comes from the fact that if one plots the $k+1$ points $(i, \sum_{j=0}^{i}\alpha_j)$, $i = 0, 1, 2, \ldots, k$, with $\alpha_0 = 0$, in the plane, then the slopes of the lines joining $(0, 0)$ to each of the consecutive points are nonincreasing. This implies that one can 'see' each of these points from $(0, 0)$ without one's sight being 'blocked' by one of the line segments joining these points. A star has this property in that one can see each of the tips of the star from the center. It should be noted that if $p_1 \geqslant p_2 \geqslant \cdots \geqslant p_k$, then $\mathbf{p}$ is lower star-shaped.

The property of being star-shaped is closely related to a convexity property. Note that if the function joining these $k+1$ points is concave then it is star-shaped. The converse is not true. The problem of estimating a regression function subject to the restriction that the estimate is concave or convex is an important practical problem (cf. Hildreth, 1954; Dent, 1973; Hanson and Pledger, 1976; Hudson, 1969). However, theory for a convexity restriction analogous to that developed in this section for a star-shaped restriction has not been developed at this time. The convexity problem seems to be difficult and one might use the methods developed for the star-shaped ordering when in fact the actual restriction is one of convexity.

Consider likelihood ratio tests for H_0 against H_1 and for H_1 against H_2–H_1. The first problem is to find the maximum likelihood estimate of $\mathbf{p}$ subject to the restriction H_1. To this end, define a one-to-one transformation of the parameter space by introducing new parameters $\theta_1, \theta_2, \ldots, \theta_{k-1}$ defined by

$$\theta_i = \frac{\sum_{j=1}^{i} p_j}{\sum_{j=1}^{i+1} p_j} \qquad \text{for } i = 1, 2, \ldots, k-1. \tag{5.3.3}$$

Thus $p_1 = \prod_{j=1}^{k-1}\theta_j$, $p_i = (1-\theta_{i-1})\prod_{j=i}^{k-1}\theta_j$, $i = 2, 3, \ldots, k-1$, and $p_k = (1-\theta_{k-1})$. In terms of the θ_i's the hypotheses are $H_0{:}\theta_i = i/(i+1)$, $i = 1, \ldots, k-1$, and $H_1{:}\theta_i \geqslant i/(i+1)$, $i = 1, \ldots, k-1$. Consider first the estimation of $\boldsymbol{\theta}$. The likelihood function can be written

$$L(\boldsymbol{\theta}) = \prod_{i=1}^{k-1} \theta_i^{n\sum_{j=1}^{i}\hat{p}_j}(1-\theta_i)^{n\hat{p}_{i+1}} \qquad \text{for } 0 \leqslant \theta_i \leqslant 1, \tag{5.3.4}$$

where $\hat{p}_i$ is the relative frequency of the event having probability p_i, $i = 1, 2, \ldots, k$. It is easy to find the maximum of the function $\theta^a(1-\theta)^b$ subject to $\theta \geqslant c (0 \leqslant \theta \leqslant 1)$. This maximum is attained at $\bar{\theta} = \{a/(a+b)\} \vee c$, where $\vee$ denotes the larger of the two numbers. It follows that the maximum likelihood estimates which satisfy H_1 are given by

$$\bar{\theta}_i = \hat{\theta}_i \vee (i/i+1), \qquad i = 1, 2, \ldots, k-1, \tag{5.3.5}$$

where $\hat{\theta}_i = (\sum_{j=1}^{i}\hat{p}_j)/(\sum_{j=1}^{i+1}\hat{p}_j)$. Evaluation of $\mathbf{p}$ at $\boldsymbol{\theta} = \bar{\boldsymbol{\theta}}$ gives the MLE of $\mathbf{p}$ under the restriction specified in (5.3.1).

The estimates given by (5.3.5) and the corresponding estimate of $\mathbf{p}$ are simple compared to the difficulty involved in finding an estimate of $\mathbf{p}$ subject to a restriction that it is isotonic with respect to a partial order. Moreover, this simplicity carries over to the distribution theory derived below for the LRTs. An explanation for this simplicity is that the restriction (5.3.1) on $\mathbf{p}$ is equivalent to requiring that $\mathbf{p}$ belongs to an orthant in R^k. In order to see this let $\mathscr{S}$ be the set of all vectors, $\mathbf{x}$, in R^k satisfying (5.3.2). The restriction (5.3.2) is equivalent to the following k restrictions:

$$x_1 \geqslant x_2, \qquad \frac{x_1 + x_2}{2} \geqslant x_3, \ldots, \frac{x_1 + x_2 + \cdots + x_k}{k} \geqslant 0. \tag{5.3.6}$$

The set $\mathscr{S}$ is a polyhedral cone with generators

$$\begin{aligned}
\mathbf{g}_1 &= (1, -1, 0, 0, \ldots, 0)\\
\mathbf{g}_2 &= (1, 1, -2, 0, \ldots, 0)\\
&\vdots\\
\mathbf{g}_{k-1} &= (1, 1, 1, \ldots, 1, -k+1)\\
\mathbf{g}_k &= (1, 1, 1, \ldots, 1, 1).
\end{aligned}$$

In order to see that these vectors generate $\mathscr{S}$, first note that each $\mathbf{g}_i$ satisfies (5.3.2) so that if $\mathbf{x}=\sum_{i=1}^{k}\alpha_i\mathbf{g}_i$ where $\alpha_i \geqslant 0$ then $\mathbf{x}\in\mathscr{S}$. Conversely, suppose $\mathbf{x}\in\mathscr{S}$ and let

$$\alpha_i = i^{-1}\sum_{j=1}^{i} x_j - (i+1)^{-1}\sum_{j=1}^{i+1} x_j \qquad \text{for } i=1,2,\ldots,k-1$$

$$\alpha_k = k^{-1}\sum_{j=1}^{k} x_j.$$

The quantities, α_i, are nonnegative since $\mathbf{x}\in\mathscr{S}$ and it is a straightforward exercise to show that

$$\mathbf{x}=\sum_{i=1}^{k}\alpha_i\mathbf{g}_i.$$

Thus, $\mathscr{S}$ is a polyhedral cone with generators $\mathbf{g}_i$, $i=1,2,\ldots,k$. Note that $\mathbf{g}_i$ and $\mathbf{g}_j$ are orthogonal for $i\neq j$ so that $\mathscr{S}$ is an orthant.

Turning to the testing problem, let Λ_{01} denote the likelihood ratio test statistic for testing H_0 against H_1–H_0 and let $T_{01}=-2\ln\Lambda_{01}$. Then

$$T_{01}=2\sum_{i=1}^{k-1}\left(n\sum_{j=1}^{i}\hat{p}_j\{\ln\bar{\theta}_i-\ln[i/(i+1)]\}\right.$$
$$\left.+\,n\hat{p}_{i+1}\{\ln(1-\bar{\theta}_i)-\ln[1/(i+1)]\}\right).$$

Expanding $\ln\bar{\theta}_i$ and $\ln[i/(i+1)]$ about $\hat{\theta}_i$ and $\ln(1-\bar{\theta}_i)$ and $\ln[1/(i+1)]$ about $1-\hat{\theta}_i$ via Taylor's theorem with a second-degree remainder term, one obtains

$$T_{01}=2\sum_{i=1}^{k-1}\left[-\frac{n\sum_{j=1}^{i}\hat{p}_j}{2\alpha_i^2}(\bar{\theta}_i-\hat{\theta}_i)^2+\frac{n\sum_{j=1}^{i}\hat{p}_j}{2\beta_i^2}\left(\hat{\theta}_i-\frac{i}{i+1}\right)^2\right.$$
$$\left.-\frac{n\hat{p}_{i+1}}{2v_i^2}(\hat{\theta}_i-\bar{\theta}_i)^2+\frac{n\hat{p}_{i+1}}{2\gamma_i^2}\left(\hat{\theta}_i-\frac{i}{i+1}\right)^2\right], \tag{5.3.7}$$

where α_i is between $\bar{\theta}_i$ and $\hat{\theta}_i$, β_i is between $\hat{\theta}_i$ and $i/(i+1)$, v_i is between $(1-\bar{\theta}_i)$ and $(1-\hat{\theta}_i)$, and γ_i is between $(1-\hat{\theta}_i)$ and $1/(i+1)$. The law of large numbers implies that, under H_0, $\hat{\theta}_i$ converges to $i/(i+1)$.

Let $\theta_i=\sum_{j=1}^{i}p_j/\sum_{j=1}^{i+1}p_j$, $i=1,2,\ldots,k-1$, so that under H_0, $\theta_i=i/(i+1)$. By conditioning on $n\sum_{j=1}^{i+1}\hat{p}_j$ it can be shown that $E(\hat{\theta}_i)=\theta_i$. A straightforward application of the central limit theorem (CLT) to $\sqrt{n}(\hat{\mathbf{p}}-\mathbf{p})$ and then use of the delta method (Serfling, 1980, p. 122) applied to $\sqrt{n}(\hat{\boldsymbol{\theta}}-\boldsymbol{\theta})$, which can be written as $\sqrt{n}[h(\hat{\mathbf{p}})-h(\mathbf{p})]$ for the appropriate function h, shows that $\sqrt{n}(\hat{\boldsymbol{\theta}}-\boldsymbol{\theta})$ has an asymptotic normal distribution with mean vector $\mathbf{0}$ and variance-covariance, $\mathbf{M}=(m_{ij})$, which is a diagonal matrix with

$$m_{ij}=\frac{\theta_i(1-\theta_i)}{\sum_{j=1}^{i+1}p_j} \qquad \text{for } i=1,2,\ldots,k-1.$$

Therefore, $\sqrt{n}\hat{\theta}_1, \sqrt{n}\hat{\theta}_2, \ldots, \sqrt{n}\hat{\theta}_k$ are asymptotically independent.

Let $\mathbf{U} = (U_1, U_2, \ldots, U_{k-1})$ have this normal distribution so that $\sqrt{n}(\hat{\boldsymbol{\theta}} - \boldsymbol{\theta}) \xrightarrow{D} \mathbf{U}$. By the law of large numbers, $\hat{p}_i \to p_i$, $\alpha_i \to \theta_i$, $\beta_i \to \theta_i$, $v_i \to (1-\theta_i)$, and $\gamma_i \to (1-\theta_i)$, all with probability one. Thus, if H_0 is satisfied, so that $\theta_i = i/(i+1)$, $i = 1, 2, \ldots, k-1$, then (recall that $a \vee b$ and $a \wedge b$ denote the larger and smaller of a and b, respectively)

$$\begin{aligned} T_{01} &\xrightarrow{D} \sum_{i=1}^{k-1} \left\{ -\frac{\sum_{j=1}^{i} p_j}{\theta_i^2} [(U_i \vee 0) - U_i]^2 + \frac{\sum_{j=1}^{i} p_j}{\theta_i^2} U_i^2 \right. \\ &\qquad \left. - \frac{p_{i+1}}{(1-\theta_i)^2} [(U_i \vee 0) - U_i]^2 + \frac{p_{i+1}}{(1-\theta_i)^2} U_i^2 \right\} \\ &= \sum_{i=1}^{k-1} (U_i \vee 0)^2 \left[\frac{\sum_{j=1}^{i} p_j}{\theta_i^2} + \frac{p_{i+1}}{(1-\theta_i)^2} \right] \\ &= \sum_{i=1}^{k-1} (Z_i \vee 0)^2, \end{aligned}$$

where $Z_1, Z_2, \ldots, Z_{k-1}$ are independent standard normal variables. This argument proves the following theorem.

Theorem 5.3.1 *Under H_0, the likelihood ratio statistic T_{01} is distributed asymptotically as*

$$L = \sum_{i=1}^{k-1} (Z_i \vee 0)^2,$$

where $Z_1, Z_2, \ldots, Z_{k-1}$ are independent standard normal variables.

Suppose I is a subset of $\{1, 2, \ldots, k-1\}$ and let E_I be the event $[Z_i \geqslant 0; i \in I$ and $Z_i < 0; i \notin I]$. Then applying Lemma B used in the proof of Theorem 2.3.2, for any real number t,

$$\begin{aligned} P[L \geqslant t, E_I] &= P\left[\sum_{i \in I} Z_i^2 \geqslant t, Z_i \geqslant 0; i \in I, Z_i < 0; i \notin I \right] \\ &= P\left[\sum_{i \in I} Z_i^2 \geqslant t, Z_i \geqslant 0; i \in I \right] P[Z_i < 0; i \notin I] \\ &= P\left[\sum_{i \in I} Z_i^2 \geqslant t \,\middle|\, Z_i \geqslant 0; i \in I \right] (\tfrac{1}{2})^{k-1} \\ &= P[\chi_m^2 \geqslant t] (\tfrac{1}{2})^{k-1}, \end{aligned}$$

where m is the number of elements in I. Partitioning the event $[L \geqslant t]$ by intersecting it with all such events, E_I, one obtains the following result.

Theorem 5.3.2 *If H_0 is true then*

$$\lim_{n\to\infty} P[T_{01} \geqslant t] = \bar{\chi}^2_{k-1}(t) \stackrel{D}{=} \sum_{l=0}^{k-1} \binom{k-1}{l} (\tfrac{1}{2})^{k-1} P[\chi_l^2 \geqslant t] \tag{5.3.8}$$

for all real $t(\chi_0^2 = 0)$.

The level probabilities, $P(l, k) = \binom{k-1}{l-1}(\tfrac{1}{2})^{k-1}$, have this simple form because the set of star-shaped vectors is an orthant. Approximate critical values for $k = 3, 4, \ldots, 15$ and for $\alpha = 0.10, 0.05, 0.01$ are given in Table 5.3.1.

Now consider the problem of testing H_1 as a null hypothesis when 'not H_1' is the alternative. Since the unrestricted maximum likelihood estimate of θ_i is equal to $\hat{\theta}_i$, it follows directly, by writing the likelihood ratio in terms of $\hat{\boldsymbol{\theta}}$ and $\bar{\boldsymbol{\theta}}$ and expanding $\ln \bar{\theta}_i$ and $\ln(1 - \bar{\theta}_i)$ about $\hat{\theta}_i$ and $(1 - \hat{\theta}_i)$, respectively, that the test statistic can be written as

$$T_{12} = -2 \ln \Lambda_{12} = n \sum_{i=1}^{k-1} \left[\frac{\sum_{j=1}^{i} \hat{p}_j}{\alpha_i^2} + \frac{\hat{p}_{i+1}}{v_i^2} \right] (\hat{\theta}_i - \bar{\theta}_i)^2,$$

where α_i is between $\bar{\theta}_i$ and $\hat{\theta}_i$ (and thus converges a.s. to θ_i) and v_i is between $1 - \bar{\theta}_i$ and $1 - \hat{\theta}_i$ (and thus converges a.s. to $1 - \theta_i$).

Table 5.3.1 The $100(1-\alpha)$ percentiles for the $\bar{\chi}^2$ distributions with $\bar{\chi}^2_{k-1}(t) = \sum_{l=0}^{k-1} \binom{k-1}{l} (\tfrac{1}{2})^{k-1} P[\chi_l^2 \geqslant t]$

	α		
k	0.10	0.05	0.01
3	2.95	4.23	7.28
4	4.01	5.44	8.77
5	4.95	6.50	10.02
6	5.84	7.48	11.18
7	6.67	8.41	12.26
8	7.48	9.29	13.31
9	8.26	10.15	14.29
10	9.02	10.99	15.29
11	9.76	11.79	16.21
12	10.49	12.59	17.12
13	11.22	13.38	18.01
14	11.93	14.15	18.91
15	12.63	14.91	19.78

By employing arguments similar to those used in Theorem 5.3.2 one concludes that

$$T_{12} \xrightarrow{D} \sum_{i=1}^{k-1} \left[\frac{\sum_{j=1}^{i} p_j}{\theta_i^2} + \frac{p_{i+1}}{(1-\theta_i)^2} \right] [U_i - (U_i \vee \delta_i)]^2$$

where $\delta_i = \lim_{n\to\infty} \sqrt{n}(i/(i+1) - \theta_i)$ is 0 if $\theta_i = i/(i+1)$ and is $-\infty$ if $\theta_i > i/(i+1)$, provided $\mathbf{p} \in H_1$ (that is $\theta_i \geq i/(i+1)$, $i = 1, 2, \ldots, k-1$). The following theorem is a consequence.

Theorem 5.3.3 *If* $\mathbf{p} \in H_1$ *then* T_{12} *is asymptotically distributed as*

$$V = \sum_{i=1}^{k-1} [Z_i - (Z_i \vee \delta_i)]^2$$

where $Z_1, Z_2, \ldots, Z_{k-1}$ *are independent standard normal variables and* $\delta_i = 0$ *or* $-\infty$ *according as* $\theta_i = i/(i+1)$ *or* $\theta_i > i(i+1)$. *Consequently, with* m *equal to the number of* i *for which* $\theta_i = i/(i+1)$,

$$\lim_{n\to\infty} P[T_{12} \geq t] = \bar{\chi}_m^2(t),$$

where $\bar{\chi}_m^2(t)$ *is given by* (5.3.8) *with* $k-1$ *replaced by* m. *Thus*

$$\sup_{\mathbf{p}\in H_1} \lim_{n\to\infty} P_{\mathbf{p}}[T_{12} \geq t] = \lim_{n\to\infty} P_0[T_{12} \geq t] = \bar{\chi}_{k-1}^2(t),$$

where as before $P_0[T_{12} \geq t]$ *is the probability of the event* $[T_{12} \geq t]$ *computed under* H_0.

Sampling from Poisson or normal populations

Maximum likelihood estimates and distribution theory for likelihood ratio tests have been studied under the star-shaped restrictions, assuming independent samples from Poisson and normal populations. These results are contained in Shaked (1979) and Dykstra and Robertson (1982b, 1983).

5.4 LIKELIHOOD RATIO TESTS FOR AND AGAINST A STOCHASTIC ORDERING

One-sample estimates and tests

Consider the sampling situation described in Example 1.7.1. Specifically, assume that $\mathbf{q} = (q_1, q_2, \ldots, q_k)$ is a known possibility for $\mathbf{p}(q_i \geq 0, \sum_{i=1}^{k} q_i = 1)$ and let H_i, $i = 0, 1, 2$, denote the following hypotheses regarding $\mathbf{p}$:

$$H_0: \mathbf{p} = \mathbf{q},$$

$$H_1: \mathbf{p} \gg \mathbf{q} \left(\text{that is } \sum_{j=1}^{i} p_j \geq \sum_{j=1}^{i} q_j,\ i = 1, 2, \ldots, k \right),$$

and H_2 places no restrictions on $\mathbf{p}$ other than $p_i \geq 0$ and $\sum_{i=1}^{k} p_i = 1$. Let Λ_{01}

and Λ_{12} denote the likelihood ratios for testing H_0 against H_1–H_0 and for testing H_1 against H_2–H_1. Assuming that $\hat{p}_i > 0$, $i = 1, 2, \ldots, k$, and using Theorem 1.7.5, the MLE of $\mathbf{p}$ subject to H_1 is given by

$$\bar{\mathbf{p}} = \hat{\mathbf{p}} P_{\hat{\mathbf{p}}}(\mathbf{q}/\hat{\mathbf{p}} \mid \mathscr{A})$$

where $\mathscr{A} = \{\mathbf{y} = (y_1, y_2, \ldots, y_k) : y_1 \geqslant y_2 \geqslant \cdots \geqslant y_k\}$.

Theorem 5.4.1 *If* $\mathbf{p}$ *satisfies* H_1 *then* $\bar{\mathbf{p}}$ *converges almost surely to* $\mathbf{p}$ *as* $n \to \infty$.

Proof: By the strong law of large numbers, $\hat{\mathbf{p}} \to \mathbf{p}$ almost surely as $n \to \infty$. Moreover, by the upper sets algorithm, $P_{\mathbf{w}}(\mathbf{y} \mid \mathscr{A})$ is continuous in both $\mathbf{w}$ and $\mathbf{y}$ so that $\bar{\mathbf{p}} \to \mathbf{p}\, P_{\mathbf{p}}(\mathbf{q}/\mathbf{p} \mid \mathscr{A})$ a.s. Using the maximum upper sets algorithm to compute $P_{\mathbf{p}}(\mathbf{q}/\mathbf{p}) \mid \mathscr{A})$, one sees that since $\mathbf{p} \gg \mathbf{q}$, $\mathrm{Av}(\{1, 2, \ldots, i\}) \leqslant 1, i = 1, 2, \ldots, k$, with equality for $i = k$. Thus $P_{\mathbf{p}}(\mathbf{q}/\mathbf{p} \mid \mathscr{A})_i = 1$, $i = 1, 2, \ldots, k$, and $\mathbf{p}\, P_{\mathbf{p}}(\mathbf{q}/\mathbf{p} \mid \mathscr{A}) = \mathbf{p}$. □

Let $T_{01} = -2 \ln \Lambda_{01} = -2n \sum_{i=1}^{k} \hat{p}_i (\ln q_i - \ln \bar{p}_i)$.

Theorem 5.4.2 *If* H_0 *is true then for any real number* t,

$$\lim_{n \to \infty} P[T_{01} \geqslant t] = \sum_{l=1}^{k} P_S(l, k; \mathbf{q}) P[\chi^2_{k-l} \geqslant t],$$

where the $P_S(l, k; \mathbf{q})$ *are the level probabilities for the usual simple order with weight vector* $\mathbf{q}$.

Proof: Expanding $\ln q_i$ and $\ln \bar{p}_i$ about the point $\hat{p}_i$, T_{01} can be expressed as follows:

$$T_{01} = \sum_{i=1}^{k} \hat{p}_i \alpha_i^{-2} [\sqrt{n}(\hat{p}_i - q_i)]^2 - \sum_{i=1}^{k} \hat{p}_i \beta_i^{-2} [\sqrt{n}(\bar{p}_i - \hat{p}_i)^2],$$

where α_i is between $\hat{p}_i$ and q_i and β_i is between $\bar{p}_i$ and $\hat{p}_i$. Under H_0, the random vector $\sqrt{n}(\hat{\mathbf{p}} - \mathbf{q})$ converges in distribution to $(q_1(U_1 - \bar{U}), q_2(U_2 - \bar{U}), \ldots, q_k(U_k - \bar{U}))$ where, as in Theorem 5.2.1, $U_1, U_2, \ldots, U_k$ are independent normal variables which are centered at their expectations and have variances q_1^{-1}, $q_2^{-1}, \ldots, q_k^{-1}$ and $\bar{U} = \sum_{i=1}^{k} q_i U_i$. Using Theorem 4.4 of Billingsley (1968), it follows that, under H_0, T_{01} converges in law to

$$\sum_{i=1}^{k} q_i (U_i - \bar{U})^2 - \sum_{i=1}^{k} q_i [P_{\mathbf{q}}(\bar{U} - \mathbf{U} \mid \mathscr{A})_i]^2. \tag{5.4.1}$$

Noting that $P_{\mathbf{q}}(\bar{U} - \mathbf{U} \mid \mathscr{A}) = \bar{U} + P_{\mathbf{q}}(-\mathbf{U} \mid \mathscr{A})$, squaring the binomials in (5.4.1), combining terms, and using (1.4.4), (5.4.1) can be rewritten as

$$\sum_{i=1}^{k} q_i [P_{\mathbf{q}}(\mathbf{W} \mid \mathscr{A})_i - W_i]^2,$$

where $W_i = -U_i$. Theorem 2.3.1 gives the desired conclusion. □

Next, consider the one-sample likelihood ratio test of H_1 versus H_2-H_1. The test rejects for large values of the statistic

$$T_{12} = -2\ln \Lambda_{12} = -2n \sum_{i=1}^{k} \hat{p}_i[\ln \bar{p}_i - \ln \hat{p}_i].$$

Theorem 5.4.3 *For any* $\mathbf{p}$ *satisfying* H_1 *(that is* $\mathbf{p} \gg \mathbf{q}$*) and for all* t,

$$\lim_{n\to\infty} P_{\mathbf{p}}[T_{12} \geqslant t] \leqslant \lim_{n\to\infty} P_{\mathbf{q}}[T_{12} \geqslant t] \tag{5.4.2}$$

and

$$\lim_{n\to\infty} P_{\mathbf{q}}[T_{12} \geqslant t] = \sum_{l=1}^{k} P_S(l, k; \mathbf{q}) P[\chi^2_{l-1} \geqslant t] \tag{5.4.3}$$

where the level probabilities, $P_S(l, k; \mathbf{q})$, *are defined in Theorem 5.4.2.*

Proof: The proof of Theorem 5.4.3 is similar to the proof of Theorem 5.4.2 and so most of the details are ommitted. (Detailed arguments can be found in Robertson and Wright, 1981.) Writing a second-order Taylor expansion for $\ln \bar{p}_i$ about $\hat{p}_i$, T_{12} can be written

$$T_{12} = \sum_{i=1}^{k} \hat{p}_i v_i^{-2}[\sqrt{n}(\bar{p}_i - \hat{p}_i)^2]$$

where v_i is between $\bar{p}_i$ and $\hat{p}_i$ and therefore converges almost surely to p_i. Assuming that $\mathbf{p} \gg \mathbf{q}$, let $0 = \eta_0 < \eta_1 < \cdots < \eta_A = k$, where $p_1 + p_2 + \cdots + p_i = q_1 + q_2 + \cdots + q_i$ for $i = \eta_1, \eta_2, \ldots, \eta_A$ and $p_1 + p_2 + \cdots + p_i > q_1 + q_2 + \cdots + q_i$ for $i \neq \eta_1, \eta_2, \ldots, \eta_A$. Using the law of large numbers and the asymptotic normality of $\sqrt{n}(\hat{\mathbf{p}} - \mathbf{p})$, T_{12} converges in law to

$$\sum_{i=1}^{k} p_i[P_{\mathbf{p}}(\mathbf{W}|\mathscr{A}')_i - \bar{W}]^2 \tag{5.4.4}$$

where

$$\mathscr{A}' = \{\mathbf{y} \in \mathscr{A} : y_1 = \cdots = y_{\eta_1}, y_{\eta_1+1} = \cdots = y_{\eta_2}, \ldots, y_{\eta_{A-1}+1} = \cdots = y_{\eta_A}\}$$

and $W_1, W_2, \ldots, W_k$ are independent normals with zero means and variances $p_1^{-1}, p_2^{-1}, \ldots, p_k^{-1}$ $(\bar{W} = \sum_{i=1}^{k} p_i W_i)$.

Let $T_1, T_2, \ldots, T_k$ be independent normal variables with zero means and variances $q_1^{-1}, q_2^{-1}, \ldots, q_k^{-1}$, respectively. In computing $P_{\mathbf{p}}(\mathbf{W}|\mathscr{A}')$, one first computes the averages $W_j^* = \sum_{i=\eta_{j-1}+1}^{\eta_j} p_i W_i / \sum_{i=\eta_{j-1}+1}^{\eta_j} p_i$ for $j = 1, 2, \ldots, A$ and then projects the A-dimensional vector, $\mathbf{W}^* = (W_1^*, W_2^*, \ldots, W_A^*)$, onto the antitonic cone, $\mathscr{A}_A = \{\mathbf{y} \in R^A; y_1 \geqslant y_2 \geqslant \cdots \geqslant y_A\}$, with weights $P_j = \sum_{i=\eta_{j-1}+1}^{\eta_j} p_i = \sum_{i=\eta_{j-1}+1}^{\eta_j} q_i$. The vector $\mathbf{W}^*$ has the same distribution as the A-dimensional random vector, T^*, where $T_j^* = \sum_{i=\eta_{j-1}+1}^{\eta_j} q_i T_i / \sum_{i=\eta_{j-1}+1}^{\eta_j} q_i$. Thus,

$$\sum_{i=1}^{k} p_i[P_{\mathbf{p}}(\mathbf{W}|\mathscr{A}') - \bar{W}]^2 = \sum_{j=1}^{A} P_j[P_{\mathbf{P}}(\mathbf{W}^*|\mathscr{A}_A) - \bar{W}]^2$$

$$\sim \sum_{j=1}^{A} Q_j[P_{\mathbf{Q}}(\mathbf{T}^*|\mathscr{A}_A) - \bar{T}]^2 = \sum_{i=1}^{k} q_i[P_{\mathbf{q}}(\mathbf{T}|\mathscr{A}') - \bar{T}]^2 \tag{5.4.5}$$

where $Q_j = \sum_{i=\eta_{j-1}+1}^{\eta_j} q_i = P_j$ and $\bar{T} = \sum_{i=1}^{k} q_i T_i = \sum_{j=1}^{A} Q_j T_j^*$. The right-hand side of (5.4.5) depends upon the vector $\mathbf{p}$ only through $\mathscr{A}'$ and this random variable is largest when $\mathscr{A}'$ is largest as a set. Moreover, $\mathscr{A}'$ is largest when $\mathscr{A}' = \mathscr{A}$ and this is implied by $\mathbf{p} = \mathbf{q}$. This yields (5.4.2). The conclusion, (5.4.3), is a consequence of Theorem 2.3.1. □

If $\mathbf{q}$ is given then the hypothesis $H_1': \mathbf{q} \gg \mathbf{p}$ is equivalent to $\mathbf{p}' \gg \mathbf{q}'$ where $p_i' = p_{k-i+1}$ and $q_i' = q_{k-i+1}$, $i = 1, 2, \ldots, k$. Thus, Theorems 5.4.2 and 5.4.3 give the distribution theory for likelihood ratio tests of H_0 against $H_1' - H_0$ and H_1' against $H_2 - H_1'$.

Two-sample estimates and tests

Suppose $\hat{\mathbf{p}}$ and $\hat{\mathbf{q}}$ are the relative frequencies of successes corresponding to independent random samples of sizes m and n from the $\mathbf{p}$ and $\mathbf{q}$ populations, respectively. Let $N = m + n$ and let $(\bar{\mathbf{p}}, \bar{\mathbf{q}})$ be the MLE of $(\mathbf{p}, \mathbf{q})$ subject to H_1.

Lemma *The estimates $\bar{\mathbf{p}}$ and $\bar{\mathbf{q}}$ satisfy*

$$\sum_{j=1}^{i} \bar{p}_j \geqslant \sum_{j=1}^{i} \frac{m\hat{p}_j + n\hat{q}_j}{N} \geqslant \sum_{j=1}^{i} \bar{q}_j \tag{5.4.6}$$

for $i = 1, 2, \ldots, k-1$.

Proof: The proof is by induction on k. First suppose $k = 2$. In this case, one wishes to find the maximum of

$$L(p_1, q_1) = m\hat{p}_1 \ln p_1 + m\hat{p}_2 \ln(1 - p_1) + n\hat{q}_1 \ln q_1 + n\hat{q}_2 \ln(1 - q_1)$$

subject to the constraint that the point (p_1, q_1) lies in the triangular-shaped region bounded by $p_1 = q_1, p_1 = 1$, and $q_1 = 0$ (see Figure 5.4.1). The theorem asserts that the maximum must lie in the shaded rectangle. If $\hat{p}_1 \geqslant \hat{q}_1$ then $(\bar{\mathbf{p}}, \bar{\mathbf{q}}) = (\hat{\mathbf{p}}, \hat{\mathbf{q}})$ and this point lies in the shaded region since $\hat{p}_1 \geqslant (m\hat{p}_1 + n\hat{q}_1)/N \geqslant \hat{q}_1$. If $\hat{p}_1 < \hat{q}_1$, then two cases are considered. If the point $(\bar{\mathbf{p}}, \bar{\mathbf{q}})$ lies in the region to the left of the shaded rectangle (i.e. region 1), then one can increase L by moving vertically to the line $p_1 = q_1$ and then along this line to the point $((m\hat{p}_1 + n\hat{q}_1)/N, (m\hat{p}_1 + n\hat{q}_1)/N)$. If the point $(\bar{\mathbf{p}}, \bar{\mathbf{q}})$ lies in region 2, then one can increase L by

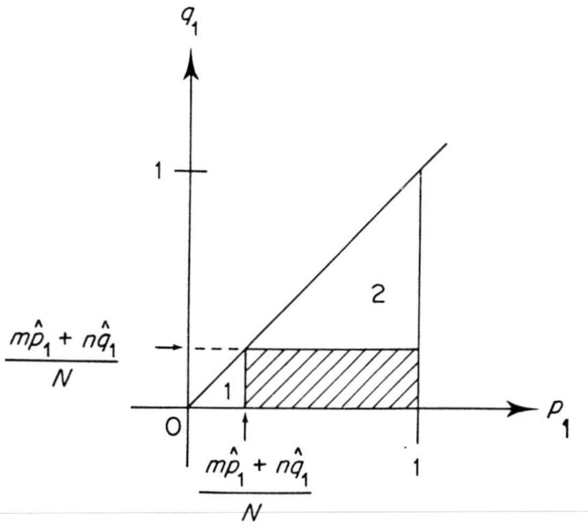

Figure 5.4.1 The constrained region for stochastic ordering with $k = 2$

moving first horizontally to the line $p_1 = q_1$ and then down this line to the point $((m\hat{p}_1 + n\hat{q}_1)/N, (m\hat{p}_1 + n\hat{q}_1)/N)$. This completes the proof for $k = 2$.

Suppose the result holds for $k < k_0$ and k_0 is the common dimension of $\mathbf{p}$ and $\mathbf{q}$. First observe that if all of the constraints hold strictly in the solution $(\bar{\mathbf{p}}, \bar{\mathbf{q}})$ then $(\bar{\mathbf{p}}, \bar{\mathbf{q}}) = (\hat{\mathbf{p}}, \hat{\mathbf{q}})$ and (5.4.6) holds. In order to see this suppose $\sum_{j=1}^{i} \bar{p}_j > \sum_{j=1}^{i} \bar{q}_j$, $i = 1, 2, \ldots, k_0 - 1$, and consider the function h, defined on $[0, 1]$ by $h(\lambda) = L(\lambda(\hat{\mathbf{p}}, \hat{\mathbf{q}}) + (1 - \lambda)(\mathbf{p}, \mathbf{q}))$. By differentiation, h is seen to be concave on $[0, 1]$ and since $(\hat{\mathbf{p}}, \hat{\mathbf{q}})$ provides the unconstrained maximum to L, the slope of the chord going from $(0, h(0))$ to $(1, h(1))$ is positive. Therefore, $h'(0)$ is at least as large as this slope. Thus $h'(0) > 0$ and one can increase L by moving from $(\bar{\mathbf{p}}, \bar{\mathbf{q}})$ in the direction of $(\hat{\mathbf{p}}, \hat{\mathbf{q}})$. Since the constraints hold strictly this yields the desired contradiction.

Now suppose $\sum_{j=1}^{i} \bar{p}_j = \sum_{j=1}^{i} \bar{q}_j$ for some $i < k_0$. The result (5.4.6) is then established by a straightforward argument, using the induction hypothesis and considering the two parts corresponding to $\{1, 2, \ldots, i\}$ and $\{i + 1, i + 2, \ldots, k_0\}$ separately. □

Theorem 5.4.4 *If $\hat{p}_i, \hat{q}_i > 0$, $i = 1, 2, \ldots, k$, then the maximum likelihood estimate of $(\mathbf{p}, \mathbf{q})$ subject to H_1 is given by $(\bar{\mathbf{p}}, \bar{\mathbf{q}})$ where*

$$\bar{\mathbf{p}} = \hat{\mathbf{p}} P_{\hat{\mathbf{p}}}\left(\frac{m\hat{\mathbf{p}} + n\hat{\mathbf{q}}}{N\hat{\mathbf{p}}} \middle| \mathscr{A}\right)$$

and

$$\bar{\mathbf{q}} = \hat{\mathbf{q}} P_{\hat{\mathbf{q}}}\left(\frac{m\hat{\mathbf{p}} + n\hat{\mathbf{q}}}{N\hat{\mathbf{q}}} \middle| \mathscr{I}\right).$$

Proof: The solution for $\bar{\mathbf{p}}$ is obtained by first applying the lemma and then the solution to the one-sample problem with $q_i = (m\hat{p}_i + n\hat{q}_i)/N$. The solution for $\mathbf{q}$ is obtained by relabeling: $q_i' = q_{k-i+1}$, $i = 1, 2, \ldots, k$. □

The next two theorems give the asymptotic distribution of the likelihood ratio statistics for testing H_0 against $H_1 - H_0$ and for testing H_1 against $H_2 - H_1$. Let $T_{01} = -2 \ln \Lambda_{01}$ where Λ_{01} is the likelihood ratio statistic for testing H_0 against $H_1 - H_0$. Then T_{01} can be expressed as

$$T_{01} = 2m \sum_{i=1}^{k} \hat{p}_i(\ln \bar{p}_i - \ln p_i^0) + 2n \sum_{i=1}^{k} \hat{q}_i(\ln \bar{q}_i - \ln q_i^0)$$

where $(\bar{\mathbf{p}}, \bar{\mathbf{q}})$ is given by Theorem 5.4.4 and

$$p_i^0 = q_i^0 = \frac{m\hat{p}_i + n\hat{q}_i}{N} \qquad i = 1, 2, \ldots, k.$$

Theorem 5.4.5 *If* $\mathbf{p} = \mathbf{q}$ *then for each real* t,

$$\lim_{m,n \to \infty} P[T_{01} \geqslant t] = \sum_{l=1}^{k} P_S(l, k; \mathbf{p}) P[\chi^2_{k-l} \geqslant t], \tag{5.4.7}$$

where the level probabilities, $P_S(l, k; \mathbf{p})$, *are defined in Theorem* 5.4.2. *Furthermore,*

$$\sup_{\mathbf{p}=\mathbf{q}} \lim_{m,n \to \infty} P[T_{01} \geqslant t] = \tfrac{1}{2}[P(\chi^2_{k-1} \geqslant t) + P(\chi^2_{k-2} \geqslant t)]. \tag{5.4.8}$$

Proof: The proof of (5.4.7) is similar to the proofs of the preceding section. It uses a Taylor expansion, the asymptotic normality of $\sqrt{m}(\hat{\mathbf{p}} - \mathbf{p})$ and $\sqrt{n}(\hat{\mathbf{q}} - \mathbf{q})$, together with properties of $P_{\mathbf{w}}(\cdot|\mathscr{A})$ and $P_{\mathbf{w}}(\cdot|\mathscr{I})$ which allow moving quantities in and out of the projection operator. Details can be found in Robertson and Wright (1981). The conclusion (5.4.8) follows from (5.4.7) and the bounds given in Theorem 3.6.1. □

Next, consider a likelihood ratio test of H_1 against $H_2 - H_1$. The test statistic, $T_{12} = -2 \ln \Lambda_{12}$, can be written as

$$T_{12} = -2m \sum_{i=1}^{k} \hat{p}_i(\ln \bar{p}_i - \ln \hat{p}_i) - 2n \sum_{i=1}^{k} \hat{q}_i(\ln \bar{q}_i - \ln \hat{q}_i).$$

Theorem 5.4.6 *If* $P_{\mathbf{p},\mathbf{q}}(E)$ *denotes the probability of the event* E *computed under the assumption that* $\mathbf{p}$ *and* $\mathbf{q}$ *are the values of the parameter vectors, then for each real* t,

$$\sup_{\mathbf{p} \gg \mathbf{q}} \lim_{m,n \to \infty} P_{\mathbf{p},\mathbf{q}}(T_{12} \geqslant t) = \sup_{\mathbf{p}=\mathbf{q}} \lim_{m,n \to \infty} P_{\mathbf{p},\mathbf{q}}(T_{12} \geqslant t) \tag{5.4.9}$$

and

$$\sup_{\mathbf{p}=\mathbf{q}} \lim_{m,n\to\infty} P_{\mathbf{p},\mathbf{q}}(T_{12} \geqslant t) = \sum_{l=1}^{k} \binom{k-1}{l-1} 2^{-k+1} P(\chi^2_{l-1} \geqslant t). \qquad (5.4.10)$$

Proof: The proof of (5.4.9), which is similar to the proofs given earlier in this section, uses a Taylor expansion and the asymptotic normality of $\sqrt{m}(\hat{\mathbf{p}} - \mathbf{p})$ and $\sqrt{n}(\hat{\mathbf{q}} - \mathbf{q})$ to argue that if $\mathbf{p} \gg \mathbf{q}$ then

$$T_{12} \xrightarrow{D} \sum_{i=1}^{k} q_i[P_{\mathbf{q}}(\mathbf{W} \mid \mathscr{A}')_i - \bar{W}]^2$$

where $\mathscr{A}'$ is defined as in the proof of Theorem 5.4.3 and $W_1, W_2, \ldots, W_k$ are independent normal random variables whose means are zero and whose variances depend upon $\mathbf{p}, \mathbf{q}$, and the way in which m and n go to infinity. The conclusion (5.4.9) follows as in the proof of Theorem 5.4.3 because this distribution is largest when $\mathscr{A}'$ is the largest (that is $\mathbf{p} = \mathbf{q}$). The conclusion (5.4.10) follows from Theorem 2.3.1 and the bound given in Theorem 3.6.1. □

It is of interest to observe that the results in Theorems 5.4.5 and 5.4.6 are valid as long as $m, n \to \infty$ in such a way that $m/N \to a \in [0, 1]$ and, in fact, a could be either 0 or 1 ($N = m + n$).

5.5 LIKELIHOOD RATIO TESTS FOR MULTINOMIAL PROBLEMS IN WHICH BOTH THE NULL AND ALTERNATIVE IMPOSE ORDER RESTRICTIONS

This is one of the areas of order restricted inference which has not been widely researched at this time. However, some progress has been made in that several examples have been studied and there is some similarity in the nature of the results in these various examples. The problem considered in this section involves two properties which play important roles in statistical theory, namely symmetry and unimodality.

Suppose one has m multinomial trials with parameter vector $\mathbf{p} = (p_1, p_2, \ldots, p_k)$ and, in order to be specific, assume that k is odd. Analogous results hold when k is even. Define the two subsets $\mathscr{S}$ and $\mathscr{U}$ of R^k by

$$\mathscr{S} = \{\mathbf{y}: y_i = y_{k-i+1}; i = 1, 2, \ldots, k\}$$

and

$$\mathscr{U} = \{\mathbf{y}: y_1 \leqslant y_2 \leqslant \cdots \leqslant y_{(k+1)/2} \geqslant \cdots \geqslant y_k\}.$$

Consider the likelihood ratio tests of H_0 against $H_1 - H_0$ and against $H_1' - H_0$

where

$$\begin{aligned} &H_0: \mathbf{p} \in \mathscr{S} \cap \mathscr{U} \\ &H_1: \mathbf{p} \in \mathscr{U} \\ &H_1': \mathbf{p} \in \mathscr{S}. \end{aligned} \tag{5.5.1}$$

Membership in all three of the sets $\mathscr{S} \cap \mathscr{U}$, $\mathscr{U}$, and $\mathscr{S}$ can be thought of as constraining the point to be isotonic with respect to an appropriate quasi-order so that the MLEs of $\mathbf{p}$ under H_0, H_1, and H_1' can be found by taking the isotonic regression of $\mathbf{p}$, with equal weights, onto $\mathscr{S} \cap \mathscr{U}$, $\mathscr{U}$, and $\mathscr{S}$, respectively. The projection, $P(\hat{\mathbf{p}} \mid \mathscr{S})$, is relatively easy to compute since $\mathscr{S}$ is a linear subspace of R^k. In fact,

$$P(\hat{\mathbf{p}} \mid \mathscr{S})_i = \frac{\hat{p}_i + \hat{p}_{k-i+1}}{2} \qquad \text{for } i = 1, 2, \ldots, k.$$

Hence, if $\hat{\mathbf{p}}'$ is defined by $\hat{p}_i' = \hat{p}_{k-i+1}$ then $P(\hat{\mathbf{p}} \mid \mathscr{S}) = (\hat{\mathbf{p}} + \hat{\mathbf{p}}')/2$. The most convenient way to compute $P(\hat{\mathbf{p}} \mid \mathscr{U})$ is via the upper sets algorithm. The upper sets for $\mathscr{U}$ are the sets of consecutive integers containing $(k+1)/2$, namely

$$\left\{\frac{k+1}{2}\right\}, \quad \left\{\frac{k+1}{2}, \frac{k+1}{2} + 1\right\}, \quad \left\{\frac{k+1}{2} - 1, \frac{k+1}{2}\right\},$$

$$\left\{\frac{k+1}{2} - 1, \frac{k+1}{2}, \frac{k+1}{2} + 1\right\}, \quad \text{etc.}$$

Find the upper set which gives the largest average value of $\hat{\mathbf{p}}$ over upper sets, say $\{a, a+1, \ldots, b\}$. The values of $P(\hat{\mathbf{p}} \mid \mathscr{U})$ on $\{1, 2, \ldots, a-1\}$ and $\{b+1, b+2, \ldots, k\}$ can now be computed independently using the upper sets algorithm or the PAVA. The projection, $P(\hat{\mathbf{p}} \mid \mathscr{S} \cap \mathscr{U})$, can be computed iteratively via the following theorem.

Theorem 5.5.1 *If $\mathscr{T}$ is a linear subspace of R^k and if $\mathscr{K}$ is a closed, convex cone in R^k and if for each $\mathbf{y} \in R^k$, $P(P(\mathbf{y} \mid \mathscr{T}) \mid \mathscr{K}) \in \mathscr{T} \cap \mathscr{K}$, then $P(\mathbf{y} \mid \mathscr{T} \cap \mathscr{K}) = P(P(\mathbf{y} \mid \mathscr{T}) \mid \mathscr{K})$.*

Proof: It is a straightforward exercise to verify that $P(P(\mathbf{y} \mid \mathscr{T}) \mid \mathscr{K})$ has the two properties characterizing $P(\mathbf{y} \mid \mathscr{T} \cap \mathscr{K})$ in Theorem 1.3.2. □

The subset $\mathscr{S}$ is a linear subspace in R^k and $\mathscr{U}$ is a closed, convex cone and it is clear from the maximum upper sets algorithm that $P(P(\hat{\mathbf{p}} \mid \mathscr{S}) \mid \mathscr{U}) \in \mathscr{S} \cap \mathscr{U}$ so that $P(\hat{\mathbf{p}} \mid \mathscr{S} \cap \mathscr{U})$ can be computed using Theorem 5.5.1. Incidentally, it is easy to construct examples to show that $P(\mathbf{y} \mid \mathscr{S} \cap \mathscr{U}) \neq P(P(\mathbf{y} \mid \mathscr{U}) \mid \mathscr{S})$.

Let $T_{01} = -2 \ln \Lambda_{01}$, where Λ_{01} is the likelihood ratio for testing H_0 against

H_1–H_0. A proof similar to that of Theorem 5.2.2 yields, for any $\mathbf{p}\in H_0$,

$$T_{01} \xrightarrow{D} \sum_{i=1}^{k} \{[P(\mathbf{Z}|\mathscr{S}\cap\mathscr{U}(\mathbf{p}))_i - Z_i]^2 - [P(\mathbf{Z}|\mathscr{U}(\mathbf{p}))_i - Z_i]^2\}$$

where $Z_1, Z_2, \ldots, Z_k$ are independent standard normal variables and, as in the proof of Theorem 5.2.2, $\mathscr{U}(\mathbf{p})$ imposes the order restrictions in $\mathscr{U}$ only over those subscripts where $\mathbf{p}$ is constant. For example, if $k=7$ and $p_1 = p_2 < p_3 = p_4 = p_5 > p_6 = p_7$ then $\mathscr{U}(\mathbf{p}) = \{\mathbf{y}\colon y_1 \leqslant y_2,\ y_3 \leqslant y_4 \geqslant y_5,\ y_6 \geqslant y_7\}$. These limiting distributions depend upon $\mathbf{p}$ through $\mathscr{U}(\mathbf{p})$ and the following theorem identifies the largest of these distributions in a stochastic sense.

Theorem 5.5.2 *Assume that $\mathscr{K} \subset R^k$ is a closed, convex cone having the properties that $\|\mathbf{y} - P_w(\mathbf{y}|\mathscr{K})\|_w^2 = \|\mathbf{y}' - P_w(\mathbf{y}'|\mathscr{K})\|_w^2$ for all $\mathbf{y}\in R^k (y_i' = y_{k-i+1})$ and $P_w(P_w(\mathbf{y}|\mathscr{S})|\mathscr{K}) \in \mathscr{S}\cap\mathscr{K}$ for all $\mathbf{y}\in R^k$. If $L(\mathbf{y}) = \|\mathbf{y} - P_w(\mathbf{y}|\mathscr{S}\cap\mathscr{K})\|_w^2 - \|\mathbf{y} - P_w(\mathbf{y}|\mathscr{K})\|_w^2$ then $L(\mathbf{y}) \leqslant \|\mathbf{y} - P_w(\mathbf{y}|\mathscr{S})\|_w^2$ for all $\mathbf{y}\in R^k$.*

Proof:

$$\begin{aligned} L(\mathbf{y}) = {}& \|\mathbf{y} - P_w(\mathbf{y}|\mathscr{S})\|_w^2 + 2\langle \mathbf{y} - P_w(\mathbf{y}|\mathscr{S}), P_w(\mathbf{y}|\mathscr{S}) - P_w(\mathbf{y}|\mathscr{S}\cap\mathscr{K})\rangle_w \\ & + \|P_w(\mathbf{y}|\mathscr{S}) - P_w(\mathbf{y}|\mathscr{S}\cap\mathscr{K})\|_w^2 - \|\mathbf{y} - P_w(\mathbf{y}|\mathscr{K})\|_w^2. \end{aligned}$$

The inner product is zero since $\mathscr{S}$ is a linear subspace so that it suffices to argue that $\|P_w(\mathbf{y}|\mathscr{S}) - P_w(\mathbf{y}|\mathscr{S}\cap\mathscr{K})\|_w^2 \leqslant \|\mathbf{y} - P_w(\mathbf{y}|\mathscr{K})\|_w^2$. However,

$$\begin{aligned} \|P_w(\mathbf{y}|\mathscr{S}) - P_w(\mathbf{y}|\mathscr{S}\cap\mathscr{K})\|_w^2 &= \tfrac{1}{4}\|\mathbf{y} + \mathbf{y}' - P_w(\mathbf{y}+\mathbf{y}'|\mathscr{K})\|_w^2 \\ &\leqslant \tfrac{1}{4}\|\mathbf{y} + \mathbf{y}' - (P_w(\mathbf{y}|\mathscr{K}) + P_w(\mathbf{y}'|\mathscr{K}))\|_w^2 \\ &\leqslant \tfrac{1}{4}\{2[\|\mathbf{y} - P_w(\mathbf{y}|\mathscr{K})\|_w^2 + \|\mathbf{y}' - P_w(\mathbf{y}'|\mathscr{K})\|_w^2]\} \\ &= \|\mathbf{y} - P_w(\mathbf{y}|\mathscr{K})\|_w^2. \end{aligned}$$

The first inequality follows because $P_w(\mathbf{y}+\mathbf{y}'|\mathscr{K})$ minimizes $\|\mathbf{y}+\mathbf{y}'-\mathbf{z}\|_w$ for $\mathbf{z}\in\mathscr{K}$ and $P_w(\mathbf{y}|\mathscr{K}) + P_w(\mathbf{y}'|\mathscr{K}) \in \mathscr{K}$ and the second inequality follows from the easily verified identity

$$\|\mathbf{u}+\mathbf{v}\|_w^2 \leqslant \|\mathbf{u}+\mathbf{v}\|_w^2 + \|\mathbf{u}-\mathbf{v}\|_w^2 = 2(\|\mathbf{u}\|_w^2 + \|\mathbf{v}\|_w^2). \qquad (5.5.2)$$

Now, in the limiting distribution of T_{01}, $\mathscr{U}(\mathbf{p})$ is a closed, convex cone satisfying the assumption placed on $\mathscr{U}$ in Theorem 5.5.2. Thus for any real number t and any $\mathbf{p}\in\mathscr{S}\cap\mathscr{U}$,

$$\lim_{n\to\infty} P_{\mathbf{p}}[T_{01} \geqslant t] \leqslant P[\|\mathbf{Z} - P(\mathbf{Z}|\mathscr{S})\|^2 \geqslant t].$$

Moreover, if

$$p_1 < p_2 < \cdots < p_{(k+1)/2} > p_{(k+3)/2} > \cdots > p_k,$$

then $\mathscr{U}(\mathbf{p}) = R^k$ so that $P[\|\mathbf{Z} - P(\mathbf{Z}|\mathscr{S})\|^2 \geqslant t]$ is actually the supremum. Since

$P(\mathbf{Z}|\mathscr{S}) = (\mathbf{Z} + \mathbf{Z}')/2$ it is straightforward to show that $\|\mathbf{Z} - P(\mathbf{Z}|\mathscr{S})\|^2$ is the sum of $(k-1)/2$ squares of independent standard normal variables. A similar result holds when k is even. □

Theorem 5.5.3 *If $T_{01} = -2\ln\Lambda_{01}$, where Λ_{01} is the likelihood ratio for testing H_0 against H_1–H_0, then*

$$\sup_{\mathbf{p}\in H_0}\lim_{n\to\infty} P_{\mathbf{p}}[T_{01} \geqslant t] = P[\chi^2_{(k-1)/2} \geqslant t]$$

if k is odd and

$$\sup_{\mathbf{p}\in H_0}\lim_{n\to\infty} P_{\mathbf{p}}[T_{01} \geqslant t] = P[\chi^2_{k/2} \geqslant t]$$

if k is even.

Testing H_0 against H_1'–H_0

Consider testing the null hypothesis, H_0, against the alternative H_1'–H_0 with H_1': $\mathbf{p}\in\mathscr{S}$. Let Λ_{01}' be the likelihood ratio and let $T_{01}' = -2\ln\Lambda_{01}'$. Proceeding as in the first problem, if the true parameter vector, $\mathbf{p}$, satisfies H_0 then T_{01}' converges in distribution to $\|\mathbf{Z} - P(\mathbf{Z}|\mathscr{S}\cap\mathscr{U}(\mathbf{p}))\|^2 - \|\mathbf{Z} - P(\mathbf{Z}|\mathscr{S})\|^2$, where $\mathbf{Z} = (Z_1, Z_2, \ldots, Z_k)$ is a vector of independent standard normal variables. Now

$$\begin{aligned}\|\mathbf{Z} - P(\mathbf{Z}|\mathscr{S}\cap\mathscr{U}(\mathbf{p}))\|^2 = {}&\|\mathbf{Z} - P(\mathbf{Z}|\mathscr{S})\|^2 \\ &+ 2\langle \mathbf{Z} - P(\mathbf{Z}|\mathscr{S}), P(\mathbf{Z}|\mathscr{S}) - P(\mathbf{Z}|\mathscr{S}\cap\mathscr{U}(\mathbf{p}))\rangle \\ &+ \|P(\mathbf{Z}|\mathscr{S}) - P(\mathbf{Z}|\mathscr{S}\cap\mathscr{U}(\mathbf{p}))\|^2\end{aligned}$$

and the inner product term is zero so that T_{01}' converges in distribution to $\|P(\mathbf{Z}|\mathscr{S}) - P(\mathbf{Z}|\mathscr{S}\cap\mathscr{U}(\mathbf{p}))\|^2$. Let $H = (k+1)/2$ and define the H-dimensional random vector $\mathbf{W}$ by $W_i = (Z_i + Z_{k-i+1})/2$, $i = 1, 2, \ldots, H$. Define $\mathscr{I}(\mathbf{p})$ as $\{\mathbf{y}\in R^H: y_i \leqslant y_{i+1}$ if $p_i = p_{i+1}\}$. For example, if $p_1 = p_2 = \cdots = p_H$ then $\mathscr{I}(\mathbf{p}) = \{\mathbf{y}\in R^H: y_1 \leqslant y_2 \leqslant \cdots \leqslant y_H\}$. Using the fact that $P(\mathbf{Z}|\mathscr{S}\cap\mathscr{U}(\mathbf{p})) = P(P(\mathbf{Z}|\mathscr{S})|\mathscr{U}(\mathbf{p}))$ and the maximum upper sets algorithm it follow that $P(\mathbf{Z}|\mathscr{S}\cap\mathscr{U}(\mathbf{p}))_i = P_{\mathbf{w}}(\mathbf{W}|\mathscr{I}(\mathbf{p}))_i$, $i = 1, 2, \ldots, H$, where $\mathbf{w}$ is the H-dimensional weight vector $\mathbf{w} = (2, 2, \ldots, 2, 1)$. Thus

$$\|P(\mathbf{Z}|\mathscr{S}) - P(\mathbf{Z}|\mathscr{S}\cap\mathscr{U}(\mathbf{p}))\|^2 = \sum_{i=1}^{H} w_i(W_i - P_{\mathbf{w}}(\mathbf{W}|\mathscr{I}(\mathbf{p})))^2.$$

The right-hand side is just the weighted distance in R^H from $\mathbf{W}$ to $\mathscr{I}(\mathbf{p})$ and for all $\mathbf{p}\in\mathscr{S}\cap\mathscr{U}$, $\mathscr{I}(\mathbf{p}) \supset \mathscr{I}_H = \{\mathbf{y}: y_1 \leqslant y_2 \leqslant \cdots \leqslant y_H\}$. Moreover, $W_1, W_2, \ldots, W_H$ are independent and $W_i \sim \mathscr{N}(0, \frac{1}{2})$, $i = 1, 2, \ldots, H-1$, and $W_H \sim \mathscr{N}(0, 1)$. The following theorem is a consequence of this discussion and Theorem 2.3.1.

Theorem 5.5.4 *Let $T_{01}' = -2\ln\Lambda_{01}'$, where Λ_{01}' is the likelihood ratio for testing H_0 against H_1'–H_0 with H_1' given in* (5.5.1). *Then*

$$\sup_{\mathbf{p}\in H_0}\lim_{n\to\infty} P_{\mathbf{p}}[T_{01}' \geqslant t] = \sum_{l=1}^{H} P_S(l, H; \mathbf{w})P[\chi^2_{H-l} \geqslant t]$$

where $\mathbf{w}$ *is the* H*-dimensional weight vector* $(2, 2, \ldots, 2, 1)$ *if* k *is odd. If* k *is even then* $H = (k/2)$ *and* $\mathbf{w} = (2, 2, \ldots, 2)$.

A caveat As we have seen in several problems, the LRT statistic in a multinomial setting has a limiting behavior which is the same as the behavior of the LRT statistic for the analogous normal means problem. Consider the normal means problems which are analogous to the problems considered in this section. Specifically, suppose one has independent random samples from k normal populations with means $\mu_1, \mu_2, \ldots, \mu_k$ and one wants to test $H_0: \boldsymbol{\mu} \in \mathscr{S} \cap \mathscr{U}$ against $H_1 - H_0$ with $H_1: \boldsymbol{\mu} \in \mathscr{U}$, where $\mathscr{U}$ and $\mathscr{S}$ are the subsets of R^k defined in (5.5.1). Assume that the population variances, $\sigma_1^2, \sigma_2^2, \ldots, \sigma_k^2$, are known and let the sample size and sample means be denoted by $n_1, n_2, \ldots, n_k$ and $\bar{Y}_1, \bar{Y}_2, \ldots, \bar{Y}_k$. The maximum likelihood estimates of $\boldsymbol{\mu}$ subject to $\boldsymbol{\mu} \in \mathscr{S} \cap \mathscr{U}$ and $\boldsymbol{\mu} \in \mathscr{U}$ are given by $P_{\mathbf{w}}(\bar{\mathbf{Y}} | \mathscr{S} \cap \mathscr{U})$ and $P_{\mathbf{w}}(\bar{\mathbf{Y}} | \mathscr{U})$ where $w_i = n_i/\sigma_i^2$. These projections may be computed in the same manner as described for the projection of $\hat{\mathbf{p}}$ onto these sets.

If $T_{01} = -2 \ln \Lambda_{01}$, where Λ_{01} is the likelihood ratio, then $T_{01} = \sum_{i=1}^k \{w_i [\bar{Y}_i - P_{\mathbf{w}}(\bar{\mathbf{Y}} | \mathscr{S} \cap \mathscr{U})_i]^2 - w_i [\bar{Y}_i - P_{\mathbf{w}}(\bar{Y}_i - P_{\mathbf{w}}(\bar{\mathbf{Y}} | \mathscr{U}_i)]^2\}$. If the weights are symmetric (that is $w_i = w_{k-i+1}$) then one can apply Theorem 5.5.2 to obtain

$$P_{\boldsymbol{\mu}}[T_{01} \geq t] \leq P_{\boldsymbol{\mu}}\left[\sum_{i=1}^{k} w_i [\bar{Y}_i - P_{\mathbf{w}}(\bar{\mathbf{Y}} | \mathscr{S})_i]^2 \geq t\right]$$

$$= P_0\left[\sum_{i=1}^{k} w_i [\bar{Y}_i + \mu_i - P_{\mathbf{w}}(\bar{\mathbf{Y}} + \boldsymbol{\mu} | \mathscr{S})_i]^2 \geq t\right]$$

$$= P_0\left[\sum_{i=1}^{k} w_i [\bar{Y}_i - P_{\mathbf{w}}(\bar{\mathbf{Y}} | \mathscr{S})_i]^2 \geq t\right]$$

since if $\boldsymbol{\mu} \in \mathscr{S} \cap \mathscr{U}$ then $P_{\mathbf{w}}(\bar{\mathbf{Y}} + \boldsymbol{\mu} | \mathscr{S}) = P_{\mathbf{w}}(\bar{\mathbf{Y}} | \mathscr{S}) + \boldsymbol{\mu}$. Moreover if k is odd and we take $\boldsymbol{\mu} = (\delta, 2\delta, \ldots, [(k+1)/2]\delta, [(k-1)/2]\delta, \ldots, \delta)$ and let $\delta \to \infty$, we find that $P_{\boldsymbol{\mu}}[T_{01} > t]$ converges to this upper bound. This establishes the following result.

Theorem 5.5.5 *Under the assumptions described above,*

$$\sup_{\boldsymbol{\mu} \in \mathscr{S} \cap \mathscr{U}} P_{\boldsymbol{\mu}}[T_{01} \geq t] = \begin{cases} P[\chi^2_{(k+1)/2} \geq t] & \text{if } k \text{ is odd} \\ P[\chi^2_{k/2} \geq t] & \text{if } k \text{ is even.} \end{cases}$$

Theorem 5.5.5 has some negative implications concerning the likelihood ratio tests of a null hypothesis of symmetry and unimodality against an alternative of unimodality but not symmetry. Let $T_{02} = -2 \ln \Lambda_{02}$, where Λ_{02} is the likelihood ratio for testing the null hypothesis that $\boldsymbol{\mu} \in \mathscr{S}$ against all alternatives ($T_{02} = \sum_{i=1}^k w_i [\bar{Y}_i - P(\bar{\mathbf{Y}} | \mathscr{S})_i]^2$), and consider using T_{02} to test H_0 against $H_1 - H_0$. Then for any $\boldsymbol{\mu} \in \mathscr{S} \cap \mathscr{U}$, $P_{\boldsymbol{\mu}}[T_{02} \geq t] = P[\chi^2_{(k-1)/2} \geq t]$ so that T_{01} and T_{02} have the same critical values and moreover by Theorem 5.5.2, $T_{01} \leq T_{02}$ so that T_{02} is uniformly more powerful than T_{01} for testing H_0 against $H_1 - H_0$. This is a rather surprising result since a test based upon T_{02} neglects much information about $\boldsymbol{\mu}$, namely that $\boldsymbol{\mu}\in\mathscr{U}$. The result is a consequence of the least favorable configuration for T_{01} ($\lim_{\delta\to\infty}(\delta, 2\delta, \ldots, \delta)$) and the fact that the power function of T_{02} is constant over H_0. Intuition tells us that we should be able to do better than T_{02}, but the authors do not know how to accomplish this. Perhaps the critical value for T_{01} should be adjusted to account for the fact that in most problems $\boldsymbol{\mu}$ is bounded. An alternative to the LRT might be to conduct a conditional test, an idea which is now receiving more attention.

The above remarks do not apply to the problem of testing the null hypothesis of symmetry and unimodality against the alternative of symmetry but not unimodality. However, this same phenomenon does occur in other testing problems where the null hypothesis constrains the parameter vector to the intersection of a linear space and a cone and the alternative requires this parameter vector to be in the cone but not in the linear space. One of these testing problems is motivated by a problem in psychiatry and is discussed in Warrack and Robertson (1984) and in Robertson and Warrack (1985). The other problem that the caveat applies to is the problem of testing symmetry and unimodality in a contingency table against the alternative of unimodality but not symmetry. This problem will be discussed briefly in the next section. Additional material and details of the results described here on the problems of testing symmetry and unimodality can be found in Robertson (1986).

5.6 CONTINGENCY TABLES

The area of statistics involving contingency tables is rich with problems in which restrictions on the parameter space can be exploited in statistical analyses. Of course, if the restriction is such that the parameters are required to be isotonic with respect to a quasi-order on some appropriate set indexing the parameter space then all the theory developed in the first five sections of this chapter applies. There is a large body of literature concerning the analysis of categorical data for which the row or column variables are ordinal measurements. It may or may not be appropriate to assume that in these problems the parameters themselves satisfy certain order restrictions such as the matrix order (cf. Examples 1.3.1 and 1.5.2). Attention in this section is focused on inference problems in which the parameters are subject to order restrictions.

At the time of the writing of this monograph other types of restrictions are beginning to be explored but, in fact, much of this research has not yet appeared in the literature. Virtually all of this theory is for two-way tables and so attention will be restricted to that setting. Assume n multinomial trials and assume that the parameter vector $\mathbf{p}$ is an $R\times C$ matrix denoted by $\mathbf{p}=(p_{ij})_{i,j=1}^{R,C}=(p_{ij})$.

Assume that $R=C$. The set of symmetric $R\times R$ matrices is defined to be $\mathscr{S}'=\{\mathbf{y}=(y_{ij}):y_{ij}=y_{R-j+1,R-i+1}\}$ and the set of unimodal matrices is defined to be $\mathscr{U}'=\{\mathbf{y}=(y_{ij}):y_{ij}\leqslant y_{i'j'}$ if $i\leqslant i'$, $j\leqslant j'$, $i+j\leqslant R+1$, $i'+j'\leqslant R+1$ and

$y_{i,j} \geqslant y_{i',j'}$ if $i \leqslant i', j \leqslant j', i+j \geqslant R+1, i'+j' \geqslant R+1\}$. The restriction that $\mathbf{y} \in \mathscr{S}'$ requires $\mathbf{y}$ to be symmetric about the diagonal running from lower left to upper right. Membership in $\mathscr{U}'$ requires that $\mathbf{y}$ is nondecreasing as one moves down or to the right and is above that diagonal and nonincreasing on the other side of the diagonal.

Let $R_{01} = -2\ln\Lambda_{01}$, where Λ_{01} is the likelihood ratio statistic for testing the null hypothesis

$$H_0 : \mathbf{p} \in \mathscr{S}' \cap \mathscr{U}'$$

against the alternative $H_1 - H_0$ where

$$H_1 : \mathbf{p} \in \mathscr{U}'.$$

The maximum likelihood estimates subject to $\mathbf{p} \in \mathscr{S}' \cap \mathscr{U}'$ and $\mathbf{p} \in \mathscr{U}'$ are again least squares projections, $P(\hat{\mathbf{p}} | \mathscr{S}' \cap \mathscr{U}')$ and $P(\hat{\mathbf{p}} | \mathscr{U}')$. Also, by the maximum upper sets algorithm for $P(\cdot | \mathscr{U}')$, $P(P(\mathbf{y} | \mathscr{S}') | \mathscr{U}') \in \mathscr{S}' \cap \mathscr{U}'$, so by Theorem 5.5.1, $P(\mathbf{y} | \mathscr{S}' \cap \mathscr{U}') = P(P(\mathbf{y} | \mathscr{S}') | \mathscr{U}')$ and it is clear that $P(\mathbf{y} | \mathscr{S}') = (\mathbf{y} + \mathbf{y}')/2$ where $y'_{ij} = y_{R-j+1,R-i+1}$, $i, j = 1, 2, \ldots, R$.

The techniques developed in Section 5.5 carry over to this problem so that under H_0, $R_{01} \xrightarrow{D} \| \mathbf{Z} - P(\mathbf{Z} | \mathscr{S}' \cap \mathscr{U}'(\mathbf{p})) \|^2 - \| \mathbf{Z} - P(\mathbf{Z} | \mathscr{U}'(\mathbf{p})) \|^2$ where $\mathscr{U}'(\mathbf{p})$ imposes the constraints in $\mathscr{U}'$ only over subscripts where $\mathbf{p}$ is constant. Theorem 5.5.2 carries over to this new setting with $y'_{ij} = y_{R-j+1,R-i+1}$ so that $\lim_{n\to\infty} P_{\mathbf{p}}[R_{01} \geqslant t] \leqslant P[\| \mathbf{Z} - P(\mathbf{Z} | \mathscr{S}') \|^2 \geqslant t]$ with Z_{ij}, $i, j = 1, 2, \ldots, R$, independent standard normal variables. Moreover, for any $\mathbf{p} \in H_0$ for which none of the values p_{ij} are equal, $\mathscr{U}'(\mathbf{p}) = R^{k^2}$ and one obtains the following result.

Theorem 5.6.1 *Let $R_{01} = -2\ln\Lambda_{01}$, where Λ_{01} is the likelihood ratio for testing $H_0 : \mathbf{p} \in \mathscr{S}' \cap \mathscr{U}'$ against $H_1 - H_0$ with $H_1 : \mathbf{p} \in \mathscr{U}'$. Then*

$$\sup_{\mathbf{p} \in H_0} \lim_{n\to\infty} P_{\mathbf{p}}[R_{01} \geqslant t] = P[\| \mathbf{Z} - P(\mathbf{Z} | \mathscr{S}') \|^2 \geqslant t]$$

$$= P[\chi^2_{(k^2-k)/2} \geqslant t].$$

The caveat which was discussed regarding Theorems 5.5.4 and 5.5.5 also applies to Theorem 5.6.1.

Now consider the likelihood ratio test of H_0 against $H'_1 - H_0$ where $H'_1 : \mathbf{p} \in \mathscr{S}'$. If $R'_{01} = -2\ln\Lambda'_{01}$ then for any $\mathbf{p} \in \mathscr{S}' \cap \mathscr{U}'$, R'_{01} converges in law to $\| \mathbf{Z} - P(\mathbf{Z} | \mathscr{S}' \cap \mathscr{U}'(\mathbf{p})) \|^2 - \| \mathbf{Z} - P(\mathbf{Z} | \mathscr{U}'(\mathbf{p})) \|^2 = \| P(\mathbf{Z} | \mathscr{S}') - P(P(\mathbf{Z} | \mathscr{S}') | \mathscr{U}'(\mathbf{p})) \|^2$. Now $P(\mathbf{Z} | \mathscr{S}')$ and $P(P(\mathbf{Z} | \mathscr{S}') | \mathscr{U}'(\mathbf{p}))$ are both symmetric about the diagonal and it is clear from the maximum upper sets algorithm that $P(P(\mathbf{Z} | \mathscr{S}') | \mathscr{U}'(\mathbf{p}))$ can be computed by restricting one's attention to the points above the diagonal, computing $P[(\mathbf{Z} + \mathbf{Z}')/2 | \mathscr{U}'(\mathbf{p})]$ on these points with regard to the order required by $\mathscr{U}'(\mathbf{p})$ on that set and then to obtain values on the other side by symmetry. Now if $W_{ij} = (Z_{ij} + Z_{R-j+1,R-i+1})/2$ for $1 \leqslant i \leqslant j \leqslant R$ then the W's are independent normal variables, $W_{ij} \sim \mathcal{N}(0, \frac{1}{2})$ if $i \neq j$ and $W_{ij} \sim \mathcal{N}(0, 1)$ if $i = j$, and it follows from Theorem 2.3.1 (see also the subsection before Theorem 2.3.1) that the least favorable, asymptotic, null hypothesis distribution of R'_{01} is the chi-bar-square distribution associated with the $(k^2 + k)/2$-dimensional random vector $\mathbf{W}$. Think of the $(k^2 + k)/2$-dimensional vectors as 'half-matrices' and let $\mathscr{I}$ be the subset $\mathscr{I} = \{\mathbf{y} = (y_{ij}): 1 \leqslant i \leqslant j \leqslant R;\ y_{ij} \leqslant y_{i',j'}$ whenever $i \leqslant i'$ and $j \leqslant j'\}$. Then one has the following result.

Theorem 5.6.2 *If $R'_{01} = -2 \ln \Lambda'_{01}$, where Λ'_{01} is the likelihood ratio for testing $H_0: \mathbf{p} \in \mathscr{S}' \cap \mathscr{U}'$ against $H_1 - H_0$ where $H'_1: \mathbf{p} \in \mathscr{S}'$, then*

$$\sup_{\mathbf{p} \in H_0} \lim_{n \to \infty} P[R'_{01} \geqslant t] = \sum_{l=1}^{(k^2+k)/2} P\left(l, \frac{k^2+k}{2}; \mathbf{w}\right) P[\chi^2_{(k^2+k)/2-l} \geqslant t]$$

where the $P(l, (k^2 + k)/2; \mathbf{w})$ are the level probabilities for the partial order considered on the half-matrices with weight vector $\mathbf{w}$. The weights, w_{ij}, are equal to 2 if $i \neq j$ and 1 if $i = j$.

Grove (1980) considers $2 \times C (R = 2)$ contingency tables and the LRT of the null hypothesis of independence:

$$H_0: p_{ij} = p_{i\cdot} p_{\cdot j} \qquad \text{for } i = 1, 2; \quad j = 1, 2, \ldots, C$$

against the alternative $H_1 - H_0$ with

$$H_1: \theta_q \geqslant 1, \qquad q = 1, 2, \ldots, C, \tag{5.6.1}$$

where

$$\theta_q = \frac{[\sum_{j=1}^{q} p_{1j}][\sum_{j=q+1}^{C} p_{2j}]}{[\sum_{j=1}^{q} p_{2j}][\sum_{j=q+1}^{C} p_{1j}]} \tag{5.6.2}$$

is the cross-product ratio in a 2×2 table formed by pooling adjacent columns of the original $2 \times C$ table. If the rows and columns are regarded as defining a pair of variables, then the alternative hypothesis, H_1, expresses both quadrant and regression dependence, according to the definitions of Lehmann (1966).

The alternative hypothesis, H_1, in (5.6.1) is equivalent to

$$p_{1\cdot}^{-1} \sum_{j=1}^{q} p_{1j} \geqslant p_{2\cdot}^{-1} \sum_{j=1}^{q} p_{2j} \qquad \text{for } q = 1, 2, \ldots, C-1, \tag{5.6.3}$$

which is a stochastic ordering condition between the two sets of conditional probabilities $\{p_{1j}/p_{1\cdot}\}$ and $\{p_{2j}/p_{2\cdot}\}$. Moreover, the likelihood is proportional to

$$\prod_{i=1}^{2} \prod_{j=1}^{C} p_{ij}^{n\hat{p}_{ij}} \tag{5.6.4}$$

and is maximized subject to the null hypothesis of independence when $p_{ij} = \hat{p}_{i\cdot} \hat{p}_{\cdot j}$.

The likelihood, (5.6.4), can be rewritten as

$$\left(\prod_{i=1}^{2} p_{i\cdot}^{np_{i\cdot}}\right)\left(\prod_{i=1}^{2}\prod_{j=1}^{C}\phi_{ij}^{np_{ij}}\right) \tag{5.6.5}$$

where $\phi_{ij} = p_{ij}/p_{i\cdot}$. The restriction, (5.6.3), involves p_{ij} only through ϕ_{ij} so that the two factors in the likelihood, (5.6.5), can be maximized separately to find the MLEs under the alternative hypothesis, H_1. The restrictions become $\sum_{j=1}^{q}\phi_{ij} \geqslant \sum_{j=1}^{q}\phi_{2j}$, $q = 1, 2, \ldots, C-1$, together with $\phi_{ij} \geqslant 0$ and $\sum_{j=1}^{C}\phi_{ij} = 1$, $i = 1, 2$. This problem was solved in Section 5.4 and the restricted estimates are given by Theorem 5.4.4. The asymptotic distribution of the likelihood ratio statistic is given in Theorem 5.4.5.

Patefield (1982) considers $R \times C$ tables and tests of hypotheses which involve the parameters

$$\phi_{ij} = \frac{p_{i,j}p_{i+1,j+1}}{p_{i,j+1}p_{i+1,j}} \qquad \text{for } i = 1, 2, \ldots, R-1; \quad j = 1, 2, \ldots, C-1.$$

These parameters are referred to as local odds ratios. Their values measure 'local' association in the table. Several tests, including a LRT, are considered for testing the null hypothesis of no association between the row and column categories:

$$H_0\colon \phi_{ij} = 1 \qquad \text{for all } i, j.$$

The alternative hypothesizes a positive trend in the association between the row and column categories and is quantified by stipulating H_1–H_0 where

$$H_1\colon \phi_{ij} \geqslant 1 \qquad \text{for all } i, j.$$

The LRT statistic is calculated using numerical routines and the attained significance level is found by summing all of the null hypothesis probabilities of these tables with a test statistic value greater than or equal to the observed test statistic value. The LRT is computationally impractical for all but rather small tables.

Agresti *et al.* (1981) consider models of the form

$$\log(np_{ij}) = \mu + \lambda_i^X + \lambda_j^Y + \beta u_i v_j. \tag{5.6.6}$$

This model is called the linear-by-linear association model when the $\{\mu_i\}$ and $\{v_j\}$ are fixed scores. It is referred to as the row effects model when the $\{\mu_i\}$ are unknown parameters, the column effects model when the $\{v_j\}$ are unknown parameters, and the row and column effects model when both sets are unknown. For the model, (5.6.6), the local log odds ratio is equal to

$$\log\phi_{ij} = \beta(\mu_{i+1} - \mu_i)(v_{j+1} - v_j).$$

Thus, assumptions about association between the row and column categories are equivalent to assumptions about the signs of the differences $(\mu_{i+1} - \mu_i)$ and

$(v_{j+1} - v_j)$. Agresti *et al.* (1985) consider fitting the model (5.6.6) subject to restrictions of the form $\mu_1 \leqslant \mu_2 \leqslant \cdots \leqslant \mu_R$. (See Chapter 6, Section 3) They also consider tests of fit for such models.

Lemke (1983) considers two models for two-way contingency tables. The first model is for square tables whose row and column response variables have the same ordinal categories. This model restricts the ratio, p_{ij}/p_{ji} of the cross-diagonal cell probabilities to be at least one. The second class of models are log-linear for rectangular tables. The restrictions placed on these log-linear models are that (1) the modal cell is known and (2) the probabilities are monotone within each row and column (i.e. the matrix order). Maximum likelihood estimates are obtained for each model, and LRT statistics, which have asymptotic chi-bar-square distributions, are investigated.

5.7 COMPLEMENTS

Maximum likelihood estimation of multinomial parameters subject to a simple order restriction was first considered by Chacko (1966), who also obtained the asymptotic null hypothesis distribution of a chi-square-type statistic for testing homogeneity of these multinomial parameters against the simple order alternative. This chi-square-type statistic is an approximation to a likelihood ratio statistic so that Theorem 5.2.1 is implicit in his work. Robertson (1965) found the maximum likelihood estimates of multinomial parameters subject to a partial order restriction. The problem of estimating multinomial parameters is, of course, closely related to the problem of density estimates discussed in Sections 7.2 and 7.3. Robertson (1978) generalized Chacko's result to the context given in Theorem 5.2.1 and considered the LRT of an order restriction as a null hypothesis. The chi-square tests discussed at the end of Section 5.2 were developed by C.I.C. Lee (1987a).

Shaked (1979) considered the estimation of a star-shaped sequence of Poisson and normal means. The theory given in Section 5.3 for multinomial parameters is contained in Dykstra and Robertson (1982b).

In the context of sampling from continuous distributions, maximum likelihood estimates of stochastically ordered distributions were first obtained by Brunk *et al.* (1966). Using Fenchel duality, Barlow and Brunk (1972) derived the two-sample maximum likelihood estimates given in Theorem 5.4.4. The authors consider both of these contributions to be seminal. The testing theory contained in Section 5.4 is from Robertson and Wright (1981). Lucas and Wright (1986) extend the work of Section 5.4 to a multivariate stochastic ordering.

Raubertas, Lee, and Nordheim (1986) consider testing problems in which the parameter vector is constrained to lie in one cone under the null hypotheses and constrained to lie in the difference between this cone and a larger cone under the alternative. The material on testing symmetry and unimodality in Sections 5.5

and 5.6 is taken from Robertson (1986). Warrack and Robertson (1984) and Robertson and Warrack (1985) consider another problem in which the parameter vector is constrained to lie in the intersection of a linear space and a cone under the null hypothesis and constrained to lie in this same cone under the alternative. This is another problem where the LRT is dominated by another test which ignores information in the model (cf. the caveats following Theorems 5.5.4 and 5.6.1).

CHAPTER 6

Duality

6.1 INTRODUCTION

As indicated in Section 1.7, duality considerations are useful in order restricted inference. For one thing, it is often possible to use duality concepts to expand the collection of problems for which one has solutions. A problem which is dual to a stated problem may appear to be very different, yet in reality be equivalent to the original problem. For example, the problem of finding the simply ordered isotonic regression is closely associated with least squares problems dealing with stochastic ordering restrictions. It is also the case that duality will often provide an alternative approach to a problem that may be more tractable, or provide additional insight into the problem. Clearly the more tools that are available for solving a problem, the better the possibility of attaining success. Finally, duality opens the door to the use of a well-developed theory. This theory allows us to employ powerful tools for working on problems in order restricted inference, and gain insight into the behavior of some challenging statistical problems.

The results stated in Section 1.7 indicate the close connection between a closed, convex cone $\mathscr{K}$ and its dual. (Really there are a whole collection of duals depending upon the set of weights that are chosen for the inner product norm. If no indication of the weights is given, the equal weights case will be assumed.) The dual cone is a special case of a more general type of duality, known as *Fenchel duality*. The natural context in which to discuss duality for normed linear spaces is the dual space determined by the collection of all continuous linear functionals on the original space. However, since the well-behaved finite-dimensional space R^k with a sum of squares norm is a Hilbert space, the dual space will be isomorphic to the original space. Thus the dual space will not be emphasized here, although it is important for theoretical considerations (for example, see Luenberger, 1969).

Fenchel duality is closely associated with a second type of duality known as *Lagrangian duality*. Lagrangian duality is sometimes referred to as *Kuhn–Tucker theory*, although Kuhn–Tucker theory is usually associated with optimization problems when inequality constraints are involved. Lagrangian duality lends itself to an asymptotic development of some problems in order restricted inference and will be discussed further in upcoming sections.

6.2 FENCHEL DUALITY

Assume that f is a proper convex function taking values in the extended reals where f is defined over some convex subset $S \subset R^k$. Proper means that $f(\mathbf{x}) < +\infty$ for at least one $\mathbf{x}$ and $f(\mathbf{x}) > -\infty$ for all $\mathbf{x} \in S$. The *effective domain* of f, which is denoted by dom f, is the set $\{\mathbf{x} \in S: f(\mathbf{x}) < +\infty\}$. The function f can be assumed to be defined over all of R^k by defining it to be $+\infty$ for all undefined values. If no particular domain is specified, we will assume that this has been done. A function h, also defined on R^k, is proper concave if $-h$ is proper convex. The usual unweighted inner product is denoted by $\langle \cdot, \cdot \rangle$ and is defined by $\langle \mathbf{x}, \mathbf{y} \rangle = \sum_{i=1}^k x_i y_i$.

The *epigraph* of an extended real value function, f, with domain $S \subset R^k$ is the set in R^{k+1} given by

$$\text{epi } f = \{(\mathbf{x}, u): \mathbf{x} \in S, u \in R, u \geqslant f(\mathbf{x})\}.$$

while the effective domain of a concave function h with domain $S \subset R^k$ is the set given by dom $h = \{\mathbf{x} \in S, h(\mathbf{x}) > -\infty\}$. The *relative interior* of a convex set $\mathscr{C}$ contained in R^k, denoted by ri $\mathscr{C}$, is defined as the interior that results when $\mathscr{C}$ is considered as a subset of its *affine hull*, the smallest linear variety containing $\mathscr{C}$ (see Luenberger, 1969).

For example, if $\mathscr{C} = \{\mathbf{x} \in R^k: \sum_{i=1}^k x_i = 0, x_1 \leqslant x_2\}$, the relative interior of $\mathscr{C}$ is $\{\mathbf{x} \in R^k: \sum_{i=1}^k x_i = 0, x_1 < x_2\}$, whereas the interior of $\mathscr{C}$ is the empty set (since for each point in $\mathscr{C}$, you can always find a point not in $\mathscr{C}$ that is arbitrarily close).

A set H is a *hyperplane* of R^k if

$$H = \{\mathbf{x} \in R^k: \langle \mathbf{x}, \mathbf{b} \rangle = \beta\}$$

for a nonzero $\mathbf{b} \in R^k$ and some $\beta \in R$. A hyperplane $H = \{\mathbf{x} \in R^k: \langle \mathbf{x}, \mathbf{b} \rangle = \beta\}$ is a *supporting hyperplane* for the set $\mathscr{C}$ if $\langle \mathbf{x}, \mathbf{b} \rangle \leqslant \beta$ for all $\mathbf{x} \in \mathscr{C}$ and $\langle \mathbf{x}, \mathbf{b} \rangle = \beta$ for some $\mathbf{x}$ on the boundary of $\mathscr{C}$.

A vector $\mathbf{b}$ is a *subgradient* of a convex function f at the vector $\mathbf{x}$ if

$$f(\mathbf{y}) \geqslant f(\mathbf{x}) + \langle \mathbf{b}, \mathbf{y} - \mathbf{x} \rangle \quad \text{for all } \mathbf{y}.$$

Thus if f is finite at $\mathbf{x}$, the graph of the affine function $g(\mathbf{y}) = f(\mathbf{x}) + \langle \mathbf{b}, \mathbf{y} - \mathbf{x} \rangle$ is a supporting hyperplane to the convex set epi f at the point $(\mathbf{x}, f(\mathbf{x}))$.

It is a well-known fact that a function defined on a convex subset of R^k is convex if and only if its epigraph is a convex set in R^{k+1}. Similarly, such a function is concave if and only if the negative of its epigraph is a convex set in R^{k+1}. An important property of convex functions is that they are totally determined by the collection of supporting hyperplanes for the epigraph of the function. This means that many properties of convex functions can be stated in terms of these supporting hyperplanes, leading to equivalences that are often not apparent. This partially motivates the definition of the *convex conjugate* of f, defined by

$$f^*(\mathbf{y}) = \sup_{\mathbf{x}} [\langle \mathbf{x}, \mathbf{y} \rangle - f(\mathbf{x})].$$

Thus $f^*(\mathbf{y})$ represents the amount that the hyperplane $\{\mathbf{x} \in R^{k+1} : \sum_{i=1}^{k} x_i y_i - x_{k+1} = 0\}$ with fixed gradient $\mathbf{y}$ must be lowered to make it a supporting hyperplane of the epigraph of f (see Figure 6.2.1).

The concave conjugate of the concave function h is given by

$$h^*(\mathbf{y}) = \inf_{\mathbf{x}} [\langle \mathbf{x}, \mathbf{y} \rangle - h(\mathbf{x})],$$

and has an analogous interpretation to that of the convex conjugate. Note that it is not the negative of the convex conjugate of $-h$.

If f and h are proper convex and proper concave, respectively, then so are f^* and h^*. The fact that there exist close connections between the convex function f and the concave function h and their conjugates is the subject of an important duality theorem.

The extreme values of convex and concave functions are often of interest. However, what is usually of more interest is locating those points in the domain where extreme values are obtained. We will abuse notation by allowing a problem

$$\inf_{\mathbf{x} \in C} g(\mathbf{x}) \left(\sup_{\mathbf{x} \in C} g(\mathbf{x}) \right)$$

to refer to either the problem of obtaining the value $\inf_{\mathbf{x} \in C} g(\mathbf{x}) (\sup_{\mathbf{x} \in C} g(\mathbf{x}))$ or the problem of identifying values $\mathbf{x}_0 \in C$ such that $g(\mathbf{x}_0) = \inf_{\mathbf{x} \in C} g(\mathbf{x}) (g(\mathbf{x}_0) = \sup_{\mathbf{x} \in C} g(\mathbf{x}))$ as the particular situation requires.

With this in mind, we will define the primal problem for a specified convex f and concave h as

$$\text{(I)} \quad \inf_{\mathbf{x}} [f(\mathbf{x}) - h(\mathbf{x})]$$

while the dual problem is

$$\text{(II)} \quad \sup_{\mathbf{y}} [h^*(\mathbf{y}) - f^*(\mathbf{y})].$$

A proof of the following theorem is given in Rockafellar (1970, p. 327).

Theorem 6.2.1 (*Fenchel's duality theorem for* R^k) *Let f and h be proper convex and proper concave functions, respectively, both defined on R^k. Then*

$$\inf_{\mathbf{x}} [f(\mathbf{x}) - h(\mathbf{x})] = \sup_{\mathbf{y}} [h^*(\mathbf{y}) - f^*(\mathbf{y})]$$

provided that the relative interiors of dom f *and* dom h *have at least one point in common. In this event, the supremum is attained.*

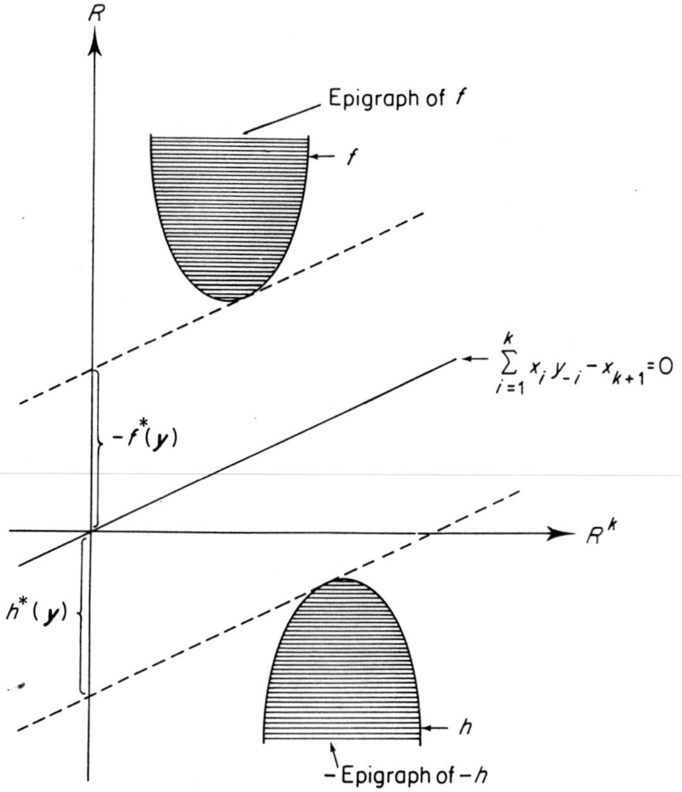

Figure 6.2.1 The graphs of a convex function f and a concave function h, their epigraphs, and tangential hyperplanes

A hueristic reason why this theorem might be true is apparent from Figure 6.2.1.

One can think of $h^*(\mathbf{y}) - f^*(\mathbf{y})$ as being the vertical distance between parallel hyperplanes which are tangential to epi f and epi h and have a gradient of $\mathbf{y}$. Clearly $f(\mathbf{x}) - h(\mathbf{x}) \geqslant h^*(\mathbf{y}) - f^*(\mathbf{y})$ for all values of $\mathbf{x}$ and $\mathbf{y}$. However, by varying $\mathbf{y}$, one can also vary the point of support for the hyperplanes. Hueristically, the theorem says that the minimum distance between two disjoint convex sets is the same as the maximum distance between parallel hyperplanes separating the sets.

An important special case of this theorem arises when

$$h(\mathbf{x}) = \begin{cases} -\infty, & \mathbf{x} \notin \mathscr{K} \\ 0, & \mathbf{x} \in \mathscr{K} \end{cases}$$

for a convex cone $\mathscr{K}$. It is straightforward to see that

$$h^*(\mathbf{y}) = \inf_{\mathbf{x}} [\langle \mathbf{x}, \mathbf{y} \rangle - h(\mathbf{x})] = \inf_{\mathbf{x} \in \mathscr{K}} \langle \mathbf{x}, \mathbf{y} \rangle = \begin{cases} 0, & -\mathbf{y} \in \mathscr{K}^* \\ -\infty, & -\mathbf{y} \notin \mathscr{K}^*. \end{cases}$$

In this event, the primal problem (I) becomes

$$\inf_{\mathbf{x}} [f(\mathbf{x}) - h(\mathbf{x})] = \inf_{\mathbf{x} \in \mathscr{K}} f(\mathbf{x}) \tag{6.2.1}$$

while the dual problem (II) is

$$\sup_{\mathbf{y}} [h^*(\mathbf{y}) - f^*(\mathbf{y})] = \sup_{\mathbf{y} \in \mathscr{K}^*} - f^*(-\mathbf{y}). \tag{6.2.2}$$

These values must be equal if the relative interiors of dom f and $\mathscr{K}$ have a point in common. Moreover, it is shown in Rockafellar (1970, p. 335) that $\hat{\mathbf{x}} \in \mathscr{K}$ and $\hat{\mathbf{y}} \in \mathscr{K}^*$ are respective solutions of (6.2.1) and (6.2.2) in that

$$f(\hat{\mathbf{x}}) = \inf_{\mathbf{x} \in \mathscr{K}} f(\mathbf{x}) = \sup_{\mathbf{y} \in \mathscr{K}^*} - f^*(-\mathbf{y}) = - f^*(-\hat{\mathbf{y}}) \tag{6.2.3}$$

if and only if

(a) $\hat{\mathbf{x}} \in \mathscr{K}$
(b) $\hat{\mathbf{y}} \in \mathscr{K}^*$
(c) $\langle \hat{\mathbf{x}}, \hat{\mathbf{y}} \rangle = 0$ (6.2.4)

and

(d) $f(\mathbf{x}) \geqslant f(\hat{\mathbf{x}}) - \langle \hat{\mathbf{y}}, \mathbf{x} - \hat{\mathbf{x}} \rangle$ for all $\mathbf{x}$, that is
$-\hat{\mathbf{y}}$ is a subgradient of f at the point $\hat{\mathbf{x}}$.

Conditions (6.2.4) are very useful, because they may allow us to identify solutions to both the primal and dual problems in an expeditious manner. For example, suppose $f(x) = (x-3)^2$ and $\mathscr{K} = \{x : x \geqslant 0\}$. Then $\mathscr{K}^* = \{y : y \leqslant 0\}$ and

$$\begin{aligned} f^*(y) &= \sup_x [xy - (x-3)^2] \\ &= \sup_x - \left[x^2 - x(6+y) + \frac{(6+y)^2}{4} \right] + \frac{(6+y)^2}{4} - 9 \\ &= \sup_x - \left[x - \left(\frac{6+y}{2} \right) \right]^2 + \tfrac{1}{4}(y^2 + 12y) = \tfrac{1}{4}(y^2 + 12y). \end{aligned}$$

By (6.2.4) the respective solutions $\hat{x}$ and $\hat{y}$ to the problems

$$\inf_{x \geqslant 0} (x-3)^2 \quad \text{and} \quad \sup_{y \leqslant 0} - \tfrac{1}{4}(y^2 - 12y)$$

must satisfy (a) $\hat{x} \geqslant 0$, (b) $\hat{y} \leqslant 0$, (c) either $\hat{x}$ or $\hat{y}$ equals zero, and (d) $-\hat{y} = 2(\hat{x} - 3)$ (since the only subgradient of f at $\hat{x}$ is the derivative of f at $\hat{x}$). Condition (d) implies that a solution to one problem determines the solution to the other.

Conditions (c) and (d) imply that $(\hat{x}, \hat{y})$ is either (0, 6) or (3, 0). Finally, the solutions must be (3, 0) since the other pair violates condition (b). The extremal values of both problems is zero. Note that the solutions were determined solely by conditions (6.2.4).

A mirrored development is of course possible when h is an arbitrary proper concave function and f is defined by

$$f(\mathbf{x}) = \begin{cases} \infty, & \mathbf{x} \notin \mathscr{K} \\ 0, & \mathbf{x} \in \mathscr{K}. \end{cases}$$

In this case, the primal problem becomes

$$\inf_{\mathbf{x} \in \mathscr{K}} - h(\mathbf{x}) \tag{6.2.5}$$

and the dual problem is

$$\sup_{\mathbf{y} \in \mathscr{K}^*} h^*(\mathbf{y}). \tag{6.2.6}$$

Then for $\hat{\mathbf{x}}$ and $\hat{\mathbf{y}}$ to be simultaneous solutions to problems (6.2.5) and (6.2.6), condition (d) of (6.2.4) must be modified to read '$-\hat{\mathbf{y}}$ is a subgradient of $-h$ at $\hat{\mathbf{x}}$'.

Condition (d) of (6.2.4) states that the hyperplane with gradient $-\hat{\mathbf{y}}$ which passes through the point $(\hat{\mathbf{x}}, \mathbf{f}(\hat{\mathbf{x}}))$ must in fact be a supporting hyperplane of the epigraph of f, as indicated in Figure 6.2.2.

Suppose now that f is differentiable at $\hat{\mathbf{x}}$ and convex and that the cone version of the Fenchel theorem is applicable. Then the collection of subgradients at $\hat{\mathbf{x}}$ consists of a single vector, namely the gradient of f at $\hat{\mathbf{x}}$. If $\hat{\mathbf{x}}$ solves the primal problem, then, by Theorem 6.1.1, the extremal value of the dual problem is attained at some point in $\mathscr{K}^*$ and, by condition (d) of (6.2.4) this point can only be the negative of the gradient of f at the point $\hat{\mathbf{x}}$. Thus the dual problem can often be easily solved if the primal problem can be solved, and vice versa.

Given a set $\mathscr{K}_1, \ldots, \mathscr{K}_t$ of closed convex cones, there are two common ways of constructing new convex cones. In particular, both

$$\mathscr{K}_1 \cap \mathscr{K}_2 \cap \cdots \cap \mathscr{K}_t = \{\mathbf{x} \in R^k : \mathbf{x} \in \mathscr{K}_i \text{ for all } i\}$$

and

$$\mathscr{K}_1 + \mathscr{K}_2 + \cdots + \mathscr{K}_t = \{\mathbf{y}_1 + \cdots + \mathbf{y}_t : \mathbf{y}_i \in \mathscr{K}_i\}$$

are convex cones. Although $\mathscr{K}_1 + \mathscr{K}_2 + \cdots + \mathscr{K}_t$ need not be closed even if $\mathscr{K}_1, \ldots, \mathscr{K}_t$ are closed, such occurrences are unusual. (See Hestenes, 1975, p. 196, for such an example.) There is an intimate relationship between these new cones and cone duality as indicated in the following theorem. This theorem is proven in Rockafellar (1970, p. 146).

Theorem 6.2.2 *If $\mathscr{K}_1, \ldots, \mathscr{K}_t$ are nonempty convex cones in R^k, then*

$$(a)\ (\mathscr{K}_1 + \cdots + \mathscr{K}_t)^* = \mathscr{K}_1^* \cap \cdots \cap \mathscr{K}_t^*$$

$$(b)\ (\mathscr{K}_1 \cap \cdots \cap \mathscr{K}_t)^* = \overline{(\mathscr{K}_1^* + \cdots + \mathscr{K}_t^*)}$$

where the overbar indicates the closure of a set.

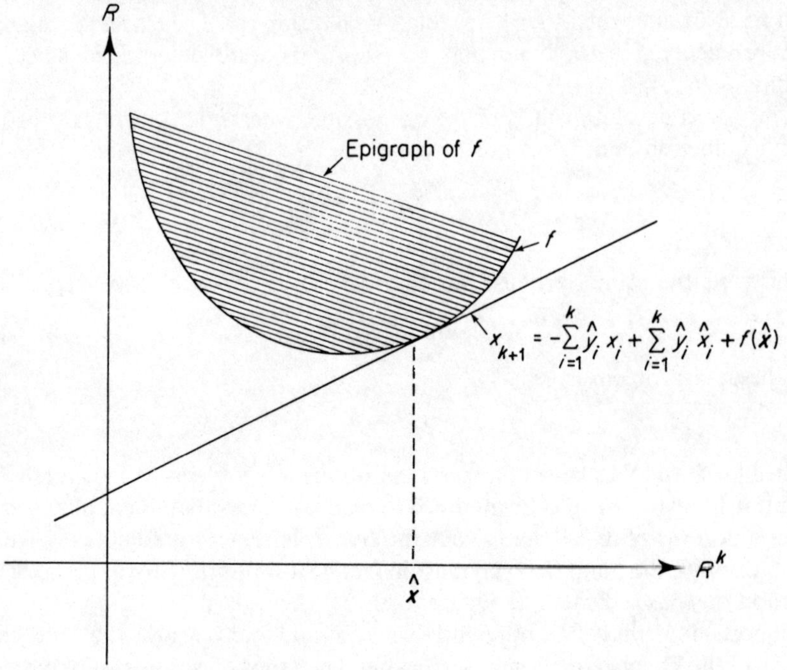

Figure 6.2.2 The convex function f, its epigraph, and the tangential hyperplane with gradient $-\hat{\mathbf{y}}$

If the $\mathscr{K}_i$'s are closed, the overbars may be omitted if the relative interiors of the $\mathscr{K}_i$ have a point in common or if the $\mathscr{K}_i^*$'s are finitely generated (see Hestenes, 1975, p. 198, and Rockafellar, 1972, p. 146). Finitely generated cones are discussed in Chapter 1.

The relationship expressed in Theorem 6.2.2 coupled with the earlier duality results associated with displays (6.2.1) and (6.2.2) often enable one to express an order restricted inference problem involving multiple constraints in a form that bears little resemblence to the original problem.

Fenchel duality in least squares problems with cone restrictions

Example 6.2.1 Consider the isotonic regression problem discussed in Chapter 1 where the function to be minimized is $f(\mathbf{x}) = \frac{1}{2}\sum_{i=1}^{k}(x_i - g_i)^2 w_i$, $\mathbf{g}$ is an arbitrary fixed vector, $\mathbf{w}$ is a set of positive weights, and $\mathscr{K}$ is an arbitrary closed, convex cone. In the present context, the primal problem becomes

$$\inf_{\mathbf{x}\in\mathscr{K}} f(\mathbf{x}).$$

The convex conjugate is given by

$$f^*(\mathbf{y}) = \sup_{\mathbf{x}}[\langle \mathbf{x}, \mathbf{y}\rangle - f(\mathbf{x})]. \tag{6.2.7}$$

The supremum in (6.2.7) is unconstrained and an easy differentiation yields that the maximal value of (6.2.7) occurs at the vector with the ith component $x_i = g_i + (y_i/w_i)$. Evaluating (6.2.7) for this $\mathbf{x}$ yields

$$f^*(\mathbf{y}) = \frac{1}{2}\sum_{i=1}^{k}\left(\frac{y_i}{w_i} + g_i\right)^2 w_i - \frac{1}{2}\sum_{i=1}^{k} g_i^2 w_i$$

The problem dual to $\inf_{\mathbf{x}\in\mathscr{K}} f(\mathbf{x})$ is then

$$\sup_{\mathbf{y}\in\mathscr{K}^*} -f^*(-\mathbf{y}) = \sup_{\mathbf{y}\in\mathscr{K}^*}\left[-\frac{1}{2}\sum_{i=1}^{k}\left(-\frac{y_i}{w_i} + g_i\right)^2 w_i + \frac{1}{2}\sum_{i=1}^{k} g_i^2 w_i\right]. \tag{6.2.8}$$

We can express the solution to (6.2.8) in terms of $P_{\mathbf{w}}(\mathbf{g}|\mathscr{K})$, the solution to $\inf_{\mathbf{x}\in\mathscr{K}} f(\mathbf{x})$, by the following argument. Since f is everywhere differentiable, the only subgradient of f at $P_{\mathbf{w}}(\mathbf{g}|\mathscr{K})$ is the gradient of f evaluated at $P_{\mathbf{w}}(\mathbf{g}|\mathscr{K})$, which must be the negative of the solution to (6.2.8). It follows that the ith coordinate of the solution to (6.2.8) must be $-(P_{\mathbf{w}}(\mathbf{g}|\mathscr{K})_i - g_i)w_i$. The change of variables $x_i = y_i/w_i$, $i = 1,\ldots,k$, then reveals that the solution to

$$\inf_{\mathbf{x}\in\mathscr{K}^{*\mathbf{w}}} \frac{1}{2}\sum_{1}^{k}(x_i - g_i)^2 w_i$$

is given by $\mathbf{g} - P_{\mathbf{w}}(\mathbf{g}|\mathscr{K})$. This result was proven by other methods in Theorem 1.7.3.

Generalized isotonic regression as a duality problem

Example 6.2.2 It will be convenient in this example to adopt the convention that a real valued function of a real argument evaluated at a vector is the vector that results from applying the function to each coordinate of the vector. Thus if $\mathbf{x}$ is a vector, $\psi(\mathbf{x})$ would be $(\psi(x_1), \psi(x_2),\ldots,\psi(x_k))$.

It was established in Section 1.7 that if Φ is an arbitrary proper convex function on R and if $\mathscr{K}$ is a cone of isotonic vectors, then the solution to

$$\inf_{\mathbf{g}-\mathbf{x}\in\mathscr{K}^{*\mathbf{w}}} \sum_{i=1}^{k} \Phi(x_i)w_i \tag{6.2.9}$$

is given by $P_{\mathbf{w}}(\mathbf{g}|\mathscr{K})$ independent of Φ. An inspection of the proof will show that the result still holds for an arbitrary closed, convex cone $\mathscr{K}$ if $\phi(P_{\mathbf{w}}(\mathbf{g}|\mathscr{K}))$ is in $\mathscr{K}$ where $\phi(x)$ is a subgradient of Φ at x; that is $\Phi(z) \geqslant \Phi(x) + \phi(x)(z-x)$ for all real z. If we make the change of variables $z_i = (g_i - x_i)w_i$, $i = 1,\ldots,k$, we may write (6.2.9) as

$$\inf_{\mathbf{z}\in\mathscr{K}^*} \sum_{i=1}^{k} \Phi\left(g_i - \frac{z_i}{w_i}\right) w_i \tag{6.2.10}$$

and the solution must then be given by the vector with coordinates $(g_i - P_w(\mathbf{g}|\mathscr{K})_i)w_i$. If f is defined on R^k by $f(\mathbf{z}) = \sum_{i=1}^{k} \Phi(g_i - z_i/w_i)w_i$, then the convex conjugate of f is

$$f^*(\mathbf{y}) = \sup_{\mathbf{z}} \left[\sum_{i=1}^{k} y_i z_i - \sum_{i=1}^{k} \Phi\left(g_i - \frac{z_i}{w_i}\right) w_i \right]. \tag{6.2.11}$$

Suppose now that Φ is a strictly convex, differentiable function defined on R. Then the supremum in (6.2.11) occurs at $z_i = (g_i - \phi^{-1}(-y_i))w_i$. Substituting this value of $\mathbf{z}$ into (6.2.11) results in

$$f^*(\mathbf{y}) = \sum_{i=1}^{k} g_i y_i w_i - \sum_{i=1}^{k} y_i \phi^{-1}(-y_i) w_i - \sum_{i=1}^{k} \Phi(\phi^{-1}(-y_i)) w_i.$$

However, the convex conjugate of Φ is given by $\Phi^*(y) = y\phi^{-1}(y) - \Phi(\phi^{-1}(y))$, so that we can write

$$f^*(\mathbf{y}) = \sum_{i=1}^{k} g_i y_i w_i + \sum_{i=1}^{k} \Phi^*(-y_i) w_i.$$

The problem dual to (6.2.10) is of the form $\sup_{\mathbf{y}\in\mathscr{K}^{**}} - f^*(-\mathbf{y})$, which can be written as

$$\inf_{\mathbf{y}\in\mathscr{K}} \sum_{i=1}^{k} [\Phi^*(y_i) - g_i y_i] w_i, \tag{6.2.12}$$

since $\mathscr{K}$ being closed implies that $\mathscr{K}^{**} = \mathscr{K}$. By condition (d) of (6.2.4), (6.2.12) can only be solved by the negative of the gradient of $\sum_{i=1}^{k} \Phi(g_i - z_i/w_i)w_i$ evaluated at the solution to (6.2.10). Moreover, since Φ is differentiable with increasing derivative ϕ, this derivative is an inverse of the derivative of Φ^*. This suggests that a solution to (6.2.12) should be given by $\psi^{-1}(P_w(\mathbf{g}|\mathscr{K}))$, where ψ is the derivative of Φ^*. This conjecture can be shown to be true by showing that the conditions of (6.2.4) hold. An important point is that the solution to problem (6.2.12) can now be given for a convex function Φ^* without making reference to its convex conjugate $\Phi^{**} = \Phi$ (which need not have a nice form). Thus the two corollaries to Theorem 1.7.4 actually give solutions to problems that are dual to each other in the Fenchel sense.

It was shown in Chapter 1 that the problems defined in (6.2.9) and (6.2.12) greatly extend the class of problems which can be solved by an isotonic regression, and are important tools to keep in mind when considering order restricted problems. It is possible to extend these results to any convex function, Φ, whose effective domain contains the values of the coordinates of $\mathbf{g}$.

Example 6.2.3 As an example of (6.2.9), consider the following problem. An amount y_i of goods is to be manufactured during the ith time period. The cost of manufacturing an amount y is given by $\Phi(y)$ where Φ is a convex function. The demand for goods in the ith time period is given as r_i, and it is desired to minimize the total production cost, $\sum_{i=1}^{k}\Phi(y_i)$, subject to the constraints that demand never exceeds supply and total production equals total demand. These constraints are expressible as

$$\sum_{i=1}^{l}(y_i - r_i) \geqslant 0, l = 1, \ldots, k-1; \qquad \sum_{i=1}^{k}(y_i - r_i) = 0.$$

If $\mathscr{A}$ is defined as $\{\mathbf{x}\in R^k: x_1 \geqslant x_2 \geqslant \cdots \geqslant x_k\}$, then since $\mathscr{A}^* = \{\mathbf{x}\in R^k: \sum_{j=1}^{i} x_j \leqslant 0,$ $i = 1, 2, \ldots, k-1$ and $\sum_{j=1}^{k} x_j = 0\}$, the previous constraints are equivalent to requiring that $\mathbf{r} - \mathbf{y} \in \mathscr{A}^*$. However, the problem

$$\inf_{\mathbf{r}-\mathbf{y}\in\mathscr{A}^*} \sum_{i=1}^{k} \Phi(y_i)$$

is of the form of (6.2.9) and hence is solved by $P(\mathbf{r}|\mathscr{A})$.

From the previous discussion, the dual problem corresponding to (6.2.12) would be

$$\inf_{\mathbf{y}\in\mathscr{A}} \sum_{i=1}^{k} [\Phi^*(y_i) - r_i y_i]$$

where Φ^* is the convex conjugate of Φ and would be solved by $\psi^{-1}(P(\mathbf{r}|\mathscr{A}))$ where ψ is a determination of the derivative of Φ^*.

Duality in *I*-Projection problems

Example 6.2.4 Consider the problem of finding a probability vector $\mathbf{x}$ which attains the extremal value

$$\inf_{\mathbf{x}\in\mathscr{C}} \sum_{i=1}^{k} x_i \ln\left(\frac{x_i}{r_i}\right) \qquad (6.2.13)$$

where $\mathscr{C}$ is a convex subset of the collection of probability vectors, $P = \{\mathbf{x}: \sum_{i=1}^{k} x_i = 1, x_i \geqslant 0; i = 1, 2, \ldots, k\}$, and the r_i are nonnegative and not all zero. (The value of $0\ln(0/r)$ is defined to be zero for all nonnegative r.) A solution to this problem is often referred to as an *I-projection of* $\mathbf{r}$ *onto* $\mathscr{C}$ if (6.2.13) is a finite number and is important in a wide variety of problems. In particular, minimization problems of the form (6.2.13) play a key role in the information–theoretic approach to statistics (e.g. Kullback, 1959, and Good, 1963) and also occur in other areas such as the theory of large deviations (Sanov, 1957) and maximization of entropy (Rao, 1965, and Jaynes, 1957). This type of problem also occurs in maximum likelihood estimation problems for log-linear models through duality considerations, and will be discussed further in Section 6.3.

Let $\mathbf{r}$ be considered fixed and f and $\mathscr{K}_{\mathscr{C}}$ be defined by

$$f(\mathbf{x}) = \begin{cases} \sum_{i=1}^{k} x_i \ln\left(\frac{x_i}{r_i}\right) & \text{for } \mathbf{x} \in P \\ \infty, & \text{elsewhere} \end{cases}$$

and

$$\mathscr{K}_{\mathscr{C}} = \{\alpha\mathbf{x} : \alpha \geqslant 0, \quad \mathbf{x} \in \mathscr{C}\}.$$

Since f is a proper convex function and $\mathscr{K}_{\mathscr{C}}$ is a convex cone, (6.2.13) can be stated in the form of (6.2.1) for the above function f and the cone $\mathscr{K}_{\mathscr{C}}$. To state the dual problem (6.2.2), we will need to find

$$f^*(\mathbf{y}) = \sup_{\mathbf{x}} \left[\sum_{i=1}^{k} x_i y_i - f(\mathbf{x}) \right]$$

or, equivalently,

$$f^*(\mathbf{y}) = \sup_{\mathbf{x}: \sum_{i=1}^{k} x_i = 1} \left[\sum_{i=1}^{k} x_i \left(y_i - \ln\left(\frac{x_i}{r_i}\right)\right)\right], \tag{6.2.14}$$

where the log of a nonpositive number is taken to be $-\infty$. To obtain $f^*(\mathbf{y})$, write $x_k = 1 - \sum_{i=1}^{k-1} x_i$ and set the partial derivative with respect to x_i of the function in (6.2.14) equal to zero; $i = 1, \ldots, k-1$. This will yield the equations

$$y_i - \ln\left(\frac{x_i}{r_i}\right) = y_k - \ln\left(\frac{x_k}{r_k}\right) \qquad \text{for } i = 1, \ldots, k,$$

or, equivalently,

$$x_i = r_i \exp(y_i) \left[\left(\frac{x_k}{r_k}\right) \exp(-y_k) \right] \qquad \text{for } i = 1, \ldots, k.$$

By summing both sides, we obtain

$$1 = \sum_{i=1}^{k} x_i = \left(\frac{x_k}{r_k}\right) \exp(-y_k) \sum_{i=1}^{k} r_i \exp(y_i)$$

so that the vector with coordinates

$$x_i = \frac{r_i \exp(y_i)}{\sum_{j=1}^{k} r_j \exp(y_j)} \qquad \text{for } i = 1, \ldots, k$$

solves the above equations. Evaluating (6.2.14) at its solution, we have that $f^*(\mathbf{y}) = \ln[\sum_{i=1}^{k} r_i \exp(y_i)]$. Note that dom f^* equals R^k, even though dom f is contained in P, the set of probability vectors. Thus the dual problem to (6.2.13) is equivalent to

$$\inf_{\mathbf{y} \in \mathscr{K}_{\mathscr{C}}^*} \sum_{i=1}^{k} r_i \exp(-y_i). \tag{6.2.15}$$

Of key interest is the relationship between the solutions to (6.2.13) and (6.2.15). Once again, the key is condition (d) of (6.2.4) and the fact that f^* is everywhere differentiable. By using the earlier duality relationships of (6.2.1) and (6.2.2) with $f(\mathbf{x})$ and $\mathscr{K}_{\mathscr{C}}$ replaced by $f^*(-\mathbf{x})$ and $\mathscr{K}_{\mathscr{C}}^*$ (nothing is lost if $\mathscr{C}$ is closed since $f^{**} = f$ and $\mathscr{K}_{\mathscr{C}}^{**} = \mathscr{K}_{\mathscr{C}}$), we know that the solution $\mathbf{p}^*$ of (6.2.13) as a function of the solution $\hat{\mathbf{y}}$ of (6.2.15) must be given by the gradient of f^* evaluated at $-\hat{\mathbf{y}}$. This is the vector with the ith coordinate

$$p_i^* = \frac{r_i \exp(-\hat{y}_i)}{\sum_{j=1}^{k} r_j \exp(-\hat{y}_j)}.$$

This is of course a probability vector since the constraint region $\mathscr{C}$ is contained in P. One can also go the other way and express the solution to (6.2.15) as a function of $\mathbf{p}^*$. It can be shown that if $\mathbf{p}^*$ solves (6.2.13), for a given region $\mathscr{C}$, the vector with the ith coordinate

$$\hat{y}_i = \ln\left(\frac{r_i}{p_i^*}\right) + \sum_{j=1}^{k} p_j^* \ln\left(\frac{p_j^*}{r_j}\right)$$

will solve (6.2.15). The latter problem (6.2.15) is often easier to solve that the original problem (6.2.13) for two reasons. First, if $\mathscr{C}$ is a region of high dimensionality (cannot be contained in a low-dimensional subspace), $\mathscr{K}_{\mathscr{C}}^*$ will often be a low-dimensional set (is contained in a low-dimensional subspace). Second, the latter problem does not involve the additional constraint that the solution be a probability vector.

I-Projections with linear inequality constraints

Example 6.2.5 As an example of problem (6.2.13), consider the situation where $\mathscr{C}$ consists of probability vectors that satisfy simultaneous linear inequality constraints and $\mathbf{r} \neq \mathbf{0}$ is a given nonnegative vector. Initially, we will suppose that $\mathscr{C}$ consists of a single linear inequality constraint. Since the solution must be in P and we are going to use the dual problem formulation, we may without loss of generality take $\mathscr{C}$ to be

$$\mathscr{C} = \left\{ \mathbf{x} \colon \sum_{i=1}^{k} c_i x_i \leqslant b, \sum_{i=1}^{k} x_i = 1 \right\}$$

for some fixed vector $\mathbf{c}$ and fixed scalar b, where we may assume that $\mathscr{C}$ contains at least one probability vector. The corresponding cone $\mathscr{K}_{\mathscr{C}}$ can be expressed as

$$\mathscr{K}_{\mathscr{C}} = \left\{ \mathbf{x} : \sum_{i=1}^{k} (c_i - b)x_i \leqslant 0 \right\} \cap \left\{ \mathbf{x} : \sum_{i=1}^{k} x_i > 0 \text{ or } \mathbf{x} = \mathbf{0} \right\}$$

$$= \mathscr{K}_1 \cap \mathscr{K}_2.$$

Since $\mathscr{K}_1^* = \{\alpha(\mathbf{c} - b\mathbf{e}) : \alpha \geqslant 0\}$ (where $\mathbf{e} = (1, \ldots, 1)$ is the vector with 1's at each coordinate) and $\mathscr{K}_2^* = \{\beta\mathbf{e} : \beta \leqslant 0\}$, we may use Theorem 6.2.2 to write

$$\mathscr{K}_{\mathscr{C}}^* = \{\alpha(\mathbf{c} - b\mathbf{e}) + \beta\mathbf{e} : \alpha \geqslant 0, \beta \leqslant 0).$$

The dual problem (6.2.15) is then equivalent to

$$\inf_{\alpha \geqslant 0, \beta \geqslant 0} \left\{ \sum_{i=1}^{k} r_i \exp[-\alpha(c_i - b)] \right\} e^{\beta}.$$

Clearly the value of β which minimizes this expression is zero since $\mathbf{r}$ is nonnegative. By setting the partial derivative with respect to α equal to zero, the solution for α must be the unique root of the equation

$$\sum_{i=1}^{k} (c_i - b) r_i \exp[-\alpha(c_i - b)] = 0, \tag{6.2.16}$$

say $\hat{\alpha}$, if this root is nonnegative. If $\hat{\alpha}$ is negative (it could be $-\infty$), the convexity of the function to be minimized implies that the solution for α must be zero. Thus if $\hat{\alpha}^+ = \max(\hat{\alpha}, 0)$, the solution to (6.2.15) will be $\hat{\alpha}^+(\mathbf{c} - b\mathbf{e})$, and the solution to (6.2.13) must be the vector with components

$$p_i^* = \frac{r_i \exp[-\hat{\alpha}^+(c_i - b)]}{\sum_{j=1}^{k} r_j \exp[-\hat{\alpha}^+(c_j - b)]} \qquad \text{for } i = 1, \ldots, k.$$

Although (6.2.16) need not have a closed form solution, it is very easy to solve using Newton's method, seldom needing more than three or four iterations to obtain adequate convergence.

If $\mathscr{C}$ consists of those vectors that satisfy t simultaneous linear inequality constraints, say

$$\langle \mathbf{x}_j, \mathbf{c} \rangle \leqslant b_j \qquad \text{for } j = 1, \ldots, t,$$

then the corresponding cone $\mathscr{K}_{\mathscr{C}}$ can be taken to be

$$\mathscr{K}_{\mathscr{C}} = \textstyle\bigcap_{j=1}^{t} \{\mathbf{x} : \langle \mathbf{x}, \mathbf{c}_j - b_j\mathbf{e} \rangle \leqslant 0\} \cap \{\mathbf{x} : \langle \mathbf{x}, \mathbf{e} \rangle > 0 \text{ or } \mathbf{x} = \mathbf{0}\}.$$

By Theorem 6.2.2, the dual cone will be

$$\mathscr{K}_{\mathscr{C}}^* = \left\{ \sum_{j=1}^{t} \alpha_j(\mathbf{c}_j - b_j\mathbf{e}) + \beta\mathbf{e} : \alpha_j \geqslant 0, j = 1, \ldots, t; \beta \leqslant 0 \right\}.$$

By replacing the $\mathbf{y}$ in (6.2.15) with an expression of this form, the dual problem is seen to be a restricted convex minimization problem in $t+1$ dimensions (really t dimensions since $\hat{\beta}$ is easily seen to be zero). The minimization region is thus really over the cone with generators $\mathbf{c}_j - b_j\mathbf{e}$, $j = 1, 2, \ldots, t$. Although the feasible region for the α_j's is just the nonnegative orthant, this problem can prove difficult to solve unless t is quite small.

Dykstra (1985a) has given an iterative procedure for finding I-projections onto the intersection of convex sets which only requires projecting onto the individual sets. (See Section 6.3 for a further discussion on this procedure.) This procedure seems to work very well for finding I-projections subject to simultaneous linear inequality constraints, since the aforementioned single constraint problem can be nicely solved in terms of its dual. Dykstra and Wollan (1987) have published a Fortran program which uses this procedure to solve linear-inequality constrained I-projection problems.

I-Projections onto cones of isotonic vectors

In the event that the convex cone $\mathscr{K}_{\mathscr{C}}$ is actually a cone of isotonic vectors, problem (6.2.13) becomes even more tractable in the sense that it can be solved by a least squares projection onto the isotonic cone. This means that the computational algorithms described in Section 1.4 can be implemented to find the solution. To see that this is the case, recall that the dual problem is specified by

$$\inf_{\mathbf{y}\in\mathscr{K}^*_{\mathscr{C}}} \sum_{i=1}^{k} r_i \exp(-y_i). \tag{6.2.17}$$

Equivalently, if the change of variables $z_i = -y_i + \ln r_i$ is made, $i = 1, \ldots, k$, the problem becomes

$$\inf_{\ln(\mathbf{r})-\mathbf{z}\in\mathscr{K}^*_{\mathscr{C}}} \sum_{i=1}^{k} \exp(z_i).$$

However, since $\exp(x)$ is a strictly convex function, this problem is of the form (6.2.9) in Example 6.2.2 so that the solution must be given by $\hat{\mathbf{z}} = P(\ln(\mathbf{r})|\mathscr{K}_{\mathscr{C}})$. Thus $\hat{\mathbf{y}} = \ln(\mathbf{r}) - \hat{\mathbf{z}}$, and since we now know $\hat{\mathbf{y}}$, we can use the duality relationship between solutions to state that the vector with components

$$\hat{x}_i = \frac{\exp[P(\ln(\mathbf{r})|\mathscr{K}_{\mathscr{C}})_i]}{\sum_{j=1}^{k} \exp[P(\ln(\mathbf{r})|\mathscr{K}_{\mathscr{C}})_j]} \qquad \text{for } i = 1, \ldots, k$$

must solve the original problem.

Recapitulating, an I-projection of $\mathbf{r}$ onto a cone of isotonic vectors is solved by finding the least squares projection of $\ln(\mathbf{r})$ with equal weights onto $\mathscr{K}_{\mathscr{C}}$, evaluating the exponential function at this projection, and then norming to obtain a probability vector.

6.3 FENCHEL DUALITY IN LOG-LINEAR MODELS

A common approach to categorical data problems is to assume that the logarithms of parameters in question follow a linear model, i.e. must lie within a linear subspace. A thorough treatment is given in the excellent book by Bishop, Fienberg, and Holland (1975). The use of order restrictions on the parameters of these models has received relatively little attention, although recent work by Lemke (1983), Agresti *et al.* (1987), and Goodman (1985) point toward further development in this area. Fenchel duality is a useful tool in these types of problems, as it can often be used to rephrase problems in another guise that may be more tractable.

In particular, let $(n_1,\dots,n_k)$ be the observed value of $(N_1,\dots,N_k)$, a random vector which has a multinomial distribution with a parameter vector $\mathbf{p}=(p_1,\dots,p_k)$ lying in P, the set of all k-dimensional probability vectors. The basic problem will be to find a maximum likelihood estimate of $\mathbf{p}$ subject to the constraints that $\ln(\mathbf{p})\in\mathscr{K}$, where $\mathscr{K}$ is a closed, convex cone containing the constant vectors lying in

$$\bar{R}^k=\{(x_1,\dots,x_k):x_i\in R\cup\{-\infty\}\}.$$

If, in fact, $\mathscr{K}$ is actually a subspace, this is the classical log-linear model problem. However, the increased generality which follows from using cones rather than subspaces proves very useful, since it allows inequality constraints to be incorporated into the problem.

Of course, if $\mathscr{K}$ is a cone of isotonic vectors, then the constraints $\ln(\mathbf{p})\in\mathscr{K}$ and $\mathbf{p}\in\mathscr{K}$ are equivalent since $\ln(x)$ is a strictly increasing function.

The addition of the point $-\infty$ to R causes a bit of a problem, but it is necessary since we want to allow the possibility of zero probabilities. To facilitate this difficulty, we will denote a cone $\mathscr{K}$ in $\bar{R}^k$ restricted to R^k by $\tilde{\mathscr{K}}$, that is $\tilde{\mathscr{K}}=\mathscr{K}\cap R^k$.

Specifically, the stated problem we consider is that of finding a probability vector which solves the problem

$$\sup_{\mathbf{p}:\mathbf{p}\in P,\ln(\mathbf{p})\in\mathscr{K}} \prod_{i=1}^{k} p_i^{n_i} \tag{6.3.1}$$

where $\mathscr{K}$ is a closed, convex cone containing the constant vectors. Equivalently, if $\hat{\mathbf{p}}=(\hat{p}_1,\dots,\hat{p}_k)$ where $\hat{p}_i=n_i/\sum_{i=1}^k n_j$ and $q_i=\ln(p_i)$ (where we take $\ln(0)=-\infty$ and $0\ln(0)=0$), we wish to solve the problem

$$\sup_{\mathbf{q}:\mathbf{q}\in\mathscr{K},\sum_{j=1}^k \exp(q_j)\leqslant 1} \sum_{i=1}^{k} \hat{p}_i q_i. \tag{6.3.2}$$

Since the $\hat{p}_i$ are nonnegative and $\mathscr{K}$ is a closed, convex cone, this supremum can only be attained if $\sum_{i=1}^k \exp(q_i)=1$, since if $\sum_{i=1}^k \exp(q_i)<1$, we could multiply $\mathbf{q}$ by some positive number less than one and still satisfy the constraints, and this would increase $\sum_{i=1}^{k}\hat{p}_i q_i$. Thus (6.3.2) is equivalent to the problem

$$\sup_{\mathbf{q}\in\mathscr{K}} f(\mathbf{q}) \tag{6.3.3}$$

where

$$f(\mathbf{q}) = \begin{cases} \sum_{i=1}^{k} \hat{p}_i q_i, & \sum_{i=1}^{k} e^{q_i} \leqslant 1 \\ -\infty, & \sum_{i=1}^{k} e^{q_i} > 1. \end{cases} \tag{6.3.4}$$

If each $\hat{p}_i > 0$, then the supremum could only be attained within $\tilde{\mathscr{K}}$, so that we could restrict f to this region. Since f is a proper, concave function, $-f$ is proper convex and the Fenchel duality result discussed in Section 6.2 must hold (see equations (6.2.1) and (6.2.2)), so that

$$-\sup_{\mathbf{q}\in\tilde{\mathscr{K}}} f(\mathbf{q}) = \sup_{\mathbf{y}\in\tilde{\mathscr{K}}^*} f^*(\mathbf{y}). \tag{6.3.5}$$

We next address the form of f^*.

Lemma *If f is defined as in (6.3.4), then its concave conjugate is given by*

$$f^*(\mathbf{y}) = \begin{cases} -\sum_{i=1}^{k} (\hat{p}_i - y_i)\ln\left[(\hat{p}_i - y_i)\Big/\sum_{j=1}^{k}(\hat{p}_j - y_j)\right], & \text{if } y_i \leqslant \hat{p}_i \text{ for all } i \\ -\infty, & \text{if } y_i > \hat{p}_i \text{ for some } i. \end{cases} \tag{6.3.6}$$

Proof: By definition

$$\begin{aligned} f^*(\mathbf{y}) &= \inf_{\mathbf{q}} [\langle \mathbf{q}, \mathbf{y}\rangle - f(\mathbf{q})] \\ &= \inf_{\mathbf{q}:\sum_{i=1}^{k}\exp(q_i)\leqslant 1} \langle \mathbf{y} - \hat{\mathbf{p}}, \mathbf{q}\rangle \\ &= \begin{cases} \inf_{\mathbf{q}:\sum_{1}^{k}\exp(q_i)=1} \langle \mathbf{y} - \hat{\mathbf{p}}, \mathbf{q}\rangle, & y_i \leqslant p_i \text{ for all } i \\ -\infty, & y_i > p_i \text{ for some } i. \end{cases} \end{aligned}$$

The above infimum is found by writing $q_k = \ln[1 - \sum_{i=1}^{k-1}\exp(q_i)]$, setting the partial derivatives with respect to $q_1, \ldots, q_{k-1}$ equal to zero, solving the resulting equations, and substituting these values into $\langle \mathbf{y} - \hat{\mathbf{p}}, \mathbf{q}\rangle$. This results in (6.3.6). □

If we now make the change of variables $p_i = \hat{p}_i - y_i$, $i = 1, \ldots, k$, we can express

$$\sup_{\mathbf{y}\in\tilde{\mathscr{K}}^*} f^*(\mathbf{y}) = -\inf_{\substack{\mathbf{y}:\mathbf{y}\in\tilde{\mathscr{K}}^* \\ y_i\leqslant p_i \text{ for all } i}} \sum_{i=1}^{k} (\hat{p}_i - y_i)\ln\left[\frac{(\hat{p}_i - y_i)}{\sum_{j=1}^{k}(\hat{p}_j - y_j)}\right] \tag{6.3.7}$$

$$= - \inf_{\substack{\mathbf{p}:\hat{\mathbf{p}}-\mathbf{p}\in\tilde{\mathscr{K}}^* \\ p_i \geq 0 \text{ for all } i}} \sum_{i=1}^{k} p_i \ln\left(\frac{p_i}{\sum_{j=1}^{k} p_j}\right). \tag{6.3.8}$$

However, since $\mathscr{K}$ contains the constant vectors, $\mathbf{x}\in\tilde{\mathscr{K}}^*$ implies that $\sum_{i=1}^{k} x_i = 0$. Moreover, since $\sum_{i=1}^{k} \hat{p}_i = 1$, the constraint region in (6.3.8) is contained in the set of probability vectors, P. Thus (6.3.8) is really an I-projection problem as discussed in Example 6.2.4 with $\mathbf{r}$ taken to be the constant one vector.

In the event that $\mathbf{p}$ and $\mathbf{q}$ are both probability vectors, the value of $\sum_{i=1}^{k} p_i \ln(p_i/q_i)$ is called the *I-divergence of* $\mathbf{p}$ *with respect to* $\mathbf{q}$ and will be denoted by $I(\mathbf{p}|\mathbf{q})$($+\infty$ is a possible value). Problem (6.3.8) can be expressed in the form

$$\inf_{\mathbf{p}\in\hat{\mathbf{p}}-\tilde{\mathscr{K}}^*} I(\mathbf{p}|\mathbf{u}) \tag{6.3.9}$$

where $\hat{\mathbf{p}} - \tilde{\mathscr{K}}^* = \{\hat{\mathbf{p}} - \mathbf{z}:\mathbf{z}\in\tilde{\mathscr{K}}^*\}$, $\mathbf{u} = (1/k, 1/k, \ldots, 1/k)$ is the uniform probability vector, and it is tacitly agreed that the domain of an I-divergence is always the set of probability vectors. The following relationship between a restricted supremum of a multinomial problem and a restricted infimum of an I-divergence problem can now be established.

Theorem 6.3.1 *If each coordinate of* $\hat{\mathbf{p}} = (\hat{p}_1, \ldots, \hat{p}_k)$ *is positive and* $\tilde{\mathscr{K}}$ *is a closed, convex cone containing the constant functions, then*

$$\sup_{\substack{\ln\mathbf{p}\in\tilde{\mathscr{K}} \\ \mathbf{p}\in P}} \prod_{i=1}^{k} p_i^{\hat{p}_i} = k^{-1} \exp\left[\inf_{\mathbf{p}\in\hat{\mathbf{p}}-\tilde{\mathscr{K}}^*} I(\mathbf{p}|\mathbf{u})\right].$$

Proof: By (6.3.5),

$$\sup\left[\sum_{i=1}^{k} \hat{p}_i q_i : \mathbf{q}\in\tilde{\mathscr{K}}^* \text{ and } \sum_{i=1}^{k} \exp(q_i) \leq 1\right]$$

$$= \inf_{\substack{\mathbf{p}\in\hat{\mathbf{p}}-\tilde{\mathscr{K}}^* \\ \mathbf{p}\in P}} \left[\sum_{i=1}^{k} p_i \ln p_i + \sum_{i=1}^{k} p_i \ln(k) - \ln(k)\right]$$

$$= \inf_{\mathbf{p}\in\hat{\mathbf{p}}-\tilde{\mathscr{K}}^*} I(\mathbf{p}|\mathbf{u}) - \ln(k).$$

The result now follows by evaluating the exponential function on each side. □

While the relationship between the extremal values of the multinomial maximum likelihood and the I-divergence expressions given in Theorem 6.3.1 is rather surprising, even more remarkable is the fact that a vector solves the multinomial problem (6.3.1) if and only if it solves the I-projection problem (6.3.9). Thus these two very different looking problems are actually equivalent. This is stated in Theorem 6.3.4, whose proof hinges upon verification of conditions (6.2.4) of the previous section.

However, it will be useful to discuss the subject of I-projections a bit more before establishing this equivalence. In many ways, I-projections exhibit properties that are analogous to least squares projections. In particular, if **p** and **q** are probability vectors, $I(\mathbf{p}|\mathbf{q})$ acts like a distance between **p** and **q** in the sense that it is always nonnegative and it is equal to zero if and only if $\mathbf{p}=\mathbf{q}$. However, $I(\mathbf{p}|\mathbf{q})$ is not a metric, nor can it easily be made into one. This is true even for the symmetrized version $\tilde{I}(\mathbf{p}|\mathbf{q})=\frac{1}{2}[I(\mathbf{p}|\mathbf{q})+I(\mathbf{q}|\mathbf{p})]$ discussed by Jeffreys (1948). Csiszar (1975) has developed an appealing geometric approach to the theory of I-projections, and has proven the following theorem which nicely characterizes I-projections.

Theorem 6.3.2 (*Csiszar*) *For a fixed probability vector* **r**, *the I-divergence of* **q** *with respect to* **r**, *where* **q** *lies in a convex set* $\mathscr{E}$ *of probability vectors, attains the value*

$$\inf_{\mathbf{p}\in\mathscr{E}} I(\mathbf{p}|\mathbf{r})=\inf_{\mathbf{p}\in\mathscr{E}}\sum_{i=1}^{k} p_i \ln\left(\frac{p_i}{r_i}\right)<\infty,$$

and hence is the I-projection of **r** *onto* $\mathscr{E}$, *if and only if*

$$\begin{aligned}&(a)\quad I(\mathbf{p}|\mathbf{r})\geqslant I(\mathbf{p}|\mathbf{q})+I(\mathbf{q}|\mathbf{r})\ \textit{for all}\ \mathbf{p}\in\mathscr{E}\ \textit{and}\\ &(b)\quad I(\mathbf{q}|\mathbf{r})<\infty.\end{aligned}\tag{6.3.10}$$

I-projections can be defined for general probability measures (Csiszar, 1975), and Theorem 6.3.2 remains valid in these more general settings. For probability vectors in R^k, part (a) of (6.3.10) is equivalent to

$$\sum_{i=1}^{k}(p_i-q_i)\ln\left(\frac{q_i}{r_i}\right)\geqslant 0 \text{ for all } \mathbf{p}\in\mathscr{E}.\tag{6.3.11}$$

If in fact **q** is an algebraic inner point of $\mathscr{E}$, i.e. for every $\mathbf{p}\in\mathscr{E}(\mathbf{p}\neq\mathbf{q})$ there exists $0<\alpha<1$ and $\mathbf{p}'\in\mathscr{E}$ such that $\mathbf{q}=\alpha\mathbf{p}+(1-\alpha)\mathbf{p}'$, then equality must hold in part (a) of (6.3.10) and (6.3.11). In particular, Csiszar defines $\mathscr{E}$ to be a linear set of probability vectors if $\mathbf{p},\mathbf{p}'\in\mathscr{E}$ implies $\alpha\mathbf{p}+(1-\alpha)\mathbf{p}'\in\mathscr{E}$ for *every* real number α for which it is a probability vector. If $\mathscr{E}$ is a linear set, then the inequality sign in part (a) of (6.3.10) and (6.3.11) may be replaced by an equality sign.

Note the similarity in the statement of Theorem 6.3.2 and least squares projections discussed in Chapter 1. In particular, if $\|\cdot\|$ denotes a least squares norm, then **q** is the least squares projection of **r** onto the convex set $\mathscr{E}$ if and only if

$$\|\mathbf{p}-\mathbf{r}\|^2\geqslant\|\mathbf{p}-\mathbf{q}\|^2+\|\mathbf{q}-\mathbf{r}\|^2 \text{ for all } \mathbf{p}\in\mathscr{E}$$

and $\mathbf{q}\in\mathscr{E}$ (see Theorem 1.2). If $\mathscr{E}$ is a subspace, then the inequality may be replaced by an equality in the above expression.

Thus in some sense, the role of $I(\mathbf{p}|\mathbf{q})$ in I-projections is analogous to that of $\|\mathbf{p}-\mathbf{q}\|^2$ in least squares projections, and linear sets of probability vectors are analogous to subspaces. Some least squares projection properties also hold for I-projections. For example, (6.3.10) can be used to prove an iterative projection property for I-projections (cf. Theorem 6.3.3) which has an analogue for least squares projections onto convex sets which are contained in subspaces.

Theorem 6.3.3 *Suppose $\mathbf{p}_{\mathscr{C}}$ denotes the I-projection of $\mathbf{r}$ onto the convex set of probability vectors $\mathscr{C}$, where $\mathscr{C}$ is contained in the linear set of probability vectors $\mathscr{E}$. If $\mathbf{p}_{\mathscr{C}}$ denotes the I-projection of $\mathbf{r}$ onto $\mathscr{E}$, $\mathbf{p}_{\mathscr{C}}$ is also the I-projection of $\mathbf{p}_{\mathscr{E}}$ onto $\mathscr{C}$.*

Proof: If one lets $\mathbf{q}\in\mathscr{C}\subset\mathscr{E}$, two applications of (6.3.10) give

$$I(\mathbf{q}|\mathbf{p}_{\mathscr{E}})+I(\mathbf{p}_{\mathscr{E}}|\mathbf{r})=I(\mathbf{q}|\mathbf{r})\geqslant I(\mathbf{q}|\mathbf{p}_{\mathscr{C}})+I(\mathbf{p}_{\mathscr{C}}|\mathbf{r}).$$

However, $\mathbf{p}_{\mathscr{C}}$ is in $\mathscr{C}\subset\mathscr{E}$ so that a third use of (6.3.10) yields

$$I(\mathbf{p}_{\mathscr{C}}|\mathbf{r})=I(\mathbf{p}_{\mathscr{C}}|\mathbf{p}_{\mathscr{E}})+I(\mathbf{p}_{\mathscr{E}}|\mathbf{r}).$$

By substituting this in the first expression and subtracting $I(\mathbf{p}_{\mathscr{E}}|\mathbf{r})$ from both sides, one obtains

$$I(\mathbf{q}|\mathbf{p}_{\mathscr{E}})\geqslant I(\mathbf{q}|\mathbf{p}_{\mathscr{C}})+I(\mathbf{p}_{\mathscr{C}}|\mathbf{p}_{\mathscr{E}})$$

which proves the desired result by one last application of (6.3.10). □

We are now ready to prove the equivalence of the multinomial problem, (6.3.1), and the I-projection problem, (6.3.9). The notation used earlier is still employed.

Theorem 6.3.4 *Suppose $(\hat{p}_1,\ldots,\hat{p}_k)$ is positive in each coordinate and $\tilde{\mathscr{K}}=\mathscr{K}\cap R^k$ is a closed, convex cone containing the constant functions. Then $\mathbf{p}^*$ solves the problem* (6.3.9), *if and only if it solves the problem* (6.3.1).

Proof: Assume $\mathbf{p}^*$ solves (6.3.9), or equivalently $\hat{\mathbf{p}}-\mathbf{p}^*$ solves (6.3.7). Note also that $\mathbf{p}^*$ solves (6.3.1) if and only if $\ln(\mathbf{p}^*)$ solves (6.3.2). Since (6.3.2) and (6.3.7) are dual problems of the form (6.2.5) and (6.2.6), it will suffice to show that $\ln(\mathbf{p}^*)$ and $\hat{\mathbf{p}}-\mathbf{p}^*$ satisfy the four conditions of (6.2.4) (modified for concave functions). These conditions will be verified in the order (d), (b), (c), (a).

(d) To show that $-(\hat{\mathbf{p}}-\mathbf{p}^*)$ is a subgradient of $-f$ (defined in equations (6.3.4)) at $\ln(\mathbf{p}^*)$, we must show

$$f(\mathbf{z})-f(\ln(\mathbf{p}^*))\leqslant\langle\hat{\mathbf{p}}-\mathbf{p}^*,\mathbf{z}-\ln(\mathbf{p}^*)\rangle\quad\text{for all }\mathbf{z}\in R^k. \tag{6.3.12}$$

If $\sum_{i=1}^k\exp(z_i)>1$, the left side is $-\infty$. If $\sum_{i=1}^k\exp(z_i)\leqslant 1$, (6.3.12) can be expressed as

$$\sum_{i=1}^k\hat{p}_iz_i-\sum_{i=1}^k\hat{p}_i\ln(p_i^*)\leqslant\sum_{i=1}^k\hat{p}_iz_i-\sum_{i=1}^k\hat{p}_i\ln(p_i^*)-\sum_{i=1}^k p_i^*z_i+\sum_{i=1}^k p_i^*\ln(p_i^*)$$

or, equivalently,

$$0 \leqslant \sum_{i=1}^{k} p_i^* \ln(p_i^*) - \sum_{i=1}^{k} p_i^* \ln[\exp(z_i)]$$

$$= \sum_{i=1}^{k} p_i^* \ln\left[p_i^* \sum_{j=1}^{k} \frac{\exp(z_j)}{\exp(z_i)} \right] - \sum_{i=1}^{k} p_i^* \ln\left[\sum_{j=1}^{k} \exp(z_j) \right]. \tag{6.3.13}$$

The first sum of (6.3.13) is nonnegative since it is an I-divergence, and the second summation is nonpositive since $\sum_{j=1}^{k} \exp(z_j) \leqslant 1$.

(b) We know $\hat{\mathbf{p}} - \mathbf{p}^* \in \tilde{\mathscr{K}}^*$ by assumption.

(c) We must show $\langle \hat{\mathbf{p}} - \mathbf{p}^*, \ln(\mathbf{p}^*) \rangle = 0$. If $\hat{\mathbf{p}} = \mathbf{p}^*$, the result is trivial, so assume otherwise. Since $\hat{\mathbf{p}} - \mathbf{p}^* \in \tilde{\mathscr{K}}^*$, $\mathbf{p}^* = \hat{\mathbf{p}} - \mathbf{v}$ for some $\mathbf{v} \in \hat{\mathscr{K}}^*$. Consider the set of probability vectors

$$\mathscr{E}_0 = \{\hat{\mathbf{p}} - \alpha \mathbf{v} : \alpha \text{ real}\} \cap P.$$

Clearly, $\mathscr{E}_0$ will be a linear set of probability vectors as defined earlier. If $\mathbf{p}_\alpha = \hat{\mathbf{p}} - \alpha \mathbf{v}$ is strictly positive,

$$I(\mathbf{p}_\alpha | \mathbf{u}) = \sum_{i=1}^{k} (\hat{p}_i - \alpha v_i) \ln(\hat{p}_i - \alpha v_i) + \ln(k)$$

is differentiable with respect to α. The first two derivatives are

$$\frac{\mathrm{d}}{\mathrm{d}\alpha} I(\mathbf{p}_\alpha | \mathbf{u}) = -\sum_{i=1}^{k} v_i \ln(\hat{p}_i - \alpha v_i)$$

(since $\sum_1^k v_i = 0$), and

$$\frac{\mathrm{d}^2}{\mathrm{d}\alpha^2} I(\mathbf{p}_\alpha | \mathbf{u}) = \sum_{i=1}^{k} \frac{v_i^2}{\hat{p}_i - \alpha v_i} > 0.$$

Thus, $I(\mathbf{p}_\alpha | \mathbf{u})$ is strictly convex as a function of α, and must have a minimum at $\alpha = 1$ (since $\tilde{\mathscr{K}}^*$ is a cone). Thus $\mathbf{p}^*$ is the I-projection of $\mathbf{u}$ onto $\mathscr{E}_0$, and since $\hat{\mathbf{p}} \in \mathscr{E}_0$, which is a linear space,

$$0 = \sum_1^k (\hat{p}_i - p_i^*) \ln\left(\frac{p_i^*}{u_i} \right) = \langle \hat{\mathbf{p}} - \mathbf{p}^*, \ln(\mathbf{p}^*) \rangle$$

by (6.3.11) and subsequent discussion.

(a) We must show $\ln(\mathbf{p}^*) \in \tilde{\mathscr{K}}$, or, equivalently, that $\langle \ln(\mathbf{p}^*), \mathbf{v} \rangle \leqslant 0$ for every $\mathbf{v} \in \tilde{\mathscr{K}}^*$. For a fixed $\mathbf{v} \in \tilde{\mathscr{K}}^*$, note that $\hat{p}_i > 0$ for all i implies there exists a positive α such that $\hat{p}_i - \alpha v_i \geqslant 0$ for all i, and hence $\mathbf{h} = \hat{\mathbf{p}} - \alpha \mathbf{v}$ is a probability vector. Now by condition (c),

$$\begin{aligned}\langle \ln(\mathbf{p}^*), (\alpha \mathbf{v}) \rangle &= \langle \ln(\mathbf{p}^*), \hat{\mathbf{p}} \rangle - \langle \ln(\mathbf{p}^*), \mathbf{h} \rangle = \langle \ln(\mathbf{p}^*), \mathbf{p}^* \rangle - \langle \ln(\mathbf{p}^*), \mathbf{h} \rangle \\ &= \langle \mathbf{p}^* - \mathbf{h}, \ln(\mathbf{p}^*) \rangle = \langle \mathbf{p}^* - \mathbf{h}, \ln(\mathbf{p}^*/\mathbf{u}) \rangle \leqslant 0\end{aligned}$$

by (6.3.11) and the fact that $\mathbf{h} \in \hat{\mathbf{p}} - \tilde{\mathscr{K}}^*$.

For the 'if' part of the proof, note that both problems (6.3.1) and (6.3.9) must have unique solutions under the conditions given in the theorem. This fact coupled with the 'only if' part of the proof will imply the desired result. □

The requirement that the $\hat{p}_i > 0$ for all i is unpleasant, but the stated Fenchel duality theorem applies only to functions defined on R^k. On the other hand, cells with zero counts have often caused problems in multinomial settings and it would be advantageous to be able to deal with this situation. Fortunately, the problem of observed zeros can be handled in the following manner if $\mathscr{K}$ is a *continuous cone*. By a continuous cone we mean that $\mathscr{K}$ has the property that $\mathbf{x} \in \mathscr{K}$ implies that there exists a sequence of vectors $\mathbf{x}_n \in \tilde{\mathscr{K}}$ such that $x_{n,i} \to x_i$ as $n \to \infty$ for all i. It can be shown that if $\mathbf{p}_n^*(\mathbf{p}^*)$ denote the solutions to the problem (6.3.9) corresponding to $\hat{\mathbf{p}}_n(\hat{\mathbf{p}})$, and if $\mathbf{p}_n'(\mathbf{p}')$ denote the solutions to (6.3.1) corresponding to $\hat{\mathbf{p}}_n(\hat{\mathbf{p}})$, then $\hat{\mathbf{p}}_n \to \hat{\mathbf{p}}$ implies that $\mathbf{p}_n^* \to \mathbf{p}^*$ and $\mathbf{p}_n' \to \mathbf{p}'$ (where the convergence is with respect to the usual Euclidian distance in R^k).

Thus to establish Theorem 6.3.4 for a $\hat{\mathbf{p}}$ which may contain some zeros, we need only select a strictly positive sequence $\hat{\mathbf{p}}_n \to \hat{\mathbf{p}}$ and use Theorem 6.3.4 and the above continuity result to argue that $\mathbf{p}^* = \mathbf{p}'$. We refer the interested reader to Lemke and Dykstra (1984) for details. This argument leads to the following corollary.

Corollary 6.3.1 *If* $(\hat{p}_1, \ldots, \hat{p}_k) = (n_1/\sum_{i=1}^k n_i, \ldots, n_k/\sum_{i=1}^k n_i)$ *is a nonnegative vector (not identically zero) and* $\mathscr{K}$ *is a closed, convex, continuous cone in* R^k *containing the constant functions, then* $\mathbf{p}^*$ *solves (6.3.9) if and only if it solves (6.3.1).*

The linear space analogue to this result, i.e. when $\mathscr{K}$ is a subspace containing the constant functions, has been known for quite some time, although we have not been able to identify the person who first discovered this relationship. We can use our duality relationship to characterize the common solution to (6.3.9) and (6.3.1) in the following manner, which we state as another corollary.

Corollary 6.3.2 *If* $\mathscr{K}$ *is a closed, convex, continuous cone containing the constant functions, then the solution to (6.3.9) and (6.3.1) is the unique vector* $\mathbf{p}^*$ *satisfying*

(*a*) $\ln(\mathbf{p}^*) \in \mathscr{K}$
(*b*) $\hat{\mathbf{p}} - \mathbf{p}^* \in \tilde{\mathscr{K}}^*$
(*c*) $\langle \hat{\mathbf{p}} - \mathbf{p}^*, \ln(\mathbf{p}^*) \rangle = 0$.

(*Note that* (c) *is redundant if* $\tilde{\mathscr{K}}$ *is a subspace, since then* $\tilde{\mathscr{K}}^*$ *becomes* $\tilde{\mathscr{K}}^{\perp}$, *the orthogonal complement of* $\tilde{\mathscr{K}}$.)

Proof: Assume $\hat{\mathbf{p}}$ is positive everywhere. Then the fourth condition of (6.2.4) always holds, and (a), (b), and (c) are just the other three conditions. The general case follows by continuity. □

Dual problem for simple independence

Example 6.3.1 Although the problem of finding maximum likelihood estimates for the parameters in a two-dimensional, independent, multinomial contingency table is not difficult, it is enlightening to consider the problem through its dual formulation.

Let us assume that we have an $r \times c$ rectangular array of cells. We will also assume that we have simple multinomial sampling, that p_{ij} is the probability of falling in the (i, j) cell, and that $\hat{p}_{ij}$ denotes the relative frequency of observations for the (i, j) cell. Since the independence criterion is equivalent to writing the probability vector p in the form $p_{ij} = a_i b_j$, the primal maximum likelihood problem can be written as

$$\sup_{\substack{\ln(\mathbf{p}) \in \mathcal{M}_1 + \mathcal{M}_2 \\ \mathbf{p} \in P}} \prod_{j=1}^{c} \prod_{i=1}^{r} p_{ij}^{\hat{p}_{ij}} \tag{6.3.14}$$

where $\mathcal{M}_1$ and $\mathcal{M}_2$ are defined to be the respective linear spaces

$$\mathcal{M}_1 = \{\mathbf{x}: x_{i1} = x_{i2} = \cdots = x_{ic}, i = 1, \ldots, r\}$$

and

$$\mathcal{M}_2 = \{\mathbf{x}: x_{1j} = x_{2j} = \cdots = x_{rj}, j = 1, \ldots, c\}.$$

Since $\mathcal{M}_1$ and $\mathcal{M}_2$ are linear spaces, the dual cones for $\mathcal{M}_1$ and $\mathcal{M}_2$ will be the orthogonal complements

$$\mathcal{M}_1^{\perp} = \left\{\mathbf{x}: \sum_{j=1}^{c} x_{ij} = 0, i = 1, \ldots, r\right\}$$

and

$$\mathcal{M}_2^{\perp} = \left\{\mathbf{x}: \sum_{i=1}^{r} x_{ij} = 0, j = 1, \ldots, c\right\}.$$

Since $\mathcal{M}_1 + \mathcal{M}_2$ contains the constant functions, Corollary 6.3.1 ensures that the solution to (6.3.14) is exactly the same as the solution to the dual problem. Thus, recalling that $\mathbf{u}$ is the uniform probability vector and that the domain of an I-divergence is the set of probability vectors, we consider

$$\inf_{\mathbf{p} \in \hat{\mathbf{p}} - (\mathcal{M}_1 + \mathcal{M}_2)^{\perp}} I(\mathbf{p} \mid \mathbf{u}) = \inf_{\mathbf{p} \in \hat{\mathbf{p}} - (\mathcal{M}_1^{\perp} \cap \mathcal{M}_2^{\perp})} I(\mathbf{p} \mid \mathbf{u}) = \inf_{\mathbf{p} \in \bigcap_{i=1}^{2} (\hat{\mathbf{p}} - \mathcal{M}_i^{\perp})} I(\mathbf{p} \mid \mathbf{u}). \tag{6.3.15}$$

Theorem 6.3.3 states that (6.3.15) can be solved by first finding the I-projection of $\mathbf{u}$ onto $\hat{\mathbf{p}} - \mathcal{M}_1^{\perp}$ and then projecting this array onto $\bigcap_{i=1}^{2} (\hat{\mathbf{p}} - \mathcal{M}_i^{\perp})$. The condition that $\mathbf{p} \in \hat{\mathbf{p}} - \mathcal{M}_1^{\perp}$ is equivalent to stating that $\mathbf{p}$ and $\hat{\mathbf{p}}$ have the same row sums, while $\mathbf{p} \in \hat{\mathbf{p}} - \mathcal{M}_2^{\perp}$ means that $\mathbf{p}$ and $\hat{\mathbf{p}}$ have the same column sums.

To project $\mathbf{u}$ onto $\hat{\mathbf{p}} - \mathcal{M}_1^{\perp}$, one must minimize

$$\sum_{j=1}^{c} \sum_{i=1}^{r} p_{ij} \ln(p_{ij}) \text{ subject to } \sum_{j=1}^{c} p_{ij} = \sum_{j=1}^{c} \hat{p}_{ij}, \quad i = 1, \dots, r.$$

If one writes $p_{ic} = \sum_{j=1}^{c} \hat{p}_{ij} - \sum_{j=1}^{c-1} p_{ij}$ and sets the partial derivatives with respect to p_{ij}, $i = 1, \dots, r$, $j = 1, \dots, c - 1$, equal to zero, the resulting equations are

$$\ln p_{ij} = \ln p_{ic} \qquad \text{for } i = 1, \dots, r; \quad j = 1, \dots, c - 1.$$

By taking the exponential function of both sides and summing over j, we obtain

$$c p_{ic} = \sum_{j=1}^{c} p_{ij} = \sum_{j=1}^{c} \hat{p}_{ij} \qquad \text{for } i = 1, \dots, r,$$

so that

$$\tilde{p}_{ij} = \frac{\sum_{j=1}^{c} \hat{p}_{ij}}{c}$$

is the I-projection of $\mathbf{u}$ onto $\hat{\mathbf{p}} - \mathcal{M}_1^{\perp}$.

In a similar manner, we can project $\tilde{\mathbf{p}}$ onto $\hat{\mathbf{p}} - \mathcal{M}_2^{\perp}$. Straightforward calculations will show that this projection is given by

$$p_{ij}^* = \left(\sum_{i=1}^{r} \hat{p}_{ij} \right) \left(\sum_{j=1}^{c} \hat{p}_{ij} \right) \qquad \text{for } i = 1, \dots, r; \quad j = 1, \dots, c.$$

However, since $\mathbf{p}^* \in \bigcap_{i=1}^{2} (\hat{\mathbf{p}} - \mathcal{M}_i^{\perp})$, this must also be the projection of $\tilde{\mathbf{p}}$ onto this region, and hence $\mathbf{p}^*$ is the solution to both the dual problem (6.3.15) and the original problem (6.3.14).

While the solution to problem (6.3.14) is easy to obtain directly without appealing to the dual problem, more complicated linear models may require the extra duality structure. Moreover, the fact that these duality results are valid in the much more general setting of cones rather than subspaces has important implications for order restricted inference in log-linear models.

Convergence of cyclic, iterated I-projections for linear sets

In the previous example, we examined a log-linear model maximum likelihood problem which was equivalent to an I-projection onto the intersection of linear sets. The structure of I-projection geometry can often be used to solve the problem of finding I-projections onto such an intersection. Csiszar (1975) has used this structure to prove that an iterated I-projection algorithm must converge correctly in a very general setting.

To elaborate on Csiszar's algorithm, let $\mathscr{E} = \bigcap_1^t \mathscr{E}_i$ denote the nonempty intersection of t *linear* sets of probability vectors (linear sets of probability vectors are always closed since that may be taken to be the intersection of a subspace with a k-dimensional simplex). The goal is to find the I-projection of

$\mathbf{r}$ onto $\mathscr{E}$ when one is able to find I-projections onto the individual $\mathscr{E}_i$'s. It is assumed that there exists a $\mathbf{q} \in \mathscr{E}$ such that $I(\mathbf{q}|\mathbf{r}) < \infty$, so that the solution to the problem exists uniquely. Csiszar's method is to sequentially cycle through the constraint regions, at each step finding the I-projection of the previous step's projection.

To be specific, let $\mathbf{q}_{nj}$ denote the jth I-projection (onto $\mathscr{E}_j$) during the nth cycle. Thus $\mathbf{q}_{11}$ would be the I-projection of $\mathbf{r}$ onto $\mathscr{E}_1$, $\mathbf{q}_{12}$ would be the I-projection of $\mathbf{q}_{11}$ onto $\mathscr{E}_2$, $\mathbf{q}_{13}$ would be the I-projection of $\mathbf{q}_{12}$ onto $\mathscr{E}_3$, etc. After projecting onto all t constraint sets, one should begin cycling through the constraint sets again. Thus $\mathbf{q}_{21}$ is the I-projection of $\mathbf{q}_{1t}$ onto $\mathscr{E}_1$, etc. This procedure is repeated until the amount of change between successive cycles is sufficiently small.

By using the structure developed for I-projections onto linear sets, in particular (6.3.10), Csiszar has proven that this sequence of I-projections must converge to the I-projection of $\mathbf{r}$ onto $\mathscr{E}$.

In fact, Csiszar's proof implies that the sequential I-projections approach the entire set $\mathscr{E} = \bigcap_1^t \mathscr{E}_i$ in a monotone manner in terms of I-divergence; i.e. if $n' > n$, or if $n' = n$, $j' > j$, then

$$I(\mathbf{p}|\mathbf{q}_{nj}) \geqslant I(\mathbf{p}|\mathbf{q}_{n'j'}) \qquad \text{for all } \mathbf{p} \in \mathscr{E}.$$

As might be anticipated,

$$I(\mathbf{p}|\mathbf{q}_{nj}) \to I(\mathbf{p}|\mathbf{q}^*) \quad \text{as } n \to \infty$$

for all $\mathbf{p} \in \mathscr{E}$ where $\mathbf{q}^*$ is the I-projection of $\mathbf{r}$ onto $\mathscr{E}$.

The often used iterated proportional fitting procedure (IPFP), discussed in Bishop, Fienberg, and Holland (1975) among other places, is really an application of Csiszar's cyclic, iterated I-projection procedure. An explanation for the fact that these proportional fittings (which are really I-projections) must converge correctly to the multinomial maximum likelihood solution is due to the duality structure which we have been discussing.

Failure of cyclic, iterated *I*-projections for nonlinear sets

The situation is much different, however, if the $\mathscr{E}_i$ are not linear sets, and the cyclic iterated projection procedure proposed by Csiszar need not converge to the correct solution in this situation. For a simple example to show that Csiszar's procedure does not work for general convex sets, consider the following example:

$$\mathscr{E}_1 = \left\{ \begin{pmatrix} p_{11} & p_{12} \\ p_{21} & p_{22} \end{pmatrix} : p_{11} \geqslant p_{12}, p_{21} \geqslant p_{22} \right\}$$

$$\mathscr{E}_2 = \left\{ \begin{pmatrix} p_{11} & p_{12} \\ p_{21} & p_{22} \end{pmatrix} : p_{11} \geqslant p_{21}, p_{12} \geqslant p_{22} \right\}$$

and

$$r = \begin{pmatrix} \frac{1}{16} & \frac{3}{16} \\ \frac{7}{16} & \frac{5}{16} \end{pmatrix}.$$

Thus $\mathscr{E}_1$ corresponds to arrays with nonincreasing rows, $\mathscr{E}_2$ corresponds to arrays with nonincreasing columns, and $\mathscr{E}_1 \cap \mathscr{E}_2$ is a version of the matrix order discussed in Chapter 1.

Csiszar's procedure converges to $(\frac{9}{32}\frac{7}{32}, \frac{9}{32}\frac{7}{32})$, whereas $(\frac{1}{4}\frac{1}{4}, \frac{1}{4}\frac{1}{4})$ is really the correct solution.

Dykstra's method for nonlinear sets

Dykstra (1985a) has proposed a modification of the cyclic iterated I-projection scheme to solve the aforementioned problem when the $\mathscr{E}_i$ are arbitrary, closed convex sets of probability vectors. As before, it is assumed that there exists a $\mathbf{q} \in \mathscr{E} = \bigcap_1^t \mathscr{E}_i$ such that $I(\mathbf{q}|\mathbf{r}) < \infty$, so that the unique solution exists. Dykstra's scheme is similar to Csiszar's, except his approach calls for a multiplicative adjustment to be made at each step. This is somewhat similar to the additive adjustments used in Theorem 1.4.6.

To be precise, let $\mathbf{s}_{nj}$ be the vector that is projected on the jth step of the nth cycle and let $\mathbf{q}_{nj}$ be the I-projection of $\mathbf{s}_{nj}$ onto $\mathscr{E}_j$. To begin, set $\mathbf{s}_{11} = \mathbf{r}$ and find $\mathbf{q}_{11}$. The algorithm then proceeds by the following rules where initially $n = 1$, $j = 2$:

1. Set $\mathbf{s}_{n,j} = \mathbf{q}_{n,j-1} \cdot \mathbf{s}_{n-1,j}/\mathbf{q}_{n-1,j}$, if $2 \leqslant j \leqslant t$, and $\mathbf{s}_{n,j} = \mathbf{q}_{n-1,t} \cdot \mathbf{s}_{n-1,1}/\mathbf{q}_{n-1,1}$, if $j = 1$ (where vector multiplication and division are done in a coordinatewise manner, 0/0 is taken to be 1, and $\mathbf{s}_{0j}/\mathbf{q}_{0j} = \mathbf{1}$, $j = 1, \ldots, r$).
2. Find $\mathbf{q}_{nj}$, the I-projection of $\mathbf{s}_{nj}$ onto $\mathscr{E}_j$, and then perform rule 1 for the indices $(n, j + 1)$ if $j < t$, or for $(n + 1, 1)$ if $j = t$.

If $\mathscr{E}_j$ is a linear set (and $j > 1$) and if $\mathbf{q} \in \mathscr{E}_j$, then

$$\langle \mathbf{q}, \ln(\mathbf{q}/\mathbf{s}_{nj}) \rangle = \langle \mathbf{q}, \ln(\mathbf{q}/\mathbf{q}_{n,j-1}) \rangle + \langle \mathbf{q}, \ln(\mathbf{q}_{n-1,j}/\mathbf{s}_{n-1,j}) \rangle. \qquad (6.3.16)$$

However, since $\mathbf{q}_{n-1,j}$ is the I-projection of $\mathbf{s}_{n-1,j}$ onto $\mathscr{E}_j$, and since $\mathbf{q} \in \mathscr{E}_j$, we may use (6.3.11) for linear sets to conclude

$$\langle \mathbf{q}, \ln(\mathbf{q}_{n-1,j}/\mathbf{s}_{n-1,j}) \rangle = \langle \mathbf{q}_{n-1,j}, \ln(\mathbf{q}_{n-1,j}/\mathbf{s}_{n-1,j}) \rangle.$$

Thus the second part of (6.3.16) does not depend upon $\mathbf{q}$, and hence minimizing (6.3.16) for $\mathbf{q} \in \mathscr{E}_j$ is equivalent to the problem

$$\inf_{\mathbf{q} \in \mathscr{E}_j} I(\mathbf{q}|\mathbf{q}_{n,j-1}).$$

It easily follows that Dykstra's procedure reduces to Csiszar's method if the $\mathscr{E}_i$ are all linear sets.

Dykstra conjectures that his scheme must always converge correctly, but is only able to prove correct convergence by requiring an extra condition on the $\mathbf{q}_{nj}$. However, he does show that if the $\mathbf{q}_{nj}$ do not converge correctly as $n \to \infty$, then $\sup \|\mathbf{s}_{nj}\|$ must go to ∞ (where $\|\cdot\|$ is the L_2 norm). A step can be put into computer routines to monitor $\|\mathbf{s}_{nj}\|$. If these values do not become large, the procedure must converge correctly.

Examples of Dykstra's procedure for convex, nonlinear sets

Example 6.3.2 Consider the case where (p_{kj}) is a square $n \times n$ array of probabilities. We denote the corresponding marginal probabilities by

$$p_{k+} = \sum_{j=1}^{n} p_{kj}, \quad p_{+j} = \sum_{k=1}^{n} p_{kj} \qquad \text{for } k, j = 1, \ldots, n.$$

We now consider the problem of finding the I-projection of a fixed nonnegative array (r_{kj}) subject to the condition that the marginal probability distributions are stochastically ordered, that is

$$\sum_{l=1}^{i} p_{l+} \geqslant \sum_{l=1}^{i} p_{+l} \qquad \text{for all } i.$$

(Kullback, 1971, has given an iterative procedure for finding I-projections where equality is forced to hold for all i. This condition is known as *marginal homogeneity*). Equivalently, we want to find the I-projection of $\mathbf{r}$ onto

$$\mathscr{E} = \bigcap_{1}^{n-1} \mathscr{E}_i \quad \text{where } \mathscr{E}_i = \left\{ \mathbf{p} \in P : \sum_{k=1}^{i} \sum_{j=1}^{n} p_{kj} \geqslant \sum_{j=1}^{i} \sum_{k=1}^{n} p_{kj} \right\}.$$

Note that the $\mathscr{E}_i$ are closed convex sets of probability vectors which are not linear sets because of the inequality. The I-projection of an arbitrary vector $\mathbf{v}$ on $\mathscr{E}_i$, denoted by $\hat{p}_i(\mathbf{v})$, can be found by forcing equality in the ith constraint if $\mathbf{v}$ violates that constraint.

To express $\hat{p}_i(\mathbf{v})$, we let

$$A_i = \{(l, m) : 1 \leqslant l \leqslant i, i + 1 \leqslant m \leqslant n\}$$
$$B_i = \{(l, m) : 1 \leqslant m \leqslant i, i + 1 \leqslant l \leqslant n\}$$
$$C_i = \{(l, m) : l, m = 1, \ldots, i\} \cup \{(l, m) : l, m = i + 1, \ldots, n\}$$

denote subblocks of the index matrix, and define $\hat{v}_i$ as

$$\hat{v}_i = \left(\sum_{A_i} v_{lm} \sum_{B_i} v_{lm} \right)^{1/2}.$$

If $\mathbf{v}$ satisfies the constraint of $\mathscr{E}_i$, namely $\sum_{A_i} v_{lm} \geqslant \sum_{B_i} v_{lm}$ since common terms may be subtracted out, then $\hat{p}_i(\mathbf{v})$ is proportional to $\mathbf{v}$. Thus

$$\hat{p}_i(\mathbf{v}) = \frac{\mathbf{v}}{\sum_{l=1}^{n}\sum_{m=1}^{n} v_{lm}}.$$

If $\mathbf{v}$ does not satisfy the constraint imposed by $\mathscr{E}_i$, that is $\sum_{A_i} v_{lm} < \sum_{B_i} v_{lm}$, then a closed form solution for $\hat{p}_i(\mathbf{v})$ is given in Dykstra (1985b) as

$$\hat{p}_i(\mathbf{v})_{kj} = \begin{cases} v_{kj}\left(\dfrac{\sum_{B_i} v_{lm}}{\sum_{A_i} v_{lm}}\right)^{1/2}\left(2\hat{v}_i + \sum_{C_i} v_{lm}\right)^{-1}, & (k, j) \in A_i \\ v_{kj}\left(\dfrac{\sum_{A_i} v_{lm}}{\sum_{B_i} v_{lm}}\right)^{1/2}\left(2\hat{v}_i + \sum_{C_i} v_{lm}\right)^{-1}, & (k, j) \in B_i \\ v_{kj}\left(2\hat{v}_i + \sum_{C_i} v_{lm}\right)^{-1}, & (k, j) \in C_i. \end{cases}$$

The key point is that finding an I-projection onto $\mathscr{E}_i$ is quite easy (and easily programmed), while finding an I-projection onto $\mathscr{E} = \bigcap_1^{n-1} \mathscr{E}_i$ is very difficult. However, Dykstra's algorithm enables one to find the latter I-projection using only the ability to find I-projections onto the individual $\mathscr{E}_i$.

To illustrate our example with some data, we consider some rather famous data from Stuart (1953) concerning grades of unaided distance vision for left and right eyes. If one wished to estimate the probabilities of falling into the various categories, subject to the provision that right eye vision is at least as good as left eye vision, one might find the I-projection of the data in Table 6.3.1 onto $\bigcap_1^{n-1} \mathscr{E}_i$. Using Dykstra's algorithm, we have essentially obtained convergence to the true I-projection by three cycles. These values are listed in Table 6.3.2 (with the relative frequencies given in parentheses).

Table 6.3.1 Unaided distance vision. (From Kendall, 1974)

	LEFT EYE				
RIGHT EYE	HIGHEST GRADE	SECOND GRADE	THIRD GRADE	LOWEST GRADE	TOTALS
Highest grade	821	112	85	35	1053
Second grade	116	494	145	27	782
Third grade	72	151	583	87	893
Lowest grade	43	34	106	331	514
Totals	1052	791	919	480	3242

Table 6.3.2 *I*-Projection of data in Table 6.3.1. (Values in parentheses are Table 6.3.1 values normed to sum to unity)

	LEFT EYE				
RIGHT EYE	HIGHEST GRADE	SECOND GRADE	THIRD GRADE	LOWEST GRADE	TOTALS
Highest grade	0.2534	0.0344	0.0262	0.0120	0.3260
	(0.2532)	(0.0345)	(0.0262)	(0.0108)	(0.3247)
Second grade	0.0358	0.1525	0.0447	0.0092	0.2422
	(0.0358)	(0.1524)	(0.0447)	(0.0083)	(0.2412)
Third grade	0.0222	0.0466	0.1799	0.0298	0.2785
	(0.0222)	(0.0466)	(0.1798)	(0.0268)	(0.2754)
Lowest grade	0.0120	0.0095	0.0295	0.1022	0.1532
	(0.0133)	(0.0105)	(0.0327)	(0.1021)	(0.1586)
Totals	0.3234	0.2430	0.2803	0.1532	0.9999
	(0.3245)	(0.2440)	(0.2834)	(0.1480)	(0.9999)

Order restricted score parameters

Example 6.3.3 As in Example 6.3.1, assume that $\hat{p}_{ij}$ is the relative frequency of falling in the (i, j)th cell of an $r \times c$ rectangular array under simple multinomial sampling and that p_{ij} is the true probability associated with the (i, j)th cell. It is assumed that the cells of the array are classified according to two ordinal variables, say X and Y.

The *local odds ratios* (also discussed in Section 6 of Chapter 5)

$$\theta_{ij} = \frac{p_{ij}p_{i+1,j+1}}{p_{i,j+1}p_{i+1,j}} \qquad \text{for } i = 1, \ldots, r-1; \quad j = 1, \ldots, c-1$$

are often used for describing properties of association between X and Y. In particular, independence between the classifications holds if and only if

$$\theta_{ij} = 1 \qquad \text{for } i = 1, \ldots, r-1; \quad j = 1, \ldots, c-1,$$

while $\theta_{ij} \geqslant 1$ for all i, j indicates a positive association between the classifications.

A particularly simple and appealing model is one proposed by Goodman (1979, 1981) of the form (see also Chapter 5, Section 6)

$$\ln(p_{ij}) = \mu + \lambda_i^X + \lambda_j^Y + \beta u_i v_j \qquad (6.3.17)$$

(where the superscripts are merely labels identifying the parameter and not numerical values). For the special case $\{u_i = i\}$, $\{v_j = j\}$, this is called the uniform association model. In this event $\ln(\theta_{ij}) = \beta$ for all i, j. In the general case where

$\{u_i\}$ and $\{v_j\}$ are unspecified parameters,

$$\ln(\theta_{ij}) = \beta(u_{i+1} - u_i)(v_{j+1} - v_j)$$

and the model is called the multiplicative row and column effects model. When the $\{u_i\}$ are parameters and the $\{v_i\}$ are predetermined monotone scores, the above model is known as the row effects model. Similarly, in the column effects model the $\{v_i\}$ are parameters and the $\{u_i\}$ are predetermined monotone scores.

It is customary to require

$$\sum_1^r \lambda_i^X = \sum_1^c \lambda_j^Y = 0$$

in model (6.3.17). Since location and scale changes in the score parameters $\{u_i\}$ and $\{v_i\}$ do not alter the general form of row, column, and row–column models, in certain situations it is plausible to require $u_1 = v_1 = 1$, $u_r = r$ and $v_c = c$ to make the scores comparable in value to the fixed integer scores of the uniform association model. Moreover, since

$$\sum_{j=1}^{c-1} \sum_{i=1}^{r-1} \ln(\theta_{ij}) = (u_r - u_1)(v_c - v_1)\beta,$$

in general these assumptions give the simple interpretation that β is the average of the $(r-1)(c-1)$ local log-odds ratios.

As Agresti and Chuang (1986) point out, the standard fits of these models do not fully utilize the ordinal nature of the variables in the sense that the same maximum likelihood estimates are obtained if the levels are permuted in any way. In many applications, one would expect the orderings of the categories to be manifested in an association that is monotone, in some sense.

One possible way to require monotonicity would be to require that the local log-odds ratios be uniformly nonnegative (or uniformly nonpositive). In a more general context, Lehmann (1966) referred to this condition as positive (or negative) likelihood ratio dependence. If one assumes model (6.3.17) where the $\{v_j\}$ are fixed increasing scores, then positive likelihood ratio dependence is equivalent to requiring that the parameters $\{u_i\}$ be nondecreasing.

Let us consider finding maximum likelihood estimates of the p_{ij} subject to the assumptions that were just stated. We can express this problem as

$$\sup_{\substack{\ln(\mathbf{p}) \in \mathcal{M}_1 + \mathcal{M}_2 + \mathcal{K}_1 \\ \mathbf{p} \in P}} \prod_{j=1}^{c} \prod_{i=1}^{r} p_{ij}^{\hat{p}_{ij}} \tag{6.3.18}$$

where

$$\begin{cases} \mathcal{M}_1 = \{\mathbf{x} : x_{i1} = x_{i2} = \cdots = x_{ic};\ i = 1, \ldots, r\} \\ \mathcal{M}_2 = \{\mathbf{x} : x_{ij} = x_{2j} = \cdots = x_{rj};\ j = 1, \ldots, c\} \\ \mathcal{K}_1 = \{\mathbf{x}\mathbf{V} : \mathbf{x} \in \mathcal{K}_0\} \end{cases}$$

where

$$\mathscr{K}_0 = \{\mathbf{x}: x_{i1} = x_{i2} = \cdots = x_{ic}; i = 1, \ldots, r, x_{11} \leqslant x_{21} \leqslant \cdots \leqslant x_{r1}\}$$

and $\mathbf{V}$ is the $c \times c$ diagonal array with elements v_i and the multiplication in $\mathscr{K}_1$ is usual matrix multiplication.

Since $\mathscr{M}_1 + \mathscr{M}_2 + \mathscr{K}_1$ contains the constant vectors, we may use Corollary 6.3.1 to say that (6.3.18) has the same solution as the I-projection problem

$$\inf I(\mathbf{p}|\mathbf{u}) \quad \text{subject to } \mathbf{p} \in \bigcap_{i=1}^{2} (\hat{\mathbf{p}} - \mathscr{M}_i^{\perp}) \cap (\hat{\mathbf{p}} - \mathscr{K}_1^*). \tag{6.3.19}$$

The condition that $\mathbf{p} \in \hat{\mathbf{p}} - \mathscr{M}_1^{\perp}$ is equivalent to requiring

$$\hat{\mathbf{p}} - \mathbf{p} \in \mathscr{M}_1^{\perp} = \left\{\mathbf{x}: \sum_{j=1}^{c} x_{ij} = 0; i = 1, \ldots, r\right\}$$

or, in other words, that $\mathbf{p}$ has the same row sums as $\hat{\mathbf{p}}$. Similarly, $\mathbf{p} \in \hat{\mathbf{p}} - \mathscr{M}_2^{\perp}$ is equivalent to $\mathbf{p}$ having the same column sums as $\hat{\mathbf{p}}$. The third region $\hat{\mathbf{p}} - \mathscr{K}_1^*$ is a bit more difficult. For $\mathbf{p}$ to lie in this region is equivalent to having $\hat{\mathbf{p}} - \mathbf{p} \in \mathscr{K}_1^*$. Note that

$$\begin{aligned} \mathscr{K}_1^* &= \left\{\mathbf{y}: \sum_{j=1}^{c} \sum_{i=1}^{r} y_{ij} z_{ij} \leqslant 0, \mathbf{z} \in \mathscr{K}_1\right\} \\ &= \left\{\mathbf{y}: \sum_{j=1}^{c} \sum_{i=1}^{r} y_{ij} v_j x_{ij} \leqslant 0, \mathbf{x} \in \mathscr{K}_0\right\} \\ &= \{\mathbf{y}\mathbf{V}^{-1}: \mathbf{y} \in \mathscr{K}_0^*\}. \end{aligned}$$

However,

$$\mathscr{K}_0^* = \left\{\mathbf{y}: \sum_{i=1}^{l} \sum_{j=1}^{c} y_{ij} \geqslant 0; l = 1, \ldots, r-1, \sum_{i=1}^{r} \sum_{j=1}^{c} y_{ij} = 0\right\}.$$

Thus $\hat{\mathbf{p}} - \mathbf{p} \in \mathscr{K}_1^*$ is equivalent to $(\hat{\mathbf{p}} - \mathbf{p})\mathbf{V} \in \mathscr{K}_0^*$, so that this last constraint region is

$$\sum_{i=1}^{l} \sum_{j=1}^{c} \hat{p}_{ij} v_j \geqslant \sum_{i=1}^{l} \sum_{j=1}^{c} p_{ij} v_j \qquad \text{for } l = 1, \ldots, r-1$$

$$\sum_{i=1}^{r} \sum_{j=1}^{c} \hat{p}_{ij} v_j = \sum_{i=1}^{r} \sum_{j=1}^{c} p_{ij} v_j.$$

Note that the entire constraint region of (6.3.19) can be written as a finite intersection of simple linear inequality (equality) constraint regions. The problem of finding I-projections onto these types of regions was discussed in Example 6.2.5. While these regions are not all linear sets, they are convex, so that Dykstra's algorithm can be used to solve the I-projection problem (6.3.19), and hence the original problem (6.3.18).

Nonnegative local log-odds ratios

Example 6.3.4 Assume the notation of the previous example. As mentioned there, an appealing model would often require that all local log-odds ratios be nonnegative. This, in conjunction with the row association model, was the reason for assuming the $\{u_i\}$ of Example 5.3.3 to be nondecreasing. Let us now examine the problem of finding maximum likelihood estimates for the p_{ij} only requiring that the local log-odds ratios are nonnegative and making no other model assumptions.

If we define the closed convex cones

$$\mathcal{K}_{ij} = \{\mathbf{x}: x_{ij} + x_{i+1,j+1} - x_{i+1,j} - x_{i,j+1} \geqslant 0\} \qquad \text{for } i = 1, \ldots, r-1; \; j = 1, \ldots, c-1,$$

then the problem we wish to consider is

$$\sup \prod_{j=1}^{c} \prod_{i=1}^{r} P_{ij}^{\hat{p}_{ij}} \text{ subject to } \ln(\mathbf{p}) \in \bigcap_{j=1}^{c-1} \bigcap_{i=1}^{r-1} \mathcal{K}_{ij} \quad \text{and} \quad \mathbf{p} \in P. \tag{6.3.20}$$

Since $\bigcap_{j=1}^{c-1} \bigcap_{i=1}^{r-1} \mathcal{K}_{ij}$ contains the constant functions, this has the same solution as the problem

$$\inf_{\mathbf{p} \in \hat{\mathbf{p}} - (\bigcap_{j=1}^{c-1} \bigcap_{i=1}^{r-1} \mathcal{K}_{ij})^*} I(\mathbf{p} \mid \mathbf{u}), \tag{6.3.21}$$

by the corollary to Theorem 6.3.4. While $(\bigcap_{j=1}^{c-1} \bigcap_{i=1}^{r-1} \mathcal{K}_{ij})^*$ can be expressed as the direct sum

$$\mathcal{K}_{11}^* + \mathcal{K}_{12}^* + \cdots + \mathcal{K}_{r-1,c-1}^*,$$

we would prefer to express it as an intersection of 'nice' regions so that Dykstra's algorithm could be used to solve (6.3.21).

If $B(i, j)$ is the matrix with a one in the (i, j)th and $(i+1, j+1)$th cells, a negative one in the $(i, j+1)$th and $(i+1, j)$th cells and zeros elsewhere, then for fixed i and j,

$$\mathcal{K}_{ij} = \left\{ \mathbf{x}: \sum_{m=1}^{c} \sum_{k=1}^{r} x_{km} B(i, j)_{km} \geqslant 0 \right\},$$

so that

$$\mathcal{K}_{ij}^* = \{\alpha B(i, j): \alpha \leqslant 0\}.$$

Let us depict an arbitrary element of $\mathcal{K}_{ij}^*$ by $\alpha_{ij} B(i, j)$ where $\alpha_{ij} \leqslant 0$. Since each $B(i, j)$ has nonzero entries (1 or -1) only in the cells (i, j), $(i+1, j)$, $(i, j+1)$, and $(i+1, j+1)$, the direct sum of all $(r-1)(c-1)$ dual cones would have

$$\alpha_{ij} - \alpha_{i,j-1} + \alpha_{i-1,j-1} - \alpha_{i-1,j} \tag{6.3.22}$$

in the (i, j)th cell (where any α with a subscript outside the region $\{(k, m): 1 \leqslant k \leqslant r-1, 1 \leqslant m \leqslant c-1\}$ is taken to be zero).

If we sum these entries over the upper left rectangle

$$A_{ij} = \{(k, m): 1 \leqslant k \leqslant i, 1 \leqslant m \leqslant j\},$$

they telescope to just the nonpositive value α_{ij} if $i \leqslant r-1$ and $j \leqslant c-1$ and zero otherwise. This implies the relationship

$$\left(\bigcap_{j=1}^{c-1} \bigcap_{i=1}^{r-1} \mathscr{K}_{ij}\right)^* \subset \bigcap_{j=1}^{c} \bigcap_{i=1}^{r} \mathscr{H}_{ij}$$

where

$$\mathscr{H}_{ij} = \begin{cases} \left\{\mathbf{x}: \sum_{A_{ij}} x_{km} \leqslant 0\right\} & \text{if } 1 \leqslant i \leqslant r-1; \quad 1 \leqslant j \leqslant c-1 \\ \left\{\mathbf{x}: \sum_{A_{ij}} x_{km} = 0\right\}, & \text{if } i = r \quad \text{or} \quad j = c. \end{cases}$$

Conversely, if $\mathbf{x} \in \bigcap_{j=1}^{c} \bigcap_{i=1}^{r} \mathscr{H}_{ij}$, we may express x_{ij} in the form of (6.3.22) for nonpositive α's by setting $\alpha_{ij} = \sum_{A_{ij}} x_{km}$, so that $\mathbf{x} \in (\bigcap_{j=1}^{c-1} \bigcap_{i=1}^{r-1} \mathscr{K}_{ij})^*$. Thus

$$\left(\bigcap_{j=1}^{c-1} \bigcap_{i=1}^{r-1} \mathscr{K}_{ij}\right)^* = \bigcap_{j=1}^{c} \bigcap_{i=1}^{r} \mathscr{H}_{ij}.$$

The dual problem (6.3.21) can therefore be expressed as

$$\inf_{\mathbf{p} \in \bigcap_{j=1}^{c} \bigcap_{i=1}^{r} (\hat{\mathbf{p}} - \mathscr{H}_{ij})} I(\mathbf{p} | \mathbf{u}), \tag{6.3.23}$$

so that Dykstra's iterative algorithm can be used. Moreover, it is quite straightforward to show that an I-projection problem of the form

$$\inf_{\mathbf{p} \in P, \sum_{j \in A} p_j \leqslant b} \sum_{1}^{k} p_j \ln\left(\frac{p_j}{r_j}\right)$$

(for $0 \leqslant b \leqslant 1$) can be solved in closed form by

$$\hat{p}_i = \begin{cases} r_i \left(\sum_{j=1}^{k} r_j\right)^{-1}, & \text{if } \sum_{\ell \in A} r_\ell \left(\sum_{j=1}^{k} r_j\right)^{-1} \leqslant b \\ r_i b \left(\sum_{j \in A} r_j\right)^{-1}, & \text{if } i \in A \quad \text{and} \quad \sum_{l \in A} r_l \left(\sum_{j=1}^{k} r_j\right)^{-1} < b \\ r_i (1-b) \left(\sum_{j \in A^c} r_j\right)^{-1}, & \text{if } i \in A^c \quad \text{and} \quad \sum_{l \in A} r_l \left(\sum_{j=1}^{k} r_j\right)^{-1} < b. \end{cases}$$

Thus the individual projections needed for using Dykstra's algorithm can be easily programmed to provide a fairly simple procedure for solving problem (6.3.21) as well as problem (6.3.20). Actually the duality result between problems (6.3.20) and (6.3.23) is rather surprising because it relates two very different concepts of positive association.

In particular, the bivariate distribution P is positive quadrant dependent (Lehmann, 1966) if

$$P[(-\infty,x]\times(-\infty,y]]\geqslant P[(-\infty,x]\times R]P[R\times(-\infty,y]] \qquad \text{for all } x,y.$$

Similarly, one says that Q is at least as positive quadrant dependent as P if

$$Q[(-\infty,x]\times(-\infty,y]]\geqslant P[(-\infty,x]\times(-\infty,y]] \qquad \text{for all } x,y.$$

Suppose now that the row and column labels for our rectangular array are numerical and increasing. Then a probability vector $\mathbf{p}$ can be thought of as the probability function for a bivariate distribution. Our duality result says that for observed relative frequencies $\hat{\mathbf{p}}$, the multinomial maximum likelihood estimate of $\mathbf{p}$ subject to the constraints that all local log-odds ratios are nonnegative is precisely the same vector which is closest (in the I-projection sense) to the uniform distribution, under the constraint that it is at least as positive quadrant dependent as $\hat{\mathbf{p}}$ and has the same marginal distributions as $\hat{\mathbf{p}}$.

If one distribution is more positive quadrant dependent than another and they have the same marginal distributions, then it is said to be more concordant. This concept is discussed in Tchen (1980) and other places.

Multinomial MLEs as I-projections

Consider again the general multinomial maximum likelihood estimation problem

$$\sup_{\substack{\ln(\mathbf{p})\in\mathscr{K}\\ \mathbf{p}\in P}} \prod_{i=1}^{k} p_i^{\hat{p}_i} \tag{6.3.24}$$

where $\hat{\mathbf{p}}$ is the vector of relative frequencies and $\mathscr{K}$ is a closed, convex cone containing the constant functions. Since $\sum_1^k \hat{p}_i \ln(\hat{p}_i)$ is free of $\mathbf{p}$, we could write this problem in the form

$$\inf_{\substack{\ln(\mathbf{p})\in\mathscr{K}\\ \mathbf{p}\in P}} \sum_{i=1}^{k} \hat{p}_i \ln(\hat{p}_i) - \sum_{i=1}^{k} \hat{p}_i \ln(p_i) = \inf_{\substack{\ln(\mathbf{p})\in\mathscr{K}\\ \mathbf{p}\in P}} \sum_{i=1}^{k} \hat{p}_i \ln\left(\frac{\hat{p}_i}{p_i}\right) = \inf_{\ln(\mathbf{p})\in\mathscr{K}} I(\hat{\mathbf{p}}|\mathbf{p})$$

(where once again the domain of I-divergence is assumed to be P).

This looks similar to previous I-projection problems, but there is a major difference. Previously, we held the second argument fixed and minimized over the first argument. Now we are fixing the first argument and minimizing over the second argument. These problems appear to be quite different, since I-divergences do not exhibit symmetry in their arguments. However, the duality results which we have derived establish a tie between these two types of problems; i.e. the problem

$$\inf_{\ln(\mathbf{p})\in\mathscr{K}} I(\hat{\mathbf{p}}|\mathbf{p}) \tag{6.3.25}$$

and

$$\inf_{\mathbf{p}\in\hat{\mathbf{p}}-\tilde{\mathscr{K}}^*} I(\mathbf{p}|\mathbf{u}) \qquad \text{(where } \mathbf{u} \text{ is the uniform distribution)} \tag{6.3.26}$$

have precisely the same solution if $\mathscr{K}$ is a closed, continuous, convex cone containing the constant functions. Moreover, some of the geometric development and characterizations developed by Csiszar for first argument I-projections carry over to second argument I-projections by our duality relationships. The following theorem should be compared to Theorem 6.3.4.

Theorem 6.3.5 *Suppose $\mathscr{K}$ is a closed, continuous, convex cone in $\bar{R}^k$ containing the constant functions. Then $\mathbf{p}^*$ is the unique solution to the problem*

$$\inf_{\ln(\mathbf{p})\in\mathscr{K}} I(\hat{\mathbf{p}}|\mathbf{p})$$

if and only if

(*a*) $\ln(\mathbf{p}^*)\in\mathscr{K}$

and

(*b*) $I(\hat{\mathbf{p}}|\mathbf{p})\geqslant I(\hat{\mathbf{p}}|\mathbf{p}^*)+I(\mathbf{p}^*|\mathbf{p})$ *for every* $\mathbf{p}\in P$ *such that* $\ln(\mathbf{p})\in\mathscr{K}$.

If $\tilde{\mathscr{K}}=\mathscr{K}\cap R^k$ is actually a subspace, the inequality in (b) may be replaced by equality.

Proof: We have shown that $\mathbf{p}^*$ solves (6.3.25) if and only if it solves (6.3.26), and that this common solution is characterized as the only vector which satisfies

$$\text{(a) } \ln(\mathbf{p}^*)\in\mathscr{K}, \quad \text{(b) } \hat{\mathbf{p}}-\mathbf{p}^*\in\tilde{\mathscr{K}}^*, \quad \text{and} \quad \text{(c) } \langle\hat{\mathbf{p}}-\mathbf{p}^*,\ln(\mathbf{p}^*)\rangle=0$$

(see Corollary 6.3.2).

Thus if $\mathbf{p}^*$ solves (6.3.25), $\hat{\mathbf{p}}-\mathbf{p}^*\in\tilde{\mathscr{K}}^*$ and hence $\langle\hat{\mathbf{p}}-\mathbf{p}^*,\ln(\mathbf{p})\rangle\leqslant 0$ if $\ln(\mathbf{p})\in\mathscr{K}$. This coupled with (c) above implies that $\langle\hat{\mathbf{p}}-\mathbf{p}^*, \ln(\mathbf{p}^*/\mathbf{p})\rangle\geqslant 0$, which is equivalent to (b) of Theorem 6.3.5.

Suppose now that $\mathbf{p}^*$ satisfies (a) and (b) of Theorem 6.3.5. Then clearly $\mathbf{p}^*$ is the unique solution to problem (6.3.25) since I-divergences are always nonnegative and equal zero if and only if both arguments are identical. The argument when $\tilde{\mathscr{K}}$ is a subspace will lead to equality in (b) of Theorem 6.3.5. □

The characterization given in Theorem 6.3.5 is useful in many ways. For example, the following iterative projection property is easily proven.

Theorem 6.3.6 *Let $\mathscr{K}$ be a closed, convex cone that contains the constant vectors and lies in the subspace $\mathscr{M}$. Then if $\mathbf{p}^*_{\mathscr{M}}$ is the solution to the problem*

$$\inf_{\ln(\mathbf{p})\in\mathscr{M}} I(\hat{\mathbf{p}}|\mathbf{p})$$

and $\mathbf{p}^*_{\mathscr{K}}$ *is the solution to*

$$\inf_{\ln(\mathbf{p})\in\mathscr{K}} I(\mathbf{p}^*_{\mathscr{M}}|\mathbf{p}),$$

it must also be the case that $\mathbf{p}^*_{\mathscr{K}}$ *is the solution to the problem*

$$\inf_{\ln(\mathbf{p})\in\mathscr{K}} I(\hat{\mathbf{p}}|\mathbf{p}).$$

Proof: Let $\mathbf{p}$ be a probability vector such that $\ln(\mathbf{p})\in\mathscr{K}\subset\mathscr{M}$. Then three applications of Theorem 6.3.5 yield

$$I(\hat{\mathbf{p}}|\mathbf{p}) = I(\hat{\mathbf{p}}|\mathbf{p}^*_{\mathscr{M}}) + I(\mathbf{p}^*_{\mathscr{M}}|\mathbf{p})$$

$$I(\hat{\mathbf{p}}|\mathbf{p}^*_{\mathscr{M}}) = I(\hat{\mathbf{p}}|\mathbf{p}^*_{\mathscr{K}}) - I(\mathbf{p}^*_{\mathscr{M}}|\mathbf{p}^*_{\mathscr{K}}) \qquad (\text{since } \ln(\mathbf{p}^*_{\mathscr{K}})\in\mathscr{M})$$

and

$$I(\mathbf{p}^*_{\mathscr{M}}|\mathbf{p}) \geqslant I(\mathbf{p}^*_{\mathscr{M}}|\mathbf{p}^*_{\mathscr{K}}) + I(\mathbf{p}^*_{\mathscr{K}}|\mathbf{p}) \qquad (\text{since } \mathbf{p}^*_{\mathscr{K}} \text{ solves the second problem}).$$

Substituting the last two lines into the first implies

$$I(\hat{\mathbf{p}}|\mathbf{p}) \geqslant I(\mathbf{p}|\mathbf{p}^*_{\mathscr{K}}) + I(\mathbf{p}^*_{\mathscr{K}}|\mathbf{p}),$$

which yields the desired conclusion by one last application of Theorem 6.3.5. □

Theorem 6.3.6 proves useful if one wishes to find multinomial maximum likelihood estimates subject to some type of independence condition together with some type of order restriction on the parameters.

Since independence restrictions can usually be written in the form $\ln(\mathbf{p})\in\mathscr{M}$ for some subspace $\mathscr{M}$, these types of problems can often be solved by first finding the maximum likelihood estimates under only the independence conditions, treating these estimates as if they were the data ($\hat{\mathbf{p}}$), and then finding the order restricted estimates for these new values.

Another nice application Theorem 6.3.5 is for proving an interated projection convergence theorem which is useful for certain types of multinomial maximum likelihood problems. The proof is similar to the one given by Csiszar (1975) for the iterated I-projection convergence theorem for linear sets, which has been discussed previously.

Theorem 6.3.7 *Let* $\mathscr{M}_1, \mathscr{M}_2, \ldots, \mathscr{M}_t$ *be subspaces of* $\bar{R}^k$ *containing the constant vectors. Let* $\mathbf{q}_n, n = 1, 2, \ldots,$ *denote the probability vector which solves*

$$\inf_{\ln(\mathbf{q})\in\mathscr{M}_n} I(\mathbf{q}_{n-1}|\mathbf{q}),$$

where $\mathbf{q}_0 = \hat{\mathbf{p}}$ *and* $\mathscr{M}_n = \mathscr{M}_j$ *if* $n = lt + j,\ 1 \leqslant j \leqslant t$. *Then* $\mathbf{q}_n$ *converges to the probability vector* $\mathbf{q}$ *which solves*

$$\inf_{\ln(\mathbf{p})\in\bigcap_{j=1}^{t}\mathscr{M}_j} I(\hat{\mathbf{p}}|\mathbf{p}),$$

and $\mathbf{q}$ *is also the solution to*

$$\inf_{\substack{\ln(\mathbf{p})\in\bigcap_{j=1}^{t}\mathcal{M}_j \\ \mathbf{p}\in P}} \prod_{i=1}^{k} p_i^{\hat{p}_i}.$$

Proof: In view of Theorem 6.3.5,

$$I(\mathbf{q}_n|\mathbf{p}) = I(\mathbf{q}_n|\mathbf{q}_{n+1}) + I(\mathbf{q}_{n+1}|\mathbf{p}) \qquad \text{for all } \mathbf{p}\in P \text{ such that } \ln(\mathbf{p})\in\mathcal{M}_{n+1},$$

so that

$$I(\hat{\mathbf{p}}|\mathbf{p}) = \sum_{i=0}^{n} I(\mathbf{q}_i|\mathbf{q}_{i+1}) + I(\mathbf{q}_{n+1}|\mathbf{p}) \qquad \text{if } \ln(\mathbf{p})\in\bigcap_{j=1}^{t}\mathcal{M}_j.$$

Setting $\mathbf{p}=\mathbf{q}$, we have

$$\infty > I(\hat{\mathbf{p}}|\mathbf{q}) = \sum_{i=0}^{n} I(\mathbf{q}_i|\mathbf{q}_{i+1}) + I(\mathbf{q}_{n+1}|\mathbf{q}) \qquad \text{for all } n.$$

Thus, $I(\mathbf{q}_n|\mathbf{q}_{n+1})\to 0$ as $n\to\infty$, and since in general

$$I(\mathbf{p}|\mathbf{p}') \geqslant \left(\tfrac{1}{2}\sum_{1}^{k}|p_i - p_i'|\right)^{1/2}$$

(see Csiszar, 1967), the vectors $\mathbf{q}_n$ and $\mathbf{q}_{n+k}$ become arbitrarily close as $n\to\infty$ for a fixed k. Thus, if a convergent subsequence is chosen, which converges say to $\mathbf{q}'$, $\mathbf{q}'$ must belong to $\bigcap_{j=1}^{t}\mathcal{M}_j$ by the closure properties of each $\mathcal{M}_i$.

By Theorem 6.3.6, $\mathbf{q}$ also solves the problem

$$\inf_{\ln(\mathbf{p})\in\bigcap_{j=1}^{t}\mathcal{M}_j} I(\mathbf{q}_1|\mathbf{p}).$$

In fact, by an induction argument and repeated use of Theorem 6.3.6, $\mathbf{q}$ solves the problem

$$\inf_{\ln(\mathbf{p})\in\bigcap_{j=1}^{t}\mathcal{M}_j} I(\mathbf{q}_n|\mathbf{p}) \qquad \text{for every } n.$$

Thus

$$I(\mathbf{q}_n|\mathbf{q}') = I(\mathbf{q}_n|\mathbf{q}) + I(\mathbf{q}|\mathbf{q}')$$

(replacing $\mathbf{p}$ by $\mathbf{q}'$ in (b) of Theorem 6.3.5). However, $\mathbf{q}_{n_j}\to\mathbf{q}'$ implies that $I(\mathbf{q}_{n_j}|\mathbf{q}')\to 0$, so that $I(\mathbf{q}|\mathbf{q}')=0$ or $\mathbf{q}'=\mathbf{q}$ which implies the desired result. □

While this theorem for second argument I-projections is analogous to the convergence result of Csiszar for cyclic iterated I-projections onto linear sets, we are not aware of a similar analogue to Dykstra's procedure when the sets are not subspaces.

Isotonic cone constraints for log-linear models

Previously in this section we have extensively discussed the importance of the duality relationships between

$$\inf_{\substack{\ln(\mathbf{p})\in\mathscr{K}\\ \mathbf{p}\in P}} \prod_{i=1}^{k} p_i^{\hat{p}_i}$$

and

$$\inf_{\mathbf{p}\in\hat{\mathbf{p}}-\tilde{\mathscr{K}}^*} I(\mathbf{p}|\mathbf{u})$$

and the fact that these problems are solved by the same vector.

If $\mathscr{K}$ is composed of a direct sum (intersection) of cones, then the region $\hat{\mathbf{p}}-\tilde{\mathscr{K}}^*$ will be an intersection (direct sum) of convex regions. Solving problems of the form

$$\inf_{\mathbf{p}\in\hat{\mathbf{p}}-\tilde{\mathscr{K}}^*} I(\mathbf{p}|\mathbf{r}) \tag{6.3.27}$$

is thus often very useful, as it can be used in the individual steps of Csiszar's and Dykstra's algorithms. If, in fact, $\mathscr{K}$ is actually a cone of isotonic vectors, problem (6.3.27) can be solved in closed form by a least squares isotonic regression, and hence the computational techniques discussed in Chapter 1 are pertinent. The following theorem should be contrasted with the solution to display (6.2.17).

There we were trying to find the I-projection subject to the probability vector lying in an isotonic cone. Here we are trying to find an I-projection where the constraint region is a translate of the negative of the dual of an isotonic cone. While both of these problems are solved by an isotonic regression, the forms of the solutions are very different.

Theorem 6.3.8 *Suppose $\mathscr{K}$ is a cone of isotonic functions, $\mathbf{r}\neq\mathbf{0}$ is a nonnegative vector, and $\hat{\mathbf{p}}$ is a probability vector which is absolutely continuous with respect to $\mathbf{r}$. Then a solution to the problem*

$$\inf_{\mathbf{p}\in\hat{\mathbf{p}}-\tilde{\mathscr{K}}^*} I(\mathbf{p}|\mathbf{r})$$

is given by $\mathbf{p}^*=\mathbf{r}P_{\mathbf{r}}(\hat{\mathbf{p}}/\mathbf{r}|\tilde{\mathscr{K}})$, where vector multiplication and division is done coordinatewise. (If a coordinate has zero weight, the corresponding term is removed from the sum of squares to be minimized, and any value for the coordinate may be chosen which satisfies the order restrictions.)

Proof: The problem can be expressed as

$$\inf_{\substack{\mathbf{p}\in\hat{\mathbf{p}}-\tilde{\mathscr{K}}^*\\ \mathbf{p}\geqslant 0}} \sum_{i=1}^{k} p_i \ln\left(\frac{p_i}{r_i}\right). \tag{6.3.28}$$

If $r_i = 0$, then p_i^* not zero implies that $I(\mathbf{p}^*|\mathbf{r}) = \infty$. However, this contradicts the fact that $I(\hat{\mathbf{p}}|\mathbf{r}) < \infty$ and $\hat{\mathbf{p}} \in \hat{\mathbf{p}} - \mathscr{K}^*$. Thus, those indices where $r_i = 0$ can be removed from the sum in (6.3.28). For the remaining indices, make the change of variables $y_i = p_i/r_i$, so that the problem (6.3.28) is equivalent to

$$\inf_{\mathbf{y} \in \hat{\mathbf{p}}/\mathbf{r} - \tilde{\mathscr{K}}_0^{*\mathbf{r}}} \sum_{i=1}^{k_0} (y_i \ln y_i) r_i \tag{6.3.29}$$

where $\tilde{\mathscr{K}}_0$ is the reduced cone with all coordinates removed for which $r_i = 0$ and $\tilde{\mathscr{K}}_0^{*\mathbf{r}}$ is the weighted dual cone with weights $\mathbf{r}$ (see Section 1.7). Since $y \ln(y)$ is a convex function, (6.3.29) is of the same form as Example 6.2.2, and hence is solved by the isotonic regression $E_{\mathbf{r}}(\hat{\mathbf{p}}/\mathbf{r}|\tilde{\mathscr{K}}_0)$. Transforming back to p gives the desired result. Note tha $\mathbf{p}^*$ must be a probability vector by Theorem 1.3.3. □

6.4 LAGRANGIAN DUALITY

The primary applications of the Fenchel duality theorem in previous sections was to phrase minimization problems over convex sets and cones into related, but different forms. A different type of duality is often of value when the restriction regions may be defined in terms of smooth convex functions.

The basic problem under consideration will be to minimize a convex function f_0 defined over a region $\mathscr{C}$ subject to a mixture of equality and inequality constraints. The constraints will be denoted by

$$f_1(\mathbf{x}) \leqslant 0, \ldots, f_r(\mathbf{x}) \leqslant 0, \quad f_{r+1}(\mathbf{x}) = 0, \ldots, f_t(\mathbf{x}) = 0 \tag{6.4.1}$$

where $\mathscr{C}$ is a convex set in R^k, $f_1, \ldots, f_r$ are convex functions defined on $\mathscr{C}$, and $f_{r+1}, \ldots, f_t$ are affine functions. It is customary to refer to f_0 as the objective function and $f_i, i = 1, \ldots, t$, as the constraint functions. Without loss of generality, the functions $f_0, f_1, \ldots, f_t$ will be taken to be finite over $\mathscr{C}$. A vector $\mathbf{x}$ is said to be *feasible* if $\mathbf{x} \in \mathscr{C}$ and satisfies all the constraints in (6.4.1). The *optimal value* of the problem is the infimum of f_0 over the set of feasible points. Those feasible points which attain the optimal value are called *optimal solutions*.

A vector $\tilde{\lambda} = (\tilde{\lambda}_1, \ldots, \tilde{\lambda}_t)$ is said to be a vector of *Kuhn–Tucker coefficients* (or just a Kuhn–Tucker vector) for the initial problem if the infimum of the proper convex function

$$f_0 + \tilde{\lambda}_1 f_1 + \cdots + \tilde{\lambda}_t f_t \qquad \text{(defined over } \mathscr{C}\text{)}$$

is finite and equal to the optimal value of the initial problem. A key theoretical point is that if a vector of Kuhn–Tucker coefficients were known, one could minimize this new objective function over $\mathscr{C}$ without concerning oneself about the additional constraints given in (6.4.1).

Kuhn–Tucker vectors exist under very mild conditions. For example, if the optimal value of the problem is not $-\infty$, then the existence of a feasible vector lying in the relative interior of $\mathscr{C}$ which strictly satisfies all the inequality constraints in (6.4.1), that is

$$f_1(\mathbf{x}) < 0, \quad f_2(\mathbf{x}) < 0, \ldots, f_r(\mathbf{x}) < 0,$$

is enough to guarantee the existence of a Kuhn–Tucker vector (see Rockafellar, 1972, pp. 278–279). The importance of Kuhn–Tucker vectors for the problem stated initially hinges on the following important fact: Kuhn–Tucker coefficients and optimal solutions to the initial convex programming problem can be characterized in terms of 'saddle-point' extrema of a particular convex–concave function defined on $R^t \times R^k$. This function is called the Lagrangian of the problem and is defined by

$$L(\boldsymbol{\lambda}, \mathbf{x}) = \begin{cases} f_0(\mathbf{x}) + \sum_{i=1}^{t} \lambda_i f_i(\mathbf{x}), & \lambda_i \geqslant 0 \quad \text{for all } 1 \leqslant i \leqslant r, \quad \mathbf{x} \in \mathscr{C} \\ -\infty, & \lambda_i < 0 \quad \text{for some } i \leqslant r, \quad \mathbf{x} \in \mathscr{C} \\ +\infty, & \mathbf{x} \notin \mathscr{C}. \end{cases}$$

(The variable λ_i is often called the Lagrange multiplier associated with the ith constraint.) It is easy to see that $L(\boldsymbol{\lambda}, \mathbf{x})$ is convex in $\mathbf{x}$ for each $\boldsymbol{\lambda}$ and concave in $\boldsymbol{\lambda}$ for each $\mathbf{x}$.

A vector pair $(\boldsymbol{\lambda}^*, \mathbf{x}^*)$ is said to be a saddle point of L (with respect to maximization in $\boldsymbol{\lambda}$ and minimization in $\mathbf{x}$) if

$$L(\boldsymbol{\lambda}, \mathbf{x}^*) \leqslant L(\boldsymbol{\lambda}^*, \mathbf{x}^*) \leqslant L(\boldsymbol{\lambda}^*, \mathbf{x}) \qquad \text{for all } \boldsymbol{\lambda} \text{ and } \mathbf{x}.$$

The key theorem is the following one which is extensively discussed in Rockafellar (1972, Chap. 28). The essence of the theorem is that one can remove constraints by considering an extremal problem with additional variables.

Theorem 6.4.1 *Let $\boldsymbol{\lambda}^*$ and $\mathbf{x}^*$ be vectors in R^t and R^k, respectively. Then $(\boldsymbol{\lambda}^*, \mathbf{x}^*)$ is a saddle point of the Lagrangian if and only if $\boldsymbol{\lambda}^*$ is a Kuhn–Tucker vector and $\mathbf{x}^*$ is an optimal solution of the problem. Moreover, this will occur if and only if the following conditions hold:*

(*a*) $\lambda_i^* \geqslant 0, \quad f_i(\mathbf{x}^*) \leqslant 0, \qquad$ *and* $\qquad \lambda_i^* f_i(\mathbf{x}^*) = 0, \qquad i = 1, \ldots, r$
(*b*) $f_i(\mathbf{x}^*) = 0,\ i = r+1, \ldots, t$
(*c*) *The vector* $\mathbf{0}$ *is a subgradient of* $L(\boldsymbol{\lambda}^*, \mathbf{x}) = f_0(\mathbf{x}) + \lambda_1^* f_1(\mathbf{x}) + \cdots + \lambda_t^* f_t(\mathbf{x})$ *at the point* $\mathbf{x}^*$, *or, equivalently,* $L(\boldsymbol{\lambda}^*, \mathbf{x}) \geqslant L(\boldsymbol{\lambda}^*, \mathbf{x}^*)$ *for all* $\mathbf{x} \in R^k$.

Thus condition (c) implies that $\mathbf{x}^*$ is a minimizing value of

$$f_0 + \lambda_1^* f_1 + \cdots + \lambda_t^* f_t.$$

If the functions $f_0, f_1, \ldots, f_t$ are all differentiable, then condition (c) is equivalent to

$$\nabla f_0(\mathbf{x}^*) + \lambda_1^* \nabla f_1(\mathbf{x}^*) + \cdots + \lambda_t^* \nabla f_t(\mathbf{x}^*) = 0, \tag{6.4.2}$$

where ∇f_i denotes the gradient (vector of partial derivatives) of f_i. The conditions in (a), (b), and (c) are often referred to as the Kuhn–Tucker conditions.

It may be the case that one can use (6.4.2) to express $\boldsymbol{\lambda}^*$ in terms of $\mathbf{x}^*$. (In particular, this is often true if the $f_i, i = 1, \ldots, t$, should be affine functions.)

This suggests a method for verifying whether a vector in $\mathscr{C}$ which is a candidate for being an optimal solution actually is one. This method is to solve (6.4.2) for $\boldsymbol{\lambda}$, a solution which will be a function of the candidate. One then checks whether this $\boldsymbol{\lambda}$ and the candidate satisfy the restrictions in (a) and (b) of the previous theorem. If so, the candidate must actually be an optimal solution.

The saddle-point property leads to some other observations as well, and implies an often used duality relationship. The key point is made in the following theorem which is proved in Rockafellar (1972).

Theorem 6.4.2 *If $\boldsymbol{\lambda}^*$ is a Kuhn–Tucker vector and $\mathbf{x}^*$ is an optimal solution, then the quantity $L(\boldsymbol{\lambda}^*, \mathbf{x}^*)$ is the optimal value of the problem. More generally, $\boldsymbol{\lambda}^*$ is a Kuhn–Tucker vector if and only if*

$$-\infty < \inf_{\mathbf{x}} L(\boldsymbol{\lambda}^*, \mathbf{x}) = \sup_{\boldsymbol{\lambda}} \inf_{\mathbf{x}} L(\boldsymbol{\lambda}, \mathbf{x}) = \inf_{\mathbf{x}} \sup_{\boldsymbol{\lambda}} L(\boldsymbol{\lambda}, \mathbf{x}).$$

Thus, if the function g is defined by

$$g(\boldsymbol{\lambda}) = \inf_{\mathbf{x}} L(\boldsymbol{\lambda}, \mathbf{x})$$

and h is defined by

$$h(\mathbf{x}) = \sup_{\boldsymbol{\lambda}} L(\boldsymbol{\lambda}, \mathbf{x}) = \begin{cases} f_0(\mathbf{x}), & \mathbf{x} \text{ feasible} \\ +\infty, & \mathbf{x} \text{ not feasible,} \end{cases}$$

then both g and h have the same extreme value in the sense that

$$\inf_{\mathbf{x}} h(\mathbf{x}) = \sup_{\boldsymbol{\lambda}} g(\boldsymbol{\lambda}).$$

By the definition of a Kuhn–Tucker vector this common extreme value is also the optimal value of the original constrained problem and the optimal solutions of h are the same as the optimal solutions of the original problem. The original problem is called the primal problem, and the problem of maximizing g is called the dual problem. Solutions to these two problems must satisfy the so-called Kuhn–Tucker conditions stated in Theorem 6.4.1. This duality approach will be explored and exploited in future sections.

The dual problem for linear programming

Consider the problem of minimizing a linear function of $\mathbf{x}$ subject to affine constraints. To be precise, the problem to be considered is defined as

$$\text{minimize } \mathbf{b}'\mathbf{x} \text{ subject to the constraints } \mathbf{A}'\mathbf{x} \geqslant \mathbf{c},\ \mathbf{x} \geqslant \mathbf{0} \qquad (6.4.3)$$

where $\mathbf{x}$ and $\mathbf{b}$ are $k \times 1$ real vectors, $\mathbf{A}$ is a $k \times t$ real matrix, and the real vector $\mathbf{c}$ is $t \times 1$. The Lagrangian corresponding to this problem is given by

$$L(\lambda, \mathbf{x}) = \begin{cases} \mathbf{b}'\mathbf{x} + \boldsymbol{\lambda}'(\mathbf{c} - \mathbf{A}'\mathbf{x}) = (\mathbf{b}' - \boldsymbol{\lambda}'\mathbf{A}')\mathbf{x} + \mathbf{c}'\boldsymbol{\lambda}, & \lambda_j \geqslant 0,\ x_i \geqslant 0 \quad \text{for all } i, j \\ -\infty, & x_i \geqslant 0 \quad \text{for all } i,\ \lambda_j < 0 \quad \text{for some } j \\ +\infty, & x_i < 0 \quad \text{for some } i. \end{cases}$$

It easily follows that

$$g(\boldsymbol{\lambda}) = \inf_{\mathbf{x}} L(\boldsymbol{\lambda}, \mathbf{x}) = \begin{cases} -\infty, & \lambda_j < 0 \text{ for some } j \text{ or } (\mathbf{A}\boldsymbol{\lambda} - \mathbf{b})_j > 0 \text{ for some } j \\ \mathbf{c}'\boldsymbol{\lambda}, & \lambda_j \geqslant 0 \text{ for all } j,\ (\mathbf{A}\boldsymbol{\lambda} - \mathbf{b})_j \leqslant 0 \text{ for all } j. \end{cases}$$

Thus the dual problem described in Theorem 6.4.2 which corresponds to (6.4.3) is maximize $g(\boldsymbol{\lambda})$ or, equivalently,

$$\text{maximize } \mathbf{c}'\boldsymbol{\lambda} \quad \text{subject to } \mathbf{A}\boldsymbol{\lambda} \leqslant \mathbf{b},\quad \boldsymbol{\lambda} \geqslant \mathbf{0}. \qquad (6.4.4)$$

Note that the dual problem is very similar to the primal problem. However, $\mathbf{x}$ is of dimension k while $\boldsymbol{\lambda}$ is of dimension t and hence one problem may be much easier to solve than the other. This duality relationship is a fundamental result in linear programming.

Example 6.4.1 Consider the linear programming problem

minimize $x_1 + 3x_2$

subject to $2x_1 + x_2 \geqslant 8, x_1 + x_2 \geqslant 6, x_1 + 2x_2 \geqslant 9, x_1 \geqslant 0,$ and $x_2 \geqslant 0.$

In the notation of (6.4.3), we would have

$$\mathbf{b} = \begin{pmatrix} 1 \\ 3 \end{pmatrix}, \qquad \mathbf{A}' = \begin{pmatrix} 2 & 1 \\ 1 & 1 \\ 1 & 2 \end{pmatrix}, \qquad \text{and} \qquad \mathbf{c} = \begin{pmatrix} 8 \\ 6 \\ 9 \end{pmatrix}.$$

The dual problem (6.4.4) would then be

maximize $8\lambda_1 + 6\lambda_2 + 9\lambda_3$

subject to $2\lambda_1 + \lambda_2 + \lambda_3 \leqslant 1, \lambda_1 + \lambda_2 + 2\lambda_3 \leqslant 3, \lambda_1 \geqslant 0, \lambda_2 \geqslant 0,$ and $\lambda_3 \geqslant 0.$

The duality nature of the problem becomes apparent from the following display:

Minimization problem	Maximization problem		
$2x_1 + 1x_2 \geqslant 8$	$2\lambda_1$	$1\lambda_1$	$8\lambda_1$
	$+$	$+$	$+$
$1x_1 + 1x_2 \geqslant 6$	$1\lambda_2$	$1\lambda_2$	$6\lambda_2$
	$+$	$+$	$+$
$1x_1 + 2x_2 \geqslant 9$	$1\lambda_3$	$2\lambda_3$	$9\lambda_3$
	$\leqslant$	$\leqslant$	$\|$
$1x_1 + 3x_2 = y$	1	3	z

From earlier discussion, we know that the minimum y must equal the maximum z. In this example the optimal value is 12. We shall not discuss techniques for solving this type of problem although much has been written on the subject. Two good sources for further discussion would be Dantzig (1963) and Campbell (1965).

Quadratic Minimization Problems

Example 6.4.2 Minimization problems involving quadratic objective functions are very important in statistical applications as well as in many other areas. Their intimate relationship with the normal distribution is especially fortuitous since this opens the door to many results in asymptotic distribution theory.

Consider the quadratic minimization problem

$$\text{minimize } \tfrac{1}{2}\mathbf{x}'\boldsymbol{\Psi}\mathbf{x} - \mathbf{y}'\mathbf{x} \text{ subject to } \mathbf{A}_1'\mathbf{x} \leqslant \mathbf{c}_1, \quad \mathbf{A}_2'\mathbf{x} = \mathbf{c}_2 \tag{6.4.5}$$

where $\boldsymbol{\Psi}$ is a symmetric $k \times k$ nonnegative definite matrix, $\mathbf{y}$ is a fixed $k \times 1$ vector, $\mathbf{A} = (\mathbf{A}_1, \mathbf{A}_2)$ is a $k \times t$ matrix (where $\mathbf{A}_1$ is $k \times r$ and $\mathbf{A}_2$ is $k \times t - r$), and $\mathbf{c} = \begin{pmatrix} \mathbf{c}_1 \\ \mathbf{c}_2 \end{pmatrix}$ and is a $t \times 1$ vector.

Of course if $\mathbf{y}' = \mathbf{z}'\boldsymbol{\Psi}$, (6.4.5) is equivalent to the Mahalanobis distance problem

$$\text{minimize } \tfrac{1}{2}(\mathbf{z} - \mathbf{x})'\boldsymbol{\Psi}(\mathbf{z} - \mathbf{x}) \text{ subject to } \mathbf{A}_1'\mathbf{x} \leqslant \mathbf{c}_1 \text{ and } \mathbf{A}_2'\mathbf{x} = \mathbf{c}_2. \tag{6.4.6}$$

By Theorem 6.4.2, the optimal value of (6.4.5) will be given by

$$\sup_{\lambda_i \geqslant 0, i=1,\ldots,r} g(\boldsymbol{\lambda}), \tag{6.4.7}$$

where $g(\boldsymbol{\lambda}) = \inf_{\mathbf{x}}[\tfrac{1}{2}\mathbf{x}'\boldsymbol{\Psi}\mathbf{x} - \mathbf{y}'\mathbf{x} + \boldsymbol{\lambda}'(\mathbf{A}'\mathbf{x} - \mathbf{c})]$. For a fixed, feasible $\boldsymbol{\lambda}$, the infimum over $\mathbf{x}$ will be finite if and only if there is an $\mathbf{x}$ satisfying

$$\boldsymbol{\Psi}\mathbf{x} - \mathbf{y} + \mathbf{A}\boldsymbol{\lambda} = \mathbf{0}. \tag{6.4.8}$$

The dual problem, (6.4.7), can be expressed as

$$\sup_{\lambda_i \geqslant 0, i=1,\ldots,r} \inf_{\mathbf{x}} (\mathbf{x}'\boldsymbol{\Psi} - \mathbf{y}' + \boldsymbol{\lambda}'\mathbf{A}')\mathbf{x} - \tfrac{1}{2}\mathbf{x}'\boldsymbol{\Psi}\mathbf{x} - \boldsymbol{\lambda}'\mathbf{c},$$

or, equivalently, as

$$\text{maximize} -\tfrac{1}{2}\mathbf{x}'\boldsymbol{\Psi}\mathbf{x} - \boldsymbol{\lambda}'\mathbf{c} \text{ subject to } \boldsymbol{\Psi}\mathbf{x} - \mathbf{y} + \mathbf{A}\boldsymbol{\lambda} = \mathbf{0}, \qquad \lambda_i \geqslant 0; i = 1,\ldots,r.$$

If $\boldsymbol{\Psi}$ is positive definite, then (6.4.8) implies that the minimizing $\mathbf{x}$ must be $\mathbf{x} = \boldsymbol{\Psi}^{-1}(\mathbf{y} - \mathbf{A}\boldsymbol{\lambda})$ for a given $\boldsymbol{\lambda}$, $\lambda_i \geqslant 0$, $i = 1,\ldots,r$. In this case, the dual problem, (6.4.7), can be written as

$$\sup_{\lambda_i \geqslant 0, i=1,\ldots,r} [-\tfrac{1}{2}\boldsymbol{\lambda}'\boldsymbol{\Xi}\boldsymbol{\lambda} - \mathbf{u}'\boldsymbol{\lambda} - \tfrac{1}{2}\mathbf{y}'\boldsymbol{\Psi}\mathbf{y}], \tag{6.4.9}$$

where $\boldsymbol{\Xi} = \mathbf{A}'\boldsymbol{\Psi}^{-1}\mathbf{A}$ and $\mathbf{u} = \mathbf{c} - \mathbf{A}'\boldsymbol{\Psi}^{-1}\mathbf{y}$. Note that this is the same type of quadratic programming problem as the primal problem (6.4.5). However, this problem may be substantially easier to solve since (1) the constraint region, $\lambda_i \geqslant 0$, $i = 1,\ldots,r$, is simpler, and (2) the dimension of the problem is smaller if $t < k$.

It is very helpful to realize that a Kuhn–Tucker vector $\boldsymbol{\lambda}^*$ and a primal problem solution must satisfy the equation (6.4.8). Thus if $t \leqslant k$ and $\mathbf{A}$ is of full column rank, a Kuhn–Tucker vector can only be given by

$$\boldsymbol{\lambda}^* = (\mathbf{A}'\mathbf{A})^{-1}\mathbf{A}'(\mathbf{y} - \boldsymbol{\Psi}\mathbf{x}^*) \tag{6.4.10}$$

where $\mathbf{x}^*$ is the solution to the primal problem. In general, a Kuhn Tucker vector could only be of the form

$$\boldsymbol{\lambda}^* = \mathbf{A}^-(\mathbf{y} - \boldsymbol{\Psi}\mathbf{x}^*),$$

where $\mathbf{A}^-$ is any generalized inverse of $\mathbf{A}(\mathbf{A}\mathbf{A}^-\mathbf{A} = \mathbf{A})$. Equivalently, a Kuhn–Tucker vector must be of the form

$$\boldsymbol{\lambda}^* = \mathbf{A}_0^-(\mathbf{y} - \boldsymbol{\Psi}\mathbf{x}^*) + (\mathbf{I} - \mathbf{A}_0^-\mathbf{A})\mathbf{z}$$

where $\mathbf{z}$ is an arbitrary vector and $\mathbf{A}_0^-$ is some fixed generalized inverse.

If $t \leqslant k$ and $\mathbf{A}$ is of full column rank, then a proposed candidate $\tilde{\mathbf{x}}$ for a solution to the primal problem (6.4.5) will actually be a solution if and only if $\tilde{\mathbf{x}}$ and $\tilde{\boldsymbol{\lambda}} = (\mathbf{A}'\mathbf{A})^{-1}\mathbf{A}'(\mathbf{y} - \psi\tilde{\mathbf{x}})$ satisfy conditions (a) and (b) of Theorem 6.4.1. Of course, for this problem $f_i(\mathbf{x})$ is the ith element of $\mathbf{A}'\mathbf{x} - \mathbf{c}$.

This duality relationship for a quadratic minimization problem is very important and will be discussed further in future sections and examples.

Example 6.4.3 As a special case of the previous example, consider the standard isotonic regression problem

$$\text{minimize} \sum_{i=1}^{k} (g_i - x_i)^2 w_i \text{ subject to } \mathbf{x} \in \mathscr{I}, \tag{6.4.11}$$

where $\mathscr{I} = \{\mathbf{x} \in R^k : x_1 \leqslant x_2 \leqslant \cdots \leqslant x_k\}$. To put this problem in the form of (6.4.5), we take $\boldsymbol{\Psi} = \text{diag}(w_1, \ldots, w_k)$, $\mathbf{y} = \boldsymbol{\Psi}\mathbf{g}$, $\mathbf{c} = \mathbf{0}$, and

$$\mathbf{A}' = \begin{pmatrix} 1 & -1 & 0 & \cdot & \cdot & \cdot & 0 \\ 0 & 1 & -1 & 0 & \cdot & \cdot & 0 \\ \cdot & \cdot & \cdot & \cdot & \cdot & \cdot & \cdot \\ \cdot & \cdot & \cdot & \cdot & \cdot & \cdot & \cdot \\ \cdot & \cdot & \cdot & \cdot & \cdot & \cdot & 0 \\ 0 & 0 & \cdot & \cdot & \cdot & 1 & -1 \end{pmatrix}$$

as the $(k-1) \times k$ constraint matrix.

Methodology for obtaining the solution $\mathbf{g}^*$ of (6.4.11) has been extensively discussed in Chapter 1. We note that $\mathbf{A}$ is of full column rank, so that the Kuhn–Tucker vector must be given by

$$\boldsymbol{\lambda}^* = (\mathbf{A}'\mathbf{A})^{-1}\mathbf{A}'\boldsymbol{\Psi}(\mathbf{g} - \mathbf{g}^*).$$

Since the generalized inverse, $(\mathbf{A}'\mathbf{A})^{-1}\mathbf{A}'$, of $\mathbf{A}$ is given by

$$(\mathbf{A}'\mathbf{A})^{-1}\mathbf{A}' = \begin{pmatrix} 1 & 0 & \cdot & \cdot & \cdot & \cdot & 0 \\ 1 & 1 & 0 & \cdot & \cdot & \cdot & 0 \\ 1 & 1 & 1 & 0 & \cdot & \cdot & \cdot \\ \cdot & \cdot & \cdot & \cdot & \cdot & \cdot & \cdot \\ \cdot & \cdot & \cdot & \cdot & \cdot & \cdot & \cdot \\ 1 & 1 & 1 & \cdot & \cdot & 1 & 0 \end{pmatrix},$$

it follows that the Kuhn–Tucker vector must have for its ith component

$$\lambda_i^* = \sum_{j=1}^{i} (g_j - g_j^*) w_j \qquad \text{for } i = 1, \ldots, k-1.$$

Note that λ_i^* is nonnegative since $\mathbf{g} - \mathbf{g}^* \in \mathscr{I}^{*w}$ (see Theorem 1.7.2). Of course $\boldsymbol{\lambda}^*$ solves a different quadratic optimization problem, namely the one corresponding to (6.4.9).

As another particularly nice example, consider problem (6.4.11) where the constraints are changed to require that $\bar{x}_1 \leqslant \bar{x}_2 \leqslant \cdots \leqslant \bar{x}_n \leqslant 0$ where $\bar{x}_i = \sum_{j=1}^{i} x_j/i$. Vectors which satisfy these constraints are sometimes referred to as being 'increasing on the average' or 'negatively star shaped' (see Dykstra and Robertson, 1982b and 1983, and Shaked, 1979, and Section 5.3). We can write these constraints as $\mathbf{A}'\mathbf{x} \leqslant 0$ where $\mathbf{A}'$ is the $n \times n$ matrix with elements $a_{ij} = [(i+1)i]^{-1/2}$ for $j = 1, 2, \ldots, i$, $a_{i,i+1} = -[i/(i+1)]^{1/2}$, $a_{ij} = 0$ for $j > i+1$ provided $i = 1, 2, \ldots, n-1$, and $a_{nj} = 1/\sqrt{n}$ for $j = 1, 2, \ldots, n$. In this instance, $\mathbf{A}$ is orthogonal so that the Kuhn–Tucker vector must be given by

$$\boldsymbol{\lambda}^* = \mathbf{A}'\boldsymbol{\Psi}(\mathbf{g} - \mathbf{g}^*).$$

We know that $\boldsymbol{\lambda}^*$ solves a different quadratic optimization problem, namely the one specified in (6.4.9).

Constrained *I*-projection problems

Example 6.4.4 Suppose, as in example, 6.2.5, one wishes to find an *I*-projection subject to linear inequality constraints; i.e. consider the problem

$$\text{minimize} \sum_{j=1}^{k} p_j \ln\left(\frac{p_j}{r_j}\right) \text{ subject to } \mathbf{A}'\mathbf{p} \leqslant \mathbf{c}, \tag{6.4.12}$$

where $\mathbf{A}' = (a_{ij})$ is a $t \times k$ matrix and $\mathbf{p}$ is required to be a k-dimensional probability vector. If P denotes the set of probability vectors, the Lagrangian for this problem will be given by

$$L(\boldsymbol{\lambda}, \mathbf{x}) = \begin{cases} \sum_{j=1}^{k} p_j \ln\left(\frac{p_j}{r_j}\right) + \boldsymbol{\lambda}'(\mathbf{A}'\mathbf{p} - \mathbf{c}), & \boldsymbol{\lambda} \geqslant \mathbf{0}, \quad \mathbf{p} \in P \\ -\infty, & \mathbf{p} \in P, \quad \lambda_i < 0 \text{ for some } i \\ +\infty, & \mathbf{p} \notin P. \end{cases}$$

To find the dual problem, note that

$$\begin{aligned} g(\boldsymbol{\lambda}) &= \inf_{\mathbf{p}} L(\boldsymbol{\lambda}, \mathbf{p}) \\ &= \begin{cases} -\infty, & \lambda_i < 0 \text{ for some } i \\ \inf_{\mathbf{p} \in P} \sum_{j=1}^{k} p_j \ln\left\{\frac{p_j}{r_j \exp\left[-\sum_{i=1}^{t} \lambda_i (a_{ij} - c_i)\right]}\right\}, & \boldsymbol{\lambda} \geqslant \mathbf{0} \end{cases} \end{aligned}$$

If we set $p_k = 1 - \sum_{m=1}^{k-1} p_m$, we may set the partial derivatives of the last expression with respect to p_m equal to zero, $m = 1, \ldots, k-1$, and solve the resulting equations to obtain a closed form for the infimum. This results in the function

$$g(\boldsymbol{\lambda}) = \begin{cases} -\infty, & \lambda_i < 0 \text{ for some } i \\ -\ln\left\{\sum_{j=1}^{k} r_j \exp\left[-\sum_{i=1}^{t} \lambda_i (a_{ij} - c_i)\right]\right\}, & \boldsymbol{\lambda} \geqslant \mathbf{0}. \end{cases}$$

Note that an equivalent problem to the dual problem $\sup_{\boldsymbol{\lambda} \geqslant 0} g(\boldsymbol{\lambda})$ is

$$\inf_{\boldsymbol{\lambda} \geqslant \mathbf{0}} \sum_{j=1}^{k} r_j \exp\left[-\sum_{i=1}^{t} \lambda_i (a_{ij} - c_i)\right]$$

which has as its constraint region the nonnegative orthant in t-dimensional space and which has been discussed in Example 6.2.5.

6.5 RELATIONSHIP BETWEEN FENCHEL AND LAGRANGIAN DUALITY UNDER AFFINE CONSTRAINTS

As is to be expected, there are some intimate connections between Fenchel duality and Lagrangian duality, particularly under affine constraints. Consider the problem

$$\inf_{\mathbf{A}'\mathbf{x} \leqslant \mathbf{0}} f(\mathbf{x}) \tag{6.5.1}$$

where f is convex, $\mathbf{x}$ is a $k \times 1$ vector, and $\mathbf{A}'$ is an $r \times k$ matrix. We remark that if the constraints were of the form $\mathbf{A}'\mathbf{x} \leqslant \mathbf{b}$ and if the equations $\mathbf{A}'\mathbf{y} = \mathbf{b}$ are consistent (have a solution), then $\mathbf{A}'(\mathbf{A}')^{-}\mathbf{b} = \mathbf{b}$ for any generalized inverse $(\mathbf{A}')^{-}$. In this situation, the constraints can be expressed as

$$\mathbf{A}'(\mathbf{x} - (\mathbf{A}')^{-}\mathbf{b}) \leqslant \mathbf{0},$$

and it is often possible to make a change of variables in the problem and express it in the general form of (6.5.1).

The Lagrangian corresponding to problem (6.5.1) would be

$$L(\boldsymbol{\lambda}, \mathbf{x}) = \begin{cases} f(\mathbf{x}) + \boldsymbol{\lambda}'\mathbf{A}'\mathbf{x}, & \boldsymbol{\lambda} \geqslant \mathbf{0} \\ -\infty, & \lambda_i < 0 \text{ for some } i, \end{cases}$$

assuming f is defined over all of R^k. However,

$$\mathbf{g}(\boldsymbol{\lambda}) = \inf_{\mathbf{x}} [f(\mathbf{x}) + \boldsymbol{\lambda}'\mathbf{A}'\mathbf{x}] = -\sup_{\mathbf{x}} [-\boldsymbol{\lambda}'\mathbf{A}'\mathbf{x} - f(\mathbf{x})] = -f^*(-\mathbf{A}\boldsymbol{\lambda})$$

where f^* is the convex conjugate of f. The dual problem is then

$$\sup_{\boldsymbol{\lambda} \geqslant \mathbf{0}} g(\boldsymbol{\lambda}) = \sup_{\boldsymbol{\lambda} \geqslant \mathbf{0}} -f^*(-\mathbf{A}\boldsymbol{\lambda}). \tag{6.5.2}$$

However, note that the convex cone

$$\mathscr{K} = \{\mathbf{x} : \mathbf{A}'\mathbf{x} \leqslant \mathbf{0}\}$$

can be written as $\bigcap_{i=1}^{r} \mathscr{K}_i$ where $\mathscr{K}_i = \{\mathbf{x} : \mathbf{a}_i'\mathbf{x} \leqslant 0\}$ and $\mathbf{a}_i$ is the ith column of $\mathbf{A}$. The dual cone $\mathscr{K}^*$ is expressible as

$$\mathscr{K}^* = \mathscr{K}_1^* + \mathscr{K}_2^* + \cdots + \mathscr{K}_r^*$$

where $\mathscr{K}_i^* = \{\lambda \mathbf{a}_i : \lambda \geqslant 0\}$. Thus (6.5.2) can be rephrased as

$$\sup_{\mathbf{y} \in \mathscr{K}^*} -f^*(-\mathbf{y})$$

which is precisely (6.2.2). Thus the Kuhn–Tucker vector associated with the original problem (6.5.1) is precisely the weights assigned to the columns of $\mathbf{A}$ in the solution of the dual problem (6.2.2). This provides some insight into the role of Kuhn–Tucker vectors. Those values of i where λ_i^* are large are precisely the vectors $\mathbf{a}_i$ that are most important (contribute the most) in the solution of the dual problem (6.2.2). If $\lambda_i^* = 0$, the coefficient of $\mathbf{a}_i$ in the solution to the dual problem is zero.

There is another interpretation of the role of Kuhn–Tucker vectors in the primal problem. Suppose the Lagrangian is modified by shifting the location of the constraint functions by a small amount. To be precise, for the initial problem discussed in the beginning of Section 6.4 define a modified Lagrangian by

$$\tilde{L}(\boldsymbol{\lambda}, \mathbf{x}, \mathbf{b}) = \begin{cases} f_0(\mathbf{x}) + \sum_{i=1}^{t} \lambda_i (f_i(\mathbf{x}) + b_i), & \mathbf{x} \in \mathscr{C}, \quad \lambda_i \geqslant 0, \quad i = 1, \ldots, r \\ -\infty, & \mathbf{x} \in \mathscr{C}, \quad \lambda_i < 0 \quad \text{for some } i \leqslant r \\ +\infty, & \mathbf{x} \notin \mathscr{C}. \end{cases}$$

Of course $\tilde{L}(\boldsymbol{\lambda}, \mathbf{x}, \mathbf{b})$ would be the Lagrangian if the constraint $f_i(\mathbf{x}) \leqslant 0$ were modified to be $f_i(\mathbf{x}) \leqslant -b_i$, $i = 1, \ldots, r$. Suppose now that $\mathbf{x}^*$ is a solution to the original problem and that $\boldsymbol{\lambda}^*$ is a Kuhn–Tucker vector, so that $L(\boldsymbol{\lambda}^*, \mathbf{x}^*) = \tilde{L}(\boldsymbol{\lambda}^*, \mathbf{x}^*, \mathbf{0})$. However,

$$\frac{\partial}{\partial b_i} \tilde{L}(\boldsymbol{\lambda}^*, \mathbf{x}^*, \mathbf{b}) \bigg|_{\mathbf{b}=\mathbf{0}} = \lambda_i^*,$$

so that the rate of change of $\tilde{L}(\boldsymbol{\lambda}^*, \mathbf{x}^*, \mathbf{b})$, when the location value of the ith constraint is slightly changed, is the ith coordinate of the Kuhn–Tucker vector. In other words, λ_i^* can be interpreted as a measure of the sensitivity of the optimal value to the ith constraint when all other constraints are in force. Thus those indices i where λ_i^* is large will index which constraints are having a large impact on the realized optimal value. Thus, the λ_i^*'s can be used as diagnostics to attempt to pinpoint the important constraints. From this approach, it is clear that the Kuhn–Tucker vector λ_i^* may prove valuable in the construction of a test statistic when parameter constraints are involved. This type of test, often called a Lagrangian test, was initially proposed by Silvey (1959). It also has much in common with scores tests (or Rao's test), proposed even earlier by Rao (1948).

Recently some work has been done in modifying Lagrangian tests for the purpose of testing inequality constraints. This seems like a fruitful avenue of work, and substantial contributions will probably be forthcoming in the near future. Some existing work on this topic is briefly discussed in Section 6.8 of this chapter.

6.6 CONSTRAINED LINEAR REGRESSION

Consider the multiple linear regression model

$$\mathbf{Y} = \mathbf{X}\boldsymbol{\beta} + \boldsymbol{\varepsilon}$$

where $\mathbf{Y}$ is an $n \times 1$ vector of observations, $\mathbf{X}$ is an $n \times k$ known design matrix of rank k, $\boldsymbol{\beta}$ is a $k \times 1$ vector of parameters, and $\boldsymbol{\varepsilon}$ is an $n \times 1$ random vector which has a normal distribution with mean $\mathbf{0}$ and known positive definite covariance matrix $\boldsymbol{\Psi}$.

The three maximum likelihood estimation problems where (1) $\boldsymbol{\beta}$ is unconstrained, (2) $\boldsymbol{\beta}$ must satisfy the constraints $\mathbf{A}'\boldsymbol{\beta} = \mathbf{c}$, and finally (3) $\boldsymbol{\beta}$ must satisfy the inequality constraints $\mathbf{A}'\boldsymbol{\beta} \leqslant \mathbf{c}$ where $\mathbf{A}$ is a $k \times t$ matrix and $\mathbf{c}$ an $r \times 1$ vector will be examined. The objective function can be taken as the negative of the logarithm of the likelihood of $\mathbf{Y}$,

$$\tfrac{1}{2}(\mathbf{y} - \mathbf{X}\boldsymbol{\beta})'\boldsymbol{\Psi}^{-1}(\mathbf{y} - \mathbf{X}\boldsymbol{\beta}) \tag{6.6.1}$$

or, equivalently, as

$$\tfrac{1}{2}\boldsymbol{\beta}'\mathbf{X}'\boldsymbol{\Psi}^{-1}\mathbf{X}\boldsymbol{\beta} - \mathbf{y}'\boldsymbol{\Psi}^{-1}\mathbf{X}\boldsymbol{\beta}. \tag{6.6.2}$$

Note that this is precisely of the same form as (6.4.5) when $\boldsymbol{\Psi}$ is replaced by $\mathbf{X}'\boldsymbol{\Psi}^{-1}\mathbf{X}$ and $\mathbf{y}'$ is replaced by $\mathbf{y}'\boldsymbol{\Psi}^{-1}\mathbf{X}$. The three aforementioned maximum likelihood problems (in the notation of (6.4.5)) correspond to

(1) $t = 0$,
(2) $r = 0$, $\mathbf{A}_2' = \mathbf{A}'$, $\mathbf{c}_2 = \mathbf{c}$,

and

(3) $r = t$, $\mathbf{A}_1' = \mathbf{A}'$, $\mathbf{c}_1 = \mathbf{c}$.

Of course in the unconstrained case, (1), the Lagrangian is free of $\boldsymbol{\lambda}$ and hence nothing is lost if $\boldsymbol{\lambda}$ is assumed to be zero. This will be done for completeness.

If we let $(\boldsymbol{\beta}_i^*, \boldsymbol{\lambda}_i^*)$, $i = 1, 2, 3$, denote the maximum likelihood estimators and Kuhn–Tucker vectors for the cases (1), (2), and (3), respectively, in all cases these pairs of vectors must satisfy equation (6.4.8). Thus

$$\boldsymbol{\beta}_i^* = (\mathbf{X}'\boldsymbol{\Psi}^{-1}\mathbf{X})^{-1}(\mathbf{X}\boldsymbol{\Psi}^{-1}\mathbf{y} - \mathbf{A}\boldsymbol{\lambda}_i^*) = (\mathbf{X}'\boldsymbol{\Psi}^{-1}\mathbf{X})^{-1}(\mathbf{X}'\boldsymbol{\Psi}^{-1}\mathbf{y}) - (\mathbf{X}'\boldsymbol{\Psi}^{-1}\mathbf{X})^{-1}\mathbf{A}\boldsymbol{\lambda}_i^*.$$

Of course in the unrestricted case, $\boldsymbol{\beta}_1^*$ is just the usual unrestricted MLE of the regression parameters $\boldsymbol{\beta}$. It clearly follows that in all three cases,

$$\boldsymbol{\beta}_1^* - \boldsymbol{\beta}_i^* = (\mathbf{X}'\boldsymbol{\Psi}^{-1}\mathbf{X})^{-1}\mathbf{A}\boldsymbol{\lambda}_i^*. \tag{6.6.3}$$

When $i = 2$, we may multiply both sides by the matrix $\mathbf{A}'$ to obtain

$$\mathbf{A}'\boldsymbol{\beta}_1^* - \mathbf{A}'\boldsymbol{\beta}_2^* = \mathbf{A}'\boldsymbol{\beta}_1^* - \mathbf{c} = \mathbf{A}'(\mathbf{X}'\boldsymbol{\Psi}^{-1}\mathbf{X})^{-1}\mathbf{A}\boldsymbol{\lambda}_2^*$$

which is important for relating $\boldsymbol{\lambda}_2^*$ to the unrestricted MLE. By evaluating (6.6.3) for $i = 2$ and $i = 3$ and subtracting, it follows that

$$\boldsymbol{\beta}_2^* - \boldsymbol{\beta}_3^* = (\mathbf{X}'\boldsymbol{\Psi}^{-1}\mathbf{X})^{-1}\mathbf{A}(\boldsymbol{\lambda}_3^* - \boldsymbol{\lambda}_2^*).$$

Thus the difference in the regression coefficients for the equality constraints and the inequality constraints situations are a linear combination of the differences of the associated Kuhn–Tucker vectors. We will exploit this relationship in Section 6.8.

6.7 DUALITY IN STOCHASTICALLY ORDERED DISTRIBUTIONS

A problem of substantial interest is that of estimating stochastically ordered distributions. These types of constraints occur quite naturally in many different situations. For example, if a mechanical device is improved in some manner, even minutely, the probability of the improved device surviving past any given time should not be less than that for the original device.

As another example, survival times for medical patients should often be stochastically ordered depending upon the status of concomitant variables. However, the inherent randomness of the observations may lead to usual estimates which do not satisfy this type of ordering. We will consider the problem from a nonparametric maximum likelihood approach. We will allow our observations to be censored on the right, although we do assume that our censoring times are either fixed or are independent of our observations. We will use the vernacular of reliability and call exact observations failures and censored observations losses. It will prove convenient to work with survival functions rather than CDFs. (The survival function of a random variable X is defined to be $P(t) = P(X > t)$).

Assume that one has independent random samples, possibly with right-censored observations, from N discrete populations. The $N \geqslant 2$ populations are assumed to be stochastically ordered, so the corresponding survival functions satisfy

$$P_1 \overset{\text{st}}{\geqslant} P_2 \overset{\text{st}}{\geqslant} \cdots \overset{\text{st}}{\geqslant} P_N. \tag{6.7.1}$$

(By $P_i \overset{\text{st}}{\geqslant} P_j$ we mean that $P_i(t) \geqslant P_j(t)$ for all t.) The problem is to find nonparametric MLEs of the survival functions subject to the constraints in (6.7.1). We will assume that $P_1(0) = P_2(0) = \cdots = P_N(0) = 1$, so all observations are positive.

Assume that exact observations (failures) occur on a subset of the times $S_1 < S_2 < \cdots < S_m (S_0 = 0, S_{m+1} = \infty)$. The number of failures from the ith population that occurs at time S_j is denoted by d_{ij}. The number of losses (right-censored observations) in the interval $[S_j, S_{j+1})$ from the ith population is denoted l_{ij}. We assume that the l_{ij} losses occur at times $L_r^{(ij)}$, $r = 1, \ldots, l_{ij}$, where these censoring times are fixed. (The same MLEs would obtain for random censoring times that are independent of the times of failure.) Let $n_{ij} = \sum_{r=j}^{m} (d_{ir} + l_{ir})$ denote the number of items from the ith population surviving just prior to S_j. We assume that the survival functions are discrete, but this is not really necessary, as one may argue in the context of generalized maximum likelihood (see Johansen, 1978) that the estimates need place probability only on those times at which observations occur.

According to our earlier notation, the problem is to find survival functions $P_1, \ldots, P_N$ that maximize the likelihood

$$\prod_{i=1}^{N} \left(\prod_{r=1}^{l_{i0}} P(L_r^{(i0)}) \prod_{j=1}^{m} \left\{ [P_i(S_j -) - P_i(S_j)]^{d_{ij}} \times \prod_{r=1}^{l_{ij}} P_i(L_r^{(ij)}) \right\} \right), \tag{6.7.2}$$

subject to the constraints

$$P_i(t) \geqslant P_{i+1}(t) \qquad \text{for all } t, \quad i = 1, \ldots, N-1.$$

For a given set of survival functions, $P_i,\ldots,P_N$, satisfying (6.7.1), the likelihood cannot be decreased and (6.7.1) cannot be violated if we replace $P_1(t)$ by a discrete $P_1'(t)$, which has possible jumps only at $S_1,\ldots,S_m$ and is such that $P_1'(S_j)=P_1(S_j)$. If we now replace $P_2(t)$ by $P_2'(t)$, defined to have possible jumps only at $S_1,\ldots,S_m$, and $P_2'(S_j)=P_2(S_j)$, the likelihood cannot decrease and (6.7.2) cannot be violated.

Continuing this reasoning, we see that it will suffice to maximize the expression

$$\prod_{i=1}^{N}\prod_{j=1}^{m}[P_i(S_{j-1})-P_i(S_j)]^{d_{ij}}P_i(S_j)^{l_{ij}},$$

subject to $P_i(S_j)\geqslant P_{i+1}(S_j)$ for all j and $i=1,\ldots,N-1$, among those survival functions that place probability only at the points $S_1, S_2,\ldots,S_m$. (Note that the MLEs need not be uniquely defined if the last observation from a population is a loss. We avoid this ambiguity by requiring the tails of our MLEs of the survival functions to go to zero in specific ways.) Equivalently, we wish to maximize

$$\prod_{i=1}^{N}\prod_{j=1}^{m}\left[1-\frac{P_i(S_j)}{P_i(S_{j-1})}\right]^{d_{ij}}\left[\frac{P_i(S_{j-1})}{P_i(S_{j-2})}\cdots P_i(S_1)\right]^{d_{ij}}\left[\frac{P_i(S_j)}{P_i(S_{j-1})}\cdots P_i(S_1)\right]^{l_{ij}}$$

or, letting $p'_{ij}=P_i(S_j)/P_i(S_{j-1})$, to maximize

$$\prod_{i=1}^{N}\prod_{j=1}^{m}(1-p'_{ij})^{d_{ij}}(p'_{ij})^{l_{ij}}\prod_{r<j}(p'_{ir})^{d_{ij}+l_{ij}},$$

subject to $\prod_{r=1}^{j}p'_{ir}\geqslant\prod_{r=1}^{j}p'_{i+1,r}$, $j=1,\ldots,m$; $i=1,\ldots,N-1$.

Finally, recalling that $n_{ij}=\sum_{r=j}^{m}(d_{ir}+l_{ir})$, letting $p_{ij}=\ln p'_{ij}$, and considering the log of the likelihood, it will suffice to minimize

$$f(\mathbf{p}_1,\ldots,\mathbf{p}_N)=-\sum_{i=1}^{N}\sum_{j=1}^{m}[d_{ij}\ln(1-e^{p_{ij}})+(n_{ij}-d_{ij})p_{ij}] \tag{6.7.3}$$

subject to the constraints

$$\sum_{j=1}^{r}p_{ij}\geqslant\sum_{j=1}^{r}p_{i+1,j},\qquad r=1,\ldots,m;\quad i=1,\ldots,N-1,\quad\text{and}\quad 0\geqslant p_{ij}\geqslant-\infty$$

$$\text{for all } i,j. \tag{6.7.4}$$

The problem has been reduced to minimizing a convex function subject to linear inequality constraints and hence is amenable to the use of Theorem 6.4.1.

By associating λ_{ir} with the constraint $\sum_{j=1}^{r}p_{ij}\geqslant\sum_{j=1}^{r}p_{i+1,j}$, the Lagrangian takes the form

$$L(\boldsymbol{\lambda},\mathbf{p})=\begin{cases}-\displaystyle\sum_{i=1}^{N}\sum_{j=1}^{m}[d_{ij}\ln(1-e^{p_{ij}})+(n_{ij}-d_{ij})p_{ij}]+\sum_{i=1}^{N-1}\sum_{r=1}^{m}\lambda_{ir}\sum_{j=1}^{r}(p_{i+1,j}-p_{ij}),\\ \qquad\text{if } 0\geqslant p_{ij}\geqslant-\infty,\quad \lambda_{ij}\geqslant 0,\quad i=1,\ldots,N-1;\quad j=1,\ldots,m\\ -\infty,\quad\text{if } \lambda_{ij}<0\ \text{ for some } (i,j) \text{ and } p_{ij}\leqslant 0\ \text{ for all } (i,j)\\ +\infty,\quad\text{if } p_{ij}>0\ \text{ for some } (i,j).\end{cases}$$

(6.7.5)

By regrouping the first line of (6.7.5), we may express $g(\boldsymbol{\lambda}) = \inf_{\mathbf{p}} L(\boldsymbol{\lambda}, \mathbf{p})$ as

$$\inf_{p_{ij} \leqslant 0} - \sum_{i=1}^{N} \sum_{j=1}^{m} [d_{ij} \ln(1 - e^{p_{ij}}) + (n_{ij} - d_{ij} + c_{ij}) p_{ij}] \tag{6.7.6}$$

where $c_{ij} = -\sum_{r=j}^{m} (\lambda_{ir} - \lambda_{i-1,r})$, $i = 1, \ldots, N$; $j = 1, \ldots, m$ and $\lambda_{0,r} = 0$ for all r.

One can minimize over each p_{ij} individually. The equations which result by setting the first derivative of (6.7.6), with respect to p_{ij}, equal to zero are

$$-\frac{d_{ij} e^{p_{ij}}}{1 - e^{p_{ij}}} + (n_{ij} - d_{ij} + c_{ij}) = 0. \tag{6.7.7}$$

These equations are easily solved by

$$p_{ij} = \ln(n_{ij} - d_{ij} + c_{ij}),$$

which, when substituted back into the initial function, yield

$$g(\boldsymbol{\lambda}) = \sum_{i=1}^{N} \sum_{j=1}^{m} \Phi\left(\frac{d_{ij}}{n_{ij} + c_{ij}}\right)(n_{ij} + c_{ij}),$$

where c_{ij} is defined in (6.7.6) and $\Phi(x)$ is the concave function defined by

$$\Phi(x) = -x \ln x - (1 - x) \ln(1 - x), \qquad 0 \leqslant x \leqslant 1.$$

The dual problem is then given by $\sup_{\boldsymbol{\lambda} \geqslant \mathbf{0}} g(\boldsymbol{\lambda})$. Equivalently, one can express the dual problem in terms of the c_{ij}'s as

$$\sup \sum_{i=1}^{N} \sum_{j=1}^{m} \Phi\left(\frac{d_{ij}}{n_{ij} + c_{ij}}\right)(n_{ij} + c_{ij})$$

$$\text{subject to } \sum_{l=1}^{i} (c_{lj} - c_{l,j+1}) \leqslant 0 \quad \text{for all } i, j. \tag{6.7.8}$$

In certain situations (say if there is no censoring and $N = 2$), it is possible to formulate (6.7.2) in such a way that Corollary A to Theorem 1.7.4 is applicable (or Example 6.2.2) and hence the stochastic ordering problem can be solved by an isotonic regression (see Barlow and Brunk, 1972).

In more general settings, Dykstra (1982) has given an algorithm for solving the stochastic ordering problem ($N = 2$ and right censoring), and Feltz and Dykstra (1985) have given an iterated procedure for solving this problem with more than two populations.

Duality becomes important in this problem for verifying that the Feltz and Dykstra procedure must converge correctly to the true MLEs. Feltz and Dykstra first prove that their procedure must converge. They are then able to exhibit a Kuhn–Tucker vector which, with the limiting value of their procedure, satisfies the conditions of Theorem 6.4.1. The theorem then guarantees that their limiting value must be the solution to the problem.

The example given in Table 6.7.1 and Figures 6.7.1 and 6.7.2 is taken from

Table 6.7.1 Survival times (in days) of patients with squamous carcinoma in the oropharynx with various degrees of lymph node deterioration

$i=1$	$i=2$	$i=3$	
216	105	94	363
324	222	99	407
338	279	112	413†
347	395	127	459
599	465	134	517
763	546	147	532
929	915	192	544
1086†	918	219	672
1092	1058†	255	696
1317†	1455†	262	800
1609†	1644†	274	914
		293	1312†
		307	1446†
		327	1472†
		334	

Note: The degree of deterioration (i) equals N-stage tumor classification in Kalbfleisch and Prentice (1980). † denotes a censored observation.

Feltz and Dykstra (1985). In this example, survival times (some censored) of patients with squamous carcinoma of the oropharynx are examined. The patients were placed in three groups depending upon the degree of lymph node deterioration. It seems inherently plausible that the disease is further advanced in those patients with more lymph node deterioration, and hence their survival times would generally be shorter. Thus it seems reasonable that the survival functions of the three populations which correspond to the three lymph node categories should be stochastically ordered according to $P_1 \geqslant P_2 \geqslant P_3$. This is generally borne out in Figure 6.7.1 where the maximum likelihood estimates (which were derived by Kaplan and Meier (1958)) of the survival functions are shown. (The Kaplan–Meier estimates are obtained by minimizing (6.7.3) with the only constraints being that $-\infty \leqslant p_{ij} \leqslant 0$.)

However, there are some reversals present in Figure 6.7.1 which are probably due to sample variation. Figure 6.7.2 shows the maximum likelihood estimates when the stochastic ordering constraints are imposed. Even though the reversals were of a fairly minor nature, the estimates with the constraints imposed are substantially changed.

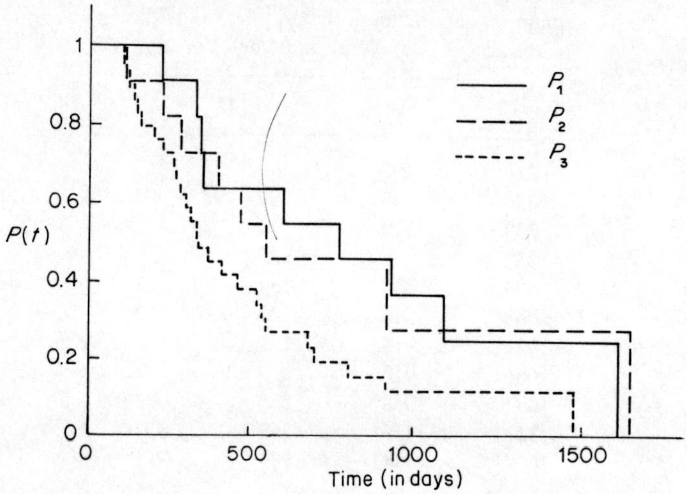

Figure 6.7.1 Kaplan–Meier MLEs of survival functions. Note that the ordering constraints are broken between P_2 and P_3 at 100 days and are also broken between P_1 and P_2 at around 400 days and again at around 1000 days. Our assumption is that these orderings actually exist and that the violations are due to sampling variation

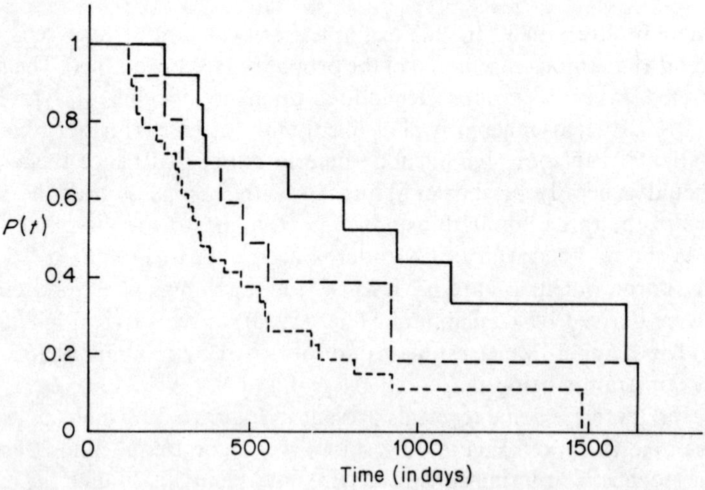

Figure 6.7.2 MLEs of survival functions under order restrictions $P_1 \overset{st}{\geqslant} P_2 \overset{st}{\geqslant} P_3$. These survival functions were smoothed using the approach given in Feltz and Dykstra (1985)

6.8 GENERAL CLASSES OF TESTS

There have been several approaches proposed for general hypothesis testing situations. Certainly the approach that is used most generally is that of likelihood ratio tests (LRT). This approach has been used extensively in earlier chapters of this book, and hence will receive little special attention here.

Competitors to the LRT are Wald's (1943) test, Rao's (1948) efficient scores test, and the closely related Lagrange multiplier test. Wald's test and Rao's test are discussed in Serfling (1980) and Rao (1973) among other places. The Lagrange multiplier test (LMT) is discussed initially in Aitchison and Silvey (1958) and Silvey (1959). The majority of papers on LMTs seem to occur in econometric literature.

Background and notation

To give a brief background on these types of tests, we will need to establish some notation.

Suppose $Y_1, Y_2, \ldots, Y_n$ are independently, identically distributed random variables (or random vectors) with common probability density $f(y; \boldsymbol{\theta})$ depending upon a q-dimensional parameter $\boldsymbol{\theta}' = (\theta_1, \theta_2, \ldots, \theta_q)$. The log likelihood function will be denoted by

$$l(\boldsymbol{\theta}) = \log f(y_1; \boldsymbol{\theta}) + \log f(y_2; \boldsymbol{\theta}) + \cdots + \log f(y_n; \boldsymbol{\theta}).$$

The gradient of the log likelihood will be denoted by

$$\mathbf{V}(\boldsymbol{\theta})' = \left(\frac{\partial l(\boldsymbol{\theta})}{\partial \theta_1}, \frac{\partial l(\boldsymbol{\theta})}{\partial \theta_2}, \ldots, \frac{\partial l(\boldsymbol{\theta})}{\partial \theta_q} \right). \tag{6.8.1}$$

Some authors term $V(\boldsymbol{\theta})/\sqrt{n}$ the efficient scores of the likelihood function.

The information matrix $I(\boldsymbol{\theta})$ of a single observation Y is the matrix with the (i, j)th component

$$E\left[\frac{\partial}{\partial \theta_i} \log f(Y; \boldsymbol{\theta}) \frac{\partial}{\partial \theta_j} \log f(Y; \boldsymbol{\theta}) \right]. \tag{6.8.2}$$

Under quite general conditions, the vector of efficient scores $V(\boldsymbol{\theta})/\sqrt{n}$ will have an asymptotic normal distribution with mean $\mathbf{0}$ and covariance matrix $I(\boldsymbol{\theta})$.

Suppose now that we wish to test as a null hypothesis that $\boldsymbol{\theta}$ is expressible as q smooth functions of r other parameters, $\tau_1, \tau_2, \ldots, \tau_r$; that is

$$\theta_i = g_i(\tau_1, \tau_2, \ldots, \tau_r) \qquad \text{for } i = 1, 2, \ldots, q.$$

Equivalently, and more tractably, we may assume as our null hypothesis that $\boldsymbol{\theta}$ is subject to $k = q - r$ restrictions

$$\mathbf{H}(\boldsymbol{\theta})' = (h_1(\boldsymbol{\theta}), h_2(\boldsymbol{\theta}), \ldots, h_k(\boldsymbol{\theta})) = \mathbf{0}. \tag{6.8.3}$$

As our alternative hypothesis, we will allow $\boldsymbol{\theta}$ to be unrestricted. If we denote our restricted and unrestricted MLEs by $\tilde{\boldsymbol{\theta}}$ and $\hat{\boldsymbol{\theta}}$, respectively, the LRT would reject for large values of

$$Q_{\text{LR}} = 2[l(\hat{\boldsymbol{\theta}}) - l(\tilde{\boldsymbol{\theta}})]. \tag{6.8.4}$$

The reasoning behind Wald's procedure is the following: (1) $\mathbf{H}(\boldsymbol{\theta}) = \mathbf{0}$ for the true parameter $\boldsymbol{\theta}$ if the null hypothesis is true, (2) $\hat{\boldsymbol{\theta}}$ is usually a good estimator for $\boldsymbol{\theta}$, and hence should be close to $\boldsymbol{\theta}$, and thus (3) $H(\hat{\boldsymbol{\theta}})$ should be close to zero if the null hypothesis is true.

Moreover, under suitable regularity conditions, $\sqrt{n}\hat{\boldsymbol{\theta}}$ will be asymptotically normal with mean vector $\boldsymbol{\theta}$ and covariance matrix $\mathbf{I}(\boldsymbol{\theta})^{-1}$ (see Serfling, 1980). Thus if $\mathbf{H}(\boldsymbol{\theta})$ is smooth (differentiable), $\sqrt{n}\mathbf{H}(\hat{\boldsymbol{\theta}})$ will be asymptotically normal with mean $\mathbf{H}(\boldsymbol{\theta})$ and covariance matrix

$$\mathbf{T}(\boldsymbol{\theta})'\mathbf{I}(\boldsymbol{\theta})^{-1}\mathbf{T}(\boldsymbol{\theta}) \quad \text{where} \quad T(\boldsymbol{\theta})_{q\times k} = \left(\frac{\partial h_j(\boldsymbol{\theta})}{\partial \theta_i}\right). \tag{6.8.5}$$

Since under the null hypothesis $\mathbf{H}(\boldsymbol{\theta}) = \mathbf{0}$, one would expect larger values of

$$Q_{\text{W}} = n\mathbf{H}(\hat{\boldsymbol{\theta}})'[\mathbf{T}(\hat{\boldsymbol{\theta}})'\mathbf{I}(\hat{\boldsymbol{\theta}})^{-1}\mathbf{T}(\hat{\boldsymbol{\theta}})]^{-1}\mathbf{H}(\hat{\boldsymbol{\theta}}) \tag{6.8.6}$$

if the null hypothesis is false than if it is true. This statistic is known as the Wald statistic and large values of Q_{W} are grounds for rejection of the null hypothesis. Note that the Wald test does not require computation of the restricted MLE.

The efficient scores test of Rao (Rao's test) evaluates the efficient scores and the information matrix at the restricted MLE, $\tilde{\boldsymbol{\theta}}$, and rejects for large values of

$$Q_{\text{R}} = \frac{1}{n}\mathbf{V}(\tilde{\boldsymbol{\theta}})'I(\tilde{\boldsymbol{\theta}})^{-1}\mathbf{V}(\tilde{\boldsymbol{\theta}}). \tag{6.8.7}$$

The rationale is that since the efficient scores are zero at the unrestricted MLE which should be close to the true parameter, $\boldsymbol{\theta}$, the magnitude of the gradient at the restricted MLE is an indication of the distance from the true parameter to the null parameter space.

The Lagrange multiplier test is closely associated with Rao's efficient scores test. In order to discuss this test, assume the log likelihood function $l(\boldsymbol{\theta})$ and the set of constraints $\mathbf{H}(\boldsymbol{\theta})$ are sufficiently nice so that $(\tilde{\boldsymbol{\theta}}, \tilde{\boldsymbol{\lambda}})$ is the unique saddle point of the Lagrangian

$$L(\boldsymbol{\theta}, \boldsymbol{\lambda}) = l(\boldsymbol{\theta}) + \boldsymbol{\lambda}'\mathbf{H}(\boldsymbol{\theta}).$$

If the unrestricted MLE $\hat{\boldsymbol{\theta}}$ satisfies the constraints $\mathbf{H}(\boldsymbol{\theta}) = \mathbf{0}$, then $\tilde{\boldsymbol{\lambda}} = \mathbf{0}$ and $\hat{\boldsymbol{\theta}} = \tilde{\boldsymbol{\theta}}$. Moreover, since $\hat{\boldsymbol{\theta}}$ is usually a good estimator of the true parameter $\boldsymbol{\theta}$, this is evidence supporting the null hypothesis. On the other hand, if $\hat{\boldsymbol{\theta}}$ is not close to satisfying $\mathbf{H}(\boldsymbol{\theta}) = \mathbf{0}$, $\tilde{\boldsymbol{\lambda}}$ will probably not be near $\mathbf{0}$. Thus the magnitude of $\tilde{\boldsymbol{\lambda}}$ is in some sense a measure of the evidence against H_0.

Aitchison and Silvey (1958) were able to show that under appropriate regularity conditions $\tilde{\boldsymbol{\lambda}}/\sqrt{n}$ is asymptotically distributed as a normal random vector with mean $\mathbf{0}$ and covariance matrix $[\mathbf{T}(\boldsymbol{\theta})'\mathbf{I}(\boldsymbol{\theta})^{-1}\mathbf{T}(\boldsymbol{\theta})]^{-1}$ when $\boldsymbol{\theta}$ is in the null hypothesis parameter region. They proposed using

$$Q_{\text{LM}} = \frac{1}{n}\tilde{\boldsymbol{\lambda}}'\mathbf{T}(\tilde{\boldsymbol{\theta}})'\mathbf{I}(\tilde{\boldsymbol{\theta}})^{-1}\mathbf{T}(\tilde{\boldsymbol{\theta}})\tilde{\boldsymbol{\lambda}} \tag{6.8.8}$$

as a test statistic, rejecting the null hypothesis for large values of Q_{LM}.

However, note that condition (c) of Theorem 6.4.1 (assuming a well-behaved Lagrangian) will have the property that

$$\nabla l(\tilde{\boldsymbol{\theta}}) + \sum_{i=1}^{k} \tilde{\lambda}_i \nabla h_i(\tilde{\boldsymbol{\theta}}) = \mathbf{0}$$

(where ∇ denotes the gradient operator) or, equivalently, that

$$\mathbf{V}(\tilde{\boldsymbol{\theta}}) = -\mathbf{T}(\tilde{\boldsymbol{\theta}})\tilde{\boldsymbol{\lambda}}.$$

Making this substitution in (6.8.7) shows that Rao's test is identical to the Lagrange multiplier test in this setting.

Under appropriate regularity conditions, all the described tests have an asymptotic chi-square distribution with $k = q - r$ degrees of freedom (see Rao, 1973, or Serfling, 1980). If the log likelihood is quadratic and the constraints are linear, all these tests are actually identical.

A natural question that arises is whether these tests are applicable to the situation when the equality constraints in (6.8.3) are changed to the inequality constraints

$$h_i(\boldsymbol{\theta}) \leqslant 0 \qquad \text{for } i = 1, 2, \ldots, k.$$

The resulting tests would depend upon whether these inequality constraints represent the null or the alternative hypothesis.

First, let us discuss the testing problem where the hypotheses of interest are

$$H_0: h_i(\boldsymbol{\theta}) = 0, \quad i = 1, 2, \ldots, k, \quad \text{versus} \quad H_1: h_i(\boldsymbol{\theta}) \leqslant 0, \quad i = 1, 2, \ldots, k.$$

We will denote the MLE and the corresponding Lagrange multipliers (Kuhn–Tucker vectors) under H_0 by $(\tilde{\boldsymbol{\theta}}, \tilde{\boldsymbol{\lambda}})$ and under H_1 by $(\boldsymbol{\theta}^*, \boldsymbol{\lambda}^*)$. Then reasonable analogues of the previously discussed tests would be the following:

1. Likelihood ratio: Reject H_0 for large values of

$$2[l(\boldsymbol{\theta}^*) - l(\tilde{\boldsymbol{\theta}})].$$

2. Wald test: Reject H_0 for large values of

$$n\mathbf{H}(\boldsymbol{\theta}^*)'[\mathbf{T}(\boldsymbol{\theta}^*)'\mathbf{I}(\boldsymbol{\theta}^*)^{-1}\mathbf{T}(\boldsymbol{\theta}^*)]^{-1}\mathbf{H}(\boldsymbol{\theta}^*).$$

3. Rao test: Reject H_0 for large values of

$$\frac{1}{n}[\mathbf{V}(\boldsymbol{\theta}^*) - \mathbf{V}(\tilde{\boldsymbol{\theta}})]'\mathbf{I}(\boldsymbol{\theta}^*)^{-1}[\mathbf{V}(\boldsymbol{\theta}^*) - \mathbf{V}(\tilde{\boldsymbol{\theta}})].$$

4. Lagrange multiplier (Kuhn–Tucker) test: Reject H_0 for large values of

$$\frac{1}{n}(\boldsymbol{\lambda}^* - \tilde{\boldsymbol{\lambda}})'\mathbf{T}(\tilde{\boldsymbol{\theta}})'\mathbf{I}(\tilde{\boldsymbol{\theta}})^{-1}\mathbf{T}(\tilde{\boldsymbol{\theta}})(\boldsymbol{\lambda}^* - \tilde{\boldsymbol{\lambda}}).$$

Note that if the constraint functions $h_i(\boldsymbol{\theta})$, $i = 1, 2, \ldots, k$, are linear, then Rao's test and the Lagrange multiplier test are identical.

If the H_1 region is taken to be the null hypothesis and the alternative region is that where no restrictions are imposed, then the corresponding test statistics would be:

1. LRT: $2[l(\hat{\boldsymbol{\theta}}) - l(\boldsymbol{\theta}^*)]$
2. Wald test: $n(\mathbf{H}(\hat{\boldsymbol{\theta}}) - \mathbf{H}(\boldsymbol{\theta}^*))'[\mathbf{T}(\hat{\boldsymbol{\theta}})'\mathbf{I}(\hat{\boldsymbol{\theta}})^{-1}\mathbf{T}(\hat{\boldsymbol{\theta}})]^{-1}[\mathbf{H}(\hat{\boldsymbol{\theta}}) - \mathbf{H}(\boldsymbol{\theta}^*)]$
3. Rao test: $\frac{1}{n}\mathbf{V}(\boldsymbol{\theta}^*)'\mathbf{I}(\boldsymbol{\theta}^*)^{-1}\mathbf{V}(\boldsymbol{\theta}^*)$
4. Lagrange multiplier (Kuhn–Tucker) test:

$$\frac{1}{n}\boldsymbol{\lambda}^{*\prime}\mathbf{T}(\boldsymbol{\theta}^*)'\mathbf{I}(\boldsymbol{\theta}^*)^{-1}\mathbf{T}(\boldsymbol{\theta}^*)\boldsymbol{\lambda}^*$$

where $\hat{\boldsymbol{\theta}}$ is the unrestricted MLE.

If one has a normal likelihood with known covariance structure and linear constraint functions on the mean vector, **r**, the four tests are again identical for both testing situations and possess chi-bar-square distributions.

Wollan (1985) has shown that under suitable regularity conditions, all four of these tests have asymptotic chi-bar-square distributions in either testing situation under the condition that $h_i(\boldsymbol{\theta}) = 0$ for all i for the true parameter vector $\boldsymbol{\theta}$.

Of course, very little is known regarding the form of the weights of these asymptotic distributions and so at the current time they have limited application.

There is some speculation (Buse, 1982) that these tests may in general be improved by replacing the information matrix by the Hessian of the log likelihood divided by n. There is some evidence that this is beneficial in the equality constraint case (Efron and Hinkley, 1978).

Some work comparing these tests has been done in a linear regression model context. To elaborate, suppose one has the model

$$\mathbf{Y} = \mathbf{X}\boldsymbol{\beta} + \boldsymbol{\varepsilon}, \qquad \boldsymbol{\varepsilon} \sim N(\mathbf{0}, \boldsymbol{\Omega})$$

where $\mathbf{Y}$ is an $n \times 1$ vector of observations, $\mathbf{X}$ is a known $n \times k$ design matrix, and $\boldsymbol{\beta}$ is a $k \times 1$ vector of parameters where $\boldsymbol{\Omega}$ depends upon a finite number of parameters independent of $\boldsymbol{\beta}$. Let $\mathbf{R}$ be an $m \times k$ matrix of rank m and $\mathbf{r}$ an $m \times 1$

vector. Suppose one wishes to test as a null hypothesis

$$H_0 : \mathbf{R}\boldsymbol{\beta} = \mathbf{r}$$

against the alternative

$$H_1 : \mathbf{R}\boldsymbol{\beta} \leqslant \mathbf{r}.$$

In this event, the Lagrange multiplier (Kuhn–Tucker) test is identical to Rao's test. Gourieroux, Holly, and Monfort (1982) have shown that the three test statistics are actually ordered in the following manner:

$$Q_{\mathrm{LM}} \leqslant Q_{\mathrm{LR}} \leqslant Q_{\mathrm{W}}$$

even though they have the same asymptotic distribution.

In the event that $\mathbf{\Omega} = \sigma^2\mathbf{I}$, Farebrother (1986) has shown that these tests are equivalent (though not identical) and related their exact null hypothesis distributions.

6.9 COMPLEMENTS

The notion of duality and dual operations for order restricted inference problems seems to be of rather recent origin, with most of the work in the area occurring since 1970. However, there appears to be an appreciation developing for the elegance and power of these approaches, and it appears likely that there will be further results in the future.

Of course the least squares duality theorems for linear subspaces has existed for a long time, and has served as a cornerstone for a substantial portion of linear model theory for many years. However, the realization that many of these duality results in statistical inference problems can be extended to cones (and hence order restrictions) seems to have been a fairly recent observation.

The duality structure inherent when dealing with normal distributions has proven useful on many occasions. Not only has it been possible to restate and approach problems from a different tack, but results which relate the weights associated with the chi-bar-square distributions for dual problems have been found. An overview of some of these types of results is given in Shapiro (1988).

Certainly, much of the general theory of convex sets and convex functions goes back much further. Many of the initial developments were done around the turn of the century, chiefly by Minkowski (1910, 1911). In particular, the early separation theorems were investigated by him. An appreciation of the importance of the convex conjugate and a development of much of the theory involving it is attributed to Fenchel (1949). We are not sure who first noted the elegant results that occur when the Fenchel duality theorem is applied to the situation where one function is an arbitrary convex function and the other is a zero–negative infinity indicator function over a convex set. However, a statement and proof of this result is given in Rockafellar (1970). Steinitz (1913) noted the importance of dual, convex cones at an early date.

The importance of dual cones in statistical problems seems to have been noted when it was observed that the dual cone to the simple linear ordering cone is closely connected with the concept of stochastic ordering. This correspondence was used in Barlow *et al.* (1972) to find maximum likelihood estimators of two stochastically ordered distributions.

Fenchel duality was recognized as being very important in order restricted statistical inference problems in an important paper by Barlow and Brunk (1972). In particular, in this paper they defined the concept of 'generalized isotonic regression', and used the Fenchel duality theorem to identify least squares projections onto isotonic cones as solutions of many general types of problems. The Barlow and Brunk duality results were extensively used by Robertson and Wright (1981) for construction of likelihood ratio tests when testing stochastic ordering hypotheses in multinomial problems.

Maximum likelihood problems in log-linear models has been an important topic in statistics during the last two decades. Methods for solving these problems, such as the 'iterated proportional fitting procedure', are discussed, for example, in Bishop, Feinberg, and Holland (1975) and have received a great deal of attention. Other procedures are discussed in Haberman (1974) and Goodman (1985). What is interesting is that some of these procedures are actually based on solving a dual problem (in terms of the Fenchel duality theorem stated in Theorem 6.2.1). This duality approach converts many multinomial-type maximum likelihood problems whose constraint region is a direct sum of subspaces into an optimization problem where the constraint region is an intersection of subspaces and hence better suited to iterative procedures. The value of this correspondence becomes even more important when considering log-linear model problems that involve order restrictions. The material in this chapter that involves duality for log-linear models with order restrictions seems to be new.

Of course, the Lagrange multiplier tests are innately connected with Lagrangian duality. Work in this area seems to largely have been centered in the econometrics literature where an active investigation has been taking place. Sources of work in this area are Gourieroux, Holly, and Monfort (1982), Farebrother (1986), and Wolak (1987).

CHAPTER 7

Inferences Regarding Distributions Subject to 'Shape' Restrictions

7.1 INTRODUCTION

The problem of estimating a distribution is fundamental to statistics and has been widely researched from a number of points of view. One familiar approach is to assume that the distribution belongs to a parametric family, such as the normal family, which is indexed by a low-dimensional vector of parameters, and then use the data to estimate the parameter values. Parametric modeling is generally an efficient approach to the problem of estimating a distribution but can be misleading if the true distribution is not a member of the parametric family. On the other extreme, the empirical distribution function (EDF) is a convenient and natural nonparametric estimator of the underlying distribution. The obvious difficulty with the EDF is that it is discrete and, in the absence of further assumptions about the distribution, it does not provide information about the underlying distribution except at its observation points.

The statistical literature contains numerous attempts to find good estimates of a distribution which lie somewhere between a low-dimensional parametric approach and the completely nonparametric approach based upon the EDF. One of the best known and oldest approaches uses the *Pearson system of generalized frequency curves.* Pearson generated twelve families of parametric frequency functions by solving a differential equation. This differential equation was motivated by a family of frequency polygons associated with the hypergeometric series. Pearson's large family of frequency functions was intended to contain a member which would agree rather closely with any observational frequency function. The idea was to use prior information about the shape of the population distribution or to use the sample moments of the data, to select one of the twelve families of frequency curves, and then to use the method of moments to fit a particular member of that family to the data. The idea has great appeal but the difficulties involved in the numerical computations were clearly indicated in Pearson's original work.

With the exception of the EDF, all of the methods which have been proposed for estimating a distribution attempt to make use of prior information regarding the *shape* or *smoothness* of the underlying distribution. This includes kernel-type estimates, orthogonal series estimates, histograms, histosplines, and penalized likelihood estimates. The literature on these estimates is too vast to try to summarize here. For more information on the subject the reader could consult the books by Tapia and Thompson (1978), Prakasa Rao (1983), or Devroye and Györfi (1985). The first two give reviews of the literature. The review of the latter two books by Silverman (1985) give a nice brief overview of the subject of density estimation.

The problem of nonparametric density estimation is much harder than the problem of nonparametric distribution function estimation. This is related to the fact that the density may be changed considerably over a short range without affecting the probability substantially. The *histogram* is one of the oldest and most useful solutions to this problem. The histogram estimates densities without making any parametric assumptions, just as the EDF estimates distributions nonparametrically. One difficulty with histograms is in the choice of the intervals over which it is constant, and those intervals are often chosen in an arbitrary fashion. One way of attacking this defect is by assuming additional smoothness conditions such as unimodality or decreasing density and obtaining the MLE. The MLE of the density under these assumptions turns out to be a histogram where the intervals have endpoints depending on the data. They can also be viewed as isotonic estimators with respect to a '*naive*' *estimator* based on the EDF. Properties of isotonic estimators provide useful tools for developing properties of the MLE in each case. In Sections 7.2 and 7.3 the maximum likelihood estimation problem is considered for *unimodal and decreasing densities*.

In Section 7.4 distributions with *monotone failure rate* are discussed. The MLE for the failure rate function again turns out to be an isotonic estimator with respect to the natural actuarial estimator of the failure rate which in turn is based upon the EDF.

In Section 7.5 an ingenious attempt to improve on the MLEs of Sections 7.2, 7.3, and 7.4 is discussed. The MLE suggests a promising class of so-called *isotonized* '*window estimators*'. By proper choice of the windows the MLE can be improved upon asymptotically.

Section 7.6 contains a brief introduction to an important testing problem: tests for *exponentiality against a monotone failure rate*. This problem has many connections with the problem of testing homogeneity of a collection of parameters against a monotone alternative. A more extensive discussion of restricted inferences regarding failure rates can be found in Barlow *et al.* (1972) and Barlow and Proschan (1975).

All of the methods for inference about a distribution discussed in this chapter make use of prior information regarding the 'shape' of the distribution. This prior information may be modeled in terms of the distribution function, the density or frequency function, or the failure rate.

7.2 ESTIMATION OF MONOTONE DENSITIES

One of the difficulties with nonparametric maximum likelihood estimation of a probability density function (PDF) is the problem of '*spiking*'; i.e. unless some assumption is made about the shape of the density, the likelihood function can be made infinitely large by making the density large in a small neighborhood around the observation points (i.e. 'spiking' the density). One solution to this problem is to *penalize* the estimate for a lack of smoothness (cf. Good and Gaskins, 1971, 1972). Another approach is to restrict the estimate to some family of distributions which have the property that if one makes the estimate large at an observation point then one must make the estimate large over a range of values. The fact that $\int f(x)\,dx = 1$ will then preclude 'spiking'.

Suppose $0 < y_1 < y_2 < \cdots < y_n$ are the order statistics of a random sample of size n from a population with distribution function F. Assume that F belongs to the family, $\mathscr{D}$, of all distributions on $[0, \infty)$ having densities which are nonincreasing on $[0, \infty)$. Note that for any distribution $F \in \mathscr{D}$ having density f, if one makes f large at y_i then one must also make f large over the interval $[0, y_i]$. This property of members of $\mathscr{D}$ precludes the problem of spiking and implies that if one restricts the estimate to belong to $\mathscr{D}$ then a MLE exists.

The first step in finding the MLE is to observe that the restricted estimate must be constant on the intervals (y_{i-1}, y_i), $i = 1, 2, \ldots, n (y_0 = 0)$, must be continuous from the left on $(0, \infty)$ (and thus constant on $(y_{i-1}, y_i])$, and must be zero on (y_n, ∞). All of these properties of the estimate follow by noting that if f does not have the property then one can construct another member of $\mathscr{D}$ which has a larger likelihood. For example, if f is not constant on the intervals (y_{i-1}, y_i) then one can replace f on $(y_{i-1}, y_i]$ by $(y_i - y_{i-1})^{-1} \int_{y_{i-1}}^{y_i} f(t)\,dt$. This new density is constant on (y_{i-1}, y_i) and has a larger likelihood. The left continuity and the fact that f must be zero on (y_n, ∞) follow similarly.

Hence, the problem of finding the distribution in $\mathscr{D}$ which maximizes the likelihood, $\prod_{i=1}^{n} f(y_i)$, reduces to the problem of finding the values, f_i, of the density on the n intervals, $(y_{i-1}, y_i]$, $i = 1, 2, \ldots, n$. In other words, one wants to maximize $\prod_{i=1}^{n} f_i$ subject to

$$\text{(a)}\ f_1 \geqslant f_2 \geqslant \cdots \geqslant f_n \qquad \text{and} \qquad \text{(b)}\ \sum_{i=1}^{n} (y_i - y_{i-1}) f_i = 1.$$

This is one of the problems which was solved in Example 1.5.7. If we let $X = \{1, 2, \ldots, k\}$ with the usual ordering and $\mathscr{A}$ be the collection of antitonic functions on X (i.e. satisfying (a)) then the solution is given by the antitonic regression, $P_{\mathbf{w}}(\mathbf{g} | \mathscr{A})$, of $\mathbf{g} = (g_1, g_2, \ldots, g_k)$ with weights $\mathbf{w} = (w_1, w_2, \ldots, w_k)$,

where

$$g_i = \frac{1}{n(y_i - y_{i-1})} \qquad \text{and} \qquad w_i = (y_i - y_{i-1}) \qquad \text{for } i = 1, 2, \ldots, k.$$

(Note that a common factor of n has been dropped from the expression for the weights. Such a factor has no effect on the antitonic regression.)

There are several interesting ways to view the solution, $\hat{f}$. If we let F_n be the empirical distribution, that is

$$F_n(x) = \begin{cases} 0, & x < y_1 \\ \dfrac{i}{n}, & y_i \leqslant x < y_{i+1}, \quad 1 \leqslant i \leqslant n-1 \\ 1, & x \geqslant y_n, \end{cases}$$

then it follows from Theorem 1.4.4 that

$$\hat{f}_i = \min_{s \leqslant i-1} \max_{t \geqslant i} \frac{F_n(y_t) - F_n(y_s)}{y_t - y_s}.$$

It follows that the MLE for the density of $F \in \mathscr{D}$ is given by

$$\hat{f}_n(x) = \begin{cases} \hat{f}_i, & y_{i-1} < x \leqslant y_i, \quad i = 1, 2, \ldots, n \\ 0, & \text{otherwise.} \end{cases}$$

Note that $\hat{f}_n$ is left continuous.

Geometrically, the solution $\hat{f}_i$ is the left slope of the least concave majorant of F_n at y_i. Figure 7.2.1 illustrates the solution. If the data are as in Figure 7.2.1 then

$$\hat{f}_n(x) = \begin{cases} \dfrac{2}{5y_2}, & 0 < x \leqslant y_2 \\ \dfrac{3}{5(y_5 - y_2)}, & y_2 < x \leqslant y_5 \\ 0, & \text{otherwise.} \end{cases}$$

Using the extended notion of isotonic regression developed in Chapter 8 the solution can also be represented as the isotonic regression of the following histogram-type estimator:

$$f_n(x) = \begin{cases} \dfrac{1}{n(y_i - y_{i-1})}, & y_{i-1} < x \leqslant y_i, \quad i = 1, 2, \ldots, n \\ 0, & \text{otherwise.} \end{cases}$$

Note that f_n is a density on $[0, \infty)$ which is a, possibly very ragged, histogram-type estimate of the underlying density. The estimator f_n is sometimes called a '*naive*' *estimator* of f. It is not a good estimator in that it is not even consistent.

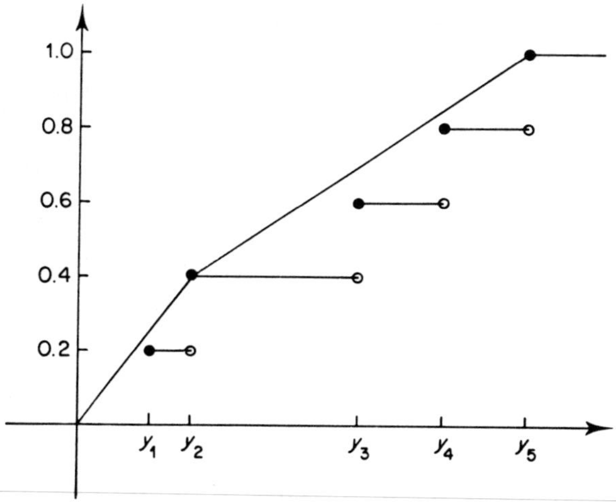

Figure 7.2.1 Empirical CDE and least concave majorant

The estimator $\hat{f}_n$ is sometimes referred to as a smoothed version of f_n. The estimate $\hat{f}_n$ can be badly spiked close to 0. This causes some technical difficulties and a discussion of possible solutions to these difficulties will be given later.

Consistency

Let $\hat{F}_n$ be the distribution function corresponding to the density estimate $\hat{f}_n$. Keep in mind that $\hat{F}_n$ is the least concave majorant of the EDF, F_n, on $[0, \infty)$. One approach to showing the consistency of $\hat{f}_n$ and $\hat{F}_n$ would be to try to verify the regularity conditions which imply the strong consistency of approximate maximum likelihood estimators. Such regularity conditions have been investigated by many researchers, notably Wald (1949). The classical regularity conditions cannot be verified in this problem because the *log-odds ratio* is not locally dominated. Wang (1985) develops new conditions which imply strong consistency in this problem.

We take a more direct approach and show that, in a sense, $\hat{F}_n$ is closer to the underlying distribution function, F, than is F_n so that consistency properties of $\hat{F}_n$ will follow from consistency properties of F_n. The well-known Glivenko–Cantelli theorem states that

$$P\left[\lim_{n\to\infty} \sup_x |F_n(x) - F(x)| = 0\right] = 1. \tag{7.2.1}$$

A stronger but less familiar result due to Smirnov (cf. Csaki, 1968) is a *law of the iterated logarithm for this sup-distance.* Specifically,

$$P\left[\limsup_{n\to\infty} (n/\log\log n)^{1/2} \sup_x |F_n(x) - F(x)| = 2^{-1/2}\right] = 1. \qquad (7.2.2)$$

Let $\varepsilon_n = \sup_x |F_n(x) - F(x)|$. Then the function $F + \varepsilon_n$ is concave on $[0, \infty)$ and majorizes F_n. Thus, for all $x \in [0, \infty)$, $F_n(x) \leqslant \hat{F}_n(x) \leqslant F(x) + \varepsilon_n$ so that $-\varepsilon_n \leqslant F_n(x) - F(x) \leqslant \hat{F}_n(x) - F(x) \leqslant \varepsilon_n$ and

$$\sup_x |\hat{F}_n(x) - F(x)| \leqslant \sup_x |F_n(x) - F(x)|. \qquad (7.2.3)$$

The inequality (7.2.3) is called *Marshall's lemma.* It can be thought of as a 'reduction of error' property analogous to the results in Section 1.6.

Theorem 7.2.1 *Suppose F_n is the EDF of a random sample of size n from a continuous population with distribution function F such that $F(0) = 0$ and F is concave down on $[0, \infty)$. If $\hat{F}_n$ is the LCM of F_n on $[0, \infty)$ then*

$$P\left[\limsup_{n\to\infty} (n/\log\log n)^{1/2} \sup_x |\hat{F}_n(x) - F(x)| = 2^{-1/2}\right] = 1. \qquad (7.2.4)$$

Proof: Using (7.2.2) and (7.2.3) one obtains

$$P\left[\limsup_{n\to\infty} (n/\log\log n)^{1/2} \sup_x |\hat{F}_n(x) - F(x)| \leqslant 2^{-1/2}\right] = 1.$$

Now choose x_0 such that $F(x_0) = \frac{1}{2}$. Using Bernoulli's law of the iterated logarithm for $F_n(x_0)$ together with the fact that $\hat{F}_n(x_0) \geqslant F_n(x_0)$ ($\hat{F}$ is the LCM of F_n), we conclude that

$$\begin{aligned} &P\left[\limsup_{n\to\infty} (n/\log\log n)^{1/2} \sup_x |\hat{F}_n(x) - F(x)| \geqslant 2^{-1/2}\right] \\ &\geqslant P\left[\limsup_{n\to\infty} (n/\log\log n)^{1/2} [\hat{F}_n(x_0) - F(x_0)] \geqslant 2^{-1/2}\right] \\ &\geqslant P\left[\limsup_{n\to\infty} (n/\log\log n)^{1/2} [F_n(x_0) - F(x_0)] \geqslant 2^{-1/2}\right] = 1. \quad \square \qquad (7.2.5) \end{aligned}$$

Note that (7.2.4) implies that if a_n is any sequence such that $a_n = o((n/\log\log n)^{1/2})$ then

$$P\left[\lim_{n\to\infty} a_n \sup_x |\hat{F}_n(x) - F(x)| = 0\right] = 1.$$

In particular this result implies a Glivenko–Cantelli-type result for $\hat{F}_n$ ($a_n = 1$) and holds for $a_n = n^\delta$ where $0 < \delta < \frac{1}{2}$.

One can use this Glivenko–Cantelli-type result to prove a strong consistency result for the density estimate, $\hat{f}_n$.

Lemma *If H and H_n, $n = 1, 2, \ldots$, are defined and concave on an open interval I and if*

$$\lim_{n\to\infty} H_n(x) = H(x)$$

uniformly in $x \in I$ then

$$H^-(x) \geqslant \overline{\lim_{n\to\infty}}\ H_n^-(x) \geqslant \underline{\lim_{n\to\infty}}\ H_n^+(x) \geqslant H^+(x) \tag{7.2.6}$$

for all $x \in I$. One uses $H^-(H^+)$ to denote the left (right) derivative of H.

Proof: Since H_n is concave, it follows that for all $x \in I$ and for all sufficiently small $\Delta > 0$,

$$\frac{H_n(x) - H_n(x-\Delta)}{\Delta} \geqslant H_n^-(x) \geqslant H_n^+(x) \geqslant \frac{H_n(x+\Delta) - H_n(x)}{\Delta}.$$

Letting $n \to \infty$ and Δ decrease to zero, one can obtain (7.2.6). □

Applying this lemma with $H_n(x) = F_n(x)$ and the fact that $F_n(x) \to F(x)$ uniformly over R with probability one, gives the following result.

Theorem 7.2.2 *Under the conditions of Theorem 7.2.1, if $\hat{f}_n$ is the maximum likelihood estimate of the density f (that is $\hat{f}_n$ is the left derivative of $\hat{F}_n$ on $(0, \infty)$) then, for $x > 0$,*

$$f(x-) \geqslant \overline{\lim}\ \hat{f}_n(x-) \geqslant \lim \hat{f}_n(x+) \geqslant f(x+)$$

for all $x > 0$ with probability one.

Prakasa Rao (1969) has shown, under regularity conditions, that

$$n^{1/3}[\tfrac{1}{2}f(x)f'(x)]^{-1/3}[\hat{f}_n(x) - f(x)]$$

has a limiting distribution which is related to the *heat equation*. This result is somewhat disappointing in that the convergence rate is like $n^{-1/3}$ rather than $n^{-1/2}$, which is the case in many other problems in statistics. This convergence rate can be improved by smoothing over wider bands of observation points. A discussion of these *isotonic window estimators* will be presented in Section 7.5. First we mention three important studies concerning the asymptotic properties of $\hat{F}_n$ and $\hat{f}_n$.

Kiefer and Wolfowitz (1976) study the sup-norm distance between $\hat{F}_n$ and F_n. Under certain assumptions on F they argue that for large n, $\hat{F}_n$ and F_n are quite close so that these two estimators of F have similar asymptotic properties. A statement of their two principal theorems are given without proof. The reader is reminded that F_n is the EDF of a random sample from a continuous population with distribution function F such that $F(0)=0$ and F is concave on $[0,\infty)$.

Theorem 7.2.3 *If*

(*a*) $\alpha_1(F)=\inf\{x:F(x)=1\}<\infty$,

(*b*) *F is twice continuously differentiable on* $(0,\alpha_1(F))$,

(*c*) $\gamma(F)=\dfrac{\sup_{0<x<\alpha_1(F)}|f'(x)|}{\inf_{0<x<\alpha_1(F)}f^2(x)}<\infty$,

and

(*d*) $\beta(F)=\inf\limits_{0<x<\alpha_1(F)}\left|\dfrac{-f'(x)}{f^2(x)}\right|>0$

then for sufficiently large n (depending only on $\beta(F)$ *and* $\gamma(F)$*)*

$$P_F\left[\sup_x|\hat{F}_n(x)-F_n(x)|>n^{-2/3}(\log n)^{5/6}\right]<2n^{-2},$$

so that

$$P_F\left[\lim_{n\to\infty}\left(\frac{n^{2/3}}{\log n}\right)\sup_x|\hat{F}_n(x)-F_n(x)|=0\right]=1.$$

Note that this result combined with Smirnov's law of the iterated logarithm (7.2.2) yields Theorem 7.2.1 under the conditions stated here.

The second result of Keifer and Wolfowitz gives the *asymptotic minimax character of* $\hat{F}_n$ as an estimate of F. Let $\mathscr{F}$ be the collection of distribution functions F such that $F(0)=0$, F being continuous and concave on $[0,\infty)$. Suppose W is any nondecreasing function on the nonnegative reals which is not identically zero and $\int_0^\infty W(t)t\,e^{-2t^2}\,dt<\infty$. A nonrandomized estimator G_n of F has the risk function

$$r_n(F;G_n)=E_F\left[W(n^{1/2}\sup_x|G_n(x)-F(x)|)\right].$$

(Note that Marshall's lemma holds for all concave distribution functions and not just F.) Using Marshall's lemma and the asymptotic minimax properties of F_n, Kiefer and Wolfowitz show the following theorem.

Theorem 7.2.4 *Under the above assumptions*

$$\lim_{n\to\infty}\frac{\sup_{F\in\mathscr{F}}r_n(F;\hat{F}_n)}{\inf_{G_n\in\mathscr{F}}\sup_{F\in\mathscr{F}}r_n(F;G_n)}=1.$$

As noted earlier, Prakasa Rao (1969) found the *asymptotic distribution* for $\hat{f}_n$ at a particular point. This result is contained in the following theorem.

Theorem 7.2.5 *Assume that f is a decreasing density on $[0, \infty)$ and has a nonzero derivative $f'(t)$ at a point $t \in (0, \infty)$. If $\hat{f}_n$ is the restricted MLE of f then*

$$n^{1/3} |\tfrac{1}{2} f(t) f'(t)|^{-1/3} [\hat{f}_n(t) - f(t)] \to 2Z,$$

where Z is distributed as the location of the maximum of the process $\{W(u) - u^2 : u \in (-\infty, \infty)\}$ with W the standard Wiener process on $(-\infty, \infty)$ originating from zero (that is $W(0) = 0$).

Groeneboom and Pyke (1983) and Groeneboom (1985) study the *asymptotic distributions* of statistics which are more global measures of the discrepancy between $\hat{f}_n$ and f. The L_1 distance between $\hat{f}_n$ and f, when f is concentrated on a bounded interval, is studied in Groeneboom (1985) and the L_2 distance is studied in Groeneboom and Pyke (1983). The following result is contained in Groeneboom and Pyke (1983). Suppose F_n is the EDF of a random sample of size n from a uniform distribution on $[0, 1]$. Let

$$L_n = \| \hat{f}_n - 1 \|_2^2 = \int_0^1 [\hat{f}_n(x) - 1]^2 \, dx.$$

Theorem 7.2.6 *As $n \to \infty$, $(3 \log n)^{-1/2} [nL_n - n - \log n]$ converges in distribution to the standard normal distribution.*

7.3 ESTIMATION OF UNIMODAL DENSITIES

Suppose $y_1 < y_2 < \cdots < y_n$ are the ordered items of a random sample of size n from a continuous population with distribution function F and density f. We assume that f is *unimodal* in the sense that there exists a number $M \in (-\infty, \infty)$, called a mode of f, such that f is nondecreasing on $(-\infty, M]$ and nonincreasing on $[M, \infty)$. The term unimodal is not quite appropriate for a distribution with such a density as it could, in fact, have more than one mode. If there exist two modes, then any point in between is also a mode so that in this case there exists an interval of values which we would call a *modal interval* of f. In such a case the height of f at each of these modes must be the same. For the lack of a better word, both the distribution function, F, and the density function, f, will be said to be unimodal at M when f has this property.

If F is unimodal with mode M then F must be convex on $(-\infty, M]$ and concave on $[M, \infty)$ because its derivative, f, is increasing and decreasing, respectively, on these intervals. Conversely, if F is a distribution function and if F is convex on $(-\infty, M]$ and concave on $[M, \infty)$, then F must be left and right differentiable (and thus continuous) on $(-\infty, M) \cup (M, \infty)$. Technically, the function F could have a discontinuity at M. However, it is assumed that the distribution is continuous and thus this possibility is ruled out. Therefore any determination of the derivative of F on $(-\infty, M)\cup(M,\infty)$ would be unimodal at M.

Now consider the problem of finding maximum likelihood estimates of F and f subject to the restriction that they are unimodal. As before, the problem of spiking must be dealt with, which is only a problem at the mode M itself because if we make the estimate large at a point away from the mode it must be made large over the interval joining this point to the mode.

The known mode case

First consider the case where the value of the mode M is known. We deal with the problem of spiking in this case by assuming that none of the observations is equal to M. Thus, making the density estimate large at an observation makes it large over an interval of values and this precludes spiking. The assumption is justified by observing that since the population is continuous, the event that all observations are different from M has probability one. Suppose a is such that $y_a < M < y_{a+1}$. As in the previous section, it is easy to see that our density estimate must be zero on $(-\infty, y_1)\cup(y_n,\infty)$ and constant on each of the intervals $[y_{j-1}, y_j)$ for $j=2,\ldots,a$, $[y_a, M)$, $(M, y_{a+1}]$ and on each of the intervals $(y_j, y_{j+1}]$ for $j=a+1,\ldots,n-1$. The problem then reduces to finding these $n+1$ nonzero values of our density estimate. (Of course, if all observations are to the left or right of M we only have n nonzero values. This problem was solved in the previous section.) If one arranges these $n+1$ intervals so that their endpoints are increasing and if we let $f_1, f_2,\ldots, f_{n+1}$ be the value of a generic unimodal density on these $n+1$ intervals, then the problem is to find maximum $\prod_{i=1}^{n+1} f_i$ subject to

$$f_1 \leqslant f_2 \leqslant \cdots \leqslant f_a, f_{a+1} \geqslant \cdots \geqslant f_{n+1}$$

and

$$\sum_{j=1}^{a-1} f_j(y_{j+1}-y_j) + f_a(M-y_a) + f_{a+1}(y_{a+1}-M) + \sum_{j=a+2}^{n+1} f_j(y_j - y_{j-1}) = 1.$$

As in the problem discussed in the preceding section, this problem was solved in Example 1.5.7. If we let $X=\{1,2,\ldots,n+1\}$ and define the partial order, $\lesssim$, on X by $1\lesssim 2\lesssim\cdots\lesssim a$, $a+1\gtrsim\cdots\gtrsim n+1$ then the solution is the isotonic regression, with respect to $\lesssim$, of the function $(g_1, g_2,\ldots, g_{n+1})$ with weights $(w_1, w_2,\ldots, w_{n+1})$ where

$$w_i = (y_{i+1}-y_i);\quad i=1,2,\ldots,a-1,\quad w_a = M - y_a,$$
$$w_{a+1} = y_{a+1} - M,\quad w_i = y_i - y_{i-1};\quad i=a+2,\ldots,n+1$$

and $g_i = (nw_i)^{-1}$ for $i=1,2,\ldots,n+1$.

Again there are several interesting ways to view the solution. If F_n is the EDF then the solution distribution function, $\hat{F}_n$, is the greatest convex minorant of F_n on $(-\infty, M]$ and the least concave majorant of F_n on $[M,\infty)$. The corresponding density estimate, $\hat{f}_n$, is the right derivative of $\hat{F}_n$ on $(-\infty, M)$ and the left derivative of $\hat{F}_n$ on (M, ∞). A second representation for $\hat{f}_n$ is as an isotonic regression in the extended notion of that concept developed in Chapter 8 of a naive density estimator whose value on each of these $n+1$ intervals is the reciprocal of n multiplied by the length of the interval. The appropriate partial order $\lesssim$ is defined on $(-\infty, \infty)$ by defining $x \lesssim y$ if $x \leqslant y \leqslant M$ or if $x \geqslant y \geqslant M$. Asymptotic properties of $\hat{F}_n$ and $\hat{f}_n$, as estimators of F and f, analogous to Theorem 7.2.3 were developed in Robertson (1967) and Prakasa Rao (1969).

Unknown mode

A more interesting problem, both from a mathematical and a practical point of view, is the problem of estimating a unimodal distribution when the value of the mode is unknown. Here the problem of spiking becomes more serious because it is possible to take the mode of the estimate at one of the observations. This value of the estimate can then be 'spiked' at that observation so that the likelihood product is unbounded and technically a MLE does not exist. The problem is complicated even further as is seen in the following observation. Suppose g is any unimodal density with mode M such that $y_a < M < y_{a+1}$. If one constructs a new unimodal density by replacing g on $[y_a, M)$ and $[M, y_{a+1})$ by $(M - y_a)^{-1}\int_{[y_a,M]} g(x)\,dx$ and $(y_{a+1} - M)^{-1}\int_{[M,y_{a+1}]} g(x)\,dx$, respectively, then this new density is unimodal with modal interval $[y_a, M)$ or $[M, y_{a+1})$ and has a likelihood product at least as large as that of g. Thus, in a sense, the MLE must be selected from the collection of all unimodal densities which have modes at one of the observations, making it difficult to avoid the problem of spiking.

One approach to this difficulty is to modify the likelihood by ignoring the value of an estimate at its mode in the likelihood. In other words, if f is a candidate for the MLE and if one is calling M the mode of f then the 'likelihood' associated with f would be $\prod_{y_i \neq M} f(y_i)$. A 'MLE' could then be found by computing the 'MLE' with mode at each of the observations using the technique described in the known mode case and then picking the one of these n estimates which has the largest modified likelihood.

One advantage to this approach is that it yields simultaneous estimates of the mode and the density. This is in contrast to other approaches which first find a *direct estimate of the mode* such as the ones proposed by Chernoff (1964) and Venter (1967) and then an indirect estimate of the density, say using the techniques described in the known mode case. Another possibility is to first construct an estimate of the density and then use it to construct an *indirect estimate of the mode*. The latter approach was suggested by Parzen (1962a) but he had a different notion of the mode of a density in mind, as his density estimate is not necessarily unimodal in the sense used here.

Another way to rationalize the above 'MLE' of a unimodal density would be to place a bound on the height of the density. If the bound is sufficiently large a density estimate would have this bound as its value at its mode. Thus each of the n estimates with modes at the n observations would have the same value at their modes. This has the same effect as neglecting the value of the estimate at the mode in the likelihood product and leads to the same estimate as the one described in the above paragraph.

A related approach was studied by Wegman (1970a, 1970b) which eliminated spiking near the mode, at least partially, by requiring the estimate to have a *modal interval* of length ε, where ε is a fixed positive number. Note that this implies that the values of the estimate are no more than ε^{-1}.

Suppose L and R are fixed points such that $R - L = \varepsilon$ and consider the problem of finding the MLE of f when the estimate is constrained to be constant on the interval $[L, R]$, nondecreasing on $(-\infty, L]$, and nonincreasing on $[R, \infty)$ (i.e. unimodal with a modal interval $[L, R]$). Reasoning as in the previous problems it is seen that the constrained estimate must be constant on the intervals

$$A_1 = [y_1, y_2), \ldots, A_{l(n)} = [y_{l(n)}, L), \quad A_{l(n)+1} = [L, R],$$
$$A_{l(n)+2} = (R, y_{r(n)}], \ldots, A_k = (y_{n-1}, y_n]$$

where $y_1 < y_2 < \cdots < y_n$ are the order statistics of our sample, $y_{l(n)}$ is the largest item less than L, and $y_{r(n)}$ is the smallest item greater than R. The problem is to find the value of the estimate on these k intervals, and again the solution is provided by Example 1.5.7. The algorithm described below is a combination of the maximum upper sets algorithm and the graphical representation given in Theorem 1.2.1.

Define a *naive density estimator* g_n by

$$g_n = \sum_{i=1}^{k} n_i [n\lambda(A_i)]^{-1} I_{A_i} \qquad \left(G_n(x) = \int_{(-\infty, x)} g_n(t)\,\mathrm{d}t \right)$$

where $\lambda(A_i)$ is the length of the interval A_i, I_{A_i} is the indicator of A_i, and n_i is the number of observations in A_i. The MLE of f constrained to be unimodal with modal interval $[L, R]$ can be computed in terms of g_n. The procedure for computing $\hat{f}_n$ from g_n will be used again in the sequel so that it will be convenient to have some notation for it. The notation comes from Chapter 8, where conditional expectations given σ-lattices are discussed in general. Suppose g is a density function on $(-\infty, \infty)$ with corresponding distribution function G. The function $P(g \mid \mathscr{L}[L, R])$ on $(-\infty, \infty)$ is defined as follows. First find the interval $[a, b]$ containing $[L, R]$ such that $[G(b) - G(a)]/(b - a)$ is maximized. On $[a, b]$ the value of $P(g \mid \mathscr{L}[L, R])$ is this ratio. To the left of a, $P(g \mid \mathscr{L}[L, R])$ is the right slope of the GCM of G and to the right of b, $P(g \mid \mathscr{L}[L, R])$ is the left slope of the LCM of G. The estimate $\hat{f}_n$ of f is given by

$$\hat{f}_n = P(g_n \mid \mathscr{L}[L, R])$$

The max–min formulas given in Theorem 1.4.4 provide another representation for $\hat{f}_n$ (cf. Barlow *et al.*, 1972).

The above gives an estimate of a unimodal density when the estimate is constrained to be unimodal and to have a predetermined modal interval of length ε. Now suppose one wishes to estimate a unimodal density and the location of the mode is not known, but in order to avoid spiking, one is willing to accept an estimate constrained to be unimodal and have modal interval of length ε. Wegman (1970a) proves that if $\hat{f}_n$ is the MLE constrained in this way then it may be assumed that one of the endpoints of the modal interval of $\hat{f}_n$ lies in the set $\{y_1, y_2, \ldots, y_n\}$. There are at most $2n$ intervals of the form $[L, R]$ where either L or R lies in this set. The MLE is found by computing these $2n$ estimates using the technique described above and choosing the one which maximizes the likelihood product. An estimate of the mode of f is taken to be the midpoint M_n of the modal interval of $\hat{f}_n$.

Consistency

Because of the constraint that $\hat{f}_n$ is constant over an interval of length ε, this estimate does not converge to f unless, of course, f has a modal interval of length ε. The problem is that the indirect mode estimate, M_n, may not converge to the true mode, M, of f. For each real number ν define the density f^ν by

$$f^\nu = P\left(f \,\middle|\, \mathscr{L}\left[\nu - \frac{\varepsilon}{2}, \nu + \frac{\varepsilon}{2}\right]\right).$$

Now choose ν to maximize

$$\int [\log f^\nu(x)] f(x)\,\mathrm{d}x$$

and call this value ν_0. Under certain regularity conditions on f, Wegman (1970a) shows that M_n converges to ν_0 and that $\hat{f}_n \to P(f \mid \mathscr{L}[\nu_0 - \varepsilon/2, \nu_0 + \varepsilon/2])$.

Another possibility is to use a direct estimate of the mode of f which is consistent and then take one of our procedures for estimating the density using this direct estimate of the mode. Direct estimates of the mode of the density are invariably based upon the fact that observations should tend to cluster and be most dense around the mode M of f. A number of such estimates have been proposed. Chernoff (1964) fixes a length and studies an estimate which is a midpoint of the interval of this length which contains the most observations. Consistency of such an estimate is obtained by assuming the interval length goes to zero as the sample size goes to infinity. Venter (1967) fixes the number of observations and studies estimates of M which are chosen from the smallest interval containing this number of observations. Robertson and Cryer (1974) study an iterative procedure for estimating the mode which also yields a density estimate.

The proof of the following theorem may be found in Wegman (1970a).

Theorem 7.3.1 *Suppose f is a continuous, unimodal density with unique mode M and M_n is a sequence of estimates of M converging to m, not necessarily M. Let*

$$L_n = M_n - \frac{\varepsilon}{2} \quad \text{and} \quad R_n = M_n + \frac{\varepsilon}{2}$$

where $\varepsilon > 0$ and let $\hat{f}_n$ be the MLE of f, described previously, with modal interval $[L_n, R_n]$. If

$$f^* = P\left(f \mid \mathscr{L}\left[m - \frac{\varepsilon}{2}, m + \frac{\varepsilon}{2}\right]\right)$$

then with probability one $\hat{f}_n$ converges to f^* except possibly at L or R or both.

If M_n is one of the direct estimates of M which is strongly consistent then the $\hat{f}_n$ in Theorem 7.3.1 converges to $P(f \mid \mathscr{L}[M - \varepsilon/2, M + \varepsilon/2])$, which should be close to f if ε is small. On the other hand, if M_n is the center of the modal interval of the MLE having modal interval of length ε then M_n converges to the value of v which maximizes $\int [\log f^v(x)] f(x)\,dx$, which may not be equal to M. In this case $\hat{f}_n$ converges to $P(f \mid \mathscr{L}[v_0 - \varepsilon/2, v_0 + \varepsilon/2])$, whose closeness to f depends upon, among other things, the closeness of v_0 to M.

Multivariate unimodal density estimation

The next logical step is to consider the estimation of a multivariate unimodal density. The complexity of the problems increases significantly for multivariate densities and it is not even clear how to define unimodal in this context. The reader has already been reminded of the slow convergence rates of density estimates in general, and these rates diminish even further for multivariate problems. Sager (1986a, b) describes an interesting idea for using restricted univariate density estimates to overcome this '*curse of dimensionality*' when the investigator has information regarding the shape of the underlying density. The interested reader is referred to Sager (1983) for a good overview of research on the problem of estimating multivariate modes and the related problem of estimating restricted densities.

7.4 MAXIMUM LIKELIHOOD ESTIMATION FOR DISTRIBUTIONS WITH MONOTONE FAILURE RATE

In many statistical studies involving failure data, biometric mortality data and actuarial mortality data, the *failure rate*, $r(x) = f(x)/[1 - F(x)]$ for $F(x) < 1$ (corresponding to a lifetime distribution with density f and distribution F), is of prime importance. If F is the life distribution of a person, an aircraft engine, a light bulb, a machine tool, etc., then the failure rate, $r(x)$, describes the way in which the item in question wears out. If the item is aging very rapidly the failure rate function will tend to increase rather steeply. The shape of the failure rate function in the case of equipment suggests when appropriate maintenance actions should be taken. A problem of considerable interest, therefore, is the estimation of the failure rate function from a sample of n independent identically distributed lifetimes. Many observed failure rate function estimators seem first to decrease and then increase (are U-shaped) or simply increase or decrease. The U-shaped case is analogous to the case of unimodal density. For simplicity, attention will be confined to monotone failure rate functions. The U-shaped case can be treated by suitably modifying the monotonic estimators (see, for example, Bray, Crawford, and Proschan, 1967).

Although most physical applications of the failure rate function correspond to distributions with support on $[0, \infty)$, the failure rate function $r(x) = f(x)/[1 - F(x)]$ (assumed right continuous) is well defined for $F(x) < 1$, if F has a density at x. Note that $r(x)\,\mathrm{d}x$ is, heuristically, the conditional probability of failure in $(x, x + \mathrm{d}x)$ given survival time x. The normal distribution, for example, has an increasing failure rate. The reciprocal of its failure rate, called *Mill's ratio*, is sometimes used in applications. The IFR (for *increasing failure rate*) class of distributions is now formally defined. (It is understood that '*increasing*' is written for 'nondecreasing' and '*decreasing*' for 'nonincreasing'.)

Definition 7.4.1 The distribution function F is IFR if $-\log[1 - F(x)]$ is convex on the support of F (i.e. the points of increase of F), an interval contained in $[0, \infty)$.

It can be shown that if F is IFR then F is absolutely continuous except for the possibility of a jump at the right-hand endpoint of its interval of support. As will be seen, the MLE for the failure rate function always jumps to plus infinity at the largest observation, corresponding to a jump in the estimate of the distribution function.

The estimator

Let $0 = y_0 < y_1 < y_2 < \cdots < y_n$ be the order statistics obtained by ordering a random sample from an unknown distribution $F \in \mathscr{F}$, the class of IFR distributions. Since $f(y_n)$ can be chosen arbitrarily large, it is not possible to obtain a maximum likelihood estimator for $F \in \mathscr{F}$ directly by maximizing

$$\prod_{i=1}^{n} f(y_i).$$

Consequently, we first consider the subclass $\mathscr{F}^M$ of distributions $F \in \mathscr{F}$ with corresponding failure rates bounded by M, obtaining

$$\sup_{F \in \mathscr{F}^M} \prod_{i=1}^{n} f(y_i) \leqslant M^n$$

since $f(x) \leqslant r(x) \leqslant M$. As will be seen, there is a unique distribution, $\hat{F}_n^M \in \mathscr{F}^M$ at which the supremum is attained. As $M \to \infty$, $\hat{F}_n^M$ converges in distribution to an estimator $\hat{F}_n \in \mathscr{F}$ which is called the *maximum likelihood estimator* for $F \in \mathscr{F}$.

For convenience, if F is IFR we define $r(x) = \infty$ for all such x that $F(x) = 1$. It is not difficult to see that for any distribution F and any x for which r is finite on $[0, x)$,

$$1 - F(x) = \exp\left[- \int_0^x r(u)\, du \right].$$

Hence, for $F \in \mathscr{F}^M$, the log likelihood $L = L(F)$ is given by

$$L = \sum_{i=1}^n \log r(y_i) - \sum_{i=1}^n \int_0^{y_i} r(u)\, du. \tag{7.4.1}$$

We now show that the maximum likelihood estimator for r is a step function. To do this, let $F \in \mathscr{F}^M$ have failure rate r and let F^* be the distribution with failure rate

$$r^*(x) = \begin{cases} 0, & x < y_1 \\ r(y_i), & y_i \leqslant x < y_{i+1}; \quad i = 1, 2, \ldots, n-1 \\ r(y_n), & x \geqslant y_n. \end{cases} \tag{7.4.2}$$

(Note that r^* is right continuous.) Then $F^* \in \mathscr{F}^M$ and $r(x) \geqslant r^*(x)$, so that

$$- \int_0^{y_i} r(u)\, du \leqslant - \int_0^{y_i} r^*(u)\, du$$

for all i. Hence, $L(F) \leqslant L(F^*)$. Thus, L may be replaced by the function

$$\sum_{i=1}^n \log r(y_i) - \sum_{i=1}^{n-1} (n-i)(y_{i+1} - y_i) r(y_i). \tag{7.4.3}$$

Since we wish to maximize (7.4.3) subject to $r(y_1) \leqslant r(y_2) \leqslant \cdots \leqslant r(y_n) \leqslant M$, clearly $r(y_n) = M$ must be chosen. Let $r(y_i) = r_i$. Then our problem can be stated as

$$\begin{aligned} &\text{maximize} \sum_{i=1}^{n-1} [\log r_i - (n-i)(y_{i+1} - y_i) r_i] \\ &\text{subject to } 0 \leqslant r_1 \leqslant r_2 \leqslant \cdots \leqslant r_{n-1} \leqslant M. \end{aligned} \tag{7.4.4}$$

Since M may be chosen arbitrarily large, (7.4.4) can be identified with the problem solved in Example 1.5.7 (cf. the discussion following (1.5.17)). As was seen in

Example 1.5.7, the isotonic regression solves the Poisson extremum problem:

$$\text{maximize} \sum_x [g(x)\log f(x) - f(x)]w(x)$$
$$\text{subject to } f \text{ isotonic.} \tag{7.4.5}$$

Rewrite (7.4.4) as

$$\text{maximize} \sum_{i=1}^{n-1} \{[(n-i)(y_{i+1}-y_i)]^{-1}\log r_i - r_i\}(n-i)(y_{i+1}-y_i)$$
$$\text{subject to } 0 \leqslant r_1 \leqslant r_2 \leqslant \cdots \leqslant r_{n-1} \leqslant M.$$

We recognize this as (7.4.5) where now $X=\{y_1, y_2, \ldots, y_{n-1}\}$, $w(y_i)=(n-i)(y_{i+1}-y_i)$, $g(y_i)=[(n-i)(y_{i+1}-y_i)]^{-1}$, and $f(y_i)=r_i$. The solution can be expressed as

$$\hat{r}_i = \hat{r}_n(y_i) = \min_{i \leqslant t \leqslant n-1} \max_{1 \leqslant s \leqslant i} \frac{t-s+1}{\sum_{j=s}^{t}(n-j)(y_{j+1}-y_j)}. \tag{7.4.6}$$

The resulting estimator is

$$\hat{r}_n^M(x) = \begin{cases} 0, & x < y_1 \\ \hat{r}_i, & y_{i-1} \leqslant x < y_i; \quad i = 2, 3, \ldots, n \\ M, & x \geqslant y_n, \end{cases} \tag{7.4.7}$$

assuming $M > \max_{1 \leqslant i \leqslant n-1}\{1/[(n-i)(y_{i+1}-y_i)]\}$. The MLE is obtained by letting $M \to \infty$, so that $\hat{r}_n(x) = +\infty$ for $x \geqslant y_n$. From Theorem 1.3.1 (equation (1.3.4)) with r_n the natural step function corresponding to g, it follows that

$$\sum_{i=1}^{n-1} [r_n(y_i) - r(y_i)]^2 w(y_i)$$
$$\geqslant \sum_{i=1}^{n-1} [r_n(y_i) - \hat{r}_n(y_i)]^2 w(y_i) + \sum_{i=1}^{n-1} [\hat{r}_n(y_i) - r(y_i)]^2 w(y_i)$$

where r is the true failure rate. Hence, in a least squares sense, $\hat{r}_n$ *is closer to r than r_n is to r*. Also $\hat{r}_n$ is the closest increasing estimator to r_n in a least squares sense.

The total time on test transformation

The MLE can also be interpreted in terms of the following transform. Let

$$H_F^{-1}(t) = \int_0^{F^{-1}(t)} [1 - F(u)]\,du \qquad \text{for } 0 \leqslant t \leqslant 1. \tag{7.4.8}$$

For F IFR, F is strictly increasing on its support, an interval. It follows that the inverse function F^{-1} is uniquely defined on $[0, 1]$ by $F^{-1}(t) = \inf\{x: F(x) \geqslant t\}$,

$0 < t \leqslant 1$, and $F^{-1}(0) = \sup\{x: F(x) = 0\}$. Note that $F^{-1}(1)$ could be $+\infty$. Also note that $H_F^{-1}(0) = 0$ and

$$H_F^{-1}(1) = \int_0^\infty [1 - F(u)]\,\mathrm{d}u = \mu,$$

where μ is the mean of F. (If F is IFR then $\mu < \infty$.) Hence H_F is a distribution with support on $[0, \mu]$ since H_F^{-1} (the inverse of H_F) is strictly increasing on $[0, 1]$. If $G(x) = 1 - \mathrm{e}^{-x}$ for $x \geqslant 0$, then $H_G^{-1}(t) = t$ and H_G is the uniform distribution on $[0, 1]$. When it is clear from the context which distribution is being transformed, H^{-1} will be written for H_F^{-1}.

Note that

$$\frac{\mathrm{d}}{\mathrm{d}t} H_F^{-1}(t)|_{t=F(x)} = \frac{1 - F(x)}{f(x)} = \frac{1}{r(x)}.$$

If F is IFR, then r is increasing and $H_F^{-1}(t)$ is a concave function of $t \in [0, 1]$ or H_F is a convex distribution on $[F^{-1}(0), \mu]$ where

$$\mu = \int_0^\infty x\,\mathrm{d}F(x).$$

If F is replaced by F_n, then one defines

$$H_n^{-1}\left(\frac{i}{n}\right) \overset{\text{def}}{=} H_{F_n}^{-1}\left(\frac{i}{n}\right) = \int_0^{y_i} [1 - F_n(u)]\,\mathrm{d}u \tag{7.4.9}$$

for $1 \leqslant i \leqslant n$. The function $H_n^{-1}(t)$ for $0 \leqslant t \leqslant 1$ is defined by linear interpolation. For example,

$$H_n^{-1}(t) = ny_1 t \qquad \text{for } 0 \leqslant t \leqslant \frac{1}{n}$$

while

$$H_n^{-1}(t) = \frac{1}{n}[ny_1 + \cdots + (n - i + 1)(y_i - y_{i-1})] + \left(t - \frac{i}{n}\right)(n - i)(y_{i+1} - y_i)$$

for $i/n \leqslant t \leqslant (i + 1)/n$ and $1 \leqslant i \leqslant n - 1$. To clarify these ideas, Figure 7.4.1 is a graph of H_n^{-1} for $n = 3$. Note that H_n^{-1} is continuous and that $H_n^{-1}(1)$ is the sample mean. In general,

$$H_n^{-1}\left(\frac{i}{n}\right) = \frac{1}{n}[ny_1 + (n - 1)(y_2 - y_1) + \cdots + (n - i + 1)(y_i - y_{i-1})]$$

for $1 \leqslant i \leqslant n$. The expression in square brackets can be interpreted as the *total time*

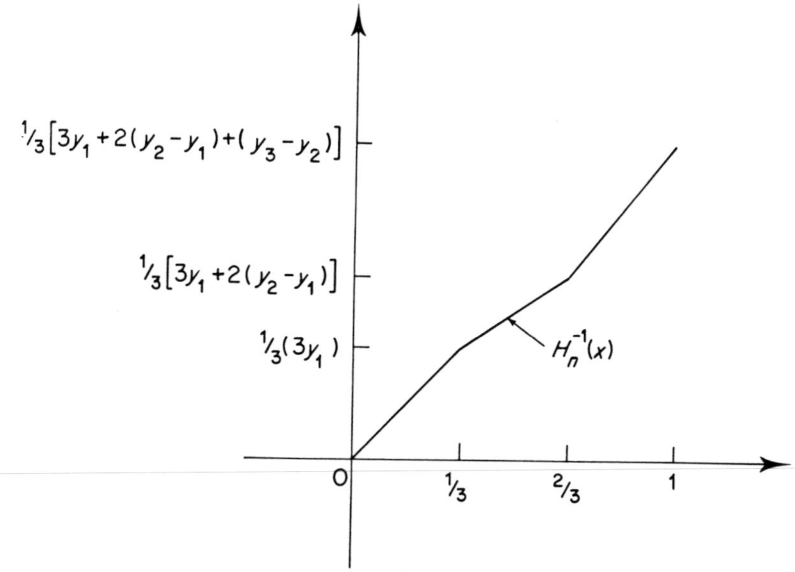

Figure 7.4.1 Graph of H_n^{-1} with three observations

on test up to the ith observation. If n items are placed on life test at time 0, then n items survive up to time y_1, $n-1$ items survive through the interval $[y_1, y_2)$, etc., while $(n-i+1)$ items survive the interval $[y_{i-1}, y_i)$. The total time on test statistic is a fundamental tool in solving life testing problems.

It is easy to verify that

$$H_n^{-1}\left(\frac{t}{n}\right) - H_n^{-1}\left(\frac{s}{n}\right) = \frac{1}{n}\sum_{j=s}^{t-1} (n-j)(y_{j+1} - y_j). \tag{7.4.10}$$

Hence (7.4.6) can be rewritten as

$$\hat{r}_n(y_i) = \min_{t \geq i+1} \max_{s \leq i} \frac{F_n(y_t) - F_n(y_s)}{H_n^{-1}[F_n(y_t)] - H_n^{-1}[F_n(y_s)]} \tag{7.4.11}$$

for $1 \leq i \leq n-1$. It is seen that $1/\hat{r}_n(x)$ is the slope (from the right) of the concave majorant of $\{H_n^{-1}(y): y \geq 1/n\}$ evaluated at the point $y = F_n(x)$ (cf. Mykytyn and Santner, 1981).

Consistency

In the proof of consistency of the MLE, H_n^{-1} plays an analogous role to that played by F_n in the proof of consistency of a monotone density in Section 7.2. In fact both results can be proved at once using the generalization in th next section. However, for clarity in exposition and because of the importanc density and failure rate estimation, both have been treated separately at the expense of conciseness.

Lemma A *If $F(0)=0$, F is strictly increasing on its support, and*

$$\mu = \int_0^\infty x\,\mathrm{d}F(x) < \infty,$$

then $H_n^{-1}(t) \to H^{-1}(t)$ uniformly in $t \in [0,1]$ with probability one.

Proof: By the strong law,

$$H_n^{-1}(1) = n^{-1}\sum_{j=1}^{n} Y_j \xrightarrow{\text{a.s.}} \mu = H^{-1}(1)$$

and clearly $H_n^{-1}(1/n) = Y_1 \xrightarrow{\text{a.s.}} F^{-1}(0) = H^{-1}(0)$. For $t \in (0,1)$,

$$H_n^{-1}\left(\frac{[nt]}{n}\right) = \frac{1}{n}\left[\sum_{i=1}^{[nt]} Y_i + (n-[nt])Y_{[nt]+1}\right] \qquad \text{(the Winsorized mean)}$$

$$\xrightarrow{\text{a.s.}} \int_0^{F^{-1}(t)} u\,\mathrm{d}F(u) + (1-t)F^{-1}(t) = H^{-1}(t)$$

by convergence of percentiles and the strong law of large numbers, where we have used integration by parts in the last step. Since H^{-1} is continuous and nondecreasing and H_n^{-1} is nondecreasing, the uniform convergence follows as in the proof of the Glivenko–Cantelli theorem. □

The above lemma and its proof are modifications of the ones in Barlow *et al.* (1972) and was communicated to us by H. Mukerjee (cf. Mukerjee and Wang, 1987).

In proving consistency of the MLE for $r(x)$, it is convenient to consider the estimator for $1/r(x)$. The basic or '*naive*' estimate for $1/r(x)$ is

$$\frac{1}{r_n(x)} = \frac{1-F_n(y_i)}{f_n(y_i)} \qquad \text{for } y_i \leqslant x < y_{i+1} \quad (1 \leqslant i < n-1)$$

or

$$\frac{1}{r_n(x)} = \frac{[1-(i/n)][y_{i+1}-y_i]}{F_n(y_{i+1})-F_n(y_i)} \qquad \text{for } y_i \leqslant x < y_{i+1}.$$

If we let

$$g_n(i) = \frac{1}{r_n(y_i)} \qquad \text{and} \qquad w_i = F_n(y_{i+1}) - F_n(y_i),$$

the cumulative sum function is

$$G_n(i)=\sum_{j=1}^{i} g_n(j)w_j=\sum_{j=1}^{i}\frac{1}{n}(n-j)(y_{j+1}-y_j)=H_n^{-1}\left(\frac{i}{n}\right).$$

Hence the transform, H_n^{-1} can be interpreted as the cumulative sum function with respect to the maximum likelihood estimator for $1/r$.

Lemma B *If F is IFR so that*

$$\int_0^\infty x\,\mathrm{d}F(x)<\infty$$

and $\hat{H}_n^{-1}$ is the least concave majorant of H_n^{-1}, then

$$\hat{H}_n^{-1}(t)\to H^{-1}(t)$$

uniformly in $t\in[0,1]$ with probability one.

Proof: Clearly

$$\sup_{0\leqslant t\leqslant 1}[\hat{H}_n^{-1}(t)-H^{-1}(t)]\leqslant \sup_{0\leqslant t\leqslant 1}[H_n^{-1}(t)-H^{-1}(t)]$$

by Marshall's lemma (7.2.3). The result follows from this fact and the previous lemma. □

The strong consistency theorem can now be stated.

Theorem 7.4.1 *If F is IFR, then for every x_0,*

$$r(x_0^-)\leqslant \underline{\lim}\,\hat{r}_n(x_0)\leqslant \overline{\lim}\,\hat{r}_n(x_0)\leqslant r(x_0^+) \tag{7.4.12}$$

with probability one.

Proof: By definition, $\hat{H}_n^{-1}(t)$ is concave in $t(0\leqslant t\leqslant 1)$. Hence, for every $\delta>0$,

$$\frac{\hat{H}_n^{-1}[F_n(x_0)]-\hat{H}_n^{-1}[F_n(x_0)-\delta]}{\delta}\geqslant\frac{1}{\hat{r}_n(x_0^-)}\geqslant\frac{1}{\hat{r}_n(x_0^+)}$$
$$\geqslant\frac{\hat{H}_n^{-1}[F_n(x_0)+\delta]-\hat{H}_n^{-1}[F_n(x_0)]}{\delta} \tag{7.4.13}$$

since

$$\frac{1}{\hat{r}_n(x_0^-)}\quad\text{and}\quad\frac{1}{\hat{r}_n(x_0^+)}$$

are the left- and right-hand derivatives, respectively, of $\hat{H}_n^{-1}(t)$ evaluated at $t=F_n(x_0)$.

Since F is IFR,

$$F(0)=0, \qquad \int_0^\infty x\,\mathrm{d}F(x)<\infty, \qquad \text{and} \qquad \hat{H}_n^{-1}(t)\to H^{-1}(t)$$

uniformly in $t(0\leqslant t\leqslant 1)$ with probability one by Lemma B. Letting $n\to\infty$ in (7.4.13) one sees that

$$\frac{\hat{H}^{-1}[F_n(x_0)]-\hat{H}^{-1}[F_n(x_0)-\delta]}{\delta}\geqslant \overline{\lim}\frac{1}{\hat{r}_n(x_0^-)}\geqslant \underline{\lim}\frac{1}{\hat{r}_n(x_0^+)}$$
$$\geqslant\frac{\hat{H}^{-1}[F_n(x_0)+\delta]-\hat{H}^{-1}[F_n(x_0)]}{\delta}$$

with probability one for all δ such that $0<\delta<\min[F(x_0), 1-F(x_0)]$. Letting $\delta\to 0$,

$$\frac{1}{r(x_0^-)}\geqslant\overline{\lim}\frac{1}{\hat{r}_n(x_0^-)}\geqslant\underline{\lim}\frac{1}{\hat{r}_n(x_0^+)}\geqslant\frac{1}{r(x_0^+)}$$

with probability one. This implies (7.4.12). □

Corollary *If r is increasing and continuous on $[a,b]$, then for $t\in[a,b]$,*

$$\lim_{n\to\infty}[\hat{r}_n(t)-r(t)]=0$$

with probability one.

Proof: This follows from the same methods as in the usual proof of the Glivenko–Cantelli theorem and Theorem 7.4.1. □

Prakasa Rao (1970) has shown, under regularity conditions, that

$$\left[\frac{2nf(x)}{r'(x)r^2(x)}\right]^{1/3}[\hat{r}_n(x)-r(x)] \tag{7.4.14}$$

has density $\frac{1}{2}\psi(u/2)$ when ψ is the density of the minimum value of $W(t)+t^2$ and $W(t)$ is a two-sided Wiener–Levy process with mean 0, variance 1 per unit t, and $W(0)=0$. Because of the relatively slow rate of convergence indicated by (7.4.14), isotonic 'window' estimators are considered in the next section.

Maximum likelihood estimator for the distribution function

Since

$$1-F(x)=\exp\left[-\int_{-\infty}^{x} r(u)\,\mathrm{d}u\right],$$

the maximum likelihood estimator for the distribution function is of course

$$\hat{F}_n(x) = 1 - \exp\left[-\int_{-\infty}^{x} \hat{r}_n(u)\,du\right].$$

Strong consistency of $\hat{F}_n$ is an almost immediate consequence of Theorem 7.4.1.

Theorem 7.4.2 *If F is IFR and r is continuous, then for all x,*

$$\lim_{n\to\infty} \hat{F}_n(x) = F(x)$$

with probability one.

Proof: It is sufficient to prove the theorem for x satisfying $F(x) < 1$, in which case $\hat{F}_n(x) < 1$ for sufficiently large n. By the corollary to Theorem 7.4.1,

$$\lim_{n\to\infty} \hat{r}_n(u) = r(u)$$

uniformly on $[0, x]$ with probability one. For $u \in [0, x]$, $\hat{r}_n(u) < \infty$, and by the Lebesgue dominated convergence theorem,

$$\lim_{n\to\infty} \int_0^x \hat{r}_n(u)\,du = \int_0^x r(u)\,du$$

with probability one. Hence

$$\lim_{n\to\infty} [1 - \hat{F}_n(x)] = 1 - F(x)$$

with probability one. □

Monte Carlo experiments indicate that for small sample sizes $\hat{F}_n$ *is not a good estimator for F in the tails of the distribution.* In fact, the empirical distribution seems to be preferable in the tails. This is perhaps to be expected since, of course, $\hat{r}_n$ will also behave badly in the tails of distribution.

Distributions with decreasing failure rate

Decreasing failure rate (DFR) distributions frequently arise in applications. They can, for example, arise as mixtures of exponential distributions.

Definition 7.4.2 A distribution F is said to be DFR if the support of F is of the form $[\alpha, \infty)$, $\alpha > -\infty$, and if $\log[1 - F(x)]$ is convex on $[\alpha, \infty)$.

Note that a DFR distribution may only have a jump at α. It is easy to check that the derivative of the absolutely continuous part must be decreasing on (α, ∞). Thus, DFR is, in a sense, a strengthening of the decreasing density assumption. For convenience it is supposed that α is known. Suppose that $\alpha = y_1 = \cdots = y_k < y_{k+1} < \cdots < y_n$. In the case $k = 0$, define $y_0 = \alpha$. It can be shown that the maximum likelihood estimator is

$$\hat{r}_n(\alpha) = \frac{k}{n} = \hat{F}(\alpha^+)$$

and

$$\hat{r}_n(x) = \hat{r}_n(y_i) \qquad \text{for } y_{i-1} < x \leqslant y_i; \quad i = 1, 2, \ldots, n,$$

where

$$\hat{r}_n(y_i) = \max_{t \geqslant i} \min_{s \leqslant i-1} \frac{t-s}{(n-s)(y_{s+1} - y_s) + \cdots + (n-t+1)(y_t - y_{t-1})}$$

and $y_0 = \alpha$ in the case $k = 0$.

Note that the estimator is left continuous in this case. Also, unlike the IFR case, this DFR estimator is not unique. It is determined by the likelihood equation only for $x \leqslant y_n$, and may be extended beyond y_n in any manner that preserves the DFR property.

7.5 ISOTONIC WINDOW ESTIMATORS FOR THE GENERALIZED FAILURE RATE FUNCTION

If we let $G(x) = 1 - e^{-x}$ and $g(x) = e^{-x}$ for $x \geqslant 0$ and note that $G^{-1}F(x) = -\log[1 - F(x)]$, then the failure rate function can be written as

$$r(x) = \frac{d}{dx} G^{-1}F(x) = \frac{f(x)}{g[G^{-1}F(x)]} = \frac{f(x)}{1 - F(x)}. \tag{7.5.1}$$

If $G(x) = x$ for $0 \leqslant x \leqslant 1$, then

$$r(x) = \frac{d}{dx} G^{-1}F(x) = \frac{f(x)}{g[G^{-1}F(x)]} = f(x) \tag{7.5.2}$$

is the density of F on the support of F.

We call

$$r(x) = \frac{f(x)}{g[G^{-1}F(x)]} \tag{7.5.3}$$

the *generalized failure rate function.* If $r(x)$ is nondecreasing then $G^{-1}F(x)$ is convex on the support of F. This motivates the notion of a partial ordering on the space of distributions called convex ordering or simply *c-ordering.* We say that $F \underset{c}{<} G$ or F *c-precedes* G if $G^{-1}F$ is convex on the support of F. If $F \underset{c}{<} G$ and G is the uniform distribution, then F has an increasing density. If $F \underset{c}{<} G$ and G is the exponential distribution then F is IFR. As another example, the gamma distributions are c-ordered with respect to the shape parameter. Let

$$f_\alpha(x) = \frac{\lambda^\alpha x^{\alpha-1} e^{-\lambda x}}{\Gamma(\alpha)} \qquad \text{for } x \geqslant 0; \quad \alpha, \lambda > 0$$

and

$$F_\alpha(x) = \int_0^x f_\alpha(u)\,du.$$

Zwet (1964) showed that $\alpha_1 > \alpha_2 > 0$ implies

$$F_{\alpha_1} \underset{c}{<} F_{\alpha_2}.$$

Also, $F < G$ implies that $F(x)$ crosses $G(\theta x)$ at most once, and from below if at all, for all $\theta^c > 0$. In this sense, F has 'lighter' tails than G. For example, if G were chosen to be the normal distribution then $F \underset{c}{<} G$ implies that F has lighter tails than the normal distribution.

Our objective is to consider a wide class of isotonic estimators for $r(x)$ which will include, as special cases, the maximum likelihood estimators considered in Sections 7.2, 7.3, and 7.4. It will be shown that, asymptotically, the maximum likelihood estimators can be improved upon. It is not assumed as in Section 7.4 that $F(0) = 0$.

It will be convenient to let $\mathscr{F}$ be the class of absolutely continuous distribution functions F on $(0, \infty)$ with positive, right (or left) continuous density f on the interval where $0 < F < 1$. It follows that the inverse function F^{-1} is uniquely defined on $(0, 1)$. The values $F^{-1}(0)$ and $F^{-1}(1)$ are taken to be equal to the left- and right-hand endpoints of the support of F (possibly $-\infty$ or $+\infty$).

Isotonic estimators

The function G will be assumed specified, $F \underset{c}{<} G$, $F, G \in \mathscr{F}$ and an ordered sample $y_1 \leqslant y_2 \leqslant \cdots \leqslant y_n$ from F available. Let F_n be the empirical distribution corresponding to our sample. Let ρ_n be an initial or *basic estimator* for r based on $y_1 \leqslant y_2 \leqslant \cdots \leqslant y_n$. For each n, a *grid* is defined on $(-\infty, \infty)$, i.e. a finite or infinite sequence $\cdots < t_{n,0} < t_{n,1} < \cdots < t_{n,i} < \cdots$. In each window $[t_{n,j}, t_{n,j+1})$ a point $t_{n,j} \leqslant x_{n,j} \leqslant t_{n,j+1}$ is chosen and to each point $x_{n,j}$ is assigned a nonnegative weight $w(x_{n,j})$. For convenience, $x_{n,j}$ is abbreviated to x_j. We call

$$r_n(x) = \min_{s \geqslant i+1} \max_{r \leqslant i} \left\{ \frac{\sum_{j=r}^{s-1} \rho_n(x_j) w(x_j)}{\sum_{j=r}^{s-1} w(x_j)} \right\} \tag{7.5.4}$$

for $t_{n,i} \leqslant x < t_{n,i+1}$, the monotonic regression or more specifically the isotonic regression of ρ_n with respect to the discrete measure w. Note that r_n is a nondecreasing step function and that $r_n = \rho_n$ whenever $\rho_n(x_j)$ happens to be nondecreasing in j.

If $y_1 \leqslant t_{n,i} < t_{n,i+1} \leqslant y_n$, then $gG^{-1}F_n(x_i) > 0$ since $F, G \in \mathscr{F}$ and $r(x)$ may be estimated for $t_{n,i} \leqslant x < t_{n,i+1}$ by

$$\hat{\rho}_n(x) = \frac{f_n(x_i)}{gG^{-1}F_n(x_i)} = \frac{F_n(t_{n,i+1}) - F_n(t_{n,i})}{gG^{-1}F_n(x_i)(t_{n,i+1} - t_{n,i})} \tag{7.5.5}$$

the *naive estimator* for r. With $\hat{\rho}_n$ for the naive estimator and weights

$$w(x_j) = gG^{-1}F_n(x_j)(t_{n,j+1} - t_{n,j}),$$

the *isotonic estimator* $\hat{r}_n$ is defined for

$$y_1 \leqslant t_{n,i} \leqslant x < t_{n,i+1} < y_n$$

by

$$\hat{r}_n(x) = \min_{\substack{s \geqslant i+1 \\ t_{n,s} \leqslant y_n}} \max_{\substack{r \leqslant i \\ t_{n,r} \geqslant y_1}} \frac{F_n(t_{n,s}) - F_n(t_{n,r})}{\sum_{j=r}^{s-1} gG^{-1}F_n(x_j)(t_{n,j+1} - t_{n,j})}. \tag{7.5.6}$$

Note that the conditions on $t_{n,i}, t_{n,i+1}, t_{n,r}$, and $t_{n,s}$ originate from the fact that $\hat{\rho}_n(x_j)$ is defined only if $y_1 \leqslant t_{n,j} < t_{n,j+1} \leqslant y_n$. An interesting special case arises when one considers the random grid $t_{n,j} = y_j$, $j = 1, 2, \ldots, n$, determined by the order statistics. For this grid and $x_j = y_j$, (7.5.6) becomes

$$\hat{r}_n(x) = \min_{i+1 \leqslant s \leqslant n} \max_{1 \leqslant r \leqslant i} \frac{s - r}{n\sum_{j=r}^{s-1} gG^{-1}(j/n)(y_{j+1} - y_j)} \tag{7.5.7}$$

for $y_i \leqslant x < y_{i+1}$, $i = 1, 2, \ldots, n-1$. This is the MLE in the case when G is the exponential or uniform distribution. Note that if G is the exponential distribution

$$ngG^{-1}\left(\frac{j}{n}\right)(y_{j+1} - y_j) = (n - j)(y_{j+1} - y_j)$$

and hence the weights $w(x_j)$ in (7.5.4) are proportional to the total time on test between y_j and y_{j+1}. Since the total time on test for an interval is a measure of our information over the interval, this choice of weights is intuitively appealing in this case. We shall call

$$gG^{-1}\left(\frac{j}{n}\right)(y_{j+1} - y_j)$$

the *total time on test weights* for general G also.

An alternative estimator r_n^* is obtained using the same basic estimator $\rho_n^* = \hat{\rho}_n$ with weights $w(x_j) = t_{n,j+1} - t_{n,j}$; thus

$$r_n^*(x) = \min_{\substack{s \geqslant i+1 \\ t_{n,s} \leqslant y_n}} \max_{\substack{r \leqslant i \\ t_{n,r} \geqslant y_1}} \frac{\sum_{j=r}^{s-1} [F_n(t_{n,j+1}) - F_n(t_{n,j})]/[gG^{-1}F_n(x_j)]}{t_{n,s} - t_{n,r}} \tag{7.5.8}$$

for $y_1 \leqslant t_{n,i} \leqslant x < t_{n,i+1} \leqslant y_n$.

If the second member of

$$r(x) = \frac{\mathrm{d}}{\mathrm{d}x} G^{-1} F(x) = \frac{f(x)}{gG^{-1}F(x)} \tag{7.5.9}$$

is considered rather than the third one, a logical choice for the basic estimator is the *graphical* estimator

$$\tilde{r}_n(x) = \min_{\substack{s \geqslant i+1 \\ t_{n,s} \leqslant y_n}} \max_{\substack{r \leqslant i \\ t_{n,r} \geqslant y_1}} \frac{G^{-1}F_n(t_{n,s}) - G^{-1}F_n(t_{n,r})}{t_{n,s} - t_{n,r}} \tag{7.5.10}$$

for $y_1 \leqslant t_{n,i} \leqslant x < t_{n,i+1} \leqslant y_n$. Notice that if $G^{-1}(1) = \infty$ and $t_{n,i+1} = y_n$, then $\tilde{\rho}_n(x) = \tilde{r}_n(x) = \infty$ for $t_{n,i} \leqslant x < t_{n,i+1}$. When G is the uniform distribution, $\hat{r}_n = r_n^* = \tilde{r}_n$.

Attention will be concentrated on the estimator $\hat{r}_n$ because of its intuitive appeal and because it generalizes the maximum likelihood estimator. We shall attempt to improve upon the MLE through an appropriate choice of the grid.

Consistency of nonrandom window estimators

For the grid consisting of order statistics, we note that the basic estimator $\hat{\rho}_n$ is not consistent even though the isotonized estimator is consistent. The grid based on order statistics will be called the *narrow grid*. By choosing a fixed grid $\{t_{n,j}\}_{j=1}^{\infty}$ with sufficiently wide windows the MLEs considered in Sections 7.2, 7.3, and 7.4 can be improved asymptotically.

In practical situations, the observed data are often heavily censored. In many life testing situations only the number of failures within specified intervals are recorded. Hence, it is important to consider estimators based on grids more general than that provided by order statistics.

For general, but nonrandom, grids $\{t_{n,j}\}$ we define

$$F_n^*(x) = F_n(x_{n,j}) \qquad \text{for } t_{n,j} \leqslant x < t_{n,j+1} \tag{7.5.11}$$

and

$$K_{F_n^*,\xi}(x) = \int_{\xi}^{x} gG^{-1}F_n^*(u)\,\mathrm{d}u \qquad \text{for } t_{n,1} \leqslant x \leqslant t_{n,\infty},$$

$$F^{-1}(0) \vee t_{n,1} < \xi < t_{n,\infty} \wedge F^{-1}(1), \tag{7.5.12}$$

where F_n is the usual empirical distribution. Strong consistency of $\hat{r}_n$ will follow from the strong uniform consistency of $K_{F_n^*,\xi}$. In order to prove uniform consistency of $K_{F_n^*,\xi}$ it is necessary to make some additional assumptions concerning the grid. The grid $\{t_{n,j}\}$ is said to *become dense* in an interval I if, for any pair $x_1 < x_2$ in I, there exists an integer N such that for $n \geqslant N$ a gridpoint $t_{n,j}$ exists with $x_1 < t_{n,j} < x_2$. It will be said that $\{t_{n,j}\}$ becomes *subexponential* *in the right tail* of F, if numbers N, $z>0$, and $c>1$ exist such that for $n \geqslant N, z \leqslant t_{n,j} < t_{n,j+1} < F^{-1}(1)$ implies $t_{n,j+1} \leqslant ct_{n,j}$. Note that if the support of F is bounded on the right, then any grid trivially becomes subexponential in the right tail of F and z may be chosen arbitrarily large. In the following lemma, a property of grids will be proved that will be needed in the sequel. Let U_n and V_n denote the infimum and the supremum of the gridpoints $t_{n,j}$ in the interval $[y_1, y_n]$.

Lemma *Let $\{t_{n,j}\}$ become dense in the support of F and subexponential in the right tail of F. Then numbers N, $z_0 \geqslant 0$, and $c>1$ exist such that for $n \geqslant N$ and $z_0 \leqslant x < V_n$, $F_n^*(x) \geqslant F_n(x/c)$.*

Proof: If $F^{-1}(1) < \infty$, the lemma is trivially true for $z_0 \geqslant F^{-1}(1)(\geqslant V_n)$. Assume $F^{-1}(1) = \infty$. By definition, $N^1, z>0$ and $c>1$ exist such that $t_{n,j+1} > t_{n,j} \geqslant z$ and $n \geqslant N^1$ imply $t_{n,j+1} \leqslant ct_{n,j}$. Choose $z_0 > z$ with $z_0 > F^{-1}(0)$. Since $\{t_{n,j}\}$ becomes dense in the support of F, an integer $N \geqslant N^1$ exists such that for $n \geqslant N$ a grid point between z and z_0 exists. Together with the definition of V_n, this implies that for $n \geqslant N$ and for every $x \in [z_0, V_n)$, gridpoints $t_{n,j}$ and $t_{n,j+1}$ exist with $0 < z \leqslant t_{n,j} \leqslant x < t_{n,j+1}$, and hence

$$F_n^*(x) = F_n(x_j) \geqslant F_n(t_{n,j+1}/c) \geqslant F_n(x/c). \quad \square$$

We will make use of the transform

$$K_{F,\xi}(x) = \int_{\xi}^{x} gG^{-1}F(u)\,du.$$

Theorem 7.5.1 *Let $F, G \in \mathscr{F}$, let*

$$EX^+ = \int_0^\infty x\,dF(x) < \infty,$$

and let gG^{-1} be uniformly continuous on $(0,1)$. Assume that either $F^{-1}(1) < \infty$ or $gG^{-1}(y)/(1-y)$ is bounded on $(0,1)$. Then, for any fixed $\xi \in (F^{-1}(0), F^{-1}(1))$, $K_{F,\xi}(F^{-1}(1)) < \infty$. If, moreover, the grid $\{t_{j,n}\}$ becomes dense in the support of F and subexponential in the right tail of F, then for $n \to \infty$,

$$\sup_{\xi \leqslant x \leqslant V_n} |K_{F_n^*,\xi}(x) - K_{F,\xi}(x)| \to 0$$

almost surely.

Proof: If $F^{-1}(1) < \infty$, then $K_{F,\xi}(F^{-1}(1))$ is an integral of a bounded function over a finite interval and, therefore, finite. If $gG^{-1}(y) \leqslant a(1-y)$ on $(0, 1)$, then

$$K_{F,\xi}(F^{-1}(1)) = \int_{\xi}^{F^{-1}(1)} gG^{-1}F(x)\,dx \leqslant a \int_{\xi}^{F^{-1}(1)} [1 - F(x)]\,dx$$

$$= aE(X - \xi)^+ < \infty.$$

Note that U_n and V_n tend a.s. to $F^{-1}(0)$ and $F^{-1}(1)$ since we assumed that the grid becomes dense in the support of F. Hence, $K_{F_n^*,\xi}(x)$ is a.s. defined for any fixed $x \in [\xi, F^{-1}(1))$ for sufficiently large n. Since both $K_{F,\xi}$ and $K_{F_n^*,\xi}$ are nondecreasing, $K_{F,\xi}(\xi) = 0$ and $K_{F,\xi}(F^{-1}(1)) < \infty$, it is obviously sufficient to show that $K_{F_n^*,\xi}$ converges pointwise to $K_{F,\xi}$ on $[\xi, F^{-1}(1))$ and that, for any $\varepsilon > 0$, $x_0 \in (\xi, F^{-1}(1))$ can be chosen in such a way that

$$\limsup_n \int_{x_0}^{V_n} gG^{-1}F_n^*(x)\,dx < \varepsilon \tag{7.5.13}$$

almost surely. (The method of proof of the Glivenko–Cantelli theorem can be used to complete the argument.) The assumption that the grid becomes dense guarantees pointwise a.s. convergence of F_n^* to F on the interval where $0 < F < 1$, even though F_n^* may be defective for all n (e.g. if no gridpoint $t_{j,n} > y_n$ exists for any n). As F is a continuous probability distribution, F_n^* is nondecreasing and $0 \leqslant F_n^* \leqslant 1$, a standard argument shows that the pointwise a.s. convergence implies that

$$\sup_x |F_n^*(x) - F(x)| \xrightarrow{\text{a.s.}} 0.$$

Hence,

$$\int_{\xi}^{x} [gG^{-1}F_n^*(u) - gG^{-1}F(u)]\,du \to 0$$

almost surely by the uniform continuity of gG^{-1} on $(0, 1)$.

It remains to prove (7.5.13), viz. for any $\varepsilon > 0$, $x_0 \in (\xi, F^{-1}(1))$ exists such that

$$\limsup_n \int_{x_0}^{V_n} gG^{-1}F_n^*(x)\,dx < \varepsilon \tag{7.5.14}$$

almost surely. It is only necessary to consider the case when $F^{-1}(1) = \infty$ and $gG^{-1}(y) \leqslant a(1-y)$ on $(0, 1)$. By the above lemma, N, $z_0 > 0$, and $c > 1$ exist such that $F_n^*(x) \geqslant F_n(x/c)$ for $n \geqslant N$ and $x \in [z_0, V_n)$. Choose $x_0 \geqslant z_0$ in such a way that

$$\int_{x_0/c}^{\infty} [1 - F(x)]\,\mathrm{d}x < \frac{\varepsilon}{ac}.$$

Then

$$\begin{aligned}
\limsup_n \int_{x_0}^{V_n} gG^{-1} F_n^*(x)\,\mathrm{d}x &\leqslant \limsup_n a \int_{x_0}^{V_n} [1 - F_n^*(x)]\,\mathrm{d}x \\
&\leqslant a \limsup_n \int_{x_0}^{V_n} \left[1 - F_n\left(\frac{x}{c}\right)\right] \mathrm{d}x \\
&\leqslant ac \limsup_n \int_{x_0/c}^{\infty} [1 - F_n(x)]\,\mathrm{d}x \\
&= ac \int_{x_0/c}^{\infty} [1 - F(x)]\,\mathrm{d}x < \varepsilon,
\end{aligned}$$

by the strong law. □

The result for the left tail that is needed in the strong consistency Theorem 7.5.1 is much weaker. Consider the special case $F = G$. Note that, for $x < \xi$, $K_{G,\xi}, K_{G_n,\xi}$, and $K_{G_n^*,\xi}$ are negative and that $K_{G,\xi}(x) = -[G(\xi) - G(x)]$.

Theorem 7.5.2 *Let $G \in \mathscr{F}$ and let gG^{-1} be uniformly continuous on $(0, 1)$. If, moreover, the grid $\{t_{n,j}\}$ becomes dense in the support of G, then for any fixed $\xi \in (G^{-1}(0), G^{-1}(1))$,*

$$\liminf_n \left[\inf_{U_n \leqslant x \leqslant \xi} (-K_{G_n^*,\xi}(x) + K_{G,\xi}(x)) \right] \geqslant 0$$

almost surely.

Proof: Let $x_0 < \xi$ with $0 < G(x_0) < \varepsilon$. Then for all $U_n \leqslant x \leqslant \xi$,

$$\begin{aligned}
-K_{G_n^*}(x) + K_G(x) &= \int_x^{\xi} [gG^{-1} G_n^*(u) - gG^{-1} G(u)]\,\mathrm{d}u \\
&\geqslant -\int_{x_0}^{\xi} |gG^{-1} G_n^*(u) - gG^{-1} G(u)|\,\mathrm{d}u - G(x_0) \xrightarrow{\text{a.s.}} -G(x_0) \geqslant -\varepsilon
\end{aligned}$$

since

$$\sup_x |G_n^*(x) - G(x)| \xrightarrow{\text{a.s.}} 0$$

and by the uniform continuity of gG^{-1} on $(0, 1)$ and the fact that $[x, \xi]$ is a.s. contained in the interval $[U_n, V_n]$, for sufficiently large n. Since $\varepsilon > 0$ may be taken arbitrarily small, the result for $K_{G_n^*,\xi}$ follows. □

From the definition of F_n^*, (7.5.11), and from (7.5.6) one sees that $\hat{r}_n(x)$ can be rewritten as

$$\hat{r}_n(x) = \min_{\substack{s \geqslant i+1 \\ t_{n,s} \leqslant y_n}} \max_{\substack{r \leqslant i \\ t_{n,r} \geqslant y_1}} \frac{F_n(t_{n,s}) - F_n(t_{n,r})}{K_{F_n^*}(t_{n,s}) - K_{F_n^*}(t_{n,r})} \tag{7.5.15}$$

for $y_1 \leqslant t_{n,i} \leqslant x < t_{n,i+1} < y_n$. It is now possible to prove the general strong consistency for isotonic estimators of the generalized failure rate function.

Theorem 7.5.3 *For $F, G \in \mathscr{F}$, let $F \underset{c}{<} G$, let*

$$\int_0^\infty x \, dG(x) < \infty,$$

and let gG^{-1} be uniformly continuous on $(0, 1)$. Assume that either $G^{-1}(1) < \infty$ or $gG^{-1}(y)/(1 - y)$ is bounded on $(0, 1)$. Let $\{t_{n,j}\}$ be a nonrandom grid that becomes dense in $(-\infty, \infty)$ and subexponential in the right tail in G. Then for $\hat{r}_n$ defined by (7.5.15), for every fixed x_0 with $0 < F(x_0) < 1$,

$$r(x_0^-) \leqslant \underline{\lim}\ \hat{r}_n(x_0) \leqslant \overline{\lim}\ \hat{r}_n(x_0) \leqslant r(x_0^+)$$

with probability one.

Proof: Let ξ be an arbitrary point satisfying $F^{-1}(0) < \xi < x_0$ and let $t_{n,a_n} \geqslant \xi$ be a sequence of gridpoints converging to ξ for $n \to \infty$. With probability 1, there exists an integral N such that for $n \geqslant N$, $\xi \leqslant t_{n,a_n} \leqslant x_0 < V_n$, and hence

$$\hat{r}_n(x_0) = \inf_{x_0 \leqslant t_{n,s} \leqslant y_n} \sup_{y_1 \leqslant t_{n,r} \leqslant x_0} \frac{F_n(t_{n,s}) - F_n(t_{n,r})}{\sum_{j=r}^{s-1} gG^{-1} F_n(x_j)(t_{n,j+1} - t_{n,j})}$$

$$\geqslant \inf_{x_0 \leqslant x \leqslant V_n} \frac{F_n(x) - F_n(t_{n,a_n})}{K_{F_n^*,\xi}(x) - K_{F_n^*,\xi}(t_{n,a_n})}.$$

Since $F \underset{c}{<} G$, it is easily verified that

$$\int_0^\infty x \, dG(x) < \infty$$

implies that

$$\int_0^\infty x\,\mathrm{d}F(x) < \infty.$$

Also $F^{-1}(1) < \infty$ if $G^{-1}(1) < \infty$, and it follows that if the grid becomes subexponential in the right tail of G, then it will also become subexponential in the right tail of F. Hence, the conditions on Theorem 7.5.1 are satisfied and as a result $K_{F_n^*,\xi}$ converges a.s. uniformly on $[\xi, V_n]$ to $K_{F,\xi}$. Since also F_n converges a.s. uniformly to F by the Glivenko–Cantelli theorem, $t_{n,a_n} \to \xi$ and $x_0 - \xi > 0$,

$$\underline{\lim}\, \hat{r}_n(x_0) \geqslant \inf_{x_0 \leqslant x < F^{-1}(1)} \frac{F(x) - F(\xi)}{K_{F,\xi}(x)} \geqslant r(\xi)$$

almost surely, where the second inequality follows from

$$K_{F,\xi}(x) = \int_\xi^x gG^{-1}F(u)\,\mathrm{d}u = \int_\xi^x \frac{f(u)}{r(u)}\mathrm{d}u \leqslant \frac{1}{r(\xi)}[F(x) - F(\xi)]$$

for $x > \xi$ since r is nondecreasing. Since $\xi \in (F^{-1}(0), x_0)$ was arbitrary, this proves the left-hand inequality of the theorem.

To prove the right-hand inequality, let ξ be an arbitrary point satisfying $x_0 < \xi < F^{-1}(1)$. Let $t_{n,b_n} \leqslant \xi$ be a sequence of gridpoints converging to ξ for $n \to \infty$. With probability one, there exists an integer N such that for $n \geqslant N$, $V_n < x_0 < t_{n,b_n} \leqslant \xi$, and hence

$$\hat{r}_n(x_0) \leqslant \sup_{U_n \leqslant t_{n,r} \leqslant x_0} \frac{F_n(t_{n,b_n}) - F_n(t_{n,r})}{\sum_{j=r}^{b_n-1} gG^{-1}F_n(x_j)(t_{n,j+1} - t_{n,j})}.$$

Consider the transformed order statistics $w_1 < w_2 < \cdots < w_i$ defined by $w_i = G^{-1}F(y_i)$. These are one order statistics of a random sample from G. The empirical distribution corresponding to $w_1 < w_2 < \cdots < w_n$ is $G_n = F_n F^{-1} G$. Define $\tilde{x}_0 = G^{-1}F(x_0)$, $\tilde{\xi} = G^{-1}F(\xi)$, $\tilde{t}_{n,j} = G^{-1}F(t_{n,j})$, $\tilde{x}_j = G^{-1}F(x_j)$, and $\tilde{U}_n = G^{-1}F(U_n)$. Note that $0 < G(\tilde{x}_0) < G(\tilde{\xi}) < 1$ and that $\tilde{U}_n$ is the infimum of the transformed gridpoints $\tilde{t}_{n,j}$ in the interval $[w_1, w_n]$. Also $F_n(t_{n,j}) = G_n(\tilde{t}_{n,j})$, $F_n(x_j) = G_n(\tilde{x}_j)$, and for $t_{n,j+1} \leqslant \xi$ one has

$$\tilde{t}_{n,j+1} - \tilde{t}_{n,j} = \int_{t_{n,j}}^{t_{n,j+1}} r(x)\,\mathrm{d}x \leqslant r(\xi)(t_{n,j+1} - t_{n,j})$$

$$\hat{r}_n(x_0) \leqslant r(\xi) \sup_{\tilde{U}_n \leqslant \tilde{t}_{n,r} \leqslant \tilde{x}_0} \frac{G_n(\tilde{t}_{n,b_n}) - G_n(\tilde{t}_{n,r})}{\sum_{j=r}^{b_n-1} gG^{-1}G_n(\tilde{x}_j)(\tilde{t}_{n,j+1} - \tilde{t}_{n,j})}$$

$$\leqslant r(\xi) \sup_{\tilde{U}_n \leqslant x \leqslant \tilde{x}_0} \frac{G_n(\tilde{t}_{n,b_n}) - G_n(x)}{K_{G_n^*,\xi}(\tilde{t}_{n,b_n}) - K_{G_n^*,\xi}(x)},$$

where $G_n^*(x) = G_n(\tilde{x}_j)$ for $\tilde{t}_{n,j} \leqslant x < \tilde{t}_{n,j+1}$. Thus, the problem has been transformed to the case when $F = G$. If $\{t_{n,j}\}$ becomes dense in $(-\infty, \infty)$, then $\{\tilde{t}_{n,j}\}$ becomes dense in the support of G. If $t_{n,j} = y_j$, then $\tilde{t}_{n,j} = w_j$. Hence, Theorem 7.5.2, the Glivenko–Cantelli theorem, the convergence of $\tilde{t}_{n,b_n}$ to $\tilde{\xi}$, and the fact that $\xi - x_0 > 0$ yields

$$\limsup_n \hat{r}_n(x_0) \leqslant r(\xi) \sup_{G^{-1}(0) < x \leqslant \tilde{x}_0} \frac{G(\xi) - G(x)}{K_{G,\xi}(x)} = r(\xi)$$

almost surely, since $K_{G,\tilde{\xi}}(x) = G(\tilde{\xi}) - G(x)$. Since $\xi \in (x_0, F^{-1}(1))$ was arbitrary, the proof of the theorem is completed. □

Asymptotic distributions

In order to obtain asymptotic comparisons, grids with spacings of the type $t_{n,i+1} - t_{n,i} = cn^{-\alpha}$ for $c > 0$ and $0 < \alpha < 1$ are considered. Although the isotonic estimator is defined for all $x \in [y_1, y_n]$, it will be mathematically convenient to consider the asymptotic distribution of $\hat{r}_n(x)$ for fixed x only. Also, for mathematical convenience, it will be assumed that $x_i = (t_{n,i+1} + t_{n,i})/2$; i.e. that x is the fixed (midpoint) of a grid spacing.

It can be shown that the 'naive' estimator, $\hat{\rho}_n(x)$, given by (7.5.5), is asymptotically normal with mean

$$r(x) + \frac{r(x) f''(x)}{24 f(x)} c^2 n^{-2\alpha} + O(n^{-3\alpha}) \tag{7.5.16}$$

and variance

$$\frac{r^2(x)}{f(x)} n^{\alpha - 1} + O(n^{-1}). \tag{7.5.17}$$

Hence, the mean square error (MSE) of $\hat{\rho}_n(x)$ is approximately

$$\text{MSE}[\hat{\rho}_n(x)] \approx \frac{r^2(x) n^{\alpha - 1}}{c f(x)} + \left[\frac{c^2 r(x) f''(x)}{24 f(x)}\right]^2 n^{-4\alpha}. \tag{7.5.18}$$

For $\frac{1}{5} < \alpha < 1$,

$$\frac{[c f(x)]^{1/2}}{r(x)} n^{(1-\alpha)/2} [r_n(x) - r(x)] \tag{7.5.19}$$

has asymptotically a $N(0, 1)$ distribution, while for $\frac{1}{7} < \alpha < \frac{1}{5}$,

$$\frac{[c f(x)]^{1/2}}{r(x)} n^{(1-\alpha)/2} \left[r_n(x) - r(x) - \frac{c^2 n^{-2\alpha} r(x)}{24} f''(x)\right] \tag{7.5.20}$$

is asymptotically $N(0, 1)$. These results follow from the asymptotic normality of related multinomial random variables (cf. Barlow and Zwet, 1971).

It will be shown that in the 'wide window' case (i.e. grid spacings like $cn^{-\alpha}$ where $0<\alpha<\frac{1}{3}$) $\hat{\rho}_n(x)$ and $\hat{r}_n(x)$ are asymptotically equivalent. Hence $\hat{r}_n(x)$ properly normalized will have, asymptotically, a normal distribution. The asymptotic distribution of $\hat{r}_n(x)$ in the narrow window case (that is $\frac{1}{3}\leqslant\alpha\leqslant 1$) is vastly more complicated. It will be argued later, in fact, that we should choose $\alpha=\frac{1}{5}$ in order to minimize mean square error and also to ensure obtaining an asymptotically normal distribution.

Theorem 7.5.4 *If* $F, G\in\mathscr{F}$ *and*

(*a*) $$r(x)=\frac{f(x)}{g[G^{-1}F(x)]}$$ *is nondecreasing in* $x\geqslant 0$;

(*b*) r *is continuously differentiable and* f'' *exists in a neighborhood of* x;

(*c*) $r'(x)>0$;

(*d*) $t_{n,i+1}-t_{n,i}=cn^{-\alpha}$ *and* $0<\alpha<\frac{1}{3}$;

then

$$\lim_{n\to\infty} P[\hat{r}_n(x)=r_n(x)]=1.$$

Proof: Only the heuristic basis for the proof will be given; the reader is referred to Barlow and Zwet (1971) for a rigorous proof.

The mean square error in estimating $r(x)$, using the naive estimator $\hat{\rho}_n(x)$, is

$$\mathrm{MSE}[\hat{\rho}_n(x)]=\mathrm{Var}[\hat{\rho}_n(x)]+\mathrm{Bia}[\hat{\rho}_n(x)]^2=O(n^{\alpha-1})+O(n^{-4\alpha}),$$

where $\mathrm{Bia}[\hat{\rho}_n(x)]$ is the bias term. If $(\mathrm{MSE}[\hat{\rho}_n(x)])^{1/2}$ is asymptotically smaller than the window size (which is $O(n^{-\alpha})$), then it is intuitively clear that asymptotically $\hat{\rho}_n(t_{n,i})<\hat{\rho}_n(t_{n,i+1})$ since $|r(t_{n,i+1})-r(t_{n,i})|=O(n^{-\alpha})$ (recall that r has a positive first derivative in a neighbourhood of x). Hence, asymptotically, one does not isotonize and $\hat{\rho}_n(x)$ and $\hat{r}_n(x)$ will have the same asymptotic distribution in this case. Clearly,

$$(\mathrm{MSE}[\hat{\rho}_n(x)])^{1/2}=O(n^{-\alpha}) \text{ if}$$
$$\sigma[\hat{\rho}_n(x)]=(\mathrm{Var}[\hat{\rho}_n(x)])^{1/2}=O(n^{(\alpha-1)/2})=O(n^{-\alpha});$$

i.e. if $(\alpha-1)/2<-\alpha$ or $\alpha<\frac{1}{3}$. The rigorous proof is, of course, more complicated but this argument is at the heart of the proof. □

Theorem 7.5.5 *Assume the conditions of Theorem 7.5.4. If* $\frac{1}{5}<\alpha<\frac{1}{3}$, *then*

$$\frac{[cf(x)]^{1/2}}{r(x)}n^{(1-\alpha)/2}\{\hat{r}_n(x)-r(x)\}$$

is asymptotically $N(0,1)$.

For $\frac{1}{7} < \alpha < \frac{1}{5}$,

$$\frac{[cf(x)]^{1/2}}{r(x)} n^{(1-\alpha)/2} \left[\hat{r}_n(x) - r(x) - \frac{c^2 n^{-2\alpha}}{24} \frac{f''(x) r(x)}{f(x)} \right]$$

is asymptotically $N(0, 1)$.

Proof: The proof follows immediately from (7.5.16), (7.5.17), and Theorem 7.5.4.

The situation in the narrow window case (that is $\frac{1}{3} \leqslant \alpha \leqslant 1$) is considerably more complicated. The asymptotic distribution of the MLE in the case of monotone densities and monotone failure rate functions was obtained by Prakasa Rao (1970). His result also holds in the generalized failure rate case for isotonic window estimators when the grid spacing satisfy $t_{n,i+1} - t_{n,i} = cn^{-\alpha}$ and $\frac{1}{3} < \alpha \leqslant 1$. Window estimates have been studied in regression problems (cf. Wright, 1982). The results given there are similar to those contained in this section.

Theorem 7.5.6 *If* $F, G \in \mathscr{F}$ *and*

(*a*) $r(x) = \dfrac{f(x)}{g[G^{-1}F(x)]}$ is nondecreasing in $x \geqslant 0$;

(*b*) *r is continuously differentiable and* f'' *exists in a neighborhood of* x;

(*c*) $r'(x) > 0$;

(*d*) $t_{n,i+1} - t_{n,i} = cn^{-\alpha}$ *and* $\frac{1}{3} < \alpha \leqslant 1$;

(*e*) $\hat{r}_n(x) = \min\limits_{s \geqslant i+1} \max\limits_{r \leqslant i} \dfrac{F_n(t_{n,s}) - F_n(t_{n,r})}{\sum_{j=r}^{s} gG^{-1}F_n(x_j)(t_{n,j+1} - t_{n,j})}$

where $t_{n,i} \leqslant x < t_{n,i+1}$, *then the asymptotic distribution of*

$$\left[\frac{2nf(x)}{r'(x)r^2(x)}\right]^{1/3} [\hat{r}_n(x) - r(x)]$$

has density $\frac{1}{2}\psi(u/2)$ *where* ψ *is the density of the minimum value of* $W(t) + t^2$ *and* $W(t)$ *is a two-sided Weiner–Levy process with mean* 0 *and variance* 1 *per unit* t *and* $W(0) = 0$.

Proof: A proof may be found in Barlow and Zwet (1971).

Recommendations for the window size

If one is willing to assume sufficient smoothness in the generalized failure rate function, say the existence of a second derivative, then recommendation on window size can be made in terms of the mean square error. For $\alpha < \frac{1}{3}$,

$$\text{MSE}[\hat{r}_n(x)] \approx \frac{r^2(x)n^{\alpha-1}}{cf(x)} + \left[\frac{r(x)f''(x)}{24}\right]^2 c^4 n^{-4\alpha}$$

$$= \frac{A}{nh} + Bh^4$$

where $h = n^{-\alpha}$. The minimum as a function of h is achieved for $h = n^{-1/5}$. Hence, $\alpha = \frac{1}{5}$ should be chosen. However, this is not helpful since the optimum choice of c in the window size still depends on the true generalized failure rate function which is, of course, unknown. The MLE takes care of this problem by choosing $t_{i,n} = y_i$, the ith order statistic from a sample size of n. The MLE might be modified by choosing $t_{i,n} = y_{[in^\beta]}$ where [ ] denotes the greatest integer of the quantity within the brackets and $\beta = 1 - \alpha$. Hence,

$$y_{[(i+1)n]} - y_{[in]} = O_p\left(\frac{n^\beta}{n}\right) = O_p(n^{-\alpha})$$

where O_p means 'big O' in probability, and by chooisng $\alpha = \frac{1}{5}$ the recommended requirement is realized.

7.6 TESTS FOR EXPONENTIALITY AGAINST A MONOTONE FAILURE RATE

A considerable part of the literature in applied statistics and operations research is based on the exponential distribution. The question of validating an exponential distribution has inspired a large number of papers on omnibus tests for exponentiality. In this section a brief discussion is given of a test statistic called the *cumulative total time on test statistic* for testing the null hypothesis of exponentiality when the alternative is that of monotone failure rate. This statistic has a very old history and has a convenient distribution under the null hypothesis of exponentiality. The test based upon this statistic has the desirable properties of *unbiasedness* and *isotonic power* with respect to an ordering on the alternative called *star-ordering*. It is also possible to prove an *asymptotic minimax property* for this test. These properties will not be proved here but the interested reader will be referred to Barlow *et al.* (1972) for more details. Again in this section, 'increasing' is used for 'nondecreasing' and 'decreasing' is used for 'nonincreasing'. In Section 7.4, we considered the problem of estimating the failure rate function,

$$r(t) = \frac{f(t)}{1 - F(t)},$$

which is assumed increasing. In this section some of the statistics and theory developed in Section 7.4 are used to test the hypothesis of constant versus increasing failure rate. In Barlow and Doksum (1972) this is generalized to the problem of testing $F \underset{c}{=} G$ versus $F \underset{c}{<} G$, where $\underset{c}{<}$ stands for c-ordering and G is specified. However, throughout this section it is assumed that $G(x) = 1 - \mathrm{e}^{-x}$ for $x \geqslant 0$.

Suppose that the first k ordered observations out of a sample of n test items from a distribution F are obtained:

$$Y_1 \leqslant Y_2 \leqslant \cdots \leqslant Y_k \qquad \text{for } 1 \leqslant k \leqslant n.$$

Let $D_{i:n} = (n - i + 1)(Y_i - Y_{i-1})$ (where $Y_0 = 0$) be the normalized sample spacings. It is well known that if $F(x) = 1 - \mathrm{e}^{-\lambda x}$ for $x \geqslant 0$, then $D_{i:n}(i = 1, 2, \ldots, n)$ are independent and exponentially distributed random variables with mean $1/\lambda$. If F is IFR, then

$$D_{1:n} \overset{\text{st}}{\geqslant} D_{2:n} \overset{\text{st}}{\geqslant} \cdots \overset{\text{st}}{\geqslant} D_{n:n}$$

will be shown to be true where $\underset{\text{st}}{\geqslant}$ indicates stochastic (or probabilistic) ordering (see the Corollary to Theorem 7.6.1). This is the basis for many tests for exponentiality versus increasing (or decreasing) failure rate. (Note that the isotonic regression $D^*_{i:n}$ of $D_{i:n}$ with unit weights and the ordering $1 \gtrsim 2 \gtrsim 3 \gtrsim \cdots \gtrsim n$ gives the maximum likelihood estimate for $1/r(t)$ when $r(t)$ is assumed increasing, Section 7.4.)

It will be convenient to use the notation $\bar{F}(x)$ for $1 - F(x)$. The following useful lemma will be needed.

Lemma *Let μ be a signed measure of bounded variation on $[a, b]$, $-\infty \leqslant a < b \leqslant \infty$ such that*

$$\int_a^b \mathrm{d}\mu(u) < \infty.$$

Then

$$\int_a^b \phi(\mu)\,\mathrm{d}\mu(u) \geqslant 0$$

for all bounded increasing ϕ if and only if

$$\int_x^b \mathrm{d}\mu(u) \geqslant 0 \quad \textit{for all } x \in [a, b] \qquad \textit{and} \qquad \int_a^b \mathrm{d}\mu(u) = 0.$$

Proof: Suppose first that

$$\int_a^b \phi(x)\,\mathrm{d}\mu(x) \geqslant 0$$

for all ϕ increasing. Then with $\phi = H_z$, where $H_z(x) = 0$ for $x \leqslant z$ and $H_z(x) = 1$ for $x > z$, it follows that

$$\int_z^b d\mu(u) \geqslant 0.$$

Also $\phi(x) = 1$ and $\phi(x) = -1$ imply

$$\int_a^b d\mu(u) = 0.$$

Next suppose

$$\int_x^b d\mu(u) \geqslant 0$$

for all $x \in [a, b]$ and suppose in addition $\phi \geqslant 0$. Approximate ϕ by an increasing sequence $\{\phi_k\}$ of increasing step functions. Since

$$\phi_k(z) = \sum_{i=1}^{n} a_i H_{x_i}(z)$$

and $a_i > 0$ it is concluded that

$$\int_a^b \phi_k(u)\, d\mu(u) \geqslant 0.$$

The lemma follows from the Lebesgue monotone convergence theorem. If $\phi \leqslant 0$,

$$\int_a^b -\phi(u)\, d\mu(u) \leqslant 0$$

is obtained by an identical argument. Finally, note that any increasing ϕ can be decomposed into $\phi_1 + \phi_2$ where $\phi_1 \geqslant 0, \phi_2 \leqslant 0$, and ϕ_i is increasing $(i = 1, 2)$.

□

Theorem 7.6.1 *If F is IFR (DFR), then $D_{i:n} = (n - i + 1)(Y_i - Y_{i-1})$ is stochastically increasing (decreasing) in $n \geqslant i$ for fixed i.*

Proof: Assume F is IFR. Let $F_{i:n}(x) = P[Y_i \leqslant x]$ and $F_u(x) = [F(x + u) -$

$F(u)]/\bar{F}(u)$. Then

$$P[(n-i)(Y_{i+1}-Y_i)>x]=\int_0^\infty \left[\bar{F}_u\left(\frac{x}{n-i}\right)\right]^{n-i} \mathrm{d}F_{i:n}(u)$$

$$\leqslant \int_0^\infty \left[\bar{F}_u\left(\frac{x}{n+1-i}\right)\right]^{n+1-i} \mathrm{d}F_{i:n}(u) \qquad (7.6.1)$$

since $[\bar{F}(t)]^{1/t}$ is decreasing in t for F IFR. (F IFR implies $\log \bar{F}(t)$ concave or $[\log \bar{F}(t)]/t$ is decreasing since $\log \bar{F}(0)=0$.) Also, since $\bar{F}_u(x)$ is decreasing in u for F IFR and $F_{i:n}(x) \leqslant F_{i:n+1}(x)$ for all F,

$$\int_0^\infty \left[\bar{F}_u\left(\frac{x}{n+1-i}\right)\right]^{n+1-i} \mathrm{d}F_{i:n}(u) \leqslant \int_0^\infty \left[\bar{F}_u\left(\frac{x}{n+1-i}\right)\right]^{n+1-i} \mathrm{d}F_{i:n+1}(u)$$

$$= P[(n+1-i)(Y_{i+1}-Y_i)>x], \qquad (7.6.2)$$

by the above lemma. (Let $\mu(u)=F_{i:n}(u)-F_{i:n+1}(u)$ and $-\phi(u)=\{\bar{F}_u[x/(n+1-i)]\}^{n+1-i}$ and note that

$$\int_x^\infty \mathrm{d}\mu(u)=\bar{F}_{i:n}(x)-\bar{F}_{i:n+1}(x)\geqslant 0.)$$

Inequalities (7.6.1) and (7.6.2) are reversed when F is DFR. □

Corollary *If F is IFR (DFR), then $D_{i:n}=(n-i+1)(Y_i-Y_{i-1})$ is stochastically decreasing (increasing) in $i=1,2,\ldots,n$ for fixed n.*

Proof: Assume F is IFR. First it will be shown that

$$(n-1)(Y_2-Y_1)\overset{\text{st}}{\leqslant} nY_1.$$

Given Y_1, Y_2-Y_1 is the minimum of $n-1$ random variables each stochastically less than Y_1. Hence,

$$Y_2-Y_1\overset{\text{st}}{\leqslant} Y_1'$$

where Y_1' is the smallest item of a sample of size $n-1$. By Theorem 7.6.1,

$$(n-1)Y_1'\overset{\text{st}}{\leqslant} Y_1,$$

so that

$$(n-1)(Y_2-Y_1)\overset{\text{st}}{\leqslant} nY_1.$$

The result follows by repeated conditioning. An analogous argument applies in the DFR case. □

Since under the exponential assumption

$$P[D_{i:n} > D_{j:n}] = \tfrac{1}{2}, \qquad i \neq j,$$

while

$$D_{1:n} \overset{\text{st}}{\geqslant} D_{2:n} \overset{\text{st}}{\geqslant} \cdots \overset{\text{st}}{\geqslant} D_{n:n}$$

under the IFR assumption, many tests have been suggested based on reversals in the assumed ordering. (See Bickel and Doksum, 1969, and Proschan and Pyke, 1967.) The fact that, under the IFR assumption, $D_{i:n}$ tends to be larger than $D_{j:n}$ for $i < j$ should be reflected by a positive slope in the linear regression of the $D_{n-i+1:n}$ on the values $i - 1$. Under exponentiality, of course, this regression should have zero slope. When the slope is nonzero, it will be sensitive to changes in scale. In order to make the statistic scale invariant, the slope is divided by the average of the $D_{i:n}$. The resulting statistic is linearly equivalent to

$$V_n \overset{\text{def}}{=} \frac{n^{-1}\sum_{i=1}^{n}(i-1)D_{n-i+1:n}}{n^{-1}\sum_{i=1}^{n} D_{i:n}} \tag{7.6.3}$$

which is the same as

$$V_n \overset{\text{def}}{=} \frac{n^{-1}\sum_{i=1}^{n-1}\sum_{j=1}^{i} D_{j=n}}{n^{-1}\sum_{i=1}^{n} D_{i:n}} \tag{7.6.4}$$

if the numerator of the right-hand side of (7.6.3) is summed by parts. Thus

$$T_n(Y_i) = \sum_{j=1}^{i} D_{j:n}$$

is recognized as the *total time on test statistic* and for this reason the following definition is made.

Definition 7.6.1 Given the first $k(1 \leqslant k \leqslant n)$ ordered observations out of a sample of n test items,

$$V_k = \frac{k^{-1}\sum_{i=1}^{k-1}\left(\sum_{j=1}^{i} D_{j:n}\right)}{k^{-1}\sum_{i=1}^{k} D_{i:n}} \tag{7.6.5}$$

is called the *cumulative total time on test statistic*.

If $k = n$, then (7.6.5) agrees with (7.6.4). Under exponentiality the distribution of V_k will depend only on k, the observed number of failures, and not on n, the sample size.

We recommend the statistic V_k for the problem of testing

$$H_0{:}F \text{ exponential}$$

versus

$$H_1{:}F \text{ IFR and not exponential.}$$

Definition 7.6.2 Let ψ be the test corresponding to V_k where

$$\psi(Y_1, Y_2, \ldots, y_k) = \begin{cases} 1, & \text{if } V_k \geqslant c_{k,1-\alpha} \\ 0, & \text{otherwise,} \end{cases}$$

where $P[V_k \geqslant c_{k,1-\alpha}] = \alpha$ under H_0 and $G(x) = 1 - \exp(-x)$ for $x \geqslant 0$.

The following well-known result is stated without proof.

Theorem 7.6.2 *If $F(x) = 1 - \exp(-\lambda x)$ for $x \geqslant 0$, then*

$$V_k \overset{\text{st}}{=} U_1 + U_2 + \cdots + U_{k-1} \tag{7.6.6}$$

where $\overset{\text{st}}{=}$ indicates stochastic equivalence and $U_i (i = 1, 2, \ldots, k-1)$ are independent uniform random variables on $[0, 1]$.

It follows that

$$[12(k-1)]^{1/2}[(k-1)^{-1}V_k - \tfrac{1}{2}] \to N(0, 1)$$

as $k \to \infty$, when $F(x) = 1 - \exp(-\lambda x)$ for $x \geqslant 0$.

Table 7.6.1 provides critical numbers for performing the test, ψ. If the observations are from an IFR distribution then V_k will tend to be large. The table provides percentage points such that if true distribution is exponential then

$$P[V_k \leqslant c_{k,1-\alpha}] = 1 - \alpha$$

where $c_{k,1-\alpha}$ is the number tabulated. For example, if $1 - \alpha = 0.90$ and $k = 5$

Table 7.6.1 Percentiles $c_{k,1-\alpha}$ of the cumulative total time on test statistic, V_k, under H_0

$k-1$	$1-\alpha$				
	0.900	0.950	0.975	0.990	0.995
2	1.553	1.684	1.776	1.859	1.900
3	2.157	2.331	2.469	2.609	2.689
4	2.753	2.953	3.120	3.300	3.411
5	3.339	3.565	3.754	3.963	4.097
6	3.917	4.166	4.367	4.610	4.762
7	4.489	4.759	4.988	5.244	5.413
8	5.056	5.346	5.592	5.869	6.053
9	5.619	5.927	6.189	6.487	6.683
10	6.178	6.504	6.781	7.097	7.307
11	6.735	7.077	7.369	7.702	7.924
12	7.289	7.647	7.953	8.302	8.535

k = number of failures observed in incomplete sample.

one must look in row $k-1=4$ and find $c_{4,0.90}=2.753$. If the computed statistic V_k exceeds the number 2.753 it is concluded that the true distribution is IFR. There is, of course, a 10 percent chance that an error has been made; that is

$$P[V_k \geqslant 2.753]=0.10$$

when the true distribution is exponential and $k=5$.

Example 7.6.1 (Air conditioning failures) In Table 7.6.2, the times between air conditioner failures are listed on selected aircraft. After roughly 2000 hours of service the planes received major overhaul; the failure interval containing major overhaul is omitted from the listing since the length of that failure interval may have been affected by major overhaul.

One wants to determine if the intervals between failures have an exponential distribution or if there is a wearout trend as the equipment ages. In the event

Table 7.6.2 Intervals between failures of air conditioning equipment on jet aircraft

7907	7908	7915	7916	8044
194	413	359	50	487
15	14	9	254	18
41	58	12	5	100
29	37	270	283	7
33	100	603	35	98
181	65	3	12	5
	9	104		85
	169	2		91
	447	438		43
	184			230
	36			3
	201			130
	118			
	34[a]			
	31			
	18			
	18			
	67			
	57			
	62			
	7			
	22			
	34			

[a] Major overhaul before this observation.

Table 7.6.3

PLANE	SAMPLE SIZE k	STATISTIC V_k	CONCLUSION
7907	6	$V_6 = 2.243$	Exponential, that is V_6 not significant at the 10% level
7908	23	$V_{23} = 9.127$ $Z = -1.384$	Exponential, that is V_{23} not significant at the 10% level
7915	9	$V_9 = 2.80$	Exponential, that is V_9 not significant at the 10% level
7916	6	$V_6 = 1.67$	Exponential, that is V_6 not significant at the 10% level
8044	12	$V_{12} = 4.22$	Exponential, that is V_{12} not significant at the 10% level

that there is a wearout trend maintenance should be scheduled accroding to equipment age rather than the present policy.

The computations associated with the above data are given in Table 7.6.3. Computations for the data from a given plane were made as if they were independent observations on different aircraft of the same type.

Since the sample size for plane 7908 exceeds the range of Table 7.6.1, one uses the fact that

$$Z = [12(k-1)]^{1/2}[(k-1)^{-1}V_k - \tfrac{1}{2}]$$

is approximately normally distributed with mean 0 and variance 1 when k is large and observations are from an exponential distribution.

It is interesting to note that all of the values of the cumulative total time on test statistics in Table 7.6.3 are below their expected values under the null hypothesis of exponentiality. If we were testing against the alternative of DFR we would use the fact that the distribution of

$$V_k \stackrel{\text{st}}{=} U_1 + U_2 + \cdots + U_{k-1}$$

is symmetric about $(k-1)/2$ so that

$$P[V_k - (k-1)/2 \geqslant x] = P[V_k - (k-1)/2 \leqslant -x].$$

The data from planes 7908, 7915, and 8044 give significant values at the 10% level while the data from planes 7907 and 7916 do not give significant values.

7.7 COMPLEMENTS

The nonparametric MLE of a monotone density was first derived by Grenander (1956). The derivation of the MLE for unimodal densities with known mode was done independently by Prakasa Rao (1966, 1969) and Robertson (1966a, 1967). The latter attributes the original idea to Ron Pyke. The idea of using the lemma preceding Theorem 7.2.2 to prove consistency is due to Albert W. Marshall. The theory developed here for the unknown mode case is due to Wegman (1968, 1970a, 1970b). Robertson (1966a, 1967) also considered estimating multivariate density functions subject to isotonic restrictions.

The MLE for IFR distributions (Section 7.4) was first obtained by Grenander (1956). This was extended and consistency was proved by Marshall and Proschan (1965). The derivation of the MLE given here follows Marshall and Proschan. The consistency proof here is a combination of ideas due to Robertson (1967), Barlow and Zwet (1970), and Albert W. Marshall. The material in Sections 7.4, 7.5, and 7.6 was taken from Barlow *et al.* (1972). Hari Mukerjee provided a corrected version of Lemmas A and B for Theorem 7.4.1.

The notion of convex ordering (Section 7.5) on the space of distribution functions is due to Zwet (1964).

Isotonic window estimators for the generalized failure rate function were first considered in Barlow and Zwet (1970, 1971). Optimal choice for the window size for nonparametric estimators of the density and failure rate functions have been investigated by Parzen (1962a), Watson and Leadbetter (1964), and Weiss and Wolfowitz (1967).

The *total time on tests transformation* first appeared in Marshall and Proschan (1965). However, they make no use of it!

The problem of testing the hypothesis that F is a negative exponential distribution with unknown scale parameter against the alternative that F has monotone increasing nonconstant failure rate (Section 7.6) has been studied by a number of authors, namely Proschan and Pyke (1967), Nadler and Eilbott (1967), Barlow (1968a), Bickel and Doksum (1969), and Bickel (1969). Bickel and Doksum show that the test proposed by Proschan and Pyke is asymptotically inadmissible. They then take an essentially parametric approach to the problem. In particular, they obtain the studentized asymptotically most powerful linear spacings tests for selected parametric families of distributions which are IFR when the parameter $\theta > 0$ and exponential when $\theta = 0$. Bickel (1969) proves that these tests are actually asymptotically equivalent to the level α tests which are most powerful among all tests which are similar and level α (for the associated parametric problems).

The analogue of the cumulative total time on test statistic for tests for trends in a series of events has been described as the oldest known statistical test, having been put forward by Laplace in 1773 to test whether comets originate in the solar system. Cox (1955) and Cox and Lewis (1966) have discussed this statistic and its optimality with respect to an interesting parametric alternative.

Epstein (1960a) proposed the test in Definition 7.6.2, among others, for life testing problems. Properties of normalized spacings from IFR distributions can be found in Barlow and Proschan (1966). The interpretation of V_n as the slope of the linear regression of the normalized spacings is due to Nadler and Eilbott (1967). The data in Example 7.6.1 was taken from Proschan (1963).

CHAPTER 8

Conditional Expectation Given a σ-lattice: Projections in a More General Setting

8.1 INTRODUCTION

In Chapter 1, the isotonic regression problem was characterized as that of finding the function f in a certain class of functions, $\mathscr{I}$, defined on a finite set X which minimizes

$$\sum_{x \in X} [g(x) - f(x)]^2 w(x), \tag{8.1.1}$$

where $\mathscr{I}$ is the collection of functions isotonic with respect to a quasi-order, $\lesssim$, on X. The function g was given and the weight function w was assumed positive. Since X was finite, real valued functions on X could be interpreted as points in a finite-dimensional Euclidean space and the weighted sum of squares as the square of a weighted Euclidean distance. Thus, the regression problem considered in Chapter 1 has a geometric interpretation of seeking the closest point in $\mathscr{I}$ to g. In this chapter a more general formulation of the regression problem is given. The setting will be the Hilbert space consisting of the square integrable functions on a measure space with the usual L_2 inner product.

Specifically, assume that $(\Omega, \mathscr{S}, \lambda)$ is a measure space. We will assume that $\mathscr{S}$ is complete in the sense that if two sets differ by a subset of a set of measure zero and one is in $\mathscr{S}$, then the other is also in $\mathscr{S}$. Although λ need not be a probability measure it will be convenient to speak of $\mathscr{S}$-measurable, real valued functions on Ω as random variables. We let $L_2(\Omega, \mathscr{S}, \lambda)$ denote the collection of square integrable random variables on $(\Omega, \mathscr{S}, \lambda)$ and denote this collection by L_2 when $(\Omega, \mathscr{S}, \lambda)$ is understood (where two random variables are regarded as identical if they differ by a set of measure zero; i.e. we really consider equivalence classes defined by the equivalence relation of equality except on a set of measure zero). The L_2 regression problem is that of finding the random variable, f, in a given subset of L_2 which minimizes

$$\int [g(\omega) - f(\omega)]^2 \lambda(\mathrm{d}\omega) \tag{8.1.2}$$

over this subset. The random variable g in L_2 is presumed to be given. In the general isotonic regression problem, the subset of interest is the collection of L_2 functions which are isotonic with respect to a quasi-order on Ω. If $\mathscr{A}$ is this subcollection and the solution exists uniquely then it is denoted by $g^* = P_\lambda(g|\mathscr{A})$ (or $P(g|\mathscr{A})$ if λ is understood).

The collection, $\mathscr{A}$, can arise in several different ways. The properties of $P(g|\mathscr{A})$ given in this chapter are derived under assumptions regarding the geometric structure of $\mathscr{A}$ as a subset of L_2. The primary concept is taken to be that of a σ-lattice of measurable subsets of Ω. Then $\mathscr{A}$ is taken to be the collection of members of L_2 which are measurable with respect to this σ-lattice. This approach is analogous to the usual approach to conditioning through conditional expectations given σ-fields. (We defined a lattice of functions in Chapter 1. A lattice of sets is a similar concept with $\cup(\cap)$ playing the role of $\vee$ ($\wedge$).)

Definition 8.1.1 A *σ-lattice* of subsets of Ω is a collection of subsets of Ω which contains both the empty set, ϕ, and Ω, and is closed under countable unions and countable intersections. It is called a *complete lattice* if it is closed under arbitrary unions and intersections.

Note that any σ-field of subsets of Ω is also a σ-lattice. Examples of σ-lattices of subsets of $(-\infty, \infty)$ which are not σ-fields are

(a) the collection of all intervals of the form $(-\infty, a)$ or $(-\infty, a]$, where $-\infty \leqslant a \leqslant \infty$,

and

(b) the collection of all intervals (open, closed, or half open and half closed) containing a given fixed point such as 0.

In general, the collection $\mathscr{A}$ will be the collection of all random variables which are measurable with respect to some σ-lattice such as those described in (a) and (b).

Definition 8.1.2 Let $\mathscr{U}$ be a σ-lattice of subsets of a space Ω. An extended real valued function f on Ω is *$\mathscr{U}$-measurable* if

$$[f > a] = \{\omega \in \Omega : f(\omega) > a\} \in \mathscr{U}$$

for all real a. It is equivalent to require that $[f \geqslant a] \in \mathscr{U}$ for all real a. The class of all extended real valued $\mathscr{U}$-measurable functions on Ω is denoted by $R(\mathscr{U})$ and $L_2(\mathscr{U})$ is defined to be $R(\mathscr{U}) \cap L_2$. (To avoid certain technical complications, we will assume that a σ-lattice $\mathscr{U}$ is complete with respect to λ in the sense that if two sets differ by a subset of a set of measure zero, and one set is in $\mathscr{U}$, then so is the other. This will imply that the property of being $\mathscr{U}$-measurable must hold for equivalence classes.)

If $f, g \in R(\mathscr{U})$, then for each real a,

$$[f+g>a]=\bigcup_r\{[f>r]\cap[g>a-r]\}\in\mathscr{U}$$

where the union is taken over all rational numbers r. Thus, $R(\mathscr{U})$ (and hence $L_2(\mathscr{U})$) is closed under addition. Also, if $f\in R(\mathscr{U})$ and $k\geqslant 0$ then $kf\in R(\mathscr{U})$. These two results imply that $R(\mathscr{U})$ and $L_2(\mathscr{U})$ are convex cones. Moreover, if $f, g\in R(\mathscr{U})$, then

$$[f\vee g>a]=[f>a]\cup[g>a]\in\mathscr{U}$$

and

$$[f\wedge g>a]=[f>a]\cap[g>a]\in\mathscr{U}$$

for all real a so that $f\vee g, f\wedge g\in R(\mathscr{U})$. Thus $R(\mathscr{U})$, and hence $L_2(\mathscr{U})$, is a lattice. If $f_n\leqslant f_{n+1}, f_n\in R(\mathscr{U})$, $n=1,2,\ldots$, and if

$$\lim_{n\to\infty} f_n(\omega)=f(\omega)\qquad \text{for } \omega\in\Omega,$$

then

$$[f>a]=\bigcup_{n=1}^{\infty}[f_n>a]=\lim_n[f_n>a]\in\mathscr{U},$$

so that $f\in R(\mathscr{U})$. Similarly, if $f_n\geqslant f_{n+1}, f_n\in R(\mathscr{U})$, then $\lim_n f_n\in R(\mathscr{U})$ if $\lim_n f_n(\omega)$ exists. These properties are summarized by saying that $R(\mathscr{U})$ is a σ-lattice.

In the generalized isotonic regression problem, a function g in L_2 and a σ-lattice of subsets, $\mathscr{U}$, are given. The set $\mathscr{A}$ is taken to be the collection, $L_2(\mathscr{U})$, of square integrable, $\mathscr{U}$-measurable random variables on $(\Omega, \mathscr{S}, \lambda)$. The function which minimizes (8.1.2) according to previous notation is $P_\lambda(g|L_2(\mathscr{U}))$, which is also denoted by $E_\lambda(g|\mathscr{U})$ for simplicity. This function is referred to in various ways including 'the conditional expectation of g given $\mathscr{U}$' (the notation $E_\lambda(g|\mathscr{U})$ is common in the literature), 'the least squares or L_2 projection of g onto $L_2(\mathscr{U})$', 'the regression of g given $\mathscr{U}$' and, if $\mathscr{U}$ is the collection of upper sets with respect to a quasi-order on Ω, 'the isotonic regression of g given $\mathscr{U}$' (the collection of upper sets is discussed in detail after the next theorem). The existence and uniqueness of $E_\lambda(g|\mathscr{U})$ follows from the fact that $L_2(\mathscr{U})$ is a closed, convex subset of L_2 (cf. Theorem 8.2.1).

Theorem 8.1.1 *If $\mathscr{U}$ is a σ-lattice of subsets of Ω then $L_2(\mathscr{U})$ is a closed, convex cone and a σ-lattice in L_2.*

Proof: It has already been verified that $L_2(\mathscr{U})$ is a convex cone and a σ-lattice. In order to see that $L_2(\mathscr{U})$ is closed, suppose $\{f_n\}$ is a sequence of members of $L_2(\mathscr{U})$ and f_n converges to f in L_2. Specifically,

$$\int[f_n(\omega)-f(\omega)]^2\lambda(\mathrm{d}\omega)\to 0\quad \text{as } n\to\infty.$$

Then there is a subsequence $\{f_{n_k}\}$ converging almost everywhere (λ) to f, so that $[f \geqslant a]$ differs by a set of measure zero from

$$\bigcap_{s=1}^{\infty} \bigcup_{r=1}^{\infty} \bigcap_{k=r}^{\infty} \left[f_{n_k} > a - \left(\frac{1}{s}\right) \right] \in \mathscr{U}$$

for each real a. Thus $f \in L_2(\mathscr{U})$ and $L_2(\mathscr{U})$ is closed in the L_2 metric. □

In a statistical inference problem the σ-lattices considered here are usually determined by a partial or quasi-order defined on Ω. Suppose $\lesssim$ is a quasi-order on Ω (cf. Definition 1.3.1) and define the collection $\mathscr{U}$ of *upper sets* corresponding to $\lesssim$ as follows: a subset U of Ω is a member of $\mathscr{U}$ if and only if $U \in \mathscr{S}$ and $x \lesssim y$ and $x \in U$ imply that $y \in U$. (Of course, we then enlarge $\mathscr{U}$ to make it complete with respect to λ.) It is a straightforward exercise to verify that the collection of upper sets corresponding to a quasi-order on Ω is a σ-lattice. In fact, if the requirement, $U \in \mathscr{S}$, were omitted the collection of upper sets would be closed under arbitrary unions and intersections and as a result would be called a complete lattice. However, since $\mathscr{S}$ is presumed to be only a σ-field, we can only conclude that $\mathscr{U}$ is a σ-lattice. Another σ-lattice of subsets of Ω corresponding to $\lesssim$ is the collection of lower sets. A subset L of Ω is a *lower set* corresponding to $\lesssim$ if and only if $L \in \mathscr{S}$ and $x \lesssim y$ and $y \in L$ imply that $x \in L$. It is a straightforward exercise to see that a set is a lower set if and only if its complement is an upper set. The proof of the following theorem is also a straightforward exercise.

Theorem 8.1.2 *Given the quasi-order $\lesssim$ on Ω, a random variable f is isotonic with respect to $\lesssim$ (cf. Definition 1.3.2) if and only if f is measurable with respect to the collection of upper sets corresponding to $\lesssim$.*

Another way these types of constraints may arise is somewhat analogous to the usual concept of the conditional expectation of one random variable given another random variable. In this setting, in addition to the random variable g, one is given a measurable mapping h from $(\Omega, \mathscr{S})$ into a measurable space $(X, \mathscr{B})$. The order restriction occurs on the image space, $(X, \mathscr{B})$, of h. Specifically, suppose $\ll$ is a quasi-order on X and let $\mathscr{U}' \subset \mathscr{B}$ be the collection of all upper subsets of X with respect to this quasi-order. In this context the isotonic regression problem is to find the real valued, measurable function f on $(X, \mathscr{B})$ which satisfies (a) $f(h) \in L_2$, (b) f is isotonic with respect to $\ll$, and (c) f minimizes

$$\int [g(\omega) - f(h(\omega))]^2 \lambda(\mathrm{d}\omega) \tag{8.1.3}$$

subject to (b).

This problem is actually equivalent to the problem described above. To see this let

$$\mathscr{U} = h^{-1}(\mathscr{U}') = \{h^{-1}(B): B \in \mathscr{U}'\}.$$

Then it is easy to verify that $\mathscr{U}$ is a σ-lattice of measurable subsets of Ω. Let T be the collection of all random variables on $(\Omega, \mathscr{S})$ of the form $f(h)$ where f is a real valued, measurable function on $(X, \mathscr{B})$ which is isotonic with respect to $\ll$. The following theorem is proved in Barlow *et al.* (1972).

Theorem 8.1.3 *The collection T is equal to $L_2(\mathscr{U})$.*

Thus the problem, (8.1.3), reduces to (8.1.2) with $\mathscr{U}$ defined as above.

A common approach to conditional expectations given a σ-field is through the theory of the Lebesgue–Radon–Nikodym (LRN) derivative. A somewhat more general LRN theory yields the conditional expectation given a σ-lattice as an operator on L_1. This operator coincides on $L_1 \cap L_2$ with the projection operator studied above. This LRN theory is developed in Section 7.4 of Barlow *et al.* (1972) and is based upon the work of Brunk and Johansen (1970). More recent research generalizing these notions of conditional expectations given σ-lattices to L_p-spaces is contained in Brunk (1975) and Landers and Rogge (1981b, 1984).

8.2 PROPERTIES OF CONDITIONAL EXPECTATION GIVEN A σ-LATTICE

Let H be a complete, real, inner product space (and thus a complete metric space with distance between elements f and g of H given by $\|f - g\|$ where $\|h\| = \langle h, h\rangle^{1/2}$ for $h \in H$ and $\langle \cdot, \cdot \rangle$ denotes the inner product). Of particular interest is the space $L_2(\Omega, \mathscr{S}, \lambda)$ discussed in Section 8.1 with the inner product given by

$$\langle f, g \rangle = \int f(\omega) g(\omega) \lambda(d\omega).$$

If $f, g \in H$, then seg fg will denote the segment joining f and g:

$$\text{seg}\, fg = \{(1-\alpha) f + \alpha g : 0 \leqslant \alpha \leqslant 1\}.$$

A subset of H is said to be *closed* if it is closed in the topology of the metric $\|f - g\|$. A subset $\mathscr{A}$ of H is *convex* if $f, g \in \mathscr{A}$ imply seg $fg \subset \mathscr{A}$.

The theorems in this section regarding projections onto closed, convex subsets have application to more general kinds of regression than isotonic regression. Suppose, for example, that g is a random variable on $(\Omega, \mathscr{S}, \lambda)$ and that $\mathscr{A}$ consists of functions $f(g) \in L_2$ such that f is convex on the range of g. Then $\mathscr{A}$ is a closed, convex cone in L_2. For another example, let $\phi_1, \phi_2, \ldots, \phi_k$ be prescribed real functions on the range of g such that $\phi_i(g) \in L_2$, $i = 1, 2, \ldots, k$, and let $\mathscr{A}$ consists of linear combinations

$$f(g) = \sum_{i=1}^{k} c_i \phi_i(g)$$

where some or all (or none) of the coefficients $c_1, c_2, \ldots, c_k$ are required to be nonnegative. Again, $\mathscr{A}$ is a closed, convex cone in L_2.

The existence, stated in Theorem 8.2.1, of the projection of a given point of H onto a given closed, convex subset of H is well known and easily verified using the identity

$$\| f + g \|^2 + \| f - g \|^2 = 2(\| f \|^2 + \| g \|^2). \tag{8.2.1}$$

This identity is easily verified by expanding the left-hand side in terms of the inner product. A special case of Theorem 8.2.1, discussed in Chapter 1, is given in Theorem 1.3.1.

Theorem 8.2.1 *If $\mathscr{A}$ is a closed, convex subset of H and $g \in H$, then there exists a unique closest point $g^* = P(g|\mathscr{A})$ of $\mathscr{A}$ to g.*

Proof: Let

$$v = \inf_{f \in \mathscr{A}} \| g - f \|$$

and let $\{f_n\}$ be a sequence of points in $\mathscr{A}$ such that

$$\lim_n \| f_n - g \| = v.$$

Using (8.2.1),

$$\tfrac{1}{4} \| f_m - f_n \|^2 = \tfrac{1}{2} \| f_m - g \|^2 + \tfrac{1}{2} \| f_n - g \|^2 - \| \tfrac{1}{2}(f_m + f_n) - g \|^2. \tag{8.2.2}$$

Since $\mathscr{A}$ is convex, $\frac{1}{2}(f_m + f_n) \in \mathscr{A}$ and $\| \frac{1}{2}(f_m + f_n) - g \|^2 \geqslant v^2$ for all positive integers m and n. However, $\| f_m - g \|^2 \to v^2$ and $\| f_n - g \|^2 \to v^2$ as $m, n \to \infty$ so that, using (8.2.2), $\{f_n\}$ is a Cauchy sequence and has a limit which is denoted by $g^* = P(g|\mathscr{A})$. The point g^* is in $\mathscr{A}$ since $\mathscr{A}$ is closed.

To prove uniqueness, suppose g_1^* and g_2^* are two closest points of $\mathscr{A}$ to g ($\| g_i^* - g \| = v$). Applying (8.2.2) with f_m and f_n replaced by g_1^* and g_2^*, respectively, one concludes that $\| g_1^* - g_2^* \| = 0$ and $g_1^* = g_2^*$. □

Corollary *If $g \in L_2(\Omega, \mathscr{S}, \lambda)$ and if $\mathscr{U}$ is a σ-lattice of measurable subsets of Ω, then there exists a unique (up to λ-equivalence) $\mathscr{U}$-measurable function $g^* = E(g|\mathscr{U})$ in the class $L_2(\mathscr{U})$ minimizing the integral $\int (g - f)^2 \, d\lambda$.*

The next theorem characterizes g^* and is given for the special setting discussed in Chapter 1 in Theorem 1.3.1. It has the interpretation that $h = P(g|\mathscr{A})$ if and only if the angle between $g - h$ and $h - f$ is obtuse for all $f \in \mathscr{A}$ (cf. Figure 1.3.1).

Theorem 8.2.2 *If $g \in H$ and if $\mathscr{A}$ is a closed, convex subset of H then an element h of $\mathscr{A}$ is equal to $P(g|\mathscr{A})$ if and only if*

$$\langle g-h, h-f\rangle \geqslant 0 \qquad \text{for all } f \in \mathscr{A}. \tag{8.2.3}$$

Proof: For every $f \in \mathscr{A}$, the point $g^* = P(g|\mathscr{A})$ is also the closest point of seg fg^* to g. Thus, the following function of $\alpha(0 \leqslant \alpha \leqslant 1)$ achieves a minimum at $\alpha = 0$:

$$\begin{aligned}\|g-(1-\alpha)g^* - \alpha f\|^2 &= \|(g-g^*) - \alpha(f-g^*)\|^2 \\ &= \|g-g^*\|^2 + 2\alpha\langle g-g^*, g^*-f\rangle + \alpha^2\|f-g^*\|^2.\end{aligned}$$

Thus the derivative of this function of α evaluated at $\alpha = 0$ must be nonnegative. This shows that (8.2.3) is necessary.

To prove sufficiency, suppose $h \in \mathscr{A}$ has property (8.2.3). Then for every $f \in \mathscr{A}$, h is the closest point of seg hf to g. In particular, h is the closest point of seg hg^* to g. Since the closest point is unique by Theorem 8.2.1, $h = g^*$. □

Theorem 8.2.3 *If $g \in H$ and $\mathscr{A}$ is a closed, convex subset of H then $-\mathscr{A} = \{-f : f \in \mathscr{A}\}$ is a closed, convex subset of H and $P(-g|-\mathscr{A}) = -P(g|\mathscr{A})$.*

Proof: Set $g^* = P(g|\mathscr{A})$. Clearly, $-g^* \in -\mathscr{A}$. One verifies that $-g^* = P(-g|-\mathscr{A})$ by verifying (8.2.3) for $-g$, $-\mathscr{A}$, and $-g^*$. If $f \in -\mathscr{A}$ then $\langle -g-(-g^*), (-g^*)-f\rangle = \langle g-g^*, g^*-(-f)\rangle \geqslant 0$. By Theorem 8.2.2, $-g^* = P(-g|-\mathscr{A})$. □

A corollary for conditional expectations given σ-lattices is obtained by noting that $-L_2(\mathscr{U})$ is the subset of L_2 consisting of all random variables which are measurable with respect to the σ-lattice, $\mathscr{U}^c$, consisting of all complements of elements of $\mathscr{U}$. Thus, if $\mathscr{U}$ is the collection of all upper sets with respect to a quasi-order on Ω, $\mathscr{U}^c$ is the collection of all lower sets.

Corollary *If $g \in L_2(\Omega, \mathscr{S}, \lambda)$ and $\mathscr{U}$ is a σ-lattice of measurable subsets of Ω then $E(-g|\mathscr{U}^c) = -E(g|\mathscr{U})$.*

The term *operator* is used in this section to denote a mapping from H into itself; it need not be linear. Thus, projection onto a closed, convex set is an operator in that it associates an element of H, namely $P(g|\mathscr{A})$, with each element g of H $(g \to P(g|\mathscr{A}))$. A *unitary* operator is by definition a *linear* operator which preserves the inner product (and hence distance):

$$\langle Tf, Tg\rangle = \langle f, g\rangle.$$

Loosely speaking, the following theorem states that projections are preserved under stretchings and under distance preserving transformations.

Theorem 8.2.4 *If $g \in H$, if $\mathscr{A}$ is a closed, convex subset of H, and if T is a unitary operator on $\mathscr{A}$, then $T\mathscr{A}$ is a closed, convex set and $P(Tg|T\mathscr{A}) = TP(g|\mathscr{A})$. If k is real then $k\mathscr{A} = \{kf : f \in \mathscr{A}\}$ is a closed, convex set and $P(kg|k\mathscr{A}) = kP(g|\mathscr{A})$. If a is real and $ka > 0$ then*

$$P(ag + (k-a)P(g|\mathscr{A})|k\mathscr{A}) = kP(g|\mathscr{A}). \tag{8.2.4}$$

Proof: The conclusion that $T\mathscr{A}$ is convex is a straightforward consequence of the linearity of T. In order to see that $T\mathscr{A}$ is closed, suppose u_n is a sequence of points in $T\mathscr{A}$ and $u_n \to u$ in L_2. Let $u_n = Tg_n$ where $g_n \in \mathscr{A}$. Then $\{u_n\}$ is a Cauchy sequence and since T is distance preserving, $\{g_n\}$ is also a Cauchy sequence. Let $g_n \to g$ in L_2. Then

$$\| Tg - u \| = \lim_{n \to \infty} \| Tg_n - u \| = 0$$

so that $u = Tg$ and $u \in T\mathscr{A}$. Thus $T\mathscr{A}$ is closed and the proof that $k\mathscr{A}$ is closed is similar. Let $g^* = P(g|\mathscr{A})$. If $h \in \mathscr{A}$ then $\langle Tg - Tg^*, Tg^* - Th \rangle = \langle g - g^*, g^* - h \rangle \geqslant 0$ so that, by Theorem 8.2.2, $Tg^* = P(Tg|T\mathscr{A})$. Equation (8.2.4) follows similarly, and the equality preceding equation (8.2.4) in the statement of Theorem 8.2.4 is obtained by setting $a = k$. □

Corollary *If $g \in L_2(\Omega, \mathscr{S}, \lambda)$, if $\mathscr{U}$ is a σ-lattice of measurable subsets of Ω, and if $k \geqslant 0$, then $E(kg|\mathscr{U}) = kE(g|\mathscr{U})$.*

Proof: If $\mathscr{A} = L_2(\mathscr{U})$, then $k\mathscr{A} = \mathscr{A}$ and the corollary follows. □

Theorem 8.2.5 *Projection onto a closed, convex set is a distance reducing, and therefore continuous, operator.*

Proof: Suppose $\mathscr{A}$ is a closed, convex subset of H and suppose $g_1, g_2 \in H$ with $g_i^* = P(g_i|\mathscr{A})$, $i = 1, 2$. Then

$$\begin{aligned} \| g_2 - g_1 \|^2 &= \| g_2 - g_2^* + g_2^* - g_1^* + g_1^* - g_1 \|^2 \\ &= \| g_2 - g_2^* + g_1^* - g_1 \|^2 + \| g_2^* - g_1^* \|^2 + 2\langle g_1 - g_1^*, g_1^* - g_2^* \rangle \\ &\quad + 2\langle g_2 - g_2^*, g_2^* - g_1^* \rangle \\ &\geqslant \| g_2^* - g_1^* \|^2 \end{aligned}$$

by Theorem 8.2.2. □

Corollary *If $g_1, g_2 \in L_2(\Omega, \mathscr{S}, \lambda)$ and if $\mathscr{U}$ is a σ-lattice of measurable subsets of Ω, then*

$$\int (g_2^* - g_1^*)^2 \, d\lambda \leqslant \int (g_2 - g_1)^2 \, d\lambda,$$

where $g_i^* = E(g_i|\mathscr{U})$, $i = 1, 2$.

Definition 8.2.1 An operator T on a Hilbert space H is *strictly monotonic at* g if $Tf \neq Tg$ implies $\langle g-f, Tg-Tf\rangle > 0$. It is *strictly monotonic* if it is strictly monotonic at each $g \in H$ (cf. Vainburg, 1973).

Theorem 8.2.6 *If* $\mathscr{A}$ *is a closed, convex subset of* H, *then* $P(\cdot|\mathscr{A})$ *is strictly monotonic. Indeed, if* $g_i \in H$ *and* $g_i^* = P(g_i|\mathscr{A})$, $i = 1, 2$, *then*

$$\langle g_1 - g_2, g_1^* - g_2^* \rangle \geqslant \|g_1^* - g_2^*\|. \tag{8.2.5}$$

Proof: Using Theorem 8.2.2,

$$\begin{aligned}\langle g_1 - g_2, g_1^* - g_2^* \rangle &= \langle g_1 - g_1^*, g_1^* - g_2^* \rangle + \langle g_1^*, g_1^* - g_2^* \rangle - \langle g_2 - g_2^*, g_1^* - g_2^* \rangle \\ &\quad - \langle g_2^*, g_1^* - g_2^* \rangle \\ &= \langle g_1 - g_1^*, g_1^* - g_2^* \rangle + \langle g_2 - g_2^*, g_2^* - g_1^* \rangle + \|g_1^* - g_2^*\|^2 \\ &\geqslant \|g_1^* - g_2^*\|^2 \geqslant 0. \quad \square\end{aligned}$$

In passing, it is noted that an alternative proof of the distance reducing property of projections given in Theorem 8.2.5 is furnished by (8.2.5) and the Schwarz inequality:

$$\|g_1^* - g_2^*\|^2 \leqslant \langle g_1 - g_2, g_1^* - g_2^* \rangle \leqslant \|g_1 - g_2\| \, \|g_1^* - g_2^*\|.$$

Properties of projections on closed, convex cones

Theorem 8.2.7 *If* $g \in H$ *and if* $\mathscr{A}$ *is a closed, convex cone in* H, *then* $h = P(g|\mathscr{A})$ *if and only if* $h \in \mathscr{A}$ *and*

$$\langle g - h, h \rangle = 0 \tag{8.2.6}$$

and

$$\langle g - h, f \rangle \leqslant 0 \tag{8.2.7}$$

for all $f \in \mathscr{A}$. *Also, if* $g^* = P(g|\mathscr{A})$ *and* $a \geqslant -1$, *then*

$$P(g + ag^*|\mathscr{A}) = (a+1)g^*. \tag{8.2.8}$$

Proof: Suppose first that $h = g^* = P(g|\mathscr{A})$. Applying Theorem 8.2.2 with $f = kh$ in the two cases $0 < k < 1$ and $k > 1$ yields (8.2.6). Then (8.2.7) follows from (8.2.3) and (8.2.6).

Conversely, if (8.2.6) and (8.2.7) are satisfied for some $h \in \mathscr{A}$, then inequality (8.2.3) follows so that $h = g^*$.

Finally, suppose $g^* = P(g|\mathscr{A})$ and $a \geqslant -1$. Then clearly $(a+1)g^* \in \mathscr{A}$ and it is straightforward to verify (8.2.6) and (8.2.7) with $g + ag^*$ in place of g and $(a+1)g^*$ in place of h. Thus (8.2.8) is verified. $\square$

Corollary A *If* $g \in L_2(\Omega, \mathscr{S}, \lambda)$ *and if* $\mathscr{U}$ *is a* σ*-lattice of measurable subsets of* Ω, *then*

$h = g^* = E(g|\mathscr{U})$ *if and only if* $h \in L_2(\mathscr{U})$,

$$\int (g - h)h \, d\lambda = 0 \tag{8.2.9}$$

and

$$\int (g - h)f \, d\lambda \leqslant 0 \tag{8.2.10}$$

for all $f \in L_2(\mathscr{U})$. *Also,* $E(g + ag^*|\mathscr{U}) = (a + 1)g^*$ *for all* $a \geqslant -1$.

It is an immediate consequence of (8.2.8) that if $\mu \in H$ and $\mathscr{A}$ is a closed, convex cone in H, then the distance from μ to $\mathscr{A}$ is the same as the distance from $\mu + a\mu^*$ to $\mathscr{A}$ for any $a \geqslant -1$ (cf. Figure 8.2.1). Thus, it might not be unreasonable to expect power functions of tests such as those discussed in Chapters 2 and 4 to have certain monotonicity properties when regarded as functions of $a (a \geqslant -1)$ with μ fixed. This is, indeed, the case and these monotonicity properties have been used to find extreme values of such power functions (cf. Mukerjee *et al.*, 1986a, and Section 2.6).

If $(\Omega, \mathscr{S}, \lambda)$ is a probability space and if $\mathscr{A}$ contains all constant functions, then (8.2.9) and (8.2.10) can be expressed in terms of variances and covariances. We must first show that under these circumstances, $E(g) = E(g^*)$.

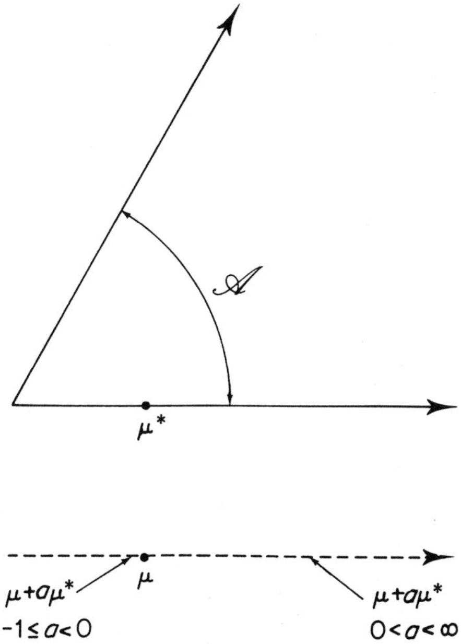

Figure 8.2.1 A closed convex cone $\mathscr{A}$, a point μ, its projection onto $\mathscr{A}$, μ^*, and the line $\{\mu + a\mu^* : a \geqslant -1\}$

Corollary B *If $g \in L_2(\Omega, \mathscr{S}, \lambda)$, if $\lambda(\Omega) = 1$, if $\mathscr{A}$ is a closed, convex cone in L_2 containing the constant functions, and if $g^* = P(g|\mathscr{A})$, then*

$$\int g \, d\lambda = \int g^* \, d\lambda.$$

Proof: Apply (8.2.10) with f equal to the constant function 1 and then f equal to the constant function -1. □

Corollary C *If $(\Omega, \mathscr{S}, \lambda)$ is a probability space, if $\mathscr{A}$ is a closed, convex cone in L_2 containing the constant functions, and if $g \in L_2$, then a function h in $\mathscr{A}$ is equal to $g^* = P(g|\mathscr{A})$ if and only if*

$$E(g) = E(h) \tag{8.2.11}$$

$$\operatorname{cov}(h, f) \geqslant \operatorname{cov}(g, f) \tag{8.2.12}$$

for all $f \in \mathscr{A}$, and

$$\operatorname{Var} h = \operatorname{cov}(g, h). \tag{8.2.13}$$

Further, for each real a, $P(g + a|\mathscr{A}) = P(g|\mathscr{A}) + a$.

Proof: The proof is a straightforward consequence of Corollaries A and B once the conditions (8.2.11), (8.2.12), and (8.2.13) are written in terms of integrals. □

Corollary D *If the assumptions of Corollary C are satisfied then $\operatorname{Var} g = E(g - g^*)^2 + \operatorname{Var} g^*$ so that $\operatorname{Var} g^* \leqslant \operatorname{Var} g$ with equality if and only if $g = g^*$.*

Corollary E *If the assumptions of Corollary C are satisfied then*

(a) *$g^* = P(g|\mathscr{A})$ is a degenerate (constant) random variable having value $E(g)$ if and only if all nondegenerate random variables in $\mathscr{A}$ have nonpositive correlations with g, that is $g \in \mathscr{A}^*$ (cf. Section 1.6)*

and

(b) *if there exists a nondegenerate random variable in $\mathscr{A}$ having nonnegative correlation with g, then $g^* = P(g|\mathscr{A})$ is the random variable in $\mathscr{A}$ most highly correlated with g.*

Proof: First assume that g^* is constant and that f is a nondegenerate random variable in $\mathscr{A}$. Then $\operatorname{cov}(g, f) = E[(g - E(g))(f - E(f))] = E[(g - g^*)(f - E(f))] \leqslant 0$ by Corollary B and (8.2.10). Conversely, suppose all nondegenerate random variables in $\mathscr{A}$ have nonpositive correlation with g. Then by (8.2.13), $\operatorname{Var} g^* = \operatorname{cov}(g, g^*) \leqslant 0$ so that $\operatorname{Var} g^* = 0$ and g^* must be constant.

In order to prove (b), we first show that if $f \in \mathscr{A}$, $E(f) = E(g)$ and $\operatorname{Var} f = \operatorname{cov}(g, f)$ then $\operatorname{Var} g^* \geqslant \operatorname{Var} f$. Let $\mu_g = E(g) = E(g^*)$. Let $\mathscr{A}_1$ denote the subset of $\mathscr{A}$ consisting of functions f such that $E(f) = E(g)$ and $\operatorname{Var} f = \operatorname{cov}(g, f)$. The random variable g^* is the closest point of $\mathscr{A}$ to g and $g^* \in \mathscr{A}_1$ so that g^* is the closest point of $\mathscr{A}_1$ to g. However, if $f \in \mathscr{A}_1$, then $E(g-f)^2 = \operatorname{Var} g - 2\operatorname{cov}(g,f) + \operatorname{Var} f = \operatorname{Var} g - \operatorname{Var} f$. Thus, $\operatorname{Var} g - \operatorname{Var} f \geqslant \operatorname{Var} g - \operatorname{Var} g^*$, which implies that $\operatorname{Var} f \leqslant \operatorname{Var} g^*$ for all $f \in \mathscr{A}$.

Suppose now that there exist nondegenerate random variables in $\mathscr{A}$ which have nonnegative correlations with g. Since $\operatorname{cov}(g, g^*) = \operatorname{Var} g^* \geqslant 0$ it suffices to prove (b) for random variables in $\mathscr{A}$ which have positive correlation with g. Suppose f is nondegenerate, $f \in \mathscr{A}$, and $\operatorname{cov}(f,g) > 0$. Set $f' = \mu_g + k[f - E(f)]$ where $k = \operatorname{cov}(g,f)/\operatorname{Var} f$. Then $f' \in \mathscr{A}_1$ and

$$\frac{\operatorname{cov}(g,g^*)}{\sigma_g \sigma_{g^*}} = \frac{\operatorname{Var} g^*}{\sigma_g \sigma_{g^*}} = \frac{\sigma_{g^*}}{\sigma_g} \geqslant \frac{\sigma_{f'}}{\sigma_g}$$

$$= \frac{\operatorname{Var} f'}{\sigma_{f'}\sigma_g} = \frac{\operatorname{cov}(g,f')}{\sigma_{f'}\sigma_g} = \frac{\operatorname{cov}(g,f)}{\sigma_f \sigma_g}$$

as was to be proved. □

The space $L_2 = L_2(\Omega, \mathscr{S}, \lambda)$ is a Hilbert space which also possesses a lattice structure, i.e. if f, $g \in L_2$ then so are $f \vee g$ and $f \wedge g$. Two useful properties involving this lattice structure are:

$$(f,g) = \int fg \, \mathrm{d}\lambda = \int (f \wedge g)(f \vee g) \, \mathrm{d}\lambda = \langle f \wedge g, f \vee g \rangle$$

and

$$f \geqslant 0, \quad g \geqslant 0 \quad \text{implies} \quad \langle f, g \rangle = \int fg \, \mathrm{d}\lambda \geqslant 0.$$

For the next theorem we assume such a lattice structure for the Hilbert space H. A real vector space which has a lattice structure which is compatible with the vector space structure is called a *Riesz space* (cf. Luxemburg and Zaanen, 1971). By this compatibility, we mean that there is a partial order $\lesssim$ on H such that for every pair f, g of elements of H there exists an infimum $f \wedge g$ and a supremum $f \vee g$ in H with the property that $f \wedge g \lesssim f, g \lesssim f \vee g$, and that

$$f \lesssim g \text{ implies } f + h \lesssim g + h \qquad \text{for all } h \in H, \tag{8.2.14}$$

and

$$f \lesssim g \text{ implies } kf \lesssim kg \qquad \text{for all } k \geqslant 0 \tag{8.2.15}$$

It can then be shown that

$$f + g = (f \wedge g) + (f \vee g) \qquad \text{for all } f,\ g \in H. \tag{8.2.16}$$

In addition to these properties, we assume that

$$(f,g) = \langle f \wedge g, f \vee g \rangle \tag{8.2.17}$$

and

$$f \gtrsim 0, \quad g \gtrsim 0 \quad \text{implies} \quad \langle f, g \rangle \geqslant 0 \tag{8.2.18}$$

for all f, $g \in H$.

Marshall, Walkup, and Wets (1967) study quasi-orders which are defined in terms of cones (cone orderings). A cone ordering on H (induced by a convex cone C) is a partial order relation $\lesssim$ on H defined by

$$f \lesssim g \quad \text{if and only if } g - f \in C.$$

It is straightforward to verify that any cone ordering is a quasi-ordering and satisfies (8.2.14) and (8.2.15). The central theorem of Marshall, Walkup, and Wets (1967) gives necessary and sufficient conditions for functions on H to be isotonic with respect to cone orderings.

The concept of majorization, which has important applications in probability and statistics, is a cone ordering. A function which is monotonic with respect to the quasi-ordering defined by majorization is called a Schur function. In addition to the paper by Marshall *et al.*, important results and applications involving Schur functions and majorization are found in Proschan and Sethuraman (1977).

Robertson and Wright (1982a, 1984) study several cone orderings in order to quantify the concept that one parameter point conforms to a partial ordering more than does another parameter point. Their cone ordering '$\gg$' gives an example of a quasi-order which does not satisfy (8.2.18).

Theorem 8.2.8 *If $g_i \in H, i = 1, 2$, if $\mathscr{A}$ is a closed, convex cone and a lattice in H, if $g_i^* = P(g_i | \mathscr{A}), i = 1, 2$, and if $g_2 \gtrsim g_1$, then $g_2^* \gtrsim g_1^*$.*

Proof: Set $\underline{g} = g_1^* \wedge g_2^*$ and $\bar{g} = g_1^* \vee g_2^*$ and note that, since $\underline{g} \in \mathscr{A}$,

$$\langle g_1 - g_1^*, g_1^* - \underline{g} \rangle \geqslant 0$$

by (8.2.3). Also, from (8.2.18),

$$\langle g_2 - g_1, g_1^* - \underline{g} \rangle \geqslant 0.$$

Adding these inequalities yields $\langle g_2 - g_1^*, g_1^* - \underline{g} \rangle \geqslant 0$ and, since $g_1^* = \underline{g} + \bar{g} - g_2^*$ by (8.2.16), $\langle g_2, \bar{g} - g_2^* \rangle \geqslant \langle g_1^*, g_1^* - \underline{g} \rangle$. This latter is equivalent to

$$-2\langle g_2, g_2^* \rangle \geqslant 2\| g_1^* \| - 2\langle g_2, \bar{g} \rangle - 2\langle g_1^*, \underline{g} \rangle. \tag{8.2.19}$$

Also, a straightforward manipulation, using (8.2.16) and (8.2.17), implies that

$$\| g_1^* \|^2 + \| g_2^* \|^2 = \| \underline{g} \|^2 + \| \bar{g} \|^2,$$

so that by adding $\| g_2 \|^2 + \| g_2^* \|^2$ to each member of (8.2.19), one obtains

$$\| g_2 - g_2^* \|^2 \geqslant \| g_2 - \bar{g} \|^2 + \| g_1^* - \underline{g} \|^2.$$

Since $\bar{g} \in \mathscr{A}$ and $h = g_2^*$ minimizes $\| g_2 - h \|$ in $\mathscr{A}$, this last inequality implies that $g_1^* = \underline{g}$ and $g_2^* = \bar{g}$. □

Corollary A *If $g_i \in L_2(\Omega, \mathscr{S}, \lambda)$, $i = 1, 2$, if $\mathscr{U}$ is a σ-lattice of measurable subsets of Ω, if $g_i^* = E(g_i | \mathscr{U})$, and if $g_2 \gtrsim g_1$, then $g_2^* \gtrsim g_1^*$.*

Corollary B *If $g \in L_2(\Omega, \mathscr{S}, \lambda)$, if $\mathscr{U}$ is a σ-lattice of measurable subsets of Ω, if $h \in L_2(\mathscr{U})$, and if $g \lesssim h$ ($g \gtrsim h$), then $g^* \lesssim h$ ($g^* \gtrsim h$).*

Proof: Corollary B follows from Corollary A on noting that $h^* = h$ since $h \in L_2(\mathscr{U})$. □

8.3 A CHARACTERIZATION OF CONDITIONAL EXPECTATION GIVEN A σ-LATTICE

What conditions on an operator $T: L_2 \to L_1$ imply the existence of a σ-lattice, $\mathscr{U}$, of measurable subsets of Ω such that $T = P(\cdot | \mathscr{U})$? This section gives an answer due to Dykstra (1970a). First a characterization of projection on a closed, convex cone is given. An operator T is *positively homogeneous* if $g \in H, a \geqslant 0$ imply $T(ag) = aT(g)$. It is *distance reducing* if $f, g \in H$ imply $\| Tf - Tg \| \leqslant \| f - g \|$. The operator T is *idempotent* if $T(T(g)) = T(g)$ for all $g \in H$.

Theorem 8.3.1 *Let T be a positively homogeneous, idempotent, distance reducing operator on a Hilbert space H which is strictly monotonic at $\mathbf{0}$. Then $\mathscr{A} = T(H)$, the range of T, is convex and $T = P(\cdot | \mathscr{A})$.*

Remark *If $T(\mathbf{0}) = \mathbf{0}$ then T is strictly monotonic at $\mathbf{0}$ if and only if for each $g \in H$ there is a positive $a = a(g)$ such that $\| g \|^2 \geqslant \| g - aTg \|^2$.*

Proof: If the latter holds then $\langle g, Tg \rangle \geqslant (a/2) \| Tg \|^2$, so that T is strictly monotonic at $\mathbf{0}$. Conversely, if T is strictly monotonic at $\mathbf{0}$ and if $Tg \neq \mathbf{0}$ then $\| g \|^2 \geqslant \| g - aTg \|^2$ where $0 < a \leqslant 2\langle g, Tg \rangle / \| Tg \|^2$. □

Thus, strict monotonicity at $\mathbf{0}$ has an obvious geometric interpretation: if $T(\mathbf{0}) = \mathbf{0}$, T is strictly monotonic at $\mathbf{0}$ if and only if for each $g \in H$ some positive multiple of Tg is at least has close to g as is $\mathbf{0}$.

Proof of Theorem 8.3.1: First we argue that $\mathscr{A}$, the range of T, is convex. Suppose that it is not. Then there exist $f', g' \in \mathscr{A}$ and $\alpha \in [0, 1]$ such that $h' = \alpha f' + (1 - \alpha) g' \notin \mathscr{A}$ (cf. Figure 8.3.1). Using the triangle inequality and the assumptions that T is idempotent and distance reducing it is possible to write

$$
\begin{aligned}
\| f' - g' \| &\leqslant \| f' - Th' \| + \| Th' - g' \| \\
&= \| Tf' - Th' \| + \| Th' - Tg' \| \\
&\leqslant \| f' - h' \| + \| h' - g' \| \\
&= \| f' - \alpha f' - (1 - \alpha) g' \| + \| \alpha f' + (1 - \alpha) g' - g' \| \\
&= \| f' - g' \|.
\end{aligned}
$$

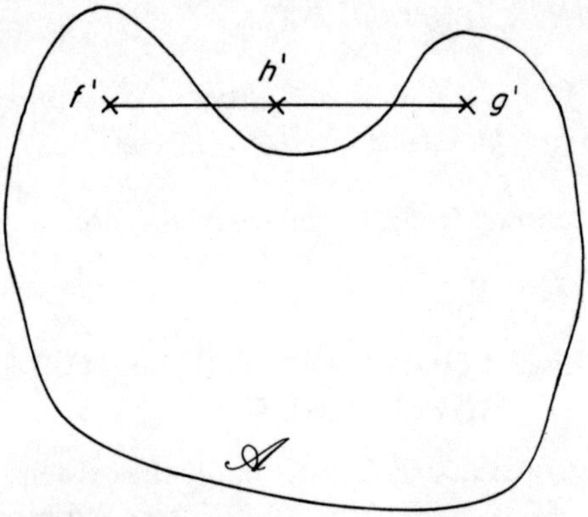

Figure 8.3.1 A set which is not convex

Thus, $\| f' - g' \| = \| f' - Th' \| + \| Th' - g' \|$ and it follows from the case of equality in the triangle inequality (cf. Smiley, 1963) that $f' - Th'$ and $Th' - g'$ are linearly dependent and thus that Th' is on the line segment joining f' and g'. If Th' is in the direction of g' from h' (cf. Figure 8.3.1), then the distance from Th' to $f' = Tf'$ is larger than the distance from f' to h', contradicting the distance reducing property of T. The case in which Th' is in the direction of f' yields a similar conclusion so that $Th' = h'$ and $h' \in \mathscr{A}$, which is a contradiction. Thus $\mathscr{A}$ is convex.

Since T is distance reducing and idempotent, $\mathscr{A}$ is closed, for if $f_n \in \mathscr{A}$, $n = 1, 2, \ldots$, and $\| f_n - g \| \to 0$ as $n \to \infty$, then $\| f_n - Tg \| = \| Tf_n - Tg \| \leqslant \| f_n - g \| \to 0$ as $n \to \infty$, and hence $g = Tg \in \mathscr{A}$. Since T is positively homogeneous, $\mathscr{A}$ is a cone containing the origin.

It will now be shown that $g \in H$, $f \in \mathscr{A}$ imply

$$\langle g - Tg, f \rangle \leqslant 0. \tag{8.3.1}$$

This follows since if $c > 0$, $\| Tg - cf \|^2 = \| Tg - T(cf) \|^2 \leqslant \| g - cf \|^2$ so that

$$\| Tg \|^2 \leqslant \| g \|^2 - 2c \langle g - Tg, f \rangle.$$

Since this holds for arbitrary positive c and since $\| Tg \| < \infty$, one concludes that $\langle g - Tg, f \rangle \leqslant 0$. Now set $T_1 = P(\cdot \mid \mathscr{A})$ and observe that

$$T_1 g = \mathbf{0} \text{ implies that } Tg = \mathbf{0}. \tag{8.3.2}$$

To see this, suppose $T_1 g = \mathbf{0}$. Then by Theorem 8.2.7, $0 \geqslant \langle g - T_1 g, Tg \rangle = \langle g, Tg \rangle$. Since T is strictly monotonic at $\mathbf{0}$ and $T(\mathbf{0}) = \mathbf{0}$ (T is positively homogeneous), it follows that $Tg = \mathbf{0}$.

Now from Theorem 8.2.7, (8.3.1), and the Schwarz inequality, we have for $g \in H$,

$$\| T_1 g \|^2 = \langle g, T_1 g \rangle \leqslant \langle Tg, T_1 g \rangle \leqslant \| Tg \| \, \| T_1 g \|.$$

Hence

$$\| T_1 g \| \leqslant \| Tg \| \tag{8.3.3}$$

for $g \in H$ with equality holding if and only if there is a real number a such that $T_1 g = aTg$. However, from Theorem 8.2.7, $T_1(g - T_1 g) = \mathbf{0}$, so that by (8.3.2), $T(g - T_1 g) = \mathbf{0}$. Then, since T is distance reducing, $\| T_1 g \|^2 = \| g - (g - T_1 g) \|^2 \geqslant \| Tg - T(g - T_1 g) \|^2 = \| Tg \|^2$. Thus equality holds in (8.3.3) and $T_1 g = aTg$. Clearly $a^2 = 1$. If $T_1 g = -Tg$ then, from (8.3.1),

$$0 \geqslant \langle g - Tg, T_1 g \rangle = \langle g - T_1 g + 2T_1 g, T_1 g \rangle = 2 \| T_1 g \|^2$$

by Theorem 8.2.7. In this case $\| T_1 g \| = \| Tg \| = 0$ and $T_1 g = Tg$. Thus in any case $T_1 g = Tg$. □

Definition 8.3.1 Let $(\Omega, \mathscr{S}, \lambda)$ be a probability space. An operator T on $L_2(\Omega, \mathscr{S}, \lambda)$ is *isotonic* if $g_i \in L_2$, $i = 1, 2$, and $g_1 \lesssim g_2$ imply $Tg_1 \lesssim Tg_2$. It is *expectation invariant* if $E(Tg) = E(g)$ for all $g \in L_2$.

Theorem 8.3.2 *Let $(\Omega, \mathscr{S}, \lambda)$ be a probability space and let T be a positively homogeneous, isotonic, expectation invariant, idempotent, distance reducing operator on $L_2(\Omega, \mathscr{S}, \lambda)$. If $\mathscr{A}$ is the range of T and $\mathscr{M} = \{[f > a]; f \in \mathscr{A}, a \in R\}$, then $\mathscr{M}$ is a σ-lattice of measurable subsets of Ω and $\mathscr{A} = L_2(\mathscr{M})$.*

Proof (1): *$\mathscr{A}$ is a lattice: $f, g \in \mathscr{A}$ implies $f \vee g$, $f \wedge g \in \mathscr{A}$.*
To prove (1), let $f, g \in \mathscr{A}$. Since T is isotonic and idempotent, $T(f \vee g) \geqslant Tf \vee Tg = f \vee g$. This implies that

$$\int [T(f \vee g) - f]^2 \, d\lambda \geqslant \int [(f \vee g) - f]^2 \, d\lambda.$$

However, since T is distance reducing,

$$\int [(f \vee g) - f]^2 \, d\lambda \geqslant \int [T(f \vee g) - Tf]^2 \, d\lambda = \int [T(f \vee g) - f]^2 \, d\lambda,$$

and hence

$$\int [T(f \vee g) - f]^2 \, d\lambda = \int [(f \vee g) - f]^2 \, d\lambda.$$

Since $T(f \vee g) \geqslant f \vee g \geqslant f$, it follows that $T(f \vee g) = f \vee g$ and $f \vee g \in \mathscr{A}$. Similarly, it can be argued that $f \wedge g \in \mathscr{A}$.

(2): *If $f \in \mathscr{A}$ and if c is a constant function then $c \in \mathscr{A}$, $f + c \in \mathscr{A}$, and $T(f + c) = Tf + c$.*
To prove (2), note first that by the Schwarz inequality

$$\int [T(f + c) - Tf]^2 \, d\lambda \geqslant \left\{ \int [T(f + c) - Tf] \, d\lambda \right\}^2 = \left(\int c \, d\lambda \right)^2$$

since T is expectation invariant. Thus

$$\int [T(f+c)-Tf]^2 \,\mathrm{d}\lambda \geqslant \int c^2 \,\mathrm{d}\lambda = \int [(f+c)-f]^2 \,\mathrm{d}\lambda.$$

However, the reverse inequality holds also since T is distance reducing, and this implies there exists a real number a such that $T(f+c)-Tf = a[(f+c)-f] = ac$. But since T is expectation invariant, we have $T(f+c)-Tf = c$. The first part of (2) follows by taking $f = \mathbf{0}$ ($\mathbf{0} \in \mathscr{A}$ since T is positively homogeneous). To obtain the second part, note that if $f \in \mathscr{A}$ ($f = Tg$), then $f + c = T(g+c)$.

(3): *$f \in \mathscr{A}$ implies $I_{[f>a]} \in \mathscr{A}$, where $I_{[f>a]}$ denotes the indicator function of the set $[f>a]$.*
To see this, set $f_n = [n(f-a) \vee 0] \wedge 1$ for $n = 1, 2, \ldots$. Using the assumption that T is positively homogeneous and (1) and (2), it follows that $f_n \in \mathscr{A}$, $n = 1, 2, \ldots$. Moreover, $f_n \uparrow I_{[f>a]}$ and hence

$$\int [f_n - I_{[f>a]}]^2 \,\mathrm{d}\lambda \to 0.$$

Since the assumptions that T is idempotent and distance reducing imply that it is closed, it follows that $I_{[f>a]} \in \mathscr{A}$.

(4): *$\mathscr{M}$ is a σ-lattice.*
Suppose $B_n \in \mathscr{M}$, $n = 1, 2, \ldots$. Then by (3), $I_{B_n} \in \mathscr{A}$, $n = 1, 2, \ldots$. By (1),

$$I_{\bigcup_{n=1}^{k} B_n} = \bigvee_{n=1}^{k} I_{B_n} \in \mathscr{A}$$

and

$$I_{\bigcap_{n=1}^{k} B_n} = \bigwedge_{n=1}^{k} I_{B_n} \in \mathscr{A}.$$

Since $\mathscr{A}$ is closed and

$$I_{\bigcup_{n=1}^{k} B_n} \uparrow I_{\bigcup_{n=1}^{\infty} B_n},$$

one concludes that $I_{\bigcup_{n=1}^{\infty} B_n} \in \mathscr{A}$ and $\bigcup_{n=1}^{\infty} B_n = [I_{\bigcup_{n=1}^{\infty} B_n} > 0] \in \mathscr{M}$. Similarly,

$$I_{\bigcap_{n=1}^{k} B_n} \downarrow I_{\bigcap_{n=1}^{\infty} B_n}$$

and $\bigcap_{n=1}^{\infty} B_n \in \mathscr{M}$.

(5): *$\mathscr{A} = L_2(\mathscr{M})$.*
By definition, $\mathscr{A} \subset L_2(\mathscr{M})$. Conversely, suppose $f \in L_2(\mathscr{M})$. For $i = 1, 2, \ldots, n2^{n+1}$, $n = 1, 2, \ldots$, set

$$B_{ni} = \left[f \geqslant -n + \frac{i}{2^{n+1}} \right].$$

Then $B_{ni} \in \mathscr{M}$ by the definition of $L_2(\mathscr{M})$. By the definition of $\mathscr{M}$, there exist $g_{ni} \in \mathscr{A}$ and real a_{ni} such that $B_{ni} = [g_{ni} > a_{ni}]$. By (3), $I_{B_{ni}} \in \mathscr{A}$ so that by (1) and (2), $f_n \in \mathscr{A}$ with f_n defined below.

$$f_n = \bigvee_{i=1}^{n2^{n+1}} \left(\frac{i}{2^n}\right) I_{[f \geqslant -n + i/2^n]} - n.$$

However $f_n \to f$ and $|f_n| \leqslant |f| + 1$, $n = 1, 2, \ldots$, so that by the dominated convergence theorem, $\int (f_n - f)^2 \, d\lambda \to 0$. Since $\mathscr{A}$ is closed, $f \in \mathscr{A}$. □

Theorem 8.3.3 *Let $(\Omega, \mathscr{S}, \lambda)$ be a probability space and let T be a positively homogeneous, isotonic, expectation invariant, idempotent, distance reducing operator on $L_2 = L_2(\Omega, \mathscr{S}, \lambda)$ which is strictly monotonic at $\mathbf{0}$. Let $\mathscr{M} = \{[Ty > a] : y \in L_2, a \in R\}$. Then $T = E(\cdot \mid \mathscr{M})$.*

Proof: This is an immediate consequence of Theorems 8.3.1 and 8.3.2. □

The following example in which T satisfies all the hypotheses of Theorem 8.3.3 except that it is not strictly monotonic at $\mathbf{0}$, and in which the conclusion fails, is from Dykstra (1968). Let $\Omega = \{1, 2\}$, $\mathscr{S} = 2^\Omega$, and $\lambda(1) = \lambda(2) = \frac{1}{2}$. Each function in L_2 can be represented as a point in the plane whose coordinates are its values at 1 and 2, respectively. If $y = (y_1, y_2)$, $y_1 < y_2$ set $Ty = (y_2, y_1)$; if $y_1 \geqslant y_2$ set $Ty = y$. Then T is positively homogeneous, isotonic, expectation invariant, idempotent, and distance reducing. However, if $y = (-1, 1)$ then $\langle y, Ty \rangle = (-1)(1)(\frac{1}{2}) + (1)(-1)(\frac{1}{2}) = -1$ so that T is not strictly monotonic at $\mathbf{0}$. Since T is not a projection on its range, it is not a conditional expectation given a σ-lattice.

8.4 COMPLEMENTS

Hilbert space projections onto convex sets occur in many places and in many different settings. Some standard sources which discuss this topic are Luenberger (1969), Kantorovich and Akilov (1964), Taylor (1958), and Dunford and Schwartz (1958).

A discussion of the relationship between partial orders and σ-lattices is given in Brunk (1961, Sec. 6). Brunk (1965) also observed that if $\mathscr{U}$ is a σ-lattice, $z \in R(\mathscr{U})$ and Q is nondecreasing, then $Q(z)$ must be in $R(\mathscr{U})$. Necessary and sufficient conditions for a family of random variables to be $R(\mathscr{U})$ for some σ-lattice are given in Brunk (1963). He also gives necessary and sufficient conditions for identifying when a family of random variables is $L_2(\mathscr{U})$ for some σ-lattice $\mathscr{U}$. Dykstra (1970b) has further refined some of Brunk's results for characterizing such families.

The phrase 'conditional expectation given a σ-lattice' is of course borrowed from the more common exposition of 'conditional expectation given a σ-field'. The observation that conditional expectations given a σ-field may be regarded as

Hilbert space projections (assuming finite variances) goes back at least as far as Blackwell (1947).

Brunk (1961) seems to have introduced the σ-lattice conditional expectations and certainly did much of the early work. Initially, he considered L_2 spaces for totally finite measures, but later extended the concept to L_1 spaces and infinite measures (Brunk, 1963).

For conditional expectations given a σ-field, an approach based on Lebesgue–Radon–Nikodym derivatives is possible, and the same thing is true for conditional expectations given a σ-lattice. This development is begun in Johansen (1967) and extended in Brunk and Johansen (1970). For reasons of space, this approach has not been included in Chapter 8. However, it naturally leads to some nice extensions of the min-max formulations discussed in Section 1.4. Darst (1970) has extended the Lebesgue–Radon–Nikodym approach to encompass additive (not necessarily σ-additive) set functions defined on lattices of sets.

A version of Jensen's inequality for conditional expectations given σ-lattices is also possible. In particular, suppose $(\Omega, \mathscr{S}, \lambda)$ is a finite measure space and $\mathscr{U}$ is a sub-σ-lattice of $\mathscr{S}$. Let $y \in L_1$ and Φ be a nondecreasing, convex function defined on an interval J containing the range of y. If $y\Phi'(E(y|\mathscr{U}))$, $E(y|\mathscr{U})\Phi'(E(y|\mathscr{U}))$, and $\Phi(E(y|\mathscr{U}))$ are all in L_1 (where Φ' is a determination of the derivative of Φ), then

$$E(\Phi(y)|\mathscr{U}) \gtrsim \Phi(E(y|\mathscr{U})),$$

as shown in Barlow *et al.* (1972).

Strictly monotonic operators are discussed in Vainburg (1973), Browder (1965), and Minty (1965). Theorem 8.2.6 appears in Nashed (1968, p. 784).

Corollary C of Theorem 8.2.7 generalizes remarks of Brillinger (1966) and Rao (1965) that the regression function $E(y|x)$ and the linear regression function are the functions in their respective classes most highly correlated with y. A version of Theorem 8.2.8 is proved in Brunk (1965) who attributes the argument to David Hanson.

Section 8.3 deals with characterizations of a conditional expectation given a σ-lattice treated as an operator on an L_2 space. Characterizations for conditional expectations given a σ-field are given in Moy (1954), Bahadur (1955), Sidák (1957), Olson (1965), Rota (1960), Douglas (1965), and Pfanzagl (1967). Of course, a conditional expectation given a σ-field is a linear operator. However, a conditional expectation with respect to a σ-lattice is not linear and hence lacks some of the other's structure. Additivity of conditional expectations given a σ-lattice has been studied in Kuenzi (1969).

Boyle (1985) has employed some of the more general structures associated with order restrictions in infinite-dimensional spaces in work dealing with constrained smoothing splines. Wollan (1985) has generalized some computational techniques developed for restricted Hilbert space problems to more general settings.

CHAPTER 9

Complements

9.1 ORDER RESTRICTED OPTIMIZATION WITH NORMS OTHER THAN L_2

The theory of conditional expectations developed in Chapter 8, the normal means estimation problem discussed in Example 1.2 and the associated isotonic regression problem discussed in Chapter 1 have been generalized from several different points of view. For example, rather than assuming random samples from normal populations, one might assume the random samples are from bilateral exponential (or Laplacian) distributions. This problem is not covered by the theory developed in Chapter 1. Specifically, assume that $\lesssim$ is a quasi-order on $X = \{x_1, x_2, \ldots, x_k\}$ and that one has independent random samples of respective sizes n_1, $n_2, \ldots, n_k$ from each of k bilateral exponential distributions with medians $\theta(x_i)$, $i = 1, 2, \ldots, k$. The problem is to find the MLE of the function θ on X when the estimate is required to be isotonic with respect to the quasi-order $\lesssim$.

Recall that the bilateral exponential density is given by

$$p(y; \theta) = \tfrac{1}{2} e^{-|y-\theta|} \qquad \text{for } -\infty < y < \infty,$$

where θ is the median (also the mean) of this population. Thus the problem is to find an isotonic function f on X which maximizes

$$\prod_{i=1}^{k} \prod_{j=1}^{n_i} \tfrac{1}{2} e^{-|y_{ij} - f(x_i)|}$$

among the collection, $\mathscr{I}$, of all isotonic functions on X. Here y_{ij}, $j = 1, 2, \ldots, n_i$, represent the values of the random sample from the ith population. Equivalently, one wants to solve

$$\min_{f \in \mathscr{I}} \sum_{i=1}^{k} \sum_{j=1}^{n_i} |y_{ij} - f(x_i)|. \qquad (9.1.1)$$

The following straightforward observations are important:

1. The same reasoning which led to the Corollary to Theorem 1.3.4 shows that

the problem reduces to minimizing a continuous function over a compact subset of Euclidean k-space so that a solution must exist.

2. By considering the case $k = 1$ and n even, and using the well-known fact that any number between the middle two items provides a minimum, one sees that the solution is not necessarily unique.

Robertson and Waltman (1968) solved (9.1.1) for a simple order restriction, say $x_1 \lesssim x_2 \lesssim \cdots \lesssim x_k$. They obtained both a min-max formula analogous to Theorem 1.4.4 and a pool adjacent violators algorithm (PAVA) and gave consistency properties of these estimators. The PAVA method proceeds as follows. Compute the medians, $M_1, M_2, \ldots, M_k$, of the k samples. (We adopt the usual convention of averaging the middle two items if the sample size is even.) If $M_1 \leqslant M_2 \leqslant \cdots \leqslant M_k$, then the medians provide the restricted MLEs of $\theta(x_1)$, $\theta(x_2), \ldots, \theta(x_k)$ since they are the unrestricted estimates. On the other hand, if $M_i > M_{i+1}$ for some i, then the ith and the $(i+1)$th samples are pooled and the median, $M(i, i+1)$, of the pooled sample is found. This yields a smaller set of $k-1$ medians. The procedure is repeated until a nondecreasing set of medians is found. Then, for each i, the restricted MLE of $\theta(x_i)$ is the median of the final set to which the ith sample contributed. Note the similarity of this procedure to the PAVA for the normal means problem described in Section 1.2, which uses the means of the samples (pooled and otherwise) rather than medians.

Actually, modifications of the above procedure can be applied to a wide class of problems involving various types of parent populations (cf. Eeden, 1957a, 1957b, and Robertson and Waltman, 1968). One formulation goes as follows. Suppose we have independent random samples, $Y_{i1}, Y_{i2}, \ldots, Y_{in_i}$, from each of k populations each having density functions belonging to the family of functions $\{p(y; \theta): \theta \in \Theta\}$. Assumptions are made which guarantee:

1. MLE solutions exist for all the one-sample problems. MLE solutions also exist when a common value of the parameter is required for any subset of the k populations.
2. The estimate computed from the pooled samples must be between the two estimates computed from the individual samples (i.e. in the notation of previous examples, $M_i \leqslant M(i, i+1) \leqslant M_{i+1}$ when $M_i \leqslant M_{i+1}$).

For example, suppose we wish to estimate $\theta_1, \theta_2, \ldots, \theta_k$ subject to the simple order $\theta_1 \leqslant \theta_2 \leqslant \cdots \leqslant \theta_k$. A PAVA would first compute the unrestricted estimates $\hat{\theta}_1, \hat{\theta}_2, \ldots, \hat{\theta}_k$ using the likelihood function involving $p(\cdot; \theta_i)$, $i = 1, 2, \ldots, k$. If these estimates satisfy the order restriction, then they are also the restricted estimates. Otherwise $\hat{\theta}_i > \hat{\theta}_{i+1}$ for some i. Replace $\hat{\theta}_i$ and $\hat{\theta}_{i+1}$ by the unrestricted estimate computed from the pooled ith and $(i+1)$th samples. This procedure is repeated until a nondecreasing set of estimates is found and the restricted estimate of θ_i is the estimate in this final set to which the ith sample contributed. Min-max formulas proceed in the obvious fashion using estimates computed from pooled samples rather than averages of unrestricted estimates.

Property 2 is referred to as the *Cauchy mean value* property. Leurgans (1981) has shown that the only linear combinations of the order statistics having the Cauchy mean value property are the arithmetic mean, the percentiles (including the median), and the weighted midrange. The Cauchy mean value property is said to hold *strictly* if the estimate computed from the pooled sample is always strictly between the estimates computed from the individual samples. Sample means have the strict Cauchy mean value property while medians are not strict Cauchy means. Whether or not the Cauchy mean value property holds strictly in a particular problem turns out to be an important factor in the form of the upper (lower) sets algorithms. These algorithms are rather more complicated if the *strict* Cauchy mean value property is not present (cf. Robertson and Wright, 1980a).

The above-mentioned theory has been extended to partial orders, but for obvious reasons, PAVA-type algorithms do not exist in this setting. However, min-max formulas and lower (upper) sets algorithms have been developed for partial orders and norm reducing and consistency properties have also been studied. Eeden authored a pioneering series of papers (1956, 1957a, 1957b, 1957c, 1957d) on this subject. The PAVA for a simple order is implicit in her results. She also studied the bounded regression problem in which the regression function is required not only to be isotonic but also to satisfy

$$a(x) \leqslant f(x) \leqslant b(x) \qquad \text{for } x \in X$$

where a and b are given functions on X. Additional research on isotonic median estimation and generalizations of this problem are contained in Cryer *et al.* (1972) and Robertson and Wright (1973, 1974b, 1975, 1980). Since this work deals largely with consistency properties, we will discuss it in more detail in the next section. Casady and Cryer (1976) consider isotonic percentile estimates.

Recall that in the normal means problem, finding the solution to

$$\min_{f \in \mathscr{I}} \sum_{i=1}^{k} \sum_{j=1}^{n_i} \frac{[y_{ij} - f(x_i)]^2}{\sigma_i^2}$$

is equivalent to the isotonic regression problem:

$$\min_{f \in \mathscr{I}} \sum_{i=1}^{k} [\bar{y}_i - f(x_i)]^2 \left(\frac{n_i}{\sigma_i^2} \right).$$

The analogous reduction does not occur in the least absolute deviations problem associated with the bilateral exponential distribution, or for other problems involving L_p norms ($p \geqslant 1$); i.e. the problem of solving (9.1.1) is not equivalent to minimizing $\sum_{i=1}^{k} |\hat{\theta}_i - f(x_i)| w_i$ for some basic estimates $\hat{\theta}_1, \hat{\theta}_2, \ldots, \hat{\theta}_k$ such as the sample medians. Restricted estimates which solve

$$\min_{f \in \mathscr{I}} \left[\sum_{i-1}^{k} |g(x_i) - f(x_i)|^p \right]^{1/p}$$

for a given g on X and $1 \leqslant p \leqslant \infty$ are studied in Barlow and Ubhaya (1971) and

Ubhaya (1974a, 1974b). The solutions are not, in general, unique, and explicit solutions exist only for $p = 1, 2, \infty$. A solution for $p = \infty$ is given by an isotonized midrange and min-max formulas are given which are analogous to Theorem 1.4.4 for means and the min-max formulas for medians given in the previously mentioned references.

Brunk and Johansen (1970) generalize the conditional expectation theory described in Chapter 8 by developing a generalized Lebesgue–Radon–Nikodym (LRN) theory. They identify the solutions to L_p minimization problems with LRN functions and point out that these LRN functions provide solutions for a wide class of minimization problems with objective functions of the form

$$\sum_i k(i, f(x_i))w_i$$

under regularity conditions on $k(\cdot, \cdot)$.

9.2 LIMIT THEOREMS

There is a large literature on limit theorems in order restricted inference as virtually every paper on the theory of the subject contains a section on consistency. It is impossible to discuss all of this research in this section, so we will only briefly survey the nature of such results and try to point the interested reader to the appropriate literature for further study. Consistency results for isotonic density estimators are discussed in Chapter 7 and a law of the iterated logarithm for the multinomial sampling situation is given at the beginning of Chapter 5.

We assume, initially, that we have random samples from populations indexed by a set, X, with corresponding distribution functions $\{F_x : x \in X\}$ and that for each of these populations we wish to estimate some characteristic, $\theta(x)$, of that population such as the mean $(\theta(x) = \int y \, dF_x(y))$ or the median $(\theta(x) = \inf\{y : F_x(y) \geqslant \frac{1}{2}\})$, subject to the restriction that θ is isotonic with respect to a quasi-order, $\lesssim$, on X.

A finite index set

If the set X is fixed and finite and the number of observations at each point of X goes to infinity, then consistency properties for an isotonic estimator of θ usually follow from norm reducing properties, such as the ones discussed in Section 1.6, together with well-known consistency properties of standard unrestricted estimators of θ. For example, consider the bioassay problem discussed in Example 1.7 in which we are interested in maximum likelihood estimation of ordered binomial parameters. Specifically, assume that the population indexed by x_i is Bernoulli with parameter $\theta(x_i)$ and that $Y_{i1}, Y_{i2}, \ldots, Y_{in_i}, i = 1, 2, \ldots, k$, are independent random samples from these populations. The unrestricted MLE of $\theta(x_i)$ is $\bar{Y}(x_i) = n_i^{-1} \sum_{j=1}^{n_i} Y_{ij}$ and the restricted MLE is the isotonic regression $\bar{\mathbf{Y}}^*$ of $\bar{\mathbf{Y}}$ with weights $w(x_i) = n_i$. Since $\bar{Y}(x_i)$ is a sample mean, there exists a large literature on consistency properties for $\bar{Y}(x_i)$ as an estimate of $\theta(x_i)$. The norm reducing property can be used to obtain rates at which $\max_{1\leqslant i\leqslant k}|\bar{Y}^*(x_i)-\theta(x_i)|$ converges weakly to zero by using known rates at which $P[\max_{1\leqslant i\leqslant k}|\bar{Y}(x_i)-\theta(x_i)|\geqslant\varepsilon]$ converges to zero. In a similar manner, almost sure rates of convergence to zero can be obtained. For example, using Kolmogorov's law of the iterated logarithm together with the norm reducing property contained in Corollary B to Theorem 1.6.1 and an argument similar to that given at the beginning of Chapter 5 for (1.5.2), Robertson and Wright (1974b) prove the following theorem.

Theorem 9.2.1 *If $0<\theta(x_i)<1$, $i=1,2,\ldots,k$, and if $n=n_1=n_2=\cdots=n_k$ then*

$$P\left[\limsup_{n\to\infty}(n/\log\log n)^{1/2}\max_{1\leqslant i\leqslant k}|\bar{Y}^*(x_i)-\theta(x_i)|\right.$$
$$\left.=\max_{1\leqslant i\leqslant k}\{2\theta(x_i)[1-\theta(x_i)]\}^{1/2}\right]=1.$$

Clearly, similar results can be obtained for other parent populations such as those discussed in the examples in Section 1.5 and those associated with the L_p problems discussed in Section 9.1. The necessary ingredients are consistency properties for the unrestricted estimators and a suitable norm reducing property. Such norm reducing properties hold in a wide variety of settings (cf. Robertson and Wright, 1974b, and Robertson, 1985).

An infinite index set: consistency properties

From a technical point of view, consistency properties for isotonic estimates are more interesting when the set X of possible observation points is infinite (such as a subset of the real line) and we do not assume that the number of observations at each observation point gets large. In these cases the unrestricted estimate is often not consistent. However, if the set of observation points becomes dense in X in a strong way, the monotonicity restriction makes it possible for the restricted estimate to be consistent. Thus, isotonicity plays a more fundamental role in establishing consistency for this situation than it does for the one hypothesized in Theorem 9.2.1. Such sampling situations are also interesting from a practical point of view and some interesting examples are described in Ayer *et al.* (1955).

We assume that the experiment is modeled by a set, $(x_1, Y(x_1))$, $(x_2, Y(x_2)),\ldots,(x_k, Y(x_k))$, of k pairs. The first partner in each pair, the observation point, is a specified point in a set X and the second partner in the pair, the observation, is a random variable whose distribution depends upon $\theta(x_i)$, where $\theta(\cdot)$ is a function on the set X. In their fundamental paper, Ayer *et al.* (1955) prove a weak convergence result for an isotonic estimator in this type of model. To elaborate, assume that $Y(x_1)$, $Y(x_2),\ldots,Y(x_k)$ are independent Bernoulli variables and that $Y(x_i)$ has parameter $\theta(x_i)$, where θ is nondecreasing on $(-\infty, \infty)$. Let $u_{k1} < u_{k2} < \cdots < u_{kl(k)}$ be the distinct values among $x_1, x_2, \ldots, x_k$, and for each u_{ki}, let $n_{ki} = \text{card}\{j \in \{1, 2, \ldots, k\} : x_j = u_{ki}\}$, and let $\hat{\theta}(u_{ki}) = \sum_{x_j = u_{ki}} Y(x_j)/n_{ki}$ be the unrestricted MLE of $\theta(u_{ki})$. Note that $\hat{\theta}$ may not be nondecreasing on $\{u_{k1}, u_{k2}, \ldots, u_{kl(k)}\}$ and this is, in fact, quite probable if the number of observations at each of these observation points is small. Let $\hat{\theta}^*(u_{ki})$, $i = 1, 2, \ldots, l(k)$, be the isotonic regression of $\hat{\theta}(u_{ki})$, $i = 1, 2, \ldots, l(k)$, with weights $w(u_{ki}) = n_{ki}$, so that $\hat{\theta}^*(u_{ki})$, $i = 1, 2, \ldots, l(k)$, is the nondecreasing MLE of $\theta(u_{ki})$, $i = 1, 2, \ldots, l(k)$. Now form an estimator, $\bar{\theta}$, of θ on $(-\infty, \infty)$ such that $\bar{\theta}(u_{ki}) = \hat{\theta}^*(u_{ki})$ and $\bar{\theta}$ is defined arbitrarily at other points of $(-\infty, \infty)$ as long as $\bar{\theta}$ is a nondecreasing function. A proof of the following theorem can be found in Ayer *et al.* (1955).

Theorem 9.2.2 *Let x_0 be a continuity point of θ and let ε and ζ be arbitrary positive numbers. Let x', x'' be chosen so that $x' < x_0 < x''$ and $|\theta(x) - \theta(x_0)| < \varepsilon/2$ for $x' \leqslant x \leqslant x''$. Then*

$$P[|\bar{\theta}(x_0) - \theta(x_0)| < \varepsilon] > 1 - \zeta$$

provided that at least N trials are made between x' and x_0 and at least N trials are made between x_0 and x'', where N is chosen so that

$$\sum_{j=N}^{\infty} j^{-2} + (4N)^{-1} < \frac{\varepsilon^2 \zeta}{32}.$$

It is important to note that in this theorem the number of trials made at x_0 need not become infinite and, in fact, there may be no observations at x_0.

Brunk (1958, 1970) improved on Theorem 9.2.2 by obtaining almost sure uniform convergence for mean functions under the assumption that the variances are bounded. (He notes in his 1958 paper that the bounded variances assumption can be relaxed by assuming a Kolmogrov-type condition.) The proof uses a strong law of large numbers for maxima of sums of permutations of random variables which has come to be known as Brunk's law of large numbers.

Suppose each $x \in [0, 1]$ indexes a population with mean $\theta(x)$. Let $(x_1, Y(x_1))$, $(x_2, Y(x_2)), \ldots$ be a sequence such that $x_1, x_2, \ldots \in [0, 1]$; $Y(x_i)$ has the distribution associated with $x_i(E(Y(x_i)) = \theta(x_i))$; and $Y(x_1)$, $Y(x_2), \ldots$ are independent. Let $a(\cdot)$ be a given bounded positive function on $[0, 1]$ which is bounded away from 0 ($a(x_i)$ is to be interpreted as a weighting factor to be applied to $Y(x_i)$). For each k let $u_{k1} < u_{k2} < \cdots < u_{kl(k)}$ be the distinct values among x_1, $x_2, \ldots, x_k$ and let $\hat{\theta}(u_{ki}) = \sum_{x_j = u_{ki}} Y(x_j)/n_{ki}$ be the mean of the observations at u_{ki}. Let $\hat{\theta}^*(u_{ki})$, $i = 1, 2, \ldots, l(k)$, be the isotonic regression of $\hat{\theta}(u_{ki})$, $i = 1, 2, \ldots, l(k)$, with weights $w(u_{ki}) = a(u_{ki})n_{ki}$ and define $\bar{\theta}_k$ on $[0, 1]$ so that $\bar{\theta}(u_{ki}) = \hat{\theta}^*(u_{ki})$ and $\bar{\theta}_k$ is nondecreasing on $[0, 1]$. For use in the following theorem, let $N_k(J)$ denote the number of points among $x_1, x_2, \ldots, x_k$ which lie in the subset J of $[0, 1]$.

Theorem 9.2.3 *Let θ be continuous and nondecreasing on $[0,1]$; let $\{x_i\}$ be a sequence of observation points, not necessarily distinct, such that for each subinterval J of $(0,1)$, $k/N_k(J)$ is bounded in k; and let the variances of the observed random variable $\{Y(x_i)\}$ be bounded. If $0<a<b<1$ then*

$$P\left[\lim_{k\to\infty}\sup_{a\leqslant x\leqslant b}|\bar{\theta}_k(x)-\theta(x)|=0\right]=1. \tag{9.2.1}$$

The consistency theorem in Brunk (1958) assumed only that the observation points become dense in $[0,1]$. This condition is not sufficient to guarantee the almost sure consistency in (9.2.1), and a counter example is given in Hanson, Pledger, and Wright (1973). The paper by Hanson *et al.* replaces the bounded variances assumption by a uniform integrability condition. It also contains a theorem giving weak consistency under the assumption that the observation points are dense in [0, 1], as well as results which give algebraic and exponential rates of convergence of $P[\sup_{a\leqslant x\leqslant b}|\bar{\theta}(x)-\theta(x)|\geqslant\varepsilon]$ to zero. Makowski (1973) obtained an almost sure rate for $|\bar{\theta}(x)-\theta(x)|$ to converge to zero (see also Makowski, 1974, for related results).

The paper by Hanson, Pledger, and Wright (1973) is the first to obtain this type of consistency result for a partial order. Assume the same model as that described preceding Theorem 9.2.2 with the following modifications. Suppose the sequence of observation points, $\{x_i\}$, are points in the unit square $X=[0,1]\times[0,1,\ ]$ and that the mean function, θ, on X is nondecreasing in each coordinate (that is θ is isotonic with respect to the 'matrix' partial order on X defined by $(x_1,x_2)\lesssim(z_1,z_2)$ if and only if $x_1\lesssim z_1$ and $x_2\lesssim z_2$). Let $u_{k1},u_{k2},\ldots,u_{kl(k)}$ be the distinct values among $x_1,x_2,\ldots,x_k$ and define $\hat{\theta}_k(u_{ki})$, $i=1,2,\ldots,l(k)$, and $\hat{\theta}_k^*(u_{ki})$, $i=1,2,\ldots,l(k)$, as before; that is $\hat{\theta}_k^*(u_{ki})$, $i=1,2,\ldots,l(k)$, is the isotonic regression of $\hat{\theta}_k(u_{ki})$, $i=1,2,\ldots,l(k)$, with respect to the partial order which is the restriction of $\lesssim$ to $\{u_{k1},u_{k2},\ldots,u_{kl(k)}\}$ with appropriate weights. Let $\bar{\theta}_k$ be an extension of $\hat{\theta}_k^*$ to X which is isotonic with respect to $\lesssim$. Then, under conditions on the distributions of $Y(x_i)-\theta(x_i)$ and conditions on the way the observation points become dense in X (such as those hypothesized in Theorem 9.2.3), Hanson *et al.* obtain algebraic and exponential rates for $P[\sup_{a\leqslant y_1,y_2\leqslant b}|\bar{\theta}_k(y_1,y_2)-\theta(y_1,y_2)|\geqslant\varepsilon]$ to converge to zero for $0<a<b<1$. Recall that exponential rates of convergence for such probabilities imply almost sure convergence to zero. The algebraic convergence rates are improved in Wright (1976).

Cryer *et al.* (1972) study an analogous problem for medians rather than means for the usual simple order on [0, 1] and obtain consistency results when $\bar{\theta}_k$ is an isotonic median estimator. They also obtain an exponential rate of convergence for $P[\sup_{a\leqslant t\leqslant b}|\bar{\theta}_k(t)-\theta(t)|\geqslant\varepsilon]$ to zero (Robertson and Waltman, 1968, study the analogous problem for X finite). Robertson and Wright (1973) study median estimates which are isotonic with respect to the matrix partial order on $[0,1]\times[0,1]$ and obtain consistency results including an exponential rate of convergence for $P[\sup_{a \leqslant y_1, y_2 \leqslant b} |\bar{\theta}_k(y_1, y_2) - \theta(y_1, y_2)| \geqslant \varepsilon]$ to zero. Robertson and Wright (1975) study isotonic estimators in a general setting which includes both the mean and median estimators described above. Their consistency results are obtained under a different condition on the way the observation points become dense in X which is satisfied for 'almost all' sequences of observations points when these observation points are generated randomly from certain distributions. They also study these problems when X is in the Cartesian product of an arbitrary finite number of copies of $[0, 1]$ to extend these results to the obvious generalization of the matrix partial order. Their consistency results include an exponential rate of convergence for $P[\sup_{x \in X'} |\bar{\theta}_k(x) - \theta(x)| \geqslant \varepsilon]$ to zero for subsets of X' of X which are bounded away from the boundaries of X. These papers on median regression contain an error which pertains to the MLSA (minimum lower sets algorithm). This mistake is discussed in Robertson and Wright (1980a) where a more complicated MLSA is established for Cauchy means which are not strict.

Consistency in concave regression

Hanson and Pledger (1976) study consistency properties for a least squares estimate which is constrained to be concave (convex). For each x in $[0, 1]$, let F_x be a distribution with mean $\theta(x)$. Suppose $\theta(x)$ is continuous and concave on $[0, 1]$. Let $(x_1, Y(x_1))$, $(x_2, Y(x_2))$,... be a sequence such that $x_i \in [0, 1]$, the random variables $Y(x_1)$, $Y(x_2)$,... are independent, and the distribution function of $Y(x_i)$ is F_{x_i}. Let $\theta_k(\cdot) = \theta_k(\cdot; Y(x_1), Y(x_2), \ldots, Y(x_k))$ be the concave function on $[0, 1]$ which minimizes

$$\sum_{i=1}^{k} [\theta_k(x_i) - Y_i]^2$$

over the class of concave functions. Under conditions on the tails of the distributions of the error random variables $Y_k - \theta(x_k)$ and under a condition requiring the points $\{x_k\}$ to be dense in $[0, 1]$ in a strong manner, Hanson and Pledger prove that, with probability one, $\hat{\theta}_k(x)$ converges uniformly to $\theta(x)$ on $[a, b]$ with $0 < a < b < 1$. They also establish bounds on $\hat{\theta}_k(x)$ at the ends of $[0, 1]$.

An infinite index set: asymptotic distributions

Using methods first used by Chernoff (1964) in determining the asymptotic distribution of an estimator of a mode and later used by Prakasa Rao (1969) in determining the asymptotic distribution of the maximum likelihood estimator of a unimodal density, Brunk (1970) derived the asymptotic distribution of a mean regression function at a point. Brunk assumed a slightly different model for the observation points than has been previously discussed. For each fixed k, let one observation Y_{ki} be made at each of the observation points $x_{ki} = i/k$, $i = 1, 2, \ldots, k$.

Assume that the mean and variance of Y_{ki} are $\theta(x_{ki})$ and $\phi(x_{ki})$ and that the observations satisfy Lindeberg's condition (this is a condition on the moments of the observations and is given in Brunk's paper). Assume that θ is nondecreasing on $[0, 1]$ and that $\bar{\theta}(x)$ is the slope of the greatest convex minorant of the set $\{(i/k, \sum_{j=1}^{i} Y_{kj}/k), i = 1, 2, \ldots, k\}$ evaluated at $x \in [0, 1]$.

Theorem 9.2.4 *In addition to the above assumptions, let $x_0 \in (0, 1)$, let the derivative θ' of θ be positive at x_0, and let there be a neighborhood of x_0 in which ϕ and θ' are continuous. Then $[2k/(\phi(x_0)\theta'(x_0))]^{1/3}[\bar{\theta}(x_0) - \theta(x_0)]$ converges in distribution to the slope at 0 of the greatest convex minorant of $W(t) + t^2$, where $W(\cdot)$ is the two-sided Wiener–Levy process with mean 0 and variance 1 per unit t and $W(0) = 0$.*

The probability density function of this asymptotic distribution can be expressed in terms of partial derivatives of a particular solution to the heat equation.

The fact that the norming constants are of order $k^{1/3}$ is somewhat disappointing as the usual norming constants for sample means in the central limit theorem is of order $k^{1/2}$. Parsons (1975) has shown that the norming constants are of order $k^{1/2}$ if the regression function, θ, is constant (note the assumption that $\theta'(x_0) > 0$ in Theorem 9.2.4). Wright (1981b) studied how the rate of growth of the regression function at a point influences the rate of convergence of the estimator at that point. Leurgans (1982) sharpened and extended Theorem 9.2.4 by developing some theory for slopes of greatest convex minorants of processes. Wright (1984) established similar results for monotone percentile regression estimates.

Martingale-type convergence theory

Brunk (1965) has developed some martingale-type convergence theory for projections onto convex subsets which includes conditional expectations given σ-lattices. Assume the notation of Section 8.2 so that H is a complete, real, inner product space. One might call a sequence $\{f_n, A_n\}$ a *martingale* if, for all n, A_n is a closed, convex subset of H, $f_n \in H$, $A_n \subset A_{n+1}$ and for $k \geqslant 0$, $f_n = P(f_{n+k}|A_n)$ (cf. Theorem 8.2.1). The following is a mean square convergence theorem of the martingale type.

Theorem 9.2.5 *Suppose $\{A_n\}, n = 0, \pm 1, \pm 2, \ldots,$ is a two-way sequence of closed, convex sets in H such that $A_n \subset A_{n+1}, n = 0, \pm 1, \pm 2, \ldots$. Let $\{f_n\}$ be a bounded sequence (that is $\|f_n\| \leqslant m$ for some positive constant m, $n = 0, \pm 1, \pm 2, \ldots$) of elements of H such that $f_n = P(f_{n+k}|A_n)$, $n = 0, \pm 1, \pm 2, \ldots, k \geqslant 0$. Then $\lim_{n\to -\infty} f_n = f_{-\infty}$ and $\lim_{n\to\infty} f_n = f_\infty$ exist, $f_n = P(f_\infty|A_n)$, $n = 0, \pm 1, \pm 2, \ldots,$ and $f_{-\infty} = P(f_n|\bigcap_k A_k)$ for all n.*

The conclusion of Theorem 9.2.5 is that every bounded martingale is of the form $\{P(f|A_n)\}$ for some $f \in H$. Brunk (1965) gives additional martingale-type convergence results.

9.3 BAYESIAN INFERENCE UNDER ORDER RESTRICTIONS

A key idea in order restricted inference is to incorporate prior information regarding a collection of parameters into a statistical inference, thereby improving the quality of the inference. The basic tenet of Bayesian statistics is very similar and thus the idea of using a Bayesian approach to order restricted problems is very appealing. However, the authors are not aware of a great deal of existing research in this area. Some of the results on Bayesian order restricted inference are described in this section.

Sedransk, Monahan, and Chiu (1985) consider the problem of estimating an ordered set of multinomial probabilities. Suppose we have a random sample of size n from a multinomial distribution with parameters $p_1, p_2, \ldots, p_k$ and it is known that these parameters satisfy $p_1 \leqslant p_2 \leqslant \cdots \leqslant p_m \geqslant \cdots \geqslant p_k$, where m is a given value in $\{1, 2, \ldots, k\}$. Assume that the prior distribution on $\mathbf{p}$ is given by

$$\pi(\mathbf{p}) = \begin{cases} C_k(\beta_1, \beta_2, \ldots, \beta_k) \prod_{i=1}^{k} p_i^{\beta_i - 1}, & \text{for } \mathbf{p} \in \mathscr{I}_k, \\ 0, & \text{otherwise,} \end{cases}$$

where $\beta_i > 0$, $i = 1, 2, \ldots, k$,

$$\mathscr{I}_k = \left\{ \mathbf{p} = (p_1, p_2, \ldots, p_k) : p_1 \leqslant \cdots \leqslant p_m \geqslant \cdots \geqslant p_k, 0 \leqslant p_i \leqslant 1, \sum_{i=1}^{k} p_i = 1 \right\}.$$

and

$$C_k(\beta_1, \beta_2, \ldots, \beta_k)^{-1} = \int_{\mathscr{I}_k} \prod_{i=1}^{k} p_i^{\beta_i - 1} \, d\mathbf{p}.$$

The posterior distribution of $\mathbf{p}$ is given by

$$\pi(\mathbf{p}|\mathbf{n}) = \begin{cases} C_k(n_1 + \beta_1, \ldots, n_k + \beta_k) \prod_{i=1}^{k} p_i^{n_i + \beta_i - 1}, & \mathbf{p} \in \mathscr{I}_k, \\ 0, & \text{otherwise,} \end{cases}$$

where n_i is the number of trails among the n which resulted in the event that has probability p_i. One possible estimate for the parameter $\mathbf{p}$ is the mode of this posterior distribution. This mode is the value of $\mathbf{p}$ in $\mathscr{I}_k$ which maximizes

$$\prod_{i=1}^{k} p_i^{n_i + \beta_i - 1}.$$

Finding this estimate is simply the multinomial isotonic regression problem discussed in Example 1.5.8 and in Chapters 5 and 6. The solution for

$n_i + \beta_i - 1 > 0$ is the isotonic regression, $\hat{\mathbf{q}}^*$, of the vector $\hat{\mathbf{q}} = (\hat{q}_1, \hat{q}_2, \ldots, \hat{q}_k)$ where

$$\hat{q}_i = \frac{n_i + \beta_i - 1}{\sum_{j=1}^{k} n_j + \beta_j - 1} \qquad \text{for } i = 1, 2, \ldots, k.$$

Another Bayesian estimate of $\mathbf{p}$ is the centroid of the posterior distribution of $\mathbf{p}$ on $\mathscr{I}_k$. The centroid would be in the interior of $\mathscr{I}_k$ and would not produce ties or flat spots in the estimate of $\mathbf{p}$. However, as with many Bayesian analyses, calculation of the moments of the posterior distribution requires the evaluation of high-dimensional integrals (i.e. integrals over $\mathscr{I}_k$) and is very difficult. One method suggested by Sedransk, *et al.* is to use Monte Carlo integration. The straightforward method is to repeatedly (and independently) sample $\mathbf{p}$ from a Dirichlet distribution with parameter $\mathbf{n} + \mathbf{p}$ (by transforming gamma $(n_i + \beta_i, 1)$ variables); test whether $\mathbf{p} \in \mathscr{I}_k$; and then use as an estimate of the posterior moment, the sample mean of the estimates of this moment. The posteiror probability of $\mathbf{p} \in \mathscr{I}_k$ can be so low that this method is inefficient. A more efficient approach is to use importance sampling. Details regarding these methods of computation are given in Sedransk *et al.* Other results include a reduction in posterior variance (versus not including the restrictions) and the treatment of uncertainty about the value of m.

Broffitt (1984) considers a Bayesian approach to some order restricted problems which are motivated by graduation techniques in actuarial science. Consider a family of density functions of the form

$$f(y \mid \theta) = a(y)\theta^{b(y)} \mathrm{e}^{-\theta c(y)}$$

which may represent either continuous or discrete distributions. The choice of the functions a, b, and c is arbitrary as long as f is a density. Specific choices of a, b, and c can produce familiar families such as the Poisson, the gamma with known shape parameter, and the normal with known mean.

Suppose we have random samples from k members of this family with corresponding parameter values $\theta_1, \theta_2, \ldots, \theta_k$ and it is known that $\boldsymbol{\theta} = (\theta_1, \theta_2, \ldots, \theta_k)$ is in the region

$$\mathscr{I}_k = \{\boldsymbol{\theta}: 0 < \theta_1 \leqslant \theta_2 \leqslant \cdots \leqslant \theta_k\}.$$

Denote the sample items by Y_{ij}, $j = 1, 2, \ldots, n_i$, $i = 1, 2, \ldots, k$. The prior distribution on $\theta_1, \theta_2, \ldots, \theta_k$ is chosen to be identical to the conditional distribution of independent gamma variables, $U_1, U_2, \ldots, U_k$ given that $U_1 < U_2 < \cdots < U_k$. Specifically, the prior p.d.f. is given by

$$\prod(\boldsymbol{\theta}) = C(\delta_1, \ldots, \delta_k, \varepsilon_1, \varepsilon_2, \ldots, \varepsilon_k) \prod_{i=1}^{k} \frac{\varepsilon_i^{\delta_i} \theta_i^{\delta_i - 1} \mathrm{e}^{-\varepsilon_i \theta_i}}{\Gamma(\delta_i)}$$

on its support, $\mathscr{I}_k$, where

$$C(\delta_1, \ldots, \delta_k, \varepsilon_1, \ldots, \varepsilon_k)^{-1} = \int_{\mathscr{I}_k} \prod_{i=1}^{k} \frac{\varepsilon_i^{\delta_i} \theta_i^{\delta_i - 1} \mathrm{e}^{-\varepsilon_i \theta_i}}{\Gamma(\delta_i)} \, \mathrm{d}\boldsymbol{\theta}$$

is the proper normalizing constant. The posterior distribution has the same support and form, and in fact is given by

$$\prod(\boldsymbol{\theta}|\mathbf{y}) = C(\alpha_1,\ldots,\alpha_k,\lambda_1,\ldots,\lambda_k)\prod_{i=1}^{k}\frac{\lambda_i^{\alpha_i}\theta_i^{\alpha_i-1}e^{-\lambda_i\theta_i}}{\Gamma(\alpha_i)}$$

where $\alpha_i = \delta_i + \sum_{j=1}^{n_i} b(y_{ij})$ and $\lambda_i = \varepsilon_i + \sum_{j=1}^{n_i} c(y_{ij})$. The proof of the following theorem is given in Broffitt (1984).

Theorem 9.3.1 *If $\hat{\theta}_i$ is the unrestricted Bayes estimate (under squared error loss) of θ_i, $i = 1,2,\ldots,k$. then*

$$\hat{\theta}_i = \frac{\alpha_i}{\lambda_i} \qquad i = 1,2,\ldots,k,$$

and the restricted (isotonic) Bayes estimate (under squared error loss) of θ_i is given by

$$\hat{\theta}_i' = \hat{\theta}_i \frac{C(\alpha_1,\alpha_2,\ldots,\alpha_k,\lambda_1,\ldots,\lambda_k)}{C(\alpha_1,\ldots,\alpha_{i-1},\alpha_i^{+1},\alpha_{i+1},\ldots,\alpha_k,\lambda_1,\ldots,\lambda_k)}.$$

In order to calculate these isotonic Bayes estimates it is necessary to evaluate the constants $C(\alpha_1,\ldots,\alpha_k,\lambda_1,\ldots,\lambda_k)$. For the case in which $\alpha_1,\ldots,\alpha_k$ are integers, Broffitt gives a formula for their calculation which involves a finite multidimensional sum. He also give a FORTRAN program for implementing this formula and applies the result to some actuarial data.

Barlow *et al.* (1972) describe an approach to Bayesian estimation for independent samples from members of an exponential-type family and give a theorem which yields the mode of the posterior distribution as an isotonic regression. Suppose $\boldsymbol{\theta} = (\theta_1,\theta_2,\ldots,\theta_k)$ is unknown but known to be isotonic with respect to a quasi-order, $\lesssim$, on $X = \{1,2,\ldots,k\}$ and suppose $\boldsymbol{\lambda} = (\lambda_1,\lambda_2,\ldots,\lambda_k)$ is known. Further suppose we have independent random samples of size $n_1, n_2,\ldots,n_k$ from distributions with densities of the form with strictly convex

$$f(y;\theta_i,\lambda_i) = \exp\{[\Phi(\theta_i) + [y-\theta_i]\phi(\theta_i)]\lambda_i\} \tag{9.3.1}$$

with respect to some measure $v_{\lambda_i}(dy)$. Let $\mathscr{I}$ be the collection of vectors isotonic with respect to $\lesssim$ and select $\boldsymbol{\theta}_0 \in \mathscr{I}$ which is a prior estimate of $\boldsymbol{\theta}$. Let the prior distribution on $\boldsymbol{\theta}$ have joint density with respect to the product measure

$$\pi(\boldsymbol{\theta}) = \begin{cases} C(\boldsymbol{\theta}_0,\mathbf{h})\prod_{i=1}^{k}\exp\{[\Phi(\theta_i) + [\theta_{0i}-\theta_i]\phi(\theta_i)]h_i\}, & \text{for } \boldsymbol{\theta}\in\mathscr{I}, \\ 0, & \text{otherwise,} \end{cases} \tag{9.3.2}$$

where $C(\boldsymbol{\theta}_0,\mathbf{h})$ is the constant which makes the integral one. The precision parameter vector $\mathbf{h}$ is assumed to be known. If $\mathbf{h}$ is large, the density in (9.3.2) is sharply peaked about $\boldsymbol{\theta}_0$ while $\mathbf{h} \equiv \mathbf{0}$ gives a flat prior. The following theorem is given in Barlow *et al.* (1972) in Section 2.4.

Theorem 9.3.2 *Under the above assumptions, the posterior distribution of* $\boldsymbol{\theta}$ *has a density of the same form as* (9.3.2) *with mode* $\hat{\boldsymbol{\theta}}^*$ *which is the isotonic regression, with weights,* $n_i\lambda_i + h_i$, $i = 1, 2, \ldots, k$, *of the weighted average*

$$\hat{\theta}_i = \frac{[n_i\lambda_i\bar{Y}_i + h_i\theta_{0i}]}{[n_i\lambda_i + h_i]} \qquad \text{for } i = 1, 2, \ldots, k$$

of $\bar{\mathbf{Y}}$ *and* $\boldsymbol{\theta}_0$. *In particular, if* $\mathbf{h} \equiv \mathbf{0}$ *(flat prior) the posterior mode coincides with the MLE.*

Kraft and Eeden (1964) propose a Bayesian approach to the problem of estimating a nondecreasing set of binomial parameters (i.e. the bioassay problem discussed in Example 1.5.1). Let F be a distribution function on $(-\infty, \infty)$ and let $Y_1, Y_2, \ldots, Y_k$ be a set of independent random variables such that Y_i has a binomial distribution with parameters n_i and $F(t_i)$, where the numbers $t_1 < t_2 < \cdots < t_k$ are the dosage levels. The experimenter knows the n_i and the t_i and wishes to make an inference regarding F. Assume that F is random and that the distribution on $\mathbf{Y} = (Y_1, Y_2, \ldots, Y_k)$ is conditional on F.

While the set of possible distribution functions, F, could be taken to be a parametric family, Kraft and Eeden propose a more general approach. Let $D = \{d_i\}$ be a countable dense subset of $(-\infty, \infty)$ and let a sequence of probability laws $\mathscr{L}[F(d_1)]$, $\mathscr{L}[F(d_2)|F(d_1)], \ldots, \mathscr{L}[F(d_n)|F(d_1), \ldots, F(d_{n-1})], \ldots$ be specified so that $P(F$ is a distribution function on $D) = 1$. Extending F to $(-\infty, \infty)$ in the obvious way by right continuity yields a separable process. In their examples, Kraft and Eeden use a specific construction process for the prior on $D = [0, 1]$, called z-interpolation. The main result of this paper is that the class of Bayes' procedures, in this problem, is a complete class.

Dykstra and Laud (1981) study a Bayesian nonparametric approach to the problem of estimating an increasing hazard rate (cf. Section 7.4). A stochastic process is defined whose sample paths may be assumed to be increasing hazard rates by properly choosing the parameter values of the process. Observations are then assumed to occur from a distribution with this hazard rate. The posterior distribution of the hazard rates is derived for both exact and censored data. Bayes' estimates of hazard rates and cumulative distribution functions are found under squared error type loss functions. Estimates of the hazard rate using some data from a paper by Kaplan and Meier are constructed. Unlike most nonparametric Bayesian methods, this procedure has both the prior and posterior probability spread over the class of continuous distributions.

9.4 SMOOTHING AND ISOTONIC ESTIMATES

The process of finding the isotonic regression has often been referred to as a 'smoothing process'. This phrase has several possible explanations. One explanation dates back to the connection of the isotonic regression with the concept of conditional expectation and the fact that conditioning is referred to as a smoothing process. Another and related explanation is that the values of the variables to be regressed are replaced in the conditioning process by constant values, which is a smoothing operation.

The term 'smoothing process' is somewhat ironic because one of the criticisms of these estimates is their lack of 'smoothness'. This criticism is directed at the presence of the 'flat spots' or ranges of constant values in the isotonic estimate. When we think of 'smooth', increasing functions they do not have 'flat spots' unless of course they are constant functions. There have been a number of proposals in the literature to resolve this difficulty. However, in the authors' opinion no really satisfactory solution has been proposed and, in fact, this is a good topic for further research.

To be specific, think of the problem within the following context. Suppose that for each $x \in (a, b)$ there is a normal distribution with mean $\mu(x)$ and variance σ^2, both of which are unknown but it is known that the function μ on (a, b) is increasing. Suppose $\bar{Y}_1, \bar{Y}_2, \ldots, \bar{Y}_k$ are the means of independent samples of sizes $n_1, n_2, \ldots, n_k$ taken at $x_1, x_2, \ldots, x_k$ with $x_1 < x_2 < \cdots < x_k$. The MLE of the vector $(\mu(x_1), \mu(x_2), \ldots, \mu(x_k))$ subject to $\mu(t_1) \leqslant \cdots \leqslant \mu(t_k)$ is the isotonic regression, $\bar{\mathbf{Y}}^*$, of the vector $\bar{\mathbf{Y}} = (\bar{Y}_1, \bar{Y}_2, \ldots, \bar{Y}_k)$ with weights $n_1, n_2, \ldots, n_k$. Smoothness relates to several questions with regard to this estimate, including the following:

1. If we are willing to accept this estimate of μ at the points $x_1, x_2, \ldots, x_k$, how can it be extended to (a, b) so that the resulting estimate of μ is smooth?
2. Unless $\bar{Y}_1 < \bar{Y}_2 < \cdots < \bar{Y}_k$, the above estimates of $\mu(x_1), \ldots, \mu(x_k)$ will have 'flat spots'. If we really believe that $\mu(x_1) < \mu(x_2) < \cdots < \mu(x_k)$ then the estimate may be deemed unsatisfactory even at the points $x_1, x_2, \ldots, x_k$.

One solution to 2 is to use an appropriate Bayesian estimate of $(\mu(x_1), \mu(x_2), \ldots, \mu(x_k))$ as discussed in the previous section. The mode of the posterior distribution is again likely to have flat spots, it is an isotonic regression of an unrestricted estimate. These flat spots occur because the isotonic regression is a projection and is on the boundary of the restricted set when the unrestricted estimate fails to satisfy the constraints. However, the centroid of the posterior distribution will, in fact, be interior to the restricted set since the restricted set is convex. Thus, the centroid of the posterior distribution will be strictly increasing on $x_1 < x_2 < \cdots < x_k$.

A second solution is discussed in Barlow *et al.* (1972, p. 58). Suppose the investigator can specify positive numbers ε_i for which

$$\mu(x_{i+1}) \geqslant \mu(x_i) + \varepsilon_i \qquad \text{for } i = 1, 2, \ldots, k-1. \tag{9.4.1}$$

The MLEs subject to the constraint (9.4.1) could then be used. The MLEs minimize

$$\sum_{i=1}^{k} [\bar{Y}_i - f_i]^2 n_i$$

in the class of functions $\mathscr{I}(\varepsilon_1, \varepsilon_2, \ldots, \varepsilon_{k-1})$ satisfying $f_{i+1} \geqslant f_i + \varepsilon_i, i = 1, 2, \ldots, k-1$. If we set $\varepsilon_0 = 0$,

$$g_i = \bar{Y}_i - \sum_{j=0}^{i-1} \varepsilon_j, \quad i = 1, 2, \ldots, k \qquad \text{and} \qquad f'_i = f_i - \sum_{j=0}^{i-1} \varepsilon_j, \quad i = 1, 2, \ldots, k,$$

then

$$\sum_{i=1}^{k} [\bar{Y}_i - f_i]^2 n_i = \sum_{i=1}^{k} [g_i - f'_i]^2 n_i$$

and $f \in \mathscr{I}(\varepsilon_1, \varepsilon_2, \ldots, \varepsilon_{k-1})$ if and only if $f'_1 \leqslant f'_2 \leqslant \cdots \leqslant f'_k$. Thus the MLEs subject to the constraint (9.4.1) are given by

$$g_i^* + \sum_{j=0}^{i-1} \varepsilon_j \qquad \text{for } i = 1, 2, \ldots, k,$$

where g^* is the isotonic regression of g with weights n_i for $i = 1, 2, \ldots, k$.

The obvious difficulty with the above solution is the specification of ε_1, $\varepsilon_2, \ldots, \varepsilon_{k-1}$. Prior information regarding μ in the form of (9.4.1) would not occur in most applied problems. If one simply 'cooked up' $\varepsilon_1, \varepsilon_2, \ldots, \varepsilon_{k-1}$ to resolve question 2, then small ε's would not have much effect on the estimate while larger ε's could have a dramatic, and probably undesirable, effect.

The research discussed in the remainder of this section relates to both questions 1 and 2. Wright (1978) proposed combining the ideas of linear and isotonic regression. Let $\bar{\mu}$ be the nondecreasing function on (a, b) such that $\bar{\mu}(t_i) = \bar{Y}_i^*$, $i = 1, 2, \ldots, k$, and $\bar{\mu}$ is linear on $[t_i, t_{i+1}]$, $i = 1, 2, \ldots, k-1$. Define $\bar{\mu}(t) = \bar{\mu}(t_1)$; $a < t < t_1$ and $\bar{\mu}(t) = \bar{\mu}(t_k)$; $t_k < t < b$. Clearly the function $\bar{\mu}$ on (a, b) is nondecreasing and continuous (it is piecewise linear). Let $\hat{\mu}$ be the least square linear regression line on the points $(t_i, \bar{Y}_i)$, $i = 1, 2, \ldots, k$. It is well known that $\hat{\mu}$ is generally a good estimate of μ even if μ is somewhat nonlinear. Consider an estimate of μ which is a weighted average of $\bar{\mu}$ and $\hat{\mu}$, that is

$$\mu_v(t) = v\hat{\mu}(t) + (1 - v)\bar{\mu}(t), \qquad a < t < b,$$

for $0 \leqslant v \leqslant 1$. The estimate μ_v is strictly increasing if $v > 0$ and $\hat{\mu}$ has a positive slope. Wright recommends a choice, $\hat{v}$, for v which is based upon the relative residuals of the observation about the two functions $\hat{\mu}$ and $\bar{\mu}$ and compares the mean square error, via simulation, of the three estimates $\hat{\mu}$, $\bar{\mu}$, and $\mu_{\hat{v}}$. In all cases considered, the mean square error of $\mu_{\hat{v}}$ was never larger than that of $\bar{\mu}$, and often considerably smaller. This is probably related to the Stein effect which has been observed in the estimation of a multivariate normal mean.

Friedman and Tibshirani (1984) propose a solution which is based upon a combination of the ideas of moving averages (a smoothing process) and isotonic regression. First the scatterplot $(x_1, \bar{Y}_1)$, $(x_2, \bar{Y}_2), \ldots, (x_k, \bar{Y}_k)$ is 'smoothed' by replacing the unrestricted estimate $\bar{Y}_i$ at x_i by an average of the $\bar{Y}_j$ over a window of values of x_j which includes x_i. This estimate is then made isotonic by taking its isotonic regression. Friedman and Tibshirani illustrate their technique on two data sets and comment on the alternative idea of using the isotonic regression first and then smoothing this using a moving average. This latter idea has some intuitive appeal to the authors and these two techniques should be compared.

Some work in a somewhat similar vein has been done by Mukerjee (1986). His approach is to first isotonize the data and then use a kernel estimator with a log concave density kernel. This approach generally yields strictly increasing estimates of the regression function. Under appropriate assumptions, Mukerjee has shown that this method yields estimators that are uniformly strongly consistent and asymptotically normal with norming constant of the order $N^{2/5-\varepsilon}$, where N is the total sample size.

The idea of using a spline to estimate a function smoothly has been extensively researched in the statistical literature during the past twenty years. It seems very natural to combine the ideas of monotonic regression and the theory of splines to construct a smooth, monotonic function. This idea has particular appeal since splines can also be viewed as projections on appropriate sets of smooth functions. A fundamental paper on this idea is Wright and Wegman (1980) (see also Wegman, 1984). The main thrust of this research is the existence and uniqueness of the solutions. Questions of computation are approached through quadratic programming techniques. These papers are written in the context of a very general optimization problem which includes isotonic density estimation (with possibly censored data), isotonic failure rate estimation, isotonic regression, convex regression, and the very interesting idea of penalized isotonic function estimation.

9.5 TRENDS IN NONHOMOGENEOUS POISSON PROCESSES

Nonhomogeneous Poisson processes (NHPPs) provide models for a variety of physical phenomena. For instance, if $T_1, T_2, \ldots$ are the times of occurrences of a certain event, e.g. failure of a system, an accident, pulses along a nerve fiber, etc., then they frequently can be modeled by a NHPP. If $\lambda(t)$ is a nonnegative function, which is not degenerate at 0, and $\Lambda(t) = \int_0^t \lambda(s)\,ds$, then the joint density of the first n occurrence times of a NHPP is

$$p_{T_1, T_2, \ldots, T_n}(t_1, t_2, \ldots, t_n) = \exp\{-\Lambda(t_n)\} \prod_{i=1}^{n} \lambda(t_i) \quad \text{for } 0 < t_1 \leqslant t_2 \leqslant \ldots \leqslant t_n. \tag{9.5.1}$$

The function $\lambda(t)$ is called the intensity function and plays a role similar to the intensity in a (homogeneous Poisson process. For a given value of t, $\Lambda(t)$ is the mean number of occurrences in $[0, t]$ (cf. Parzen, 1962b).

Boswell (1966) considers the maximum likelihood estimation of λ subject to the restriction that λ is a nondecreasing function. He considers sampling schemes in which the number of occurrences is fixed or the time interval over which the process is observed is fixed. The results in both of these cases are similar. The sampling scheme with fixed time is discussed here. Let N be the number of occurrences in $[0, T]$. Conditional on $N = n > 0$, the occurrence times are known to be distributed as the order statistics of a random sample of size n from the distribution with support $[0, T]$ and density $\lambda(t)/\Lambda(T)$ for $0 \leqslant t \leqslant T$; that is, conditional on $N = n$, the likelihood function is

$$n! \prod_{i=1}^{n} \frac{\lambda(t_i)}{\Lambda(T)} \qquad \text{for } 0 < t_1 \leqslant t_2 \leqslant \cdots \leqslant t_n \leqslant T.$$

Boswell shows that the conditional MLE, subject to the assumption that λ is nondecreasing, is unique up to a multiplicative constant and is given by

$$\hat{\lambda}(t) = \begin{cases} 0, & 0 \leqslant t < T_1 \\ \hat{\lambda}(T_i), & T_i \leqslant t < T_{i+1}; \quad i = 1, 2, \ldots, n, \end{cases} \tag{9.5.2}$$

where $T_{n+1} = T$ and $\hat{\lambda}(T_i) = \max_{1 \leqslant \alpha \leqslant i} \min_{i \leqslant \beta \leqslant n} (\beta - \alpha + 1)/(T_{\beta+1} - T_\alpha)$. The choice of $\hat{\lambda}$ given in (9.5.2) satisfies $\int_0^T \lambda(s) \mathrm{d}s = n$, which seems heuristically reasonable since $\Lambda(T) = \int_0^T \lambda(s)\, \mathrm{d}s$ is the average number of occurrences in $(0, T]$.

If $\lambda(t)$ is constant, then the occurrence times are distributed as a homogeneous Poisson process and hence tests of the hypothesis, $H_0 : \lambda$ is constant, are of interest. If a researcher believes that λ is nondecreasing, then it is desirable to use a test which utilizes this information. Boswell found that the conditional LRT rejects H_0 in favor of $H_1 - H_0$, with $H_1 : \lambda$ is nondecreasing, for large values of

$$W = 2\left[\sum_{i=1}^{n} \ln \hat{\lambda}(T_i) + n \ln\left(\frac{T}{n}\right)\right] \tag{9.5.3}$$

and that under H_0 (with n large)

$$P[W \geqslant w] \approx \sum_{l=1}^{n} P_S(l, n) P[\chi^2_{l+1} \geqslant w] \tag{9.5.4}$$

where the $P_S(l, n)$ are the equal-weights level probabilities for the simple order which are discussed in Chapter 2 and are given in Table A.10 in the Appendix. The techniques discussed in Section 4.1 for obtaining approximations to the null distribution of the LRT in the exponential means problem can also be applied to improve the approximation in (9.5.4) (cf. Guffey and Wright, 1986). Barlow (1968a) and Bain, Engelhardt, and Wright (1985) compare the power of W with some *ad hoc* tests as well as several other tests of H_0 versus H_1 which are optimal for parametric families for λ.

In some situations it may be appropriate to model the occurrence times by a NHPP with an intensity function which is a nondecreasing, or nonincreasing, step function with the set of points at which the 'jumps' occur a subset of a collection of known points. For instance, if a system is modified at known times $0 = s_0 < s_1 < s_2 < \ldots < s_m$ in an attempt to improve it, then one might model its failures by a NHPP with $\lambda(t) = \lambda_i$ for $s_{i-1} < t \leqslant s_i$, $i = 1, 2, \ldots, m$, and $\lambda_1 \geqslant \lambda_2 \geqslant \cdots \geqslant \lambda_m$, provided a constant intensity over each $(s_{i-1}, s_i]$ seems reasonable. Magel and Wright (1984a) show that the LRT of $\lambda_1 = \lambda_2 = \cdots = \lambda_m$ versus $\lambda_1 \leqslant \lambda_2 \leqslant \cdots \leqslant \lambda_m$ with $\lambda_1 < \lambda_m$, conditional on $n > 0$ occurrences in $(0, T]$ (which is equivalent to the unconditional LRT), leads to the following testing problem: test $p_1/a_1 = p_2/a_2 = \cdots = p_m/a_m$ versus $p_1/a_1 \leqslant p_2/a_2 \leqslant \cdots \leqslant p_m/a_m$ with $p_1/a_1 < p_m/a_m$ where $\mathbf{p} = (p_1, p_2, \ldots, p_m)$ is a multinomial probability vector and $a_1, a_2, \ldots, a_m$ are known, positive constants. They study the LRT for this multinomial testing problem, as well as its dual, and find that it is closely related to the multinomial problem studied by Robertson (1978). (See also Chapter 5.)

If the assumption that $\lambda(t)$ is nondecreasing is in question, then one might wish to test H_1 as the null hypothesis. In some reliability problems, it is hypothesized that a component has a burn-in time, i.e. a time in which the intensity of failures decreases, followed by wear-out with the failure intensity increasing. Evidence in favor of this hypothesis could be obtained by rejecting $H_1{:}\lambda$ is nondecreasing. Following Boswell's work, Guffey and Wright (1986) study the conditional LRT for this testing situation.

9.6 CONFIDENCE INTERVALS

Statistical confidence interval procedures for problems involving order restrictions have been somewhat slow in developing. This is primarily due to the general intractability of these types of problems. However, some recent progress and developments lend hope to the premise that tractable, efficient confidence region procedure will become commonplace in order restricted settings.

One of the earliest efforts in this direction is due to Bohrer (1967), with further refinements occurring in Bohrer and Francis (1972). The basic problem is to find simultaneous confidence bounds for linear combinations of normal means when the coefficients are known to be nonnegative. In particular, assume that $\hat{\beta}_1, \hat{\beta}_2, \ldots, \hat{\beta}_n$, the components of $\hat{\boldsymbol{\beta}}$, are independent normal random variables with respective means $\beta_1, \beta_2, \ldots, \beta_n$ and respective, known variances $\sigma_1^2, \sigma_2^2, \ldots, \sigma_n^2$. Let $X^+ = \{\mathbf{x}: x_p \geqslant 0 \text{ for } 1 \leqslant p \leqslant n\}$ denote the nonnegative orthant in R^n. The goal is to find the constant $c = c(\alpha, n)$ such that

$$P[\hat{\boldsymbol{\beta}}'\mathbf{x} - c\sigma(\mathbf{x}) \leqslant \boldsymbol{\beta}'\mathbf{x} \leqslant \hat{\boldsymbol{\beta}}'\mathbf{x} + c\sigma(\mathbf{x})\text{: all } \mathbf{x} \text{ in } X^+] = 1 - \alpha, \tag{9.6.1}$$

where $\sigma(\mathbf{x}) = [\sum_{p=1}^n x_p^2 \sigma_p^2]^{1/2}$ is the standard deviation of $\hat{\boldsymbol{\beta}}'\mathbf{x}$, since this will give simultaneous $100(1 - \alpha)$ percent confidence intervals for $\boldsymbol{\beta}'\mathbf{x}$ for every $\mathbf{x}$ in X^+. Of course, if $\mathbf{x}$ were allowed to range over all of R^n, a Scheffe-type procedure would be appropriate, and c could be chosen from the χ^2 tables. If the variances were unknown, but equal, a Scheffe-type approach is still appropriate and c can be chosen from the F tables.

The key question is the effect that restricting $\mathbf{x}$ to X^+ has on c. Bohrer (1967) shows that the left side of (9.6.1) can be expressed as

$$\sum_{s=0}^{n} 2^{-n}\binom{n}{s}P(\chi_s^2 \leqslant c^2)P(\chi_{n-s}^2 \leqslant c^2). \tag{9.6.2}$$

This expression can then be used to find appropriate values of $c = c(\alpha, n)$ for particular values of α and n. Bohrer tables some values of $c(\alpha, n)$, and also shows that if $c_0(\alpha, n)$ is defined by

$$P[\chi_n^2 \leqslant c_0(\alpha, n)] = 1 - \alpha,$$

then $c_0(\alpha, n)/c(\alpha, n) \to \sqrt{2}$ as $n \to \infty$. Thus, for large n, the length of the simultaneous confidence intervals for $\boldsymbol{\beta}'\mathbf{x}$ when $\mathbf{x}$ is restricted to the positive will be approximately 30 percent shorter than if Scheffe's procedure is used.

Bohrer and Francis (1972) have extended these developments to the case where $\hat{\beta}_1, \hat{\beta}_2, \ldots, \hat{\beta}_n$ are not independent, but the covariance matrix is known except for scale. They also handle one-sided simultaneous confidence bounds for $\boldsymbol{\beta}'\mathbf{x}$, $\mathbf{x} \geqslant \mathbf{0}$.

Marcus and Peritz (1976) also develop methodology for finding simultaneous confidence intervals for linear combinations of normal means with certain restrictions on the coefficients. They obtain the Bohrer–Francis confidence intervals as a special case of this procedure. To illustrate, suppose $\mathbf{Y}^{k \times 1}$ is $\mathcal{N}(\boldsymbol{\eta}, \boldsymbol{\Sigma})$ and $\Omega \subset R^k$ is closed under multiplication by a positive scalar. Assume $\boldsymbol{\Sigma}$ is nonsingular and known, and that $\hat{\boldsymbol{\eta}}$ is the maximum likelihood estimate of $\boldsymbol{\eta}$ over Ω. Then Marcus and Peritz show

$$\max_{\mathbf{a}:\boldsymbol{\Sigma}\mathbf{a}\in\Omega, \mathbf{a}'\boldsymbol{\Sigma}\mathbf{a}=1} \mathbf{a}'\mathbf{Y} = \frac{\hat{\boldsymbol{\eta}}'\boldsymbol{\Sigma}^{-1}\mathbf{Y}}{(\hat{\boldsymbol{\eta}}'\boldsymbol{\Sigma}^{-1}\hat{\boldsymbol{\eta}})^{1/2}}.$$

It follows that

$$P[\mathbf{a}'\boldsymbol{\eta} > \mathbf{a}'\mathbf{Y} - t_\alpha(\mathbf{a}'\boldsymbol{\Sigma}\mathbf{a})^{1/2} \text{ for all } \mathbf{a} \text{ with } \boldsymbol{\Sigma}\mathbf{a}\in\Omega] = 1 - \alpha \tag{9.6.3}$$

if t_α is the upper 100α percentile of the distribution of $\hat{\boldsymbol{\eta}}'\boldsymbol{\Sigma}^{-1}\mathbf{Y}/(\hat{\boldsymbol{\eta}}'\boldsymbol{\Sigma}^{-1}\hat{\boldsymbol{\eta}}^{1/2})$ when $\boldsymbol{\eta} = \mathbf{0}$. For the ordered ANOVA case where X_i are independent $\mathcal{N}(\mu_i, n_i^{-1})$ random variables, $i = 1, 2, \ldots, k$, with $\mu_i \leqslant \mu_{i+1}$ for $i = 1, 2, \ldots, k-1$, (9.6.3) becomes

$$P\left[\sum_1^k n_i c_i \mu_i > \sum_1^k n_i c_i X_i - d_\alpha\left(\sum_1^k n_i c_i^2\right)^{1/2} \text{ for all } \mathbf{c} \text{ with } c_i \leqslant c_{i+1}, \sum_1^k n_i c_i = 0\right] = 1 - \alpha,$$

where d_α is the upper 100α percentile of the null distribution of the square root of Bartholomew's likelihood ratio statistic. They also show that the bounds can be improved in certain cases without altering the confidence level $1 - \alpha$.

Williams (1977) obtains some similar results for normal means. He obtains a limiting distribution for the estimated maximum and range of a set of monotonically ordered normal means when in fact all means are equal. He then uses percentiles of Studentized distributions in forming simultaneous confidence limits for contrasts between normal means when the contrast coefficients are monotonically ordered.

Marcus (1978b) is able to improve the Bohrer–Francis bounds when *a priori* information is available on the parameters. Schoenfeld (1986) has taken a different approach. He assumes that $Y_1, Y_2, \ldots, Y_k$ are normal random variables with ordered means $\mu_1 \leqslant \mu_2 \leqslant \cdots \leqslant \mu_k$. He seeks confidence intervals for each individual μ_i and also simultaneous confidence intervals for all the means. For a given mean μ_i, his upper and lower bounds are the maximum and minimum values of x such that the hypotheses that $x < \mu_i$ and $\mu_i < x$ are accepted by their respective likelihood ratio tests. Schoenfeld's method for finding simultaneous confidence intervals is based on an idea by Lee (1984).

Appendix

Tables are provided to facilitate the use of the $\bar{\chi}^2$ and $\bar{E}^2$ tests described in Chapter 2. Critical values for the $\bar{\chi}^2_{01}$, $\bar{\chi}^2_{12}$, $S_{01}(v)$, and $S_{12}(v)$ statistics are given for the simple order and the simple tree ordering. The equal weights and some limiting $P(l, k; w)$ are given for these two orderings. The equal-weights level probabilities are also given for some unimodal and matrix orders.

Tables A.1, A.2, A.3 give the upper 10, 5, 2.5, 1, 0.5, and 0.1 percent points for $\bar{\chi}^2_{01}$ and $\bar{\chi}^2_{12}$ under homogeneity for the simple order, $k = 3$ and 4, and arbitrary weights. For $k = 3$ compute $-\rho = \{w_1 w_3/[(w_1 + w_2)(w_2 + w_3)]\}^{1/2}$ and, if necessary, use linear interpolation between the values given. For $k = 4$ compute $-\rho_{12} = -\rho$ as in the $k = 3$ case and $-\rho_{23} = \{w_2 w_4/[(w_2 + w_3)(w_3 + w_4)]\}^{1/2}$ and, if necessary, use linear interpolation between the values given.

For $k = 3$ with the simple tree ordering, $\mu_0 \leqslant [\mu_1, \mu_2]$ or $\mu_0 \geqslant [\mu_1, \mu_2]$, one may use Table A.1; however, $\bar{\chi}^2_{01}$ and $\bar{\chi}^2_{12}$ need to be interchanged and $-\rho = \{w_1 w_2/[(w_0 + w_1)(w_0 + w_2)]\}^{1/2}$.

Table A.4 gives the same upper percentage points for the equal-weights distributions of $\bar{\chi}^2_{01}$ and $\bar{\chi}^2_{12}$ under homogeneity for the simple order, $\mu_1 \leqslant \mu_2 \leqslant \cdots \leqslant \mu_k$, with $3 \leqslant k \leqslant 24$. Table A.5 gives the corresponding values for the simple tree ordering, $\mu_0 \leqslant [\mu_1, \mu_2, \ldots, \mu_k]$, with $2 \leqslant k \leqslant 23$.

The $\bar{\chi}^2$ statistics are used in the normal means problems when σ^2 is known, as an approximation for one-parameter problems for exponential families and for certain rank tests. If the common variance is not known in a normal means problem, then the LRT statistics are $\bar{E}^2_{01}(v)$ and $\bar{E}^2_{12}(v)$. However, the null distribution of $S_{01}(v) = v\bar{E}^2_{01}(v)/[1 - \bar{E}^2_{01}(v)]$ and $S_{12}(v) = v\bar{E}^2_{12}(v)/[1 - \bar{E}^2_{12}(v)]$ are tabled. (Recall, as $v \to \infty$, that $S_{01}(v)$ and $S_{12}(v)$ approach $\bar{\chi}^2_{01}$ and $\bar{\chi}^2_{12}$, respectively.) For equal weights, the simple order, $\alpha = 0.1, 0.05, 0.01$, the critical values are given for $S_{01}(v)$ and $S_{12}(v)$ in Tables A.6 and A.7. The corresponding values for the simple tree ordering are given in Tables A.8 and A.9. The equal-weights values are given in the rows labeled $w = 1$. The limiting values as the first weight, w_1, for the simple order and w_0, for the simple tree ordering, approaches infinity, with the other weights fixed, is given in the rows labeled $w = \infty$. Interpolation between the $w = 1$ and $w = \infty$ values for the case $w_1 > w_2 = \cdots = w_k$

for a simple order and for $w_0 > w_1 = w_2 = \cdots = w_k$ for the simple tree ordering is discussed in Chapter 3.

Tables A.10 and A.11 give the equal-weights level probabilities for the simple order and the simple tree ordering, respectively. The limiting values, $\lim_{w \to \infty} P_S(l, k; (w, 1, \ldots, 1))$, are given in Table A.12, and for the simple tree the values $\lim_{w \to \infty} P_T(l, k+1; (1, 1, \ldots, w))$ are given in Table A.13.

In Chapter 3, two moment approximations to the equal-weights, null distributions of $\bar{\chi}^2_{01}$ and $\bar{\chi}^2_{12}$ are discussed for the simple order and the simple tree ordering. The coefficients needed to implement these approximations are given in Tables A.14 and A.15. The first two cumulants of the equal-weights, null $\bar{\chi}^2_{01}$ and $\bar{\chi}^2_{12}$ distributions are given in Tables A.16 and A.17 for the simple order and the simple tree ordering.

To approximate the null distributions of $\bar{\chi}^2_{01}$ and $\bar{\chi}^2_{12}$ distributions with unequal weights one needs the limiting cumulants of these distributions as $w \to \infty$ with weights $(w, 1, \ldots, 1)$ in the simply ordered case, and with weights $(1, 1, \ldots, 1, w)$ in the case of a simple tree ordering. These cumulants are given in Tables A.18 and A.19.

While the work in the Appendix focuses on the simple order and the simple tree order, the equal-weights level probabilities are given for some unimodal and matrix orders in Tables A.20 and A.21, respectively. The level probabilities for the unimodal order, $\mu_1 \leqslant \mu_2 \leqslant \cdots \leqslant \mu_h \geqslant \mu_{h+1} \geqslant \cdots \geqslant \mu_k$, are denoted by $P_h(l, k)$. Because of symmetry, one only needs to table these values for $1 < h \leqslant [(k+1)/2]$. The values in Table A.20 were computed using the recursive relation discussed in Chapter 2 for $k \leqslant 7$. For $8 \leqslant k \leqslant 15$, they were estimated by Monto Carlo techniques with 100 000 iterations (see Lucas and Wright, 1987).

For the index set $\{(i, j): 1 \leqslant i \leqslant R \text{ and } 1 \leqslant j \leqslant C\}$ with the matrix order, i.e. $(i, j) \lesssim (s, t)$ if and only if $i \leqslant s$ and $j \leqslant t$, the level probabilities are denoted by $P_{R \times C}(l, RC)$. The 2×2 case is a simple loop ordering which is discussed in Chapter 2. In the other cases considered, the level probabilities were estimated by Monte Carlo techniques with 10 000 iterations (see Lemke, 1983).

Table A.1 Critical values for the $\bar{\chi}^2$ tests with the simple order[a] ($\mu_1 \leqslant \mu_2 \leqslant \mu_3$ or $\mu_1 \geqslant \mu_2 \geqslant \mu_3$) and $k = 3$

$-\rho$	0.0	0.1	0.2	0.3	0.4	0.5	0.6	0.7	0.8	0.9	1.0
α					$\bar{\chi}^2_{01}$						
0.100	2.952	2.885	2.816	2.742	2.664	2.580	2.486	2.379	2.251	2.080	1.642
0.050	4.231	4.158	4.081	4.001	3.914	3.820	3.715	3.593	3.446	3,245	2.706
0.025	5.537	5.460	5.378	5.292	5.200	5.098	4.984	4.852	4.689	4.465	3.841
0.010	7.290	7.208	7.122	7.030	6.932	6.823	6.700	6.556	6.378	6.129	5.412
0.005	8.628	8.544	8.455	8.360	8.258	8.144	8.016	7.865	7.677	7.413	6.635
0.001	11.763	11.674	11.580	11.479	11.370	11.249	11.110	10.947	10.742	10.449	9.550
α					$\bar{\chi}^2_{12}$						
0.100	2.952	3.018	3.082	3.145	3.210	3.275	3.343	3.415	3.496	3.594	3.808
0.050	4.231	4.301	4.370	4.439	4.508	4.577	4.650	4.726	4.811	4.915	5.138
0.025	5.537	5.612	5.684	5.756	5.828	5.901	5.977	6.057	6.145	6.252	6.483
0.010	7.290	7.368	7.445	7.520	7.595	7.672	7.750	7.833	7.925	8.035	8.273
0.005	8.628	8.709	8.788	8.865	8.943	9.021	9.101	9.186	9.279	9.392	9.634
0.001	11.763	11.848	11.931	12.011	12.092	12.173	12.258	12.345	12.442	12.559	12.809

[a]This table can be used to obtain the critical values for the $\bar{\chi}^2$ tests for the simple tree ordering ($\mu_0 \leqslant [\mu_1, \mu_2]$ or $\mu_0 \geqslant [\mu_1, \mu_2]$). For $\bar{\chi}^2_{01}(\bar{\chi}^2_{12})$ use the critical value above for $\bar{\chi}^2_{12}(\bar{\chi}^2_{01})$ with $-\rho$ replaced by $\{w_1 w_2/[(w_0 + w_1)(w_0 + w_2)]\}^{1/2}$.

Note: $-\rho = \{w_1 w_3/[(w_1 + w_2)(w_2 + w_3)]\}^{1/2}$.

Table A.2 Critical values for the $\bar{\chi}^2_{01}$ test with the simple order ($\mu_1 \leqslant \mu_2 \leqslant \mu_3 \leqslant \mu_4$ or $\mu_1 \geqslant \mu_2 \geqslant \mu_3 \geqslant \mu_4$) and $k = 4$

		$-\rho_{12}$							
$-\rho_{23}$	α	0.0	0.1	0.2	0.3	0.4	0.5	0.6	0.7
	0.100	4.010							
	0.050	5.435							
0.0	0.025	6.861							
	0.010	8.746							
	0.005	10.171							
	0.001	13.475							
	0.100	3.952	3.891						
	0.050	5.372	5.305						
0.1	0.025	6.795	6.724						
	0.010	8.676	8.601						
	0.005	10.098	10.020						
	0.001	13.397	13.314						
	0.100	3.893	3.827	3.758					
	0.050	5.307	5.235	5.160					
0.2	0.025	6.725	6.649	6.570					
	0.010	8.602	8.522	8.437					
	0.005	10.022	9.939	9.851					
	0.001	13.316	13.227	13.133					
	0.100	3.831	3.760	3.685	3.606				
	0.050	5.239	5.162	5.080	4.993				
0.3	0.025	6.653	6.571	6.484	6.391				
	0.010	8.525	8.438	8.346	8.246				
	0.005	9.942	9.852	9.756	9.653				
	0.001	13.230	13.134	13.032	12.921				
	0.100	3.765	3.688	3.607	3.519	3.423			
	0.050	5.166	5.083	4.994	4.898	4.791			
0.4	0.025	6.575	6.486	6.392	6.289	6.174			
	0.010	8.442	8.348	8.247	8.137	8.014			
	0.005	9.855	9.758	9.653	9.539	9.411			
	0.001	13.137	13.033	12.921	12.799	12.662			
	0.100	3.695	3.610	3.521	3.423	3.313	3.187		
	0.050	5.088	4.997	4.898	4.791	4.670	4.528		
0.5	0.025	6.491	6.394	6.289	6.173	6.043	5.891		
	0.010	8.352	8.248	8.136	8.013	7.873	7.709		
	0.005	9.761	9.654	9.537	9.409	9.264	9.092		
	0.001	13.035	12.921	12.797	12.659	12.503	12.318		
	0.100	3.617	3.523	3.422	3.310	3.183	3.031	2.837	
	0.050	5.002	4.900	4.789	4.665	4.524	4.354	4.135	
0.6	0.025	6.398	6.289	6.170	6.038	5.886	5.702	5.462	
	0.010	8.251	8.135	8.008	7.867	7.703	7.504	7.244	
	0.005	9.656	9.535	9.404	9.257	9.085	8.877	8.604	
	0.001	12.921	12.793	12.652	12.494	12.310	12.085	11.786	
	0.100	3.530	3.422	3.305	3.172	3.017	2.822	2.550	1.987
	0.050	4.904	4.787	4.657	4.510	4.337	4.118	3.805	3.137
0.7	0.025	6.291	6.166	6.027	5.870	5.682	5.443	5.100	4.346
	0.010	8.135	8.002	7.854	7.684	7.482	7.223	6.846	6.000
	0.005	9.534	9.395	9.242	9.065	8.853	8.581	8.183	7.279
	0.001	12.788	12.640	12.476	12.286	12.057	11.761	11.324	10.305

Table A.2 (*continued*)

		$-\rho_{12}$							
$-\rho_{23}$	α	0.0	0.1	0.2	0:3	0.4	0.5	0.6	0.7
	0.100	3.427	3.296	3.151	2.981	2.770	2.473	1.642	
	0.050	4.787	4.644	4.483	4.294	4.056	3.715	2.706	
0.8	0.025	6.163	6.011	5.838	5.634	5.375	4.999	3.841	
	0.010	7.994	7.832	7.647	7.428	7.146	6.734	5.412	
	0.005	9.385	9.217	9.025	8.795	8.500	8.064	6.635	
	0.001	12.625	12.445	12.239	11.991	11.670	11.190	9.550	
	0.100	3.291	3.110	2.897	2.621	2.166			
	0.050	4.631	4.432	4.195	3.883	3.353			
0.9	0.025	5.990	5.778	5.523	5.182	4.591			
	0.010	7.804	7.577	7.303	6.933	6.277			
	0.005	9.183	8.948	8.661	8.273	7.576			
	0.001	12.399	12.148	11.840	11.417	10.642			
	0.100	2.952							
	0.050	4.231							
1.0	0.025	5.537							
	0.010	7.290							
	0.005	8.628							
	0.001	11.763							

Notes:
(1)

$$-\rho_{12} = \left[\frac{w_1 w_3}{(w_1 + w_2)(w_2 + w_3)}\right]^{1/2}$$

and

$$-\rho_{23} = \left[\frac{w_2 w_4}{(w_2 + w_3)(w_3 + w_4)}\right]^{1/2}.$$

(2) These critical values are symmetric in ρ_{12} and ρ_{23}.
(3) $\rho_{12}^2 + \rho_{23}^2 \leqslant 1$.

Table A.3 Critical values for the $\bar{\chi}^2_{12}$ test with the simple order ($\mu_1 \leqslant \mu_2 \leqslant \mu_3 \leqslant \mu_4$ or $\mu_1 \geqslant \mu_2 \geqslant \mu_3 \geqslant \mu_4$) and $k = 4$

		$-\rho_{12}$							
$-\rho_{23}$	α	0.0	0.1	0.2	0.3	0.4	0.5	0.6	0.7
	0.100	4.010							
	0.050	5.435							
0.0	0.025	6.861							
	0.010	8.746							
	0.005	10.171							
	0.001	13.475							
	0.100	4.067	4.126						
	0.050	5.496	5.560						
0.1	0.025	6.926	6.994						
	0.010	8.815	8.886						
	0.005	10.242	10.316						
	0.001	13.549	13.627						
	0.100	4.123	4.185	4.247					
	0.050	5.556	5.623	5.690					
0.2	0.025	6.989	7.061	7.131					
	0.010	8.882	8.957	9.032					
	0.005	10.311	10.389	10.466					
	0.001	13.622	13.704	13.785					
	0.100	4.179	4.245	4.311	4.378				
	0.050	5.616	5.687	5.758	5.831				
0.3	0.025	7.053	7.128	7.203	7.279				
	0.010	8.948	9.028	9.107	9.187				
	0.005	10.380	10.462	10.543	10.626				
	0.001	13.694	13.781	13.866	13.954				
	0.100	4.236	4.306	4.376	4.449	4.526			
	0.050	5.677	5.752	5.829	5.907	5.988			
0.4	0.025	7.117	7.196	7.276	7.359	7.444			
	0.010	9.015	9.100	9.184	9.270	9.360			
	0.005	10.448	10.536	10.623	10.711	10.804			
	0.001	13.766	13.858	13.950	14.044	14.141			
	0.100	4.294	4.369	4.445	4.524	4.609	4.701		
	0.050	5.739	5.820	5.902	5.987	6.077	6.175		
0.5	0.025	7.182	7.267	7.354	7.443	7.537	7.640		
	0.010	9.083	9.174	9.265	9.358	9.457	9.564		
	0.005	10.518	10.612	10.706	10.802	10.903	11.013		
	0.001	13.840	13.938	14.037	14.138	14.245	14.360		
	0.100	4.355	4.437	4.521	4.608	4.702	4.807	4.932	
	0.050	5.804	5.892	5.982	6.076	6.176	6.288	6.420	
0.6	0.025	7.249	7.343	7.438	7.536	7.641	7.757	7.895	
	0.010	9.154	9.253	9.353	9.456	9.565	9.687	9.831	
	0.005	10.591	10.693	10.796	10.902	11.015	11.139	11.287	
	0.001	13.915	14.023	14.132	14.243	14.362	14.492	14.646	
	0.100	4.420	4.512	4.607	4.706	4.815	4.940	5.100	5.381
	0.050	5.873	5.972	6.074	6.180	6.296	6.428	6.597	6.891
0.7	0.025	7.322	7.426	7.533	7.644	7.765	7.904	8.079	8.384
	0.010	9.229	9.340	9.452	9.569	9.695	9.840	10.022	10.337

Table A.3 (*Continued*)

		$-\rho_{12}$							
$-\rho_{23}$	α	0.0	0.1	0.2	0.3	0.4	0.5	0.6	0.7
	0.005	10.668	10.782	10.898	11.018	11.148	11.296	11.482	11.804
	0.001	13.996	14.116	14.238	14.365	14.501	14.655	14.849	15.183
	0.100	4.494	4.602	4.714	4.834	4.970	5.141	5.528	
	0.050	5.951	6.068	6.187	6.315	6.460	6.640	7.045	
0.8	0.025	7.402	7.525	7.651	7.786	7.936	8.124	8.542	
	0.010	9.313	9.443	9.576	9.716	9.873	10.068	10.501	
	0.005	10.753	10.888	11.024	11.169	11.330	11.529	11.971	
	0.001	14.084	14.227	14.371	14.522	14.690	14.898	15.355	
	0.100	4.584	4.729	4.880	5.054	5.296			
	0.050	6.046	6.201	6.364	6.548	6.803			
0.9	0.025	7.500	7.665	7.835	8.028	8.293			
	0.010	9.414	9.588	9.767	9.968	10.243			
	0.005	10.857	11.036	11.221	11.427	11.708			
	0.001	14.192	14.382	14.575	14.791	15.083			
	0.100	4.784							
	0.050	6.254							
1.0	0.025	7.715							
	0.010	9.636							
	0.005	11.083							
	0.001	14.426							

Notes:

(1)

$$-\rho_{12} = \left[\frac{w_1 w_3}{(w_1 + w_2)(w_2 + w_3)}\right]^{1/2}$$

and

$$-\rho_{23} = \left[\frac{w_2 w_4}{(w_2 + w_3)(w_3 + w_4)}\right]^{1/2}.$$

(2) These critical values are symmetric in ρ_{12} and ρ_{23}.

(3) $\rho_{12}^2 + \rho_{23}^2 \leqslant 1$.

Table A.4 Critical values for the $\bar{\chi}^2$ tests with the simple order ($\mu_1 \leqslant \mu_2 \leqslant \cdots \leqslant \mu_k$ or $\mu_1 \geqslant \mu_2 \geqslant \cdots \geqslant \mu_k$), equal weights, and k populations

$\bar{\chi}^2_{01}$

$\alpha \backslash k$	3	4	5	6	7	8	9	10	11	12	13
0.100	2.580	3.187	3.636	3.994	4.289	4.542	4.761	4.956	5.130	5.288	5.432
0.050	3.820	4.528	5.049	5.460	5.800	6.088	6.339	6.560	6.758	6.937	7.100
0.025	5.098	5.891	6.471	6.928	7.304	7.624	7.901	8.145	8.363	8.561	8.740
0.010	6.823	7.709	8.356	8.865	9.284	9.638	9.945	10.216	10.458	10.676	10.875
0.005	8.144	9.092	9.784	10.327	10.774	11.152	11.480	11.768	12.025	12.257	12.469
0.001	11.249	12.318	13.098	13.711	14.214	14.640	15.009	15.333	15.622	15.883	16.120
$\alpha \backslash k$	14	15	16	17	18	19	20	21	22	23	24
0.100	5.564	5.687	5.801	5.908	6.008	6.103	6.192	6.277	6.357	6.433	6.507
0.050	7.250	7.389	7.518	7.639	7.752	7.858	7.958	8.053	8.144	8.230	8.312
0.025	8.905	9.058	9.199	9.332	9.456	9.572	9.682	9.786	9.885	9.979	10.069
0.010	11.057	11.225	11.381	11.527	11.664	11.792	11.913	12.028	12.137	12.240	12.339
0.005	12.663	12.842	13.008	13.163	13.308	13.444	13.573	13.695	13.810	13.920	14.024
0.001	16.338	16.538	16.724	16.898	17.060	17.213	17.357	17.493	17.622	17.745	17.862

$\bar{\chi}^2_{12}$

$\alpha \backslash k$	3	4	5	6	7	8	9	10	11	12	13
0.100	3.275	4.701	6.048	7.353	8.630	9.888	11.131	12.361	13.581	14.793	15.997
0.050	4.577	6.175	7.665	9.095	10.485	11.846	13.185	14.505	15.811	17.103	18.384
0.025	5.901	7.640	9.248	10.783	12.268	13.717	15.137	16.534	17.912	19.274	20.621
0.010	7.672	9.564	11.305	12.957	14.550	16.098	17.611	19.096	20.557	21.997	23.420
0.005	9.021	11.013	12.841	14.571	16.234	17.848	19.423	20.966	22.482	23.976	25.450
0.001	12.173	14.360	16.358	18.242	20.047	21.792	23.491	25.151	26.779	28.378	29.955
$\alpha \backslash k$	14	15	16	17	18	19	20	21	22	23	24
0.100	17.194	18.385	19.570	20.751	21.927	23.099	24.267	25.432	26.593	27.752	28.907
0.050	19.655	20.918	22.172	23.419	24.660	25.894	27.123	28.347	29.566	30.780	31.991
0.025	21.956	23.279	24.592	25.897	27.193	28.481	29.762	31.036	32.305	33.568	34.826
0.010	24.827	26.221	27.602	28.971	30.331	31.680	33.022	34.355	35.681	36.999	38.312
0.005	26.906	28.346	29.772	31.186	32.589	33.981	35.362	36.735	38.099	39.454	40.803
0.001	31.511	33.047	34.565	36.068	37.560	39.034	40.498	41.952	43.393	44.829	46.250

Table A.5 Critical values for the $\bar{\chi}^2$ tests with the simple tree ordering ($\mu_0 \leqslant [\mu_1, \mu_2, \ldots, \mu_k]$ or $\mu_0 \geqslant [\mu_1, \mu_2, \ldots, \mu_k]$), equal weights, and k treatments

	$\bar{\chi}^2_{01}$										
$\alpha \backslash k$	2	3	4	5	6	7	8	9	10	11	12
0.100	3.275	4.696	6.036	7.333	8.600	9.848	11.080	12.300	13.510	14.711	15.905
0.050	4.577	6.171	7.653	9.075	10.456	11.807	13.136	14.446	15.741	17.023	18.294
0.025	5.901	7.635	9.237	10.764	12.240	13.679	15.089	16.476	17.844	19.195	20.533
0.010	7.672	9.560	11.295	12.939	14.523	16.061	17.564	19.039	20.490	21.920	23.333
0.005	9.021	11.010	12.831	14.553	16.208	17.812	19.377	20.910	22.417	23.900	25.364
0.001	12.173	14.357	16.348	18.225	20.021	21.757	23.446	25.096	26.716	28.307	29.873
$\alpha \backslash k$	13	14	15	16	17	18	19	20	21	22	23
0.100	17.092	18.273	19.449	20.620	21.786	22.949	24.108	25.264	26.417	27.566	28.713
0.050	19.555	20.808	22.052	23.290	24.521	25.747	26.967	28.182	29.392	30.598	31.800
0.025	21.857	23.171	24.475	25.770	27.056	28.335	29.608	30.874	32.133	33.388	34.637
0.010	24.731	26.115	27.486	28.846	30.197	31.538	32.870	34.195	35.512	36.822	38.126
0.005	26.810	28.241	29.658	31.063	32.456	33.839	35.212	36.576	37.932	39.279	40.620
0.001	31.419	32.944	34.455	35.948	37.429	38.897	40.352	41.797	43.231	44.655	46.069

$\bar{\chi}^2_{12}$

α/k	2	3	4	5	6	7	8	9	10	11	12
0.100	2.580	3.199	3.667	4.046	4.365	4.640	4.884	5.101	5.299	5.479	5.645
0.050	3.820	4.543	5.087	5.524	5.891	6.208	6.487	6.736	6.961	7.167	7.356
0.025	5.098	5.908	6.515	7.002	7.410	7.762	8.071	8.347	8.596	8.824	9.033
0.010	6.823	7.729	8.407	8.951	9.406	9.798	10.142	10.449	10.726	10.979	11.211
0.005	8.144	9.114	9.839	10.421	10.907	11.326	11.694	12.022	12.318	12.588	12.836
0.001	11.249	12.344	13.163	13.820	14.370	14.843	15.259	15.630	15.965	16.270	16.550

α/k	13	14	15	16	17	18	19	20	21	22	23
0.100	5.799	5.943	6.077	6.204	6.323	6.436	6.544	6.646	6.744	6.837	6.927
0.050	7.531	7.694	7.847	7.991	8.126	8.255	8.376	8.492	8.603	8.709	8.810
0.025	9.227	9.407	9.576	9.734	9.884	10.025	10.159	10.287	10.409	10.525	10.636
0.010	11.426	11.626	11.813	11.989	12.155	12.311	12.460	12.601	12.736	12.865	12.988
0.005	13.065	13.279	13.478	13.666	13.842	14.009	14.168	14.318	14.462	14.599	14.731
0.001	16.809	17.050	17.275	17.486	17.686	17.874	18.053	18.222	18.385	18.539	18.687

Table A.6 Critical values for $S_{01}(v)$ with k populations, the simple order, equal weights ($w = 1$), and limiting coefficients, $Q(l, k)$ ($w = \infty$)

		$\alpha = 0.100$									
v	$w\backslash k$	3	4	5	6	7	8	9	10	11	12
4	1	3.271	3.443	3.344	3.157	2.949	2.750	2.566	2.402	2.254	2.122
	∞	2.890	2.838	2.642	2.425	2.223	2.044	1.889	1.754	1.636	1.533
5	1	3.124	3.406	3.412	3.305	3.153	2.991	2.831	2.681	2.541	2.412
	∞	2.784	2.843	2.732	2.571	2.405	2.248	2.105	1.975	1.860	1.755
6	1	3.029	3.378	3.456	3.409	3.304	3.176	3.040	2.906	2.776	2.654
	∞	2.714	2.843	2.793	2.678	2.543	2.407	2.277	2.156	2.045	1.943
7	1	2.962	3.356	3.487	3.487	3.421	3.322	3.208	3.090	2.972	2.859
	∞	2.664	2.842	2.838	2.759	2.652	2.535	2.418	2.306	2.202	2.104
8	1	2.912	3.338	3.509	3.547	3.512	3.440	3.347	3.245	3.139	3.034
	∞	2.627	2.841	2.872	2.823	2.739	2.639	2.537	2.434	2.336	2.243
9	1	2.874	3.324	3.527	3.594	3.587	3.537	3.463	3.375	3.281	3.186
	∞	2.598	2.839	2.899	2.875	2.810	2.728	2.636	2.544	2.452	2.364
10	1	2.844	3.312	3.539	3.632	3.649	3.619	3.561	3.488	3.406	3.319
	∞	2.576	2.837	2.920	2.917	2.871	2.801	2.722	2.639	2.553	2.471
11	1	2.819	3.302	3.550	3.665	3.701	3.689	3.646	3.585	3.514	3.436
	∞	2.556	2.835	2.938	2.952	2.921	2.865	2.796	2.721	2.644	2.565
12	1	2.799	3.294	3.559	3.692	3.744	3.748	3.719	3.670	3.609	3.540
	∞	2.540	2.833	2.952	2.983	2.965	2.920	2.861	2.794	2.723	2.651
13	1	2.781	3.286	3.566	3.714	3.782	3.801	3.784	3.747	3.694	3.633
	∞	2.527	2.832	2.965	3.008	3.003	2.969	2.918	2.859	2.794	2.727
14	1	2.767	3.279	3.571	3.734	3.816	3.845	3.841	3.814	3.771	3.717
	∞	2.517	2.831	2.975	3.031	3.036	3.012	2.970	2.916	2.858	2.796
15	1	2.754	3.274	3.577	3.751	3.844	3.886	3.892	3.873	3.839	3.793
	∞	2.506	2.830	2.984	3.050	3.065	3.050	3.016	2.970	2.916	2.859

16	1	2.743	3.269	3.581	3.766	3.871	3.922	3.937	3.928	3.901	3.862
	∞	2.498	2.829	2.993	3.067	3.091	3.084	3.056	3.016	2.968	2.916
17	1	2.733	3.265	3.585	3.779	3.894	3.954	3.979	3.977	3.957	3.924
	∞	2.490	2.828	2.998	3.083	3.114	3.114	3.094	3.059	3.016	2.968
18	1	2.724	3.260	3.587	3.791	3.914	3.983	4.016	4.021	4.008	3.981
	∞	2.484	2.826	3.006	3.096	3.135	3.141	3.126	3.098	3.060	3.016
19	1	2.717	3.257	3.591	3.801	3.932	4.010	4.050	4.062	4.055	4.034
	∞	2.478	2.825	3.011	3.108	3.154	3.167	3.157	3.133	3.100	3.060
20	1	2.709	3.254	3.594	3.811	3.949	4.033	4.081	4.099	4.099	4.083
	∞	2.472	2.825	3.015	3.121	3.172	3.190	3.186	3.167	3.137	3.101
30	1	2.667	3.232	3.609	3.871	4.059	4.192	4.287	4.352	4.395	4.419
	∞	2.438	2.819	3.047	3.192	3.286	3.343	3.377	3.392	3.395	3.388
40	1	2.643	3.222	3.617	3.903	4.116	4.274	4.397	4.490	4.560	4.611
	∞	2.422	2.816	3.063	3.231	3.345	3.426	3.481	3.518	3.541	3.553
50	1	2.633	3.214	3.621	3.921	4.149	4.325	4.466	4.575	4.663	4.732
	∞	2.410	2.812	3.072	3.252	3.380	3.474	3.544	3.597	3.632	3.660
60	1	2.624	3.212	3.625	3.931	4.173	4.362	4.515	4.633	4.735	4.817
	∞	2.405	2.812	3.078	3.269	3.408	3.510	3.592	3.650	3.699	3.732
70	1	2.617	3.206	3.624	3.942	4.186	4.388	4.548	4.679	4.786	4.879
	∞	2.402	2.811	3.085	3.281	3.426	3.534	3.624	3.690	3.747	3.790
80	1	2.614	3.202	3.627	3.948	4.201	4.407	4.571	4.713	4.828	4.927
	∞	2.396	2.808	3.086	3.286	3.440	3.557	3.648	3.723	3.782	3.830
90	1	2.609	3.200	3.626	3.954	4.212	4.417	4.593	4.739	4.861	4.964
	∞	2.394	2.807	3.088	3.294	3.448	3.572	3.667	3.745	3.810	3.864
100	1	2.608	3.203	3.627	3.956	4.221	4.433	4.607	4.760	4.888	4.995
	∞	2.390	2.808	3.093	3.301	3.457	3.581	3.686	3.765	3.837	3.890
∞	1	2.580	3.187	3.636	3.994	4.289	4.542	4.761	4.956	5.130	5.288
	∞	2.371	2.804	3.110	3.344	3.535	3.695	3.832	3.952	4.058	4.154

Table A.6 (*continued*)

		$\alpha = 0.050$									
v	$w\backslash k$	3	4	5	6	7	8	9	10	11	12
4	1	5.744	5.756	5.392	4.948	4.518	4.133	3.796	3.503	3.249	3.028
	∞	5.143	4.817	4.330	3.869	3.471	3.138	2.858	2.621	2.420	2.247
5	1	5.296	5.525	5.369	5.077	4.751	4.434	4.139	3.873	3.634	3.418
	∞	4.793	4.698	4.383	4.034	3.705	3.412	3.154	2.928	2.732	2.558
6	1	5.016	5.367	5.343	5.160	4.916	4.657	4.404	4.164	3.942	3.738
	∞	4.570	4.611	4.414	4.149	3.877	3.621	3.387	3.177	2.987	2.818
7	1	4.825	5.251	5.318	5.217	5.039	4.829	4.612	4.400	4.197	4.006
	∞	4.416	4.546	4.433	4.232	4.008	3.786	3.575	3.379	3.201	3.038
8	1	4.687	5.164	5.297	5.258	5.131	4.965	4.782	4.594	4.410	4.233
	∞	4.303	4.496	4.445	4.296	4.112	3.198	3.730	3.550	3.382	3.226
9	1	4.581	5.096	5.278	5.289	5.206	5.075	4.921	4.757	4.591	4.428
	∞	4.218	4.455	4.452	4.345	4.195	4.028	3.860	3.695	3.538	3.390
10	1	4.498	5.040	5.262	5.311	5.266	5.167	5.037	4.895	4.746	4.597
	∞	4.150	4.422	4.458	4.385	4.264	4.120	3.969	3.819	3.673	3.533
11	1	4.432	4.995	5.247	5.330	5.315	5.243	5.137	5.014	4.881	4.745
	∞	4.096	4.394	4.461	4.418	4.321	4.198	4.063	3.926	3.791	3.660
12	1	4.378	4.957	5.235	5.346	5.357	5.308	5.223	5.116	4.998	4.875
	∞	4.050	4.371	4.464	4.445	4.370	4.264	4.145	4.020	3.896	3.772
13	1	4.333	4.925	5.224	5.357	5.391	5.364	5.297	5.207	5.103	4.991
	∞	4.013	4.351	4.465	4.468	4.412	4.323	4.218	4.104	3.988	3.873
14	1	4.294	4.897	5.215	5.367	5.421	5.411	5.362	5.287	5.195	5.095
	∞	3.981	4.335	4.466	4.487	4.448	4.373	4.281	4.178	4.072	3.963
15	1	4.261	4.873	5.205	5.376	5.447	5.454	5.420	5.359	5.278	5.188
	∞	3.953	4.320	4.467	4.504	4.479	4.419	4.337	4.244	4.145	4.045

16	1	4.231	4.852	5.198	5.383	5.470	5.492	5.470	5.421	5.353	5.272
	∞	3.930	4.306	4.468	4.519	4.508	4.458	4.388	4.305	4.214	4.119
17	1	4.206	4.832	5.191	5.389	5.490	5.525	5.517	5.479	5.421	5.349
	∞	3.908	4.295	4.469	4.532	4.532	4.494	4.433	4.359	4.275	4.186
18	1	4.183	4.816	5.184	5.396	5.507	5.554	5.558	5.530	5.481	5.418
	∞	3.890	4.284	4.469	4.544	4.555	4.527	4.474	4.407	4.331	4.250
19	1	4.163	4.802	5.177	5.399	5.524	5.582	5.596	5.578	5.538	5.482
	∞	3.873	4.274	4.468	4.554	4.575	4.555	4.513	4.452	4.382	4.307
20	1	4.146	4.788	5.172	5.405	5.537	5.607	5.629	5.621	5.589	5.541
	∞	3.858	4.266	4.469	4.563	4.592	4.583	4.546	4.493	4.429	4.358
30	1	4.035	4.703	5.135	5.426	5.626	5.764	5.853	5.908	5.936	5.944
	∞	3.764	4.211	4.467	4.620	4.711	4.757	4.774	4.772	4.752	4.725
40	1	3.979	4.660	5.115	5.437	5.671	5.844	5.969	6.060	6.125	6.167
	∞	3.718	4.184	4.466	4.648	4.770	4.846	4.895	4.923	4.932	4.932
50	1	3.946	4.633	5.104	5.443	5.699	5.893	6.042	6.154	6.242	6.308
	∞	3.692	4.167	4.466	4.666	4.805	4.904	4.971	5.019	5.048	5.063
60	1	3.927	4.616	5.096	5.447	5.713	5.925	6.089	6.218	6.325	6.406
	∞	3.675	4.157	4.464	4.676	4.829	4.941	5.022	5.083	5.126	5.156
70	1	3.909	4.601	5.090	5.450	5.729	5.950	6.126	6.268	6.385	6.477
	∞	3.662	4.148	4.461	4.684	4.845	4.967	5.060	5.129	5.183	5.223
80	1	3.900	4.593	5.081	5.453	5.738	5.963	6.149	6.303	6.428	6.530
	∞	3.653	4.142	4.462	4.691	4.861	4.988	5.087	5.164	5.225	5.275
90	1	3.888	4.587	5.081	5.450	5.741	5.978	6.172	6.329	6.462	6.575
	∞	3.644	4.134	4.459	4.696	4.867	5.007	5.111	5.197	5.259	5.314
100	1	3.884	4.580	5.076	5.455	5.748	5.987	6.186	6.352	6.490	6.608
	∞	3.640	4.134	4.460	4.700	4.874	5.016	5.130	5.218	5.292	5.346
∞	1	3.820	4.528	5.049	5.460	5.800	6.088	6.339	6.560	6.758	6.937
	∞	3.584	4.097	4.455	4.730	4.951	5.136	5.295	5.433	5.556	5.666

Table A.6 (*continued*)

v	$w \backslash k$	3	4	5	6	7	8	9	10	11	12
						$\alpha = 0.010$					
4	1	15.953	14.802	12.980	11.260	9.810	8.626	7.662	6.875	6.223	5.680
	∞	14.386	12.427	10.443	8.834	7.586	6.614	5.845	5.228	4.725	4.308
5	1	13.357	13.094	12.093	10.952	9.880	8.933	8.117	7.417	6.817	6.299
	∞	12.226	11.259	9.986	8.811	7.815	6.986	6.302	5.729	5.246	4.835
6	1	11.888	12.039	11.487	10.708	9.895	9.130	8.436	7.819	7.271	6.788
	∞	10.986	10.515	9.652	8.766	7.957	7.249	6.640	6.113	5.658	5.262
7	1	10.949	11.328	11.051	10.511	9.889	9.264	8.673	8.127	7.631	7.183
	∞	10.180	9.997	9.404	8.716	8.052	7.443	6.899	6.418	5.992	5.615
8	1	10.300	10.817	10.721	10.351	9.871	9.357	8.853	8.371	7.923	7.509
	∞	9.621	9.621	9.209	8.671	8.117	7.590	7.105	6.665	6.269	5.912
9	1	9.826	10.432	10.462	10.219	9.848	9.426	8.992	8.568	8.161	7.780
	∞	9.210	9.332	9.052	8.627	8.163	7.704	7.271	6.868	6.500	6.163
10	1	9.464	10.129	10.256	10.109	9.825	9.476	9.103	8.728	8.363	8.011
	∞	8.894	9.103	8.925	8.588	8.195	7.796	7.407	7.039	6.697	6.380
11	1	9.178	9.887	10.085	10.014	9.800	9.514	9.194	8.863	8.531	8.211
	∞	8.644	8.920	8.817	8.552	8.221	7.869	7.520	7.183	6.866	6.568
12	1	8.950	9.690	9.942	9.932	9.779	9.545	9.268	8.975	8.677	8.381
	∞	8.443	8.769	8.727	8.520	8.239	7.931	7.617	7.308	7.013	6.733
13	1	8.762	9.526	9.823	9.862	9.757	9.566	9.332	9.071	8.802	8.531
	∞	8.277	8.642	8.649	8.492	8.254	7.981	7.698	7.416	7.142	6.877
14	1	8.602	9.385	9.717	9.800	9.736	9.585	9.385	9.154	8.911	8.662
	∞	8.137	8.533	8.582	8.465	8.264	8.025	7.770	7.510	7.255	7.008
15	1	8.467	9.264	9.626	9.746	9.718	9.599	9.431	9.228	9.008	8.781
	∞	8.017	8.440	8.523	8.440	8.274	8.062	7.833	7.593	7.355	7.124

16	1	8.353	9.161	9.549	9.697	9.699	9.611	9.469	9.292	9.094	8.884
	∞	7.915	8.359	8.471	8.421	8.280	8.095	7.887	7.667	7.447	7.226
17	1	8.251	9.071	9.480	9.652	9.683	9.622	9.502	9.347	9.169	8.978
	∞	7.825	8.287	8.426	8.400	8.287	8.123	7.934	7.733	7.526	7.320
18	1	8.164	8.991	9.417	9.612	9.667	9.630	9.532	9.397	9.237	9.062
	∞	7.746	8.225	8.384	8.382	8.290	8.148	7.978	7.791	7.601	7.407
19	1	8.085	8.918	9.360	9.579	9.653	9.637	9.561	9.443	9.301	9.140
	∞	7.678	8.170	8.349	8.365	8.293	8.170	8.016	7.846	7.667	7.485
20	1	8.015	8.854	9.311	9.546	9.640	9.642	9.583	9.482	9.355	9.211
	∞	7.616	8.119	8.315	8.350	8.296	8.191	8.051	7.894	7.728	7.556
30	1	7.594	8.459	8.993	9.333	9.541	9.666	9.724	9.740	9.717	9.672
	∞	7.238	7.805	8.098	8.246	8.306	8.309	8.273	8.210	8.130	8.039
40	1	7.392	8.265	8.836	9.222	9.485	9.669	9.790	9.862	9.904	9.915
	∞	7.058	7.650	7.989	8.190	8.304	8.364	8.382	8.375	8.343	8.300
50	1	7.273	8.151	8.739	9.154	9.451	9.667	9.824	9.938	10.012	10.061
	∞	6.950	7.558	7.925	8.155	8.304	8.395	8.450	8.475	8.475	8.462
60	1	7.196	8.075	8.674	9.104	9.426	9.667	9.850	9.985	10.089	10.165
	∞	6.880	7.496	7.878	8.132	8.297	8.416	8.493	8.540	8.564	8.574
70	1	7.140	8.022	8.632	9.071	9.405	9.664	9.864	10.020	10.143	10.238
	∞	6.830	7.453	7.847	8.112	8.299	8.427	8.524	8.588	8.632	8.653
80	1	7.100	7.983	8.595	9.047	9.393	9.662	9.877	10.044	10.180	10.292
	∞	6.794	7.419	7.823	8.095	8.297	8.440	8.547	8.625	8.679	8.715
90	1	7.071	7.955	8.570	9.027	9.381	9.663	9.886	10.062	10.211	10.334
	∞	6.765	7.398	7.805	8.085	8.294	8.445	8.564	8.649	8.715	8.762
100	1	7.046	7.928	8.550	9.012	9.369	9.662	9.890	10.082	10.237	10.363
	∞	6.747	7.376	7.786	8.077	8.292	8.457	8.579	8.672	8.745	8.802
∞	1	6.832	7.709	8.356	8.865	9.284	9.638	9.945	10.216	10.458	10.676
	∞	6.545	7.198	7.651	7.996	8.274	8.505	8.702	8.875	9.027	9.164

Table A.7 Critical values for $S_{12}(v)$ with k populations, the simple order, equal weights ($w = 1$), and limiting coefficients, $Q(l, k)$ ($w = \infty$)

		$\alpha = 0.100$									
v	$w\backslash k$	3	4	5	6	7	8	9	10	11	12
4	1	5.434	8.520	11.673	14.895	18.174	21.505	24.877	28.276	31.705	35.161
	∞	5.768	9.151	12.568	16.032	19.528	23.053	26.603	30.178	33.762	37.361
5	1	4.874	7.504	10.161	12.853	15.591	18.356	21.152	23.983	26.820	29.690
	∞	5.154	8.027	10.902	13.791	16.704	19.634	22.578	25.539	28.512	31.498
6	1	4.541	6.910	9.280	11.673	14.093	16.539	19.004	21.494	23.994	26.513
	∞	4.793	7.374	9.934	12.497	15.070	17.656	20.253	22.858	25.472	28.092
7	1	4.323	6.522	8.707	10.907	13.122	15.359	17.609	19.875	22.157	24.452
	∞	4.555	6.946	9.304	11.656	14.011	16.370	18.741	21.113	23.490	25.876
8	1	4.168	6.249	8.306	10.369	12.443	14.531	16.631	18.747	20.867	23.005
	∞	4.387	6.646	8.861	11.066	13.269	15.471	17.678	19.891	22.107	24.328
9	1	4.053	6.047	8.011	9.972	11.941	13.920	15.910	17.910	19.916	21.936
	∞	4.262	6.425	8.537	10.631	12.718	14.808	16.899	18.988	21.084	23.178
10	1	3.965	5.891	7.781	9.665	11.554	13.451	15.352	17.268	19.187	21.110
	∞	4.166	6.253	8.285	10.296	12.299	14.296	16.296	18.295	20.293	22.300
11	1	3.894	5.767	7.600	9.425	11.249	13.080	14.915	16.759	18.606	20.461
	∞	4.090	6.118	8.086	10.031	11:963	13.891	15.817	17.746	19.669	21.599
12	1	3.837	5.667	7.454	9.229	11.002	12.776	14.556	16.344	18.134	19.930
	∞	4.027	6.009	7.925	9.815	11.695	13.565	15.430	17.299	19.161	21.029
13	1	3.789	5.585	7.332	9.066	10.797	12.528	14.263	16.000	17.741	19.491
	∞	3.975	5.917	7.793	9.638	11.469	13.294	15.112	16.929	18.740	20.555
14	1	3.749	5.514	7.230	8.930	10.623	12.318	14.012	15.714	17.415	19.120
	∞	3.932	5.840	7.679	9.488	11.283	13.063	14.843	16.614	18.386	20.156
15	1	3.714	5.454	7.144	8,813	10.477	12.139	13.803	15.467	17.136	18.802
	∞	3.894	5.775	7.585	9.360	11.121	12.869	14.611	16.349	18.083	19.818

16	1	3.684	5.402	7.067	8.713	10.350	11.984	13.619	15.254	16.893	18.531
	∞	3.862	5.718	7.501	9.251	10.982	12.701	14.413	16.120	17.823	19.523
17	1	3.658	5.358	7.002	8.625	10.239	11.850	13.459	15.068	16.681	18.295
	∞	3.833	5.669	7.429	9.157	10.860	12.554	14.241	15.921	17.598	19.269
18	1	3.635	5.318	6.945	8.548	10.142	11.731	13.319	14.905	16.495	18.086
	∞	3.809	5.626	7.368	9.072	10.754	12.427	14.089	15.745	17.397	19.042
19	1	3.614	5.283	6.893	8.479	10.056	11.626	13.194	14.761	16.328	17.899
	∞	3.786	5.588	7.311	8.996	10.660	12.313	13.954	15.587	17.216	18.845
20	1	3.597	5.252	6.847	8.419	9.979	11.533	13.084	14.633	16.182	17.732
	∞	3.768	5.553	7.260	8.930	10.577	12.209	13.832	15.450	17.057	18.667
30	1	3.483	5.060	6.567	8.045	9.506	10.958	12.400	13.841	15.274	16.711
	∞	3.646	5.342	6.952	8.522	10.060	11.582	13.091	14.588	16.081	17.566
40	1	3.432	4.966	6.432	7.866	9.277	10.679	12.073	13.457	14.840	16.213
	∞	3.588	5.242	6.806	8.325	9.813	11.282	12.735	14.173	15.607	17.035
50	1	3.398	4.912	6.355	7.761	9.146	10.518	11.880	13.232	14.578	15.924
	∞	3.555	5.182	6.718	8.209	9.667	11.105	12.522	13.933	15.330	16.718
60	1	3.379	4.876	6.303	7.691	9.056	10.411	11.752	13.086	14.408	15.732
	∞	3.531	5.143	6.663	8.132	9.569	10.985	12.386	13.771	15.147	16.510
70	1	3.360	4.849	6.263	7.641	8.994	10.333	11.660	12.978	14.289	15.598
	∞	3.516	5.114	6.620	8.080	9.504	10.900	12.287	13.656	15.014	16.365
80	1	3.350	4.833	6.234	7.606	8.950	10.279	11.591	12.903	14.198	15.495
	∞	3.504	5.092	6.593	8.036	9.454	10.842	12.210	13.573	14.914	16.253
90	1	3.341	4.818	6.216	7.579	8.914	10.232	11.543	12.839	14.132	15.412
	∞	3.495	5.075	6.569	8.007	9.414	10.794	12.155	13.503	14.843	16.171
100	1	3.333	4.807	6.200	7.552	8.882	10.200	11.497	12.787	14.075	15.352
	∞	3.483	5.063	6.552	7.985	9.383	10.759	12.108	13.451	14.778	16.104
∞	1	3.275	4.701	6.048	7.353	8.630	9.888	11.131	12.361	13.581	14.793
	∞	3.421	4.951	6.385	7.763	9.104	10.418	11.711	12.987	14.248	15.498

Table A.7 (*continued*)

v	$w\backslash k$	3	4	5	6	7	8	9	10	11	12
		$\alpha = 0.050$									
4	1	9.193	13.821	18.540	23.358	28.260	33.247	38.295	43.370	48.534	53.716
	∞	9.678	14.743	19.862	25.043	30.267	35.539	40.872	46.200	51.563	56.935
5	1	7.894	11.608	15.345	19.140	22.983	26.882	30.812	34.777	38.772	42.809
	∞	8.276	12.325	16.369	20.437	25.537	28.650	32.795	36.956	41.139	45.304
6	1	7.157	10.372	13.577	16.811	20.079	23.375	26.697	30.055	33.424	36.825
	∞	7.484	10.980	14.440	17.898	21.371	24.860	28.354	31.875	35.400	38.918
7	1	6.685	9.587	12.460	15.345	18.248	21.175	24.119	27.088	30.074	33.066
	∞	6.979	10.127	13.222	16.303	19.392	22.479	25.577	28.678	31.792	34.910
8	1	6.357	9.046	11.694	14.339	16.997	19.673	22.358	25.061	27.778	30.500
	∞	6.628	9.541	12.389	15.213	18.035	20.855	23.679	26.506	29.337	32.175
9	1	6.117	8.652	11.135	13.611	16.088	18.580	21.078	23.591	26.113	28.650
	∞	6.370	9.114	11.782	14.422	17.050	19.674	22.304	24.925	27.558	30.191
10	1	5.934	8.352	10.712	13.058	15.404	17.753	20.115	22.479	24.856	27.241
	∞	6.175	8.790	11.321	13.823	16.305	18.782	21.258	23.729	26.204	28.683
11	1	5.789	8.117	10.380	12.625	14.866	17.109	19.354	21.611	23.870	26.133
	∞	6.021	8.535	10.961	13.350	15.722	18.085	20.439	22.797	25.150	27.505
12	1	5.672	7.927	10.114	12.277	14.431	16.587	18.746	20.908	23.074	25.247
	∞	5.897	8.331	10.673	12.974	15.252	17.523	19.785	22.042	24.298	26.554
13	1	5.577	7.771	9.894	11.992	14.076	16.159	18.247	20.335	22.425	24.515
	∞	5.793	8.161	10.435	12.663	14.869	17.060	19.245	21.423	23.600	25.772
14	1	5.496	7.640	9.712	11.751	13.781	15.803	17.829	19.854	21.876	23.910
	∞	5.707	8.020	10.235	12.403	14.549	16.675	18.794	20.910	23.017	25.126
15	1	5.426	7.529	9.555	11.550	13.529	15.501	17.471	19.447	21.417	23.391
	∞	5.634	7.901	10.066	12.184	14.276	16.349	18.412	20.471	22.523	24.572

16	1	5.367	7.434	9.422	11.374	13.311	15.243	17.168	19.095	21.018	22.949
	∞	5.571	7.798	9.920	11.996	14.040	16.068	18.087	20.096	22.094	24.092
17	1	5.315	7.352	9.305	11.224	13.126	15.017	16.907	18.789	20.677	22.561
	∞	5.516	7.709	9.796	11.832	13.836	15.824	17.801	19.767	21.725	23.682
18	1	5.270	7.279	9.205	11.091	12.960	14.821	16.673	18.524	20.373	22.225
	∞	5.468	7.630	9.685	11.689	13.658	15.614	17.551	19.480	21.403	23.319
19	1	5.230	7.216	9.114	10.974	12.815	14.645	16.469	18.288	20.110	21.925
	∞	5.426	7.560	9.587	11.560	13.503	15.422	17.330	19.226	21.118	23.002
20	1	5.195	7.158	9.033	10.871	12.688	14.491	16.286	18.082	19.871	21.660
	∞	5.387	7.498	9.501	11.449	13.364	15.255	17.137	19.003	20.865	22.719
30	1	4.977	6.808	8.546	10.237	11.901	13.549	15.183	16.809	18.428	20.040
	∞	5.155	7.122	8.974	10.761	12.514	14.239	15.943	17.635	19.312	20.985
40	1	4.874	6.643	8.314	9.938	11.528	13.102	14.657	16.203	17.743	19.274
	∞	5.047	6.944	8.724	10.438	12.110	13.755	15.377	16.985	18.580	20.161
50	1	4.812	6.545	8.180	9.763	11.310	12.839	14.349	15.849	17.338	18.820
	∞	4.982	6.840	8.575	10.246	11.875	13.472	15.049	16.604	18.145	19.681
60	1	4.770	6.482	8.089	9.648	11.170	12.670	14.150	15.621	17.074	18.527
	∞	4.936	6.771	8.478	10.124	11.721	13.287	14.833	16.356	17.861	19.364
70	1	4.742	6.436	8.027	9.564	11.066	12.548	14.005	15.457	16.888	18.320
	∞	4.908	6.722	8.411	10.036	11.614	13.158	14.675	16.176	17.665	19.137
80	1	4.724	6.405	7.983	9.503	10.994	12.457	13.901	15.335	16.750	18.163
	∞	4.883	6.685	8.362	9.969	11.534	13.061	14.565	16.048	17.513	18.971
90	1	4.708	6.380	7.948	9.461	10.938	12.389	13.824	15.239	16.647	18.038
	∞	4.867	6.658	8.320	9.919	11.466	12.983	14.479	15.950	17.401	18.844
100	1	4.693	6.359	7.914	9.420	10.886	12.331	13.758	15.165	16.558	17.943
	∞	4.854	6.635	8.292	9.882	11.422	12.927	14.409	15.866	17.309	18.738
∞	1	4.577	6.175	7.665	9.095	10.485	11.846	13.185	14.505	15.811	17.103
	∞	4.732	6.440	8.020	9.525	10.981	12.399	13.789	15.156	16.503	17.834

Table A.7 (*continued*)

		$\alpha = 0.010$									
v	$w\backslash k$	3	4	5	6	7	8	9	10	11	12
4	1	25.121	36.169	47.421	58.925	70.685	82.516	94.625	106.796	119.072	131.545
	∞	26.236	38.322	50.545	62.908	75.389	87.980	100.774	113.553	126.420	139.248
5	1	19.240	26.907	34.623	42.449	50.370	58.430	66.515	74.727	82.944	91.320
	∞	19.995	28.335	36.679	45.089	53.535	62.010	70.607	79.150	87.827	96.449
6	1	16.228	22.260	28.270	34.330	40.446	46.639	52.882	59.167	65.468	71.803
	∞	16.811	23.358	29.845	36.327	42.822	49.367	55.924	62.481	69.127	75.750
7	1	14.431	19.521	24.555	29.601	34.682	39.802	44.954	50.130	55.347	60.603
	∞	14.918	20.434	25.848	31.236	36.632	42.043	47.445	52.874	58.331	63.773
8	1	13.245	17.733	22.142	26.543	30.957	35.394	39.863	44.355	48.852	53.378
	∞	13.670	18.519	23.256	27.945	32.636	37.314	41.999	46.693	51.403	56.110
9	1	12.406	16.474	20.453	24.410	28.364	32.333	36.322	40.325	44.339	48.387
	∞	12.789	17.179	21.444	25.659	29.850	34.034	38.224	42.405	46.591	50.783
10	1	11.783	15.550	19.213	22.844	26.470	30.093	33.731	37.387	41.048	44.714
	∞	12.136	16.196	20.115	23.981	27.816	31.642	35.454	39.268	43.083	46.899
11	1	11.304	14.837	18.260	21.646	25.020	28.389	31.763	35.146	38.546	41.937
	∞	11.631	15.439	19.100	22.696	26.263	29.816	33.352	36.887	40.412	43.955
12	1	10.924	14.276	17.514	20.705	23.881	27.044	30.222	33.390	36.575	39.759
	∞	11.233	14.843	18.301	21.692	25.043	28.375	31.696	35.019	38.328	41.623
13	1	10.615	13.820	16.908	19.943	22.957	25.963	28.973	31.978	34.987	37.998
	∞	10.912	14.358	17.653	20.875	24.065	27.221	30.366	33.510	36.631	39.766
14	1	10.359	13.446	16.407	19.318	22.205	25.079	27.945	30.813	33.682	36.551
	∞	10.642	13.957	17.121	20.207	23.251	26.273	29.274	32.268	35.249	38.232
15	1	10.143	13.130	15.989	18.793	21.569	24.331	27.086	29.843	32.596	35.345
	∞	10.419	13.625	16.676	19.646	22.575	25.478	28.363	31.226	34.088	36.941

16	1	9.959	12.863	15.635	18.350	21.034	23.704	26.360	29.015	31.667	34.320
	∞	10.226	13.341	16.298	19.171	22.000	24.804	27.585	30.352	33.104	35.853
17	1	9.801	12.632	15.329	17.967	20.575	23.160	25.735	28.308	30.878	33.435
	∞	10.061	13.096	15.972	18.766	21.511	24.224	26.915	29.596	32.266	34.921
18	1	9.664	12.430	15.064	17.637	20.174	22.692	25.195	27.691	30.184	32.677
	∞	9.915	12.882	15.687	18.411	21.079	23.722	26.338	28.942	31.528	34.109
19	1	9.542	12.253	14.831	17.343	19.822	22.278	24.718	27.156	29.583	32.011
	∞	9.790	12.695	15.437	18.097	20.709	23.281	25.833	28.371	30.891	33.402
20	1	9.435	12.099	14.626	17.087	19.510	21.916	24.302	26.681	29.050	31.421
	∞	9.677	12.529	15.221	17.823	20.377	22.893	25.388	27.861	30.323	32.771
30	1	8.792	11.168	13.399	15.560	17.672	19.757	21.818	23.862	25.896	27.918
	∞	9.005	11.543	13.915	16.193	18.409	20.586	22.735	24.860	26.965	29.059
40	1	8.493	10.734	12.832	14.853	16.827	18.764	20.673	22.567	24.444	26.315
	∞	8.691	11.090	13.314	15.443	17.510	19.532	21.522	23.482	25.428	27.358
50	1	8.316	10.487	12.508	14.451	16.340	18.196	20.020	21.825	23.613	25.395
	∞	8.512	10.828	12.969	15.013	16.991	18.924	20.825	22.692	24.544	26.379
60	1	8.207	10.325	12.296	14.189	16.025	17.824	19.595	21.347	23.080	24.792
	∞	8.393	10.654	12.745	14.730	16.653	18.533	20.370	22.184	23.974	25.746
70	1	8.128	10.210	12.152	14.005	15.803	17.565	19.297	21.011	22.703	24.374
	∞	8.315	10.536	12.584	14.538	16.417	18.252	20.054	21.828	23.573	25.299
80	1	8.065	10.130	12.042	13.868	15.642	17.375	19.076	20.759	22.420	24.063
	∞	8.249	10.448	12.470	14.388	16.246	18.053	19.823	21.563	23.275	24.971
90	1	8.020	10.062	11.957	13.766	15.518	17.229	18.908	20.566	22.206	23.826
	∞	8.203	10.382	12.382	14.279	16.110	17.896	19.637	21.359	23.049	24.720
100	1	7.985	10.015	11.886	13.679	15.417	17.108	18.772	20.413	22.036	23.639
	∞	8.163	10.326	12.308	14.194	16.005	17.773	19.500	21.197	22.869	24.513
∞	1	7.672	9.564	11.305	12.957	14.550	16.098	17.611	19.096	20.557	21.997
	∞	7.839	9.853	11.690	13.422	15.083	16.691	18.257	19.789	21.293	22.774

Table A.8 Critical values for $S_{01}(v)$ with k treatments, simple tree, equal weights ($w = 1$), and limiting coefficients, $R(l, k+1)(w = \infty)$

		$\alpha = 0.100$									
v	$w\backslash k$	2	3	4	5	6	7	8	9	10	11
4	1	4.714	6.723	8.563	10.298	11.955	13.554	15.104	16.619	18.092	19.536
	∞	4.010	5.030	5.701	6.157	6.470	6.688	6.840	6.945	7.015	7.060
5	1	4.380	6.277	8.033	9.706	11.320	12.885	14.415	15.911	17.373	18.817
	∞	3.774	4.829	5.583	6.139	6.558	6.880	7.130	7.323	7.474	7.592
6	1	4.172	5.990	7.689	9.316	10.889	12.425	13.930	15.405	16.859	18.294
	∞	3.623	4.695	5.497	6.118	6.610	7.006	7.328	7.592	7.810	7.990
7	1	4.029	5.792	7.445	9.035	10.578	12.088	13.570	15.028	16.470	17.892
	∞	3.519	4.598	5.430	6.097	6.643	7.094	7.473	7.792	8.064	8.295
8	1	3.925	5.647	7.265	8.825	10.341	11.831	13.293	14.738	16.163	17.573
	∞	3.444	4.525	5.379	6.078	6.662	7.157	7.581	7.945	8.261	8.538
9	1	3.846	5.536	7.125	8.660	10.156	11.625	13.073	14.501	15.914	17.310
	∞	3.386	4.469	5.338	6.061	6.676	7.205	7.665	8.066	8.421	8.733
10	1	3.785	5.447	7.014	8.529	10.006	11.460	12.891	14.307	15.706	17.097
	∞	3.339	4.423	5.304	6.046	6.685	7.242	7.731	8.165	8.550	8.896
11	1	3.736	5.376	6.923	8.422	9.885	11.323	12.743	14.145	15.537	16.914
	∞	3.303	4.386	5.275	6.034	6.692	7.272	7.786	8.245	8.659	9.030
12	1	3.694	5.317	6.849	8.331	9.781	11.208	12.614	14.008	15.388	16.759
	∞	3.272	4.355	5.252	6.022	6.695	7.295	7.830	8.312	8.749	9.146
13	1	3.660	5.268	6.785	8.256	9.694	11.109	12.506	13.891	15.261	16:621
	∞	3.246	4.328	5.230	6.011	6.699	7.313	7.868	8.369	8.826	9.246
14	1	3.631	5.224	6.731	8.189	9.617	11.023	12.412	13.788	15.151	16.504
	∞	3.224	4.305	5.213	6.001	6.701	7.330	7.899	8.419	8.893	9.331
15	1	3.606	5.188	6.684	8.134	9.553	10.950	12.329	13.699	15.053	16.401
	∞	3.205	4.287	5.196	5.992	6.702	7.343	7.926	8.460	8.952	9.406

16	1	3.584	5.157	6.643	8.084	9.494	10.882	12.256	13.619	14.966	16.310
	∞	3.189	4.269	5.182	5.985	6.703	7.355	7.950	8.498	9.003	9.472
17	1	3.565	5.129	6.607	8.040	9.442	10.826	12.191	13.545	14.892	16.223
	∞	3.175	4.254	5.170	5.978	6.704	7.365	7.970	8.531	9.049	9.533
18	1	3.548	5.104	6.575	8.001	9.397	10.773	12.135	13.482	14.821	16.151
	∞	3.161	4.240	5.159	5.972	6.705	7.374	7.989	8.560	9.090	9.584
19	1	3.534	5.082	6.546	7.965	9.358	10.728	12.081	13.425	14.758	16.082
	∞	3.151	4.228	5.149	5.965	6.705	7.380	8.005	8.586	9.124	9.632
20	1	3.519	5.063	6.521	7.934	9.319	10.686	12.036	13.374	14.703	16.023
	∞	3.140	4.218	5.139	5.960	6.705	7.388	8.020	8.607	9.160	9.674
30	1	3.436	4.938	6.359	7.735	9.086	10.413	11.730	13.035	14.331	15.615
	∞	3.076	4.149	5.080	5.923	6.701	7.425	8.107	8.753	9.364	9.950
40	1	3.394	4.877	6.278	7.633	8.964	10.275	11.573	12.858	14.133	15.405
	∞	3.044	4.113	5.050	5.902	6.696	7.440	8.148	8.822	9.467	10.087
50	1	3.370	4.842	6.231	7.575	8.891	10.193	11.476	12.752	14.013	15.272
	∞	3.028	4.092	5.033	5.889	6.694	7.449	8.171	8.861	9.528	10.171
60	1	3.354	4.817	6.196	7.533	8.843	10.134	11.413	12.675	13.932	15.182
	∞	3.013	4.081	5.018	5.881	6.690	7.454	8.189	8.891	9.569	10.225
70	1	3.342	4.801	6.172	7.505	8.810	10.092	11.365	12.625	13.876	15.115
	∞	3.006	4.071	5.011	5.874	6.686	7.458	8.197	8.907	9.597	10.266
80	1	3.334	4.784	6.155	7.483	8.781	10.062	11.330	12.588	13.827	15.066
	∞	2.996	4.061	5.004	5.873	6.685	7.460	8.208	8.926	9.619	10.292
90	1	3.330	4.775	6.141	7.469	8.762	10.041	11.306	12.553	13.795	15.029
	∞	2.994	4.056	5.001	5.866	6.683	7.463	8.209	8.934	9.636	10.314
100	1	3.320	4.767	6.131	7.454	8.745	10.022	11.278	12.531	13.766	14.996
	∞	2.989	4.049	4.995	5.864	6.684	7.468	8.213	8.940	9.647	10.334
∞	1	3.275	4.696	6.036	7.333	8.600	9.848	11.080	12.300	13.510	14.711
	∞	2.952	4.010	4.955	5.835	6.671	7.476	8.257	9.018	9.764	10.496

Table A.8 (*continued*)

		$\alpha = 0.050$									
v	$w\backslash k$	2	3	4	5	6	7	8	9	10	11
4	1	7.949	10.805	13.400	15.826	18.137	20.358	22.506	24.600	26.631	28.613
	∞	6.885	8.210	8.999	9.470	9.744	9.892	9.957	9.966	9.939	9.889
5	1	7.092	9.674	12.051	14.305	16.470	18.564	20.604	22.596	24.547	26.465
	∞	6.235	7.596	8.515	9.150	9.596	9.909	10.127	10.276	10.374	10.435
6	1	6.580	8.984	11.215	13.342	15.396	17.397	19.352	21.265	23.149	24.996
	∞	5.839	7.205	8.186	8.918	9.472	9.895	10.223	10.476	10.672	10.820
7	1	6.240	8.520	10.646	12.680	14.654	16.577	18.461	20.316	22.142	23.942
	∞	5.572	6.933	7.952	8.742	9.369	9.872	10.279	10.610	10.879	11.102
8	1	5.997	8.187	10.231	12.194	14.101	15.969	17.799	19.603	21.378	23.137
	∞	5.381	6.733	7.774	8.605	9.284	9.844	10.313	10.705	11.036	11.316
9	1	5.816	7.936	9.918	11.826	13.680	15.496	17.282	19.047	20.786	22.504
	∞	5.238	6.580	7.634	8.495	9.212	9.817	10.332	10.773	11.154	11.484
10	1	5.676	7.740	9.675	11.534	13.347	15.123	16.870	18.596	20.304	21.991
	∞	5.126	6.461	7.524	8.404	9.150	9.789	10.341	10.825	11.246	11.619
11	1	5.564	7.583	9.477	11.298	13.076	14.818	16.534	18.231	19.905	21.570
	∞	5.036	6.362	7.431	8.328	9.097	9.764	10.347	10.863	11.320	11.725
12	1	5.472	7.456	9.314	11.105	12.851	14.567	16.254	17.923	19.576	21.214
	∞	4.963	6.281	7.354	8.264	9.051	9.740	10.351	10.892	11.379	11.815
13	1	5.397	7.348	9.179	10.941	12.663	14.350	16.016	17.662	19.294	20.908
	∞	4.902	6.214	7.290	8.209	9.011	9.718	10.351	10.917	11.427	11.890
14	1	5.333	7.257	9.063	10.804	12.501	14.170	15.814	17.441	19.049	20.646
	∞	4.850	6.155	7.234	8.161	8.976	9.700	10.349	10.934	11.468	11.952
15	1	5.278	7.180	8.964	10.684	12.363	14.010	15.638	17.246	18.840	20.420
	∞	4.806	6.105	7.186	8.118	8.943	9.681	10.345	10.950	11.501	12.005

16	1	5.231	7.114	8.877	10.580	12.241	13.872	15.483	17.076	18.650	20.215
	∞	4.768	6.063	7.142	8.082	8.915	9.664	10.342	10.963	11.529	12.053
17	1	5.189	7.054	8.803	10.489	12.133	13.751	15.347	16.923	18.488	20.036
	∞	4.734	6.024	7.106	8.049	8.889	9.647	10.337	10.972	11.554	12.094
18	1	5.153	7.002	8.736	10.407	12.039	13.641	15.224	16.788	18.339	19.880
	∞	4.704	5.991	7.072	8.019	8.865	9.633	10.334	10.980	11.575	12.128
19	1	5.121	6.955	8.677	10.336	11.955	13.547	15.116	16.668	18.207	19.735
	∞	4.677	5.961	7.042	7.993	8.845	9.621	10.330	10.986	11.593	12.159
20	1	5.091	6.915	8.622	10.269	11.877	13.459	15.018	16.561	18.087	19.606
	∞	4.653	5.933	7.015	7.967	8.826	9.607	10.328	10.991	11.610	12.187
30	1	4.913	6.657	8.291	9.865	11.400	12.907	14.395	15.866	17.323	18.774
	∞	4.508	5.764	6.844	7.811	8.698	9.522	10.290	11.016	11.702	12.352
40	1	4.828	6.534	8.130	9.665	11.162	12.633	14.084	15.518	16.936	18.345
	∞	4.436	5.681	6.756	7.730	8.630	9.470	10.263	11.019	11.740	12.427
50	1	4.776	6.460	8.031	9.546	11.018	12.465	13.893	15.304	16.702	18.088
	∞	4.394	5.631	6.706	7.680	8.588	9.438	10.246	11.018	11.754	12.465
60	1	4.740	6.410	7.967	9.465	10.924	12.360	13.766	15.165	16.546	17.917
	∞	4.367	5.599	6.672	7.649	8.559	9.416	10.235	11.016	11.768	12.493
70	1	4.718	6.375	7.921	9.410	10.860	12.281	13.680	15.064	16.430	17.786
	∞	4.345	5.574	6.645	7.626	8.540	9.399	10.227	11.014	11.777	12.512
80	1	4.702	6.348	7.888	9.368	10.805	12.217	13.613	14.983	16.345	17.695
	∞	4.331	5.559	6.628	7.606	8.523	9.387	10.217	11.013	11.777	12.523
90	1	4.684	6.329	7.864	9.334	10.766	12.176	13.561	14.925	16.278	17.620
	∞	4.321	5.543	6.613	7.592	8.511	9.381	10.211	11.008	11.782	12.532
100	1	4.673	6.310	7.843	9.310	10.736	12.139	13.514	14.875	16.227	17.562
	∞	4.314	5.530	6.601	7.581	8.500	9.369	10.208	11.006	11.787	12.539
∞	1	4.577	6.171	7.653	9.075	10.456	11.807	13.136	14.446	15.741	17.023
	∞	4.231	5.435	6.498	7.480	8.407	9.295	10.152	10.985	11.799	12.596

Table A.8 (*continued*)

		$\alpha = 0.010$									
v	$w\backslash k$	2	3	4	5	6	7	8	9	10	11
4	1	21.446	27.530	32.974	38.024	42.761	47.300	51.657	55.878	59.969	63.948
	∞	18.839	21.028	21.909	22.094	21.888	21.486	20.980	20.422	19.862	19.310
5	1	17.186	22.112	26.611	30.843	34.873	38.772	42.559	46.216	49.814	53.327
	∞	15.391	17.639	18.929	19.634	19.979	20.079	20.025	19.881	19.664	19.414
6	1	14.887	19.145	23.071	26.795	30.375	33.856	37.239	40.534	43.787	46.951
	∞	13.491	15.693	17.133	18.079	18.696	19.087	19.313	19.418	19.438	19.398
7	1	13.467	17.291	20.840	24.220	27.487	30.658	33.763	36.816	39.802	42.745
	∞	12.300	14.435	15.931	17.006	17.784	18.348	18.752	19.033	19.223	19.338
8	1	12.504	16.030	19.310	22.443	25.475	28.434	31.331	34.179	36.988	39.758
	∞	11.486	13.560	15.074	16.221	17.102	17.779	18.306	18.714	19.022	19.258
9	1	11.811	15.116	18.193	21.139	23.999	26.795	29.535	32.229	34.892	37.523
	∞	10.898	12.915	14.434	15.623	16.567	17.329	17.940	18.441	18.840	19.170
10	1	11.290	14.426	17.350	20.148	22.870	25.536	28.151	30.721	33.270	35.785
	∞	10.453	12.421	13.937	15.150	16.141	16.958	17.636	18.207	18.681	19.078
11	1	10.886	13.888	16.686	19.374	21.981	24.537	27.050	29.523	31.967	34.391
	∞	10.105	12.030	13.539	14.770	15.789	16.652	17.379	18.001	18.538	18.995
12	1	10.561	13.453	16.153	18.746	21.264	23.731	26.158	28.550	30.913	33.254
	∞	9.825	11.715	13.214	14.456	15.500	16.393	17.160	17.827	18.404	18.911
13	1	10.294	13.100	15.715	18.228	20.672	23.061	25.415	27.745	30.042	32.313
	∞	9.595	11.455	12.945	14.190	15.254	16.171	16.967	17.671	18.288	18.835
14	1	10.075	12.804	15.349	17.794	20.171	22.504	24.795	27.059	29.299	31.516
	∞	9.402	11.235	12.717	13.964	15.040	15.978	16.803	17.532	18.182	18.761
15	1	9.888	12.553	15.039	17.429	19.749	22.026	24.262	26.475	28.663	30.829
	∞	9.240	11.049	12.522	13.772	14.856	15.811	16.656	17.407	18.083	18.696

16	1	9.727	12.338	14.773	17.109	19.384	21.613	23.806	25.973	28.113	30.238
	∞	9.101	10.888	12.354	13.602	14.698	15.662	16.526	17.299	17.998	18.632
17	1	9.589	12.151	14.542	16.837	19.067	21.251	23.404	25.533	27.632	29.713
	∞	8.981	10.749	12.207	13.455	14.556	15.533	16.407	17.198	17.919	18.574
18	1	9.468	11.991	14.339	16.596	18.788	20.940	23.054	25.144	27.208	29.258
	∞	8.875	10.625	12.076	13.325	14.428	15.412	16.302	17.111	17.845	18.519
19	1	9.360	11.847	14.161	16.381	18.544	20.658	22.743	24.798	26.836	28.851
	∞	8.781	10.519	11.961	13.210	14.313	15.308	16.208	17.028	17.777	18.467
20	1	9.266	11.720	14.001	16.194	18.323	20.412	22.465	24.495	26.503	28.491
	∞	8.697	10.421	11.855	13.104	14.214	15.214	16.122	16.953	17.714	18.422
30	1	8.692	10.945	13.035	15.038	16.979	18.881	20.753	22.594	24.423	26.228
	∞	8.193	9.830	11.219	12.451	13.569	14.596	15.551	16.446	17.287	18.077
40	1	8.421	10.578	12.578	14.488	16.339	18.148	19.924	21.678	23.409	25.129
	∞	7.954	9.545	10.908	12.126	13.245	14.285	15.256	16.175	17.045	17.876
50	1	8.263	10.366	12.308	14.163	15.961	17.716	19.438	21.139	22.815	24.476
	∞	7.814	9.382	10.724	11.936	13.052	14.093	15.075	16.003	16.893	17.744
60	1	8.160	10.225	12.136	13.954	15.715	17.432	19.121	20.779	22.418	24.046
	∞	7.723	9.270	10.608	11.810	12.923	13.966	14.953	15.890	16.791	17.652
70	1	8.091	10.126	12.011	13.803	15.540	17.232	18.895	20.530	22.137	23.734
	∞	7.662	9.196	10.520	11.719	12.828	13.870	14.863	15.803	16.711	17.585
80	1	8.036	10.056	11.919	13.693	15.412	17.080	18.725	20.335	21.934	23.504
	∞	7.612	9.137	10.454	11.649	12.758	13.800	14.791	15.739	16.650	17.527
90	1	7.994	10.001	11.845	13.605	15.307	16.965	18.595	20.194	21.773	23.327
	∞	7.579	9.093	10.409	11.599	12.703	13.751	14.738	15.692	16.609	17.487
100	1	7.963	9.956	11.794	13.537	15.230	16.874	18.489	20.078	21.638	23.183
	∞	7.545	9.056	10.371	11.558	12.663	13.703	14.698	15.653	16.566	17.452
∞	1	7.672	9.560	11.295	12.939	14.523	16.061	17.564	19.039	20.490	21.920
	∞	7.290	8.746	10.019	11.183	12.274	13.313	14.310	15.274	16.211	17.126

Table A.9 Critical values for $S_{12}(v)$ with k treatments, simple tree, equal weights ($w = 1$), and limiting coefficients, $R(l, k+1)$ ($w = \infty$)

		$\alpha = 0.100$									
v	$w\backslash k$	2	3	4	5	6	7	8	9	10	11
4	1	3.990	5.155	6.071	6.832	7.484	8.057	8.568	9.028	9.450	9.839
	∞	4.735	6.887	8.940	10.940	12.910	14.857	16.789	18.712	20.633	22.538
5	1	3.634	4.653	5.450	6.107	6.669	7.161	7.598	7.994	8.354	8.684
	∞	4.279	6.134	7.884	9.575	11.233	12.869	14.489	16.094	17.689	19.283
6	1	3.420	4.355	5.081	5.677	6.187	6.633	7.028	7.385	7.710	8.008
	∞	4.007	5.690	7.264	8.779	10.258	11.714	13.150	14.576	15.989	17.397
7	1	3.278	4.157	4.838	5.395	5.871	6.286	6.654	6.987	7.289	7.566
	∞	3.826	5.398	6.860	8.260	9.625	10.964	12.283	13.588	14.885	16.172
8	1	3.177	4.017	4.665	5.196	5.648	6.042	6.392	6.707	6.993	7.254
	∞	3.699	5.192	6.576	7.896	9.180	10.436	11.676	12.900	14.113	15.316
9	1	3.102	3.913	4.537	5.048	5.483	5.861	6.196	6.498	6.772	7.023
	∞	3.604	5.039	6.364	7.627	8.851	10.047	11.226	12.387	13.542	14.682
10	1	3.043	3.831	4.437	4.933	5.354	5.720	6.045	6.337	6.602	6.846
	∞	3.530	4.921	6.202	7.420	8.598	9.748	10.881	11.996	13.100	14.196
11	1	2.996	3.767	4.358	4.842	5.252	5.610	5.925	6.209	6.468	6.705
	∞	3.471	4.828	6.072	7.255	8.397	9.511	10.606	11.685	12.753	13.809
12	1	2.957	3.714	4.295	4.768	5.169	5.519	5.827	6.105	6.358	6.589
	∞	3.423	4.750	5.967	7.121	8.234	9.318	10.384	11.432	12.470	13.499
13	1	2.926	3.671	4.241	4.706	5.100	5.443	5.747	6.020	6.267	6.495
	∞	3.383	4.688	5.880	7.010	8.100	9.160	10.200	11.224	12.238	13.239
14	1	2.899	3.634	4.197	4.654	5.042	5.380	5.678	5.947	6.191	6.413
	∞	3.349	4.634	5.807	6.916	7.985	9.026	10.045	11.048	12.040	13.019
15	1	2.875	3.602	4.157	4.610	4.993	5.327	5.620	5.885	6.125	6.345
	∞	3.320	4.589	5.745	6.837	7.888	8.912	9.914	10.898	11.872	12.834

16	1	2.856	3.575	4.124	4.572	4.951	5.279	5.569	5.832	6.069	6.286
	∞	3.296	4.550	5.691	6.768	7.804	8.814	9.800	10.770	11.727	12.673
17	1	2.839	3.552	4.095	4.539	4.913	5.239	5.526	5.784	6.020	6.235
	∞	3.274	4.515	5.644	6.708	7.733	8.727	9.701	10.658	11.601	12.535
18	1	2.823	3.530	4.070	4.508	4.880	5.202	5.489	5.744	5.976	6.190
	∞	3.254	4.484	5.601	6.655	7.668	8.651	9.612	10.559	11.490	12.411
19	1	2.809	3.511	4.046	4.483	4.851	5.170	5.453	5.707	5.939	6.148
	∞	3.238	4.458	5.565	6.609	7.612	8.584	9.537	10.472	11.393	12.303
20	1	2.798	3.496	4.026	4.460	4.825	5.143	5.422	5.675	5.903	6.113
	∞	3.222	4.435	5.533	6.568	7.560	8.525	9.466	10.393	11.305	12.206
30	1	2.721	3.392	3.901	4.316	4.664	4.967	5.236	5.475	5.691	5.892
	∞	3.129	4.287	5.332	6.311	7.249	8.157	9.042	9.911	10.765	11.607
40	1	2.685	3.342	3.841	4.247	4.587	4.883	5.143	5.377	5.592	5.783
	∞	3.083	4.214	5.233	6.186	7.099	7.982	8.840	9.680	10.504	11.318
50	1	2.663	3.314	3.804	4.203	4.542	4.834	5.089	5.320	5.530	5.722
	∞	3.055	4.171	5.178	6.115	7.010	7.876	8.718	9.541	10.353	11.150
60	1	2.648	3.293	3.782	4.178	4.510	4.799	5.057	5.286	5.490	5.678
	∞	3.038	4.144	5.139	6.067	6.953	7.808	8.641	9.456	10.250	11.037
70	1	2.640	3.281	3.766	4.157	4.490	4.776	5.031	5.257	5.465	5.649
	∞	3.024	4.124	5.114	6.035	6.912	7.757	8.583	9.388	10.182	10.957
80	1	2.630	3.271	3.755	4.142	4.473	4.762	5.010	5.236	5.442	5.632
	∞	3.017	4.109	5.092	6.008	6.880	7.723	8.541	9.344	10.130	10.899
90	1	2.626	3.259	3.745	4.134	4.459	4.745	4.995	5.222	5.425	5.611
	∞	3.012	4.098	5.075	5.991	6.854	7.695	8.511	9.307	10.089	10.856
100	1	2.621	3.255	3.732	4.121	4.453	4.734	4.989	5.211	5.414	5.598
	∞	3.002	4.088	5.063	5.973	6.837	7.673	8.485	9.274	10.052	10.819
∞	1	3.199	3.667	4.046	4.365	4.640	4.884	5.101	5.299	5.479	5.645
	∞	2.952	4.010	4.955	5.835	6.671	7.476	8.257	9.018	9.764	10.496

Table A.9 (*continued*)

		$\alpha = 0.050$									
v	$w\backslash k$	2	3	4	5	6	7	8	9	10	11
4	1	7.069	8.849	10.245	11.404	12.394	13.262	14.042	14.738	15.381	15.971
	∞	8.171	11.418	14.505	17.512	20.467	23.392	26.292	29.174	32.038	34.905
5	1	6.181	7.653	8.797	9.743	10.549	11.252	11.882	12.446	12.963	13.436
	∞	7.078	9.717	12.197	14.591	16.936	19.240	21.524	23.789	26.048	28.294
6	1	5.669	6.970	7.975	8.801	9.507	10.119	10.666	11.158	11.606	12.016
	∞	6.452	8.757	10.904	12.965	14.972	16.944	18.890	20.819	22.740	24.639
7	1	5.338	6.529	7.447	8.200	8.840	9.397	9.892	10.336	10.741	11.114
	∞	6.050	8.143	10.081	11.933	13.734	15.494	17.234	18.950	20.652	22.343
8	1	5.107	6.223	7.082	7.783	8.379	8.898	9.357	9.771	10.148	10.491
	∞	5.770	7.719	9.515	11.226	12.883	14.504	16.096	17.673	19.230	20.779
9	1	4.937	5.998	6.812	7.478	8.042	8.532	8.966	9.357	9.712	10.038
	∞	5.564	7.409	9.103	10.709	12.264	13.782	15.275	16.745	18.198	19.641
10	1	4.805	5.826	6.607	7.245	7.785	8.254	8.668	9.041	9.381	9.691
	∞	5.406	7.173	8.788	10.319	11.794	13.235	14.647	16.041	17.414	18.777
11	1	4.702	5.690	6.446	7.061	7.581	8.034	8.434	8.793	9.120	9.419
	∞	5.283	6.987	8.540	10.011	11.426	12.806	14.159	15.486	16.802	18.104
12	1	4.619	5.580	6.316	6.913	7.417	7.856	8.245	8.595	8.910	9.201
	∞	5.181	6.837	8.341	9.762	11.129	12.461	13.763	15.046	16.307	17.558
13	1	4.548	5.490	6.207	6.790	7.284	7.712	8.089	8.429	8.737	9.020
	∞	5.098	6.713	8.177	9.559	10.887	12.175	13.437	14.681	15.902	17.116
14	1	4.490	5.413	6.116	6.688	7.171	7.589	7.959	8.290	8.593	8.868
	∞	5.028	6.608	8.039	9.388	10.681	11.940	13.169	14.376	15.564	16.741
15	1	4.441	5.349	6.040	6.601	7.074	7.484	7.848	8.173	8.469	8.739
	∞	4.968	6.521	7.924	9.243	10.508	11.738	12.938	14.117	15.279	16.425

16	1	4.398	5.293	5.973	6.526	6.992	7.396	7.752	8.073	8.364	8.629
	∞	4.917	6.445	7.824	9.118	10.361	11.564	12.739	13.896	15.032	16.155
17	1	4.360	5.244	5.916	6.460	6.919	7.318	7.669	7.986	8.271	8.533
	∞	4.874	6.379	7.737	9.010	10.231	11.414	12.569	13.700	14.815	15.917
18	1	4.327	5.202	5.865	6.402	6.857	7.250	7.597	7.909	8.190	8.448
	∞	4.834	6.320	7.659	8.917	10.118	11.282	12.421	13.533	14.628	15.710
19	1	4.298	5.164	5.819	6.351	6.801	7.189	7.533	7.839	8.120	8.375
	∞	4.798	6.269	7.592	8.833	10.019	11.166	12.288	13.384	14.464	15.529
20	1	4.273	5.130	5.780	6.306	6.750	7.136	7.475	7.780	8.056	8.308
	∞	4.769	6.224	7.533	8.758	9.929	11.064	12.168	13.252	14.317	15.366
30	1	4.113	4.923	5.536	6.031	6.445	6.808	7.125	7.411	7.669	7.904
	∞	4.579	5.944	7.165	8.303	9.389	10.433	11.449	12.444	13.419	14.379
40	1	4.039	4.825	5.418	5.898	6.300	6.650	6.957	7.235	7.481	7.709
	∞	4.487	5.812	6.991	8.087	9.129	10.133	11.110	12.060	12.995	13.910
50	1	3.992	4.765	5.349	5.821	6.215	6.557	6.859	7.129	7.373	7.599
	∞	4.434	5.734	6.887	7.962	8.980	9.960	10.909	11.838	12.743	13.640
60	1	3.965	4.727	5.303	5.771	6.160	6.500	6.794	7.063	7.302	7.523
	∞	4.400	5.683	6.821	7.878	8.881	9.845	10.781	11.689	12.584	13.457
70	1	3.942	4.703	5.272	5.734	6.121	6.456	6.753	7.016	7.254	7.474
	∞	4.374	5.644	6.773	7.821	8.810	9.764	10.689	11.585	12.465	13.334
80	1	3.926	4.680	5.247	5.704	6.093	6.422	6.719	6.978	7.216	7.431
	∞	4.358	5.620	6.737	7.776	8.757	9.705	10.616	11.508	12.379	13.240
90	1	3.912	4.666	5.228	5.685	6.072	6.399	6.689	6.950	7.186	7.405
	∞	4.345	5.599	6.708	7.740	8.722	9.656	10.567	11.452	12.318	13.163
100	1	3.903	4.653	5.218	5.673	6.049	6.379	6.670	6.928	7.166	7.376
	∞	4.333	5.584	6.691	7.715	8.687	9.618	10.527	11.399	12.261	13.106
∞	1	3.820	4.543	5.087	5.524	5.891	6.208	6.487	6.736	6.961	7.167
	∞	4.231	5.435	6.498	7.480	8.407	9.295	10.152	10.985	11.799	12.596

Table A.9 (*continued*)

		$\alpha = 0.010$									
v	$w \backslash k$	2	3	4	5	6	7	8	9	10	11
4	1	20.187	24.512	27.899	30.721	33.120	35.231	37.101	38.820	40.356	41.781
	∞	22.766	30.556	37.970	45.183	52.278	59.289	66.205	73.147	79.967	86.833
5	1	15.810	18.915	21.337	23.322	25.024	26.514	27.827	29.027	30.106	31.112
	∞	17.620	23.108	28.253	33.236	38.082	42.892	47.597	52.307	56.990	61.629
6	1	13.534	16.039	17.974	19.563	20.922	22.099	23.149	24.095	24.957	25.747
	∞	14.967	19.326	23.384	27.273	31.061	34.782	38.451	42.082	45.698	49.274
7	1	12.158	14.312	15.967	17.321	18.478	19.478	20.369	21.175	21.900	22.570
	∞	13.369	17.081	20.506	23.776	26.956	30.062	33.122	36.140	39.143	42.106
8	1	11.240	13.163	14.640	15.846	16.869	17.758	18.546	19.258	19.902	20.491
	∞	12.313	15.602	18.627	21.497	24.280	26.995	29.659	32.286	34.897	37.472
9	1	10.589	12.353	13.701	14.804	15.739	16.549	17.268	17.910	18.498	19.036
	∞	11.564	14.561	17.306	19.899	22.410	24.855	27.252	29.616	31.943	34.261
10	1	10.102	11.748	13.006	14.032	14.898	15.650	16.318	16.914	17.460	17.957
	∞	11.006	13.791	16.326	18.726	21.033	23.277	25.482	27.651	29.791	31.908
11	1	9.726	11.284	12.471	13.435	14.254	14.963	15.588	16.152	16.665	17.131
	∞	10.577	13.199	15.580	17.824	19.980	22.078	24.128	26.148	28.141	30.105
12	1	9.425	10.913	12.045	12.965	13.742	14.417	15.012	15.546	16.033	16.475
	∞	10.234	12.729	14.986	17.112	19.151	21.130	23.068	24.967	26.844	28.693
13	1	9.183	10.612	11.699	12.583	13.326	13.973	14.545	15.057	15.523	15.949
	∞	9.958	12.349	14.509	16.539	18.482	20.366	22.208	24.020	25.800	27.558
14	1	8.980	10.362	11.414	12.267	12.984	13.607	14.160	14.652	15.099	15.511
	∞	9.727	12.034	14.112	16.064	17.931	19.739	21.504	23.233	24.940	26.623
15	1	8.808	10.153	11.171	11.999	12.696	13.302	13.833	14.311	14.748	15.142
	∞	9.533	11.770	13.779	15.664	17.467	19.212	20.911	22.575	24.218	25.834

16	1	8.664	9.974	10.968	11.771	12.452	13.038	13.559	14.023	14.445	14.831
	∞	9.366	11.543	13.499	15.325	17.072	18.760	20.406	22.022	23.608	25.169
17	1	8.538	9.819	10.790	11.577	12.240	12.813	13.319	13.775	14.185	14.563
	∞	9.223	11.347	13.253	15.035	16.734	18.375	19.977	21.543	23.079	24.599
18	1	8.427	9.685	10.636	11.407	12.055	12.616	13.114	13.557	13.960	14.328
	∞	9.100	11.180	13.042	14.781	16.442	18.042	19.600	21.126	22.624	24.103
19	1	8.332	9.566	10.500	11.257	11.893	12.446	12.929	13.368	13.759	14.122
	∞	8.990	11.032	12.858	14.561	16.185	17.751	19.273	20.759	22.223	23.664
20	1	8.245	9.461	10.382	11.123	11.750	12.292	12.770	13.198	13.586	13.941
	∞	8.892	10.900	12.691	14.364	15.955	17.490	18.979	20.442	21.873	23.284
30	1	7.729	8.829	9.659	10.330	10.894	11.379	11.805	12.189	12.536	12.851
	∞	8.309	10.109	11.709	13.193	14.596	15.947	17.255	18.528	19.772	21.001
40	1	7.488	8.536	9.325	9.961	10.492	10.955	11.358	11.723	12.048	12.347
	∞	8.038	9.745	11.254	12.650	13.968	15.232	16.455	17.647	18.811	19.952
50	1	7.349	8.366	9.129	9.745	10.264	10.711	11.100	11.453	11.768	12.058
	∞	7.880	9.533	10.991	12.337	13.610	14.823	15.998	17.139	18.252	19.344
60	1	7.255	8.255	9.002	9.608	10.114	10.547	10.934	11.279	11.585	11.868
	∞	7.775	9.397	10.822	12.136	13.375	14.560	15.703	16.815	17.898	18.956
70	1	7.192	8.176	8.918	9.509	10.008	10.440	10.815	11.152	11.457	11.736
	∞	7.705	9.300	10.701	11.993	13.213	14.370	15.496	16.580	17.645	18.682
80	1	7.146	8.119	8.847	9.435	9.932	10.354	10.729	11.063	11.362	11.636
	∞	7.653	9.228	10.616	11.887	13.088	14.232	15.342	16.415	17.455	18.480
90	1	7.110	8.072	8.801	9.381	9.872	10.293	10.663	10.994	11.292	11.564
	∞	7.611	9.173	10.546	11.810	12.997	14.132	15.224	16.278	17.315	18.324
100	1	7.081	8.042	8.759	9.340	9.823	10.237	10.609	10.939	11.233	11.505
	∞	7.574	9.128	10.490	11.741	12.919	14.043	15.125	16.178	17.200	18.198
∞	1	6.823	7.729	8.407	8.951	9.406	9.798	10.142	10.449	10.726	10.979
	∞	7.290	8.746	10.019	11.183	12.274	13.313	14.310	15.274	16.211	17.126

Table A.10 Level probabilities for equal weights and the simple order, $P_S(l, k)$

l	$k = 3$	$k = 4$	$k = 5$	$k = 6$	$k = 7$	$k = 8$	$k = 9$	$k = 10$	$k = 11$
1	0.33333	0.25000	0.20000	0.16667	0.14286	0.12500	0.11111	0.10000	0.09091
2	0.50000	0.45833	0.41667	0.38056	0.35000	0.32411	0.30198	0.28290	0.26627
3	0.16667	0.25000	0.29167	0.31250	0.32222	0.32569	0.32552	0.32316	0.31950
4		0.04167	0.08333	0.11806	0.14583	0.16788	0.18542	0.19943	0.21068
5			0.00833	0.02083	0.03472	0.04861	0.06186	0.07422	0.08560
6				0.00139	0.00417	0.00799	0.01250	0.01744	0.02260
7					0.00020	0.00069	0.00150	0.00260	0.00395
8						0.00002	0.00010	0.00024	0.00045
9							0.00000	0.00001	0.00003
10								0.00000	0.00000
11									0.00000

l	$k = 12$	$k = 13$	$k = 14$	$k = 15$	$k = 16$	$k = 17$	$k = 18$	$k = 19$	$k = 20$
1	0.08333	0.07692	0.07143	0.06667	0.06250	0.05882	0.05556	0.05263	0.05000
2	0.25166	0.23871	0.22715	0.21677	0.20739	0.19887	0.19109	0.18395	0.17739
3	0.31507	0.31019	0.30508	0.29989	0.29469	0.28956	0.28452	0.27960	0.27482
4	0.21974	0.22708	0.23301	0.23782	0.24170	0.24482	0.24730	0.24926	0.25078
5	0.09602	0.10554	0.11422	0.12214	0.12937	0.13598	0.14203	0.14757	0.15265
6	0.02785	0.03309	0.03827	0.04333	0.04826	0.05303	0.05764	0.06208	0.06635
7	0.00551	0.00723	0.00907	0.01102	0.01304	0.01511	0.01722	0.01934	0.02148
8	0.00075	0.00111	0.00155	0.00205	0.00261	0.00322	0.00388	0.00459	0.00532
9	0.00007	0.00012	0.00019	0.00028	0.00039	0.00052	0.00067	0.00084	0.00103
10	0.00000	0.00001	0.00002	0.00003	0.00004	0.00006	0.00009	0.00012	0.00016
11[a]	0.00000	0.00000	0.00000	0.00000	0.00000	0.00001	0.00001	0.00001	0.00002

[a] For $12 \leqslant k \leqslant 20$ and $l > 11$, $P_S l k$ is 0 to five places.

Table A.11 Level probabilities for equal weights and the simple tree ordering, $P_T(l, k+1)$ with k the number of treatments

l	$k=2$	$k=3$	$k=4$	$k=5$	$k=6$	$k=7$	$k=8$	$k=9$	$k=10$
1	0.16667	0.04387	0.00978	0.00192	0.00034	0.00006	0.00001	0.00000	0.00000
2	0.50000	0.25000	0.08774	0.02446	0.00577	0.00119	0.00022	0.00004	0.00001
3	0.33333	0.45613	0.29022	0.12373	0.04060	0.01101	0.00258	0.00053	0.00010
4		0.25000	0.41226	0.30887	0.15185	0.05650	0.01714	0.00443	0.00101
5			0.20000	0.37434	0.31620	0.17352	0.07139	0.02374	0.00669
6				0.16667	0.34238	0.31731	0.19018	0.08499	0.03054
7					0.14286	0.31541	0.31492	0.20299	0.09725
8						0.12500	0.29246	0.31054	0.21282
9							0.11111	0.27274	0.30505
10								0.10000	0.25563
11									0.09091

Table A.11 (*continued*)

l	$k=11$	$k=12$	$k=13$	$k=14$	$k=15$	$k=16$	$k=17$	$k=18$	$k=19$
1	0.00000	0.00000	0.00000	0.00000	0.00000	0.00000	0.00000	0.00000	0.00000
2	0.00000	0.00000	0.00000	0.00000	0.00000	0.00000	0.00000	0.00000	0.00000
3	0.00002	0.00000	0.00000	0.00000	0.00000	0.00000	0.00000	0.00000	0.00000
4	0.00021	0.00004	0.00001	0.00000	0.00000	0.00000	0.00000	0.00000	0.00000
5	0.00164	0.00036	0.00007	0.00001	0.00000	0.00000	0.00000	0.00000	0.00000
6	0.00925	0.00243	0.00057	0.00012	0.00002	0.00000	0.00000	0.00000	0.00000
7	0.03734	0.01204	0.00336	0.00083	0.00019	0.00004	0.00001	0.00000	0.00000
8	0.10824	0.04402	0.01500	0.00442	0.00115	0.00027	0.00006	0.00001	0.00000
9	0.22035	0.11804	0.05050	0.01807	0.00558	0.00152	0.00037	0.00008	0.00002
10	0.29898	0.22609	0.12677	0.05673	0.02121	0.00684	0.00194	0.00050	0.00011
11	0.24064	0.29264	0.23041	0.13455	0.06269	0.02439	0.00817	0.00241	0.00064
12	0.08333	0.22742	0.28622	0.23360	0.14148	0.06836	0.02757	0.00957	0.00292
13		0.07692	0.21565	0.27986	0.23590	0.14766	0.07375	0.03074	0.01102
14			0.07143	0.20512	0.27363	0.23748	0.15318	0.07885	0.03387
15				0.06667	0.19563	0.26756	0.23848	0.15810	0.08368
16					0.06250	0.18704	0.26170	0.23900	0.16249
17						0.05882	0.17922	0.25604	0.23914
18							0.05556	0.17207	0.25059
19								0.05263	0.16550
20									0.05000

Table A.12 Limiting level probabilities for the simple order, $Q(l, k) = \lim_{w \to \infty} P_S(l, k; (w, 1, \ldots, 1))$ with k the number of populations

l	$k = 3$	$k = 4$	$k = 5$	$k = 6$	$k = 7$	$k = 8$	$k = 9$	$k = 10$	$k = 11$
1	0.37500	0.31250	0.27344	0.24609	0.22559	0.20947	0.19638	0.18547	0.17620
2	0.50000	0.47917	0.45833	0.43984	0.42370	0.40955	0.39704	0.38589	0.37587
3	0.12500	0.18750	0.22396	0.24740	0.26343	0.27488	0.28330	0.28962	0.29443
4		0.02083	0.04167	0.05990	0.07552	0.08894	0.10056	0.11072	0.11966
5			0.00260	0.00651	0.01096	0.01557	0.02016	0.02462	0.02893
6				0.00026	0.00078	0.00151	0.00239	0.00337	0.00444
7					0.00002	0.00008	0.00017	0.00029	0.00044
8						0.00000	0.00001	0.00002	0.00003

l	$k = 12$	$k = 13$	$k = 14$	$k = 15$	$k = 16$	$k = 17$	$k = 18$	$k = 19$	$k = 20$
1	0.16819	0.16118	0.15498	0.14945	0.14446	0.13995	0.13583	0.13206	0.12859
2	0.36680	0.35852	0.35093	0.34393	0.33745	0.33142	0.32579	0.32051	0.31555
3	0.29813	0.30099	0.30321	0.30491	0.30621	0.30719	0.30790	0.30840	0.30872
4	0.12761	0.13471	0.14111	0.14690	0.15216	0.15698	0.16139	0.16546	0.16923
5	0.03305	0.03699	0.04075	0.04433	0.04775	0.05102	0.05413	0.05711	0.05996
6	0.00555	0.00670	0.00786	0.00904	0.01021	0.01139	0.01255	0.01371	0.01485
7	0.00062	0.00083	0.00106	0.00130	0.00156	0.00183	0.00211	0.00240	0.00270
8	0.00005	0.00007	0.00010	0.00013	0.00017	0.00022	0.00026	0.00032	0.00037
9[a]	0.00000	0.00000	0.00001	0.00001	0.00001	0.00002	0.00003	0.00003	0.00004

[a] For larger l, $Q(lk)$ is 0 to five places.

Table A.13 Limiting level probabilities for the simple tree, $S(l, k+1) = \lim_{w \to \infty} P_T(l, k+1; (1, 1, \ldots, 1, w))$ with k the number of treatments

l	$k = 2$	$k = 3$	$k = 4$	$k = 5$	$k = 6$	$k = 7$	$k = 8$	$k = 9$	$k = 10$
1	0.12500	0.02704	0.00521	0.00091	0.00015	0.00002	0.00000	0.00000	0.00000
2	0.50000	0.20833	0.06250	0.01531	0.00324	0.00061	0.00010	0.00002	0.00000
3	0.37500	0.47296	0.26042	0.09738	0.02853	0.00701	0.00150	0.00029	0.00005
4		0.29167	0.43750	0.29094	0.12804	0.04315	0.01198	0.00286	0.00061
5			0.23438	0.40171	0.30726	0.15351	0.05793	0.01781	0.00467
6				0.19375	0.36872	0.31450	0.17400	0.07213	0.02417
7					0.16406	0.33946	0.31605	0.19020	0.08535
8						0.14174	0.31391	0.31409	0.20287
9							0.12451	0.29170	0.31003
10								0.11089	0.27235
11									0.09990

l	$k = 11$	$k = 12$	$k = 13$	$k = 14$	$k = 15$	$k = 16$	$k = 17$	$k = 18$	$k = 19$
1	0.00000	0.00000	0.00000	0.00000	0.00000	0.00000	0.00000	0.00000	0.00000
2	0.00000	0.00000	0.00000	0.00000	0.00000	0.00000	0.00000	0.00000	0.00000
3	0.00001	0.00000	0.00000	0.00000	0.00000	0.00000	0.00000	0.00000	0.00000
4	0.00011	0.00002	0.00000	0.00000	0.00000	0.00000	0.00000	0.00000	0.00000
5	0.00107	0.00022	0.00005	0.00000	0.00000	0.00000	0.00000	0.00000	0.00000
6	0.00686	0.00170	0.00037	0.00008	0.00001	0.00000	0.00000	0.00000	0.00000
7	0.03080	0.00937	0.00248	0.00058	0.00013	0.00002	0.00000	0.00000	0.00000
8	0.09741	0.03749	0.01212	0.00340	0.00084	0.00020	0.00003	0.00001	0.00000
9	0.21269	0.10829	0.04410	0.01505	0.00444	0.00116	0.00028	0.00005	0.00001
10	0.30475	0.22024	0.11805	0.05054	0.01810	0.00560	0.00152	0.00038	0.00008
11	0.25543	0.29881	0.22601	0.12677	0.05675	0.02123	0.00685	0.00194	0.00050
12	0.09086	0.24055	0.29254	0.23036	0.13454	0.06270	0.02440	0.00818	0.00241
13		0.08331	0.22737	0.28617	0.23357	0.14148	0.06837	0.02758	0.00958
14			0.07691	0.21563	0.27983	0.23588	0.14766	0.07375	0.03074
15				0.07142	0.20511	0.27361	0.23747	0.15317	0.07885
16					0.06666	0.19563	0.26755	0.23847	0.15809
17						0.06250	0.18704	0.26169	0.23900
18							0.05882	0.17922	0.25603
19								0.05556	0.17207
20									0.05263

Table A.14 Coefficients for the corrected two-moment approximations to the null $\bar{\chi}^2$ distributions for a simple order and equal weights with k populations

k	ρ	b	k	ρ	b	k	ρ	b
				$\bar{\chi}^2_{01}$				
5	2.31791	0.69207	17	2.60658	0.99442	29	2.68791	1.14119
6	2.37111	0.73383	18	2.61641	1.00973	30	2.69233	1.15077
7	2.41322	0.77006	19	2.62545	1.02431	31	2.69653	1.16006
8	2.44757	0.80213	20	2.63378	1.03823	32	2.70053	1.16909
9	2.47626	0.83093	21	2.64151	1.05153	33	2.70434	1.17786
10	2.50066	0.85709	22	2.64870	1.06428	34	2.70798	1.18638
11	2.52174	0.88108	23	2.65541	1.07651	35	2.71145	1.19469
12	2.54017	0.90325	24	2.66168	1.08828	36	2.71477	1.20278
13	2.55646	0.92386	25	2.66757	1.09961	37	2.71796	1.21066
14	2.57098	0.94313	26	2.67310	1.11055	38	2.72101	1.21835
15	2.58403	0.96122	27	2.67832	1.12110	39	2.72394	1.22586
16	2.59584	0.97828	28	2.68325	1.13131	40	2.72675	1.23319
				$\bar{\chi}^2_{12}$				
5	2.30174	1.18027	17	2.13655	6.34688	29	2.09388	11.95787
6	2.27003	1.56385	18	2.13128	6.80571	30	2.09163	12.43290
7	2.24530	1.96283	19	2.12646	7.26667	31	2.08950	12.90870
8	2.22537	2.37360	20	2.12203	7.72951	32	2.08748	13.38528
9	2.20891	2.79370	21	2.11795	8.19408	33	2.08556	13.86255
10	2.19505	3.22136	22	2.11417	8.66023	34	2.08374	14.34043
11	2.18318	3.65527	23	2.11066	9.12783	35	2.08200	14.81902
12	2.17290	4.09444	24	2.10739	9.59673	36	2.08035	15.29814
13	2.16388	4.53809	25	2.10434	10.06684	37	2.07876	15.77787
14	2.15589	4.98562	26	2.10147	10.53812	38	2.07726	16.25802
15	2.14876	5.43651	27	2.09879	11.01042	39	2.07581	16.73877
16	2.14235	5.89038	28	2.09627	11.48366	40	2.07442	17.21997

Table A.15 Coefficients for the corrected two-moment approximations to the null $\bar{\chi}^2$ distributions for a simple tree and equal weights with k treatments

k	ρ	b	k	ρ	b	k	ρ	b
				$\bar{\chi}^2_{01}$				
4	2.31135	1.17029	16	2.16133	6.21085	28	2.11843	11.70470
5	2.28328	1.54569	17	2.15619	6.65912	29	2.11606	12.17091
6	2.26135	1.93529	18	2.15145	7.10963	30	2.11382	12.63799
7	2.24357	2.33592	19	2.14708	7.56218	31	2.11168	13.10598
8	2.22877	2.74540	20	2.14301	8.01662	32	2.10964	13.57478
9	2.21618	3.16215	21	2.13922	8.47280	33	2.10769	14.04440
10	2.20531	3.58501	22	2.13568	8.93055	34	2.10583	14.51469
11	2.19580	4.01307	23	2.13237	9.38978	35	2.10405	14.98575
12	2.18737	4.44563	24	2.12925	9.85036	36	2.10233	15.45749
13	2.17985	4.88212	25	2.12631	10.31224	37	2.10069	15.92977
14	2.17307	5.32205	26	2.12353	10.77533	38	2.09912	16.40265
15	2.16693	5.76507	27	2.12091	11.23947	39	2.09760	16.87619
				$\bar{\chi}^2_{12}$				
4	2.32655	0.69580	16	2.67955	1.02157	28	2.80194	1.18450
5	2.38563	0.73980	17	2.69353	1.03845	29	2.80910	1.19523
6	2.43385	0.77829	18	2.70655	1.05454	30	2.81597	1.20564
7	2.47431	0.81255	19	2.71873	1.06992	31	2.82255	1.21576
8	2.50897	0.84349	20	2.73016	1.08465	32	2.82888	1.22561
9	2.53916	0.87172	21	2.74091	1.09879	33	2.83496	1.23519
10	2.56581	0.89770	22	2.75105	1.11238	34	2.84082	1.24453
11	2.58958	0.92178	23	2.76065	1.12546	35	2.84646	1.25363
12	2.61099	0.94424	24	2.76974	1.13808	36	2.85191	1.26252
13	2.63042	0.96529	25	2.77838	1.15026	37	2.85717	1.27119
14	2.64817	0.98510	26	2.78661	1.16205	38	2.86225	1.27966
15	2.66448	1.00382	27	2.79445	1.17345	39	2.86716	1.28794

Table A.16 The first two cumulants of the null $\bar{\chi}^2$ distributions for a simple order and equal weights with k populations

$\bar{\chi}^2_{01}$

k	$\kappa_1^*(k,1)$	$\kappa_2^*(k,1)$	k	$\kappa_1^*(k,1)$	$\kappa_2^*(k,1)$	k	$\kappa_1^*(k,1)$	$\kappa_2^*(k,1)$
2	0.50000	1.25000	15	2.31823	6.37425	28	2.92717	8.17166
3	0.83333	2.13889	16	2.38073	6.55784	29	2.96165	8.27392
4	1.08333	2.82639	17	2.43955	6.73085	30	2.99499	8.37281
5	1.28333	3.38639	18	2.49511	6.89443	31	3.02725	8.46854
6	1.45000	3.85861	19	2.54774	7.04956	32	3.05850	8.56132
7	1.59286	4.26677	20	2.59774	7.19706	33	3.08880	8.65131
8	1.71786	4.62615	21	2.64536	7.33765	34	3.11821	8.73868
9	1.82897	4.94714	22	2.69081	7.47194	35	3.14678	8.82358
10	1.92897	5.23714	23	2.73429	7.60049	36	3.17456	8.90614
11	2.01988	5.50160	24	2.77596	7.72375	37	3.20159	8.98649
12	2.10321	5.74466	25	2.81596	7.84215	38	3.22790	9.06474
13	2.18013	5.96951	26	2.85442	7.95606	39	3.25354	9.14101
14	2.25156	6.17869	27	2.89146	8.06580	40	3.27854	9.21539

$\bar{\chi}^2_{12}$

k	$\kappa_1^*(k,2)$	$\kappa_2^*(k,2)$	k	$\kappa_1^*(k,2)$	$\kappa_2^*(k,2)$	k	$\kappa_1^*(k,2)$	$\kappa_2^*(k,2)$
2	0.50000	1.25000	15	11.68177	25.10133	28	24.07283	50.46298
3	1.16667	2.80556	16	12.61927	27.03492	29	25.03835	52.42731
4	1.91667	4.49306	17	13.56045	28.97264	30	26.00501	54.39286
5	2.71667	6.25306	18	14.50489	30.91400	31	26.97275	56.35956
6	3.55000	8.05861	19	15.45226	32.85860	32	27.94150	58.32734
7	4.40714	9.89535	20	16.40226	34.80610	33	28.91120	60.29612
8	5.28214	11.75472	21	17.35464	36.75621	34	29.88179	62.26584
9	6.17103	13.63126	22	18.30919	38.70869	35	30.85322	64.23645
10	7.07103	15.52126	23	19.26571	40.66332	36	31.82544	66.20790
11	7.98012	17.42209	24	20.22404	42.61992	37	32.79841	68.18014
12	8.89679	19.33181	25	21.18404	44.57832	38	33.77210	70.15314
13	9.81987	21.24897	26	22.14558	46.53838	39	34.74646	72.12684
14	10.74844	23.17244	27	23.10854	48.49997	40	35.72146	74.10121

Table A.17 The first two cumulants of the null $\bar{\chi}^2$ distributions for a simple tree and equal weights with k treatments

$\bar{\chi}^2_{01}$

k	$\tau_1(k+1,1)$	$\tau_2(k+1,1)$	k	$\tau_1(k+1,1)$	$\tau_2(k+1,1)$	k	$\tau_1(k+1,1)$	$\tau_2(k+1,1)$
1	0.50000	1.25000	14	11.56520	25.13201	27	23.83796	50.55829
2	1.16667	2.80556	15	12.49250	27.07036	28	24.79554	52.52756
3	1.91226	4.49230	16	13.42367	29.01292	29	25.75440	54.49803
4	2.70495	6.25208	17	14.35830	30.95920	30	26.71448	56.46961
5	3.52925	8.05829	18	15.29604	32.90876	31	27.67569	58.44225
6	4.37638	9.89652	19	16.23661	34.86127	32	28.63798	60.41586
7	5.24081	11.75812	20	17.17974	36.81641	33	29.60128	62.39039
8	6.11885	13.63749	21	18.12522	38.77392	34	30.56554	64.36578
9	7.00791	15.53081	22	19.07284	40.73360	35	31.53070	66.34199
10	7.90607	17.43535	23	20.02246	42.69523	36	32.49673	68.31898
11	8.81189	19.34911	24	20.97390	44.65866	37	33.46358	70.29670
12	9.72425	21.27054	25	21.92704	46.62373	38	34.43121	72.27513
13	10.64226	23.19849	26	22.88177	48.59031	39	35.39957	74.25423

$\bar{\chi}^2_{12}$

k	$\tau_1(k+1,2)$	$\tau_2(k+1,2)$	k	$\tau_1(k+1,2)$	$\tau_2(k+1,2)$	k	$\tau_1(k+1,2)$	$\tau_2(k+1,2)$
1	0.50000	1.25000	14	2.43480	6.87119	27	3.16204	9.20646
2	0.83333	2.13889	15	2.50750	7.10034	28	3.20446	9.34545
3	1.08774	2.84326	16	2.57633	7.31823	29	3.24560	9.48047
4	1.29505	3.43228	17	2.64170	7.52600	30	3.28553	9.61176
5	1.47075	3.94127	18	2.70396	7.72459	31	3.32431	9.73954
6	1.62362	4.39102	19	2.76339	7.91483	32	3.36203	9.86399
7	1.75919	4.79490	20	2.82026	8.09745	33	3.39872	9.98529
8	1.88115	5.16208	21	2.87478	8.27306	34	3.43446	10.10361
9	1.99209	5.49917	22	2.92716	8.44222	35	3.46929	10.21910
10	2.09393	5.81107	23	2.97754	8.60542	36	3.50326	10.33189
11	2.18811	6.10155	24	3.02610	8.76307	37	3.53641	10.44211
12	2.27575	6.37355	25	3.07296	8.91557	38	3.56878	10.54989
13	2.35774	6.62946	26	3.11823	9.06326	39	3.60041	10.65534

Table A.18 The first two cumulants of the limiting null $\bar{\chi}^2$ distributions with k populations and coefficients $\lim_{w \to \infty} P_S(l, k; (w, 1, 1, \ldots, 1))$

$\bar{\chi}^2_{01}$

k	$\kappa'_1(k,1)$	$\kappa'_2(k,1)$	k	$\kappa'_1(k,1)$	$\kappa'_2(k,1)$	k	$\kappa'_1(k,1)$	$\kappa'_2(k,1)$
2	0.50000	1.25000	15	1.62578	4.48334	28	1.94573	5.43504
3	0.75000	1.93750	16	1.65911	4.58223	29	1.96359	5.48829
4	0.91667	2.40972	17	1.69036	4.67501	30	1.98083	5.53972
5	1.04167	2.76910	18	1.71978	4.76238	31	1.99749	5.58944
6	1.14167	3.05910	19	1.74755	4.84494	32	2.01362	5.63757
7	1.22500	3.30215	20	1.77387	4.92319	33	2.02925	5.68420
8	1.29643	3.51134	21	1.79887	4.99757	34	2.04440	5.72943
9	1.35893	3.69493	22	1.82268	5.06843	35	2.05910	5.77333
10	1.41448	3.85851	23	1.84541	5.13610	36	2.07339	5.81598
11	1.46448	4.00601	24	1.86715	5.20084	37	2.08728	5.85745
12	1.50994	4.14031	25	1.88798	5.26291	38	2.10079	5.89781
13	1.55161	4.26357	26	1.90798	5.32251	39	2.11395	5.93711
14	1.59007	4.37748	27	1.92721	5.37983	40	2.12677	5.97541

$\bar{\chi}^2_{12}$

k	$\kappa'_1(k,2)$	$\kappa'_2(k,2)$	k	$\kappa'_1(k,2)$	$\kappa'_2(k,2)$	k	$\kappa'_1(k,2)$	$\kappa'_2(k,2)$
2	0.50000	1.25000	15	12.37422	25.98022	28	25.05427	51.65213
3	1.25000	2.93750	16	13.34089	27.94578	29	26.03641	53.63395
4	2.08333	4.74306	17	14.30964	29.91355	30	27.01917	55.61641
5	2.95833	6.60243	18	15.28022	31.88327	31	28.00251	57.59947
6	3.85833	8.49243	19	16.25245	33.85472	32	28.98638	59.58308
7	4.77500	10.40215	20	17.22613	35.82771	33	29.97075	61.56721
8	5.70357	12.32562	21	18.20113	37.80209	34	30.95560	63.55183
9	6.64107	14.25922	22	19.17732	39.77771	35	31.94090	65.53691
10	7.58552	16.20057	23	20.15459	41.75447	36	32.92661	67.52242
11	8.53552	18.14807	24	21.13285	43.73226	37	33.91272	69.50834
12	9.49006	20.10055	25	22.11202	45.71099	38	34.89921	71.49464
13	10.44839	22.05715	26	23.09202	47.69059	39	35.88605	73.48131
14	11.40993	24.01721	27	24.07279	49.67099	40	36.87323	75.46832

Table A.19 The first two cumulants of the limiting null $\bar{\chi}^2$ distributions with k treatments and coefficients, $S(l, k+1) = \lim_{w\to\infty} P_T(l, k+1; (1, 1, \ldots, 1, w))$

$\bar{\chi}^2_{01}$

k	$\tau_1'(k+1,1)$	$\tau_2'(k+1,1)$	k	$\tau_1'(k+1,1)$	$\tau_2'(k+1,1)$	k	$\tau_1'(k+1,1)$	$\tau_2'(k+1,1)$
1	0.50000	1.25000	14	11.64171	25.19935	27	23.88175	50.59082
2	1.25000	2.93750	15	12.56487	27.13252	28	24.83795	52.55874
3	2.02925	4.66581	16	13.49230	29.07065	29	25.79553	54.52796
4	2.83333	6.43056	17	14.42355	31.01308	30	26.75439	56.49831
5	3.65847	8.22688	18	15.35822	32.95934	31	27.71447	58.46993
6	4.50145	10.05005	19	16.29600	34.90892	32	28.67568	60.44248
7	5.35957	11.89576	20	17.23657	36.86152	33	29.63797	62.41607
8	6.23057	13.76028	21	18.17971	38.81676	34	30.60127	64.39055
9	7.11259	15.64044	22	19.12519	40.77436	35	31.56553	66.36597
10	8.00406	17.53363	23	20.07282	42.73407	36	32.53070	68.34212
11	8.90370	19.43771	24	21.02243	44.69579	37	33.49673	70.31912
12	9.81043	21.35092	25	21.97388	46.65919	38	34.46358	72.29675
13	10.72335	23.27185	26	22.92703	48.62426	39	35.43121	74.27515

$\bar{\chi}^2_{12}$

k	$\tau_1'(k+1,2)$	$\tau_2'(k+1,2)$	k	$\tau_1'(k+1,2)$	$\tau_2'(k+1,2)$	k	$\tau_1'(k+1,2)$	$\tau_2'(k+1,2)$
1	0.50000	1.25000	14	2.35832	6.63278	27	3.11822	9.06318
2	0.75000	1.93750	15	2.43515	6.87335	28	3.16203	9.20640
3	0.97075	2.54882	16	2.50772	7.10172	29	3.20446	9.34539
4	1.16667	3.09722	17	2.57646	7.31908	30	3.24560	9.48043
5	1.34153	3.59299	18	2.64178	7.52648	31	3.28552	9.61173
6	1.49855	4.04485	19	2.70400	7.72483	32	3.32431	9.73951
7	1.64043	4.45750	20	2.76341	7.91492	33	3.36202	9.86397
8	1.76943	4.83800	21	2.82026	8.09743	34	3.39872	9.98528
9	1.88741	5.19009	22	2.87478	8.27299	35	3.43446	10.10360
10	1.99594	5.51739	23	2.92714	8.44212	36	3.46929	10.21909
11	2.09630	5.82294	24	2.97753	8.60531	37	3.50326	10.33188
12	2.18958	6.10930	25	3.02609	8.76296	38	3.53641	10.44211
13	2.27667	6.37863	26	3.07294	8.91548	39	3.56878	10.54989

Table A.20 Level probabilities[a] for equal weights and unimodal orders ($\mu_1 \leqslant \mu_2 \leqslant \cdots \leqslant \mu_h \geqslant \mu_{h+1} \geqslant \cdots \geqslant \mu_k$), $P_h(l, k)$

	$k=3$	$k=4$	$k=5$		$k=6$		$k=7$		
l^b	$h=2$	$h=2$	$h=2$	$h=3$	$h=2$	$h=3$	$h=2$	$h=3$	$h=4$
1	0.1667	0.0980	0.0690	0.0549	0.0531	0.0376	0.0432	0.0284	0.0253
2	0.5000	0.3750	0.2990	0.2646	0.2493	0.2045	0.2143	0.1671	0.1555
3	0.3333	0.4020	0.3977	0.3951	0.3781	0.3603	0.3563	0.3271	0.3170
4		0.1250	0.2010	0.2354	0.2438	0.2816	0.2681	0.2999	0.3055
5			0.0333	0.0500	0.0688	0.1021	0.0993	0.1415	0.1537
6					0.0069	0.0139	0.0176	0.0330	0.0390
7							0.0012	0.0030	0.0040

	$k=8$			$k=9$				$k=10$			
l	$h=2$	$h=3$	$h=4$	$h=2$	$h=3$	$h=4$	$h=5$	$h=2$	$h=3$	$h=4$	$h=5$
1	0.035	0.024	0.019	0.030	0.020	0.015	0.014	0.027	0.017	0.012	0.012
2	0.189	0.141	0.127	0.169	0.123	0.106	0.101	0.151	0.108	0.092	0.083
3	0.335	0.300	0.281	0.317	0.276	0.254	0.247	0.300	0.255	0.230	0.221
4	0.283	0.306	0.309	0.289	0.305	0.305	0.305	0.293	0.303	0.297	0.295
5	0.125	0.170	0.187	0.146	0.192	0.211	0.216	0.163	0.207	0.225	0.233
6	0.029	0.051	0.063	0.042	0.069	0.086	0.090	0.054	0.086	0.107	0.113
7	0.003	0.008	0.012	0.006	0.014	0.020	0.023	0.011	0.021	0.031	0.036
8	0.000	0.001	0.001	0.001	0.002	0.003	0.004	0.001	0.003	0.005	0.007
9				0.000	0.000	0.000	0.000	0.000	0.000	0.001	0.001

	$k = 11$					$k = 12$				
l	$h = 2$	$h = 3$	$h = 4$	$h = 5$	$h = 6$	$h = 2$	$h = 3$	$h = 4$	$h = 5$	$h = 6$
1	0.024	0.014	0.011	0.010	0.009	0.022	0.013	0.009	0.008	0.007
2	0.137	0.098	0.081	0.072	0.070	0.127	0.090	0.074	0.064	0.062
3	0.285	0.239	0.211	0.199	0.197	0.271	0.223	0.197	0.182	0.177
4	0.297	0.298	0.289	0.283	0.280	0.294	0.291	0.279	0.273	0.268
5	0.176	0.217	0.235	0.243	0.243	0.188	0.227	0.243	0.249	0.249
6	0.064	0.100	0.122	0.132	0.138	0.075	0.112	0.134	0.147	0.152
7	0.015	0.029	0.040	0.047	0.050	0.020	0.036	0.051	0.058	0.064
8	0.002	0.005	0.009	0.011	0.012	0.003	0.008	0.012	0.016	0.018
9	0.000	0.001	0.001	0.002	0.002	0.000	0.001	0.002	0.003	0.003
10	0.000	0.000	0.000	0.000	0.000	0.000	0.000	0.000	0.000	0.000

	$k = 13$						$k = 14$					
l	$h = 2$	$h = 3$	$h = 4$	$h = 5$	$h = 6$	$h = 7$	$h = 2$	$h = 3$	$h = 4$	$h = 5$	$h = 6$	$h = 7$
1	0.020	0.012	0.008	0.007	0.006	0.006	0.019	0.010	0.007	0.006	0.005	0.005
2	0.115	0.081	0.066	0.057	0.053	0.051	0.108	0.074	0.060	0.052	0.047	0.045
3	0.260	0.210	0.182	0.168	0.160	0.159	0.248	0.199	0.171	0.156	0.148	0.145
4	0.290	0.284	0.271	0.258	0.256	0.254	0.288	0.280	0.261	0.249	0.244	0.241
5	0.200	0.233	0.245	0.257	0.251	0.253	0.206	0.235	0.250	0.255	0.254	0.252
6	0.085	0.125	0.149	0.158	0.167	0.168	0.095	0.134	0.156	0.169	0.176	0.181
7	0.024	0.043	0.059	0.069	0.076	0.077	0.029	0.050	0.069	0.079	0.086	0.090
8	0.005	0.010	0.016	0.022	0.024	0.026	0.006	0.014	0.021	0.026	0.031	0.032
9	0.001	0.002	0.003	0.004	0.006	0.006	0.001	0.002	0.004	0.007	0.008	0.009
10	0.000	0.000	0.001	0.001	0.001	0.001	0.000	0.000	0.001	0.001	0.001	0.001

Table A.20 (*continued*)
$k = 15$

$h=2$	$h=3$	$h=4$	$h=5$	$h=6$	$h=7$	$h=8$
0.017	0.009	0.006	0.005	0.004	0.004	0.004
0.102	0.069	0.055	0.047	0.043	0.040	0.039
0.237	0.189	0.162	0.147	0.138	0.131	0.131
0.287	0.273	0.254	0.241	0.233	0.231	0.229
0.211	0.242	0.251	0.253	0.252	0.250	0.251
0.102	0.140	0.164	0.179	0.186	0.189	0.189
0.034	0.057	0.076	0.087	0.096	0.101	0.104
0.008	0.016	0.025	0.031	0.036	0.039	0.039
0.001	0.003	0.006	0.008	0.010	0.011	0.012
0.000	0.000	0.001	0.001	0.002	0.002	0.002

[a]For $8 \leqslant k \leqslant 15$ the values were estimated by Monte Carlo techniques (100 000 iterations).
[b]For larger l, the estimated $P_h(l,k)$ are zero to three decimal places.

Table A.21 Level probabilities[a] for equal weights and matrix orders ($\mu_{ij} \leqslant \mu_{st}$ if $1 \leqslant i \leqslant s \leqslant R$ and $1 \leqslant j \leqslant t \leqslant C$), $P_{R\times C}(l, RC)$

l[b]	2×2	2×3	2×4	3×3	2×5	2×6	3×4
1	0.1959	0.114	0.074	0.059	0.053	0.040	0.037
2	0.4167	0.304	0.232	0.187	0.185	0.154	0.139
3	0.3041	0.329	0.312	0.282	0.292	0.253	0.226
4	0.0833	0.191	0.232	0.254	0.244	0.256	0.258
5		0.056	0.111	0.148	0.151	0.171	0.188
6		0.007	0.033	0.056	0.058	0.086	0.100
7			0.006	0.012	0.013	0.033	0.040
8			0.000	0.002	0.004	0.006	0.012
9				0.000	0.000	0.001	0.002

[a] Except for the 2 × 2, the values are estimated by Monte Carlo techniques with 10 000 iterations.
[b] For $l > 9$ the values are 0 to three places.

Bibliography and Author Index

This compilation is both a bibliography and an author index. It contains references that are related to the topic of statistical inference under inequality constraints but are not cited in this book. However, if an entry is cited then the pages in the text on which it is discussed are given in brackets at the end of the entry.

Abelson, R. P., and V. Sermat (1962). 'Multidimensional scaling of facial expressions', *J. Experimental Psychology*, **63**, 546–54. [11]

Abelson, R. P., and J. W. Tukey (1963). 'Efficient utilization of non-numerical information in quantitative analysis: general theory and the case of the simple order', *Ann. Math. Statist.*, **34**, 1347–69. [168, 176, 178–180, 182, 183, 187, 191, 194, 197, 214, 227]

Abrahamson, I. G., (1964). 'Orthant probabilities for the quadrivariate normal distribution', *Ann. Math. Statist.*, **35**, 1685–703. [75]

Abramowitz, M., and I. S. Stegun (1965). *Handbook of Mathematical Functions*, Dover, New York. [82]

Acar, B. S., and A. N. Pettitt (1985). 'Prediction using nonparametric techniques', *J. Statist. Comp. Simul.*, **21**, 117–34.

Adegbeye, O. S., and A. K. Gupta (1985). 'On testing against ordered alternatives for Penrose model', *J. Statist. Comp. Simul.*, **22**, 147–60.

Adichie, J. N. (1976). 'Testing parallelism of regression lines against ordered alternatives', *Comm. Statist. (A)*, **5**, 985–98.

Agresti, A (1983a). 'A survey of strategies for modeling cross-classifications having ordinal variables', *J. Amer. Statist. Assoc.*, **78**, 184–98.

Agresti, A (1983b). 'A simple diagonal parameter symmetry and quasi-symmetry model', *Statist. Prob. Lett.*, **1**, 313–16.

Agresti, A (1983c). 'Testing marginal homogeneity for ordinal categorical variables', *Biometrics*, **39**, 505–10.

Agresti, A., and C. Chuang (1986). 'Bayesian and maximum likelihood approaches to order restricted inference for models from ordinal categorical data'. In R. L. Dykstra, T. Robertson, and F. T. Wright (eds.), *Advances in Order Restricted Statistical Inference*, Springer-Verlag, New York, pp. 6–27. [279, 293]

Agresti, A., C. Chuang, and A. Kezouk (1987). 'Order restricted score parameters in association models for contingency tables', *J. Amer. Statist. Assoc.*, **82**, 619–23. [262, 263]

Aitchison, J., and S. D. Silvey (1958). 'Maximum likelihood estimation of parameters subject to restraints', *Ann. Math. Statist.*, **29**, 813–28. [318, 320]

Aiyar, R. J., C. L. Guillier, and W. Albers (1979). 'Asymptotic relative efficiencies of rank tests for trend alternatives', *J. Amer. Statist. Assoc.*, **74**, 226–31. [208]

Akilov, J. T. See Kantorovich and Akilov (1964).

Alalouf, I. S. (1987). 'The chi-square test with both margins fixed', *Comm. Statist. Theory and Meth.*, **16**, 29–43.

Alam, K., and A. Mitra (1981). 'Polarization test for the multinomial distribution', *J. Amer. Statist. Assoc.*, **76**, 107–9.

Alam, K., K. M. L. Saxena, and Y. L. Tong (1973). 'Optimal confidence interval for a ranked parameter', *J. Amer. Statist. Assoc.*, **68**, 720–5.

Albers, W. See Aiyar, Guillier, and Albers (1979)

Alexander, M. J. (1970). 'An algorithm for obtaining maximum-likelihood estimates of a set of partially ordered parameters', Rocketdyne Technical Paper. [57]

Amundsen, H. T., and H. Ljøgodt (1979). 'Small sample tests against an ordered set of binomial probabilities', *Scand. J. Statist.*, **6**, 81–5.

Anderson, J. A., and R. R. Philips (1981). 'Regression, discrimination and measurement models for ordered categorical variables', *Appl. Statist.*, **30**, 22–31.

Anderson, R. L., and L. A. Nelson (1984). 'Recent developments in the use of linear plateau models to estimate response relationships'. In H. A. David and H. T. David (eds.), *Statistics: An Appraisal*, Iowa State University Press, pp. 371–392.

Anderson, T. W. (1955). 'The integral of a symmetric, unimodal function over a symmetric, convex set and some probability inequalities', *Proc. Amer. Math. Soc.*, **6**, 170–6. [104]

Andrews, F. C. (1954). 'Asymptotic behavior of some rank tests for analysis of variance', *Ann. Math. Statist.*, **25**, 724–35. [207]

Antillo, A., G. Kersting, and W. Zucchino (1982). 'Testing symmetry', *J. Amer. Statist. Assoc.*, **77**, 639–46.

Araki, T., and S. Shirahato (1981). 'Rank tests for ordered alternatives in randomized blocks', *J. Japanese Statist.*, **11**, 27–42.

Armitage, P. V. (1955). 'Tests for linear trends in proportions and frequencies', *Biometrics*, **11**, 375–86. [168, 178]

Armitage, P. V. See also Krishnaiah and Armitage (1965, 1966).

Arnold, B. C. (1970). 'An alternative derivation of a result due to Srivastava and Bancroft', *J. R. Statist. Soc. (B)*, **32**, 265–7.

Arnold, B. C. (1974). 'On estimates of the smaller of two ordered normal means which incorporate a preliminary test of significance', *Utilities Math.*, **5**, 65–74.

Arrow, K. J., S. Karlin, and H. Scarf (1958). *Studies in the Mathematical Theory of Inventory and Production*, Stanford University Press.

Arthur, K. H. See Chou, Arthur, Rosenstein, and Owen (1984).

Avenhaus, R. (1979). 'Significance thresholds of one sided tests for means of bivariate normally distributed variables', *Comm. Statist. (A)*, **8**, 223–30.

Ayer, M., H. D. Brunk, G. M. Ewing, W. T. Reid, and E. Silverman (1955). 'An empirical distribution function for sampling with incomplete information', *Ann. Math. Statist.*, **26**, 641–7. [8, 56, 58, 392, 393]

Bagai, I. See Kusum and Bagai (1987).

Bahadur, R. R. (1955). 'Measurable subspaces and subalgebras', *Proc. Amer. Math. Soc.*, **6**, 565–70. [387]

Bain, L. J., and M. E. Engelhardt (1975). 'A two-moment chi-square approximation for the statistic $\log(\bar{x}/\tilde{x})$', *J. Amer. Statist. Assoc.*, **70**, 948–50. [171, 175]

Bain, L. J., M. E. Engelhardt, and F. T. Wright (1985). 'Tests for an increasing trend in the intensity of a Poisson process: a power study', *J. Amer. Statist. Assoc.*, **80**, 419–22. [404]

Balius, M. E., and N. Hernandez (1983). 'Application and comparison of likelihood ratio tests for the one-way models of analysis of variance with order restrictions in the parameters', *Rl. Opr. Cba.*, **4**, 71–93.

Bancroft, T. A. See Srivastava and Bancroft (1967).

Baras, M. (1983a). 'Testing equality of probabilities of k mutually exclusive events against ordered alternatives', *Biometrika*, **70**, 473–8.

Baras, M. (1983b). 'Testing randomness against ordered alternatives in a multinomial experiment with grouped frequencies', *Comm. Statist. (A)*, **12**, 2575–80.

Barlow, R. E. (1968a). 'Likelihood ratio tests for restricted families of probability distributions', *Ann. Math. Statist.*, **39**, 547–60. [367, 404]

Barlow, R. E. (1968b). 'Some recent developments in reliability theory'. In *Selected Statistical Papers*, Vol. 2, Mathematical Centre, Amsterdam. pp. 49–66.

Barlow, R. E., and H. D. Brunk (1972). 'The isotonic regression problem and its dual', *J. Amer. Statist. Assoc.*, **67**, 140–7. [58, 263, 315, 323]

Barlow, R. E., and K. Doksum (1972). 'Isotonic tests for convex orderings', *Proc. 6th Berkeley Symp. Math. Statist. Probab.*, **1**, 293–323. [360]

Barlow, R. E., A. W. Marshall, and F. Proschan (1963). 'Properties of probability distributions with monotone hazard rate', *Ann. Math. Statist.*, **34**, 375–89.

Barlow, R. E., and F. Proschan (1965). *Mathematical Theory of Reliability*, John Wiley and Sons, New York.

Barlow, R. E., and F. Proschan (1966). 'Inequalities for linear combinations of order statistics from restricted families', *Ann. Math. Statist.*, **37**, 1574–92. [368]

Barlow, R. E., and F. Proschan (1967). 'Exponential life test procedures when the distribution has monotone failure rate', *J. Amer. Statist. Assoc.*, **62**, 548–60.

Barlow, R. E., and F. Proschan (1969). 'A note on tests for monotone failure rate based on incomplete data', *Ann. Math. Statist.*, **40**, 595–600.

Barlow, R. E., and F. Proschan (1975). *Statistical Theory of Reliability and Life-Testing*. Holt, Rhinehart and Winston, New York. [325]

Barlow, R. E., F. Proschan, and E. M. Scheuer (1971). 'A system debugging model'. In D. Grouchko (ed.), *Operations Research and Reliability*, Gordon and Breach, New York, pp. 401–20.

Barlow, R. E., and E. M. Scheuer (1966). 'Reliability growth during a development testing program', *Technometrics*, **8**, 53–60.

Barlow, R. E., and E. M. Scheuer (1971). 'Estimation from accelerated life tests', *Technometrics*, **13**, 145–59.

Barlow, R. E., and V. A. Ubhaya (1971). 'Isotonic approximation'. In J. S. Rustagi (ed.), *Optimizing Methods in Statistics*, Academic Press, New York. [58, 390]

Barlow, R. E., and W. R. van Zwet (1970a). 'Asymptotic properties of isotonic estimators for the generalized failure rate function. Part I: strong consistency'. In M. L. Puri (ed.) *Nonparametric Techniques in Statistical Inference*, Cambridge University Press, pp. 159–73. [367]

Barlow, R. E., and W. R. van Zwet (1971). 'Comparisons of several nonparametric estimators of the failure rate function'. *Operations Research and Reliability*. (Proceedings of the NATO Conference at Turin, 1969.) Gordon Breach, New York, 375–99. [356–358, 367]

Barlow, R. E., D. J. Bartholomew, J. M. Bremner, and H. D. Brunk (1972). *Statistical Inference under Order Restrictions*, John Wiley and Sons, London. [56, 57, 68, 79, 113, 219, 221, 226, 323, 325, 336, 343, 359, 367, 373, 387, 399]

Barlow, R. E. See also Proschan, Barlow, Madansky, and Scheuer (1968).

Bartholomew, D. J. (1956). 'Tests for randomness in a series of events when the alternative is a trend', *J. R. Statist. Soc. (B)*, **18**, 234–9. [173]

Bartholomew, D. J. (1959a). 'A test of homogeneity for ordered alternatives', *Biometrika*, **46**, 36–48. [56, 68, 112, 113, 116, 161, 167, 226]

Bartholomew, D. J. (1959b). 'A test of homogeneity for ordered alternatives II', *Biometrika*, **46**, 328–35. [68, 76, 114, 116, 161, 226]

Bartholomew, D. J. (1961a). 'A test of homogeneity of means under restricted alternatives' (with discussion), *J. R. Statist. Soc. (B)*, **23**, 239–81. [83, 91, 94, 97, 112–114, 152, 163, 196–198, 226]

Bartholomew, D. J. (1961b). 'Ordered tests in the analysis of variance', *Biometrika*, **48**, 325–32. [91, 94, 120, 207, 208, 227]

Bartholomew, D. J. (1963). 'On Chassan's test for order', *Biometrics*, **19**, 188–91.

Bartholomew, D. J. (1983a). 'Latent variable models for ordered categorical data', *J. Econmtx.*, **22**, 229–43.

Bartholomew, D. J. (1983b). 'Isotonic inference', *Encyclopedia of Statistical Sciences*, Wiley, New York.

Bartholomew, D. J. See also Barlow, Bartholomew, Bremner, and Brunk (1972).

Barton, D. E., and C. L. Mallows (1961). 'The randomization bases of the problem of the amalgamation of weighted means', *J. R. Statist. Soc. (B)*, **23**, 423–33. [80, 114]

Bassiouni, El M. Y. (1983). 'On the existence of explicit restricted maximum likelihood estimation in multivariate normal models', *Sankhya (B)*, **45**, 303–5.

Basu, A. P. (1967). 'On two *k*-sample rank tests for censored data', *Ann. Math. Statist.*, **38**, 1520–35.

Bates, D. M., and D. A. Wolf (1984). 'Non-negative regression by Givens rotations', *Comm. Statist.*, **9**, 841–50.

Bechhofer, R. E., and R. V. Kulkarni (1984). 'Closed sequential procedures for selecting the multinomial events which have the largest probabilities', *Comm. Statist. (A)*, **13**, 2997–3031.

Beckett, J. III. See Schucany and Beckett (1976).

Bee, R. H. See Liu and Bee (1984).

Bellhouse, D. R., and J. N. K. Rao (1975). 'Systematic sampling in the presence of a trend', *Biometrika*, **62**, 694–7.

Bennett, B. M. (1962). 'On an exact test for trend in binomial trials and its power function', *Metrika*, **5**, 49–53.

Bennett, B. M. (1964). 'On a test of homogeneity for samples from a negative binomial distribution', *Metrika*, **8**, 1–4. [169]

Beran, R. (1975). 'Local asymptotic power of quadratic rank tests for trend', *Ann. Statist.*, **3**, 401–12.

Berenson, M. L. (1982a). 'A comparison of several *k* sample tests for ordered alternatives in completely randomized designs', *Psychometrika*, **47**, 265–80. [207]

Berenson, M. L. (1982b). 'A study of several useful tests for ordered alternatives in the randomized block design', *Comm. Statist. (B)*, **11**, 563–81. [213, 216]

Berenson, M. L. (1982c). 'Some useful nonparametric tests for ordered alternatives in randomized block experiments', *Comm. Statist. (A)*, **11**, 1681–93.

Berger, R. L. (1984a). 'Testing whether one regression function is larger than another', *Comm. Statist. (A)*, **13**, 1793–810.

Berger, R. L. (1984b). 'Testing for the same order in several groups of means'. In T. J. Santner and A. C. Tamhane (eds.), *Designs of Experiments, Statistics*, Vol. 56, Dekker, New York, pp. 241–9.

Berger, R. L. (1985). 'Testing for stochastic ordering' (abstract), *Bull. Inst. Math. Statist.*, **14**, 266.

Berger, R. L., and D. F. Sinclair (1984). 'Testing hypotheses concerning unions of linear subspaces', *J. Amer. Statist. Assoc.*, **79**, 158–63.

Bergman, B. (1977). 'Crossings in the total time on test plot', *Scand. J. Statist.*, **4**, 171–7.

Bergman, B. (1981). 'Tests against certain ordered multinomial alternatives', *Scand. J. Statist.*, **8**, 218–26.

Bernshtein, A. V. (1980). 'On the construction of majorizing tests', *Theory of Probability and its Applications*, **25**, 16–26.

Bhattacharyya, G. K., and R. A. Johnson (1970). 'A layer rank test for ordered bivariate alternatives', *Ann. Math. Statist.*, **41**, 1296–310.

Bhattacharyya, G. K., and J. H. Klotz (1966). 'The bivariate trend of Lake Mendota', Technical Report No. 98, Department of Statistics, University of Wisconsin, Madison, Wisconsin.

Bhattacharya, P. K. (1981). 'Posterior distribution of a Dirichlet process from quantal response data', *Ann. Math. Statist.*, **9**, 803–11.

Bickel, P. J. (1969). 'Tests for monotone failure rate II', *Ann. Math. Statist.*, **40**, 1250–60. [367]

Bickel, P. J. (1981). 'Minimax estimation of the mean of a normal distribution when the parameter space is restricted', *Ann. Statist.*, **9**, 1301–9.

Bickel, P. J., and K. Doksum (1969). 'Tests for monotone failure rate based on normalized spacings', *Ann. Math. Statist.*, **40**, 1216–35. [363, 367]

Billingsley, P. (1968). *Convergence of Probability Measures*, Wiley, New York. [165, 166, 233, 249]

Birnbaum, Z. W. (1953). 'On the power of a one-sided test of fit for continuous probability functions', *Ann. Math. Statist.*, **24**, 484–9.

Birnbaum, Z. W., J. D. Esary, and A. W. Marshall (1966). 'A stochastic charcterization of wear-out for components and systems', *Ann. Math. Statist.*, **37**, 816–25.

Birnbaum, Z. W., and F. H. Tingey (1951). 'One-sided confidence contours for probability distribution functions', *Ann. Math. Statist.*, **22**, 592–6. [173]

Bishop, Y. M. M., S. E. Fienberg, and P. W. Holland (1975). *Discrete Multivariate Analysis: Theory and Practice*, MIT Press, Cambridge, Mass. [279, 288, 323]

Blackwell, D. (1947). 'Conditional expectation and unbiased estimation', *Ann. Math. Statist.*, **18**, 105–10. [387]

Bliss, C. I., and R. A. Fisher (1953). 'Fitting the negative binomial distribution to biological data and a note on efficient fitting of the negative binomial', *Biometrics*, **9**, 176–200. [235]

Blumenthal, S., and A. Cohen (1968). 'Estimation of two-ordered translation parameters', *Ann. Math. Statist.*, **39**, 517–30. [58]

Bock, M. E. (1982). 'Employing vague inequality information in the estimation of normal mean vectors'. In *Statistical Decision Theory and Related Topics III*, Vol. 1, Academic Press, pp. 169–93.

Bock, M. E. See also Judge, Yancey, Bock, and Bohrer (1984) and Yancey, Judge, and Bock (1981).

Bohrer, R. (1967). 'On sharpening Scheffe's bounds', *J. Royal Statist. Soc. B*, **29**, 110–14. [405, 406]

Bohrer, R. (1975). 'Algorithm A5 90:One sided multivariable inference', *Appl. Statist.*, **24**, 380–4. [117]

Bohrer, R., and W. Chow (1978). 'Weights for one-sided multivariate inference', *Appl. Statist.*, **27**, 100–4. [117]

Bohrer, R., and G. K. Francis (1972). 'Sharp one-sided confidence bounds for linear regression over an interval', *Biometrika*, **59**, 99–107. [117, 405–407]

Bohrer, R., and G. K. Francis (1972b). 'Likelihood ratio tests for order restrictions in exponential families', *Ann. Statist.*, **6**, 485–505.

Bohrer, R. See also Judge, Yancey, Bock, and Bohrer (1984) and Yancey, Bohrer, and Judge (1982).

Boos, D. D. (1982). 'A test for asymmetry associated with the Hodges–Lehman estimator', *J. Amer. Statist. Assoc.*, **77**, 647–51.

Boswell, M. T. (1966). 'Estimating and testing trend in a stochastic process of Poisson type', *Ann. Math. Statist.*, **37**, 1564–73. [226, 404]

Boswell, M. T., and H. D. Brunk (1969). 'Distribution of likelihood ratio in testing against trend', *Ann. Math. Statist.*, **40**, 371–80. [170, 172, 205, 226]

Bowman, A. W. See Titterington and Bowman (1985).

Boyd, M. N., and P. K. Sen (1983). 'Union-intersection rank tests for ordered alternatives in some simple linear models', *Comm. Statist. (A)*, **12**, 1737–53. [216]

Boyd, M. N., and P. K. Sen (1984). 'Union-intersection rank tests for ordered alternatives in a complete block design', *Comm. Statist. (A)*, **13**, 285–303. [216]

Boyd, M. N., and P. K. Sen (1986). 'Union-intersection rank tests for ordered alternatives in ANOCOVA', *J. Amer. Statist. Assoc.*, **81**, 526–32. [216]

Boyett, J. M., and J. J. Schuster (1977). 'Nonparametric one-sided tests in multivariate analysis with medical applications', *J. Amer. Statist. Assoc.*, **72**, 665–8. [216]

Boyle, J. P. (1984). 'Restricted Hilbert space projections with applications to smoothing splines', *Ph.D. Dissertation*, University of Missouri–Columbia.

Boyle, J. P., and R. L. Dykstra (1986). 'An approach for finding Hilbert space least squares projections onto the intersection of convex sets'. In R. L. Dykstra, T. Robertson, and F. T. Wright (eds.), *Advances in Order Restricted Statistical Inference*, Springer-Verlag, New York, pp. 28–47.

Boyle, J. P. See also Dykstra and Boyle (1987).

Boyles, R. A., A. W. Marshall, and F. Proschan (1985). 'Inconsistency of the maximum likelihood estimator of a distribution having increasing failure rate average', *Ann. Statist.*, **13**, 413–17.

Boyles, R. A., and F. J. Samaniego (1984). 'Modeling and inference for multivariate binary data with positive dependence', *J. Amer. Statist. Assoc.*, **79**, 188–93.

Bray, T. A., G. B. Crawford, and F. Proschan (1967). 'Maximum likelihood estimation of a U-shaped failure rate function', Math. Note No. 534, Math. Res. Lab., Boeing Sci. Res. Labs. [338]

Bremner, J. M. (1967). 'Problems of estimation and testing arising from ordered hypotheses concerning normal means', Unpublished M.Sc. dissertation, University of Wales. [219]

Bremner, J. M. (1978). 'Mixtures of beta distributions', *J. R. Statist. Soc. (C)*, **27**, 104–9.

Bremner, J. M. See also Barlow, Bartholomew, Bremner, and Brunk (1972).

Breslow, N. E., J. H. Lubin, P. Marek, and B. Langholz (1983). 'Multiplicative models and cohort analysis', *J. Amer. Statist. Assoc.*, **78**, 1–12.

Bril, G., R. L. Dykstra, C. Pillers, and T. Robertson (1984). 'Isotonic regression in two independent variables', *J. R. Statist. Soc. (C)*, **33**, 352–7. [28]

Brillinger, D. R. (1966). 'An extremal property of the conditional expectation.' *Biometrika*, **53**, 594–5.

British Association Mathematical Tables, Vol. 1, 3rd ed., Cambridge University Press, 1951.

Broffitt, J. D. (1984). 'A Bayes estimator for ordered parameters and isotonic Bayesian graduation', *Scand. Actuarial J.*, **1984**, 231–47. [398, 399]

Broffitt, J. D. (1986). 'Restricted Bayes estimates for binomial parameters', Technical Report, University of Iowa, Iowa City.

Browder, F. E. (1965). 'Nonlinear monotone operations and convex sets in Banach spaces', *Bull. Amer. Math. Soc.*, **71**, 780–5. [387]

Brown, B. M., and S. I. Resnick (1984). 'Rank tests for multivariate trend', *Austral. J. Statist.*, **26**, 58–67.

Brunk, H. D. (1955). 'Maximum likelihood estimates of monotone parameters', *Ann. Math. Statist.*, **26**, 607–16. [6, 56]

Brunk, H. D. (1956). 'On an inequality for convex functions', *Proc. Amer. Math. Soc.*, **7**, 817–24. [6, 56]

Brunk, H. D. (1958). 'On the estimation of parameters restricted by inequalities', *Ann. Math. Statist.*, **29**, 437–54. [57, 393, 394]

Brunk, H. D. (1960). 'On a theorem of E. Sparre Andersen and its application to tests against trend', *Mathematica Scand.*, **8**, 305–26. [206]

Brunk, H. D. (1961). 'Best fit to a random variable by a random variable measurable with respect to a σ-lattice', *Pacific J. Math.*, **11**, 785–802. [386, 387]

Brunk, H. D. (1963). 'On an extension of the concept of conditional expectation', *Proc. Amer. Math. Soc.*, **14**, 298–304. [386, 387]

Brunk, H. D. (1964). 'A generalization of Spitzer's combinatorial lemma', *Z. Wahrsch. verw. Gebiete*, **2**, 395–405.

Brunk, H. D. (1965). 'Conditional expectation given a σ-lattice and applications', *Ann. Math. Statist.*, **36**, 1339–50. [57, 386, 387, 396, 397]

Brunk, H. D. (1970). 'Estimation of isotonic regression' (with discussion by Ronald Pyke). In M. L. Puri (ed.), *Nonparametric Techniques in Statistical Inference*, Cambridge University Press, pp. 177–97. [393, 395, 396]

Brunk, H. D. (1975). 'Uniform inequalities for conditional p-means given σ-lattices', *Ann. Prob.*, **3**, 1025–30. [373]

Brunk, H. D., G. M. Ewing, and W. R. Utz (1957). 'Minimizing integrals in certain classes of monotone functions', *Pacific J. Math.*, **7**, 833–47. [56, 57]

Brunk, H. D., and S. Johansen (1970). 'A generalized Random–Nikodym derivative', *Pacific J. Math.*, **34**, 585–617. [373, 387]

Brunk, H. D., W. E. Franck, D. L. Hanson, and R. V. Hogg (1966). 'Maximum likelihood estimation of the distributions of two stochastically ordered random variables', *J. Amer. Statist. Assoc.*, **61**, 1067–80. [58, 263]

Brunk, H. D. See also Ayer, Brunk, Ewing, Reid, and Silverman (1955); Barlow and Brunk (1972); Barlow, Bartholomew, Bremner, and Brunk (1972); Boswell and Brunk (1969); and Lombard and Brunk (1963).

Bunke, O., and M. Möhner (1984). 'Minimax estimators of regression functions under normalized quadratic loss functions and inequality restrictions', *Math. Operations Forsch. Ser. Statist.*, **15**, 471–82.

Burr, E. J. (1960). 'The distribution of Kendall's score *S* for a pair of tied rankings', *Biometrika*, **47**, 151–71.

Buse, A. (1982). 'The likelihood ratio, Wald and Lagrange multiplier tests: An expository note', *Amer. Statist.*, **36**, 153–7. [321]

Cadoret, R. J., R. Woolson, and G. Winokur (1977). 'Relationship of age of onset in unipolar disorder to risk of alcoholism and depression in parents', *J. of Psychiatric Research*, **13**, 137–42.

Cani, J. S. de (1984). 'Balancing type I risk and loss of power in ordered Bonferonni procedures', *J. Educ. Psych.*, **76**, 1035–7.

Casady, R. J., and J. D. Cryer (1976). 'Monotone percentile regression', *Ann. Statist.*, **4**, 532–41. [390]

Casady, R. J. See also Cryer, Robertson, Wright, and Casady (1972).

Causey, B. D. (1983). 'Estimation of proportions for multinomial contingency tables subject to marginal constraints', *Comm. Statist. (A)*, **12**, 2581–7.

Causey, B. D. (1984). 'Estimation under generalized sampling of cell proportions for contingency tables subject to marginal constraints', *Comm. Statist. (A)*, **13**, 2487–94.

Chacko, V. J. (1963). 'Testing homogeneity against ordered alternatives', *Ann. Math. Statist.*, **34**, 945–56. [112, 114, 204–206, 210, 227]

Chacko, V. J. (1966). 'Modified chi-square test for ordered alternatives', *Sankhya (B)*, **28**, 185–90. [57, 263]

Chang, Y. C. (1984). 'A truncated maximum likelihood estimator of a constrained bivariate linear regression coefficient', *J. Amer. Statist. Assoc.*, **79**, 454–8.

Chao, M., and Glaser, R. E. (1978). 'The exact distribution of Bartlett's test statistic for homogeneity of variances with unequal sample sizes', *J. Amer. Statist. Assoc.*, **73**, 422.

Chapman, D. G. (1958). 'A comparative study of several one-sided goodness-of-fit tests', *Ann. Math. Statist.*, **29**, 655–74.

Chase, G. R. (1974). 'On testing for ordered alternatives with increased sample size for a control', *Biometrika*, **61**, 569. [125, 162, 190, 196]

Chassan, J. B. (1960). 'On a test for order', *Biometrics*, **16**, 119–21.

Chassan, J. B. (1962). 'An extension of a test for order', *Biometrics*, **18**, 245–7.

Chatterjee, S. K. (1984). 'Restricted alternatives'. In P. R. Krishnaiah and P. K. Sen (eds.), *Handbook of Statistics—Nonparametric Methods*, Vol. 4, North Holland.

Chatterjee, S. K., and N. K. De (1972). 'Bivariate nonparametric location test against restricted alternatives', *Calcutta Statist. Assoc. Bull.*, **21**, 1–20.

Chatterjee, S. K., and N. K. De (1974). 'On the power superiority of certain bivariate location tests against restricted alternatives', *Calcutta Statist. Assoc. Bull.*, **23**, 73–84.

Chatterjee, S. K., and N. K. De (1977). 'Corrigendum: On the power superiority of certain bivariate location tests against restricted alternatives', *Calcutta Statist. Assoc. Bull.*, **26**, 125.

Chen, H. J., and J. R. Pickett (1984). 'Selecting all treatments better than a control under a multivariate normal distribution and a uniform prior distribution', *Comm. Statist. (A)*, **13**, 59–80.

Chen, R. W., and E. V. Slud (1984). 'On the product of symmetric random variables', *Comm. Statist. (A)*, **13**, 611–15.

Chen, Y. Y., M. Hollander, and N. A. Langberg (1982). 'Small sample results for the Kaplan–Meier estimator', *J. Amer. Statist. Assoc.*, **77**, 141–4.

Chernoff, H. (1954). 'On the distribution of the likelihood ratio', *Ann. Math. Statist.*, **25**, 573–8. [68, 112, 226]

Chernoff, H. (1964). 'Estimation of the mode', *Ann. Inst. Statist. Math.*, **16**, 31–41. [334, 336, 395]

Chernoff, H., and I. R. Savage (1958). 'Asymptotic normality and efficiency of certain nonparametric test statistics', *Ann. Math. Statist.*, **29**, 972–97. [202, 203, 205, 209, 214]

Childs, D. R. (1967). 'Reduction of the multivariate normal integral to characteristic form', *Biometrika*, **54**, 293–300. [75]

Chinchilli, V. M., and P. K. Sen (1981a). 'Multivariate linear rank statistics and the union–intersection principle for hypothesis testing under ordered alternatives', *Sankhya (B)*, **43**, 135–51. [216]

Chinchilli, V. M., and P. K. Sen (1981b). 'Multivariate linear rank statistics and the union–intersection principle for the orthant restriction problem', *Sankhya (B)*, **43**, 152–71. [216]

Chiu, H. Y., and J. Sedransk (1986). 'A Bayesian procedure for imputing missing values in sample surveys', *J. Amer. Statist. Assoc.*, **81**, 667–76.

Chiu, H. Y. See also Sedransk, Monahan and Chiu (1985).

Choi, J. R. (1975). 'An equality involving orthant probabilities', *Comm. Statist.*, **12**, 1167–75.

Choi, J. R. See also Kudô and Choi (1975) and Kudô, Sasabuchi, and Choi (1981).

Chotai, J. (1979). 'Selection and ranking procedures based on likelihood ratio', Thesis, University of Umea, Sweden.

Chotai, J. (1980). 'Subset selection based upon likelihood from uniform and related populations', *Comm. Statist.*, **11**, 1147–64.

Chotai, J. (1983). 'Isotonic inference for populations related to the uniform distribution', *Comm. Statist. (A)*, **12**, 2109–18. [226]

Chou, Y. M., K. H. Arthur, R. B. Rosenstein, and D. B. Owen (1984). 'New representations of the noncentral chi-square density and cumulative', *Comm. Statist. (A)*, **13**, 2673–8.

Chow, W. See Bohrer and Chow (1978).

Christopeit, N., and G. Tosstorff (1987). 'Strong consistency of least squares estimators in the monotone regression model with stochastic regressors', *Ann. Statist.*, **15**, 568–86.

Chuang, C. See Agresti and Chuang (1986) and Agresti, Chuang, and Kezouh (1987).

Chung, K. L. (1968). *A Course in Probability Theory*, Harcourt, Brace and World, New York.

Church, J. D., and E. B. Cobb (1971). 'Nonparametric estimation of the mean using quantile response data', *Ann. Inst. Stat. Math.*, **23**, 105–17.

Church, J. D., and E. B. Cobb (1973). 'On the equivalence of Spearman, Karber and maximum likelihood estimates of the mean', *J. Amer. Statist. Assoc.*, **68**, 201–2.

Church, J. D. See also Cobb and Church (1983).

Ciesielski, A. (1958). 'A note on some inequalities of Jensen's type', *Ann. Pol. Math.*, **4**, 269–74.

Clarkson, D. B., and D. B. Wolfson (1983). 'An application of a displaced Poisson process', *St. Neerla.*, **37**, 21–8.

Cobb, E. B., and J. D. Church (1983). 'Small-sample quantal response methods for estimating the location parameter for a location-scale family of dose response curves', *J. Amer. Statist. Assoc.*, **78**, 99–107.

Cobb, E. B. See also Church and Cobb (1971, 1973).

Cochran, W. G. (1954). 'Some methods for strengthening the common χ^2 tests', *Biometrics*, **10**, 417–51.

Cochran, W. G. (1955). 'A test of a linear function of the deviations between observed and expected numbers', *J. Amer. Statist. Assoc.*, **50**, 377–97.

Cochran, W. G. (1966). 'Analyse des classifications d'ordre', *Revue Statist. Appl.*, **14**, 5–17.

Cohen, A., and H. B. Sackrowitz (1970). 'Estimation of the last mean of a monotone sequence', *Ann. Math. Statist.*, **41**, 2021–34. [58]

Cohen, A. See also Blumenthal and Cohen (1968).

Cohen, M. P., and L. Kuo (1985). 'The admissibility of the empirical distribution function', *Ann. Stat.*, **13**, 262–71.

Collings, B. J., B. H. Margolin, and G. W. Oehlert (1981). 'Analyses for binomial data, with application to the fluctuation test for mutagenicity', *Biometrics*, **37**, 775–94.

Conover, W. J. (1967). 'A k-sample extension of the one-sided two sample Smirnov test statistic', *Ann. Math. Statist.*, **38**, 1726–30.

Cooke, W. P. (1983). 'Surrogate geometric programming estimates of restricted multinomial probabilities', *Comm. Statist. (B)*, **12**, 291–305.

Cooper, M. M. See White, Landis, and Cooper (1982).

Corley, H. W. (1984). 'Multivariate order statistics', *Comm. Statist. (A)*, **13**, 1299–304.

Cox, C. (1984). 'An elementary introduction to maximum likelihood estimation for multinomial models: Birch's theorem and the delta method', *Amer. Statist.*, **38**, 283–7.

Cox, D. R. (1955). 'Some statistical methods connected with series of events' (with discussion). *J. R. Statist. Soc. (B)*, **17**, 129–64. [367]

Cox, D. R. (1982). 'On the role of data of possibly lowered sensitivity', *Biometrika*, **69**, 215–9.

Cox, D R., and P. A. W. Lewis (1966). *The Statistical Analysis of Series of Events*, Methuen, London. [367]

Craig, A. T. See Hogg and Craig (1978).

Cran, G. W. (1981). 'Calculation of the probabilities $\{P(l,k)\}$ for the simply ordered alternative', *J. R. Statist. Soc. (C)*, **30**, 85–91.

Crawford, G. B. (1967). 'Maximum likelihood estimation: A practical theorem on consistency of the nonparametric maximum likelihood estimates with applications', Math. Note No. 503, Math. Res. Lab., Boeing Sci. Res. Labs.

Crawford, G. B. See also Bray, Crawford, and Proschan (1967).

Crawley, P., and R. P. Dilworth (1973). *Algebraic Theory of Lattices*, Prentice-Hall, Englewood Cliffs, New Jersey. [144]

Cressie, Noel, and T. R. C. Read (1984). 'Multinomial goodness of fit models', *J. R. Statist. Soc. (B)*, **46**, 440–64. [242]

Crouse, C. F. See Steyn and Crouse (1975).

Cryer, J. D., and T. Robertson (1975). 'Isotonic estimation of the probability of extinction of a branching process', *J. Amer. Statist. Assoc.*, **70**, 905–12.

Cryer, J. D., T. Robertson, F. T. Wright, and R. J. Casady (1972). 'Monotone median regression', *Ann. Math. Statist.*, **43**, 1459–69. [390–394]

Cryer, J. D. See also Casady and Cryer (1976) and Robertson and Cryer (1974).

Csaki, E. (1968). 'An interated logarithm law for semimartingales and its application to empirical distribution functions', *Studia Sci. Math. Hungar.*, **3**, 287–92. [328]

Csiszar, I. (1967). 'Information-type measures of difference of probability distributions and indirect observation', *Studia Sci. Math. Hungar.*, **2**, 299–318. [300]

Csiszar, I. (1975). '*I*-divergence geometry of probability distributions and minimization problems', *Ann. Prob.*, **3**, 146–59. [282, 287–289, 298–301]

Csiszar, I. (1984). 'Sanov property, generalized *I*-projection and a conditional limit theorem', *Ann. Prob.*, **12**,, 768–93.

Cunningham, F., and N. Grossman (1971). 'On Young's inequality', *Amer. Math. Monthly*, **78**, 781–3.

D'Agostino, R. B. See Meeks and D'Agostino (1983).

Dahlbom, U. (1985). 'Consistency of unimodal regression', *Proc. ISI Meeting in Amsterdam.*

Daly, C. (1962). 'A simple test for trends in a contingency table', *Biometrics*, **18**, 114–9.

Daniel, W. W. See Roth and Daniel (1978).

Dantzig, G. B. (1963). *Linear Programming and Extensions*, Princeton University Press, Princeton, New Jersey. [306]

Dantzig, G. B. (1971). 'A control problem of Bellman', *Management Science*, **17**, 542–6. [ ]

Darst, R. B. (1970). 'The Lesbegue decomposition, Randon–Nikodym derivative, conditional expectation, and martingale convergence for lattices of sets', *Pacific J. Math.*, **35**, 581–600. [387]

Darst, R. B., and R. Huotari (1985a). 'Best L_1-approximation of bounded approximately continuous functions on [0, 1] by nondecreasing functions', *J. Approx. Theory*, **43**, 178–89.

Darst, R. B., and R. Huotari (1985b). 'Monotone L_1-approximation on the unit *N*-cube', *Proc. Amer. Math. Soc.*, **95**, 425–8.

Darst, R. B., and S. Sahab (1983). 'Approximation of continuous and quasi-continuous functions by monotone functions', *J. Approx. Theory*, **38**, 9–27.

Das Gupta, S. (1983). 'On Anderson's probability inequality'. In *Studies in Econometrics, Time Series, and Multivariate Statistics*, Academic press, New York, pp. 407–417.

Davis, A. W. (1977). 'A differential equation approach to linear combinations of independent χ^2', *J. Amer. Statist. Assoc.*, **72**, 212–4.

Davis, H. T. (1933, 1935). *Tables of Higher Mathematical Functions*, Vols 1 and 2, Principia Press, Bloomington, Indiana.

De, N. K. (1975). 'Rank tests for randomized blocks against ordered alternatives', *Calcutta Statist. Assoc. Bull.*, **24**, 1–27.

De, N. K. See Chatterjee and De (1972, 1974, 1977).

Deaton, L. W. (1980). 'An empirical Bayes approach to polynomial regression under order restriction', *Biometrika*, **67**, 111–7.

Demyanov, V. F. See Mitchell, Demyanov, and Malozemov (1974).

Denby, L., and Y. Vardi (1984). 'The survival curve with monotone density and a short cut method for estimation in renewal processes' (abstract), *Bull. Inst. Math. Statist.*, **13**, 75.

Denier van der Gon, J. J. (1963). 'Comparison of two tests against trend for a number of probabilities' (in Dutch), *Statist. Neerlandica*, **17**, 105–12.

Denis, J. B. (1979). 'Interactive submodels with order restrictions' (in French), *Biom. Prax.*, **19**, 49–58.

Denis, J. B. (1982). 'Tests of interaction under order restrictions' (in French), *Biom. Prax.*, **22**, 29–45.

Dent, W. T. (1973). 'A note on least squares fitting of functions constrained to be either nonnegative, nondecreasing or convex', *Management Science*, **20**, 130–2, [243]

Dent, W. T. (1980). 'On restricted alternatives in linear models', *J. Econmtx*, **12**, 49–58.

Derman, C., L. J. Gleser and I. Olkin (1973). *A Guide to Probability Theory and Application*, Holt, Rinehart and Winston, New York.

Deshpande, J. V. (1980). 'Linear combinations of two sample rank tests for ordered alternatives', *J. Indian Statist. Assoc.*, **18**, 55–61.

Deshpande, J. V. (1983). 'A class of tests for exponentiality against increasing failure rate average alternatives', *Biometrika*, **69**, 514–8.

Deshpande, J. V., and S. C. Kochar (1983). 'A test for exponentiality against IFR alternatives', *J. Indian Assoc. for Productivity, Quality and Reliability*, **8**, 1–8.

Deshpande, J. V. See also Kochar and Deshpande (1985).

Devroye, L., and L. Gyorfi (1985). *Nonparametric Density Estimation: The L_1 View*, John Wiley and Sons, New York. [325]

Dhariyal, I. D., D. Sharma, and K. Krishnamoorthy (1985). 'Non-existence of unbiased estimators of ordered parameters', *Mathematische Operations Forsch. Statist.*, **16**, 88–95.

Diaz, J. B., and F. T. Metcalf (1970). 'An analytic proof of Young's inequality', *Amer. Math. Monthly*, **77**, 603.

Dietz, E. J., and T. J. Killeen (1981). 'A nonparametric multivariate test for monotone trend with pharmaceutical applications', *J. Amer. Statist. Assoc.*, **75**, 169–74.

Dilworth, R. P. See Crawley and Dilworth (1973).

Dimitrov, Boyan (1983). 'Isotonic estimate of distributions in reliability' (Russian), *Bulgar. Acad. Sci. Sofia*, **1983**, 39–74.

Dinse, G. E. (1985). 'Testing for trend in tumor prevalence rates: I. nonlethal tumors', *Biometrics*, **41**, 751–70.

Dinse, G. E. (1986). 'Nonparametric prevalence and mortality estimators for animal experiments with incomplete cause of death data', *J. Amer. Statist. Assoc.*, **81**, 324–36.

Disch, D. (1981). 'Bayesian nonparametric inferences for effective doses in a quantal-response experiment', *Biometrics*, **37**, 713–22.

Doksum, K. (1966). 'Asymptotically minimax distribution-free procedures', *Ann. Math. Statist.*, **37**, 619–28.

Doksum, K. (1967a). 'Robust procedures for some linear models with one-observation per cell', *Ann. Math. Statist.*, **38**, 878–83. [211, 212, 215]

Doksum, K. (1967b). 'Asymptotically optimal statistics in some models with increasing failure rate averages', *Ann. Math. Statist.*, **38**, 1731–9 (for corrections to the above paper, see *Ann. Math. Statist.*, **39**, 684–5).

Doksum, K. (1969a). 'Starshaped transformations and the power of rank tests', *Ann. Math. Statist.*, **40**, 1167–76.

Doksum, K. (1969b). 'Minimax results for IFRA scale alternatives', *Ann. Math. Statist.*, **40**, 1778–83.

Doksum, K. See also Barlow and Doksum (1972) and Bickel and Doksum (1969).

Don, F. J. H. (1985). 'The use of generalized inverses in restricted maximum likelihood', *Linear Algebra Appl.*, **70**, 225–40.

Doran, J. E. See Hodson, Sneath and Doran (1966).

Douglas, R. G. (1965). 'Contractive projections on an L_p space', *Pacific J. Math.*, **15**, 433–62. [387]

Dudewicz, E. J. (1969). 'Estimation of ordered parameters', Technical Report, Department of Operations Research, Cornell University.

Dunford, N., and J. T. Schwartz (1958). *Linear Operators*, Vol. I. Interscience, New York. [386]

Dunnett, C. W. (1955). 'A multiple comparisons procedure for comparing several treatments with a control', *J. Amer. Statist. Assoc.*, **50**, 1096–121. [161, 192–194, 196, 198–200]

Dunnett, C. W., and M. Sobel (1955). 'Approximations to the probability integral and certain percentage points of a multivariate analogue of Student's t-distribution', *Biometrika*, **42**, 258–60. [83, 227]

Dupacova, J. (1984). 'On the asymptotic normality of inequality contrained optimal decisions'. In P. Mandel and M. Hustova (eds.), *Asymptotic Statistics*, Vol. 2, Elsevier, Amsterdam, pp. 249–57.

Dykstra, R. L. (1968). 'Characterizations of a conditional expectation with respect to a sigma-lattice', *Ph.D. Dissertation*, University of Iowa. [386]

Dykstra, R. L. (1970a). 'A characterization of a conditional expectation with respect to a σ-lattice', *Ann. Math. Statist.*, **41**, 698–701. [382]

Dykstra, R. L. (1970b). 'A note on a theorem by H. D. Brunk', *Proc. Amer. Math. Soc.*, **24**, 171–4. [386]

Dykstra, R. L. (1981). 'An isotonic regression algorithm', *J. Statist. Planning and Inf.*, **5**, 355–63. [57]

Dykstra, R. L. (1982). 'Maximum likelihood estimation of the survival functions of stochastically ordered random variables', *J. Amer. Statist. Assoc.*, **77**, 621–8. [315]

Dykstra, R. L. (1983). 'An algorithm for restricted least squares regression', *J. Amer. Statist. Assoc.*, **78**, 837–42. [29, 56, 57, 111]

Dykstra, R. L. (1984). 'Dual convex cones of order restrictions with applications'. In Y. L. Tong (ed.), *Inequalities in Statistics and Probability*, Institute Math. Statist., Hayward, pp. 228–35.

Dykstra, R. L. (1985a). 'An iterative procedure for obtaining I-projections onto the intersection of convex sets', *Ann. Prob.*, **13**, 975–84. [278, 289–291, 294, 296, 300, 301]

Dykstra, R. L. (1985b). 'Computational aspects of I-projections', *J. Statist. Comp. Simul.*, **21**, 265–74. [291]

Dykstra, R. L., and J. P. Boyle (1987). 'An algorithm for least squares projections onto the intersection of translated, convex cones', *J. Statist. Plann. Inf.*, **11**, 391–400.

Dykstra, R. L., and C. Feltz (1982). 'Nonparametric estimates for reliability growth'. In D. J. DePriest and R. L. Launer (eds.), *Reliability in the Acquisitions Process*, Marcel Dekker, New York.

Dykstra, R. L., and P. Laud (1981). 'A Bayesian nonparametric approach to reliability', *Ann. Statist.*, **9**, 356–67. [400]

Dykstra, R. L., and R. Madsen (1976). 'Restricted maximum likelihood estimators for Poisson parameters', *J. Amer. Statist. Assoc.*, **71**, 711–8.

Dykstra, R. L., R. Madsen, and K. Fairbanks (1983). 'A nonparametric likelihood ratio test', *J. Statist. Comp. and Simul.*, **18**, 247–64.

Dykstra, R. L., and T. Robertson (1982a). 'An algorithm for isotonic regression for two or more independent variables', *Ann. Statist.*, **10**, 708–11. [56]

Dykstra, R. L., and T. Robertson (1982b). 'Order restricted statistical tests on multinomial and Poisson parameters: The starshaped restriction', *Ann. Statist.*, **10**, 1246–52. [248, 263, 308]

Dykstra, R. L., and T. Robertson (1983). 'On testing monotone tendencies', *J. Amer. Statist. Assoc.*, **78**, 342–50. [227, 248]

Dykstra, R. L., and P. C. Wollan (1986). 'Constrained estimation using interated partial Kuhn–Tucker vectors'. In A. Basu (ed.), *Reliability and Quality Control*, North Holland, New York, pp. 133–48.

Dykstra, R. L., and P. C. Wollan (1987). 'Finding *I*-projections subject to a finite set of linear inequality constraints', *Appl. Statist.*, **36**, 377–83. [278]

Dykstra, R. L. See also Boyle and Dykstra (1987); Bril, Dykstra, Pillers, and Robertson (1984); Feltz and Dykstra (1985); Lemke and Dykstra (1984); Wollan and Dykstra (1986, 1987).

Eaton, M. L. (1970). 'A complete class theorem for multidimensional one-sided alternatives', *Ann. Math. Statist.*, **41**, 1884–8.

Ebrahimi, N., and M. Habibullah (1983). 'Testing for linearity and convexity of a potency curve', *Comm. Statist. (A)*, **12**, 693–700.

Eeden, C. van (1956). 'Maximum likelihood estimation of ordered probabilities', *Proc. K. ned. Akad. Wet (A)*, **59**/*Indag. Math.*, **18**, 444–55. [56, 57, 390]

Eeden, C. van (1957a). 'Maximum likelihood estimation of partially or completely ordered parameters. I', *Proc. K. ned. Akad. Wet. (A)*, **60**/*Indag. Math.*, **19**, 128–36. [57, 389, 390]

Eeden, C. van (1957b). 'Maximum likelihood estimation of partially or completely ordered parameters. II', *Proc. K. ned. Akad. Wet. (A)*, **60**/*Indag. Math.*, **19**, 201–11.

Eeden, C. van (1957c). 'Note on two methods for estimating ordered parameters of probability distributions', *Proc. K. ned. Akad. Wet. (A)*, **60**/*Indag. Math.*, **19**, 506–12.

Eeden, C. van (1957d). 'A least squares inequality for maximum likelihood estimates of ordered parameters', *Proc. K. ned. Akad. Wet. (A)*, **60**/*Indag. Math.*, **19**, 512–21. [390]

Eeden, C. van (1958). 'Testing and estimating ordered parameters of probability distributions', Doctoral Dissertation, University of Amsterdam, Studentendrukkerij Poortpers, Amsterdam. [57, 58, 112, 190, 227]

Eeden, C. van (1960). 'On distribution-free bio-assay', *Proc. Symp. Quantitative Methods in Pharmacology*, Leiden, pp. 206–10.

Eeden, C. van (1970). 'Efficiency-robust estimation of location', *Ann. Math. Statist.*, **41**, 172–81.

Eeden, C. van, and J. Hemelrijk (1955a). 'A test for equality of probabilities against a class of specified alternative hypotheses, including trend. I', *Proc. K. ned. Akad. Wet. (A)*, **58**/*Indag. Math.*, **17**, 191–8.

Eeden, C. van, and J. Hemelrijk (1955b). 'A test for equality of probabilities against a class of specified alternative hypotheses, including trend. II', *Proc. K. ned. Akad. Wet. (A)*, **58**/*Indag. Math.*, **17**, 301–8.

Eeden, C. van. See also Kraft and Eeden (1964).

Efron, B., and D. V. Hinkley (1978). 'Assessing the accuracy of the maximum likelihood estimator: Observed versus expected Fisher information', *Biometrika*, **65**, 457–87. [321]

Eicker, F. (1979). 'The asymptotic distribution of the suprema of the standardized empirical process', *Ann. Statist.*, **7**, 116–38.

Eilbott, J. M. See Nadler and Eilbott (1967).

Elfving, G., and J. H. Whitlock (1950). 'A simple trend test with application to erthrocyte data', *Biometrics*, **6**, 282–8.

Engelhardt, M. E. See Bain and Engelhardt (1975) and Bain, Engelhardt, and Wright (1985).

Epstein, B. (1960a). 'Tests for the validity of the assumption that the underlying distribution of life is exponential. Part I', *Technometrics*, **2**, 83–101. [367]

Epstein, B. (1960b). 'Tests for the validity of the assumption that the underlying distribution of life is exponential. Part II', *Technometrics*, **2**, 167–83.

Esary, J. D. See Birnbaum. Esary and Marshall (1966).

Escobar, L. A., and B. Skarpness (1984). 'A closed form solution for the least squares regression problem with linear inequality constraints', *Comm. Statist. (A)*, **13**, 1127–34.

Escobar, L. A., and B. Skarpness (1986). 'The bias of least squares estimator over interval constraints', *Econ. Lett.*, **20**, 331–5.

Escobar, L. A., and B. Skarpness (1987). 'Mean square error and efficiency of the least squares estimator over interval constraints', *Comm. Statist.-Theory and Meth.*, **16**, 397–406.

Ewing, G. M. See Ayer, Brunk, Ewing, Reid, and Silverman (1955) and Brunk, Ewing, and Utz (1957).

Fairbanks, K. See Dykstra, Madsen, and Fairbanks (1983).

Fairley, D., and M. Fligner (1987). 'Linear rank statistics for the ordered alternatives problems', *Comm. Statist. -Theory and Meth.*, **16**, 1–16.

Fairley, D., D. K. Pearl, and J. S. Verducci (1987). 'The penalty for assuming that a monotone regression is linear', *Ann. Statist.*, **15**, 443–8.

Fang, K. T., and S. D. Ho (1985). 'Regression models with linear constraints and nonnegative regression coefficients' (Chinese, English summary), *Math. Numer. Sinica.*, **7**, 237–46.

Farebrother, R. W. (1984). 'The restricted least squares estimator and ridge regression', *Commun. Statist. (A)*, **13**, 191–6.

Farebrother, R. W. (1986). 'Testing linear inequality constraints in the standard linear model', *Comm. Statist. (A)*, **15**, 7–31. [322, 323]

Farewell, V. T. See Matthews and Farewell (1982).

Feder, P. I. (1968). 'On the distribution of the log likelihood ratio test statistic when the true parameter is 'near' the boundaries of the hypothesis regions', *Ann. Math. Statist.*, **39**, 2044–55. [226]

Feigin, P. D., and U. Passay (1981). 'The geometric programming dual to the extinction probability problem in simple branching processes', *Ann. Prob.*, **9**, 498–503.

Feller, W. (1968). *An Introduction to Probability Theory and Its Applications*, 3rd ed., Wiley, New York. [127, 334]

Feltz, C. J., and R. L. Dykstra (1985). 'Maximum likelihood estimation of the survival functions of N stochastically ordered random variables', *J. Amer. Statist. Assoc.*, **80**, 1012–9. [315–317]

Fenchel, W. (1949). 'On conjugate convex functions', *Canad. J. Math.*, **1**, 73–7. [322]

Fernandez, J. A. M. (1984). 'Likelihood ratio tests for normal population means subject to restrictions', *Trabajos Estadist. Investigacion Oper.*, **35**, 305–318.

Fienberg, S. E. See Bishop, Fienberg, and Holland (1975).

Finkelstein, D. M., and R. A. Wolfe (1986). 'Isotonic regression for interval censored survival data using an E–M algorithm,' *Comm. Statist. (A)*, **15**, 2493–505.

Fisher, L. See Gillespie and Fisher (1979).

Fisher, R. A. See Bliss and Fisher (1953).

Fligner, M. A., and S. W. Rust (1982). 'A modification of Mood's median test for the generalized Behrens–Fisher problem', *Biometrika*, **69**, 221–6.

Fligner, M. See also Fairley and Fligner (1987).

Földes, A., and L. Rejtö (1981). 'Strong uniform consistency for nonparametric survival curve estimators from randomly censored data', *Ann. Statist.*, **9**, 122–9.

Francis, G. K. See Bohrer and Francis (1972a, 1972b).

Franck, W. E. (1984). 'A likelihood ratio test for stochastic ordering', *J. Amer. Statist. Assoc.*, **79**, 686–91.

Franck, W. E. See also Brunk, Franck, Hanson, and Hogg (1966).

Friedman, J. H., W. Stuetzle, and A Schroeder (1984). 'Projection pursuit density estimation', *J. Amer. Statist. Assoc.*, **79**, 599–608.

Friedman, J. H., and R. Tibshirani (1984). 'The monotone smoothing of scatterplots', *Technometrics*, **26**, 243–50. [402]

Friedman, M. (1937). 'The use of ranks to avoid the assumption of normality implicit in the analysis of variance', *J. Amer. Statist. Assoc.*, **32**, 675–701. [210]

Frisen, M. (1985). 'Unimodal regression', Research Report No. 3, Department of Statistics, University of Göteborg.

Frisen, M. See also Holm and Frisen (1985).

Froggatt, P. (1970). 'Application of discrete distribution theory to the study of non-communicable events in medical epidemiology', In G. P. Patil, (ed.), *Random Counts in Biomedical and Social Sciences*, Vol. 2, Pennsylvania State University Press.

Fujino, Y. (1979). 'Tests for the homogeneity of a set of variances against ordered alternatives', *Biometrika*, **66**, 133–9.

Fujisawa, H. See Kudô and Fujisawa (1964).

Gabriel, K. R. See Marcus, Peritz, and Gabriel (1976).

Gafarion, A. V. See Murthy and Gafarion (1970).

Gaskins, R. A. See Good and Gaskins (1971, 1972).

Gebhardt, F. (1970). 'An algorithm for monotone regression with one or more independent variables', *Biometrika*, **57**, 263–71. [57]

Geer, S. van de (1987). 'A new approach to least-squares estimation with applications', *Ann. Statist.*, **15**, 587–602.

Gehrlein, W. V. (1979). 'A representation for quadrivariate normal positive orthant probabilities', *Comm. Statist. (B)*, **8**, 349–58.

Gillespie, M. J., and L. Fisher (1979). 'Confidence bands for Kaplan–Meier'. *Ann. Statist.*, **7**, 920–4.

Gine, E., and J. Zinn (1984). 'Some limit theorems for empirical processes', *Ann. Prob.*, **7**, 929–89.

Glaser, R. E. See Chao and Glaser (1978).

Gleser, L. J. See Derman, Gleser and Olkin (1973).

Goldstein, A. J., and J. B. Kruskal (1976). 'Least squares fitting by monotonic functions having integer values', *J. Amer. Statist. Assoc.*, **71**, 370–3.

Gong, G., and F. J. Samaniego (1981). 'Pseudo-maximum likelihood estimation: Theory and applications', *Ann. Statist.*, **8**, 861–9.

Good, I. J. (1963). 'Maximum entropy for hypothesis formulation, especially for multidimensional contingency tables', *Ann. Math. Statist.*, **34**, 911–34. [274]

Good, I. J., and R. A. Gaskins (1971). 'Nonparametric roughness penalties for probability densities', *Biometrika*, **58**, 255–77. [326]

Good, I. J., and R. A. Gaskins (1972). 'Global nonparametric estimation of probability densities', *Virginia J. Science*, **23**, 171–93. [326]

Goodman, L. A. (1979). 'Simple models for the analysis of association in cross classifications having ordered categories', *J. Amer. Statist. Assoc.*, **74**, 537–52. [292]

Goodman, L. A. (1981). 'Association models and canonical correlation in the analysis of cross classifications having ordered categories', *J. Amer. Statist. Assoc.*, **76**, 320–34. [292]

Goodman, L. A. (1985). 'The analysis of cross-classified data having ordered and/or unordered categories: Association models, correlation models and asymmetry models for contingency tables with or without missing entries', *Ann. Statist.*, **13**, 10–69. [279, 323]

Goodman, L. A., and S. Kotz (1980). 'Hazard rates based on isoprobability contours', In C. Taillie, G. P. Patil, and B. A. Baldessari (eds.), *Statistical Distributions in Scientific Work*, D. Reidel, Hingham, MA, pp. 239–308.

Gordon, R. D. (1978). 'On monotonicity of power of two-sample rank tests when testing the alternative G is stochastically larger than F', *Comm. Statist. (A)*, **7**, 535–41.

Gore, A. P. See Rao and Gore (1984a, 1984b, 1984c).

Gourieroux, C., A. Holly, and A. Monfort (1982). 'Likelihood ratio test, Wald test and Kuhn–Tucker test in linear model with inequality constraints on the regression parameters', *Econometrica*, **50**, 63–79. [220, 226, 322, 323]

Gove, W. R. See Leik and Gove (1969).

Govindarajulu, Z., and G. D. Gupta (1978). 'Tests for homogeneity of scale against ordered alternatives', *C. Prague A.*, **8**, 235–74. [216]

Govindarajulu, Z., and H. S. Haller (1968). 'c-sample tests for homogeneity against ordered alternatives' (abstract)', *Ann. Math. Statist.*, **39**, 1089.

Govindarajulu, Z., and H. S. Haller (1977). 'c-sample tests of homogeneity against ordered alternatives'. In R. Bartosyznski, E. Fidelis, and N. Klonecki (eds.), *Proc. Symp. to Honor Jerzy Neyman*, Polish Scientific Pub., Warsaw, pp. 91–102. [216]

Govindarajulu, Z., and S. H. Mansouri-Ghiassi (1986). 'On nonparametric tests for ordered alternatives in two-way layouts'. In R. L. Dykstra, T. Robertson, and F. T. Wright (eds.), *Advances in Order Restricted Statistical Inference*, Springer-Verlag, pp. 153–68. [216]

Greenberg, I. (1985). 'A one-sided goodness-of-fit test for multinomial populations', *J. Amer. Statist. Assoc.*, **80**, 558–62.

Grenander, U. (1956). 'On the theory of mortality measurement. Part II', *Skand. Akt.*, **39**, 125–53. [56, 367]

Groeneboom, P. (1983). 'The concave majorant of Brownian motion', *Ann Prob.*, **11**, 1028–36.

Groeneboom, P. (1984). 'Brownian motion with a parabolic drift and airy functions', Report Ms-R8413, Centre for Mathematics and Computer Science, Amsterdam.

Groeneboom, P. (1985). 'Estimating a monotone density', *Proc. Berkeley Conf. in Honor of Neyman and Keifer*, Vol. II, Wadsworth, Inc., pp. 529–55. [332]

Groeneboom, P., and R. Pyke (1983). 'Asymptotic normality of statistics based on the convex minorants of empirical distribution functions', *Ann. Prob.*, **11**, 328–45. [332]

Gross, S. T. (1981). 'On asymptotic power and efficiency of tests of independence in contingency tables with ordered classifications', *J. Amer. Statist. Assoc.*, **76**, 935–41.

Grossman, N. See Cunningham and Grossman (1971).

Grotzinger, S. J., and C. Witzgall (1984). 'Projections onto order simplexes', *Appl. Math. and Optimization*, **12**, 247–70.

Grove, D. M. (1980). 'A test of independence against a class of ordered alternatives in a $2 \times C$ contingency table', *J. Amer. Statist. Assoc.*, **75**, 454–9. [116, 161, 261]

Grove, D. M. (1984). 'Positive association in a two-way contingency table: Likelihood ratio tests', *Comm. Statist. (A)*, **13**, 931–45.

Grove, D. M. (1986). 'Positive association in a two way contingency table: A numerical study', *Comm. Statist.—Simula.*, **15**, 633–48.

Guess, F., M. Hollander, and F. Proschan (1986). 'Testing exponentiality versus a trend change in mean residual life', *Ann. Statist.*, **14**, 1388–98.

Guffey, J. M., and F. T. Wright (1986). 'Testing for trends in a nonhomogeneous Poisson process', Technical Report, Department of Mathematics and Statistics, University of Missouri-Rolla. [170, 173, 404, 405]

Guillier, C. L. See Aiyar, Guillier, and Albers (1979).

Gupta, A. K., and B. K. Kim (1975). 'Two-way classification for stochastically ordered populations' (abstract), *Bull. Inst. Math. Statist.*, **4**, 197.

Gupta, A. K., and T. T. Nguyen (1987). 'A test for homogeneity of multinomial populations against restricted stochastically ordered alternatives', Technical Report, Bowling Green State University, Department of Mathematics and Statistics.

Gupta, A. K., and V. K. Rohatgi (1980). 'On estimation of restricted mean', *J. Statist. Plann. Inf.*, **4**, 369–79.

Gupta, A. K. See also Adegbeye and Gupta (1985).

Gupta, G. D. See Govindarajulu and Gupta (1978).

Gupta, S. S. (1963). 'Bibliography on the multivariate normal integrals and related topics', *Ann. Math. Statist.*, **34**, 829–38. [75]

Gupta, S. S., and W. T. Huang (1984). 'On isotonic selection rules for binomial populations better than a standard', *King Savd. Univ. Lib. Riyadh*, **1984**, 89–112.

Gupta, S. S., and L. Y. Leu (1986). 'Isotonic procedures for selecting populations better than a standard for two-parameter exponential distributions'. In A. Basu (ed.), *Reliability and Quality Control*, North Holland, Amsterdam.

Gupta, S. S., and H. M. Yang (1984). 'Isotonic procedures for selecting populations better than a control under ordering prior'. In J. K. Ghosh and J. Roy (eds.), *Statist. Appl. and New Direct. Proc. of Ind. Stat. Inst. Golden Jubilee*, Indian Statist., Inst., Calcutta, pp. 279–312.

Gyorfi, L. See Devroye and Gyorfi (1985).

Haber, M. (1981). 'On the asymptotic power and relative efficiency of the frequency χ^2 test', *J. Statist. Plann. Inference*, **5**, 299–308.

Haberman, S. J. (1974). 'Log-linear models for frequency tables with ordered classifications', *Biometrics*, **30**, 589–600. [323]

Habibullah, M. See Ebrahimi and Habibullah (1983).

Haff, L. R. (1987). 'On correcting the eigenvalue distortion in the sample convariance matrix', *Bull. IMS*, **69**.

Hahn, G. J. (1970). 'A new tabulation of percentage points for the multivariate *t*-distribution and its use in constructing prediction intervals', *Proc. Amer. Statist. Assoc. Meeting*, Detroit. [192]

Hájek, J., and Z. Sidák (1967). *Theory of Rank Tests*, Academic Press, New York. [204, 206]

Hall, P. (1982). 'Asymptotic theory of Grenander's mode estimator', *Z. Wahrsch. keitstheorie verw. Gebiete* **60**, 315–34.

Haller, H. S. See Govindarajulu and Haller (1968, 1977).

Halperin, M., and K. K. G. Lan (1987). 'A two sample ordered alternative test for means and variances', *Comm. Statist.*, **16**, 1297–1313.

Halpern, J. W. See Miller and Halpern (1980).

Hanson, D. L., and G. Pledger (1976). 'Consistency in concave regression', *Ann. Statist.*, **4**, 1038–50. [243, 395]

Hanson, D. L., G. Pledger, and F. T. Wright (1973). 'On consistency in monotonic regression', *Ann. Statist.*, **1**, 401–21. [394]

Hanson, D. L. See also Brunk, Franck, Hanson and Hogg (1966).

Hardy, G. H., J. E. Littlewood, and G. Pólya (1952). *Inequalities*, 2nd ed., Cambridge University Press.

Hartigan, J. A. (1967a). 'Distribution of the residual sum of squares in fitting inequalities', *Biometrika*, **54**, 69–84.

Hartigan, J. A. (1967b). 'Representation of similarity matrices by trees', *J. Amer. Statist. Assoc.*, **62**, 1140–58.

Hartigan, J. A. (1987). 'Estimation of a convex density contour in two dimensions', *J. Amer. Statist. Assoc.*, **82**, 267–70.

Hartigan, J. A., and P. M. Hartigan (1985). 'The dip test of unimodality', *Ann. Statist.*, **13**, 70–84.

Hartigan, P. M. (1985). 'Computation of the dip statistic to test for unimodality', *Appl. Statist.*, **34**, 320–5.

Hartigan, P. M. See also Hartigan and Hartigan (1985).

Harville, D. A., and R. W. Mee (1984). 'A mixed-model procedure for analyzing ordered categorical data', *Biometrics*, **40**, 393–408.

Haseman, J. K. See Lin and Haseman (1976).

Hemelrijk, J. (1958). 'Distribution-free tests against trend and maximum likelihood estimates of ordered parameters', *Bull. Inst. Int. Statist.*, **36**, 15–25.

Hemelrijk, J. See also Eeden and Hemelrijk (1955a, 1955b).

Herbach, L. H. (1959). 'Properties of model II-type analysis of variance tests, A: optimum nature of the *F*-test for model II in the balanced case', *Ann. Math. Statist.*, **30**, 939–59.

Hernándes, N. See Marqués and Hernándes (1981).

Hestenes, M. R. (1975). *Optimization Theory*, John Wiley, New York. [270, 271]

Hettmansperger, T. P. (1975). 'Nonparametric inference for ordered alternatives in a randomized block design', *Psychometrika*, **40**, 53–62. [210, 215]

Hettmansperger, T. P. (1987). 'Tests for patterned alternatives in *k*-sample problems', *J. Amer. Statist. Assoc.*, **82**, 292–9.

Hettmansperger, T. P., and R. M. Norton (1987). 'Tests for patterned alternatives in *k*-sample problems', *J. Amer. Statist. Assoc.*, **82**, 292–9.

Hettmansperger, T. P. See also Tryon and Hettmansperger (1983).

Hewett, J. E., and R. Madsen (1985). 'A robust test for comparing three populations relative to an ordered alternative with bivariate data', *J. Statist. Comp. Simul.*, **21**, 195–212.

Hewitt, E., and K. Stromberg (1965). *Real and Abstract Analysis*, Springer-Verlag, Berlin.

Hildreth, C. (1954). 'Point estimates of ordinates of concave functions', *J. Amer. Statist. Assoc.*, **49**, 598–619. [243]

Hill, D. (1983). 'One way analysis of variance for an ordered contingency table', *N. Znd. St.*, **18**, 33–6.

Hille, E. (1959). *Analytic Function Theory*, Vol. I, Ginn and Company, New York. [8]

Hillier, G. H. (1986). 'Joint tests for zero restrictions on nonnegative regression coefficients', *Biometrika*, **73**, 657–69.

Hinkley, D. V. See Efron and Hinkley (1978).

Hirotsu, C. (1978). 'Ordered alternatives for interaction effects', *Biometrika*, **65**, 561–70.

Hirotsu, C. (1979). 'The cumulative chi-squares method testing an ordered alternative in one way analysis of variance model', *Rep. Statist. Appl. Res. Union Sci. Engr.*, **26**, 12–21.

Hirotsu, C. (1982). 'Use of cumulative efficient scores for testing ordered alternatives in discrete models', *Biometrika*, **69** 576–7.

Hirotsu, C. (1983). 'Defining the pattern of association in two-way contingency tables', *Biometrika*, **70**, 579–89.

Hirotsu, C. (1986). 'Cumulative chi-squared statistic as a tool for testing goodness of fit', *Biometrika*, **73**, 165–73.

Hirotsu, C. See also Takeuchi and Hirotsu (1982).

Ho, S. D. See Fang and Ho (1985).

Hoadley, B. (1971). 'The Southern Bell left-in stations study I: Statistical methods for data analysis', Bell Laboratories Memorandum. [57]

Hodges, J. L., Jr., and E. L. Lehmann (1956). 'The efficiency of some nonparametric competitors of the *t*-test', *Ann. Math. Statist.*, **27**, 324–35.

Hodson, F. R., P. H. A. Sneath, and J. E. Doran (1966). 'Some experiments in the numerical analysis of archaeological data', *Biometrika*, **53**, 311–24.

Hoel, D. See Portier and Hoel (1984).

Hogg, R. V. (1965). 'On models and hypotheses with restricted alternatives', *J. Amer. Statist. Assoc.*, **60**, 1153–62. [176, 227]

Hogg, R. V., and A. T. Craig (1978). *Introduction to Mathematical Statistics*, fourth edition, Macmillan, New York. [74]

Hogg, R. V. See also Brunk, Franck, Hanson and Hogg (1966).

Holland, P. W. See Bishop, Fienberg, and Holland (1975).

Hollander, M. (1967). 'Rank tests for randomized blocks when the alternatives have an *a-priori* ordering', *Ann. Math. Statist.*, **38**, 867–77. [210–212, 215]

Hollander, M., F. Proschan, and J. Sethuraman (1977). 'Functions decreasing in transposition and their applications in ranking problems', *Ann. Statist.*, **5**, 722–33.

Hollander, M. See also Chen, Hollander, and Langberg (1982); and Guess, Hollander, and Proschan (1986); Pirie and Hollander (1972, 1975).

Holly, A. See Gourieroux, Holly, and Monfort (1982).

Holm, S., and M. Frisen (1985). 'Nonparametric regression based upon simple curve characteristics', *Proc. ISI meeting at Amsterdam.*

Holt, J. N. See Wright and Holt (1985).

Hooper, J. H., and T. J. Santner (1979). 'Design of experiments for selection from ordered families of distributions', *Ann. Statist.*, **7**, 615–43.

Horan, C. B. (1969). 'Multidimensional scaling: Combining observations when individuals have different perceptual structures', *Psychometrika*, **34**, 139–65.

Houwelingon, van J. C. (1987). 'Monotone empirical Bayes distributions using maximum likelihood estimator of a decreasing density', *Ann. Statist.*, **15**, 875–9.

Houwelingon, J. C. van, and T. Stiijnen (1983). 'Monotone empirical Bayes estimators for the continuous one-parameter exponential family', *Statistica Neerlandica*, **37**, 29–43.

Huang, W. T. (1984). 'Nonparametric isotonic selection rules under a prior ordering'. In T. J. Santner and A. C. Tamhane (eds.), *Design of Experiments, Statistics*, Dekker, New York, pp. 95–111.

Huang, W. T. See also Gupta and Huang (1984).

Hudson, D. J. (1969). 'Least squares fitting of a polynomial constrained to be either nonnegative, nondecreasing, or convex', *J. R. Statist. Soc. (B)*, **31**, 113–8. [243]

Hudson, H. M. (1986). 'Evaluation of trends in middle ear disease among Australian aborigines', *Biometrics*, **42**, 159–69.

Huotari, R. See Darst and Huotari (1985a, 1985b).

Huskova, M. (1984). 'Hypothesis of symmetry'. In P. R. Krishnaiah and P. K. Sen (eds.), *Handbook of Statistics*, Vol. 4, *Nonparametric Methods*, North Holland.

Huynh, H. (1981). 'Testing the identity of trends under the restriction of monotonicity in repeated measured designs', *Psychometrika*, **40**, 295–305.

Inutsuka, M., S. Sasabuchi, and D. D. S. Kulatunga (1983). 'A multivariate version of isotonic regression', *Biometrika*, **70**, 465–72.

Itakura, H., and Y. Nishikawa (1981). 'A class of linear multiple regression techniques for ordered alternatives', *Mem. Fac. Engrg. Kyoto Univ.*, **42**, 335–48.

Jaeschke, D. (1979). 'The asymptotic distribution of the supremum of the standardized empirical distribution function on subintervals', *Ann. Statist.*, **7**, 108–15.

James, B. R., and K. L. James (1979). 'Analogues of *R*-estimates for quantal bioassay', Technical Report no. 49, Stanford University.

James, B. R., and K. L. James (1983). 'On the influence curve for quantal bioassay', *J. Statist. Planning and Inf.*, **8**, 331–45.

James, K. L. See James and James (1979, 1983).

Jaynes, E. T. (1957). 'Information theory and statistical mechanics', *Phys. Rev.*, **106**, 620–30. [274]

Jeffreys, H. (1948). *Theory of Probability*, 2nd ed., Clarendon Press, Oxford. [282]

Jessen, B. See Sparre Andersen and Jessen (1948).

Jewell, W. S. (1975). 'Isotonic optimization in tariff construction', *ASTIN Bull.*, **1975**, 175–203.

Jha, V. D. (1978). 'An asymptotic nonparametric test of symmetry of population', *Math. Forum*, **1978**, 22–6.

Johansen, S. (1967). 'The descriptive approach to the derivative of a set function with respect to a σ-lattice', *Pacific J. Math.*, **21**, 49–58. [387]

Johansen, S. See also Brunk and Johansen (1970).

Johnson, E. G., and R. D. Routledge (1985). 'The line transect method: a nonparametric estimator based on shape restrictions', *Biometrics*, **41**, 669–79.

Johnson, M. M. (1921). 'Generalized mean value function', M. S. thesis, University of Iowa.

Johnson, N. L., and Kotz, S. (1976). *Distributions in Statistics: Continuous Multivariate Distributions*, Wiley, New York. [192]

Johnson, R. A., and K. G. Mehrota (1971). 'Some c-sample nonparametric tests for ordered alternative', *J. Indian Statist. Assoc.*, **9**, 8–23.

Johnson, R. A., and K. G. Mehrota (1972). 'Nonparametric tests for ordered alternative in the bivariate case', *J. Multivariate Anal.*, **2**, 219–29.

Johnson, R. A. See also Bhattacharyya and Johnson (1970).

Jonckheere, A. R. (1954a). 'A distribution-free *k*-sample test against ordered alternatives', *Biometrika*, **41**, 133–45. [201–203, 206, 208, 211, 215, 227]

Jonckheere, A. R. (1954b). 'A test for significance of the relation between *m* rankings and *k* ranked categories', *Br. J. Statist. Psychol.*, **7**, 93–100. [209]

Judge, G. G., T. A. Yancey, M. E. Bock, and R. Bohrer (1984). 'The non-optimality of the inequality restricted estimator under squared error loss', *J. Econmtx.*, **25**, 165–177.

Judge, G. G. See also Yancey, Bohrer and Judge (1982) and Yancey, Judge and Bock (1981).

Kabe, D. G. See Scobey and Kabe (1980).

Kalbfleisch, J., and R. L. Prentice (1980). *The Statistical Analysis of Failure Time Data*, Wiley, New York.

Kallenberg, W. C. M. (1984). 'Testing statistical hypotheses: worked solutions', math. Centre, Amsterdam.

Kamenarov, G. N. (1984). 'Ridge estimates with linear constraints' (Bulgarian; English and Russian summaries), *Godishnik Vissh. Vchebn. Zaved. Prilozhna. Mat.*, **19**, 121–30.

Kantorovich, L. V., and J. T. Akilov (1964). In A. P. Robertson (ed.), *Functional Analysis in Normed Spaces* (trans. by D. E. Brown), Pergamon Press and Macmillan, New York. [386]

Kaplan, E. L., and P. Meier (1958). 'Non parametric estimation from incomplete observations', *J. Amer. Statist. Assoc.*, **53**, 457–81. [316]

Karlin, S. See Arrow, Karlin, and Scarf (1958).

Katz, M. W. (1963). 'Estimating ordered probabilities', *Ann. Math. Statist.*, **34**, 967–72.

Kendall, M. G. (1941). 'Proof of relations connected with the tetrachoric series and its generalization', *Biometrika*, **32**, 196–8.

Kendall, M. G. (1974). *Rank Correlation Methods*, Griffin, London. [291]

Kendall, M. G., and A. Stuart (1958). *The Advanced Theory of Statistics*, Vol. 1, Griffin, London. [72]

Kepner, J. L., and D. H. Robinson (1984). 'A distribution-free rank test for ordered alternatives in randomized complete block designs', *J. Amer. Statist. Assoc.*, **79**, 212–7.

Kersting, G. See Antillo, Kersting, and Zucchino (1982).

Kezouh, A. See Agresti, Chuang, and Kezouh (1987).

Kiefer, J. (1982). 'Optimum rates for nonparametric density and regression estimates under order restrictions', *Statist. Prob. C. R. R.*, **1982**, 419–28.

Kiefer, J., and J. Wolfowitz (1976). 'Asymptotically minimax estimation of concave and convex distribution functions', *Z. Wahrsch. verw. Gebiete*, **36**, 73–85. [330, 331]

Killeen, T. J. See Dietz and Killeen (1981).

Kim, B. K. See Gupta and Kim (1975).

Klahr, D. (1969). 'A Monte Carlo investigation of the statistical significance of Kruskal's nonmetric scaling procedure', *Psychometrika*, **34**, 319–33.

Klotz, J. H. See Bhattacharyya and Klotz (1966).

Knoll, R. L. See Stenson and Knoll (1969).

Kochar, S. C. (1985). 'Testing exponentiality against monotone failure rate average', *Comm. Statist. (A)*, **14**, 381–92.

Kochar, S. C., and J. V. Deshpande (1985). 'On exponential scores statistic for testing against positive aging', *Statist. Prob. Lett.*, **3**, 71–3.

Kochar, S. C., and R. P. Gupta (1986). 'A class of distribution-free tests for testing homogeneity of variances against ordered alternatives'. In R. L. Dykstra, T. Robertson and F. T. Wright (eds.), *Advances in Order Restricted Statistical Inference*, Springer-Verlag, pp. 169–83. [216]

Kochar, S. C. See also Deshpande and Kochar (1983).

Konkin, P. R. See May and Konkin (1970).

Kopocinski, B. (1985). 'Some characterizations of IFR distributions', *Bull. Polish Acad. Sci. Math.*, **32**, 57–69.

Korkhin, A. S. (1985). 'Some properties of estimates for regression parameters with a priori inequality constraints' (Russian), *Kibernetika*, **1985**, 106–14.

Kormann, U. (1985). 'Asymptotic properties of a sequential estimator of expectation in the presence of a trend', *Statistics*, **16**, 203–11.

Korn, E. L. (1982). 'Confidence bands for isotonic dose–response curves', *Appl. Statist.*, **31**, 59–63. [115]

Kotz, S. See Johnson and Kotz (1976).

Koul, H. L. (1976). 'A test for new better than used in expectation' (abstract), *Bull. Inst. Math. Statist.*, **5**, 133.

Koziol, J. A. (1983). 'Test for symmetry about an unkonwn value based on the empirical distribution function', *Comm. Statist. (A)*, **12**, 2823–46.

Krafft, D. (1981). 'Dual optimization problems in stochastics', *Jahresbar. Deutsch Math. Verein*, **83**, 97–105.

Kraft, C. H., and C. van Eeden (1964). 'Bayesian bioassay', *Ann. Math. Statist.*, **35**, 886–90. [400]

Krishnaiah, P. R., and P. V. Armitage, (1965). 'Percentage points of the multivariate *t*-distribution', Report ARL-65-199, Aerospace Research Laboratories, Wright Patterson Air Force Base, Ohio. [192]

Krishnaiah, P. R., and P. V. Armitage (1966). 'Tables for the multivariate *t*-distribution', *Sankhya (B)*, **28**, 31–56. [192]

Krishnamoorthy, K. See Dhariyal, Sharma, and Krishnamoorthy (1985).

Kroonenberg, P. M., and A. Verbuek (1987). Comments on 'A generalization of Fisher's exact test in $P \times Q$ contingency tables using more concordant relations' (by Nguyen), *Comm. Statist. -Simul. and Comp.*, **16**, 301–6.

Kruskal, J. B. (1964a). 'Multidimensional scaling by optimizing goodness of fit to a nonmetric hypothesis', *Psychometrika*, **29**, 1–27. [11, 57]

Kruskal, J. B. (1964b). 'Nonmetric multidimensional scaling: A numerical method', *Psychometrika*, **29**, 115–29. [57]

Kruskal, J. B. (1965). 'Analysis of factorial experiments by estimating monotone transformations of the data', *J. Roy. Statist. Soc. (B)*, **27**, 251–63. [56]

Kruskal, J. B. (1971). 'Monotone regression and continuity and differentiability properties', *Psychometrika*, **36**, 57–62.

Kruskal, J. B. See also Goldstein and Kruskal (1976).

Kruskal, W. H. (1952). 'A nonparametric test for the several sample problem', *Ann. Math. Statist.*, **23**, 525–40. [205, 206]

Kudô, A. (1963). 'A multivariate analogue of the one-sided test', *Biometrika*, **50**, 403–18. [112, 115, 117, 218–220, 228]

Kudô, A., and J. R. Choi (1975). 'A generalized multivariate analogue of the one-sided test', *Mem. Fac. Sci. Kyushu Univ. (A)*, **29**, 303–28. [217–219]

Kudô, A., and H. Fujisawa (1964). 'A bivariate normal test with two sided alternative', *Mem. Fac. Sci. Kyushu Univ. (A)*, **18**, 104–8.

Kudô, A., S. Sasabuchi, and J. R. Choi (1981). 'Test of equality of normal means in the absence of independent estimator of variance', *Comm. Statist. (A)*, **10**, 659–68. [114]

Kudô, A., and J. S. Yao (1982a). 'Tables for testing ordered alternatives in an analysis of variance without replications', *Biometrika*, **69**, 237–8.

Kudô, A., and J. S. Yao (1982b). 'An approximate formula for $P(l,k)$ used in order restricted inference, and the analysis of freezing dates of Lake Mendota', *Comm. Statist. (A)*, **11**, 607–14.

Kudô, A. See also Shi and Kudô (1986) and Shiraishi and Kudô (1981).

Kuenzi, N. J. (1969). 'An investigation of some problems concerning the additivity of conditional expectation with respect to a σ-lattice', Unpublished Ph.D. thesis, University of Iowa. [387]

Kukuk, C. See Wegman, Nour, and Kukuk (1980).

Kulatunga, D. D. S. (1984a). 'Convolutions of the probabilities $P(l,k)$ used in order restricted inference', *Mem. Fac. Sci. Kyushu Univ. (A)*, **38**, 9–15.

Kulatunga, D. D. S. (1984b). 'An approximate formula for the convolutions of the probabilities $P(l,k)$', *Mem. Fac. Sci. Kyushu Univ. (A)*, **38**, 177–82.

Kulatunga, D. D. S. (1985). 'Testing homogeneity of t normal means against ordered alternatives in 2 groups having uncommon and unknown variances', *Comm. Statist. (A)*, **13**, 701–11.

Kulatunga, D. D. S., and S. Sasabuchi (1984a). 'A test of simultaneous homogeneity against ordered alternatives in multifactorial design', *Comm. Statist. (A)*, **13**, 3173–83.

Kulatunga, D. D. S., and S. Sasabuchi (1984b). 'A test of homogeneity of mean vectors against multivariate isotonic alternatives', *Mem. Fac. Sci. Kyushu Univ. (A)*, **38**, 151–61.

Kulatunga, D. D. S. See also Inutsuka, Sasabuchi, and Kulatunga (1983) and Sasabuchi and Kulatunga (1985).

Kulkarni, R. V. See Bechhofer and Kulkarni (1984).

Kullback, S. (1959). *Information Theory and Statistics*, Wiley, New York. [274]

Kumazawa, Y. (1983). 'A class of test statistics for testing whether new is better than used', *Comm. Statist. (A)*, **12**, 311–21.

Kuo, L. (1983). 'Bayesian bioassay design', *Ann. Statist.*, **11**, 886–95.

Kuo, L. See also Cohen and Kuo (1985).

Kusum, K., and I. Bagai (1987). 'A new class of distribution-free tests for testing homogeneity of variance against ordered alternatives', *Bull. IMS*, **16**, 121.

Labelle, G. (1982). 'Asymptotic relative efficiency of certain tests of monotonic tendency', *Ann. Sci. Math. Quebec*, **6**, 151–61.

Lan, K. K. G. See Halperin and Lan (1987).

Landers, D., and L. Rogge (1976). 'Convergence of conditional p-means given a σ-lattice', *Ann. Prob.*, **4**, 147–50.

Landers, D., and L. Rogge (1979a). 'Characterization of P-predictors', *Proc. of Amer. Math. Soc.*, **76**, 307–9.

Landers, D., and L. Rogge (1979b). 'On projections and monotony in L_p-spaces', *Manuscripta Math.*, **26**, 363–9.

Landers, D., and L. Rogge (1979c). 'A functional relationship between different r-means for indicator functions', *Ann. Prob.*, **7**, 166–9.

Landers, D., and L. Rogge (1979d). 'Martingale representation in uniformly convex spaces', *Proc. Amer. Math. Soc.*, **75**, 108–10.

Landers, D., and L. Rogge (1980a). 'Best approximants in L_Φ-spaces', *Z. Wahrsch. verw. Gebiete*, **51**, 215–37.

Landers, D., and L. Rogge (1980b). 'A short proof for a.e. convergence of generalized conditional expectations', *Proc. Amer. Math. Soc.*, **79**, 471–3.

Landers, D., and L. Rogge (1981a). 'Characterization of conditional expectation operators for Banach-valued functions', *Proc. Amer. Math. Soc.*, **81**, 107–10.

Landers, D., and L. Rogge (1981b). 'Isotonic approximation in L_s', *J. Approx. Theory*, **31**, 199–223. [373]

Landers, D., and L. Rogge (1984). 'Natural α-quantiles and conditional α-quantiles', *Metrika*, **31**, 99–113. [373]

Langberg, N. A., R. V. Leon, and F. Proschan (1980). 'Characterization of nonparametric classes of life distributions', *Ann. Prob.*, **8**, 1163–70.

Langberg, N. A. See also Chen, Hollander, and Langberg (1982).

Langholz, B. See Breslow, Lubin, Marek, and Langholz (1983).

La Riccia, V. N. (1984). 'Optimal weights for general L^2 distance estimates', *Statist. Probab. Lett.*, **2**, 169–73.

Larntz, K. (1978). 'Small sample comparisons of chi-squared goodness-of-fit statistics', *J. Amer. Statist. Assoc.*, **73**, 253.

Laslett, G. M. (1982). 'The survival curve under monotone density constraints with applications to two-dimensional line-segment processes', *Biometrika*, **69**, 153–60.

Lawrence, M. J. (1975). 'Inequalities for S ordered distributions', *Ann. Statist.*, **3**, 413–28.

Leach, E. B., and M. C. Sholander (1978). 'Extended mean values', *Amer. Math. Monthly*, **85**, 84–90.

Leadbetter, M. R. See Watson and Leadbetter (1964).

LeCam, L. (1966). 'Likelihood functions for large numbers of independent observations'. In F. N. David (ed.), *Research Papers in Statistics* (Festschrift for J. Neyman), Wiley, New York, pp. 167–87.

Lee, C. I. C. (1981). 'The quadratic loss of isotonic regression under normality', *Ann. Statist.*, **9**, 686–8. [58]

Lee, C. I. C. (1983). 'The min-max algorithm and isotonic regression', *Ann. Statist.*, **11**, 467–77. [57]

Lee, C. I. C. (1984). 'Truncated Bayesian confidence region and its corresponding simultaneous confidence intervals in restricted normal models', Report, Mem. Univ. of Newfoundland. [407]

Lee, C. I. C. (1986). 'The quadratic loss of order restricted estimators for several treatment means and a control mean', Report, Mem. Univ. of Newfoundland. [237]

Lee, C. I. C. (1987a). 'Chi-square tests for and against an order restriction on multinomial parameters', *J. Amer. Statist. Assoc.*, **82**, 611–8. [263]

Lee, C. I. C. (1987b). 'Maximum likelihood estimation for stochastically ordered multinomial populations with fixed and random zeros', *Proc. Symp. in Statistics and Festschrift in Honor of V. M. Joshi*, 189–73.

Lee, C. I. C., T. Robertson, and F. T. Wright (1986). 'On testing marginal homogeneity against a stochastic ordering', University of Iowa Technical Report No. 121.

Lee, C. I. C. See also Raubertas, Lee, and Nordheim (1986).

Lee, S. Y. (1984). 'Multidimensional scaling models with inequality and equality constraints', *Comm. Statist. (B)*, **13**, 127–40.

Lee, Y. J. (1977). 'Maximin tests of randomness against ordered alternatives: The multinomial distribution case', *J. Amer. Statist. Assoc.*, **72**, 673–5. [239]

Lee, Y. J. (1980). 'Test of trend in count data: Multinomial distribution case', *J. Amer. Statist. Assoc.*, **75**, 1010–14.

Lee, Y. J. (1985). 'Test of monotone trend in *K* Poisson means', *J. Quality Technology*, **17**, 44–9.

Lee, Y. J., and D. A. Wolfe (1976). 'A distribution-free test for stochastic ordering'. *J. Amer. Statist. Assoc.*, **71**, 722–727.

Lehmann, E. L. (1959). *Testing Statistical Hypotheses*, Wiley, New York. [3, 34]

Lehmann, E. L. (1964). 'Asymptotically nonparametric inference in some linear models with one observation per cell', *Ann. Math. Statist.*, **35**, 726–34. [212]

Lehmann, E. L. (1966). 'Some concepts of dependence', *Ann. Math. Statist.*, **37**, 1137–53. [216, 293, 297]

Lehmann, E. L. See also Hodges and Lehmann (1956).

Leik, R. K., and W. R. Gove (1969). 'The conception and measurement of asymmetric monotonic relationships in sociology', *Amer. J. Sociology*, **74**, 696–709.

Lemke, J. H. (1983). 'Estimation and testing for two-way contingency tables within order restricted inference parameter spaces', Ph.D. thesis, Pennsylvania State University. [85, 263, 410]

Lemke, J. H., and R. L. Dykstra (1984). 'An algorithm for multinomial maximum likelihood estimation with multiple cone restrictions', Technical Report 84–1, Department of Preventive Medicine, University of Iowa. [279, 285]

Leon, R. V. See Langberg, Leon, and Proschan (1980).

Leu, L. Y. See Gupta and Leu (1986).

Leurgans, S. (1981). 'The Cauchy mean value property and linear functions of order statistics', *Ann. Statist.*, **9**, 905–8. [390]

Leurgans, S. (1982). 'Asymptotic distributions of slope-of-greatest-convex-minorant estimators', *Ann. Statist.*, **10**, 287–96. [396]

Leurgans, S. (1986). 'Isotonic *M*-estimation'. In R. L. Dykstra, T. Robertson and F. T. Wright (eds.), *Advances in Order Restricted Statistical Inference*, Springer-Verlag, pp. 48–68.

Levin, M. (1973). 'A new nonparametric procedure for estimating location parameters of quantal response curves' (abstract), *Bull. Inst. Math. Statist.*, **2**, 241.

Levin, V. I., and S. B. Steckin (1960). 'Inequalities', *Amer. Math. Soc. Translations (2)*, **14**, 1–29.

Lewis, P. A. W., and D. W. Robinson (1973). 'Tests for a montone trend in a modulated renewal process'. In F. Proschan and R. J. Serfling (eds.), *Reliability and Biometry: Proceedings of Conference at Florida State University*, pp. 163–82.

Lewis, P. A. W. See also Cox and Lewis (1966).

Li, H. C. (1984). 'A generalized problem of least squares', *Amer. Math. Monthly*, **91**, 135–7.

Liang, K. Y. (1984). 'The asymptotic efficiency of conditional likelihood methods', *Biometrika*, **71**, 305–13.

Liang, K. Y. See also Self and Liang (1987).

Lin, F. O., and J. K. Haseman (1976). 'A modified Jonckheere test against ordered alternatives when ties are present at a single extreme value', *Biom. Zeit.*, **18**, 623–32.

Lindsay, B. G. (1983). 'The geometry of mixture likelihood: A general theory', *Ann. Statist.*, **11**, 86–94.

Littell, R. C., and W. C. Louv (1981). 'Confidence regions based on methods of combining test statistics', *J. Amer. Statist. Assoc.*, **76**, 125.

Littlewood, J. E. See Hardy, Littlewood, and Pólya (1952).

Liu, Y. W., and R. H. Bcc (1984). 'Ridge regression: An application in criminology', *Comm. Statist. (A)*, **13**, 263–71.

Ljøgodt, H. See Amundsen and Ljøgodt (1979, 1981).

Lo, S. H. (1986). 'Estimation of a unimodal distribution function', *Ann. Statist.*, **14**, 1132–8.

Lockhart, R. A., and C. G. McLaren (1985). 'Asymptotic points for a test of symmetry about a specified median', *Biometrika*, **72**, 208–10.

Loh, N. Y. (1984). 'Bounds on AREs for restricted classes of distributions defined via tail-orderings', *Ann. Statist.*, **12**, 685–701.

Loh, W. Y. (1985). 'The Cauchy mean value property for *M*-estimates', *J. Statist. Plann. and Infer.*, **12**, 265–7.

Lombard, P. B., and H. D. Brunk (1983). 'Evaluating the relation of juice composition of mandarin oranges to percent acceptance of a taste panel', *Fd. Technol.*, **17**, 113–15. [56]

Louv, W. C. See Littell and Louv (1981).

Lubin, A. (1957). 'A rank order test for trend in correlated means' (abstract), *Ann. Math. Statist.*, **28**, 524.

Lubin, J. H. See Breslow, Lubin, Marek, and Langholz (1983).

Lucas, L. A., and F. T. Wright (1986). 'Testing for and against a stochastic ordering between multivariate multinomial populations', Technical Report, Dept. of Math and Statist., University of Missouri-Rolla. [263]

Lucas, L. A., and F. T. Wright (1987). 'The level probabilities for a unimodal ordering', Technical Report, Dept. of Math and Statist., University of Missouri-Rolla. [410]

Luenberger, D. G. (1969). *Optimization by Vector Space Methods*, Wiley, New York. [265, 266, 386]

Lütkepohl, H. (1983). 'Nonlinear least squares estimation under nonlinear equality constraints', *Econom. Lett.*, **13**, 191–6.

Luxemburg, W. A. J., and A. C. Zaanen (1971). *Riesz Spaces*, North Holland, London. [380]

Lyerly, S. B. (1952). 'The average Spearman rank correlation coefficient', *Psychometrika*, **17**, 421–8.

Lynch, J., G. Mimmack, and F. Proschan (1983). 'Dispersive ordering results', *Adv. Appl. Prob.*, **15**, 889–91.

Macdonald, R. R., and P. T. Smith (1983). 'Testing differences between means with ordered hypotheses', *Br. J. Math. Statist. Prob.*, **36**, 22–35.

McFadden, J. A. (1960). 'Two expansons for the quadravariate normal integral', *Biometrika*, **47**, 325–33. [75]

Mack, G. A., and D. A. Wolfe (1981). '*k*-sample rank tests for umbrella alternatives', *J. Amer. Statist. Assoc.*, **76**, 175. [203]

McLaren, C. G. See Lockhart and McLaren (1985).

McNichols, D. T., and W. J. Padgett (1982). 'Maximum likelihood estimation of unimodal and decreasing densities based on arbitrarily right censored data', *Comm. Statist. (A)*, **11**, 2259–70.

Madansky, A. See Proschan, Barlow, Madansky, and Scheuer (1968).

Madsen, R. See Dykstra and Madsen (1976), Dykstra, Madsen, and Fairbanks (1983) and Hewett and Madsen (1985).

Magee, L. (1987). 'Asymptotic risk comparisons of restricted and unrestricted maximum likelihood estimators', *Comm. Statist.–Theory and Math.*, **16**, 545–58.

Magel, R. (1983). 'A comparison between the Jonckheere and Chacko nonparametric test statistics when sample sizes are small', Technical Report, Department of Mathematics, North Dakota State University. [206]

Magel, R., and K. Magel (1986). 'A comparison of some nonparametric tests for the simple tree alternative', *Comm. Statist. Simul.*, **15**, 435–49. [206]

Magel, R., and F. T. Wright (1984a). 'Tests for and against trends in Poisson intensities', In Proceedings of Symposium on Inequalities in Statistics and Probability, pp. 244–50.

Magel, R., and F. T. Wright (1984b). 'Robust estimates of ordered parameters', *J. Statist. Comp. and Simulation*, **14**, 47–58. [58]

Magel, K. See Magel and Magel (1986).

Makowski, G. (1973). 'Laws of the iterated logarithm for permuted random variables and regression applications', *Ann. Statist.*, **1**, 872–7. [394]

Makowski, G. (1974). 'A rate of convergence of a distribution connected with integral regression function estimation', *Ann. Statist.*, **2**, 829–32. [394]

Makowski, G. (1975a). 'A rate of convergence for a nondecreasing regression estimator', *Acad. Sinica.*, **3**, 61–4.

Makowski, G. (1975b). 'Consistency and logarithmic convergence for iterated nondecreasing regression function estimators', Unpublished manuscript.

Makowski, G. (1976). 'Convergence rates for multidimensional integral regression function estimators', *Sankhya (A)*, **38**, 304–7.

Makowski, G. (1977). 'Consistency of an estimator of a doubly nondecreasing regression function', *Z. Wahr.*, **77**, 263–8.

Mäkuläinen, T., K. Schmidt, and G. P. H. Styan (1981). 'On the existence and uniqueness of the maximum likelihood estimate of a vector-valued parameter in fixed size samples', *Ann. Statist.*, **9**, 758–67.

Mallows, C. L. See Barton and Mallows (1961).

Malozemov, V. N. See Mitchell, Demyanov, and Malozemov (1974).

Manden, J. I. (1985). 'Combining independent one-sided noncentral and normal mean tests', *Ann. Statist.*, **13**, 1535–53.

Mann, H. B. (1945). 'Nonparametric tests against trend', *Econometrica*, **13**, 245–59.

Mansouri-Ghiassi, S. H. See Govindarajulu and Mansouri-Ghiassi (1986).

Mantel, N. (1983). 'Ordered alternatives and the $1\frac{1}{2}$ tail test', *Amer. Statistn.*, **37**, 225–8.

Marcus, R. (1976a). 'The powers of some tests of the equality of exponential means against an ordered alternatives', *J. Statist. Comp. Simul.*, **5**, 37–42. [174]

Marcus, R. (1976b). 'The powers of some tests of the equality of normal means against an ordered alternative', *Biometrika*, **63**, 177–83.

Marcus, R. (1978a). 'A note on analyzing ordered alternatives', *Psychometrika*, **43**, 133–40. [115]

Marcus, R. (1978b). 'Further results on simultaneous confidence bounds in normal models with restricted alternatives', *Comm. Statist. (A)*, **7**, 573–90. [115, 210]

Marcus, R. (1980a). 'Two stage procedures for testing homogeneity of means against ordered alternatives in one-way ANOVA with common unknown variance', *Austral. J. Statist.*, **22**, 250–9. [114]

Marcus, R. (1980b). 'A two stage procedure for testing homogeneity of means against ordered alternatives in analysis of variance with unequal variances', *Comm. Statist. (A)*, **9**, 949–63. [114]

Marcus, R. (1982). 'Some results on simultaneous confidence intervals for monotone contrasts in one-way ANOVA model', *Comm. Statist. (A)*, **11**, 615–22. [115]

Marcus, R., and E. Peritz (1976). 'Some simultaneous confidence bounds in normal models with restricted alternatives', *J. R. Statist. Soc. (B)*, **38**, 157–65. [115, 406]

Marcus, R., E. Peritz, and K. R. Gabriel (1976). 'On closed testing procedures with special reference to ordered analysis of variance', *Biometrika*, **63**, 655–60.

Marcus, R., and H. Talpaz (1983). 'On testing homogeneity of t normal means against ordered alternatives in r groups', *Comm. Statist. (A)*, **12**, 2897–902.

Marden, J. I. (1982a). 'Combining independent noncentral chi-squared or F-tests', *Ann. Statist.*, **10**, 266–77.

Marden, J. I. (1982b). 'Minimal complete classes of tests of hypotheses with multivariate one-sided alternative', *Ann. Statist.*, **10**, 962–70.

Marek, P. See Breslow, Lubin, Marek, and Langholz (1983).

Margolin, B. H. See Collings, Margolin, and Oehlert (1981) and Simpson and Margolin (1986).

Margosches, E. H. See Wolfe, Roi, and Margosches (1981).

Marqués, L. L., and N. Hernándes (1981). 'Application of the likelihood ratio test to the model of analysis of variance with a single classification factor with restrictions on the order of the parameters', *Investigación Oper.* (Spanish), **2–3**, 191–240.

Marshall, A. W. (1970). Discussion on Barlow and van Zwet's paper. In M. L. Puri (ed.), *Nonparametric Techniques in Statistical Inference*, Cambridge University Press, pp. 175–6.

Marshall, A. W., and F. Proschan (1965). 'Maximum Likelihood estimation for distributions with monotone failure rate', *Ann. Math. Statist.*, **36**, 69–77. [367]

Marshall, A. W., B. W. Walkup, and R. J. B. Wets (1967). 'Order-preserving functions: Applications to majorization and order statistics', *Pacific J. Math.*, **23**, 569–84. [103, 381]

Marshall, A. W. See also Barlow, Marshall, and Proschan (1963); Birnbaum, Esary, and Marshall (1966); and Boyles, Marshall, and Proschan (1985).

Mathow, T. (1985). 'Admissible linear estimation in singular linear models with respect to a restricted parameter set', *Comm. Statist. (A)*, **14**, 491–8.

Matthews, D. E., and V. T. Farewell (1982). 'On testing for a constant hazard against a change-point alternative', *Biometrics*, **38**, 463–8.

Mau, J. (1985). 'Isotonic regression trend tests for counting processes', *Statistics and Decision Supplement*, **2**, 187–91.

May, R. B., and P. R. Konkin (1970). 'A nonparametric test of an ordered hypothesis for k independent samples', *Educ. Psychol. Measur.*, **30**, 251–7.

Mee, R. W. See Harville and Mee (1984).

Meeks, L., and R. B. D'Agostino (1983). 'A model for comparisons with ordered categorical data', *Comm. Statist. (A)*, **12**, 895–906.

Meelis, E. (1974). 'Testing for homogeneity of k independent negative binomial distributed random variables', *J. Amer. Statist. Assoc.*, **69**, 181–6. [169]

Mehrota, K. G. See Johnson and Mehrota (1971, 1972).

Meilijson, I., and A. Nádas (1979). 'Convex majorization with an application to the length of critical paths', *J. Appl. Prob.*, **16**, 671–7.

Metcalf, F. T. See Diaz and Metcalf (1970).

Mhyre, J. M. See Saunders and Mhyre (1983).

Miles, R. E. (1959). 'The complete amalgamation into blocks, by weighted means, of a finite set of real numbers', *Biometrika*, **46**, 317–27. [56, 80, 114]

Miller, D., and A. Sofer (1986). 'Least-squares regression under convexity and higher order difference constraints with applications to software reliability'. In R. L. Dykstra, T. Robertson, and F. T. Wright (eds.), *Advances in Order Restricted Statistical Inference*, Springer-Verlag, pp. 91–124.

Miller, R. G., and J. W. Halpern (1980). 'Robust estimators for quantal bioassay', *Biometrika*, **67**, 103–10.

Mimmack, G. See Lynch, Mimmack, and Proschan (1983).

Minkowski, H. (1910). *Geometrie der Zahlen*, Teubner, Leipzig. [322]

Minkowski, H. (1911). *Theorie der Konvexen Körper, Insebesondere Begründung ihres Oberflächenbegriffs*, Gesammelte Abhandlungen II, Leipzig. [322]

Minty, G. J. (1965). 'A theorem on maximal monotonic sets in Hilbert space', *J. Math. Anal. Applic.*, **11**, 434–9. [387]

Mitchell, B. F., V. F. Demyanov, and V. N. Malozemov (1974). 'Finding the point of a polyhedron closest to the origin', *SIAM J. on Control.*, **12**, 19–26.

Mitra, A. See Alam and Mitra (1981).

Mittal, Y. (1981). 'Smoothing of samples for maxima', *Ann. Statist.*, **9**, 66–77.

Mittelhammer, R. C. (1985). 'Quadratic risk domination of restricted least squres estimators via Stein-ruled auxiliary constraints', *J. Econometrics*, **29**, 289–303.

Miyamoto, T. (1981). 'On testing hypotheses represented by inequalities. Part I', *Sci. Bull. Fac. Ed. Nagasaki Univ.*, **32**, 7–14.

Miyamoto, T. (1982). 'On testing hypotheses represented by inequalities. Part II', *Sci. Bull. Fac. Ed. Nagasaki Univ.*, **33**, 11–18.

Miyazaki, H. (1987). 'Critical values for testing partially ordered alternatives in the analysis of variance', *Comm. Statist.*, **16**, 1339–41.

Möhner, M. See Bunke and Möhner (1984).

Monahan, J. See Sedransk, Monahan and Chiu (1985).

Monfort, A. See Gourieroux, Holly, and Monfort (1982).

Moore, D. S. (1968). 'An elementary proof of asymptotic normality of linear functions of order statistics', *Ann. Math. Statist.*, **39**, 263–5.

Moore, D. S. (1977). 'Generalized inverses, Wald's method and the construction of χ^2 tests of fit', *J. Amer. Statist. Assoc.*, **72**, 131.

Moore, J. R. See Thompson and Moore (1963).

Moran, P. A. P. (1966a). 'Rank correlation and product moment correlation', *Biometrika*, **35**, 203–6.

Moran, P. A. P. (1966b). 'Estimation from inequalities', *Austral. J. Statist.*, **8**, 1–8.

Moran, P. A. P. (1972). 'Maximum likelihood estimators with known incidental parameters', *Proc. Cambridge Philos. Soc.*, **72**, 233–41.

Moran, P. A. P. (1984). 'The Monte Carlo evaluation of orthant probabilities for multivariate normal distributions', *Austral. J. Statist.*, **26**, 39–44.

Moran, P. A. P. (1985). 'Calculation of multivariate normal probabilities—another special case', *Austral. J. Statist.*, **27**, 60–7.

Moy, S. C. (1954). 'Characterization of conditional expectation as a transformation on function spaces', *Pacific J. Math.*, **4**, 47–63. [387]

Mudholkar, G. S., and S. George (1977). 'A note on the estimation of two ordered Poisson parameters', *Metrika*, **24**, 87–98.

Mudholkar, G. S., and P. Subbaiah (1987). 'Testing significance of orthant restricted multivariate mean', *Bull. IMS*, March **1987**, 65.

Mukerjee, H. (1981). 'A stochastic approximation by observations on a discrete lattice using isotonic regression', *Ann. Statist.*, **9**, 1020–5. [58]

Mukerjee, H. (1986). 'Monotone nonparametric regression', Technical Report No. 79, University of California at Davis. [403]

Mukerjee, H., and T. Robertson (1986). 'From order restriced inference to geometry: Characterization of the smallest equilateral triangle(s) circumscribing an arbitrary triangle', Technical Report No. 73, University of California at Davis.

Mukerjee, H., T. Robertson, and F. T. Wright (1986a). 'Multiple contrast tests for testing against a simple tree ordering'. In R. L. Dykstra, T. Robertson, and F. T. Wright (eds.), *Advances in Order Restricted Statistical Inference*, Springer-Verlag, pp. 203–30. [187, 196, 200]

Mukerjee, H., T. Robertson, and F. T. Wright (1986b). 'A probability inequality for elliptically contoured densities with applications in order restricted inference', *Ann. Statist.*, **14**, 1545–54. [93, 103, 104, 107, 378]

Mukerjee, H., T. Robertson, and F. T. Wright (1987). 'Comparison of several treatments with a control using multiple contrasts', *J. Amer. Statist. Assoc.*, **82**, 902–10. [161, 193, 194, 227]

Mukerjee, H., and J. L. Wang (1987). 'Maximum likelihood estimator of an increasing hazard rate under unidentifiable censoring', Technical Report No. 92, University of California at Davis. [343]

Murphy, A. H., and R. L. Winkler (1984). 'Probability forecasting in meteorology', *J. Amer. Statist. Assoc.*, **79**, 489–500.

Murray, G. D. (1983). 'Nonconvergence of the minimax order algorithm', *Biometrika*, **70**, 490–1. [57]

Murthy, V. K., and A. V. Gafarion (1970). 'Limiting distributions of some variations of the chi-square statistic', *Ann. Math. Statist.*, **41**, 188–94.

Mykytyn, S. W., and T. J. Santner (1981). 'Maximum likelihood estimation of the survival function based on censored data under hazard rate assumptions', *Comm. Statist. (A)*, **10**, 1369–87. [342]

Nádas, A. See Meilijson and Nádas (1979).

Nadler, J., and J. M. Eilbott (1967). 'Testing for monotone failure rate' (unpublished). [367, 368]

Nagahata, H. (1980). 'Tests against order restrictions in binomial populations', *Mem. Fac. Sci. Kyushu Univ. (A)*, **34**, 369–78.

Nair, K. A. (1982). 'An estimator of the common mean of two normal populations', *J. Statist. Plann. Inference*, **6**, 119–22.

Nair, V. N. (1986). 'On testing against ordered alternatives in analysis of variance models', *Biometrika*, **73**, 493–9.

Nair, V. N. (1987). 'Chi-squared-type tests for ordered alternatives in contingency tables', *J. Amer. Statist. Assoc.*, **82**, 283–91.

Nandi, S. B. (1980). 'On a generalized Stirling distribution of the first kind', *Pure Appl. Math. Sci.*, **11**, 91–5.

Nashed, M. Z. (1968). 'A decomposition relative to convex sets', *Proc. Amer. Math. Soc.*, **19**, 782–8. [387]

Nelson, L. A. See Anderson and Nelson (1984).

Nelson, L. S. (1977). 'Tables for testing ordered alternatives in an analysis of variance', *Biometrika*, **64**, 335–8.

Nelson, P. L., and L. E. Toothaker (1975). 'An empirical study of Jonckheere's nonparametric test of ordered alternatives', *Br. J. Math. Statist. Prob.*, **28**, 167–76. [201]

Neumann, J. von (1940). 'On rings of operators, III', *Ann. Math.*, **41**, 94–161.

Neveu, J. (1965). *Mathematical Foundations of the Calculus of Probability* (translated by Amiel Feinstein), Holden-Day, San Francisco.

Nguyen, T. T., and A. R. Sampson (1983). 'Testing for positive quadrant dependence in ordinal contingency tables', Technical Report No. 83–10, Center for Multivariate Analysis, University of Pittsburgh.

Nguyen, T. T. See also Gupta and Nguyen (1987).

Nikitin, Y. Y. (1982). 'Asymptotic efficiency in the sense of Bahadur of integral tests for symmetry', *Zap. Nauchn. Sem. Leningrad., Otdel. Math. Inst.*, 181–94, 241–7.

Nishikawa, Y. See Itakura and Nishikawa (1981).

Nomakuchi, K. (1983). 'The likelihood ratio test of normal mean with hypothesis determined by a convex polyhedral cone and the monotonicity of its power function', *Mem. Fac. Sci. Kyushu Univ. (A)*, **37**, 195–204.

Nordheim, E. V. See Raubertas, Lee, and Nordheim (1986).

Norton, R. M. See Hettmansperger and Norton (1987).

Nour, E. S. See Wegman, Nour, and Kukuk (1980).

Nüesch, P. E. (1964). 'Multivariate tests of location for restricted alternatives', Doctoral dissertation, Swiss Federal Institute of Technology, Juris-Verlag, Zürich. [117, 218, 220, 228]

Nüesch, P. E. (1966). 'On the problem of testing location in multivariate populations for restricted alternatives', *Ann. Math. Statist.*, **37**, 113–19. [218, 220, 228]

Nüesch, P. E. (1970). 'Estimation of monotone parameters and the Kuhn–Tucker conditions' (abstract), *Ann. Math. Statist.*, **41**, 1800.

Odeh, R. E. (1971). 'On Jonckheere's k-sample test against ordered alternatives', *Technometrics*, **13**, 912–18. [20]

Odeh, R. E. (1972). 'On the power of Jonckheere's k-sample test against ordered alternatives', *Biometrika*, **59**, 467–71. [201]

Odeh, R. E. (1977). 'The exact distribution of Page's L-statistic in the two-way layout', *Comm. Statist. (B)*, **6**, 49–62. [209]

Oehlert, G. W. See Collings, Margolin, and Oehlert (1981).

Olkin, I. See Derman, Gleser and Olkin (1973).

Olson, M. P. (1965). 'A characterization of conditional probability', *Pacific J. Math.*, **15**, 971–83. [387]

Oosterhoff, J. (1985). 'The choice of cells in chi-square tests', *Statist. Neerlandica*, **39**, 115–28.

Oosterhoff, J. See Zwet and Oosterhoff (1967).

Owen, D. B. (1962). *Handbook of Statistical Tables*, Addison-Wesley, Reading, Mass. [173]

Owen, D. B. See also Chou, Arthur, Rosenstein, and Owen (1984).

Pace, L. (1984). 'Consistency of two tests for exponentiality against alternatives of monotone failure rate'. *Metron.*, **42**, 195–206.

Padgett, W. J. See McNichols and Padgett (1982).

Padmanabhan, A. R., M. L. Puri, and A. K. Saleh (1981). 'A nonparametric test for equality against ordered alternatives in the case of skewed data, with a biomedical application'. In *Statistics and Related Topics*, North-Holland, pp. 279–83.

Page, E. B. (1963). 'Order hypotheses for multiple treatments: A significance test for linear ranks', *J. Amer. Statist. Assoc.*, **58**, 216–30. [210, 211, 215]

Parsons, V. L. (1975). 'Distribution theory of the isotonic estimators', *Ph.D. Dissertation*, University of Iowa. [396]

Parsons, V. L. (1979a). 'A nonparametric test for trend', Technical Report, Department of Mathematics, University of Cincinnati. [205]

Parsons, V. L. (1979b). 'A note on the exact distribution of a nonparametric test statistic for ordered alternatives', *Ann. Statist.*, **7**, 454–8. [205]

Parsons, V. L. (1981). 'Small sample distribution for a nonparametric test for trend', *Comm. Statist. (B)*, **10**, 289–302. [205]

Parzen, E. (1962a). 'On estimation of a probability density and mode', *Ann. Math. Statist.*, **33**, 1065–76. [334, 367]

Patefield, W. M. (1978). 'Determining the precision of estimators in problems of constrained likelihood inference', *Sankhya (B)*, **40**, 316–28.

Patefield, W. M. (1982). 'Exact tests for trends in ordered contingency tables', *J. R. Statist. Soc. (C)*, **31**, 32–43. [262]

Patel, G. P. (1962). 'On homogeneity and combined estimation for the generalized power series distribution and certain applications', *Biometrics*, **18**, 365–74. [169]

Patnaik, P. B. (1949). 'The non-central χ^2-and F-distributions and their application', *Biometrika*, **36**, 202–32.

Pearl, D. K. See Fairley, Pearl and Verducci (1987).

Pearson, K. (1909). 'On a new method of determining correlation between a character A, and a character B, of which only the percentage of cases where in B exceeds (or falls short of) a given intensity is recorded for each grade of A', *Biometrika*, **7**, 96–105.

Pearson, K. (1910). 'On a new method of determining correlation, when one variable is given by an alternative and the other by multiple categories', *Biometrika*, **7**, 248–57.

Pearson, K. (1934). *Tables of the Incomplete Γ-Function*, Cambridge University Press.

Pendergast, J., and J. Broffitt (1985). 'Robust estimation in growth curve methods', *Comm. Statist. (A)*, **14**, 1919–39.

Peritz, E. See Marcus and Peritz (1976) and Marcus, Peritz, and Gabriel (1976).

Perlman, M. D. (1969). 'One-sided problems in multivariate analysis', *Ann. Math. Statist.*, **40**, 549–67 (for corrections to the above paper, see *Ann. Math. Statist.*, **42**, 1777). [101, 114, 141, 218, 222, 223, 226, 228]

Perron, O. (1910). 'Über das Verhalten der Integrale linearer Differenzegleichungen im Unendlichen', *Jber. dt. MatVerein*, **19**, 129–37.

Pettitt, A. N. See Acar and Pettitt (1985).

Pfanzagl, J. (1960a). 'Tests und Konfidenzintervalle für exponentielle Verteilungen und deren Anwendung auf einige diskrete Verteilungen', *Metrika*, **3**, 1–25.

Pfanzagl, J. (1960b). 'Über lokal optimale Rang-Tests', *Metrika*, **3**, 143–50.

Pfanzagl, J. (1967). 'Characterizations of conditional expectations', *Ann. Math. Statist.*, **38**, 415–21. [387]

Philips, P. R. See Anderson and Philips (1981).

Pickett, J. R. See Chen and Pickett (1984).

Pillers, C., T. Robertson, and F. T. Wright (1984). 'A FORTRAN program for the level probabilities of order restricted inference', *J. R. Statist. Soc. (C)*, **33**, 115–19. [135]

Pincus, R. (1975). 'Testing linear hypotheses under restricted alternatives', *Math. Operationsforsch. u. Statist.*, **6**, 733–51. [111, 112, 114]

Pirie, W. R. (1974). 'Comparing rank tests for ordered alternatives in randomized blocks', *Ann. Statist.*, **2**, 374–82. [215]

Pirie, W. R. (1983). 'Jonckheere tests for ordered alternatives', *Encyclopedia Statist. Sci.*, **4**, 315–18.

Pirie, W. R. (1985). 'Page test for ordered alternatives', *Encyclopedia Statist. Sci.*, **6**, 553–5.

Pirie, W. R., and M. Hollander (1972). 'A distribution-free normal scores test for ordered alternatives in the randomized block design', *J. Amer. Statist. Assoc.*, **67**, 855–7. [210]

Pirie, W. R., and M. Hollander (1975). 'Note on a Tukey test for ordered alternatives', *Ann. Statist. Math.*, **27**, 521–4.

Plackett, R. L. (1954). 'A reduction formula for normal multivariate integrals', *Biometrika*, **41**, 351–60.

Pledger, G. (1976). 'Consistency in integral regression estimation with a triangular array of observation points', *Ann. Statist.*, **4**, 234–6.

Pledger, G. See also Hanson and Pledger (1976) and Hanson, Pledger, and Wright (1973).

Poirer, D. J. (1973). 'Piecewise regression using cubic splines', *J. Amer. Statist. Assoc.*, **68**, 515–24.

Pollak, M. See Rinott and Pollak (1980).

Pólya, G. See Hardy, Littlewood, and Pólya (1952).

Ponnapalli, R. (1983). 'On multinomial parametric estimation', *Comm. Statist.*, **12**, 47–72.

Poon, A. H. (1980). 'A Monte Carlo Study of the power of some k-sample tests for ordered binomial alternatives', *J. Statist. Comp. Simul.*, **11**, 251–9. [168]

Portier, C., and D. Hoel (1984). 'Type I error of trend tests in proportions and in the design of cancer screens', *Comm. Statist.*, **13**, 1–13.

Potter, R. W., and G. W. Sturm (1981). 'The power of Jonckheere's test', *Amer. Statist.*, **35**, 249–50. [201]

Potthoff, R. F., and S. N. Roy (1964). 'A generalized multivariate analysis of variance model useful especially for growth curve problems', *Biometrika*, **51**, 313–26. [2]

Prakasa Rao, B. L. S. (1966). 'Asymptotic distributions in some non-regular statistical problems', Technical Report No. 9, Department of Statistics and Probability, Michigan State University. [367]

Prakasa Rao, B. L. S. (1969). 'Estimation of a unimodal density', *Sankhya (A)*, **31**, 23–36. [331, 334, 367, 395]

Prakasa Rao, B. L. S. (1970). 'Estimation for distributions with monotone failure rate', *Ann. Math. Statist.*, **41**, 507–19. [358]

Prakasa Rao, B. L. S. (1983). *Nonparametric Functional Estimation*, Academic Press. [325]

Prentice, R. L. See Kalbfleisch and Prentice (1980).

Proctor, C. H. (1971). 'Processing survey data with ordered alternative hypotheses', *J. Amer. Statist. Assoc.*, **66**, 720–4. [115]

Proschan, F. (1963). 'Theoretical explanation of observed decreasing failure rate', *Technometrics*, **5**, 375–83. [368]

Proschan, F. (1966). 'Reliability estimation under plausible assumptions', *Bull. Inst. int. Statist.*, **41**, 143–55.

Proschan, F., and R. Pyke (1967). 'Tests for monotone failure rate', *Proc. 5th Berkeley Symp. Math. Statist. Probab.*, **III**, 293–312. [363, 367]

Proschan, F., and Sethuraman, J. (1977). 'Schur function in statistics I. The preservation theorem', *Ann. Statist.*, **5**, 256–62. [381]

Proschan, F., R. E. Barlow, A. Madansky, and E. M. Schcucr (1968). 'Statistical estimation procedures for the "burn-in" process', *Technometrics*, **10**, 51–62.

Proschan, F. See also Barlow, Marshall, and Proschan (1963); Barlow and Proschan (1965, 1966, 1967, 1969, 1975); Barlow, Proschan, and Scheuer (1971); Boyles, Marshall, and Proschan (1985); Bray, Crawford, and Proschan (1967); Guess, Hollander and Proschan (1986); Hollander, Proschan, and Sethuraman (1977); Langberg, Leon, and Proschan (1980); Lynch, Mimmack, and Proschan (1983); and Marshall and Proschan (1965).

Pukelsheim, F. (1984). 'A note on nonparametric trend conformity', *Ann. Statist.*, **12**, 775–7.

Puri, M. L. (1965). 'Some distribution-free k-sample rank tests of homogeneity against ordered alternatives', *Comm. Pure Appl. Math.*, **18**, 51–63. [202, 206–208]

Puri, M. L., and P. K. Sen (1968). 'On Chernoff–Savage tests for ordered alternatives in randomized blocks', *Ann. Math. Statist.*, **39**, 967–72. [212, 215]

Puri, M. L. See also Padmanabhan, Puri, and Saleh (1981).

Pyke, R. See Brunk (1970); Groeneboom and Pyke (1983); and Proschan and Pyke (1967).

Quade, D. See Salama and Quade (1981, 1984).

Randles, R. H., and D. A. Wolfe (1979). *Introduction to the Theory of Nonparametric Statistics*, Wiley, New York. [201, 203]

Rao, C. R. (1948). 'Large sample tests of statistical hypotheses concerning several parameters with application to problems of estimation', *Proc. Cambridge Philosophical Soc.*, **44**, 50–7. [311, 318–322]

Rao, C. R. (1965). *Linear Statistical Inference and its Applications*, Wiley, New York. [274, 387]

Rao, C. R. (1973). *Linear Statistical Inference and Its Applications*, 2nd ed., Wiley, New York [318, 320]

Rao, K. S. M. (1982). 'Nonparametric tests for homogeneity of scale against ordered alternatives', *Ann. Inst. Statist. Math.*, **34**, 327–34. [216]

Rao, K. S. M., and A. P. Gore (1984a). 'Testing concurrence and parallelism of several sample regressions against ordered alternatives', *Math. Operationsfursch. Statist. Ser. Statist.*, **15**, 43–50.

Rao, K. S. M., and A. P. Gore (1984b). 'Testing against ordered alternatives in a one-way layout', *Biometrics*, **26**, 25–32.

Rao, K. S. M., and A. P. Gore (1984c). 'Some nonparametric tests for ordered alternatives in randomized block designs', *J. Indian Statist. Assoc.*, **22**, 71–80.

Raubertas, R. F., C. I. C. Lee, and E. V. Nordheim (1986). 'Hypothesis tests for normal means constrained by linear inequalities', *Comm. Statist.—Theor. Meth.*, **15**, 2809–33. [103, 109, 111, 115, 263]

Read, T. R. C. (1984). 'Small-sample comparisons for the power divergence goodness-of-fit statistics', *J. Amer. Statist. Assoc.*, **79**, 929–35.

Read, T. R. C. See Cressie and Read (1984).

Reid, W. T. (1968). 'A simple optimal control problem involving approximation by monotone functions', *J. Optimization Theory and Applications*, **2**, 365–77.

Reid, W. T. See also Ayer, Brunk, Ewing, Reid, and Silverman (1955).

Rejtö, L. See Földes and Rejtö (1981).

Resnick, S. I. See Brown and Resnick (1984).

Riley, C. (1981). 'A note on strict convexity', *Amer. Math. Monthly*, **88**, 198.

Rinott, Y., and M. Pollak (1980). 'A stochastic ordering induced by a concept of positive dependence and monotonicity of asymptotic test sizes', *Ann. Statist.*, **8**, 190–8.

Robertson, T. (1965). 'A note on the reciprocal of the conditional expectation of a positive random variable', *Ann. Math. Statist.*, **36**, 1302–5. [56, 58, 263]

Robertson, T. (1966a). 'Conditional expectation given a σ-lattice and estimation of restricted densities', Ph.D. thesis, University of Missouri-Columbia. [367]

Robertson, T. (1966b). 'A representation for conditional expectations given σ-lattices', *Ann. Math. Statist.*, **37**, 1279–83.

Robertson, T. (1967). 'On estimating a density which is measurable with respect to a σ-lattice', *Ann. Math. Statist.*, **38**, 482–93. [334, 367]

Robertson, T. (1968). 'A smoothing property for conditional expectations given σ-lattices', *Amer. Math. Monthly*, **75**, 515–18.

Robertson, T. (1978). 'Testing for and against an order restriction on multinomial parameters', *J. Amer. Statist. Assoc.*, **73**, 197–202. [263, 405]

Robertson, T. (1985). 'Monotone relationships'. In *Encyclopedia of Statistical Sciences*, Vol. 5, John Wiley and Sons, pp. 609–11. [392]

Robertson, T. (1986). 'On testing symmetry and unimodality'. In R. L. Dykstra, T. Robertson, and F. T. Wright (eds.), *Advances in Order Restricted Statistical Inference*, Springer-Verlag, pp. 231–48. [259, 263]

Robertson, T., and J. D. Cryer (1974). 'An iterative procedure for estimating the mode', *J. Amer. Statist. Assoc.*, **69**, 1012–16. [336]

Robertson, T., and P. Waltman (1968). 'On estimating monotone parameters', *Ann. Math. Statist.*, **39**, 1030–9. [389, 394]

Robertson, T., and G. Warrack (1985). 'An application of order restricted inference methodology to a problem in psychiatry', *Psychometrika*, **50**, 421–7. [259, 264]

Robertson, T., and E. J. Wegman (1978). 'Likelihood ratio tests for order restrictions in exponential families', *Ann. Statist.*, **6**, 485–505. [68, 114, 190, 226, 227]

Robertson, T., and F. T. Wright (1973). 'Multiple isotonic median regression', *Ann. Statist.*, **1**, 422–32. [390, 394]

Robertson, T., and F. T. Wright (1974a). 'On the maximum likelihood estimation of stochastically ordered random variables', *Ann. Statist.*, **2**, 528–34.

Robertson, T., and F. T. Wright (1974b). 'A norm reducing property for isotonized Cauchy mean value functions', *Ann. Statist.*, **2**, 1302–7. [58, 390, 392]

Robertson, T., and F. T. Wright (1975). 'Consistency in generalized isotonic regression', *Ann. Statist.*, **3**, 350–62. [390, 395]

Robertson, T., and F. T. Wright (1980a). 'Algorithms in order restricted statistical inference and the Cauchy mean value property', *Ann. Statist.*, **8**, 645–51. [390, 395]

Robertson, T., and F. T. Wright (1980b). 'Statistical inferences for ordered parameters: A personal view of isotonic regression since the work by Barlow, Bartholomew, Bremner, and Brunk', *Proc. Symp. Appl. Math.*, **23**, 55–71. [390]

Robertson, T., and F. T. Wright (1981). 'Likelihood ratio tests for and against stochastic ordering between multinomial populations', *Ann. Statist.*, **9**, 1248–57. [250, 253, 263, 323]

Robertson, T., and F. T. Wright (1982a). 'On measuring the conformity of a parameter set to a trend, with applications', *Ann. Statist.*, **10**, 1234–45. [381]

Robertson, T., and F. T. Wright (1982b). 'Testing for ordered alternatives with increased precision in one of the samples', *Biometrika*, **69**, 579–86. [128, 162]

Robertson, T., and F. T. Wright (1982c). 'Bounds on mixtures of distribution arising in order restricted inference', *Ann. Statist.*, **10**, 302–6.

Robertson, T., and F. T. Wright (1983). 'On approximation of the level probabilities and associated distributions in order restricted inference', *Biometrika*, **70**, 597–606. [132, 136, 162]

Robertson, T., and F. T. Wright (1984). 'A measure of the conformity of a parameter set to a trend: The partially ordered case'. In Y. L. Tong (ed.), *Inequalities, Statistics Probability*, pp. 244–50. [381]

Robertson, T., and F. T. Wright (1985). 'One-sided comparisons for treatments with a control', *Can. J. Statist.*, **13**, 109–22. [93, 131, 161, 162]

Robertson, T. See also Bril, Dykstra, Pillers, and Robertson (1984); Cryer and Robertson (1975); Cryer, Robertson, Wright, and Casady (1972); Dykstra and Robertson (1982a, 1982b, 1983); Lee, Robertson, and Wright (1986); Mukerjee and Robertson (1986); Mukerjee, Robertson, and Wright (1986a, 1986b, 1987); Pillers, Robertson, and Wright (1984); and Warrack and Robertson (1984).

Robinson, D. H. See Kepner and Robinson (1984).

Robinson, D. W. See Lewis and Robinson (1973).

Rockafellar, R. T. (1970). *Convex Analysis*, Princeton University Press. [20, 30, 267, 269–271, 303, 304, 322]

Rockafellar, R. T. (1972). 'Monotropic programming: Descent algorithms and duality'. In *Nonlinear Programming*, Vol. 4, Academic Press, New York, pp. 327–66.

Rockafellar, R. T., and R. J. B. Wets (1982). 'On the interchange of subdifferentiation and conditional expectation for convex functionals', *Stochastics*, 7, 173–82.

Rogge, L. See Landers and Rogge (1976, 1979a, 1979b, 1979c, 1979d, 1980a, 1980b, 1981a, 1981b, 1984).

Rohatgi, V. K. See Gupta and Rohatgi (1980).

Roi, L. D. See Wolfe, Roi, and Margosches (1981).

Rosenstein, R. B. See Chou, Arthur, Rosenstein, and Owen (1984).

Rosenthal, R., and D. B. Rubin (1984). 'Multiple contrasts and ordered Bonferroni procedures', *J. Educational Psychol.*, **76**, 1028–34.

Rosner, N. (1961). 'System analysis: Non-linear estimation techniques', In *Proceedings of the Seventh National Symposium on Reliability and Quality Control.* pp. 203–7.

Rota, G. C. (1960). 'On the representations of averaging operators', *Rendiconti Sem. Mat. Univ. Padova.*, **30**, 52–64. [387]

Roth, A. J. (1983). 'Robust trend tests derived and simulated analogs of the Welch and Brown–Forsythe tests', *J. Amer. Statist. Assoc.*, **78**, 972–80.

Roth, G. L., and W. W. Daniel (1978). 'Critical values for Chacko's homogeneity test against ordered alternatives', *Ed. Ps. Meas.*, **38**, 889–92.

Rothe, G. (1984). 'An efficiency property of linear nonparametric tests on trend', *Statist. Prob. Lett.*, **2**, 197–201.

Routledge, R. D. See Johnson and Routledge (1985).

Roy, S. N. See Potthoff and Roy (1964).

Ruben, H. (1954). 'On the moments of order statistics in samples from normal populations', *Biometrika*, **41**, 200–27. [83]

Rubin, D. B. See Rosenthal and Rubin (1984).

Rüschendorf, L. (1981a). 'Ordering distributions and rearrangement of functions', *Ann. Prob.*, **9**, 276–83.

Rüschendorf, L. (1981b). 'Stochastically ordered distributions and monotonicity of the OC-function of sequential probability ratio tests', *Math. Operationsforsch. Statist. Ser. Statist.*, **12**, 327–38.

Rüschendorf, L. (1983). 'Essential completeness of monotone tests and estimators', *Math. Operationsforsch. Statist. Ser. Statist.*, **13**, 243–50.

Rust, S. W. See Fligner and Rust (1982).

Ryzin, J. van. See Susarla and Ryzin (1980).

Sackrowitz, H. B. (1970). 'Estimation for ordered parameter sequences: The discrete case', *Ann. Math. Statist.*, **41**, 609–20. [58]

Sackrowitz, H. B. (1982). 'Procedures for improving the MLE for ordered binomial parameters', *J. Statist. Plann. Inf.*, **6**, 287–96. [58]

Sackrowitz, H. B., and W. E. Strawderman (1974). 'On the admissability of the MLE for ordered binomial parameters', *Ann. Statist.*, **2**, 822–8. [58, 109]

Sackrowitz, H. B. See also Cohen and Sackrowitz (1970).

Sager, T. W. (1975). 'Consistency in nonparametric estimation of the mode', *Ann. Statist.*, **3**, 698–706.

Sager, T. W. (1978). 'Estimation of a multivariate mode', *Ann. Statist.*, **6**, 802–12.

Sager, T. W. (1979). 'An iterative procedure for estimating a multivariate mode and isopleth', *J. Amer. Statist. Assoc.*, **74**, 329–39.

Sager, T. W. (1982). 'Nonparametric maximum likelihood estimation of spatial patterns', *Ann. Statist.*, **11**, 1125–36.

Sager, T. W. (1983). 'Estimating modes and isopleths', *Comm. Statist. (A)*, **12**, 529–58. [337]

Sager, T. W. (1985). 'Searching for and exploiting structure in higher dimensional density estimation', Technical Report No. 18, Center for Statistical Sciences, University of Texas at Austin.

Sager, T. W. (1986a). 'An application of isotonic regression to multivariate density estimation'. In R. L. Dykstra, T. Robertson, and F. T. Wright (eds.), *Advances in Order Restricted Statistical Inference*, Springer-Verlag, pp. 69–90. [337]

Sager, T. W. (1986b). 'Dimensionality reduction in density estimation'. In E. J. Wegman and D. J. Depriest (eds.), *Statistical Image Processing and Graphics*, Vol. 72, Marcel Dekker, Inc., pp. 307–19. [337]

Sager, T. W., and Thisted (1982). 'Maximum likelihood estimation of isotonic modal regression', *Ann. Statist.*, **10**, 690–707.

Sahab, S. See Darst and Sahab (1983).

Salama, I. A., and D. Quade (1981). 'Using weighted rankings to test against ordered alternatives in complete blocks', *Comm. Statist. (A)*, **10**, 385–99. [213]

Salama, I. A., and D. Quade (1984). 'Some simple statistics to test against ordered alternatives in block designs', *Biometrics*, **26**, 867–82.

Saleh, A. K. See Padmanabhan, Puri, and Saleh (1981).

Samaniego, F. J. See Boyles and Samaniego (1984) and Gong and Samaniego (1981).

Sampson, A. R., and L. R. Whitaker (1987). 'Estimation of multivariate distributions under stochastic ordering', Technical Report, University of Pittsburg.

Sampson, A. R. See also Nguyen and Sampson (1983).

Sandford, M. D. (1985). 'Nonparametric one-sided confidence intervals for an unknown distribution function using censored data', *Technometrics*, **27**, 41–8.

Sanov, I. N. (1957). 'On the probability of large deviations of random variables', *Mat. Sb.*, **42**, 11–44. [274]

Santner, T. J. See Mykytyn and Santner (1981) and Tenga and Santner (1984).

Sasabuchi, S. (1980). 'A test of a multivariate normal mean with composite hypotheses determined by linear inequalities', *Biometrika*, **67**, 429–39.

Sasabuchi, S., and D. D. S. Kulatunga (1985). 'Some approximations for the null distribution of the $\bar{E}^2$ statistic used in order restricted inference', *Biometrika*, **72**, 476–80. [122, 124, 161]

Sasabuchi, S. See also Inutsuka, Sasabuchi, and Kulatunga (1983); Kudô, Sasabuchi, and Choi (1981); and Kulatunga and Sasabuchi (1984a, 1984b).

Saunders, S. C., and J. M. Mhyre (1983). 'Maximum likelihood estimation for two-parameter decreasing hazard rate distributions using censored data', *J. Amer. Statist. Assoc.*, **78**, 664–73.

Savage, I. R. (1957). 'Contributions to the theory of rank order statistics- the "trend" case', *Ann. Math. Statist.*, **28**, 968–77.

Savage, I. R. See also Chernoff and Savage (1958).

Saxena, K. M. L. (1976). 'Distribution-free tolerance intervals for stochastically ordered distributions', *Ann. Statist.*, **4**, 1210–18.

Saxena, K. M. L. See also Alam, Saxena, and Tong (1973).

Scarf, H. See also Arrow, Karlin, and Scarf (1958).

Schaafsma, W. (1966). 'Hypothesis testing problems with the alternative restricted by a number of inequalities', Doctoral dissertation, University of Groningen, Noordhoff, Groningen. [226]

Schaafsma, W. (1968). 'A comparison of the most stringent and the most stringent somewhere most powerful test for certain problems with restricted alternative', *Ann. Math. Statist.*, **39**, 531–46. [227]

Schaafsma, W. (1970). 'Most stringent and maximin tests as solutions of linear programming problems', *Z. Wahrsch. verw. Gebiete* **14**, 290–307.

Schaafsma, W., and L. J. Smid (1966). 'Most stringent somewhere most powerful test against alternatives restricted by a number of linear inequalities', *Ann. Math. Statist.*, **37**, 1161–72. [176, 180, 182, 184, 187, 191, 194, 197, 226, 227]

Schectman, K. (1981). 'A test for zero point triserial correlation coefficient and its relationship with tests for homogeneity of means with ordered alternatives', *Comm. Statist. (A)*, **10**, 1167–82.

Scheffé, H. (1959). *The Analysis of Variance*, Wiley, New York.

Schell, M., and B. Singh (1986). 'Robustness of $\bar{\chi}^2$ and $\bar{E}^2$: A case for a ranked test in ordered restricted inference'. In R. L. Dykstra, T. Robertson, and F. T. Wright (eds.), *Advances in Order Restricted Statistical Inference*, Springer-Verlag, pp. 184–202.

Schervish, M. J. (1983). 'Multivariate normal probabilities with error bound', *Appl. Statist.*, **32**, 81–7.

Scheuer, E. M. See Barlow, Proschan, and Scheuer (1971); Barlow and Scheuer (1966, 1971); and Proschan, Barlow, Madansky, and Scheuer (1968).

Schlee, W. (1980). 'Nonparametric tests of the monotone and convexity of regression'. In *Nonparametric Statistical Inference*, Vol. II, Budapest, pp. 823–36.

Schmidt, K. See Mäkuläinen, Schmidt, and Styan (1981).

Schmidt, P., and M. Thomson (1982). 'A note on the comparison of the mean square error of inequality constrained least squares estimates', *Econom. Lett.*, **5**, 355–8.

Schmoyer, R. L. (1984). 'Sigmoidally constrained maximum likelihood estimation in quantal bioassay', *J. Amer. Statist. Assoc.*, **79**, 448–53.

Schmoyer, R. L. (1986a). 'An exact distribution-free analysis for accelerated life testing at several levels of a single stress', *Technometrics*, **28**, 165–75.

Schmoyer, R. L. (1986b). 'Dose–response analysis under unimodality of response-to-dose'. In R. L. Dykstra, T. Robertson and F. T. Wright (eds.), *Advances in Order Restricted Statistical Inference*, Springer-Verlag, pp. 125–152.

Schmoyer, R. L. (1986c). 'Convex sets in minimum distance estimation'. *Comm. Statist. (B)*, **15**, 3625–35.

Schoenfeld, D. (1986). 'Confidence intervals for normal means under order restrictions, with application to dose–response curves, toxicology experiments and low-dose extrapolation', *J. Amer. Statist. Assoc.*, **81**, 186–95. [115, 407]

Scholander, M. C. See Leach and Sholander (1978).

Schroeder, A. See Friedman, Stuetzle, and Schroeder (1984).

Schucany, W. R., and J. Beckett III (1976). 'Analysis of multiple sets of incomplete rankings', *Comm. Statist. (A)*, **5**, 1327–34.

Schuster, E. F. (1975). 'Estimating the distribution function of a symmetric distribution', *Biometrika*, **62**, 631.

Schuster, E. F. (1987). 'Identifying the closest symmetric distribution or density function', *Ann. Statist.*, **15**, 865–74.

Schwartz, J. T. See Dunford and Schwartz (1958).

Scobey, P., and D. G. Kabe (1980). 'On constrained least squares estimation'. In *Proceedings of Conference on Multivariate Statistical Analysis*, North Amsterdam, pp. 179–89.

Sedransk, J., J. Monahan and H. Y. Chiu (1985). 'Bayesian estimation of finite population parameters in categorical data models incorporating order restrictions'. *J. Royal Statist. Soc. (B)*, **47**, 519–27. [397, 398]

Sedransk, J. See Chiu and Sedransk (1986).

Self, S. G., and K. Y. Liang (1987). 'Asymptotic properties of maximum likelihood estimators and likelihood ratios tests under nonstandard condition', *J. Amer. Stat. Assoc.*, **82**, 605–10.

Sen, P. K. (1968). 'On a class of aligned rank order test in two-way layouts', *Ann. Math. Statist.*, **39**, 1115–24. [212, 213, 215, 216]

Sen, P. K. (1982). 'The UI-principle and LMP rank tests', *Colloq. Nonparametric Statistical Inference, Janos Bolyai Math. Soc.*, **32**, 843–58. [216]

Sen, P. K. (1984). 'Subhypothesis testing against restricted alternatives for the Cox regression model', *J. Statist. Plann. Inference*, **8**, 31–42.

Sen, P. K. (1985). 'Nonparametric testing against restricted alternatives under progressive censoring', *Sequential Analysis*, **4**, 247–73.

Sen, P. K. See also Boyd and Sen (1983, 1984, 1986); Chinchilli and Sen (1981a, 1981b); and Puri and Sen (1968).

SenGupta, A. 'On tests under order restrictions in generalized homogeneity and reduction of dimensionality'. In R. L. Dykstra, T. Robertson and F. T. Wright (eds.), *Advances in Order Restricted Statistical Inference*, Springer-Verlag, pp. 249–56.

Serfling, R. J. (1980). *Approximation Theorems of Mathematical Statistics*, Wiley, New York. [242, 245, 318 320]

Sermat, V. See Abelson and Sermat (1962).

Sethuraman, J. See Hollander, Proschan, and Sethuraman (1977).

Shaked, M. (1979). 'Estimation of star-shaped sequences of Poisson and normal means', *Ann. Statist.*, **7**, 729–41. [248, 263, 308]

Shaked, M. (1982). 'Dispersive ordering distributions', *J. Applied Probab.*, **19**, 310–20.

Shapiro, M. (1985). 'Asymptotic distribution of test statistics in the analysis of moment structures under inequality constraints', *Biometrika*, **72**, 133–44. [111, 115, 219, 220]

Shapiro, A. (1987a). 'A conjecture related to chi-bar-squared distributions', *Amer. Math. Monthly*, **94**, 46–7.

Shapiro, A. (1987b). 'On differentiability of metric projections in R^n, I: Boundary case', *Proc. Amer. Math. Soc.* (to appear).

Shapiro, A. (1988). 'Towards a unified theory of inequality constrained testing in multivariate analysis', *Internat. Statist. Rev.*, **56** (to appear). [112, 315, 322]

Sharma, D. See Dhariyal, Sharma, and Krishnamoorthy (1985).

Shepard, R. N. (1962a). 'The analysis of proximities: Multidimensional scaling with an unknown distance function. I', *Psychometrika*, **27**, 125–40.

Shepard, R. N. (1962b). 'The analysis of proximities: Multidimensional scaling with an unknown distance function. II', *Psychometrika*, **27**, 219–46. [11]

Shepard, R. N. (1966). 'Metric structures in ordinal data', *J. Math. Psych.*, **3**, 287–315. [11]

Shepp, L. A. (1982). 'The XYZ conjecture and the FKG inequality', *Ann. Prob.*, **10**, 824–7.

Shi, N. Z. (1986). 'Testing a normal mean vector against the alternative determined by a convex cone', Technical Report, Kyushu University. [226]

Shi, N. Z., and A. Kudô (1986). 'The most stringent somewhere most powerful one sided test of multivariate normal mean', Technical Report, Kyushu University. [226]

Shiller, R. J. (1984). 'Smoothness priors and nonlinear regression', *J. Amer. Statist. Assoc.*, **79**, 609–15.

Shirahata, S. (1978). 'An approach to a one-sided test in the bivariate normal distribution', *Biometrika*, **32**, 61–7.

Shirahata, S. (1979). 'Tests of homogeneity for ordered alternatives in the normal population', *Bull. math. Statist.*, **18**, 61–8.

Shirahata, S. (1980). 'Rank tests for the k-sample problem with restricted alternatives', *Comm. Statist. (A)*, **9**, 1071–86. [203, 206]

Shirahata, S. See also Araki and Shirahata (1981).

Shiraishi, T. (1982). 'Testing homogeneity against trend based on rank in one-way layout', *Comm. Statist. (A)*, **11**, 1255–68. [205, 206, 208]

Shiraishi, T. (1984). 'Rank analogues of the likelihood ratio test for an ordered alternative in a two-way layout', *Ann. Inst. St. Math.*, **36**, 223–37.

Shiraishi, T., and A. Kudô (1981). 'A nonparametric test of trend based on amalgamation', *Mem. Fac. Sci. Kyushu Univ. (A)*, **35**, 235–45.

Shirley, E. (1977). 'A nonparametric equivalent of William's test for contrasting increasing dose levels of a treatment', *Biometrics*, **33**, 386–9.

Shisha, O. (1965). 'Monotone approximation', *Pacific J. Math.*, **15**, 667–71.

Shorack, G. R. (1967). 'Testing against ordered alternatives in model I analysis of variance; Normal theory and nonparametric', *Ann. Math. Statist.*, **38**, 1740–53. [113, 115, 117, 167, 204, 206, 210, 211, 214, 215, 218, 221, 227]

Shuster, J. J., and M. C. K. Yang (1975). 'A distribution free approach to quantal response assays', *Can. J. Statist.*, **3**, 57–70.

Sidák, Z. (1957). 'On relations between strict-sense and wide-sense conditional expectations', *Theory Prob. Appl.*, **2**, 267–71. [387]

Sidák, Z. See Hájek and Sidák (1967).

Siegel, S. (1956). *Nonparametric Statistics for the Behavioural Sciences*, McGraw-Hill, New York. [237]

Sielken, R. L., Jr., and H. O. Hartley (1973). 'Two linear programming algorithms for unbiased estimation of linear models', *J. Amer. Statist. Assoc.*, **68**, 639–47.

Silverman, B. W. (1985). 'Two books on density estimation' (Book Review), *Ann. Statist.*, **13**, 1630–8. [325]

Silverman, E. See Ayer, Brunk, Ewing, Reid, and Silverman (1955).

Silvey, J. D. (1959). 'The Lagrangian multiplier test', *Ann. Math. Statist.*, **30**, 389–407. [311, 318]

Silvey, J. D. See also Aitchison and Silvey (1958).

Simpson, D. G., and B. H. Margolin (1986). 'Recursive nonparametric testing for dose–response relationships subject to downturns at high doses', *Biometrika*, **73**, 589–596.

Sinclair, D. F. See Berger and Sinclair (1984).

Singh, B., and F. T. Wright (1986a). 'Power series approximations to the null distribution of some chi-bar square statistics'. In R. L. Dykstra, T. Robertson, F. T. Wright (eds.), *Advances in Order Restricted Statistical Inference*, Springer-Verlag, pp. 257–78. [122, 123, 140, 161]

Singh, B., and F. T. Wright (1986b). 'Power series approximations to the null distributions in order restricted inference: The case of unequal weights', Technical Report, Department of Mathematics, University of Missouri-Rolla. [122, 125, 161]

Singh, B., and F. T. Wright (1987). 'Approximations to the power of some order restricted tests with slippage alternatives'. *Biometrika*, **74**, (to appear). [154, 157, 160–162]

Singh, B. See also Schell and Singh (1986).

Siskind (1976). 'Approximate probability integrals and critical values for Bartholomew's test for ordered means', *Biometrika*, **63**, 647–54.

Skarpness, B. (1986). 'A discussion of equivalent forms of the solution to the least squares regression problem with linear inequality constraints', Unpublished manuscript.

Skarpness, B. See Escobar and Skarpness (1984, 1986, 1987) and Sposito, Hand, and Skarpness (1983).

Skillings, J. H. (1978). 'Adaptively combining independent Jonckheere statistics in a randomized block design with unequal scales', *Comm. Statist. (A)*, **7**, 1027–40.

Skillings, J. H. (1980). 'On the null distribution of Jonckheere's statistic used in two-way models for ordered alternative', *Technometrics*, **22**, 431–6. [209]

Skillings, J. H., and D. A. Wolfe (1977). 'Testing for ordered alternatives by combining independent distribution-free block statistics', *Comm. Statist. (A)*, **6**, 1453–64.

Skillings, J. H., and D. A. Wolfe (1978). 'Distribution-free tests for ordered alternatives in a randomized block design', *J. Amer. Statist. Assoc.*, **73**, 427. [209, 215]

Slud, E. V. See Chen and Slud (1984).

Smid, L. J. See Schaafsma and Smid (1966).

Smiley, M. F. (1963). 'The proof of the triangle inequality'. *Amer. Math. Monthly*, **70**, 546–547. [383]

Smith, A. M. F. (1977). 'A Bayesian note on reliability growth during a development testing program', *IEEE Transactions on Reliability*, **26**, 346–7.

Smith, P. T. See Macdonald and Smith (1983).

Smith, W. (1983). 'Inequalities for bivariate distributions with $X \leqslant Y$ and marginals given', *Comm. Statist. (A)*, **12**, 1371–9.

Smythe, R. T. (1980). 'Maxima of partial sums and a monotone regression estimator', *Ann. Prob.*, **8**, 630–5.

Sneath, P. H. A. See Hodson, Sneath, and Doran (1966).

Snidjers, T. (1979). *Asymptotic Optimality Theory for Testing Problems with Restricted Alternatives*, Mathematisch Centrum, Amsterdam. [176, 227]

Snidjers, T. (1980). 'Asymptotic optimality theory for testing problems with restricted alternatives', *J. Amer. Statist. Assoc.*, **75**, 1044.

Snidjers, T. (1981). 'Rank tests for bivariate symmetry', *Ann. Statist.*, **9**, 1087–5.

Sobel, M. See Dunnett and Sobel (1955).

Sofer, A. See also Miller and Sofer (1986).

Soms, A. P. 'Permutation tests for k-sample binomial data with comparisons of exact and approximate p-levels', *Comm. Statist.*, **14**, 217–33.

Sparre Andersen, E. (1953a). 'On sums of symmetrically dependent random variables', *Skand. Akt.*, **36**, 123–8.

Sparre Andersen E. (1953b). 'On the fluctuations of sums of random variables', *Mathematica Scand.*, **1**, 263–85 (for corrections to the above paper, see *Mathematica Scand.*, **2**, 193–4).

Sparre Andersen E. (1954). 'On the fluctuations of sums of random variables II'. *Mathematica Scand.*, **2**, 195–223. [113, 206]

Sparre Andersen E., and B. Jessen (1948). 'Some limit theorems on set-functions'. *Math.-fys. Meddr.*, **25**, 8pp.

Spiegelhalter, D. J. (1983). 'Diagnostic tests of distributional shape', *Biometrika*, **70**, 401–9.

Spjøtvoll, E. (1977). 'Ordering ordered parameters', *Biometrika*, **64**, 327–34. [115]

Sposito, V. A. (1975). *Linear and Nonlinear Programming*, Iowa State University Press, Ames, Iowa.

Sposito, V. A., M. L. Hand, and B. Skarpness (1983). 'On the efficiency of using the sample kurtosis in selecting optimal L_p estimators', *Comm. Statist. Simula. Computa.*, **12**, 265–72.

Spring, D. (1985). 'On the second derivative test for constrained local extrema', *Amer. Math. Monthly*, **92**, 631–42.

Srivastava, S. R., and T. A. Bancroft (1967). 'Inferences concerning a population correlation coefficient from one or possibly two samples subsequent to a preliminary test of significance', *J. R. Statist. Soc. (B)*, **29**, 282–91.

Steckin, S. B. See Levin and Steckin (1960).

Steele, M. (1977). 'Limit properties of random variables associated with a partial order on R^d', *Ann. Prob.*, **5**, 395–403.

Stegun, I. S. See Abramowitz and Stegun (1965).

Steinitz, E. (1913). 'Bedingt konvergente reihen und konvexe systemme, I', *J. Math.*, **143**, 128–75. [322]

Stenson, H. H., and R. L. Knoll (1969). 'Goodness of fit for random rankings in Kruskal's nonmetric scaling procedure', *Psych. Bull.*, **71**, 122–6.

Steyn, A. G. W., and C. F. Crouse (1975). 'Testing the homogeneity of variances against an ordered alternative', *South African Statist. J.*, **9**, 91–108.

Stiijnen, T. See Houwelingon and Stiijnen (1983).

Stoer, J., and C. Witzgall (1970). *Convexity and Optimization in Finite Dimensions*, Springer-Verlag, New York. [109]

Strang, G. (1984). 'Duality in the classroom', *Amer. Math. Monthly*, **91**, 250–4.

Strawderman, W. E. See Sackrowitz and Strawderman (1974).

Stromberg, K. See Hewitt and Stromberg (1965).

Stuart, A. (1953). 'The estimation and comparison of strengths of association in contingency tables', *Biometrics*, **40**, 105–10. [291]

Stuart, A. (1963). 'Calculation of Spearman's rho for ordered 2-way classifications', *Amer. Statist.*, **17**, 23–4. [168]

Stuart, A. See also Kendall and Stuart (1958).

Stuetzle, W. See Friedman, Stuetzle, and Schroeder (1984).

Strum, G. W. See Potter and Strum (1981).

Styan, G. P. H. See Mäkuläinen, Schmidt, and Styan (1981).

Subbaiah, P. See Mudholkar and Subbaiah (1987).

Susarla, V., and J. van Ryzin (1980). 'Large sample theory for an estimator of the mean survival time from censored samples', *Ann. Statist.*, **8**, 1002–1016.

Takeuchi, T., and C. Hirotsu (1982). 'The cumulative chi-squares method against ordered alternatives in two way contingency tables', *Rep. Statist. Appl. Res. Union Jap. Sci. Eng.*, **29**, 1–13.

Talpaz, H. See Marcus and Talpaz (1983).

Tamhane, A. C. (1986). 'A survey of literature on quantal response curves with a view towards application to the problem of selecting a curve with the smallest q-quantile (ED100Q)', *Comm. Stat.: Theor. Meth.*, **15**, 2679–718.

Tapia, R. A., and J. R. Thompson (1978). *Nonparametric Probability Density Estimation*, John Hopkins University Press, Baltimore. [325]

Tarone, R. E. (1975). 'Tests for trend in life table analysis', *Biometrika*, **62**, 679.

Tarone, R. E. (1982a). 'The use of historical control information in testing for trend in proportions', *Biometrics*, **24**, 215–20.

Tarone, R. E. (1982b). 'The use of historical control information in testing for a trend in Poisson means', *Biometrics*, **24**, 457–62.

Taylor, A. E. (1958). *Introduction to Functional Analysis*, John Wiley, New York. [386]

Tchen, A. E. (1980). 'Inequalities for distributions with given marginals', *Ann. Prob.*, **8**, 814–27. [297]

Tenga, R., and T. J. Santner (1984). 'Testing goodness of fit to the increasing failure rate family', *Naval Res. Logist. Quart.*, **31**, 617–30.

Terpstra, T. J. (1952a). 'The asymptotic normality and consistency of Kendall's test against trend when ties are present in one ranking', *Proc. Sect. Sci. K. ned. Akad. Wet. (A)*, **55**/*Indag. Math.*, **14**, 327–33. [201–203, 206, 208, 209, 227]

Terpstra, T. J. (1953). 'The exact probability distribution of the T statistic for testing against trend and its normal approximation', *Proc. Sect. Sci. K. ned. Akad. Wet. (A)*, **56**/*Indag. Math.*, **15**, 433–7.

Terpstra, T. J. (1955). 'A generalization of Kendall's rank correlation statistic I', *Proc. Sect. Sci. K. ned. Akad. Wet. (A)*, **58**/*Indag. Math.*, **17**, 690–6. [201]

Terpstra, T. J. (1956). 'A generalization of Kendall's rank correlation statistic II', *Proc. Sect. Sci. K. ned. Akad. Wet. (A)*, **59**/*Indag. Math.*, **18**, 59–66. [201]

Terpstra, T. J. (1976). 'Asymptotic power and efficiency of a class of k-sample rank tests against trend and the determination of optimal tests' (abstract), *Bull. Inst. Math. Stat.*, **5**.

Terpstra, T. J. (1980). 'Efficiency and optimality properties of a class of k-sample rank tests against trend', *Metrika*, **27**, 225–34.

Terpstra, T. J. (1985). 'One-sided and multi-sided rank tests against trend in the case of concordant observers', *Metrika*, **32**, 109–23.

Thakur, A. K. (1984). 'A FORTAN program to perform nonparametric Terpstra–Jonckheere test', *Computer Programs in Biomedicine*, **18**, 235–40.

Theodorescu, R., and H. Wolff (1981). 'Sequential estimation of expectations in the presence of trend', *Austral. J. Statist.*, **23**, 196–203.

Thisted, R. A. See Sager and Thisted (1982).

Thomas, P. C. (1983). 'Nonparametric estimation and tests of fit for dose response relations', *Biometrics*, **39**, 263–8.

Thompson, J. R. See Tapia and Thompson (1978).

Thompson, W. A., Jr. (1962). 'The problem of negative estimates of variance components', *Ann. Math. Statist.*, **33**, 273–89. [57, 58]

Thompson, W. A., Jr., and J. R. Moore (1963). 'Non-negative estimates of variance components', *Technometrics*, **5**, 441–9.

Thomson, M. (1982). 'Some results on the statistical properties of an inequality constrained least squares estimator in a linear model with two regressions', *J. Econometrics*, **19**, 215–31.

Thomson, M. See Schmidt and Thomson (1982).

Tibshirani, R. See Friedman and Tibshirani (1984).

Tika, J. L. (1975). 'A new statistic for testing normality', *Comm. Statist. (A)*, **4**, 223–32.

Tingey, F. H. See Birnbaum and Tingey (1951).

Titterington, D. M., and A. W. Bowman (1985). 'A comparative study of smoothing procedures for ordered categorical data', *J. Statist. Comp. Simul.*, **21**, 291–312.

Tomizawa, S. (1987). 'Diagonal weighted marginal homogeneity models and decompositions for linear diagonals-parameter symmetry model', *Comm. Statist. -Theory and Meth.*, **16**, 477–88.

Tong, Y. L. (1980). *Probability Inequalities in Multivariate Distributions*, Academic Press, New York. [143]

Tong, Y. L. See Alam, Saxena, and Tong (1973).

Toothaker, L. E. See Nelson and Toothaker (1975).

Tosstorff, G. See Christopeit and Tosstorff (1987).

Tran, T. See Wright and Tran (1985).

Tryon, P. V. (1970). 'Nonparametric tests of homogeneity against restricted alternatives in a one-way classification', Unpublished Ph.D. thesis, Pennsylvania State University.

Tryon, P. V., and T. P. Hettmansperger (1973). 'A class of nonparametric tests for homogeneity against ordered alternative', *Ann. Statist.*, **1**, 1061–70. [202, 203]

Tukey, J. W. See Abelson and Tukey (1963).

Ubhaya, V. A. (1974a). 'Isotone optimization I', *J. Approx. Theory*, **12**, 146–59.

Ubhaya, V. A. (1974b). 'Isotone optimization II', *J. Approx. Theory*, **12**, 315–31.

Ubhaya, V. A. See also Barlow and Ubhaya (1971).

Umbach, D. (1981). 'A note on the median of a distribution', *Ann. Inst. Statist. Math.*, **33**, 135–40.

Ury, H. K. (1968). 'The behavior of some tests for ordered alternatives under interior slippage'. In *Selected Statistical Papers*, Vol. 2, Mathematical Centre, Amsterdam, pp. 15–25.

Utreras, F. I. (1986). 'Smoothing noisy data using monotonicity constraints', *Numerische Mathematik*, **47**, 611–25.

Utz, W. R. See Brunk, Ewing, and Utz (1957).

Vainberg, M. M. (1973). *Variational and the Method of Monotone Operators*, John Wiley, Chichester.

Vardi, Y. See Denby and Vardi (1984).

Veinott, A. F., Jr. (1971). 'Least *d*-majorized network flows with inventory and statistical applications', *Management Science (Theory)*, **17**, 547–67.

Venter, J. H. (1967). 'On estimation of the mode', *Ann. Math. Statist.*, **38**, 1446–55. [334, 336]

Verbuek, A. See Kroonenberg and Verbuek (1987).

Verducci, J. S. See Fairley, Pearl and Verducci (1987).

Villalobes, M., and G. Wahba (1987). 'Inequality constrained multivariate smoothing splines with application to the estimation of posterior probabilities', *J. Amer. Statist. Assoc.*, **82**, 239–48.

Vincent, S. E. (1961). 'A test of homogeneity for ordered variances', *J. R. Statist. Soc. (B)*, **23**, 195–206. [176]

Vitale, R. A. (1979). 'Regression with given marginals', *Ann. Statist.*, **7**, 653–8.

Vogt, H. (1983). 'Unimodality of differences', *Metrika*, **30**, 165–70.

Wahba, G. (1982). 'Constrained regularization for ill posed linear operator equations, with applications in meteorology and medicine'. In S. S. Gupta and J. O. Berger (eds.), *Statistical Decision Theory and Related Topics*, Vol. III, Academic Press, New York, pp. 383–418.

Wahba, G. (1986). 'Multivariate thin plate spline smoothing with positivity and other linear inequality constraints'. In E. J. Wegman and D. J. DePriest (eds.), *Statistical Image Processing and Graphics*, Marcel Dekker Inc., pp. 275–89.

Wahba, G. See also Villalobes and Wahba (1987).

Wakimoto, K. (1981). '*k*-multiple chart and its application to the test for homogeneity against ordered alternatives', *J. Japan. Statist. Soc.*, **11**, 1–7.

Wald, A. (1943). 'Tests of statistical hypotheses concerning several parameters when the number of observations is large', *Trans. Amer. Math. Soc.*, **54**, 426–82. [318–320]

Wald, A. (1949). 'Note on the consistency of the maximum likelihood estimate', *Ann. Math. Statist.*, **20**, 595–601. [328]

Walk, H. (1983). 'Stochastic iteration for a constrained optimization problem', *Comm. Statist. (C)*, **2**, 369–85.

Walk, H. (1985). 'On recursive estimation of the mode', *Statist. Decisions*, **3**, 337–50.

Walkup, B. W. See Marshall, Walkup, and Wets (1967).

Wallenstein, S. (1980). 'Distributions of some one-sided *k*-sample Smirnov-type statistics', *J. Amer. Statist. Assoc.*, **75**, 441–6.

Waltman, P. See Robertson and Waltman (1968).

Wang, J. L. (1985). 'Strong consistency of approximate maximum likelihood estimators with applications in nonparametrics', *Ann. Statist.*, **13**, 932–46. [328]

Wang, J. L. (1986). 'Asymptotically minimax estimates for distributions with increasing failure rate', *Ann. Statist.*, **14**, 113–31.

Wang, J. L. See also Mukerjee and Wang (1987).

Warrack, G., and T. Robertson (1984). 'A likelihood ratio test regarding two nested but oblique order restricted hypotheses', *J. Amer. Statist. Assoc.*, **79**, 881–6. [259, 264]

Warrack, G. See also Robertson and Warrack (1985).

Watson, G. S. and M. R. Leadbetter (1964). 'Hazard analysis II', *Sankhya*, **26**, 101–16. [367]

Weerahandi, S., and J. V. Zidek (1979). 'A characterization of the general mean', *Can. J. Statist.*, **7**, 83–90.

Wegman, E. J. (1968). 'On estimating a unimodal density', *Ph.D. dissertation*, University of Iowa.

Wegman, E. J. (1969a). 'Maximum likelihood histograms', University of North Carolina Mimeo Series 629.

Wegman, E. J. (1969b). 'A note on estimating a unimodal density', *Ann. Math. Statist.*, **40**, 1661–7.

Wegman, E. J. (1970a). 'Maximum likelihood estimation of a unimodal density function', *Ann. Math. Statist.*, **41**, 457–71. [335, 336, 367]

Wegman, E. J. (1970b). 'Maximum likelihood estimation of a unimodal density function, II', *Ann. Math. Statist.*, **41**, 2169–74. [335, 367]

Wegman, E. J. (1980). 'Two approaches to nonparametric regression: Splines and isotonic inference'. In *Proceedings of International Symposium Recent Developments in Statistical Inference and Data Analysis*, North Holland, Amsterdam, pp. 323–34.

Wegman, E. J. (1984). 'Optimal nonparametric function estimation', *J. Statist. Plan. Inference*, **8**, 375–87. [403]

Wegman, E. J. (1986). 'Another look at Box–Jenkins forecasting procedures', *Comm. Statist. Simulation*, **15**, 523–30.

Wegman, E. J., E. S. Nour, and C. Kukuk (1980). 'A time series approach to life table constructions', *Comm. Statist. (A)*, **9**, 1587–607.

Wegman, E. J., and I. Wright (1983). 'Splines in statistics', *J. Amer. Statist. Assoc.*, **78**, 351–65.

Wegman, E. J. See also Robertson and Wegman (1978) and Wright and Wegman (1980).

Weisberg, H. (1972). 'Upper and lower probability inferences from ordered multinomial data', *J. Amer. Statist. Assoc.*, **67**, 884–90.

Weiss, L., and J. Wolfowitz (1967). 'Estimators of a density function at a point', *Z. Wahrsch. verw. Gebiete*, **7**, 327–335. [367]

Welch, B. L. (1937). 'On the *z*-test in randomized blocks and Latin squares', *Biometrika*, **29**, 21–52.

Wets, R. J. B. See Marshall, Walkup, and Wets (1967).

Whitaker, L. R. See Sampson and Whitaker (1987).

White, A. A., J. R. Landis, and M. M. Cooper (1982). 'A note on the equivalence of several marginal homogeneity test criteria for categorical data', *Internat. Statist. Rev.*, **50**, 27–34.

Whitlock, J. H. See Elfving and Whitlock (1950).

Wilhemsen, D. R. (1976). 'A nearest point algorithm for convex polyhedral cones and applications to positive linear approximation', *Mathematics of Computation*, **30**, 48–57. [111]

Williams, D. A. (1971). 'A test for differences between treatment means when several dose levels are compared with a zero dose control', *Biometrics*, **27**, 103–17. [190, 196]

Williams, D. A. (1972). 'The comparison of several dose levels with a zero dose control', *Biometrics*, **28**, 519–31. [190]

Williams, D. A. (1977). 'Some inference procedures for monotonically ordered normal means', *Biometrika*, **64**, 9–14. [406]

Williams, D. A. (1986). 'A note on Shirley's nonparametric test for comparing several dose levels with a zero-dose control', *Biometrics*, **42**, 183–6.

Williams, G. (1976) 'On tests for order among proportions obtained from matched samples', Technical Report No. 10, Department of Biostatistics, University of Michigan.

Winkler, R. L. See Murphy and Winkler (1984).

Winokur, G. See Cadoret, Woolson, and Winokur (1977).

Witzgall, C. See Stoer and Witzgall (1970).

Winter, B. B. (1987). 'Nonparametric estimation with censored data from a distribution with a non-increasing density', *Comm. Statist. -Theory and Meth.*, **16**, 93–120.

Wolak, F. A. (1986). 'Testing inequality constraints in linear econometrics models', Technical Report.

Wolak, F. A. (1987). 'An exact test for multiple inequality and equality constraints in the linear regression model', *J. Amer. Statist. Assoc.*, **82**, 782–93. [323]

Wolfe, D. A. See Bates and Wolfe (1984); Lee and Wolfe (1976); Mack and Wolfe (1981); Randles and Wolfe (1979); and Skillings and Wolfe (1977, 1978).

Wolfe, R. A., L. D. Roi, and E. H. Margosches (1981). 'Monotonic dichotomous regression estimates: A burn case example', *Biometrics*, **37**, 157–67.

Wolfe, R. A. See also Finkelstein and Wolfe (1986).

Wolfe, P. (1976). 'Finding the nearest point in a polytope', *Math. Programming*, **11**, 128–49.

Wolff, H. See Theodorescu and Wolff (1981).

Wolfowitz, J. See Weiss and Wolfowitz (1967) and Kiefer and Wolfowitz (1976).

Wolfson, D. B. See Clarkson and Wolfson (1983).

Wollan, P. C. (1985). 'Estimation and Hypothesis Testing under Inequality Constraints', Ph.D. thesis, University of Iowa. [321, 387]

Wollan, P. C., and R. L. Dykstra (1986). 'Conditional tests with an order restriction as the null hypothesis'. In R. L. Dykstra, T. Robertson and F. T. Wright (eds.), *Advances in Order Restricted Statistical Inference*, Springer-Verlag, pp. 279–95. [114, 200]

Wollan, P. C., and R. L. Dykstra (1987). 'Minimizing linear inequality constrained Mahalanobis distances', *Appl. Statist.*, **36**, 234–240 [219]

Wollan, P. C. See also Dykstra and Wollan (1986, 1987).

Woolson, R. See Cadoret, Woolson, and Winokur (1977).

Wright, F. T. (1976). 'Algebraic rates of convergence for isotonic regression functions', *Bull. Inst. Math. Acad. Sinica.*, **4**, 281–7. [394]

Wright, F. T. (1978). 'Estimating strictly increasing regression functions', *J. Amer. Statist. Assoc.*, **73**, 636–9. [402]

Wright, F. T. (1979). 'A strong law for variables indexed by a partially ordered set with applications to isotone regression', *Ann. Prob.*, **7**, 109–27.

Wright, F. T. (1981a). 'The empirical discrepancy over lower layers and a related law of large numbers', *Ann. Prob.*, **9**, 323–9.

Wright, F. T. (1981b). 'The asymptotic behavior of monotone regression estimates', *Ann. Statist.*, **9**, 443–8. [396]

Wright, F. T. (1981c). 'Sums of random variables indexed by a partially ordered set and the estimation of integral regression functions', *Ann. Statist.*, **9**, 449–52.

Wright, F. T. (1982). 'Monotone regression estimates for grouped observations', *Ann. Statist.*, **10**, 278–86. [358]

Wright, F. T. (1984). 'The asymptotic behavior of monotone percentile regression estimates', *Can. J. Statist.*, **12**, 229–36. [396]

Wright, F. T. (1985). 'Order-restricted inference', *Encyclopedia of Statist. Sci.*, **6**, 501–3.

Wright, F. T. (1986). 'Bounds on distributions arising in order restricted inference: The partially ordered case', Technical Report, Department of Mathematics, University of Missouri-Rolla. [147, 149]

Wright, F. T., and T. Tran (1985). 'Approximating the level probabilities in order restricted inference: The simple tree ordering', *Biometrika*, **72**, 429–39. [139, 162]

Wright, F. T. See also Bain, Engelhardt, and Wright (1985); Cryer, Robertson, Wright, and Casady (1972); Guffey and Wright (1986); Hanson, Pledger, and Wright (1973); Lucas and Wright (1986, 1987); Magel and Wright (1984a, 1984b); Mukerjee, Robertson, and Wright (1986a, 1986b, 1987); Pillers, Robertson, and Wright (1984); Robertson and Wright (1973, 1974a, 1974b, 1975, 1980a, 1980b, 1981, 1982a, 1982b, 1982c, 1983, 1984, 1985); and Singh and Wright (1986a, 1986b, 1987).

Wright, I., and E. J. Wegman (1980). 'Isotonic, convex and related splines', *Ann. Statist.*, **8**, 1023–35. [403]

Wright, I. See also Wegman and Wright (1983).

Wright, S. J., and J. N. Holt (1985). 'Algorithms for nonlinear least squares with linear inequality constraints', *SIAM J. Sci. Statist. Comp.*, **6**, 1033–48.

Wu, Chien-Fu (1982). 'Some algorithms for concave and isotonic regression'. In S. H. Zanokis and J. S. Rustagi (eds.), *Optimization in Statistics*, North Holland, Amsterdam, pp. 105–16.

Wynn, H. P. (1975). 'Integrals for one-sided confidence bounds: A general result', *Biometrika*, **62**, 393–96. [219]

Yancey, T. A., G. G. Judge, and M. E. Bock (1981). 'Testing multiple equality and inequality hypotheses in economics', *Econ. Lett.*, **7**, 249–55.

Yancey, T. A., R. Bohrer, and G. G. Judge (1982). 'Power function comparisons in inequality hypothesis testing', *Econ. Lett.*, **9**, 161–7.

Yancey, T. A. See also Judge, Yancey, Bock, and Bohrer (1984).

Yang, M. C. K. See Gupta and Yang (1984) and Shuster and Yang (1975).

Yao, J. S. (1984a). 'Estimation of a noisy discrete-time *step* function: Bayes and empirical Bayes approaches', *Ann. Statist.*, **12**, 1434–7.

Yao, J. S. (1984b). 'A test of homogeneity of means against isotonic alternatives in *p* groups', *Chinese J. Math.*, **12**, 105–14.

Yao, J. S. See also Kudô and Yao (1982a, 1982b).

Yarnold, J. K. (1972). 'Asymptotic approximations for the probability that a sum of lattice random vectors lies in a convex set', *Ann. Math. Statist.*, **43**, 1566–80.

Yates, F. (1948). 'The analysis of contingency tables with groupings based on quantitative characters', *Biometrika*, **35**, 176–81.

Yuan, W. J. C. (1975). 'Asymptotically minmax rank tests for ordered alternatives' (abstract), *Bull. Inst. Math. Statist.*, **4**, 149–54.

Zaanen, A. C. See Luxemburg and Zaanen (1971).

Zalman, R. (1970). 'On polynomial *S*-type functions and approximation by monotonic polynomials', *J. Approx. Theory*, **3**, 1–6.

Zidek, J. V. See Weerahandi and Zidek (1979).

Zinn, J. See Gine and Zinn (1984).

Zucchino, W. See Antillo, Kersting, and Zucchino (1982).

Zwet, W. R. van (1964). *Convex Transformations of Random Variables*, Mathematical Centre, Amsterdam. [348, 367]

Zwet, W. R. van, and Oosterhoff, J. (1967). 'On the combination of independent test statistics', *Ann. Math. Statist.*, **38**, 659–80.

Zwet, W. R. van. See also Barlow and van Zwet (1970, 1971).

Zwick, D. (1987). 'Best approximation by convex functions', *Amer. Math. Monthly*, **94**, 528–34.

Subject Index

Applied Probability and Statistics (Continued)

HOEL • Elementary Statistics, *Fourth Edition*

HOEL and JESSEN • Basic Statistics for Business and Economics, *Third Edition*

HOGG and KLUGMAN • Loss Distributions

HOLLANDER and WOLFE • Nonparametric Statistical Methods

IMAN and CONOVER • Modern Business Statistics

JESSEN • Statistical Survey Techniques

JOHNSON • Multivariate Statistical Simulation

JOHNSON and KOTZ • Distributions in Statistics
Discrete Distributions
Continuous Univariate Distributions—1
Continuous Univariate Distributions—2
Continuous Multivariate Distributions

JUDGE, HILL, GRIFFITHS, LÜTKEPOHL and LEE • Introduction to the Theory and Practice of Econometrics, *Second Edition*

JUDGE, GRIFFITHS, HILL, LÜTKEPOHL and LEE • The Theory and Practice of Econometrics, *Second Edition*

KALBFLEISCH and PRENTICE • The Statistical Analysis of Failure Time Data

KISH • Statistical Design for Research

KISH • Survey Sampling

KUH, NEESE, and HOLLINGER • Structural Sensitivity in Econometric Models

KEENEY and RAIFFA • Decisions with Multiple Objectives

LAWLESS • Statistical Models and Methods for Lifetime Data

LEAMER • Specification Searches: Ad Hoc Inference with Nonexperimental Data

LEBART, MORINEAU, and WARWICK • Multivariate Descriptive Statistical Analysis: Correspondence Analysis and Related Techniques for Large Matrices

LINHART and ZUCCHINI • Model Selection

LITTLE and RUBIN • Statistical Analysis with Missing Data

McNEIL • Interactive Data Analysis

MAGNUS and NEUDECKER • Matrix Differential Calculus with Applications

MAINDONALD • Statistical Computation

MALLOWS • Design, Data, and Analysis by Some Friends of Cuthbert Daniel

MANN, SCHAFER and SINGPURWALLA • Methods for Statistical Analysis of Reliability and Life Data

MARTZ and WALLER • Bayesian Reliability Analysis

MIKÉ and STANLEY • Statistics in Medical Research: Methods and Issues with Applications in Cancer Research

MILLER • Beyond ANOVA, Basics of Applied Statistics

MILLER • Survival Analysis

MILLER, EFRON, BROWN, and MOSES • Biostatistics Casebook

MONTGOMERY and PECK • Introduction to Linear Regression Analysis

NELSON • Applied Life Data Analysis

OSBORNE • Finite Algorithms in Optimization and Data Analysis

OTNES and ENOCHSON • Applied Time Series Analysis: Volume I, Basic Techniques

OTNES and ENOCHSON • Digital Time Series Analysis

PANKRATZ • Forecasting with Univariate Box-Jenkins Models: Concepts and Cases

PLATEK, RAO, SARNDAL and SINGH • Small Area Statistics: An International Symposium

POLLOCK • The Algebra of Econometrics

RAO and MITRA • Generalized Inverse of Matrices and Its Applications

RÉNYI • A Diary on Information Theory

RIPLEY • Spatial Statistics

Applied Probability and Statistics (Continued)

RIPLEY • Stochastic Simulation
ROSS • Introduction to Probability and Statistics for Engineers and Scientists
ROUSSEEUW and LEROY • Robust Regression and Outlier Detection
RUBIN • Multiple Imputation for Nonresponse in Surveys
RUBINSTEIN • Monte Carlo Optimization, Simulation, and Sensitivity of Queueing Networks
SCHUSS • Theory and Applications of Stochastic Differential Equations
SEARLE • Linear Models
SEARLE • Linear Models for Unbalanced Data
SEARLE • Matrix Algebra Useful for Statistics
SPRINGER • The Algebra of Random Variables
STEUER • Multiple Criteria Optimization
STOYAN • Comparison Methods for Queues and Other Stochastic Models
TIJMS • Stochastic Modeling and Analysis: A Computational Approach
TITTERINGTON, SMITH, and MAKOV • Statistical Analysis of Finite Mixture Distributions
UPTON • The Analysis of Cross-Tabulated Data
UPTON and FINGLETON • Spatial Data Analysis by Example, Volume I: Point Pattern and Quantitative Data
VAN RIJCKEVORSEL and DE LEEUW • Component and Correspondence Analysis
WEISBERG • Applied Linear Regression, *Second Edition*
WHITTLE • Optimization Over Time: Dynamic Programming and Stochastic Control, Volume I and Volume II
WHITTLE • Systems in Stochastic Equilibrium
WILLIAMS • A Sampler on Sampling
WONNACOTT and WONNACOTT • Econometrics, *Second Edition*
WONNACOTT and WONNACOTT • Introductory Statistics, *Fourth Edition*
WONNACOTT and WONNACOTT • Introductory Statistics for Business and Economics, *Third Edition*
WOOLSON • Statistical Methods for The Analysis of Biomedical Data

Tracts on Probability and Statistics

AMBARTZUMIAN • Combinatorial Integral Geometry
BIBBY and TOUTENBURG • Prediction and Improved Estimation in Linear Models
BILLINGSLEY • Convergence of Probability Measures
DEVROYE and GYORFI • Nonparametric Density Estimation: The L_1 View
KELLY • Reversibility and Stochastic Networks
RAKTOE, HEDAYAT, and FEDERER • Factorial Designs
TOUTENBURG • Prior Information in Linear Models